Anorectal Physiology

Lucia Camara Castro Oliveira
Editor

Anorectal Physiology

A Clinical and Surgical Perspective

Springer

Editor
Lucia Camara Castro Oliveira, MD, PhD, FASCRS
Director Department of Anorectal Physiology of
Rio de Janeiro and CEPEMED
Rio de Janeiro
Brazil

ISBN 978-3-030-43810-4 ISBN 978-3-030-43811-1 (eBook)
https://doi.org/10.1007/978-3-030-43811-1

This Springer imprint is published by the registered company Springer Nature Switzerland AG
The registered company address is: Gewerbestrasse 11, 6330 Cham, Switzerland

This book is dedicated to all patients that I could help, to the ones I will help through other doctors' hands, and to all physicians interested, like me, in helping patients with colon and anorectal disorders.

Foreword 1

We live in a world of incredible change in the practice of medicine with rapid advancements, exponentially increasing knowledge, and innovative surgical procedures. Our understanding of anorectal physiology and pelvic floor disorders has rapidly advanced in the last several decades, and with this, the range of surgical options for selected patients has significantly expanded. As such, this book is an extremely valuable contribution. It is the first comprehensive textbook written from a surgical perspective dedicated to anorectal physiology and to the diagnosis, evaluation, and treatment of pelvic disorders.

Pelvic floor disorders, including pelvic organ prolapse, rectal prolapse, fecal incontinence, chronic constipation, pelvic pain, and defecatory disorders, are quite common and may affect up to 20% of the population. Optimal treatment of this heterogeneous group of disorders and challenging group of patients is predicated not only on understanding the anatomy and physiology of the region but also on an in-depth understanding of the indications for and limitations of anorectal physiology testing and radiographic testing combined with operative and non-operative techniques.

This textbook serves as a unique reference for all surgeons who treat pelvic floor disorders. The text is clear, readable, up-to-date, and written with a surgical perspective. Close to 400 figures, the majority of which are in color and are superb. Starting with the anatomy and physiology of the anorectal region and pelvic floor, detailed chapters follow on clinical evaluation and the plethora of tests including anal manometry, endoanal and endovaginal ultrasound, and defecography (including echo-, cine-, and MRI defecography), providing a complete and clinically useful catalog of current and evolving modalities for pelvic floor investigations. All modalities for treatment of the full gamut of pelvic floor disorders including fecal incontinence, rectal prolapse, and obstructed defecation are included in addition to rarer conditions such as Chagasic megacolon, and emerging techniques including the use of stem cells and posterior tibial nerve stimulation for fecal incontinence are detailed.

Dr. Oliveira is to be commended for assembling a truly outstanding group of internationally known experts to produce this comprehensive textbook. All trainees and established colon and rectal surgeons will find this book to be an invaluable, all-inclusive resource for anorectal physiology and pelvic floor disorders.

Patricia Roberts
Professor of Surgery
Tufts University School of Medicine
Boston, MA, USA

Chair Emeritus, Department of Surgery
Lahey Hospital and Medical Center
Burlington, MA, USA

Foreword 2

It is a privilege and an honor for me to write this foreword for one of my highly accomplished and internationally renowned alumni, Dr. Lucia Camara Castro Oliveira. Dr. Oliveira has established herself as one of the global authorities in anorectal physiology. Her content expertise is paralleled by her network of key opinion leaders with whom she interacts and who clearly respect her. Attestation to my reflections are attested to by perusal of the table of contents. Lucia's newest book, *Anorectal Physiology: A Clinical and Surgical Perspective*, includes 40 chapters undoubtedly making it the most comprehensive current and complete textbook of anorectal physiology. The authors of the chapters include luminaries from throughout the Americas and Europe. Every single aspect of anorectal physiology is delved into in great detail commencing with the anatomy and then moving through physiology to evaluation of anatomy and physiology. Ultimately, the etiology and clinical treatment of incontinence, constipation, rectal prolapse, and other disorders are addressed.

Each chapter is authoritatively written, clinically relevant, and comprehensively referenced. Some of the newest and most controversial areas of both evaluation and management are included in this wonderful treatise. Lesser known topics are given equal coverage to the more "common" etiologies, evaluations, and therapies which we employ in our practices. In addition to the depth and breadth of topics and the unparalleled reputations of the authors, Dr. Oliveira has skillfully managed to include a sufficient amount of overlap such that individual chapters can be read and clinically applied. I congratulate Dr. Oliveira on how she has managed to blend appropriate amounts of overlap among the chapters with sufficient non-repetitive content within each chapter to encourage the reader to devote the necessary time and attention to study the entire textbook.

Once again, I thank Dr. Oliveira for having afforded me the privilege of writing this foreword. I thank all of the authors who have helped my alumnus, Dr. Oliveira, achieve such a superlative result as this textbook of anorectal physiology: a clinical and surgical perspective. I again congratulate Dr. Oliveira upon her tremendous accomplishment in creating what I am confident will become *the* reference standard textbook of anorectal physiology.

Most important, I congratulate the reader of this textbook on having made an exceptionally wise choice in selecting it as the best, most recent, and comprehensive educational resource for anorectal physiology.

Steven D. Wexner, MD, PhD(hon), FACS
Chair, Department of Colorectal Surgery
Cleveland Clinic Florida, Digestive Disease Center
Weston, FL, USA

Preface

Anorectal physiology and pelvic floor disorders are an interesting area of coloproctology, which allow for multispecialty interaction. New insights and theories to explain the mechanisms of continence and defecation are still challenging our understanding of functional disorders of the pelvic floor. Although quite developed over the last 20 years, anorectal physiology is still an area of medicine that offers opportunities for research and development of new treatment modalities. Dynamic imaging methods, including tridimensional ultrasound, are contributing to the assessment of the entire pelvis, as a unit. Conventional and high-resolution manometry as well as other methods to assess motility and neuromuscular integrity of the pelvic floor can help surgical decision-making and guide treatment algorithms. Neuromodulation by sacral or tibial nerve stimulation is a valuable and important tool for patients with both incontinence and constipation.

Five years of surgical residency in Brazil were not sufficient time to prepare me to address often challenging pelvic floor disorders. I was fortunate for the opportunity to spend time under the unique mentorship of Dr. Steven Wexner during a fellowship at the Cleveland Clinic Florida. It was a period dedicated to learning new and innovative ways to evaluate and assess patients with fecal incontinence and constipation. After returning to Brazil, I had the opportunity to set up the first Anorectal Physiology Department in Rio de Janeiro. I was blessed to evaluate patients and to share my experience and expertise with many fellows that rotated through my department. After 24 years of helping patients with a wide array of anorectal dysfunctions, I am happy to share my experiences, along with the world-renowned colorectal surgeons who have contributed to this book. It is my hope that this comprehensive book will contribute to a better understanding of anorectal and pelvic floor disorders, its mechanisms, and best treatment practices.

Rio de Janeiro, Brazil — Lucia Camara Castro Oliveira

Acknowledgments

I am very privileged to have met many physicians with a particular interest to pelvic floor disorders. All of the authors in this book share the same compassion and devotion for this subject, and I would like to thank them for the precious time spent in the process of writing their chapters. I also like to thank Elektra McDermott, a true friend and editorial consultant, for her precious help during all phases of the process of creating this book. I am sure that without her help, I would not be able to accomplish the challenge of editing a book in English. To my friends and family who, although not directly involved in this book, are always there for support, I would like to thank all of you for your patience and encouragement. Finally, a special thank you to my dearest friend and mentor, Dr. Steven Wexner, for all of his invaluable lessons and continued support throughout the last 25 years.

Contents

Contributors

Editor

Lucia Camara Castro Oliveira, MD, PhD, FASCRS Director Department of Anorectal Physiology of Rio de Janeiro and CEPEMED, Rio de Janeiro, Brazil

Authors

Nouf Y. Akeel, MD Department of Surgery, Faculty of Medicine, King Abdulaziz University, Jeddah, Saudi Arabia

Mauro Pereira de Azevedo, MD Marcilio Dias Navy Hospital - Brazil, Rio de Janeiro, Brazil

Rio de Janeiro Society of Anesthesiology, Rio de Janeiro, Brazil

Brazilian Society of Anesthesiology, Rio de Janeiro, Brazil

Mauro Bafutto, MD, MSc, PhD School of Medicine, Universidade Federal de Goiás, Internal Medicine Department, Goiania, Brazil

Bary Berghmans, PhD MSc RPT Pelvic Care Center Maastricht, Maastricht University Medical Centre, Maastricht, The Netherlands

Mitchell A. Bernstein, MD NYU Langone Medical Center, Division of Colon and Rectal Surgery, Department of Surgery, New York, NY, USA

Anuradha R. Bhama, MD Rush University, Division of Colorectal Surgery, Chicago, IL, USA

Birgit Bittorf, MD University Hospital Erlangen, Department of Surgery, Section of Coloproctology, Erlangen, Germany

Liliana Bordeianou, MD, MPH Center for Pelvic Floor Disorders, Section of Colon and Rectal Surgery, Massachusetts General Hospital, Boston, MA, USA

Alice Brandão, MD Fonte Imagem Medicina Diagnóstica, Rio de Janeiro, Brazil

Leonardo A. Bustamante-Lopez, MD, PhD Division of Coloproctology, Department of Gastroenterology, University of São Paulo, Hospital das Clinicas, São Paulo, Brazil

Laura Cancelliere, MD Department of Obstetrics and Gynecology, Faculty of Medicine, University of Toronto, Toronto, ON, Canada

Paul Cavallaro, MD Department of General Surgery, Massachusetts General Hospital, Boston, MA, USA

Peter Christensen, MD, DMSc Aarhus University Hospital, Aarhus, Denmark

Claudio Saddy Rodrigues Coy, MD, PhD Group of Coloproctology, Discipline of Diseases of the Digestive System, Department of Surgery – Campinas State University (UNICAMP), Campinas, SP, Brazil

J. Manuel Devesa, MD Colorectal Unit, Ruber International, General and Digestive Surgery, Madrid, Spain

Hossam Elfeki, MD, PhD Aarhus University Hospital, Aarhus, Denmark

Katrine J. Emmertsen, MD, PhD Aarhus University Hospital, Aarhus, Denmark

Henrique Sarubbi Fillmann, MD Hospital Sao Lucas-PUC-RS, Department of Surgery, Porto Alegre, Brazil

Eliane Garcez da Fonseca, MD, PhD Polyclinic Piquet Carneiro, Department of Pediatrics, Rio de Janeiro, Brazil

Raul Fonseca, MD Pelvic Floor Clinic of Guadalajara, Department of Functional Disorders, Zapopan, Mexico

Rogerio de Fraga, MD, PhD Hospital das Clinicas UFPR, Department of Surgery, Curitiba, Brazil

Ilario Froehner Jr, MD Division of Coloproctology, Department of Gastroenterology, University of São Paulo, Hospital das Clinicas, São Paulo, Brazil

Ezio Ganio, MD Colorectal Surgical Division, Humanitas San PioX, Via Nava 31, Milan, Lombardia, Italy

Darren M. Gold, MSc FRCS (Eng) St. Vincent's Hospital Clinical School, Department of Surgery, Sydney, UNSW, Australia

Tracy Hull, MD Cleveland Clinic, Department of Colorectal Surgery, Cleveland, OH, USA

Lawna Hunter San Diego, CA, USA

José Marcio N. Jorge, MD, PhD Division of Coloproctology, Department of Gastroenterology, University of São Paulo, Hospital das Clinicas, São Paulo, Brazil

T. Juul, Cand.Cur.PhD Aarhus University Hospital, Aarhus, Denmark

Charles H. Knowles, MA (cantab), PhD National Bowel Research Centre, Blizard Institute, Barts and the London School of Medicine and Dentistry, London, UK

Kinshuk Kumar, MD Department of Obstetrics and Gynecology, Faculty of Medicine, University of Toronto, Toronto, ON, Canada

Sergio Larach, MD Endosurgical Center of Florida, Orlando, FL, USA

Rebeca Kisel Laska, MD Clinica del Piso Pelvico, Department of Urogynecology, Zapopan, Mexico

Soren Laurberg, MD, DMSc Aarhus University Hospital, Aarhus, Denmark

Janet T. Lee, MD, MS University of Minnesota, Department of Surgery, Saint Paul, MN, USA

Jason Lei, MD Jefferson Health, Philadelphia, PA, USA

Nucelio L. B. M. Lemos, MD Faculty of Medical Sciences, Santa Casa de São Paulo, Department of Gynecology, UNIFESP-EPM, Department of Neurodisorders in Gynecology, Sao Paulo, Brazil

Adrienne Li, MD Division of Urogynecology, Department of Obstetrics and Gynecology, Mount Sinai Hospital and Women's College Hospital, University of Toronto, Toronto, ON, Canada

José Arturo Aragón López, MD Hospital Espanol de Veracruz, Department of Surgery, Veracruz, Mexico

Robert D. Madoff, MD University of Minnesota, Department of Surgery, Minneapolis, MN, USA

Beatriz Martin-Perez, MD Hospital Clinic, Gastrointestinal Surgery, Barcelona, Spain

Klaus E. Matzel, MD University Hospital Erlangen, Department of Surgery, Section of Coloproctology, Erlangen, Germany

Anders Mellgren, MD, PhD Division of Colon & Rectal Surgery, University of Illinois at Chicago, Chicago, IL, USA

Andre da Luz Moreira, MD University Hospital Pedro Ernesto, Department of Colorectal Surgery, Rio de Janeiro, Brazil

Helio Moreira Jr, MD, PhD Surgical Department, Colorectal Unit, School of Medicine, Federal University of Goiás, Goiânia, Goiás, Brazil

Sthela M. Murad-Regadas, MD, PhD, TSBCP University Hospital, School of Medicine, Federal University of Ceara, Anorectal Physiology and Pelvic Floor Unit, Sao Carlos Hospital, Ceara, Brazil

María Margarita Murphy, MD East Cooper Regional Medical Center, Department of Colon and Rectal Surgery, Mount Pleasant, SC, USA

Isaac José Felippe Corrêa Neto, MD, MSc Department of Gastroenterology, Division of Colorectal Surgery, University of São Paulo, São Paulo, Brazil

Salustiano Gabriel Neto, MD, MSc School of Medicine, Universidade Federal de Goiás, Department of Surgery, Goiania, Brazil

Colleen Nofi, MS, MBA New York Institute of Technology College of Osteopathic Medicine, Huntington, NY, USA

Anelise Oliveira, MD Diimagem Grupo Alliar, Campo Grande, MS, Brazil

Enio Chaves de Oliveira, MD, MSc, PhD School of Medicine, Universidade Federal de Goiás, Department of Surgery, Goiania, Brazil

Paulo Palma, MD, PhD UNICAMP, Department of Urology, Sao Paulo, Brazil

Peter Petros, DSc FRCOG St. Vincent's, Department of Surgery, Sydney, NSW, Australia

Johann Pfeifer, MD Department of General Surgery, University Surgical Clinic, Medical University of Graz, Graz, Austria

Andrea Povedano, MD, PhD Universidad Federal do Estado do Rio de Janeiro, Department of General Surgery, Rio de Janeiro, Brazil

Francisco Sérgio Regadas, MD, PhD, TSBCP School of Medicine, Federal University of Ceara, Ceara, Brazil

David E. Rivadeneira, MD, MBA Northwell Health, Huntington Hospital, Hofstra School of Medicine, Huntington, NY, USA

Virginia C. Roncatti, MD Department of Urogynecology and Pelvic Reconstruction, Heliópolis Hospital, Sao Paulo, Brazil

Guillermo O. Rosato, MD Hospital Universitario Austral, Department of Colorectal Surgery, Buenos Aires, Argentina

Gonzalo Hagerman Ruíz-Galindo, MD Centro Medico ABC, Department of Colorectal Surgery/General Surgery, Mexico City, Mexico

Renato Caram Saad, MD Sirio-Libanese Hospital, Department of Surgery, São Paulo, Brazil

Mara Rita Salum, MD Sirio-Libanes and Albert Einstein Hospital, Department of Surgery, São Paulo, Brazil

Maura R. Seleme, PhD, RPT Faculdade Inspirar, Department of Pelvic Physiotherapy, Curitiba, Brazil

Corey Sermer, MSc Faculty of Medicine, University of Toronto, Department of Obstetrics and Gynecology, Mount Sinai Hospital and Women's College Hospital, Toronto, ON, Canada

Ali A. Shafik, MD, PhD Chairman and Professor of Colorectal Surgery Department, Faculty of Medicine, Cairo University, Cairo, Egypt

Giovanna Da Silva, MD Department of Colorectal Surgery, Cleveland Clinic Florida, Weston, FL, USA

Alexandre Barcelos Morais da Silveira, MD, MSc, PhD Universidade Federal de Uberlândia, Anatomy Department, Uberlandia, Brazil

Jorge Daniel Silva-Velazco, MD Centro Medico ABC, Department of Colorectal Surgery/General Surgery, Mexico City, Mexico

Anne Sirany, MD University of Minnesota, Department of Surgery, Minneapolis, MN, USA

Carlos Walter Sobrado, MD, PhD Department of Gastroenterology, Division of Colorectal Surgery, University of São Paulo, São Paulo, Brazil

Lucas Faraco Sobrado, MD Department of Gastroenterology, Division of Colorectal Surgery, University of São Paulo, São Paulo, Brazil

Scott R. Steele, MD, MBA Department of Colorectal Surgery, Digestive Disease and Surgery Institute, Cleveland Clinic Foundation, Cleveland, OH, USA

Adam Studniarek, MD Division of Colon & Rectal Surgery, University of Illinois at Chicago, Chicago, IL, USA

Silvana Uchôa, MSc, RPT Universidade Catolica de Pernambuco, Recife, Brazil

Sarah A. Vogler, MD, MBA University of Minnesota, Department of Surgery, Edina, MN, USA

James Elvis Waha, MD Department of General Surgery, University Surgical Clinic, Medical University of Graz, Graz, Austria

Steven D. Wexner, MD, PhD(hon), FACS Chair, Department of Colorectal Surgery, Cleveland Clinic Florida, Digestive Disease Center, Weston, FL, USA

Carlos Zaslavsky, MD, PhD Santa Casa of Porto Alegre, Department of Digestive Motility, Porto Alegre, Brazil

Massarat Zutshi, MD Associate Professor of Surgery Lerner College of Medicine of Case Wesetern Reserve University, Staff Surgeon, Department of Colorectal Surgery, Cleveland Clinic Foundation, Cleveland, OH, USA

1 Anatomy of the Anorectal Region and Pelvic Floor

José Marcio N. Jorge, Leonardo A. Bustamante-Lopez, and Ilario Froehner Jr

Introduction

The study of the anatomy of the rectum and anus is described since 1543 by Andreas Vesalius through anatomic dissections [1]. However, the anatomy of the rectum, anal canal, and pelvic floor is so intrinsically related to its physiology that much can be appreciated only in the living. Thus, it is a region in which the surgeon has an advantage over the anatomist through in vivo dissection, physiologic investigation, and endoscopic examination. On the other hand, anatomy of the pelvis is also challenging to the surgeon: the pelvis is a narrow space, packed with intestinal, urologic, gynecologic, vascular, and neural structures, all confined within a rigid and deep osseous–muscular cage. Whereas, detailed anatomy of this region is difficult to learn in the setting of an operating room, and it demands not only observations in vivo but also anatomy laboratory studies, including dissections of humans and animals, with in-depth descriptions and drawings and sometimes associated with physiologic evaluation. Based on these studies, some controversial concepts of the anorectal anatomy have been actually changed [2–8]. In addition, virtual reality models have been designed to improve visualization of three-dimensional structures and teach more properly anatomy, pathology, and surgery of the anorectum and pelvic floor [9].

J. M. N. Jorge (✉) · L. A. Bustamante-Lopez
I. Froehner Jr
Division of Coloproctology, Department of Gastroenterology, University of São Paulo, Hospital das Clinicas, São Paulo, Brazil

Anatomy of the Anus and Rectum

Anal Canal Structure, Anus, and Anal Verge

The anal canal is anatomically peculiar and has a complex physiology, which accounts for its crucial role in continence and, in addition, its susceptibility to a variety of diseases.

The anus or anal orifice is an anteroposterior cutaneous slit that along with the anal canal remains virtually closed at rest, as a result of tonic circumferential contraction of the sphincters and the presence of anal cushions. The edge of the anal orifice, the anal verge or margin (anocutaneous line of Hilton), marks the lowermost edge of the anal canal and is sometimes the level of reference for measurements taken during endoscopy examination. The dentate line is, however, considered a more precise landmark . The difference between the anal verge and the dentate line is usually 1–2 cm. The epithelium distal to the anal verge acquires hair follicles, glands, including apocrine glands, and other features of normal skin, and it is the source of perianal hidradenitis suppurativa, an inflammation of the apocrine glands.

L. C. C. Oliveira (ed.), *Anorectal Physiology*, https://doi.org/10.1007/978-3-030-43811-1_1

The anal sphincter is a multilayered cylindrical structure, with the innermost layer being the anal lining, with the subsequent layers: internal sphincter, the fat-containing intersphincteric space with the longitudinal layer, and subsequently the outer striated muscle layer. The latter constitutes the sling-like puborectalis muscle for the upper half and the cylindrical external sphincter for the lower half.

Anatomic Versus Surgical Anal Canal

Two definitions are found describing the anal canal (Fig. 1.1). The "anatomic" or "embryologic" anal canal is only 2.0 cm long, extending from the anal verge to the dentate line, the level that corresponds to the proctodeal membrane. The "surgical" or "functional" anal canal is longer, extending for approximately 4.0 cm (in men) from the anal verge to the anorectal ring (levator ani). This "long anal canal" concept was first introduced by Milligan and Morgan [10] and has been considered, in spite of not being proximally marked by any apparent epithelial or developmental boundary, useful both as a physiological and surgical parameter. The anorectal ring is at the level of the distal end of the ampullary part of the rectum and forms the anorectal angle and the beginning of a region of higher intraluminal pressure. Therefore, this definition correlates with digital, manometric, and sonographic examinations.

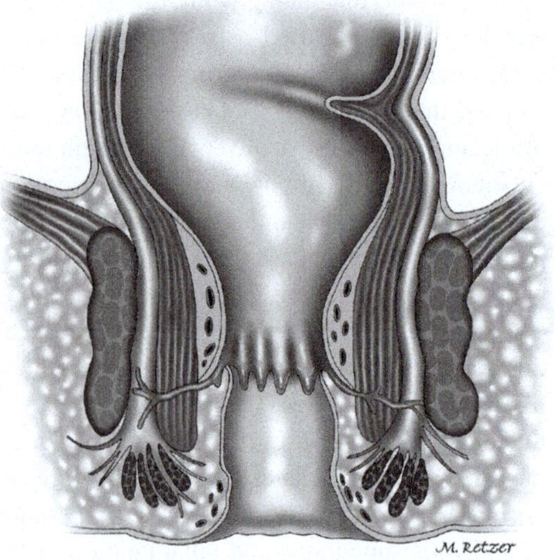

Fig. 1.1 Anal canal. In memorium Marcos Retzer (illustrator)

Anatomic Relations of the Anal Canal

Posteriorly, the anal canal is related to the coccyx and anteriorly to the perineal body and the lowest part of the posterior vaginal wall in the female, and to the urethra in the male. The ischium and the ischiorectal fossa are situated on either side. The ischiorectal fossa contains fat and the inferior rectal vessels and nerves, which cross it to enter the wall of the anal canal.

Anal Sphincter Support

General support is provided by the fibroelastic network which is present within the anal sphincter and is continuous with the network outside the sphincter traversing the perianal fat. This network arises from the connective tissue within the longitudinal layer (conjoined longitudinal coat). The network extends through the sphincters, interlacing with each other as well as with the perimysium and endomysium to the pelvic side wall to connect with the caudal levator fascia and to the perianal skin, thus anchoring the anus within the pelvic cavity.

Additional support is given anteriorly by the perineal body and its attachments and by supportive structures in the anovaginal septum in females and Denonvilliers' fascia in males.

Lateral support is given by the levator ani muscle and superficial transverse perineal muscles. Posterior support is given by the attachment of the anococcygeal ligament to the coccyx and superiorly by the continuity with the rectum.

Muscles of the Anal Canal

Internal Anal Sphincter

The internal anal sphincter represents the distal 2.5–4.0 cm condensation of the circular muscle layer of the rectum. As a consequence of both intrinsic myogenic and extrinsic autonomic neurogenic properties, the internal anal sphincter is a smooth muscle in a state of continuous maximal contraction and represents a natural barrier to the involuntary loss of stool and gas.

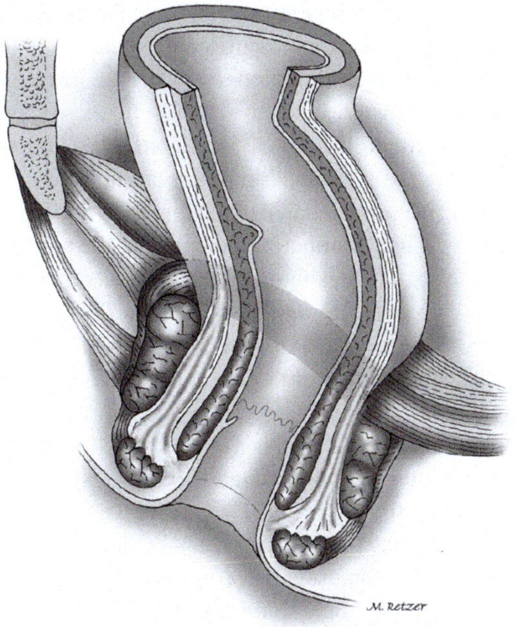

Fig. 1.2 Muscles of the anal canal. In memorium Marcos Retzer (illustrator)

The lower rounded edge of the internal anal sphincter can be felt on physical examination, about 1.2 cm distal to the dentate line. The groove between the internal and external anal sphincter, the intersphincteric sulcus, can be visualized or easily palpated. Endosonographically, the internal anal sphincter is a 2- to 3-mm-thick circular band and shows a uniform hypoechogenicity (Fig. 1.2).

External Anal Sphincter

The external anal sphincter was initially described as encompassing three divisions: subcutaneous, superficial, and deep [10]. Goligher et al. [11] described the external anal sphincter as a simple, continuous sheet that forms, along with the puborectalis and levator ani, one funnel-shaped skeletal muscle.

The external sphincter is a cylindrical striated muscle under voluntary control and comprises predominantly slow-twitch muscle fibers, capable of prolonged contraction.

This muscle envelops the entire length of the inner tube of smooth muscle, but it ends slightly more distal than the internal anal sphincter. The external sphincter has a thickness of 4 mm on endoluminal imaging. A decrease in the thickness of the external sphincter in men with age has been demonstrated. In females this is also most likely in normal aging; however, when coinciding with external sphincter defects, this may lead to incontinence. It extends approximately 1 cm beyond the internal sphincter and has posterior fibers continuous with the anococcygeal ligament. Some of the anterior fibers decussate into the superficial transverse perineal muscles and perineal body. The deepest part of the external anal sphincter is intimately related to the puborectalis muscle, which can be actually considered a component of both the levator ani and the external anal sphincter muscle complexes. Others considered the external anal sphincter as being subdivided into two parts, deep (deep sphincter and puborectalis) and superficial (subcutaneous and superficial sphincter) [6, 12, 13]. Shafik [14] proposed the three *U*-shaped loop system, but clinical experience has not supported this schema. The external anal sphincter is more likely to be one muscle unit, attached by the anococcygeal ligament posteriorly to the coccyx and anteriorly to the perineal body, not divided into layers or laminae. Nevertheless, differences in the arrangement of the external anal sphincter have been described between the sexes [15]. In the male, the upper half of the external anal sphincter is enveloped anteriorly by the conjoined longitudinal muscle, while the lower half is crossed by it. In the female, the entire external anal sphincter is encapsulated by a mixture of fibers derived from both longitudinal and internal anal sphincter muscles (Fig. 1.2).

Endosonographically, the puborectalis and the external anal sphincter, despite their mixed linear echogenicity, are both predominantly hyperechogenic, with a mean thickness of 6 mm (range, 5–8 mm). Distinction is made by position, shape, and topography. Recently, both anal endosonography and endocoil magnetic resonance imaging have been used to detail the anal sphincter complex in the living healthy subjects [16–19]. These tests provide a three-dimensional mapping of the anal sphincter; they help to study the differences in the arrangement of the external anal sphincter between the sexes and uncover sphincter disruption or defect during vaginal deliveries. In addition, there is some degree of "anatomical asymmetry" of the external anal sphincter, which accounts for both radial and longitudinal "functional asymmetry" observed during anal manometry [20].

The automatic continence mechanism is formed by the resting tone, maintained by the internal anal sphincter, magnified by voluntary reflex, and resting external anal sphincter contractile activities. In response to conditions of threatened incontinence, such as increased intra-abdominal pressure and rectal distension, the external anal sphincter and puborectalis reflexively and voluntarily contract further to prevent fecal leakage. Because of muscular fatigue, maximal voluntary contraction of the external anal sphincter can be sustained for only 30–60 seconds. However, the external anal sphincter and the pelvic floor muscles, unlike other skeletal muscles, which are usually inactive at rest, maintain unconscious resting electrical tone through a reflex arc at the cauda equina level. Histologic studies have shown that the external anal sphincter, puborectalis, and levator ani muscles have a predominance of type I fibers, which are a peculiarity of skeletal muscles connecting tonic contractile activity [21]. The external sphincter has a nerve supply by the inferior rectal branch of the pudendal nerve (S-2, S-3) and the perineal branch of the fourth sacral nerve (S-4).

Conjoined Longitudinal Muscle

Whereas the inner circular layer of the rectum gives rise to the internal anal sphincter, the outer longitudinal layer, at the level of the anorectal ring, mixes with fibers of the levator ani muscle to form the conjoined longitudinal muscle [22]. This muscle descends between the internal and external anal sphincter, and ultimately some of its fibers, referred to as the *corrugator cutis ani muscle*, traverse the lowermost part of the external anal sphincter to insert into the perianal skin. Some of these fibers may enter the fat of the ischiorectal fossa [23]. Other sources for the striated component of the conjoined longitudinal muscle include the puborectalis and deep external anal sphincter, the pubococcygeus and top loop of the external anal sphincter, and the lower fibers of the puborectalis [7, 24, 25]. In its descending course, the conjoined longitudinal muscle may give rise to medial extensions that cross the internal anal sphincter to contribute the smooth muscle of the submucosa (*musculus canalis ani, sustentator tunicae mucosae, Treitz muscle, musculus submucosae ani*) (Fig. 1.2) [26].

Possible functions of the conjoined longitudinal muscle include attaching the anorectum to the pelvis and acting as a skeleton that supports and binds the internal and external sphincter complex together [23]. Haas and Fox [27] consider that the meshwork formed by the conjoined longitudinal muscle may minimize functional deterioration of the sphincters after surgical division and act as a support to prevent hemorrhoidal and rectal prolapse. In addition, the conjoined longitudinal muscle and its extensions to the intersphincteric plane divide the adjacent tissues into subspaces and may actually play a role in the septation of thrombosed external hemorrhoids and containment of sepsis [7]. Finally, Shafik [24] ascribes to the conjoined longitudinal muscle the action of shortening and widening of the anal canal as well as eversion of the anal orifice and proposed the term *evertor ani muscle*. This is controversial. In addition to this primary function during defecation, a limited role in anal continence, specifically a potentialization effect in maintaining an anal seal, has also been proposed [24].

Histology of the Anal Canal

The lining of the anal canal consists of an upper mucosal (endoderm) and a lower cutaneous (ectoderm) segment (Figs. 1.1 and 1.2). The dentate (pectinate) line is the "saw-toothed" junction between these two distinct origins of venous and lymphatic drainage, nerve supply, and epithelial lining. Above this level, the intestine is innervated by the sympathetic and parasympathetic systems, with venous, arterial, and lymphatic drainage to and from the hypogastric vessels.

Distal to the dentate line, the anal canal is innervated by the somatic nervous system, with blood supply and drainage from the inferior hemorrhoidal system. These differences are important when the classification and treatment chosen for hemorrhoids are considered.

The pectinate or dentate line corresponds to a line of anal valves that represent remnants of the proctodeal membrane. Above each valve, there is a little pocket known as an anal sinus or crypt. These crypts are connected to a variable number of glands, in average 6 (range, 3–12) [28, 29]. The anal glands first described by Chiari [30] in 1878 are more concentrated in the posterior quadrants. More than one gland may open into the same crypt, while half the crypts have no communication. The anal gland ducts, in an outward and downward route, enter the submucosa; two-thirds enter the internal anal sphincter, and half of them terminate in the intersphincteric plane [29]. Obstruction of these ducts, presumably by accumulation of foreign material in the crypts, may lead to perianal abscesses and fistulas [31]. Cephalad to the dentate line, 8–14 longitudinal folds, known as the rectal columns (columns of Morgagni), have their bases connected in pairs to each valve at the dentate line. At the lower end of the columns are the anal papillae. The mucosa in the area of the columns consists of several layers of cuboidal cells and has a deep purple color because of the underlying internal hemorrhoidal plexus. This 0.5- to 1.0-cm strip of mucosa above the dentate line is known as the anal transition or cloacogenic zone. Cephalad to this area, the epithelium changes to a single layer of columnar cells and macroscopically acquires the characteristic pink color of the rectal mucosa.

The cutaneous part of the anal canal consists of modified squamous epithelium that is thin, smooth, pale, stretched, and devoid of hair and glands. The terms pecten and pecten band have been used to define this segment [32]. However, as pointed out by Goligher, the round band of fibrous tissue called pecten band, which is divided in the case of anal fissure (pectenotomy), probably represents the spastic internal anal sphincter [11, 33].

Anorectal Spaces

There are several spaces around the rectum and anal canal that are clinically significant.

These spaces normally contain loose areolar tissue or fat. These spaces include ischiorectal, perianal, intersphincteric, submucous, superficial postanal, deep postanal, supralevator, and retrorectal spaces.

The intersphincteric space exists between internal and external sphincter muscles and is contiguous with the supralevator space superiorly, which is covered by peritoneum. It is important in the genesis of perianal abscess, because most of the anal glands end in this space. Lateral to the external sphincter lies the triangular ischioanal space which is bordered superiorly by the levator ani muscle.

Posteriorly, the most caudal space is the superficial postanal space that terminates at the coccyx.

Above the superficial postanal space is the anococcygeal ligament, and deep to this ligament, but below the levator ani muscle, is the deep postanal space of Courtney (retrosphincteric space). This space is continuous laterally with each ischioanal space and when infected can create a large "horseshoe" abscess.

Above the levator ani, below and posterior to the rectum, and anterior and superior to the sacrum is the supralevator space that can extend into the retroperitoneum.

The ischiorectal fossa is subdivided by a thin horizontal fascia into two spaces: the perianal and ischiorectal. The ischiorectal space comprises the upper two-thirds of the ischiorectal fossa. It is pyramid-shaped, situated on both sides between the anal canal and the lower part of the rectum medially and the side wall of the pelvis laterally [34]. The apex is at the origin of the levator ani muscle from the obturator fascia; the base is the perianal space. Anteriorly, the fossa is bounded by the urogenital diaphragm and transversus perinei muscle. Posterior to the ischiorectal fossa is the sacrotuberous ligament and the inferior border of the gluteus maximus. On the superolateral wall, the pudendal nerve and the internal pudendal vessels run in the pudendal canal (Alcock's canal). The ischiorectal fossa contains fat and the inferior rectal vessels and nerves.

The perianal space surrounds the lower part of the anal canal and contains the external hemorrhoidal plexus, the subcutaneous part of the external anal sphincter, the lowest part of the internal anal sphincter, and fibers of the longitudinal muscle. This space is the typical site of anal hematomas, perianal abscesses, and anal fistula tracts. The perianal space is continuous with the subcutaneous fat of the buttocks laterally and extends into the intersphincteric space medially. The submucous space is situated between the internal anal sphincter and the mucocutaneous lining of the anal canal. This space contains the internal hemorrhoidal plexus and the muscularis submucosae ani. Above, it is continuous with the submucous layer of the rectum, and, inferiorly, it ends at the level of the dentate line.

The supralevator spaces are situated between the peritoneum superiorly and the levator ani inferiorly. Medially, these bilateral spaces are limited by the rectum and laterally by the obturator fascia. Supralevator abscesses may occur as a result of upward extension of a cryptoglandular infection or develop from a pelvic origin. The retrorectal space is located between the fascia propria of the rectum anteriorly and the presacral fascia posteriorly. Laterally are the lateral rectal ligaments and inferiorly the rectosacral ligament, and above the space is continuous with the retroperitoneum. The retrorectal space is a site for embryologic remnants and rare presacral tumors (Fig. 1.3) [35].

Rectum

Both proximal and distal limits of the rectum are controversial: the rectosigmoid junction is considered to be at the level of the third sacral vertebra by anatomists but at the sacral promontory by surgeons, and likewise, the distal limit is regarded to be the muscular anorectal ring by surgeons and the dentate line by anatomists [36, 37]. The rectum measures 12–15 cm in length and has three lateral curves: the upper and lower are convex to the right, and the middle is convex to the left [38]. These curves correspond intraluminally to the folds or valves of Houston. The rectum has two or three curves within its lumen, created by submucosal folds called the valves of Houston. The peritoneum covers the upper two-thirds of the rectum anteriorly but only the upper third laterally. The two left-sided folds are usually noted at 7–8 cm and at 12–13 cm, respectively, and the one on the right is generally at 9–11 cm. The middle valve (Kohlrausch's plica) is the most consistent in presence and location and corresponds to the level of the anterior peritoneal reflection. The rectal valves do not contain all the muscle wall layers and do not have a specific function. However, from a clinical point of view, they represent adequate locations for performing a rectal biopsy, as they are readily accessible with minimal risk for perforation [13, 39]. The valves of Houston are absent after mobilization of the rectum, and this is attributed to the 5-cm

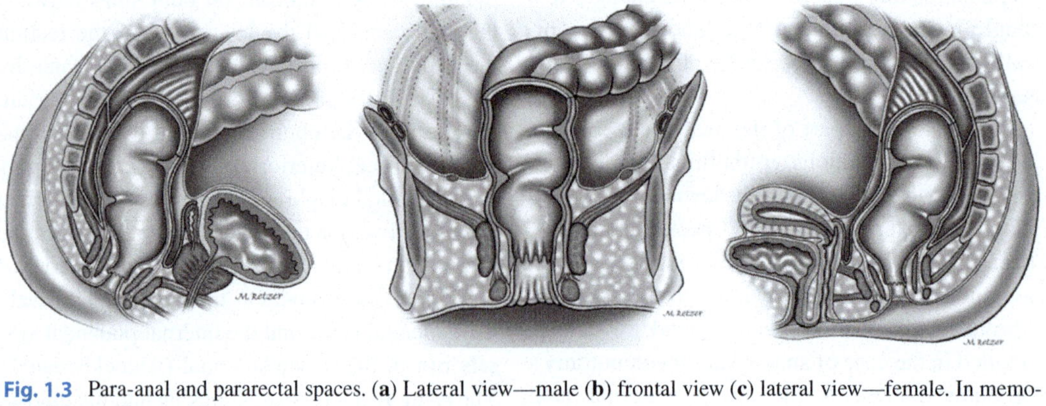

Fig. 1.3 Para-anal and pararectal spaces. (**a**) Lateral view—male (**b**) frontal view (**c**) lateral view—female. In memorium Marcos Retzer (illustrator)

length gained following complete surgical dissection. The rectal mucosa is smooth, pink, and transparent, which allows visualization of small and large submucosal vessels. This characteristic "vascular pattern" disappears in inflammatory conditions and in melanosis coli.

The rectum is characterized by its wide, easily distensible lumen and the absence of taeniae, epiploic appendices, haustra, or a well-defined mesentery. The prefix "meso," in gross anatomy, refers to two layers of peritoneum that suspend an organ. Normally the rectum is not suspended but entirely extraperitoneal on its posterior aspect and closely applied to the sacral hollow. Consequently, the term "mesorectum" is anatomically inapplicable. An exception, however, is that a peritonealized mesorectum may be noted in patients with procidentia. But, the word "mesorectum" has gained widespread popularity among surgeons to address the perirectal areolar tissue, which is thicker posteriorly, containing terminal branches of the inferior mesenteric artery and enclosed by the fascia propria [40, 41]. The "mesorectum" may be a metastatic site for a rectal cancer and is removed during surgery for rectal cancer without neurologic sequelae, as no functionally significant nerves pass through it.

The upper third of the rectum is anteriorly and laterally invested by peritoneum; the middle third is covered by peritoneum on its anterior aspect only. Finally, the lower third of the rectum is entirely extraperitoneal, as the anterior peritoneal reflection occurs at 9.0–7.0 cm from the anal verge in men and at 7.5–5.0 cm from the anal verge in women.

Relations of the Rectum

The rectum occupies the sacral concavity and ends 2–3 cm anteroinferior from the tip of the coccyx. At this point it angulates backward sharply to pass through the levators and becomes the anal canal. Anteriorly, in women, the rectum is closely related to the uterine cervix and posterior vaginal wall; in men it lies behind the bladder, vas deferens, seminal vesicles, and prostate. Posterior to the rectum lie the median sacral vessels and the roots of the sacral nerve plexus.

Fascial Relationship of the Rectum

The parietal endopelvic fascia lines the walls and floor of the pelvis and continues on the internal organs as a visceral pelvic fascia (Fig. 1.3) [42, 43]. Thus, the *fascia propria of the rectum* is an extension of the pelvic fascia, enclosing the rectum, fat, nerves, and the blood and lymphatic vessels. It is more evident in the posterior and lateral extraperitoneal aspects of the rectum.

The *lateral ligaments* or *stalks of the rectum* are distal condensations of the pelvic fascia that form a roughly triangular structure with a base on the lateral pelvic wall and an apex attached to the lateral aspect of the rectum [33]. Still subject of misconception, the lateral stalks are comprised essentially of connective tissue and nerves and that the middle rectal artery does not traverse the lateral stalks of the rectum. Branches, however, course through in approximately 25% of cases [44]. Consequently, division of the lateral stalks during rectal mobilization is associated with a 25% risk for bleeding. Although the lateral stalks do not contain important structures, the middle rectal artery and the pelvic plexus are both closely related, running, at different angles, underneath it [45]. One theoretical concern in ligation of the stalks is leaving behind lateral mesorectal tissue, which may limit adequate lateral or mesorectal margins during cancer surgery [40, 41, 46].

The *presacral fascia* is a thickened part of the parietal endopelvic fascia that covers the concavity of the sacrum and coccyx, nerves, the middle sacral artery, and presacral veins. Operative dissection deep to the presacral fascia may cause troublesome bleeding from the underlying presacral veins. Presacral hemorrhage occurs as frequently as 4.6–7.0% of resections for rectal neoplasms and, despite its venous nature, can be life-threatening [47–49]. This is a consequence of two factors: the difficulty in securing control because of retraction of the vascular stump into the sacral foramen and the high hydrostatic pressure of the presacral venous system. The presacral veins are avalvular and communicate via basivertebral veins with the internal vertebral venous system. The adventitia of the basivertebral veins adheres firmly to the sacral periosteum at the level of the ostia of the sacral foramina, mainly at the level of S-3–S-4. With the patient in the lithotomy position, the presacral veins can attain hydrostatic pressures of 17–23 cm H_2O, two to three times the normal pressure of the inferior vena cava [48].

The *rectosacral fascia* is an anteroinferior directed thick fascial reflection from the presacral fascia at the S-4 level to the fascia propria of the rectum just above the anorectal ring [49]. The rectosacral fascia, classically known as the fascia of Waldeyer, is an important landmark during posterior rectal dissection [2, 50].

The *visceral pelvic fascia of Denonvilliers* is a tough investing fascia that separates the extraperitoneal rectum anteriorly from the prostate and seminal vesicles or the vagina [50]. Therefore, three structures lie between the anterior rectal wall and the seminal vesicles and prostate: "anterior mesorectum," fascia propria of the rectum, and Denonvilliers' fascia. A consensus has generally been reached about the anatomy of the plane of posterior and lateral rectal dissection, but anteriorly, the matter is more controversial. The anterior plane of rectal dissection may not necessarily follow the same plane of posterior and lateral dissection, and the use of the terms close rectal, mesorectal, and extramesorectal has been recently suggested to describe the available anterior planes [51]. The close rectal or perimuscular plane lies inside the fascia propria of the rectum, and therefore, it is more difficult and bloody than the mesorectal plane. The mesorectal plane represents the continuation of the same plane of posterior and lateral dissection of the rectum. This is a natural anatomical plane and consequently more appropriate for most rectal cancers. Finally, the extra mesorectal plane involves resection of the Denonvilliers' fascia, with exposure of prostate and seminal vesicles, and associated with high risk of mixed parasympathetic and sympathetic injury due to damage of the periprostatic plexus.

Identification of the ureters is advisable to avoid injury to their abdominal or pelvic portions during colorectal operations. On both sides, the ureters rest on the psoas muscle in their inferomedial course; they are crossed obliquely by the spermatic vessels anteriorly and the genitofemoral nerve posteriorly. In its pelvic portion, the ureter crosses the pelvic brim in front of or a little lateral to the bifurcation of the common iliac artery and descends abruptly between the peritoneum and the internal iliac artery. Before entering the bladder in the male, the vas deferens crosses lateromedially on its superior aspect. In the female, as the ureter traverses the posterior layer of the broad ligament and the parametrium close to the side of the neck of the uterus and upper part of the vagina, it is enveloped by the vesical and vaginal venous plexuses and is crossed above and lateromedially by the uterine artery (Fig. 1.4).

Arterial Supply of the Rectum and Anal Canal

The superior hemorrhoidal artery is the continuation of the inferior mesenteric artery, once it crosses the left iliac vessels [52]. The artery descends in the sigmoid mesocolon to the level of S-3 and then to the posterior aspect of the rectum. In 80% of cases, it bifurcates into right, usually wider, and left terminal branches; multiple branches are present in 17% [53]. These divisions, once within the submucosa of the rectum, run straight downward to supply the lower rectum and the anal canal. Approximately five branches reach the level of the rectal columns and condense in capillary plexuses, mostly at the right posterior, right anterior, and left lateral positions, corresponding to the location of the major internal hemorrhoidal groups [54, 55].

The superior and inferior hemorrhoidal arteries represent the major blood supply to the anorectum. In addition, it is also supplied by the internal iliac arteries [56].

The contribution of the middle hemorrhoidal artery varies with the size of the superior hemorrhoidal artery; this may explain its controversial anatomy. Some authors report absence of the middle hemorrhoidal artery in 40–88% [57, 58], whereas others identify it in 94–100% of specimens [53]. It originates more commonly from the anterior division of the internal iliac or the pudendal arteries and reaches the rectum. The middle hemorrhoidal artery reaches the lower third of the rectum anterolaterally, close to the

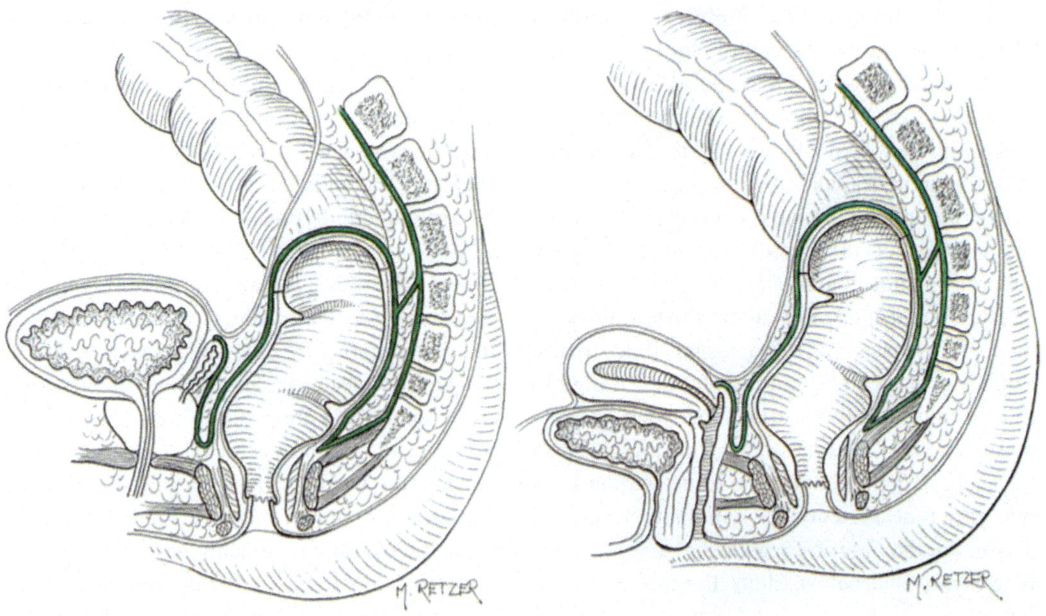

Fig. 1.4 Fascial relationships of the rectum: (**a**) male (**b**) female. In memorium Marcos Retzer (illustrator)

level of the pelvic floor and deep to the levator fascia. It therefore does not run in the lateral ligaments, which are inclined posterolaterally [2]. The middle hemorrhoidal artery is more prone to be injured during low anterior resection, when anterolateral dissection of the rectum is performed close to the pelvic floor and the prostate and seminal vesicles or upper part of the vagina are being separated [45]. The anorectum has a profuse intramural anastomotic network, which probably accounts for the fact that division of both superior and middle hemorrhoidal arteries does not result in necrosis of the rectum.

The paired inferior hemorrhoidal arteries are branches of the internal pudendal artery, which in turn is a branch of the internal iliac artery. The inferior hemorrhoidal artery arises within the pudendal canal and is throughout its course entirely extrapelvic. It traverses the obturator fascia, the ischiorectal fossa, and the external anal sphincter to reach the submucosa of the anal canal, ultimately ascending in this plane. Klosterhalfen et al. [4] performed postmortem angiographic, manual, and histologic evaluations and demonstrated that in 85% of cases, the posterior commissure was less well perfused than were the other sections of the anal canal. In addition, the blood supply could be jeopardized by contusion of the vessels passing vertically through the muscle fibers of the internal anal sphincter with increased sphincter tone. The resulting decreased blood supply could lead to ischemia at the posterior commissure, in a pathogenetic model of primary anal fissure.

Venous Drainage and Lymphatic Drainage of the Rectum and Anal Canal

The anorectum also drains, via middle and inferior hemorrhoidal veins, to the internal iliac vein and then to the inferior vena cava. Although it is still a controversial subject, the presence of communications among these three venous systems may explain the lack of correlation between portal hypertension and hemorrhoids [59]. The paired inferior and middle hemorrhoidal veins and the single superior hemorrhoidal vein originate from three anorectal arteriovenous plexuses. The external hemorrhoidal plexus, situated subcutaneously around the anal canal below the dentate line, constitutes when dilated the external hemorrhoids. The internal hemorrhoidal plexus is situated submucosally, around the upper anal canal and above the dentate line. The internal hemorrhoids originate from this plexus. The perirectal or perimuscular rectal plexus drains to the middle and inferior hemorrhoidal veins.

Lymph from the upper two-thirds of the rectum drains exclusively upward to the inferior mesenteric nodes and then to the para-aortic nodes. Lymphatic drainage from the lower third of the rectum occurs not only cephalad, along the superior hemorrhoidal and inferior mesentery arteries, but also laterally, along the middle hemorrhoidal vessels to the internal iliac nodes. Studies using lymphoscintigraphy have failed to demonstrate communications between inferior mesenteric and internal iliac lymphatics [60]. In the anal canal, the dentate line is the landmark for two different systems of lymphatic drainage: above, to the inferior mesenteric and internal iliac nodes, and below, along the inferior rectal lymphatics to the superficial inguinal nodes, or less frequently along the inferior hemorrhoidal artery. In the female, drainage at 5 cm above the anal verge in the female lymphatic may also spread to the posterior vaginal wall, uterus, cervix, broad ligament, fallopian tubes, ovaries, and cul-de-sac, and at 10 cm above the anal verge, spread seems to occur only to the broad ligament and cul-de-sac [61].

Innervation of the Rectum and Anal Canal

The sympathetic supply of the rectum and the left colon arises from L-1, L-2, and L-3 (Fig. 1.4). Preganglionic fibers, via lumbar sympathetic nerves, synapse in the preaortic plexus, and the postganglionic fibers follow the branches of the inferior mesenteric artery and superior rectal artery to the left colon and upper rectum. The lower rectum is innervated by the presacral nerves, which are formed by fusion of the aortic plexus and lumbar splanchnic nerves. Just below the sacral promontory, the presacral nerves form the hypogastric plexus (or superior hypogastric plexus). Two main hypogastric nerves, on either side of the rectum, carry sympathetic innervation from the hypogastric plexus to the pelvic plexus. The pelvic plexus lies on the lateral side of the pelvis at the level of the lower third of the rectum, adjacent to the lateral stalks.

The parasympathetic fibers to the rectum and anal canal emerge through the sacral foramen and are called the nervi erigentes (S-2, S-3, and S-4). They pass laterally, forward and upward to join the sympathetic hypogastric nerves at the pelvic plexus.

From the pelvic plexus, combined postganglionic parasympathetic and sympathetic fibers are distributed to the left colon and upper rectum via the inferior mesenteric plexus and directly to the lower rectum and upper anal canal. The periprostatic plexus, a subdivision of the pelvic plexus situated on Denonvilliers' fascia, supplies the prostate, seminal vesicles, corpora cavernosa, vas deferens, urethra, ejaculatory ducts, and bulbourethral glands. Sexual function is regulated by cerebrospinal, sympathetic, and parasympathetic components. Erection of the penis is mediated both parasympathetic (arteriolar vasodilatation) and sympathetic inflow (inhibition of vasoconstriction).

All pelvic nerves lie in the plane between the peritoneum and the endopelvic fascia and are in danger of injury during rectal dissection. Permanent bladder paresis occurs in 7–59% of patients after abdominoperineal resection of the rectum [62]; the incidence of impotence is reported to range from 15% to 45% and that of ejaculatory dysfunction from 32% to 42% [62]. The overall incidence of sexual dysfunction after proctectomy has been reported to reach 100% when wide dissection is performed for malignant disease [63–66]; however, this kind of procedure is unnecessary, and these rates are much lower for benign conditions, such as inflammatory bowel disease (0–6%) [64, 65, 67, 68]. Dissections performed for benign conditions are undertaken closer to the bowel wall, thus reducing the possibility of nerve injury [69].

Trauma to the autonomic nerves may occur at several points. During high ligation of the inferior mesenteric artery, close to the aorta, the sympathetic preaortic nerves may be injured. Division of both superior hypogastric plexus and hypogastric nerves may occur also during dissection at the level of the sacral promontory or in the presacral region. In such circumstances, sympathetic denervation with intact nervi erigentes results in retrograde ejaculation and bladder dysfunction. The nervi erigentes are located in the posterolateral aspect of the pelvis and at the point of fusion with the sympathetic nerves are closely related to the middle hemorrhoidal artery. Injury to these nerves will completely abolish erectile function [67]. The pelvic plexus may be damaged either by excessive traction on the rectum, particularly laterally, or during division of the lateral stalks when this is performed close to the lateral pelvic wall. Finally, dissection near the seminal vesicles and prostate may damage the periprostatic plexus, leading to a mixed parasympathetic and sympathetic injury. This can result in erectile impotence as well as a flaccid, neurogenic bladder. Sexual complications after rectal surgery are readily evident in men but are probably underdiagnosed in women (Fig. 1.5) [70, 71].

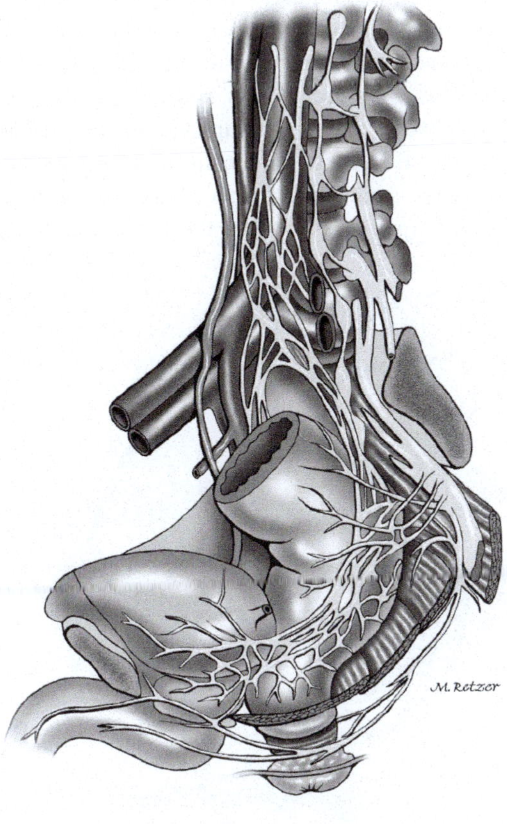

Fig. 1.5 Innervation of the colon, rectum, and anal canal. In memorium Marcos Retzer (illustrator)

Anatomy of the Pelvic Floor

Pelvic Floor Musculature

The pelvic floor is a complex interrelated structure of muscles, ligaments, and fascia with multiple functions. These functions concern support of visceral organs, maintaining continence, facilitating micturition and evacuation, as well as forming part of the birth canal. This multifunctional unit has connections to the pelvis, to organs, and to the extensive fibroelastic network in the fat containing anatomical spaces. The pelvic floor is traversed by the urethra and anal sphincters and, in women, the vagina.

Understanding the anatomic relationship of the pelvic floor muscles with the pelvic girdle, spine, and hips aids the rehabilitation provider in diagnosis, management, and appropriate referrals.

The muscles within the pelvis can be divided into three categories: (1) the anal sphincter complex, (2) pelvic floor muscles, and (3) muscles that line the sidewalls of the osseous pelvis [34]. Muscles in this last category form the external boundary of the pelvis and include the obturator internus and piriformis. These muscles, compared to the other two groups, lack clinical relevance to anorectal diseases; however, they provide an open communication for pelvic infection to reach extrapelvic spaces. For example, infection from the deep postanal space, which originated from posterior midline glands, can track along the obturator internus fascia and reach the ischiorectal fossa. The anal sphincter and pelvic floor muscles, based on phylogenetic studies, derive from two embryonic cloaca groups, respectively, sphincteric and lateral compressor [72]. The sphincteric group is present in almost all animals. In mammals, this group is divided into ventral (urogenital) and dorsal (anal) components [73]. In primates, the latter form the external anal sphincter. The lateral compressor or pelvicaudal group connects the rudimentary pelvis to the caudal end of the vertebral column. This group is more differentiated and subdivided into lateral and medial compartments only in reptiles and mammals. The homolog of the lateral compartment is the ischiococcygeus and of the medial, pelvicaudal compartment, the pubococcygeus and ileococcygeus. In addition, most primates possess a variably sized group of muscle fibers close to the inner border of the medial pelvicaudal muscle, which attaches the rectum to the pubis. In humans, the fibers are more distinct and known as the puborectalis muscle.

Levator Ani

The levator ani muscle, or pelvic diaphragm, comprises the major component of the pelvic floor. It is a pair of broad, symmetric sheets composed of three striated muscles: ileococcygeus, pubococcygeus, and puborectalis (Fig. 1.6). A variable fourth component, the ischiococcygeus or coccygeus, is rudimentary in humans and represented by only a few muscle fibers on the surface of the sacrospinous ligament. The levator ani is supplied by sacral roots on its pelvic surface (S-2, S-3, and S-4) and by the perineal branch of the pudendal nerve on its inferior surface. The puborectalis muscle receives additional innervation from the inferior rectal nerves.

The ileococcygeus muscles arise from the ischial spine and posterior part of the obturator fascia and course inferiorly and medially to insert into the lateral aspects of S-3 and S-4, the coccyx, and the anococcygeal raphe. The pubococcygeus arises from the posterior aspect of the pubis and the anterior part of the obturator fascia; it runs dorsally alongside the anorectal junction to decussate with fibers of the opposite side at the anococcygeal raphe and inserts into the anterior surface of the fourth sacral and first coccygeal segments.

The pelvic floor is "incomplete" in the midline where the lower rectum, urethra, and either the dorsal vein of the penis in men or the vagina in women pass through it. This defect is called the levator hiatus and consists of an elliptic space situated between the two pubococcygeus muscles. The hiatal ligament, originating from the pelvic fascia, keeps the intrahiatal viscera together and prevents their constriction during contraction of the levator ani. A possible (but controversial) dilator function has been attributed to the anococcygeal raphe because of its crisscross arrangement [14].

The puborectalis muscle is a strong, *U*-shaped loop of striated muscle that slings the anorectal junction to the posterior aspect of the pubis. The puborectalis is the most medial portion of the levator ani muscle. It is situated immediately cephalad to the deep component of the external sphincter. Because the junction between the two muscles is indistinct and they have similar innervation (pudendal nerve), the puborectalis has

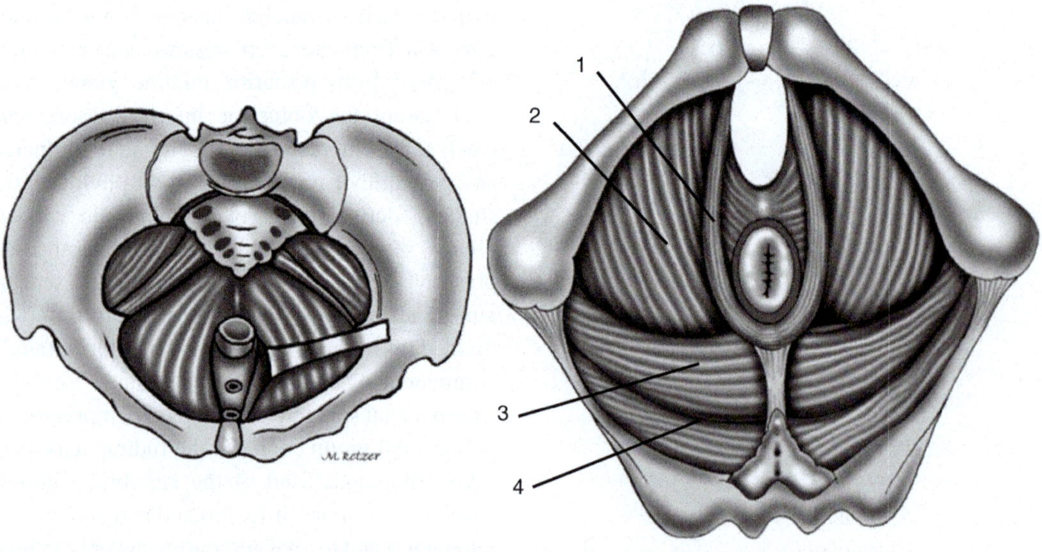

Fig. 1.6 Levator ani muscle. (**a**) Superior and (**b**) Pelvic floor muscles: (1) puborectal muscle; (2) pubococcygeus or pubovisceral; (3) iliococcygeus; (4) Ischiococcygeus. In memorium Marcos Retzer (illustrator)

been regarded by some authors as a part of the external anal sphincter and not of the levator ani complex [14, 15]. Anatomic and phylogenetic studies suggest that the puborectalis may be a part of the levator ani [73] or of the external anal sphincter [25, 72]. Embryologically, the puborectalis has a common primordium with the ileococcygeus and pubococcygeus muscles, and it is never connected with the external anal sphincter during the different stages of development [6]. In addition, neurophysiologic studies have implied that the innervation of these muscles may not be the same, because stimulation of the sacral nerves results in electromyographic activity in the ipsilateral puborectalis muscle but not in the external anal sphincter [74]. Currently, due to this controversy, the puborectalis has been considered to belong to both muscular groups, the external anal sphincter and the levator ani [75].

Anorectal Ring and the Anorectal Angle

Two anatomic structures of the junction of the rectum and anal canal are related to the puborectalis muscle: the anorectal ring and the anorectal angle. The anorectal ring, term coined by Milligan and Morgan [10], is a strong muscular ring that represents the upper end of the sphincter, more precisely the puborectalis, and the upper border of the internal anal sphincter, around the anorectal junction. Despite its lack of embryologic significance, it is an easily recognized boundary of the anal canal appreciated on physical examination, and it is of clinical relevance, as division of this structure during surgery for abscesses or fistula inevitably results in fecal incontinence.

The anorectal angle is thought to be the result of the anatomic configuration of the *U*-shaped sling of puborectalis muscle around the anorectal junction. Whereas the anal sphincters are responsible for closure of the anal canal to retain gas and liquid stool, the puborectalis muscle and the anorectal angle are designed to maintain gross fecal continence. Different theories have been postulated to explain the importance of the puborectalis and the anorectal angle in the maintenance of fecal continence. Parks et al. [76] opined that increasing intra-abdominal pressure forces the anterior rectal wall down into the upper anal canal, occluding it by a type of flap valve mechanism that creates an effective seal. Subsequently, it has been demonstrated that the flap mechanism does not occur. Instead, a continuous sphincteric occlusion-like activity that is attributed to the puborectalis is noted [77, 78].

Blood Supply

Within the abdomen, the inferior mesenteric artery branches into the left colic artery and two to six sigmoidal arteries. After crossing the left common iliac artery, it acquires the name superior hemorrhoidal artery (superior rectal artery). The sigmoidal arteries form arcades within the sigmoid mesocolon, resembling the small-bowel vasculature, and anastomose with branches of the left colic artery proximally and with the superior hemorrhoidal artery distally. The marginal artery terminates within the arcade of sigmoidal arteries. The superior hemorrhoidal artery is the continuation of the inferior mesenteric artery, once it crosses the left iliac vessels. The artery descends in the sigmoid mesocolon to the level of S-3 and then to the posterior aspect of the rectum. In 80% of cases, it bifurcates into right and left terminal branches; multiple branches are present in 17% [79]. These divisions, once within the submucosa of the rectum, run straight downward to supply the lower rectum and the anal canal.

The venous drainage basically follows its arterial supply. Blood from the right colon, via the superior mesenteric vein, and from left colon and rectum, via the inferior mesenteric vein, reaches the intrahepatic capillary bed through the portal vein.

Collateral Circulation

A potential area of discontinuity of the marginal artery is the Sudeck's critical point, situated between the lowest sigmoid and the superior hemorrhoidal arteries; however, both surgical experience and radiological studies have demonstrated adequate communications between these vessels [80]. There is also a collateral network involving middle hemorrhoidal, internal iliac, and external iliac arteries which could potentially prevent gangrene of the pelvis and even the lower extremities in case of occlusion of the distal aorta [81, 82].

Lymphatic Drainage

The submucous and subserous layers of the rectum have a rich network of lymphatic plexuses, which drain into an extramural system of lymph channels and follow their vascular supply [61]. They are more numerous in the sigmoid and are known in the rectum as the nodules of Gerota. The lymphatic drainage from all parts of the colon follows its vascular supply. Colorectal carcinoma staging systems are based on the neoplastic involvement of these various lymph node groups.

Innervation

The pelvic floor muscles receive innervation through somatic, visceral, and central pathways. Skin innervation of the lower trunk, perineum, and proximal thigh is mediated through the iliohypogastric, ilioinguinal, and genitofemoral nerves (L1–L3). The sympathetic and parasympathetic components of the autonomic innervation of the rectum closely follow the blood supply.

Perhaps the most clinically relevant nerve to this article is the pudendal nerve and its branches (Fig. 1.5). Arising from the ventral branches of S-2–S-4 of the sacral plexus, the pudendal nerve passes between the piriformis and coccygeal muscle as it traverses through the greater sciatic foramen, over the spine of the ischium, and back into the pelvis through the lesser sciatic foramen. It courses along the lateral wall of the ischiorectal fossa where it is contained in a sheath of the obturator fascia termed the pudendal (or Alcock's) canal. There are three main terminal branches of the pudendal nerve—the inferior rectal nerve (which typically originates proximal to Alcock's canal), the perineal nerve, and the dorsal nerve of the penis/clitoris. The pudendal nerve innervates the penis/clitoris, the bulbospongiosus and ischiocavernosus muscles, the perineum, the anus, the external anal sphincter, and the urethral sphincter. This nerve contributes to external genital sensation, continence, orgasm, and ejaculation. Muscles of the levator ani are thought to have direct innervation from sacral nerve roots S-3–S-5.

Summary

Anorectal and pelvic floor anatomy is complex, and the understanding of its dynamic interactions depends on the integrity of each one of its components. Classical anatomic dissections and studies are now associated to dynamic tridimensional engineering reconstructions and to high-technology image-retrieving systems, including magnetic resonance and ultrasonographic devices.

The interaction between normal anatomy and physiological events, childbirth, and variation in bowel habits pregnancy may result in a myriad of pelvic dysfunction. And by the proper recognizing of all the components of the system (anatomy) is possible to understand how it works (physiology) and, therefore, is possible to comprehend its malfunctions (pathophysiology). Understanding of the anatomy of the pelvis and its three compartments' (urological, gynecological, and anorectal) interaction is essential to fully diagnose patients' symptoms and to properly treat them.

References

1. Vesalii Bruxellensis Andreae, de Humani corporis fabrica. De Recti intestini musculis. 1st ed. 1543. p. 228.
2. Church JM, Raudkivi PJ, Hill GL. The surgical anatomy of the rectum – a review with particular relevance to the hazards of rectal mobilisation. Int J Colorect Dis. 1987;2:158–66.
3. Shafik A. A concept of the anatomy of the anal sphincter mechanism and the physiology of defecation. Dis Colon Rectum. 1987;30:970–82.
4. Klosterhalfen B, Vogel P, Rixen H, Mitterman C. Topography of the inferior rectal artery. A possible cause of chronic, primary anal fissure. Dis Colon Rectum. 1989;32:43–52.
5. Stoss F. Investigations of the muscular architecture of the rectosigmoid junction in humans. Dis Colon Rectum. 1990;33:378–83.

6. Levi AC, Borghi F, Garavoglia M. Development of the anal canal muscles. Dis Colon Rectum. 1991;34:262–6.
7. Lunniss PJ, Phillips RKS. Anatomy and function of the anal longitudinal muscle. Br J Surg. 1992;79:882–4.
8. Tjandra JJ, Milsom JW, Stolfi VM, Lavery I, Oakley J, Church JM, Fazio V. Endoluminal ultrasound defines anatomy of the anal canal and pelvic floor. Dis Colon Rectum. 1992;35:465–70.
9. Dobson HD, Pearl RK, Orsay CP, Rasmusen M, Evenhouse R, Ai Z, Blew G, Dech F, Edison MI, Silverstein JC, Abcarian H. Virtual reality: new method of teaching anorectal and pelvic floor anatomy. Dis Colon Rectum. 2003;46:349–52.
10. Milligan ETC, Morgan CN. Surgical anatomy of the anal canal: with special reference to anorectal fistulae. Lancet. 1934;2:1150–6.
11. Goligher JC, Leacock AG, Brossy JJ. The surgical anatomy of the anal canal. Br J Surg. 1955;43:51–61.
12. Garavoglia M, Borghi F, Levi AC. Arrangement of the anal striated musculature. Dis Colon Rectum. 1993;36:10–5.
13. Nivatvongs S, Gordon PH. Surgical anatomy. In: Gordon PH, Nivatvongs S, editors. Principle and practice of surgery for the colon, rectum and anus. St. Louis: Quality Medical Publishing; 1992. p. 3–37.
14. Shafik A. A new concept of the anatomy of the anal sphincter mechanism and the physiology of defecation II. Anatomy of the levator ani muscle with special reference to puborectalis. Investig Urol. 1975;12:175–82.
15. Oh C, Kark AE. Anatomy of the external anal sphincter. Br J Surg. 1972;59:717–23.
16. Bollard RC, Gardiner A, Lindow S, Phillips K, Duthie GS. Normal female anal sphincter: difficulties in interpretation explained. Dis Colon Rectum. 2002;45:171–5.
17. Fritsch H, Brenner E, Lienemann A, Ludwikowski B. Anal sphincter complex: reinterpreted morphology and its clinical relevance. Dis Colon Rectum. 2002;45:188–94.
18. Morren GL, Beets-Tan RGH, van Engelshoven JMA. Anatomy of the anal canal and perianal structures as defined by phased-array magnetic resonance imaging. Br J Surg. 2001;88:1506–12.
19. Williams AB, Bartram CI, Halligan S, Marshall MM, Nicholls RJ, Kmiot WA. Endosonographic anatomy of the normal anal compared with endocoil magnetic resonance imaging. Dis Colon Rectum. 2002;45:176–83.
20. Jorge JMN, Habr-Gama A. The value of sphincter asymmetry index in anal incontinence. Int J Color Dis. 2000;15:303–10.
21. Swash M. Histopathology of pelvic floor muscles in pelvic floor disorders. In: Henry MM, Swash M, editors. Coloproctology and the pelvic floor. London: Butterworth-Heinemann; 1992. p. 173–83.
22. Fraser ID, Condon RE, Schulte WJ, Decosse JJ, Cowles VE. Longitudinal muscle of muscularis externa in human and nonhuman primate colon. Arch Surg. 1981;116:61–3.
23. Courtney H. Anatomy of the pelvic diaphragm and anorectal musculature as related to sphincter preservation in anorectal surgery. Am J Surg. 1950;79:155–73.
24. Shafik A. A new concept of the anatomy of the anal sphincter mechanism and the physiology of defecation III. The longitudinal anal muscle: anatomy and role in sphincter mechanism. Investig Urol. 1976;13:271–7.
25. Lawson JON. Pelvic anatomy II. Anal canal and associated sphincters. Ann R Coll Surg Engl. 1974;54:288–300.
26. Roux C. Contribution to the knowledge of the anal muscles in man. Arch Mikr Anat. 1881;19:721–3.
27. Haas PA, Fox TA. The importance of the perianal connective tissue in the surgical anatomy and function of the anus. Dis Colon Rectum. 1977;20:303–13.
28. Gordon PH. Anorectal anatomy and physiology. Gastroenterol Clin N Am. 2001;30:1–13.
29. Lilius HG. Investigation of human fetal anal ducts and intramuscular glands and a clinical study of 150 patients. Acta Chir Scand Suppl. 1968;383:1–88.
30. Chiari H. Über die Nalen Divertik Fel der Rectumschleimhaut und Ihre Beziehung zu den anal fisteln. Wien Med Press. 1878;19:1482.
31. Parks AG. Pathogenesis and treatment of fistula-in-ano. BMJ. 1961;1:463–9.
32. Abel AL. The pecten band: pectenosis and pectenectomy. Lancet. 1932;1:714–8.
33. Goligher J. Surgery of the anus, rectum and colon. London: Baillière Tindall; 1984. p. 1–47.
34. Kaiser AM, Ortega AE. Anorectal anatomy. Surg Clin N Am. 2002;82:1125–38.
35. Skandalakis JE, Gray SW, Ricketts R. The colon and rectum. In: Skandalakis JE, Gray SW, editors. Embryology for surgeons. The embryological basis for the treatment of congenital anomalies. Baltimore: Williams & Wilkins; 1994. p. 242–81.
36. Mayo WJ. A study of the rectosigmoid. Surg Gynecol Obstet. 1917;25:616–21.
37. Balli R. The sphincters of the colon. Radiology. 1939;33:372–6.
38. Otis WJ. Some observations on the structure of the rectum. J Anat Physiol. 1898;32:59–63.
39. Abramson DJ. The valves of Houston in adults. Am J Surg. 1978;136:334–6.
40. Cawthorn SJ, Parums DV, Gibbs NM, A'Hern RP, Caffarey SM, Broughton CIM, Marks CG. Extent of mesorectal spread and involvement of lateral resection margin as prognostic factors after surgery for rectal cancer. Lancet. 1990;335:1055–9.
41. Heald RJ, Husband EM, Ryall RD. The mesorectum in rectal cancer surgery—the clue to pelvic recurrence? Br J Surg. 1982;69:613–6.
42. Cantlie J. The sigmoid flexure in health and disease. J Trop Med Hyg. 1915;18:1–7.

43. Ewing MR. The significance of the level of the peritoneal reflection in the surgery of rectal cancer. Br J Surg. 1952;39:495–500.
44. Boxall TA, Smart PJG, Griffiths JD. The blood-supply of the distal segment of the rectum in anterior resection. Br J Surg. 1963;50:399–404.
45. Nano M, Dal Corso HM, Lanfranco G, Ferronato M, Hornung JP. Contribution to the surgical anatomy of the ligaments of the rectum. Dis Colon Rectum. 2000;43:1592–8.
46. Quirke P, Durdey P, Dixon MF, Williams NS. Local recurrence of rectal adenocarcinoma due to inadequate surgical resection. Histopathological study of lateral tumour spread and surgical excision. Lancet. 1986;1:996–8.
47. Jorge JMN, Habr-Gama A, Souza AS Jr, Kiss DR, Nahas P, Pinotti HW. Rectal surgery complicated by massive presacral hemorrhage. Arq Bras Circ Dig. 1990;5:92–5.
48. Wang Q, Shi W, Zhao Y, Zhou W, He Z. New concepts in severe presacral hemorrhage during proctectomy. Arch Surg. 1985;120:1013–20.
49. Zama N, Fazio VW, Jagelman DG, Lavery IC, Weakley FL, Church JM. Efficacy of pelvic packing in maintaining hemostasis after rectal excision for cancer. Dis Colon Rectum. 1988;31:923–8.
50. Crapp AR, Cuthbertson AM. William Waldeyer and the rectosacral fascia. Surg Gynecol Obstet. 1974;138:252–6.
51. Tobin CE, Benjamin JA. Anatomical and surgical restudy of Denonvilliers' fascia. Surg Gynecol Obstet. 1945;80:373–88.
52. Sonneland J, Anson BJ, Beaton LE. Surgical anatomy of the arterial supply to the colon from the superior mesenteric artery based upon a study of 600 specimens. Surg Gynecol Obstet. 1958;106:385–98.
53. Michels NA, Siddharth P, Kornblith PL, Park WW. The variant blood supply to the small and large intestines: its importance in regional resections. A new anatomic study based on four hundred dissections with a complete review of the literature. J Int Col Surg. 1963;39:127–70.
54. Steward JA, Rankin FW. Blood supply of the large intestine. Its surgical considerations. Arch Surg. 1933;26:843–91.
55. Griffiths JD. Surgical anatomy of the blood supply of the distal colon. Ann R Coll Surg Engl. 1956;19:241–56.
56. Drummond H. The arterial supply of the rectum and pelvic colon. Br J Surg. 1914;1:677–85.
57. Ayoub SF. Arterial supply of the human rectum. Acta Anat. 1978;100:317–27.
58. Didio LJA, Diaz-Franco C, Schemainda R, Bezerra AJC. Morphology of the middle rectal arteries: a study of 30 cadaveric dissections. Surg Radiol Anat. 1986;8:229–36.
59. Bernstein WC. What are hemorrhoids and what is their relationship to the portal venous system? Dis Colon Rectum. 1983;26:829–34.
60. Miscusi G, Masoni L, Dell'Anna A, Montori A. Normal lymphatic drainage of the rectum and the anal canal revealed by lymphoscintigraphy. Coloproctology. 1987;9:171–4.
61. Block IR, Enquist IF. Studies pertaining to local spread of carcinoma of the rectum in females. Surg Gynecol Obstet. 1961;112:41–6.
62. Gerstenberg TC, Nielsen ML, Clausen S, Blaabgerg J, Lindenberg J. Bladder function after abdominoperineal resection of the rectum for anorectal cancer. Am J Surg. 1980;91:81–6.
63. Orkin BA. Rectal carcinoma: treatment. In: Beck DE, Wexner SD, editors. Fundamentals of anorectal surgery. New York: McGraw-Hill; 1992. p. 260–369.
64. Balslev I, Harling H. Sexual dysfunction following operation for carcinoma of the rectum. Dis Colon Rectum. 1983;26:785.
65. Danzi M, Ferulano GP, Abate S, Califano G. Male sexual function after abdominoperineal resection for rectal cancer. Dis Colon Rectum. 1983;26:665–8.
66. Weinstein M, Roberts M. Sexual potency following surgery for rectal carcinoma. A follow-up of 44 patients. Ann Surg. 1977;185:295–300.
67. Bauer JJ, Gerlent IM, Salky B, Kreel I. Sexual dysfunction following proctectomy for benign disease of the colon and rectum. Ann Surg. 1983;197:363–7.
68. Walsh PC, Schlegel PN. Radical pelvic surgery with preservation of sexual function. Ann Surg. 1988;208:391–400.
69. Lee ECG, Dowling BL. Perimuscular excision of the rectum for Crohn's disease and ulcerative colitis. A conservative technique. Br J Surg. 1972;59:29–32.
70. Metcalf AM, Dozois RR, Kelly KA. Sexual function in women after proctocolectomy. Ann Surg. 1986;204:624–7.
71. Duthie HL, Gairns FW. Sensory nerve endings and sensation in the anal region in man. Br J Surg. 1960;47:585–95.
72. Wendell-Smith CP. Studies on the morphology of the pelvic floor [Ph.D. thesis]. University of London. 1967.
73. Paramore RH. The Hunterian lectures on the evolution of the pelvic floor in non-mammalian vertebrates and pronograde mammals. Lancet. 1910;1(1393–1399):1459–67.
74. Percy JP, Swash M, Neill ME, Parks AG. Electrophysiological study of motor nerve supply of pelvic floor. Lancet. 1981;1:16–7.
75. Russell KP. Anatomy of the pelvic floor, rectum and anal canal. In: Smith LE, editor. Practical guide to anorectal testing. New York: Igaku-Shoin Medical Publishers; 1991. p. 744–7.
76. Parks AG, Porter NH, Hardcastle J. The syndrome of the descending perineum. Proc R Soc Med. 1966;59:477–82.
77. Bannister JJ, Gibbons C, Read NW. Preservation of faecal continence during rises in intra-abdominal pressure: is there a role for the flap valve? Gut. 1987;28:1242–5.

78. Bartolo DCC, Roe AM, Locke-Edmunds JC, Virjee J, Mortensen NJMcC. Flap-valve theory of anorectal continence. Br J Surg. 1986;73:1012–4.
79. Lindsey I, Guy RJ, Warren BF, Mortensen NJMcC. Anatomy of Denonvilliers' fascia and pelvic nerves, impotence, and implications for the colorectal surgeon. Br J Surg. 2000;87:1288–99.
80. Sudeck P. Über die Gefassversorgung des Mastdarmes in Hinsicht auf die Operative Gangran. Munch Med Wochenschr. 1907;54:1314.
81. Griffiths JD. Extramural and intramural blood supply of the colon. BMJ. 1961;1:322–6.
82. Lindstrom BL. The value of the collateral circulation from the inferior mesenteric artery in obliteration of the lower abdominal aorta. Acta Chir Scand. 1950;1:677–85.
83. Petros PE. The anatomy and dynamics of pelvic floor function and dysfunction. In: Petros PE, editor. The female pelvic floor—function, dysfunction and management according to the integral theory. 2nd ed. Berlin/Heidelberg/New York: Springer; 2006. p. 13–50.
84. Petros PE. The integral system. Cent European J Urol. 2011;64:110–9.
85. Petros PEP, Woodman PJ. The integral theory of continence. Int Urogynecol J. 2008;19:35–40.
86. Petros PEP. The Integral Theory System. A simplified clinical approach with illustrative case histories. Pelviperineology. 2010;29:37–51.
87. Petros PEP. Anatomy and surgical cure of descending perineal syndrome. Int Urogynecol J. 2018; https://doi.org/10.1007/s00192-018-3557-y.
88. Lipschuetz M, Cohen SM, Liebergall-Wischnitzer M, Zbedat K, Hochner-Celnikier D, Lavy Y, et al. Degree of bother from pelvic floor dysfunction in women one year after first delivery. European J Obst Gynecol Repr Biol. 2015;191:90–4.
89. Durnea CM, Khashan AS, Kenny LC, Durnea UA, Dronan JC, O'Sullivan SM, et al. What is to blame for postnatal pelvic floor dysfunction in primiparous women – pre-pregnancy or intrapartum risk factors? Eur J Obstet Gynecol Reprod Biol. 2017; 214: 36–43.
90. Fitzpatrick M, O'Herlihy C. The effects of labour and delivery on the pelvic floor. Best Pract Res Clin Obstet Gynaecol. 2001;15:63–79.
91. Rogers RG, Leeman LM, Borders N, Fullilove AM, Teaf D, Hall RJ, et al. The contribution of the second stage of labor to pelvic floor dysfunction: a prospective cohort comparison of nulliparous women. BJOG. 2014;121:1145–54.
92. Bozkurt M, Yumru AE, Sahin L. Pelvic floor dysfunction, and effects of pregnancy and mode of delivery on pelvic floor. Taiwan J Obstet Gynecol. 2014;53:452–8.
93. Vila Pouca MCP, Ferreira JPS, Oliveira DA, Parente MPL, Natal Jorge RM. Viscous effects in pelvic floor muscles during childbirth: a numerical study. Int J Numer Meth Biomed Engng. 2017:e2927.
94. Goldberg RP. Effects of pregnancy and childbirth on the pelvic floor. In: Culligan PJ, Goldberg RP, editors. Urogynecology in primary care. 1st ed. London: Springer; 2007. p. 21–33.
95. Woodman PJ, Graney DO. Anatomy and physiology of the female perineal body with relevance to obstetrical injury and repair. Clin Anat. 2002;15:321–34.
96. Farouk R, Duthie GS, Bartolo DCG. Functional anorectal disorders and physiological evaluation. In: Beck DE, Wexner SD, editors. Fundamentals of anorectal surgery. New York: McGraw-Hill, Inc.; 1992. p. 68–88.
97. O'Beirne J, editor. New views of the process of defecation and their application to the pathology and treatment of diseases of the stomach, bowels and other organs. Dublin: Hodges and Smith; 1833.
98. Kumar D, Wingate DL. Colorectal motility. In: Henry MM, Swash M, editors. Coloproctology and the pelvic floor. Oxford: Butterworth-Heinemann Ltd.; 1992. p. 72–85.
99. Haller A. The large intestine. In: Cullen W, editor. First lines of physiology. A reprint of the 1786 edition (sources of science 32). New York, London: Johnson Reprint Corporation; 1966. p. 139–40.

Physiology of Continence and Defecation

2

Lucia Camara Castro Oliveira, Mara Rita Salum, and Renato Caram Saad

Introduction

The maintenance of continence and defecation is determined by complex and multifactorial mechanisms, involving the integration of somatic and visceral functions, under the control of the central nervous system [1]. Therefore, there is an interaction between the brain, spinal cord, enteric neurons, and the muscle of the colon, the rectum, anus, and pelvic floor. Those structures are coordinated and are dependent on conscious control.

The defecation process is triggered by the arrival of feces into the rectum as a result of the peristaltic movements of the colon. As the peristaltic movements increase, the rectum receives a larger quantity of feces thus triggering the reflex of defecation. At that moment, the individual is able to control the involuntary passage of feces and gases through the voluntary contraction of the external sphincter muscle, and the puborectalis muscle, and the formation of the anorectal angle. At the appropriate time, the defecation reflex initiates the process of elimination of the rectal contents, when abdominal muscles, through the abdominal press, help the expulsion of this content and the pelvic floor relaxes with the opening of the anal canal, allowing the passage of feces. In fact, the defecation process is very complex and not very well understood.

Several theories have been proposed to explain the mechanism of anal continence [1, 2]. In 1965, Phillips and Edward [3] proposed a mechanism similar to that of the esophageal sphincter, believing that there would be a high-pressure zone created by the abdominal muscles which would prevent the passage of stools to the anal canal. However, other authors demonstrated that, in fact, the high-pressure zone was below the levator muscles at the level of the anal canal [4]. Another hypothesis postulated by Parks [5], in 1975, was that the anorectal angle would function as a valve, controlling the passage of the feces in dependence on the change of this angle by the movement of the pelvic floor muscles. However, as comparative imaging studies in incontinent patients demonstrated that many of them had normal anorectal angles, these theories were abandoned, and emphasis was placed on sphincter action [6–9]. Similarly, incontinent patients undergoing Parks' surgery, with correction of the supposed abnormal anorectal angle, remained incontinent despite the correction of this anatomy.

The coexistence of incontinence and intestinal constipation in the same patient may be related to rectal dysfunction, with fecal leakage

L. C. C. Oliveira (✉)
Director Department of Anorectal Physiology of Rio de Janeiro and CEPEMED, Rio de Janeiro, Brazil

M. R. Salum
Sirio-Libanes and Albert Einstein Hospital, Department of Surgery, São Paulo, Brazil

R. C. Saad
Sirio-Libanese Hospital, Department of Surgery, São Paulo, Brazil

L. C. C. Oliveira (ed.), *Anorectal Physiology*, https://doi.org/10.1007/978-3-030-43811-1_2

after bowel movements. A recent theory to explain defecation process was proposed by Petros and Swash: [10] the external striated muscle mechanism when stretched opens both walls of the anorectum; the puborectalis muscle (PRM) relaxes, relieving the closure pressure on the posterior rectal wall, thereby allowing the anorectal angle (ARA) to be actively opened by backward and downward vectors created by contraction of the levator plate (LP) and the longitudinal muscle of the anus (LMA). This hypothesis was tested in a group of patients with idiopathic fecal incontinence (FI) [11]. Active opening of the anal canal by the pelvic striated muscle contraction, external to the rectum, is attractive from a flow mechanical perspective. Contraction of striated musculature external to the rectum has the capacity to reduce friction within the rectum by stretching its walls, thereby reducing the resistance of the mucosal folds and, by opening the lumen, reducing the internal expulsion pressure required for evacuation. These dynamic processes are important in women during micturition, reducing micturition pressure to the fifth power for non-laminar flow at the urethral opening [12]. The notion that a similar external opening mechanism driven by striated muscles is necessary for defecation has a strong historical basis. These ideas will be further discussed in-depth in a separate chapter.

In addition, the development of new research and the understanding of neuromodulation mechanisms have brought new explanations on the mechanism of defecation. We already know that defecation is facilitated by increasing intra-abdominal pressure or by stimulation of the mucosal lining of the rectum, indicating that sensory pathways from the abdomen and rectum play a role on defecation control pathways [13].

Callaghan et al. [14] recently presented new concepts for the nerve pathways for voluntary control of defecation and fecal continence (Fig. 2.1).

Some of the factors associated with the continence and defecation mechanisms are presented as follows.

Anal Continence Mechanisms

Anal incontinence can be defined as the involuntary loss of stool or gas usually after the age of 4 years when the individual acquires sphincter control.

The main mechanisms involved in maintaining continence are listed in Table 2.1.

Anal Sphincter Muscles

Internal Anal Sphincter Muscle

The anal internal sphincter muscle has a resting tonus with cyclic variations represented by the short and ultrashort waves [15–19]. The most common finding of internal sphincter motility is the presence of short waves, with frequencies of 10–23 cycles per minute [17, 18, 20, 21]. These waves are not related to respiration, not even with the activity of the external anal sphincter muscle. In fact, they occur regardless of the state of wakefulness or feeding [15, 21, 22]. Ultrashort waves can be found in 5–90% of individuals, with frequencies of 0.5–2 cycles per minute and may decrease during sleep [15, 19, 23]. The importance of these waves is related to the reflex of accommodation or sampling reflex in the upper part of the internal anal sphincter. An intermittent relaxation of this muscle of 10–20 seconds occurs with a frequency of seven times every hour [24–26]. This relaxation leads to a balance between the pressures of the anal and rectum, thus allowing contact of the rectal contents with the sensitive mucosa of the anal canal. This reflex occurs, then, seven times an hour and, in general, is not perceived.

The internal anal sphincter muscle contributes with approximately 50% to 85% of the anal resting tonus. Nearly 15–20% of the resting pressure is represented by the puborectalis and anal external sphincter muscle together with the vascular cushions [27, 28]. This is due to intrinsic myogenic activity and extrinsic adrenergic innervation. However, when there is an increase in the intra-abdominal pressure (Valsalva maneuver),

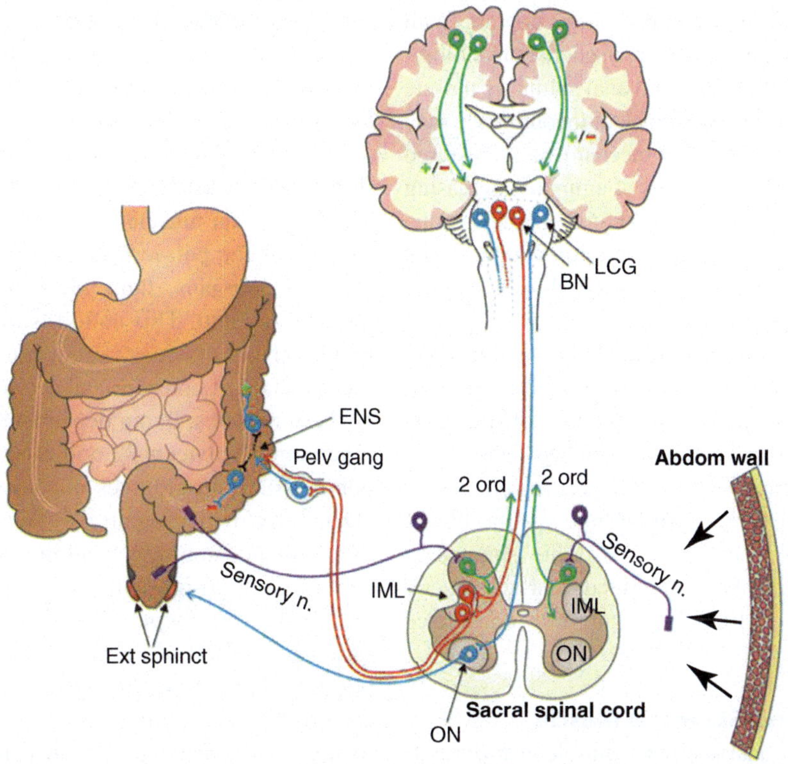

Fig. 2.1 Nerve pathways for voluntary control of defecation and fecal continence. Cortical centers that govern voluntary control provide inputs that either inhibit or enhance excitability of neurons in the brain stem, a medial nucleus (Barrington's nucleus, BN) through which autonomic pathways to the distal colon and rectum are activated and a lateral cell group (LCG) that controls the external anal sphincter. The medial group of neurons projects to the spinal defecation center in the intermediolateral column (IML) at S1 level. This in turn connects with intrinsic reflex pathways of the enteric nervous system (ENS), via the pelvic ganglia. Afferent (sensory) neurons that detect pressure and mucosal irritation in the colon contribute to urge, and neurons that sense pressure in the abdominal cavity enhance defecation. These connect to second-order neurons that make local connections in the spinal cord and provide sensory information to the pons and cortex. Descending neurons from the LCG synapse in Onuf's nucleus (ON) on motor neurons supply the external sphincter. (Reused with permission © Springer Nature [14])

Table 2.1 Mechanisms of continence

Anal sphincter muscles
High-pressure zone
Anorectal angle
Anorectal sensitivity and rectoanal inhibitory reflex
Rectal compliance, tonus, and capacity
Rectal filling and emptying
Colon and rectal motility and transit time
Vascular cushions
Stool volume and consistency

there is a greater electrical activity of these supporting muscles, which provides an additional force to the sphincter mechanism, preventing the involuntary passage of the rectal contents through the anal canal.

Complete rupture of the internal anal sphincter muscle is associated with gas leakage in up to 40% of patients. Thus, anal dilation formerly used to treat anal fissure was abandoned as a result of this important complication. Partial rupture of the internal anal sphincter muscle or partial sphincterotomy can also lead to up to 15% incontinence for gases and risk of stool leakage [29, 31]. The division or rupture of this muscle impairs the mechanism of continence of rectal contents or rectal inhibitory reflex previously described. In low anterior rectal resection and coloanal anastomosis, or when staplers with excessive anal dilatation are used, the physiological response of the internal anal sphincter muscle may be impaired, and transient incontinence may be generated until the mechanism is restored again [25, 26]. The effects of aging by increasing the thickness of the internal anal sphincter muscle and leading to its degeneration with collagen replacement have been considered to be a cause of incontinence [2].

In addition to longitudinal, the resting pressure presents radial variation, partially explained by the anatomical conformation of the anal sphincter and puborectalis muscles. Thus, in the upper and lower anal canal, the resting pressure is higher posteriorly, distributing equally in the middle anal canal [32, 33]. This sphincteric asymmetry has been described, and a sphincter asymmetry of 10–20% is currently considered normal [34].

External Anal Sphincter Muscle

The external anal sphincter muscle also presents a resting tone, even during sleep [35–38]. The tonic activity of this muscle presents variations, according to daily activities. This muscle forms a ring in continuity with the puborectalis muscle, and both act together in the voluntary maintenance of continence. The maximum contraction of the external anal sphincter muscle can be maintained for up to 1 minute; after that, fatigue occurs [39, 40]. The distension of the rectum by the arrival of gases or feces in the rectal ampulla causes a contraction of the external anal sphincter muscle for 20–30 seconds, a phenomenon known as guarding reflex constituting a low spinal reflex with cortical control that can be found even in patients submitted to coloanal anastomosis, once it is triggered by receptors located in the puborectalis region [30, 41]. The anal canal is normally closed at rest and during sleep, due to the constant activity of the internal anal sphincter muscle, reinforced by the tonic activity of the external anal sphincter and puborectalis.

The importance of the external anal sphincter muscle in maintaining continence can be demonstrated when incontinence to gas and stool is observed in up to 50% of the women who present small previous defects of obstetric origin and the satisfactory results when these patients are submitted to a sphincter repair or sphincteroplasty [42, 43].

Puborectalis and Levator Ani Muscles

The levator ani muscles are responsible for the support of the pelvic floor and the pelvic and abdominal organs, thus preventing excessive perineal descent. This action is mediated by a pelvic reflex known as postural reflex that provides a state of constant and active contraction of these muscles, being dependent on an intact innervation between S2 and S4 [36, 39]. The puborectalis muscle does not involve the anal canal fully. It participates in the maintenance of continence not only as an important component of the high-pressure zone but as responsible for the formation of the anorectal angle, a fact that resulted in the flap-valve theory proposed by Parks [44]. In addition, it is also in a state of constant tonic activity. Its contribution to the continence mechanism can be demonstrated in children who were born with anal agenesis and absence of the external and internal anal sphincters. Continence in these cases is represented by the action of the puborectalis muscle [45].

High-Pressure Zone

Resting tonus, arising from the involuntary activity of the internal anal sphincter muscle, can be perceived through digital examination and during anorectal manometry. A high-pressure zone can also be found in the anal canal, consisting of the length of the anal canal, where pressures are observed 30% larger than those found in the rectum. It occurs, in part, because of the myogenic properties of both the internal and external anal sphincter muscles. The high-pressure zone is greater in males than in females [26, 46]. The extension of the high-pressure zone usually corresponds to 2–3 cm in women and 2.5–3.5 cm in men. The high-pressure zone is one of the main factors to avoid the passage of stools.

Anorectal Angle

For many years, the anorectal angle (angle formed by the intersection of the upper limit of the anal canal with the median line of the rectum) has been believed to be the most efficient mechanism of valvular control for anal continence [5, 44]. This angle, originated mainly due to the contractile force of the puborectalis muscle, performs the traction of the rectum anteriorly, presenting in normal resting conditions, around 90 degrees. In the case of defecation, with the relaxation of the puborectalis muscle and the pelvic floor muscles, this angle is more obtuse, resulting in rectification or alignment of the rectum with respect to the anal canal. However, in some studies in patients with fecal incontinence, several authors have demonstrated normal anorectal angles, and continence restoration by surgical approach was not necessarily associated with more acute anorectal angles [8, 24, 48]. Dynamic radiological evaluation in association with other physiological studies were also not able to demonstrate the valvular mechanism [9].

Anorectal Sensitivity and rectoanal inhibitory reflex

The anal canal, especially the transition zone, has a large number of nerve endings and sensory cells responding to pressure, temperature, and friction stimuli [50]. Distension of the rectal wall by the intestinal contents directly stimulates the pressure receptors on the wall of the rectum. This is the mechanism of the rectoanal inhibitory reflex, which allows contact of the rectal contents with the sensitive area of the anal canal through relaxation of the internal anal sphincter muscle and contraction of the external anal sphincter [47–49, 51, 52]. At this time, the rectum can accomodate the fecal contents, due to a decrease in rectal pressure and its capacity of compliance, contributing decisively to anal continence. The importance of the rectoanal inhibitory reflex in the continence mechanism becomes more evident when we observe that the reflex recovery in patients submitted to ileal pouch construction decreases the incidence of nocturnal soiling [53]. Any factor that alters rectal compliance or the sensitivity of the rectum to recognize the intestinal contents would interfere directly in anal continence.

Rectal Compliance, Tonus, and Capacity

The rectum has the ability to passively accommodate distension. This ability enables the rectum to maintain rectal pressure despite sudden increases in intrarectal pressure, contributing to the maintenance of lower pressures as compared to those of the anal canal. Intrarectal volumes above 200 ml usually cause a sensation of urgency, and the maximum tolerated volume corresponds to about 400 ml. Certain conditions, such as inflammatory bowel disease and actinic or ischemic proctitis, can reduce rectal compliance, usually manifested by urgency, increased evacuation frequency, tenesmus, and incontinence episodes [47, 51]. In Hirschsprung's disease, on the other hand, compliance is increased by the presence of functional megarectum. A compliant rectum is therefore essential to the maintenance of continence. Compliance is a volumetric parameter of anorectal manometry whose concept still presents some controversies. It consists of the measurement of the change in intrarectal pressure in response to change in volume through an infusion catheter or balloon system positioned into the rectum. The variability in the techniques used for assessment and the differences in values obtained in accordance with the balloon material are the two major factors responsible for the existence of controversy.

Rectal Filling and Emptying

The rectum normally remains free of fecal contents of the sigmoid colon by the rectosigmoid angle, by the Houston valves, and by its contractile resting activity, which is believed to be larger than that of the sigmoid. Increased rectal volume results in a progressive accommodation mechanism. Approximately half of this instilled volume immediately refluxes into the sigmoid colon. These findings suggest a significant role of the sigmoid colon in the continence mechanism, helping to maintain intrarectal pressure [55].

Colon and rectal motility and emptying and transit time

Motility of the rectum is represented by contraction waves called the rectal motor complex, and more intense activities are observed after meals. The function of these rectal waves is still uncertain but is believed to contribute to perfect rectal emptying [17–23, 54]. In the anal canal, short-wave gradients are also present, with a greater frequency distally, tending to keep the contents inside the rectum. Ultrashort waves have been found in about 40% of normal continents. These waves present amplitude of up to 100 cm H2O with duration of about 30 seconds and frequency of 0.5–2 cycles per minute. They are related to high-resting pressures and reflect the activity of the internal anal sphincter muscle [17–23, 56].

Vascular Cushions

The hemorrhoidal vascular cushions and anal canal mucosa help to maintain fecal continence [57, 58]. The cushions have the ability to expand to keep the anal canal closed and prevent incontinence when anal pressure decreases. These vessels are useful for filling spaces that cannot be occupied by the musculature and can contribute in about 15% of the basal pressure at rest [59]. The importance of the cushions becomes obvious when patients presents with soiling after hemorrhoidectomy, even with normal sphincter pressure.

The mucocutaneous junction, which is in the high-pressure zone of the anal canal, also works as a barrier that prevents the loss of mucus and feces. The displacement of this junction out of the anal margin, as occurs in hemorrhoidal prolapse, can be an important cause of leakage and involuntary loss of mucus. Correction of hemorrhoidal prolapse improves continence in these cases [60]. The anal canal is rich in nerve endings which participate in the discrimination of rectal contents and promote the sampling reflex, previously described. In diabetics the reflex may not be found, and loss of the reflex and the threshold could be elevated [61].

Stool Volume and Consistency

Stool consistency definitely influences sensation, emptying, and anal continence. The time taken to expel a single solid piece of stool varies inversely with its diameter, and greater effort is necessary to eliminate hard and small stools compared to large and soft stools [62]. One study has shown that the ideal diameter for stool to be successfully eliminated is about 2 cm [60].

Mechanisms of Defecation

Although extensively studied, the mechanism of defecation remains partially understood (Fig. 2.1). It is known that the act of defecating requires the perfect coordination of the pelvic floor and anal sphincter muscles, being mediated by the neuromotor and sensorineural impulses. The propulsion of feces through the rectosigmoid junction causing the intestinal contents to reach the rectum seems to be the starting point for the stimulation of the evacuation, provoking the desire for defecation. The receptors for this sensation of urge to defecate are located in the puborectalis muscle [47, 63]. Rectal perception usually occurs with volumes between 11 and 68 ml, and the maximum tolerable volume varies between 220 and 510 ml [64]. The distension of the rectal walls by the presence of the fecal content generates an intrinsic reflex through the enteric plexus that results in the relaxation of the anal internal sphincter muscle also known as the rectoanal inhibitory reflex [65–67]. This mechanism makes possible the contact of the fecal content with the sensory cells of the anal canal making it possible to discriminate the quality of this content (gases, liquid, or solid feces). At that moment the continence is maintained by the reflex contraction of the external anal sphincter muscle. For the evacuation to occur, the intra-abdominal pressure should increase by contracting the lumbar, rectus abdominis, and diaphragm muscles (Valsalva maneuver). Then, the puborectalis muscle relaxes, along with the other sphincter muscles, allowing an increase of the anorectal angle that reaches about 130 to 140 during defecation, resulting in rectification of the rectum with respect to the anal canal and facilitating the evacuation. After the expulsion of the feces, the reflex contraction of the external sphincter muscle of the anus and the return of the pelvic floor to the normal position finally occur [9].

If defecation is an undesirable act, the process may be temporarily postponed by the voluntary contraction of the external sphincter muscle and the lifting muscles. Because of the rectal accommodation mechanism, associated with rectal compliance, intrarectal pressure returns to normal levels, the integrated action of the pelvic floor muscles displaces the feces toward the rectosigmoid junction, and gradually the tone of the sphincter muscles returns to the resting condition. The difficulty of opening the anal canal during evacuation attempts is three to four times greater in the initial phase. After opening the anal canal about 1–2 cm in diameter, the evacuation difficulty decreases, and therefore the formed and soft fecal bolus facilitates evacuation [60]. The normal frequency of evacuation corresponds to about 6.3–7.5 for 7 days varying between 3 and 11 for 7 days [68, 69].

Summary

The mechanism to maintain normal continence and defecation is complex and multifactorial. The anatomical and anorectal physiological correlations have allowed a better understanding of the diseases that involve the structures of the rectum and anus and their functional alterations.

References

1. Rasmussen OO. Anorectal function. Dis Colon Rectum. 1994;37:386–403.
2. Bartolo DCC, MacDonald ADH. Fecal continence and defecation. In: Pemberton JH, Swash M, Henry MM, editors. The pelvic floor its function and disorders. London: WB Saunders; 2002. p. 77–83.
3. Phillips SF, Edwards DAW. Some aspects of anal continence and defecation. Gut. 1965;6:396–405.
4. Duthie HL. Dynamics of the rectum and anus. Clin Gastroenterol. 1975;4:467–77.
5. Parks AG. Anorectal incontinence. Proc Roy Soc Med. 1975;68:61–90.
6. Miller R, Bartolo DC, Locke-Edmunds JC, Mortensen NJ. A prospective study of conservative and operative treatment for fecal incontinence. Br J Surg. 1988;75:101–5.
7. Miller R, Orrom WJ, Cornes H, Duthie HL, Bartolo DC. Anterior sphincter plication and levatorplasty in the treatment of faecal incontinence. Br J Surg. 1989;76:1058–60.
8. Womack NR, Morrison JFB, Williams NS. Prospective study of the effect of post anal repair in neurogenic faecal incontinence. Br J Surg. 1988;75:48–52.
9. Bartolo DC, Roe AM, Locke-Edmunds JC, Virjee J, Mortensen NJ. Flap-valve theory of anorectal continence. Br J Surg. 1986;73:1012–4.
10. Petros PE, Swash M. The musculoelastic theory of ano- 376 rectal function and dysfunction. Pelviperineology. 2008;27:89–93.
11. Petros PE, Swash M. Directional muscle forces activate anorectal continence and defecation in the female. Pelviperineology. 2008;27:94–7.
12. Bush MB, Petros PE, Barrett-Lennard BR. On the flow through the human urethra. J Biomech. 1997;30:967–9.
13. Shafik A. Recto-colic reflex: role in the defecation mechanism. Int Surg. 1996;81:292–4.
14. Callaghan B, Furness JB, Pustovit RV. Neural pathways for colorectal control, relevance to spinal cord injury and treatment: a narrative review. Spinal Cord. 2018;56(3):199–205.
15. Garavoglia M, Borghi F, Levi AC. Arrangement of the anal striated musculature. Dis Colon Rectum. 1993;36(1):10–5.
16. Garavoglia M, Borghi F, Bargoni A, Fronticelli CM. Suspensory elements of the anal canal. Anatomo-surgical considerations. Minerva Chir. 1989;44(12):1623–6.
17. Penninckx F. Morphological and physiological aspects of anal function. Leuven: Belgium ACCO; 1981.
18. Wankling WJ, Brown BH, Collins CD, Duthie HL. Basal electrical activity in the anal canal in man. Gut. 1968;9:457–60.
19. Ustach TJ, Tobon F, Hambrecht T, Bass DD, Schuster MM. Electrophysiological aspects of human sphincter function. J Clin Invest. 1970;49:41–8.

20. Gutierrez JG, Oliai A, Chey WY. Manometric profile of the internal sphincter in man. Gastroenterology. 1975;68:907.
21. Hancock BD. Measurement of anal pressure and motility. Gut. 1976;17:645–51.
22. Pedersen IK, Christiansen J. A study of anal physiological variation in anal manometry. Br J Surg. 1989;76:69–71.
23. Orkin BA, Hanson RB, Kelly KA, Phillips SF, Dent J. Human anal motility while fasting, after feeding, and during sleep. Gastroenterology. 1991;100:1016–23.
24. Kerremans R. Morphological and physiological aspects of anal continence and defecation. Bruxelles Editions Arscia SA. 1969;95:1–294.
25. Read NW, Haynes WG, Bartolo DC, et al. Use of anorectal manometry during rectal infusion of saline to investigate function in incontinent patients. Gastroenterology. 1983;85:105–13.
26. Miller R, Lewis GT, Bartolo DC, Cervero F, Mortensen NJ. Sensory discrimination and dynamic activity in the anorectum evidence of a new ambulatory technique. Br J Surg. 1988;75:1003–7.
27. Miller R, Bartolo DC, Cervero F, Mortensen NJ. Anorectal sampling a comparison of normal and incontinent patients. Br J Surg. 1988;75:44–7.
28. Waldron DJ, Kumar D, Hallan RI, Williams NS. Prolonged ambulant assessment of anorectal function in patients with prolapse hemorrhoids. Dis Colon Rectum. 1989;32:968–74.
29. Bennet LRC, Duthie HL. The functional importance of the internal sphincter. Br J Surg. 1964;51:355–7.
30. Duthie HL, Watts J. Contribution of the external anal sphincter to the pressure zone in the anal canal. Gut. 1965;6:64–8.
31. Walker WA, Rothenberger DA, Goldberg SM. Morbidity of internal sphincterotomy for anal fissure and stenosis. Dis Colon Rectum. 1985;28:832–5.
32. Lane RH, Parks AG. Function of the anal sphincters following colo-anal anastomosis. Br J Surg. 1977;64:596–9.
33. Wunderlich M, Teleky B, Schiessel R. Sphincter function nach coloanaler anastomose. Langenbecks Arch Chir. 1986;367:259–69.
34. Taylor BM, Beart RW, Phillips SF. Longitudinal and radial variation of pressure in the human anal canal. Gastroenterology. 1984;86:693–7.
35. Jorge JMN, Wexner SD. A practical guide to basic anorectal physiology. Contemp Surg. 1993;43:214–24.
36. Jorge JMN, Habr-Gama A. The value of sphincter asymmetry index in anal incontinence. Int J Color Dis. 2000;15(5-6):303–10.
37. Floyd WF, Walls EW. Electromyography of the sphincter ani externus in manm. J Physio (Lond). 1953;122:599–609.
38. Porter NH. A physiological study of the pelvic floor in rectal prolapse. Ann R Coll Surg Engl. 1962;31:379–404.
39. Taverner D, Smiddy FG. An electromyographic study of the normal function of the external anal sphincter and pelvic diaphragm. Dis Colon Rectum. 1959;2:153–60.
40. Kumar D, Waldron D, Williams NS, Browning C, Hutton MR, Wingate DL. Prolonged anorectal manometry and external anal sphincter electromyography in ambulant human subjects. Dug Dis Sci. 1990;35:641–8.
41. Parks AG, Porter NH, Melzak J. Experimental study of the reflex mechanism controlling the muscles of the pelvic floor. Dis Colon Rectum. 1962;5:407–14.
42. Gaston EA. Physiological basis for preservation of fecal continence after resection of rectum. JAMA. 1951;146:1486–9.
43. Freckner B. Function of the anal sphincters in spinal man. Gut. 1975;16:638–44.
44. Nielsen MB, Hauge C, Rasmussen O, Pedersen JF, Christiansen J. Anal endosonographic findings in the follow-up of primarily sutured sphincteric ruptures. Br J Surg. 1992;79:104–6.
45. Oliveira L, Pfeifer J, Wexner SD. Physiological and clinical outcome of anterior sphincteroplasty. Br J Surg. 1996;83:1244–51.
46. Parks AG, Porter NH, Hardcastle J. The syndrome of the descending perineum. Proc R Soc Med. 1966;59:477–82.
47. Varma KK, Stephens D. Neuromuscular reflexes of rectal continence. Aust N Z J Surg. 1972;41:263–72.
48. Bouvier M, Gonella J. Nervous control of the internal sphincter in the cat. J Physiol. 1997;310:457–69.
49. Sun WM, Read NW. Anorectal function in normal human subjects effect of gender. In J Colorectal Dis. 1989;4:188–96.
50. Bannister JJ, Abouzecky L, Read NW. Effect of aging on anorectal function. Gut. 1987;28:353–7.
51. Duthie HL, Gairns FW. Sensory nerve endings and sensation in the anal region of man. Br J Surg. 1960;47:585–95.
52. Goligher JC, Hughes ESR. Sensibility of the rectum and colon. Its role in the mechanism of anal continence. Lancet. 1951;1:543–8.
53. Oresland T, Fasth S, Kervall S, Nordgren S, Hulten L. Manovolumetric and sensory characteristics of the ileoanal pouch compared with healthy rectum. Br J Surg. 1990;77:803–7.
54. Duthie GS, Miller R, Bartolo DCC. Internal anal sphincter electromyographic frequency is related to anal canal resting pressure. Both are reduced in idiopathic faecal incontinence. Gut. 1990;31:A619.
55. O'Riordain DS, O'Connell PR. Completion proctectomy in ulcerative colitis. Br J Surg. 1987;84:136–7.
56. Bell AM, Pemberton JH, Hanson RB, Zinsmeister AR. Variations in muscle tone of the human rectum recordings with an electromechanical barostat. Am J Phys. 1991;260:G17–25.
57. Gibbons CP, Trowbridge EA, Bannister JJ, Read NW. Role of anal cushions in maintaining continence. Lancet. 1986;1:886–8.
58. Gibbons CP, Bannister JJ, Trowbridge EA, Read NW. An analysis of anal sphincter pressure and

anal compliance in normal subjects. Int J Color Dis. 1986;1:231–7.
59. Lestar B, Penninckx F, Kerremans RP. The composition of anal basal pressure. An in vivo and in vitro study in man. Int J Color Dis. 1989;4:118–22.
60. Murie JA, Sim AJ, Mackenzie I. The importance of pain, pruritus and soiling as symptoms of haemorrhoids and their response to haemorrhoidectomy or rubberband ligation. Br J Surg. 1981;68:247–9.
61. Aitchison M, Fisher BM, Carter K, McKee R, Mac Cuish AC. Impaired anal sensation and early diabetic faecal incontinence. Diabet Med. 1991;9:960–3.
62. Bannister JJ, Davison P, Timms JM, Gibbons C, Read NW. Effect of stool size and continency of defecation. Gut. 1987;28:1246–50.
63. Scarli AF, Kiesewetter WB. Defecation and continence. Dis Colon Rectum. 1970;13:81–107.
64. Varma KK, Stephens D. Neuromuscular reflexes of rectal continence. Aus N Z J Surg. 1972;41:263–72.
65. Burleigh DE, D'Mello A. Neural and pharmacologic factors affecting motility of the internal anal sphincter. Dis Colon Rectum. 1993;34:4–8.
66. Uher EM, Swash M. Sacral reflexes physiology and clinical application. Dis Colon Rectum. 1998;41:1165–677.
67. Sangwan YP, Solla JA. Internal anal sphincter: advances and insights. Dis Colon Rectum. 1998;41:1297–311.
68. Rendtorff RC, Kashgarian M. Stool patterns of healthy adult males. Dis Colon Rectum. 1967;10:222–8.
69. Martelli H, Devroede G, Arhan P, Duguay C, Dornic C, et al. Some parameters of large bowel motility in normal man. Gastroenterology. 1978;75:612–8.

3 New Concepts of Anorectal Anatomy, Physiology, and Surgery According to the Integral System

Peter Petros and Darren M. Gold

Introduction

Historically and even today, the cause of bladder and bowel dysfunction is thought to arise from the organs themselves. The *Integral Theory* (IT) of female urinary incontinence [1] is holistic and entirely anatomical theory. It is a different way of thinking. The Theory's concept is that the cause of dysfunction lies outside the organs, mainly laxity in the suspensory ligaments, because of collagen/elastin degeneration [1]. The Integral System [2] is an entirely anatomical management system based on the Integral Theory.

Diagnosis, reinforcement and repair of ligaments based on the Integral System [2] can give high rates of cure not only for pelvic organ prolapse (POP), but also can improve chronic pelvic pain, bladder and bowel dysfunctions of closure and evacuation.

New concepts for anorectal anatomy and physiology

- Ligaments and muscles work interactively, holisticallly.
- The almost identical nature of bladder and bowel control mechanisms.
- The anorectum is reflexly closed and opened by external directional pelvic striated muscles forces contracting against competent suspensory ligaments.
- Reflex control of the defecation reflex is by directional muscle forces contracting against competent ligaments.
- Voluntary control is by forward contraction of puborectalis against the symphysis.
- Role of damaged ligaments in intussusception and descending perineal syndrome.
- The critical role of internal anal resistance in evacuation disorders.
- Reversal of dysfunctions by reinforcement of damaged pelvic ligaments with tapes.

Pelvic ligaments and muscles are codependent and work together in a balanced system (Fig. 3.1). One cannot work without the other. However, it is the ligaments which are the most vulnerable to damage, especially during pregnancy, childbirth, and old age, where the collagen and elastin weaken. The striated muscles contract against the pubourethral (PUL) and uterosacral (USL) ligaments (Fig. 3.1). These muscles open and close the urethral and anal tubes and stretch the organs to support the bladder and bowel stretch receptors which control the micturition and defecation reflexes. A loose or weak ligament may weaken these muscles

Electronic supplementary material The online version of this chapter (https://doi.org/10.1007/978-3-030-43811-1_3) contains supplementary material, which is available to authorized users.

P. Petros (✉)
St. Vincent's, Department of Surgery, Sydney, NSW, Australia
e-mail: pp@kvinno.com

D. M. Gold
St. Vincent's Hospital Clinical School, Department of Surgery, Sydney, UNSW, Australia

L. C. C. Oliveira (ed.), *Anorectal Physiology*, https://doi.org/10.1007/978-3-030-43811-1_3

forces, so the patient may not be able to adequately close these tubes (incontinence), open them (evacuation difficulties), or stretch the organs sufficiently to control bladder and bowel urge incontinence.

Given the Integral System's anatomical emphasis, it is appropriate that we follow the dictums of the great Spanish anatomist and urologist Salvador Gil Vernet (1892–1987), who stated that it was not sufficient to describe an anatomical structure. He said that an answer was required to the question of "*what is it for*" [3] and

> *Precise, almost mathematical knowledge of anatomy is a highly fertile source of surgical applications, suggesting new techniques and helping perfect and simplify existing surgical methods, making them less mutilating and more benign and, in short, raising surgery to the rank of true science* [3].

Accordingly, we will outline the dynamic anatomy of relevant structures, muscle, ligaments, and nerves, then proceed to the key role of ligaments in function, role of damaged ligaments in dysfunction, then to diagnosis of damaged ligaments and surgical correction of the ligaments.

Normal Function of Bladder and Bowel

The bladder and rectum have only two functions, storage and emptying. For storage, the emptying tubes (urethra, anus) must be closed. Externally sited striated muscle forces compress these tubes to close them. Stretch receptors in the bladder and rec-

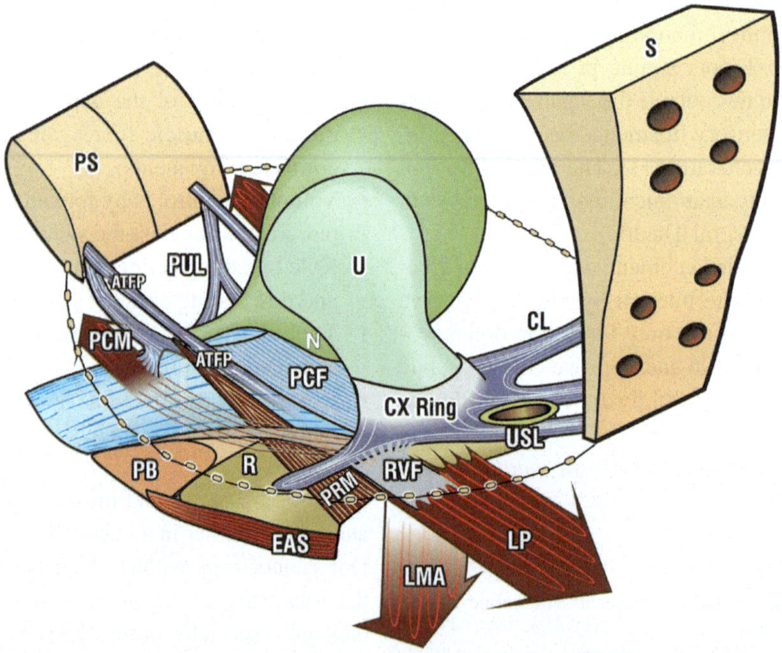

Fig. 3.1 There are 5 main ligamentous structures in the female pelvic floor. Pubourethral (PUL) inserts into midurethra and pubococcygeus muscle; Arcus tendineus fascia pelvis (ATFP): arises above PUL and inserts into ischial spine; Cardinal (CL): attaches anterior vaginal wall to the cervical ring (CX) and laterally to the skeleton; Uterosacral (USL): attaches CX to sacrum and lateral walls of the rectum); Perineal body (PB; is attached to the posterior part of descending ramus by deep transverse perineal ligaments (Fig. 3.2)). Urethra and anorectum in closed phase. Four muscles control function [4]. Three directional muscles (arrows), pubococcygeus muscle (PCM), levator plate (LP), and conjoint longitudinal muscle of the anus (LMA), contract against suspensory ligaments; the fourth muscle, puborectalis muscle (PRM), contracts only against pubic symphysis (PS); EAS = external anal sphincter. U = urethra; V = vagina; R = rectum. There are also the vaginal fascia usually known as pubocervical fascia (PCF) or rectovaginal fascia (RVF)

tum signal cortex when storage reaches capacity. If appropriate, the cortex activates the micturition and defecation reflexes. These reflexes relax the forward vectors and activate the posterior muscle forces to open out the urethra and anal tubes; the bladder and rectum contract to empty the organs (Fig. 3.5).

Ligaments have two main functions [4]:

1. *Structural*. To attach the organs and other structures to the skeleton (Fig. 3.1). Ligaments have a high content of collagen. When the ligaments are stretched, the collagen is tensioned like the ropes of a suspension bridge. The stretched collagen is very strong. Ligaments have a breaking strain of approximately 300 mg/mm [1, 5]. In contrast, because they need to expand for many of their functions, organs are far more elastic with a much lower breaking strain. For example, the vagina has a breaking strain of approximately 60 mg/mm^2.
2. *Functional*. To act as anchoring points for the three oppositely acting striated muscle force vectors deriving from PCM (pubococcygeus), LP (levator plate), and LMA (conjoint longitudinal muscle of the anus) (Fig. 3.2). These muscles force the following:
   - To open the urethral and anal tubes
   - To close the urethral and anal tubes
   - To stretch the organs to prevent the stretch receptors from firing off at a low volume

The main ligaments involved in the above functions are the pubourethral and cardinal/uterosacral complex.

Ligaments (Fig. 3.1)

There are five main ligamentous structures in the female pelvic floor:

- Pubourethral (PUL) ligament that inserts into midurethra and pubococcygeus muscle
- Arcus tendineus fascia pelvis (ATFP) ligament that arises above PUL and inserts into ischial spine

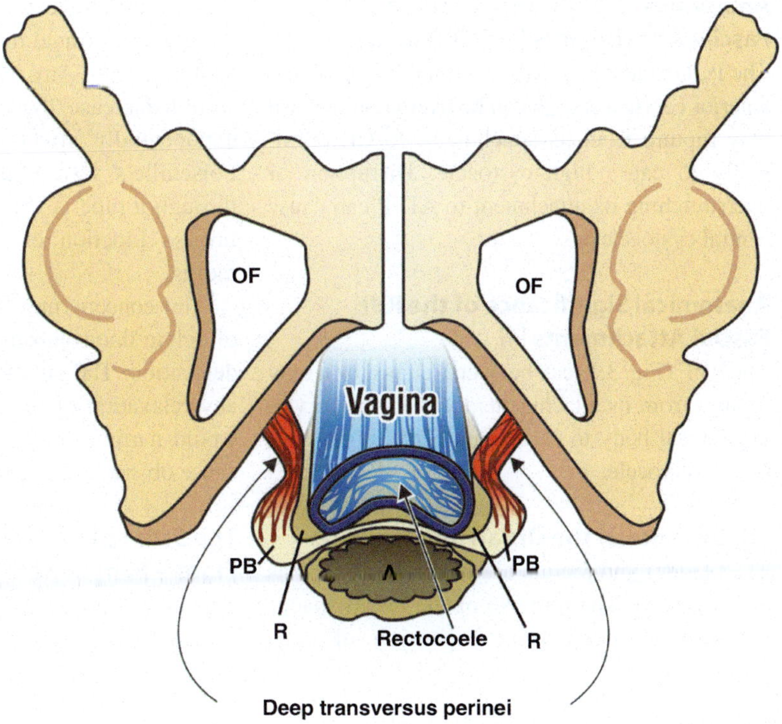

Fig. 3.2 Deep transversus perinei ligaments [4] attach perineal body "PB" to behind the descending ramus, between its upper 2/3 and lower 1/3. In this figure, the ligaments are elongated. PBs have been separated into two parts during childbirth, by stretching of their central part. This causes the rectum to protrude into the vagina as a rectocele

- Cardinal (CL) ligament that attaches anterior vaginal wall to the cervical ring (CX) and laterally to the skeleton
- Uterosacral (USL) ligament that attaches CX to sacrum and lateral walls of the rectum
- Perineal body (PB) ligament that is attached to the posterior part of descending ramus by deep transverse perineal ligaments

The Vaginal "Fascia" [4]

The vaginal fascia is usually known as pubocervical fascia (PCF) or rectovaginal fascia (RVF) and *is a fibromuscular tissue* composed of smooth muscle, collagen, elastin, nerves, and blood vessels. It forms the inner part of the vaginal wall. "Fascia" is the main structural component of the vagina. Ligaments have a similar histological structure to fascia but are far less distensile, with more collagen 1 and less elastin. PCF and RVF are attached to ligaments. Damage to PCF or RVF is invariably associated with damage to ligaments also. It is impossible for the RVF to stretch or rupture if the ligaments are intact.

Anatomical Significance of the PCF Fascial Attachments [4] PCF (Fig. 3.1)

The PCF is attached to the cardinal ligament and anterior cervical ring. It can be overstretched or it may rupture from its attachments to CL proximally to cause high cystocele. Disruption or overstretching its attachment to ATFP can cause central cystocele.

Anatomical Significance of the RVF Fascial Attachments [4]

The RVF (Fig. 3.1) can be overstretched or it may rupture from its attachments to USL, the cervix, or perineal body to cause enterocele, high, mid, or low rectocele.

Significance of the Organ Spaces

The bladder and rectum are highly distensile as is the vagina. The purpose of the organ spaces is to allow independent expansion of each structure. Especially important is the ability of the vagina to expand posteriorly with different sizes of the penis during intercourse. An elastic vagina allows the curved shape of the erect penis to stretch the apex around the curve of the sacrum. A vagina fibrosed by insertion of a mesh sheet may inhibit this action. Stretching a visceral structure anchored on one wall is an important cause of dyspareunia after mesh insertion.

The Muscles [4]

The pelvic floor muscles, smooth and striated, work in a highly coordinated holistic way with ligaments, fascia, and the organs themselves, directed by opening and closure reflexes and ultimately by the cortex itself (Fig. 3.8).

The Role of Smooth Muscle of the Urethra and Anus During Closure and Evacuation [1]

Smooth muscle must contract to create a semirigid tube. A semirigid tube allows the external striated muscle vector forces to efficiently close the urethral or anal tubes and to pull open their posterior wall. Any unstretched fold of mucosa would decrease the diameter of the tube and exponentially affect evacuation, according to Poiseuille's law, which states that the flow through a tube is exponentially determined, an inverse function of the fourth power of the radius.

This concept runs counter to existing concepts of pelvic floor relaxation during micturition and defecation. The emptying tubes would sag with any relaxation of the striated muscles, and this would narrow the diameter of the anal tube to cause obstructive defecation or urination.

The Striated Muscles of the Pelvic Floor – Functional Anatomy [4]

The pelvic floor has upper and lower striated muscle layers.

Upper Layers

The upper striated muscles of the pelvic floor consist mainly of slow-twitch fibers (approximately 80%) with 20 % fast-twitch fibers. There are three striated muscles of the pelvic floor which control all anorectal and urethral functions: pubococcygeus (PCM), levator plate (LP), and conjoint longitudinal muscle of the anus (LMA) (Fig. 3.3a, b). These three muscles (arrows) (Fig. 3.1) contract against suspensory ligaments. A fourth muscle, puborectalis (PRM), contracts only against the pubic symphysis (Fig. 3.1 and Fig. 3.3b). PRM is concerned with voluntary "squeezing," voluntary interruption of micturition, and defecation. It has an ancillary role in anorectal closure and defecation. The pelvic muscles have three main functions:

1. To support the organs from below
2. To close and open the urethra and anus
3. To stretch the organs bilaterally to prevent inappropriate activation of the micturition and defecation reflexes

The fast-twitch fibers act during stress to close the emptying tubes more tightly and to stretch the tubes open for evacuation.

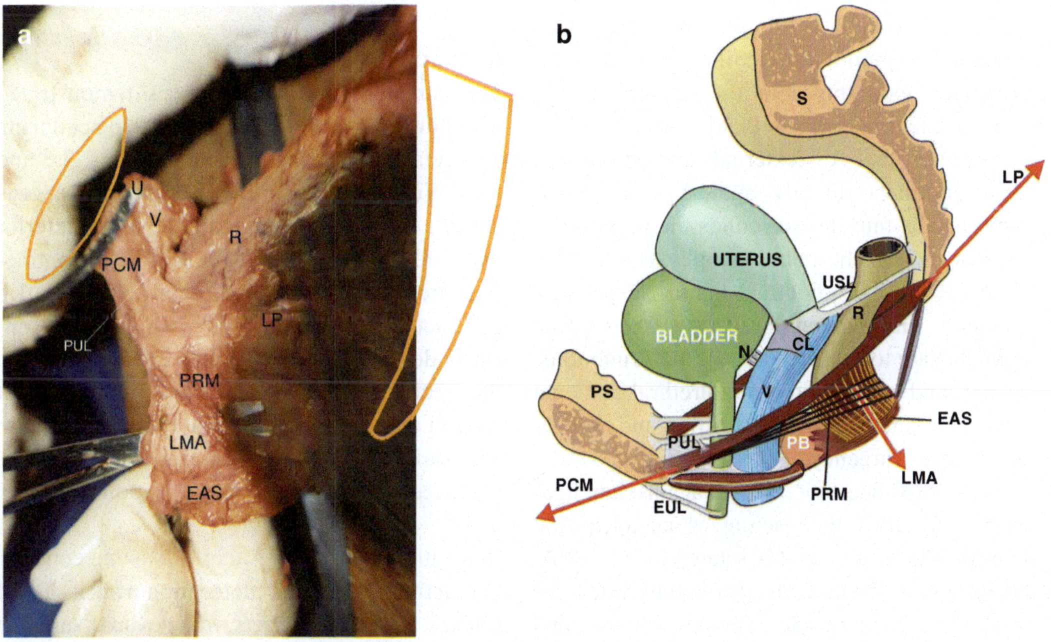

Fig. 3.3 (**a**) The pelvic floor striated muscles – cadaver specimen sagittal view. This is an anatomical specimen from a female cadaver, cut away from its bony insertions. The bladder and vagina have been excised at the level of bladder neck. Pubococcygeus muscle (PCM) inserts into the lateral wall of distal vagina (V). The PCMs sweep behind the rectum (R) and merge with the contralateral side to form part of the levator plate (LP). LP inserts into the posterior wall of the rectum. LMA connects LP to EAS (external anal sphincter). PRM (puborectalis muscle) surrounds the rectum (R) and inserts into the pubic symphysis; PUL – insertion of pubourethral ligament into lateral part of midurethra and PCM; LMA = (conjoint) longitudinal muscle of the anus; EAS = external anal sphincter. U = urethra; V = vagina. (**b**) Functional anatomy: diagrammatic sagittal analogue of the cadaver specimen. PCM attaches to lateral wall of the vagina "V." LP inserts into the posterior wall of the rectum "R"; PCM contracts forward against PUL. LP contracts backward against PUL. LMA connects LP to the external anal sphincter and contracts downward against USL

Three oppositely acting directional striated muscles control mechanical closure and evacuation of urethra and anorectum (Fig. 3.3b, arrows) [1]. The opposite stretching of tissues supports the organ stretch receptors to control the micturition and defecation reflexes. With reference to Fig. 3.1, pubococcygeus muscle (PCM) inserts into the lateral wall of distal vagina (V); it contracts forwards against PUL to close the distal part of urethra and to stabilize the perianal body and lower part of the anus; the lateral parts of PCM sweep behind the rectum (R) and merge with the contralateral side to form part of the levator plate (LP); LP inserts into the posterior wall of the rectum; LP contracts backward against PUL; because USLs are attached to the lateral rectal wall, they are also tensioned; LMA, the conjoint longitudinal muscle of the anus, is a vertically acting muscle. LMA takes longitudinal smooth muscle fibers from the longitudinal smooth muscle of the rectum and striated muscle fibers from PCM and LP; LMA connects LP to EAS (external anal sphincter); LMA contracts directly against USL; PRM (puborectalis muscle) surrounds the rectum (R) and inserts into the pubic symphysis. Because PCM, LP, and LMA contract against suspensory ligaments, their contractile force weakens if the ligaments are loose and so do all their functions, closure, and evacuation of the urethral and anal tubes and control of the micturition and defecation reflexes (urgency).

Note: PRM does not contract against any ligaments [4]. Other than acting as an anchoring point for the rectum as it is rotated by LP/LMA, it does not control the same involuntary functions as the other three muscles, opening, closure, and organ stretching to support stretch receptors. It is the muscle which contracts during “squeezing,” voluntary cutting off urine flow, or voluntary interruption of defecation.

Lower Layers (Fig. 3.4) [3]

The ischiocavernosus and bulbocavernosus attach to tissues lateral to the urethra. X-ray studies show that the distal urethra is pulled downward during closure and micturition, the latter action being important as part of the distal expansion of urethra required to reduce frictional resistance to the urine stream.

The deep transversus perinei (DTP) is actually a thick ligamentous structure containing collagen, smooth muscle, elastin, nerves, and blood vessels. It inserts into the perineal body and attaches it bilaterally immediately behind the descending ramus at the junction of the upper 2/3 and lower 1/3. The DTP stabilizes the whole perineum, anus, distal vagina, external anal sphincter, and even bulbocavernosus muscle. As such it will have an impact on many colorectal conditions, including hemorrhoids, anal mucosal prolapse, and descending perineal syndrome.

Anorectal Closure [4]

The same three directional forces demonstrated to act during urethral closure [2] also act during anorectal closure and evacuation (Figs. 3.5 and 3.6). With reference to Fig. 3.5, pubococcygeus (PCM) contraction immobilizes the anterior wall of the anus. Puborectalis (PRM) contraction immobilizes the anorectal junction. LP contracts backward against PUL. LMA contracts downward against USL. LP/LMA stretch the rectum backward/downward to rotate it around the contracted PRM to close the rectum. The anorectal angle is decreased and the cavity is closed off.

Defecation [4, 6]

On activation of the defecation reflex, PRM relaxes. LP contracts backward against PUL. LMA contracts downward. LP/LMA stretch opens the posterior wall of rectum (Fig. 3.5, *broken lines*; Fig. 3.7a, b; Video 3.1). This action reduces the internal resistance to expulsion of feces exponentially, inversely by the third power [6]. The active opening vastly reduces the work of the rectum in expelling feces.

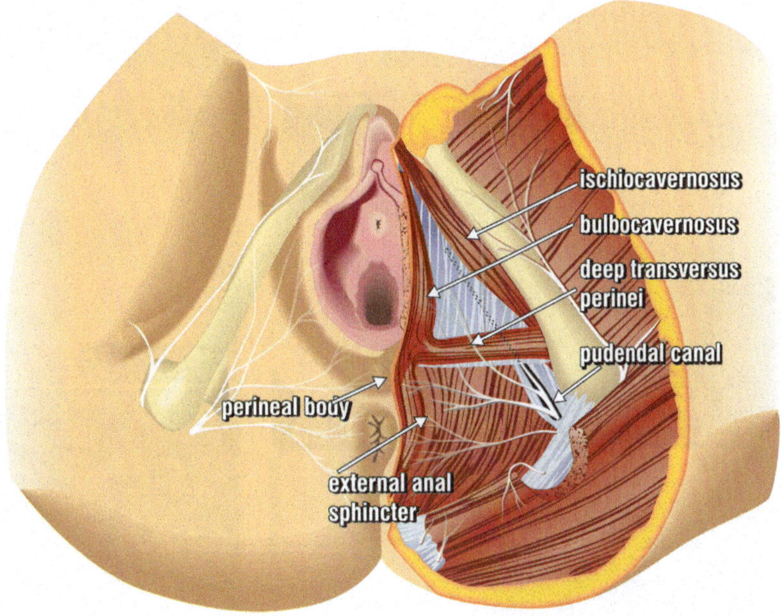

Fig. 3.4 Lower layer of muscles splint distal parts of the urethra, vagina, rectum, and perineal body. Patient supine. View into perineum. Lower-layer muscles of the pelvic floor anchor the organs distally

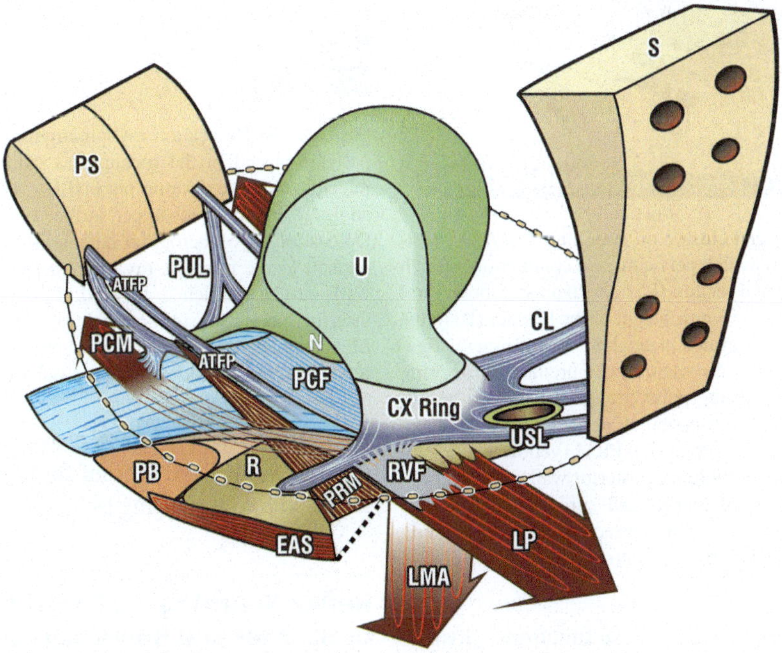

Fig. 3.5 Anorectal function – system in closed mode. *Anorectal closure* PRM contracts forward. LP contracts backward against PUL. Smooth muscles of the anorectal walls contract. LMA contracts downward against USL. The resultant forces rotate rectum around PRM to "close the anorectal angle and the cavity." *Defecation* (broken lines) PRM relaxes. LP/LMA pulls the posterior rectal wall backward/downward to open out the anorectal angle. PCM contracts forward to stiffen the anterior rectal wall. Both actions vastly decrease the internal anal resistance to passage of feces, inversely by third power of the radius. 5

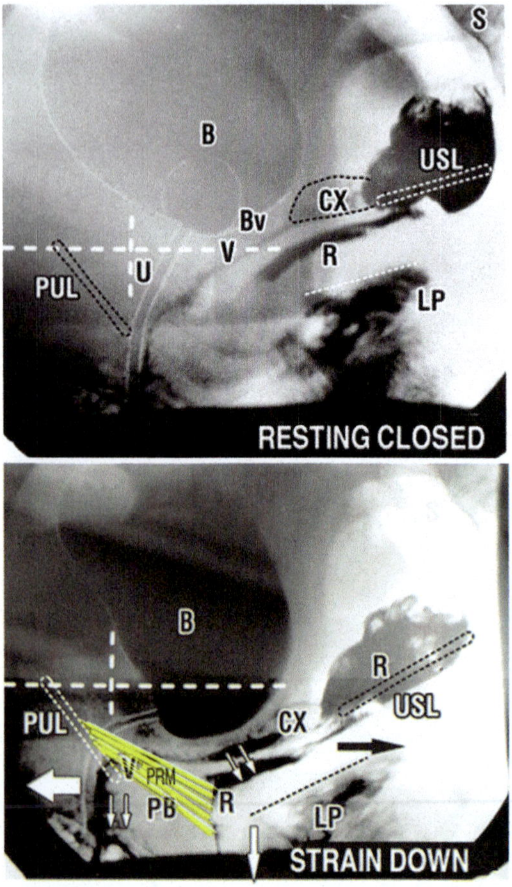

Fig. 3.6 Anorectal closure on straining. *Top figure* resting closed. Slow-twitch pelvic muscle contractions stretch the vagina (V) and rectum (R) back against pubourethral ligament (PUL). Bv = attachment of the bladder (B) to the vagina (v). LP = levator plate; USL = uterosacral ligaments; CX = cervix; S = sacrum; U = urethra; R = rectum. *Bottom figure* straining pubococcygeus (forward arrow) contracts forward to immobilize perineal body (PB) and anterior wall of the rectum (R). PRM (yellow lines) contracts forward to immobilize posterior wall of the rectum. LP/LMA (backward /downward arrows) contracts to rotate the rectum around a contracted PRM for closure. Arrows = fast-twitch muscle contractions

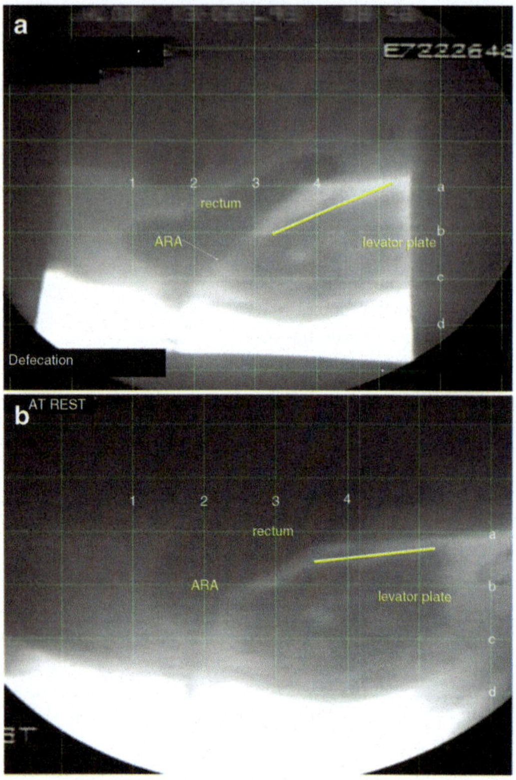

Fig. 3.7 (**a**) Defecation myoproctogram at rest. Barium paste is placed in the rectum and radio-opaque dye is injected into the levator plate (LP). ARA = anorectal angle. A line indicates upper surface of levator plate. (**b**) Defecation myoproctogram defecation starting. In comparison to the upper X-ray, the anterior rectal and anal walls are all stretched forward. Levator plate is seen inserting into the posterior rectal wall. The anterior part of LP is angulated downward. This action pulls the posterior rectal wall back and down to open out the anorectal canal

If USLs are loose, all these functions, closure, evacuation, and control of the defecation reflex may weaken to cause fecal incontinence and obstructed defecation and urge fecal incontinence (FI). Because LP also pulls against PUL anteriorly, a lax PUL may also cause FI. In this case, the FI is associated with USI "double incontinence." Clinical proof was provided by Hocking [7], who found >90% proof for both USI and FI with a midurethral sling [8].

Reflex Neurological Control of Bladder and Bowel Function

Similar mechanisms prevail for bowel control and evacuation [4]. In simple terms, both the bladder and bowel are controlled by cortical reflexes, a closure reflex for continence and micturition and defecation reflexes for evacuation (Fig. 3.8).

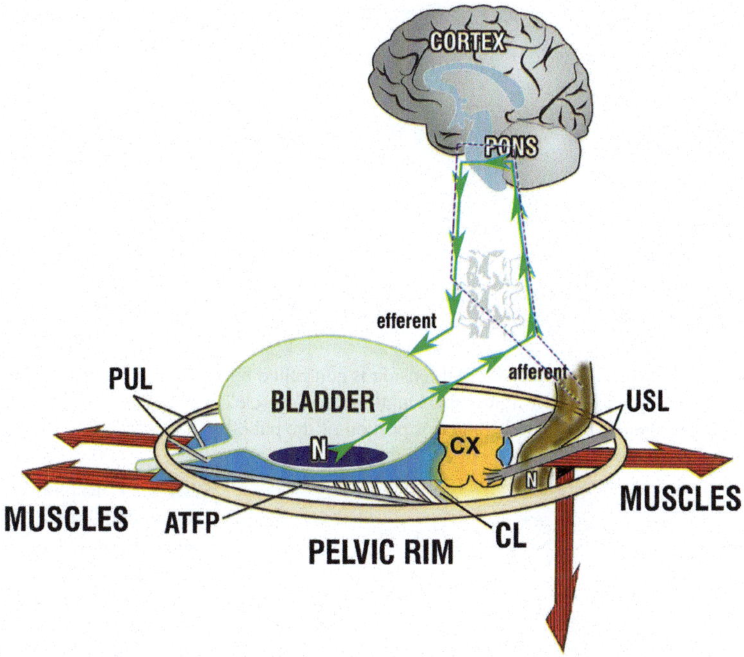

Fig. 3.8 Afferent fibers from stretch receptors "N" in the bladder and bowel transmit nerve impulses to the cortex which interprets them as "fullness." By stretching the organs bidirectionally (arrows), the muscles tension the underlying supports of the stretch receptors "N" to prevent stretch receptors from firing off prematurely, thereby controlling the micturition and defecation reflexes. At a certain point, the micturition reflex and defecation reflexes are activated. The cortex signals this as bladder or bowel urgency

The afferent nerve axis (*arrows*, Fig. 3.8) works as a sensor of organ fullness and sends afferent signals to the cortex. Once activated by the micturition or defecation reflex, the efferent nerve axis (Fig. 3.8, *arrows*) acts as a "motor" via a positive feedback loop (Fig. 3.8) [9] to coordinate the external opening of the outflow tract and evacuation by bladder or bowel contraction. A series of positive feedback loops in cats were described in detail by Barrington 100 years ago [9].

The presence of nerve endings in ligaments and vaginal fascia indicates that their smooth muscle content is also part of this holistic coordination.

Peripheral sensors activate the emptying reflexes, micturition, and defecation. In the human bladder, sensory nerves form a plexus immediately below the urothelium [10]. The urothelium, with its myelinated sensory nerves, acts as a mechanosensor. It controls the activity of the afferent nerves and, therefore, the micturition reflex, via specific neurotransmitters. It is thought that unmyelinated nerves may transmit pain and urge in pathological conditions, for example, interstitial cystitis. It is likely that a similar sensory system may initiate defecation.

Peripheral Neurological Feedback Mechanisms Balance Muscle Function

Discovery of a muscle spindle (Petros PE, Kakulas B and Swash MM, unpublished data) in the anterior portion of pubococcygeus muscle suggested that a precise feedback system controls the striated muscles which tension the connective tissue structures which support organ stretch receptors.

Pathogenesis of Loose Ligaments

There are three main causes of loose ligaments: congenital laxity, birth damage, and age [4]. Childbirth stretches the suspensory ligaments and/or distends the vagina or tears its fascial attachments to ligamentous structures (Figs. 3.9, 3.10, 3.11, and 3.12). In the younger woman, damage may be insufficient to affect structure or function. After the menopause collagen loss accelerates, and the ligaments weaken further to cause prolapse, chronic pain, bladder, and bowel dysfunctions.

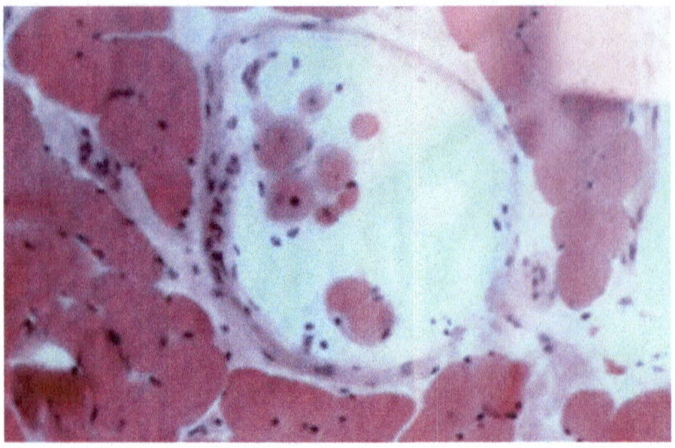

Fig. 3.9 Muscle spindle. Vaginal tension is controlled by muscle spindles which adjust the length of the muscle. Muscle spindle found in the anterior portion of the pubococcygeus muscle

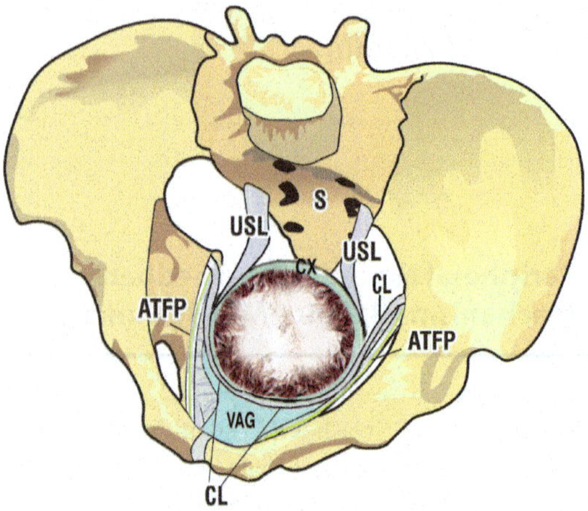

Fig. 3.10 Ligament damage at childbirth. View from above at 10 cm dilatation of the cervix. The maximal strain of the 10 cm dilatation is on the structures which attach to the cervical ring: vagina ("vag") uterosacral ligaments (USL) and cardinal (CL) ligaments. Extension or tearing of these attachments may cause apical prolapse, cystocele anteriorly, rectocele, and enterocele posteriorly. The ligamentous attachments of perineal body (deep transversus perinei) may be damaged and separated to cause low rectocele (perineocele) as the head exits the birth canal

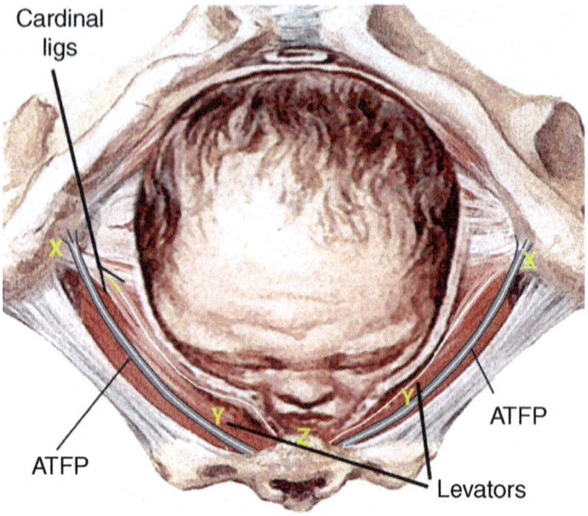

Fig. 3.11 Rupture/stretching points from cervical dilatation at childbirth [4]. The diameter of the pelvic inlet and outlet is only 12–13 cm. The smallest diameter of the flexed head is 9.4 cm. A deflexed head (posterior presentation) is 11.2 cm. ATFP is stretched and dislocates mainly at its insertion points at the ischial spine "X." The cardinal ligaments are stretched and may tear off the cervix at "Z" to be displaced laterally. Simultaneously with CL tearing, the pubocervical fascial layer of the vagina attached to CL and cervical ring at "Z" may be stretched or torn, so the vagina rotates down like a trapdoor, "high cystocele." The "levators" "Y" (pubococcygeus and puborectalis) may be stretched or torn from their insertions behind pubic symphysis

Uterosacral Ligament Laxity – A Key Element in Anorectal Dysfunction

(Fig. 3.13) [4]

The USLs are important structural components for the uterus, vaginal apex, and rectum. If loose, they can cause uterovaginal prolapse, enterocele, and rectal wall intussusception. The USLs are an important anchoring point for the backward/downward vectors. These activate urethral and anorectal closure (continence) and open the organs externally prior to evacuation. If the USLs lengthen, the striated muscle forces weaken [11]. The muscles cannot close the anus adequately, open it adequately for evacuation, or stretch the organs sufficiently to control the defecation reflex. The patient may complain of constipation, fecal leakage, or urge incontinence of feces. Often the prolapse is minor.

Gordon's law [11] (Fig. 3.14) is the ultimate pathway for understanding how pelvic ligament looseness may cause bladder and bowel dysfunction. It states "A striated muscle contracts optimally over a short length only" ("N"; Fig. 3.14, *lower Figure*). If the ligaments against which the three vector muscles contract are firm, the muscles contract efficiently over an optimal length "N." If the ligaments against which the three vector muscles contract lengthen by "E," the muscles lengthen accordingly by "E" and their contractile

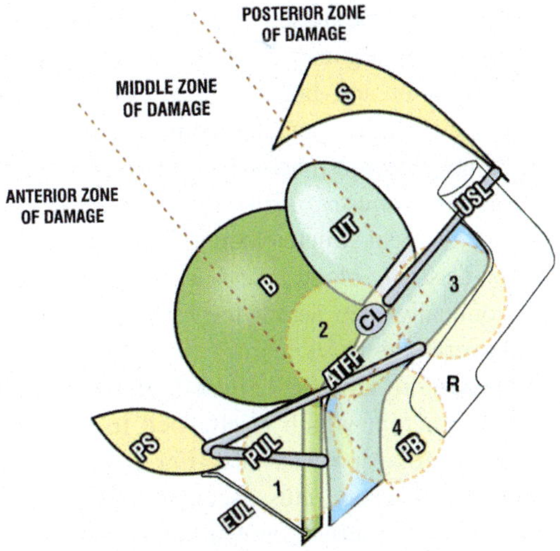

Fig. 3.12 Dysfunctions are a consequence of which ligaments are damaged. (1) Anterior-zone ligaments. External urethral ligament "EUL," pubourethral ligament "PUL," and pubovesical ligament "PVL." Anatomical dysfunction: (mainly) urinary stress incontinence. (2) Middle-zone ligaments: cardinal (CL) and ATFP. Anatomical dysfunction: cystocele. (3 and 4) Posterior-zone ligaments USL and PB (3) Uterosacral "USL" anatomical dysfunction: uterine prolapse. (4) Deep transversus perinei (DTP) as part of perineal body ("PB"). Anatomical dysfunction: low rectocele and descending perineal syndrome

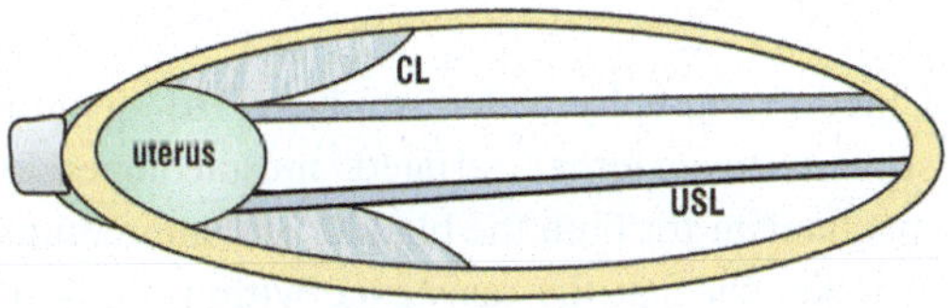

Fig. 3.13 Uterosacral (USL) and cardinal (CL) ligaments are related. Both stretch with prolapse, and both must be repaired to properly restore anatomy during surgical reconstruction. Because the fascial wall is attached to both, it too must be reattached or repaired during surgical reconstruction

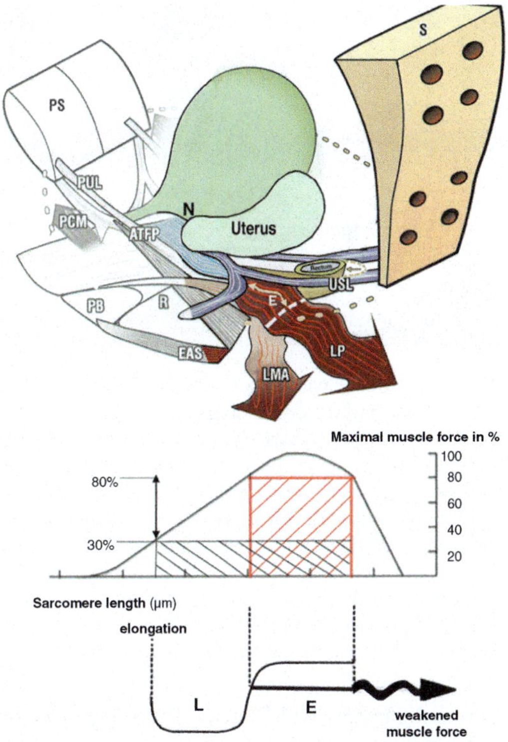

Fig. 3.14 3D view from above. The uterus has prolapsed to first degree. The USLs have elongated by "E," as have LP and LMA. The rectum also has descended, by virtue of its attachments laterally to the elongated USL. The "wavy" shape of LP and LMA indicates diminution of contractile strength. Gordon's law. A striated muscle contracts optimally over a short length only, "N," red square. If the ligament against which the pelvic muscle contracts lengthens by "E," the muscle also lengthens by "E." This results in a rapid loss of contractile force, from 80% to 30%, black rectangle

strength weakens [11], from a nominal 80% to 30%, (Fig. 3.14, lower Figure).

Perineal Body Laxity (Fig. 3.15)

Structure

The perineal body (PB) is an essential inferior supporting structure for the vagina and anorectum (Fig. 3.2). It is approximately 4 cm in length. It supports 50% of the posterior vaginal wall [12] and a significant part of the anterior wall of rectum. In the distal 2 cm, the PBs are densely adherent to the vagina and anus. PBs are attached to the descending ramus by deep transversus perinei (DTP) ligaments. At surgical dissection, DTP is distinctively whitish in its macroscopic appearance, and its structure is similar to the other ligaments, consisting of collagen, elastin, smooth muscle, nerve, and blood vessels. It has small amounts of striated muscle.

Anatomical and Surgical Significance of DTP Ligaments

It is not well known that the key anatomical components of the perineal bodies are the deep transversus

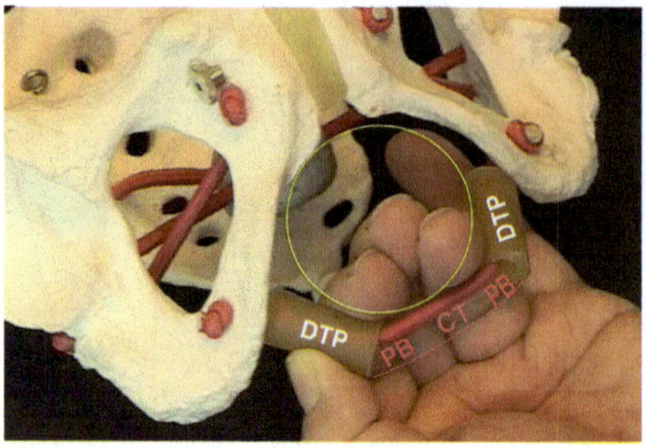

Fig. 3.15 Model pathogenesis of DTP ligaments and their causation of the descending perineal syndrome. The perineal body "PB" is attached behind the upper 2/3 and lower 1/3 of the descending ramus by deep transversus perinei ligaments (DTP). The PBs are connected in the midline by fibromuscular tissue, "CT." Childbirth may stretch PB and CT∗ and elongate and push DTP laterally. The rectum may protrude as a rectocele (fingers). The angulated DTPs are the ultimate cause of "descending perineal syndrome." ∗CT is known in some parts of Europe as the "central tendon of the perineal body"

perinei ligaments which attach PB to the skeleton [12–15]. When the rectum is dissected off the PB, the DTPs appear as whitish structures. They can elongate to cause "descending perineal syndrome" (Fig. 3.15). They can only be repaired by insertion of a tensioned tape which penetrates the DTP behind the descending ramus. It is not unusual for one DTP to be destroyed as a result of childbirth.

Perineal Body Function is Linked to USL Function

The PB and USL work as a unit. The extension of RVF from PB to cervix (Fig. 3.5) helps anchor the PBs when the backward vectors (LP/LMA) contract. Tensioning the RVF "smooths out" the rectal mucosa, and this helps facilitate rectal evacuation by reducing frictional resistance to fecal flow. Structural damage to PB, its RVF attachment, may result in a patient having to splint her perineum to adequately evacuate feces and mitigate anal mucosal prolapse or hemorrhoid descent during defecation. Stretching of RVF upward by the levator plate by slow-twitch levator plate contractions at rest helps prevent infolding and back pressure on the rectal veins which may otherwise manifest as hemorrhoids or anal mucosal prolapse. The deep transversus perinei (DTP) ligaments " (Fig. 3.15) attach PB to the skeleton and thus stabilize it. Loose DTPs may cause the descending perineal syndrome. DTPs have been erroneously called *deep transversus perinei muscles*. Histology demonstrates a typical ligament structure: smooth muscle, collagen, elastin, blood vessels, and nerves. Anatomically they are in the same position of "puboperinei muscles."

Damage to USL will cause both the vagina and the rectum which is attached to its sidewalls to prolapse downward (Fig. 3.16a). This may cause rectal intussusception and symptoms of obstructive defecation. Inability to stretch the loose rectovaginal fascia and the attached distal part of rectum upward may contribute to tissue back pressure. This may cause prolapsed rectal mucosa, hemorrhoids, and, if the backward pressure is sufficient, solitary rectal ulcer.

Anatomical Pathways to "Obstructive Micturition or Defecation" (Organ Emptying Problems)

Cinedefecography video studies [16] demonstrated an external striated muscle opening mechanism for both the bladder and anorectum. It is known that the resistance within a tube is inversely related to the radius (Poiseuille's law). A loose USL (Figs. 3.14 and 3.16a, b) may result in weakening of the urethral or anorectal LP/LMA opening forces. The bladder detrusor or rectum has to contract against an unopened tube. This is perceived by the patient as "obstructed micturition" or "obstructed defecation," with symptoms such as "feeling bladder has not emptied," "stopping and starting," multiple emptying, post-micturition dribble, raised residual urine and, for bowel, constipation or obstructive defecation (ODS). It was demonstrated that shortening and reinforcing CL/USL by TFS restore prolapse and the external opening mechanism with symptom and residual urine improvement for bladder [17].

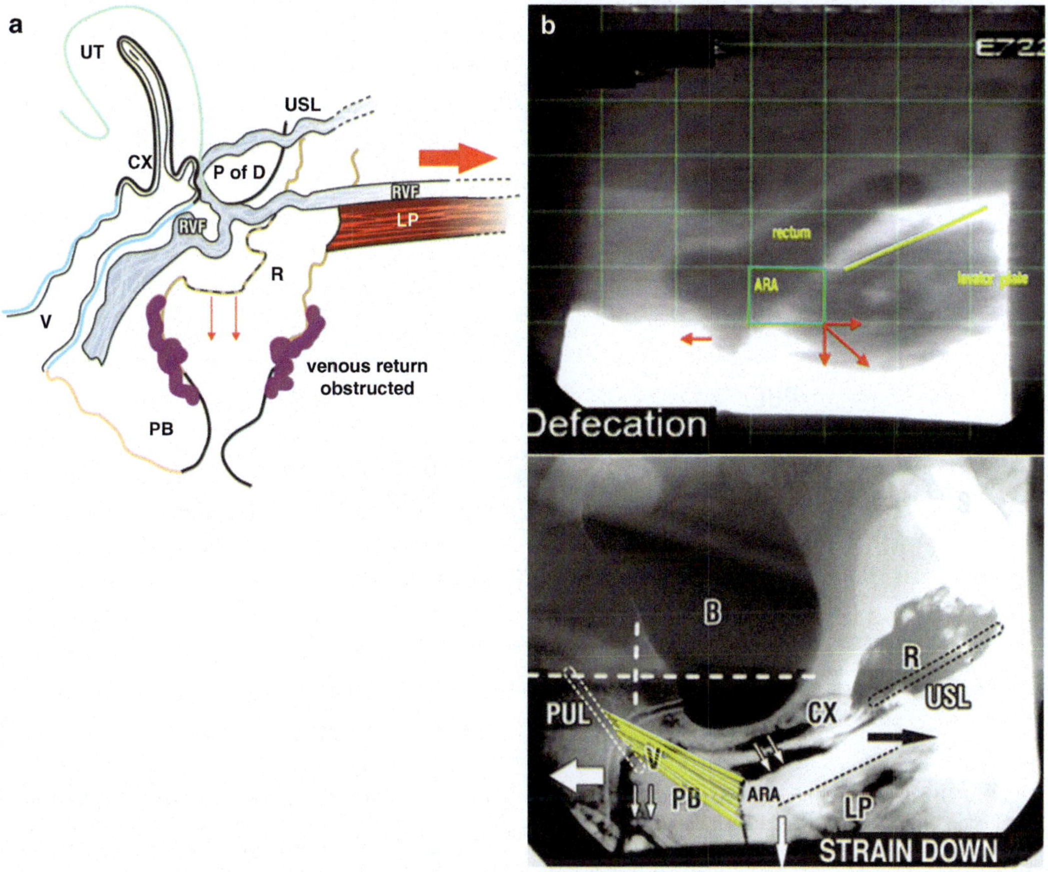

Fig. 3.16 (**a**) Structural effect of loose USLs. The rectum is held up by USLs like the apex of a tent. Loose USLs may cause anterior rectal wall intussusception ("R," downward arrows). The prolapsed rectal mucosa may cause back pressure in the rectal veins to cause hemorrhoids. (**b**) The paradox of co-occurrence of FI and ODS. *Upper figure, defecation.* A loose USL anchoring point will weaken ability to open the anorectal angle (ARA) to cause "obstructive defecation syndrome" (ODS). *Lower figure, anorectal closure.* A loose USL anchoring point will weaken ability to close the anorectal angle (ARA) to cause fecal incontinence

The Paradox of Obstructive Defecation (ODS) and Fecal Incontinence (FI) Coexistence

It is counterintuitive, but it is a fact that ODS and FI frequently coexist. A loose USL can explain both (Fig. 3.16b). Anorectal closure is achieved by the rectum being stretched backward and rotated by the backward/downward vectors (LP/LMA) around a contracted PRM, to create the anorectal angle, "ARA," lower figure "straining" (Fig. 3.16b). Relaxation of PRM allows the same backward/downward vectors to open out the anorectal angle, ARA, "defecation" (Fig. 3.16b, upper). The downward vector contracts against USL. A loose USL will weaken both mechanisms. The patient cannot adequately close the anus (FI) or open it (ODS).

Anatomical Pathway to Symptoms of Bladder and Bowel Urge Incontinence

Inability of the weakened pelvic muscles to stretch the connective tissue structures sufficiently to support bladder and rectal stretch

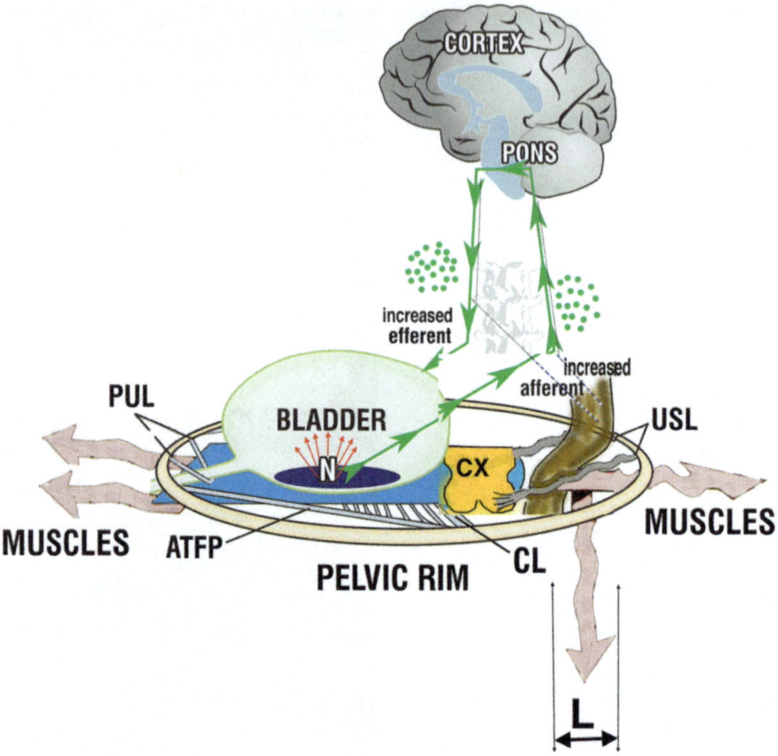

Fig. 3.17 Bladder bowel urge incontinence as interpreted by Gordon's law. The uterosacral ligaments (USL) lengthen "L" and are unable to suspend the vagina adequately. The downward/backward muscle forces (arrows) lengthen by "L" and weaken. The vagina cannot be stretched sufficiently to support the stretch receptors "N." "N" fire off increased afferent impulses at a low bladder volume, and this is perceived by the cortex as urgency. If the quantum of afferents is sufficient to activate the micturition reflex, the efferents are activated; the forward muscles relax; the backward muscles open out the urethra; the bladder contracts; the patient may uncontrollably lose urine ("urge incontinence"). The wavy form and pink color of the arrows denote weakened muscle contractile force

receptors may cause them to fire off at a low volume to stimulate the micturition and defecation reflexes. This is perceived by the patient as "urgency" or "frequency" (Fig. 3.17).

Function of Puborectalis Muscle ("Kegel" Muscle)

Puborectalis muscle (PRM) has an involuntary slow twitch action which assists anorectal closure and relaxes for defecation (Fig. 3.18). It also has a voluntary action in interrupting defecation and micturition. It is this "squeeze" function which is harnessed in Kegel pelvic floor exercises. PRM works by contracting directly against the pubic symphysis. Hence it is totally independent of the pelvic suspensory ligaments.

Diagnosis of Damaged Ligaments

Symptoms guide the surgeon as to which ligaments have been damaged (Fig. 3.19). The ligaments are checked by vaginal examination using specific diagnostic criteria [19]. "Simulated operations," supporting specific ligaments to assess diminution of USI, urge, or pain symptoms, further confirm diagnosis of which ligament has been damaged.

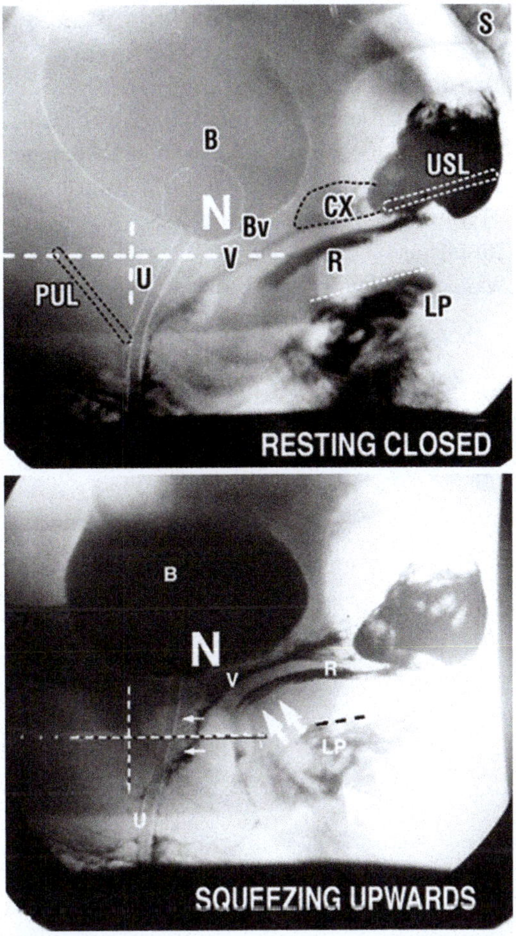

Fig. 3.18 Role of puborectalis in bladder A/R function. *Upper X-ray*. Resting Closed. Sitting lateral X-ray. U = urethra; V = vagina; B = bladder; Bv = fascial attachment of bladder base to vagina; CX = cervix; LP = levator plate. Broken white line coordinates drawn from fixed bony points. *Lower X-ray*. With reference to the horizontal and vertical coordinates, LP and all the organs have been lifted upward and forward (arrows). Lifting upward stretches and narrows the urethral and anal tubes, exponentially increasing the resistance to flow through the tubes, temporarily restoring continence. Note how the vagina has been elevated to temporarily support the bladder base and the stretch receptors "N." This explains how "squeezing" mitigates urgency

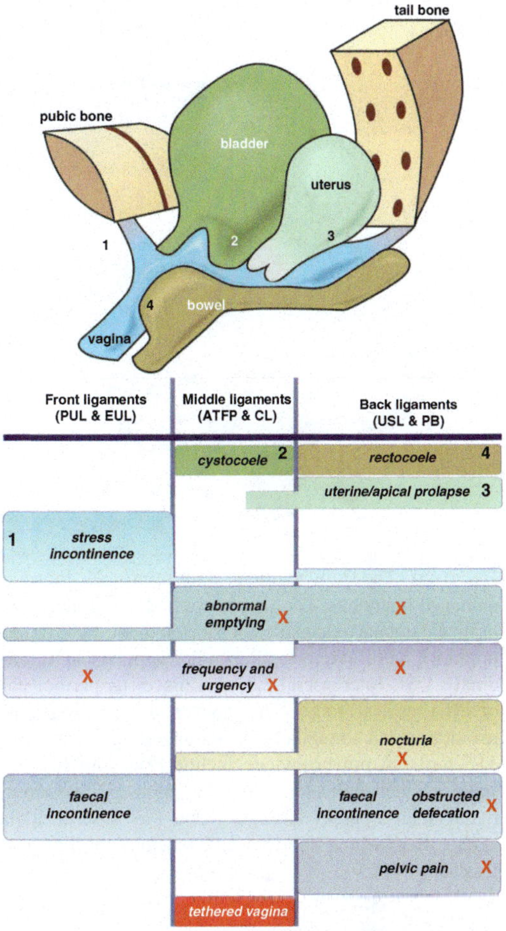

Fig. 3.19 Pictorial diagnostic algorithm. The damaged ligaments are located using the pictorial algorithm. Because symptoms such as urgency and abnormal emptying may be caused by ligament damage in more than one zone, all rectangles for that symptom should be marked. The area of the symptom rectangles indicates the estimated frequency of symptom causation occurring in each zone. The main ligaments causing the symptoms and prolapse in each zone are indicated in capital letters, two in each zone: PUL = pubourethral ligament, EUL = external urethral ligament (*anterior ligaments*); ATFP = arcus tendineus fascia pelvis, CL = cardinal ligament complex (*middle ligaments*); USL = uterosacral ligament, PB = perineal body (*posterior ligaments*). The numbers in the figure correlate with structural damage and with ligament damage: 1, USI; 2, cystocele; 3, uterine prolapse; 4, rectocele. The diagnosis for this patient is cardinal/uterosacral damage and perineal body ligament damage. Note: Major symptoms may occur with minimal ligament damage

Surgery

The damaged ligaments are shortened and repaired with tensioned strips of tape placed along the length of the damaged ligament (Fig. 3.20). PUL can be repaired using any midurethral sling method. The posterior sling can similarly be repaired by any posterior sling method (without mesh sheet attached). ATFP, PB, and CL can only be repaired using the TFS system [19].

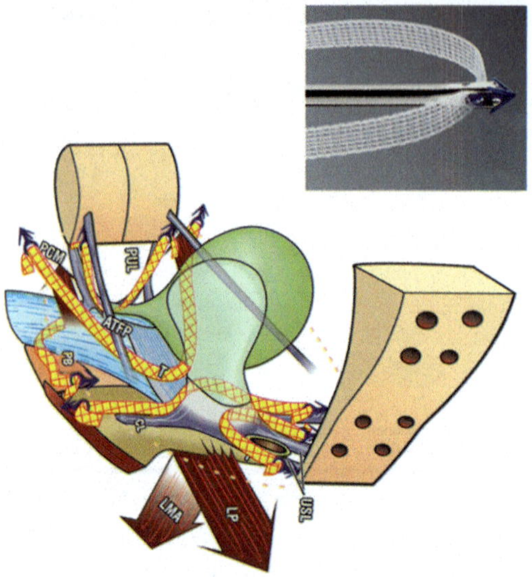

Fig. 3.20 The TFS shortens and reinforces loose or damaged ligaments. The tape is applied along the length of the ligaments to tension and shorten them: PUL (pubourethral); ATFP (arcus tendineus fascia pelvis); CL (cardinal); USL (uterosacral); deep transversus perinei part of PB (perineal body). The tape creates a collagenous reaction which strengthens the damaged ligament. Inset: TFS tool – A polypropylene anchor 11×4 mm sits on a stainless steel applicator. A lightweight macropore monofilament tape passes through a one-way system at the anchor base which shortens and tensions the damaged ligament

Results of Posterior Sling Surgery Cures Pain, Bladder, and Fecal Incontinence

Two different methods were used to repair uterine/apical prolapse ranging from stage 2 to stage 4, a total of 1420 patients, 809 with infracoccygeal sacropexy (posterior IVS) and 611 patients with TFS cardinal/uterosacral ligament repair [18]. Results are shown in Table 3.1. A 90% cure rate was achieved with both methods for POP [20].

Table 3.1 Results of posterior sling surgery cures pain, bladder, and fecal incontinence

	No. patients with symptom or condition/ total patients (%)				Confidence interval of difference between proportions[b]		Probability (two-tailed)[a]	
	Pre-PIVS	Pre-TFS	Post-PIVS	Post-TFS	Pre-PIVS vs pre-TFS	Post-PIVS vs post-TFS	Pre-PIVS vs pre = TFS	Post-PIVS vs post-TFS
Pelvic pain	405/809 (50)	194/611 (31)	131/809 (16)	42/611 (7)	0.13 to 0.23	0.06 to 0.12	<0.0001	<0.0002
Nocturia	286/809 (35)	254/611 (41)	59/809 (13)	77/611 (7)	0.011 to 0.11	0.021 to 0.085	0.017	0.0008
Urge incontinence	322/809 (40)	317/611 (52)	100/809 (12)	51/611 (8)	0.069 to 0.17	0.007 to 0.071	<0.0002	0.015
Frequency	233/549 (42)	310/611 (51)	48/549 (9)	55/611 (9)	0.025 to 0.14	−0.03 to 0.035	0.0047	0.87
Fecal incontinence	69/324 (21)	95/532 (17)	17/324 (5)	34/532 (6)	−0.015 to 0.094	0.023 to 0.042	0.167	0.49
Apical prolapse	809/809 (100)	611/611 (100)	56/809 (7)	63/611 (10)	−0.005 to 0.005	0.0047 to 0.064	#	0.022

[a]No continuity correction
[b]Z-test comparing PIVS vs TFS, before or after surgery (www.vassarstats.net)

Summary

Bladder and anorectal function are linked and conceptually similar in normal function and dysfunction. Three cortically controlled directional striated muscle vector forces pull in opposite directions against firm suspensory ligaments. *This concept is key to understanding the Integral Theory. If the ligaments are loose, the contractile force of the muscles which contract against them diminishes. All functions related to the muscles are affected: closure, evacuation, control of the micturition, and defecation reflexes.*

These actions stretch the tissues to close the urethral and anorectal tubes. Simplistically, the forward muscles close the distal part of these tubes. The backward muscles stretch the proximal part of the tubes backward and downward against firm suspensory ligaments to close them in a "kinking-type" action. The same opposite muscle forces stretch the organs to support peripheral stretch receptors in the bladder and bowel mucosa. All these three actions are dependent on firm suspensory ligaments. When the bladder and bowel are ready to evacuate, afferent impulses from their stretch receptors proceed to the cortex, which perceive them as the urge symptom. The urge symptom signals early activation of the micturition and defecation reflexes. When fully activated, these reflexes coordinate a series of events which proceed to evacuate the organs. The first event is coordinated relaxation of the forward muscles which hold closed the posterior walls of the urethral and anal tubes, PCM (pubococcygeus muscle) for the urethra and PRM (puborectalis muscle) for the anus. Relaxation of PCM or PRM allows the posterior muscle vectors LP (levator plate) and LMA (conjoint longitudinal muscles of the anus) to pull open the posterior walls of the urethra or anus. This action opens out the urethra and anus and vastly diminishes the internal resistance to evacuation of urine and feces. The smooth muscle walls of the bladder and rectum then contract in a type of spasm to empty their contents.

Age and childbirth weaken the collagen of the suspensory ligaments, collagen being the most vulnerable component of this system. Because the striated muscle vectors require firm ligamentous anchoring points, the contractile forces which close the urethral and anal tubes weaken. The outlet tubes cannot be closed adequately (incontinence). Because the same backward forces also open the urethral and anal tubes, inadequate backward stretching to open the tubes means the bladder and rectum have to open against a partly unopened tube. The patient perceives this as "obstructive symptoms," micturition difficulties, or constipation. Nor can the organs be stretched sufficiently to support the stretch receptors. These may fire off inappropriately to be perceived as urge incontinence of urine or feces. A symptom-based diagnostic algorithm guides diagnosis of damaged ligaments. The damage is confirmed on vaginal examination using specific morphological criteria. Surgery consists of shortening and reinforcing the damaged ligament along its length using precisely tensioned slings. The restored ligament reverses the cascade of events and so restores normal function.

> *Repair the structure and you will restore the function* – Integral Theory

The take-home message is that the urethra and anus are emptying tubes subject to all the physical rules of flow through a tube. No flow (closure) = continence. Full flow = normal evacuation. Flow is exponentially determined by the radius of the tube. The posterior muscles contract against USL to either close the anal tube or open it by pulling back the posterior wall of the rectum. A loose USL can invalidate both closure (FI) and evacuation (ODS). Because flow is exponentially determined, only a minor looseness of USL can cause FI, ODS, or both. Shortening and reinforcing the USLs with short strips of implanted tape have been demonstrated to reverse the cascade of dysfunctions, not only anorectal but also bladder and chronic pelvic pain.

References

1. Petros PE, Ulmsten U. An integral theory of female urinary incontinence. Acta Obstet Gynecol Scand. 1990;69(Supp. 153):1–79.
2. Petros PE. Doctor of Science Thesis University of Western Australia 2012 The Integral Theory System. A management system based on the Integral Theory: a universal theory of pelvic floor function and dysfunction in the female.
3. Gil-Vernet JM. Salvador Gil-Vernet, a pioneer in Urological Anatomy, Chaper in De Historia Urologiae Europeaeae Vol 24 2017, Eds. Van Kerrebroecke P, Schultheiss D, EAU History Office European Association of Urology. p. 13–43.
4. Petros PE. Chapter Chapter 2: The anatomy and dynamics of pelvic floor function and dysfunction. In: Petros PE, editor. The female pelvic floor. Heidelberg: Springer; 2010. p. 17–77.
5. Yamada H. Aging rate for the strength of human organs and tissues. In: Evans FG, editor. Strength of biological materials. Balt: Williams & Wilkins Co; 1970. p. 272–80.
6. Bush M, Petros P, Swash M, Fernandez M, Gunnemann A. Defecation 2: internal anorectal resistance is a critical factor in defecatory disorders. Tech Coloproctol. 2012; https://doi.org/10.1007/s10151-012-0860-3.
7. Hocking IW. Experimental study no. 9: double incontinence, urinary and fecal, cured by surgical reinforcement of the pubourethral ligaments. Pelviperineology. 2008;27:110.
8. Petros PE, Swash M. The Musculoelastic theory of anorectal function and dysfunction. J Pelviperineology. 2008;27:89–93.
9. Barrington FJ. The relation of the hind- brain to micturition. Brain. 1921;44:23–53.
10. Everaerts W, Gevaert T, Nilius B, De Ridder D. On the origin of bladder sensing: Tr(i)ps in urology. Nerourol Urodynam. 2008;27:264–73.
11. Gordon AM, Huxley AF, Julian FJ. The variation in isometric tension with sarcomere length in vertebrate muscle fibres. J Physiol. 1966;184(1):170–92.
12. Abendstein B, Petros PE, Richardson PA, Goeschen K, Dodero D. The surgical anatomy of rectocele and anterior rectal wall intussusception. Int Urogynecol J. 2008;19(5):513–7.
13. Wagenlehner FME, Del Amo E, Santoro GA, Petros P. Live anatomy of the perineal body in patients with third- degree rectocele. Color Dis. 2013;15(11):1416–22.
14. Wagenlehner FM, Del Amo E, Santoro G, Petros P. Perineal body repair in patients with 3rd degree rectocele. A critical analysis of the tissue fixation system. Color Dis. 2013;15(12):e760–5.
15. Petros PEP, Inoue H. How I do it? Transvaginal perineal body repair for low rectocoele. Tech Coloproctol. 2013;17:449–54. https://doi.org/10.1007/s10151-013-0980-4.
16. Petros PE, Bush MB. The same posterior muscle vectors act to open urethra and anus during micturition and defecation. Pelviperineology. 2017;36:35–6.
17. Petros P, Lynch W, Bush M. Surgical repair of uterosacral/cardinal ligaments in the older female using the tissue fixation system improves symptoms of obstructed micturition and residual urine. Pelviperineology. 2015;34:112–6.
18. Liedl B, Inoue H, Sekiguchi Y, et al. Is overactive bladder in the female surgically curable by ligament repair? Cent European J Urol. 2017;70:51–7.
19. Petros PE. Chapter 3 Diagnosis of connective tissue damage. In: Petros PE, editor. The female pelvic floor. Heidelberg: Springer; 2010. p. 77–117.
20. Wagenlehner F, Muller-Funogea I, Perletti G, Abendstein B, Goeschen K, Inoue H, Sekiguch Y, Sivaslioglu A, Haverfield M, Richardson P. Vaginal apical prolapse repair using two different sling techniques improves chronic pelvic pain, urgency and nocturia– a multicentre study of 1420 patients. Pelviperineology. 2016;35:99–104.

4 Clinical Evaluation of Continence and Defecation

Lucia Camara Castro Oliveira, Andrea Povedano, and Raul Fonseca

Anal continence is maintained by complex mechanisms, previously described in Chap. 2. Anal incontinence is defined as the involuntary loss of stool or gas and the inability to postpone an evacuation until socially convenient in an individual above 4 years of age [1]. The definition has been simplified by the International Continence Society recently, as to the involuntary loss of flatus, liquid, or solid stool that is a social or hygienic problem [2]. It can be related to the change in any involved mechanisms (see Table 2.1 in Chap. 2) and has multiple etiology (Table 4.1).

Therefore, clinical evaluation should be very meticulous providing important information regarding the severity of the problem and the impact on quality of life.

Table 4.1 Etiologies of anal incontinence

Anal sphincter dysfunction	Congenital anorectal malformations Radiation therapy Obstetric anal sphincter injury
	Anal surgery Perianal fistulas Sexual abuse
Rectal disorders	Inflammatory bowel disease Radiation proctitis Rectocele Rectal intussusception Rectal prolapse Fecal impaction
Neurological disorders	Spinal cord lesions Stroke Multiple sclerosis Spina bifida Diabetic neuropathy Obstetric nerve damage
	Systemic scleroderma
Rapid colorectal transit time	Chronic diarrhea Irritable bowel syndrome
Psychological	Encopresis Dementia

Clinical Evaluation of Continency

Evaluation of an incontinent patient should begin with a detailed history, focusing on defecatory habits and particular aspects of defecation, type, and frequency of incontinent episodes, as well as the associated factors, such as previous surgeries, use of medications, or trauma to the anorectal region. With a structural workup as proposed by Ruiz and Kaiser [3], the strategy is to review all the important issues in the diagnostic process and define the type and

L. C. C. Oliveira (✉)
Director Department of Anorectal Physiology of Rio de Janeiro and CEPEMED, Rio de Janeiro, Brazil

A. Povedano
Universidad Federal do Estado do Rio de Janeiro, Department of General Surgery, Rio de Janeiro, Brazil

R. Fonseca
Pelvic Floor Clinic of Guadalajara, Department of Functional Disorders, Zapopan, Mexico

L. C. C. Oliveira (ed.), *Anorectal Physiology*, https://doi.org/10.1007/978-3-030-43811-1_4

mechanism of the incontinence, providing the best treatment options to each case scenario (Table 4.2).

According to the last guideline for the treatment of fecal incontinence by the American Society of Colon and Rectal Surgeons (ASCRS), a thorough clinical history should be obtained and has a grade 1C recommendation [4]. With this initial clinical information, we should be able to better classify the type of incontinence and the basic involved mechanism.

This clinical evaluation has allowed the development of many anal incontinence scoring systems which could help us to objectively describe and measure the type, frequency of incontinent episodes, the impact on quality of life, and therefore the severity of the problem. The ASCRS and International Continence Society's (ICS) already mentioned guidelines recommend the use of those instruments because they can help to select the best treatment options as well as to measure the response to treatment over time. Moreover, the use of similar validated instruments can facilitate comparisons between different institutions.

In Table 4.3 we can appreciate the different incontinence scoring systems described in the medical literature [5–17]. Unfortunately, the initial scores left immediate gaps by way of com-

Table 4.2 Structured workup of patients with fecal incontinence

Assessment tool	Details
History	Onset Quantification: staining, soilage, seepage, accidents Qualitative assessment: passive incontinence or urge incontinence Obstetric history: pregnancies, vaginal deliveries Previous surgeries: anorectal surgery, hysterectomy, bladder surgery (colon and rectal surgery, spinal surgery Underlying diseases (diabetes, stroke, etc.) Bowel function and stool quality Incomplete evacuation Stool/gas passage through vagina Medications
Scoring instruments	Cleveland Clinic Florida Fecal Incontinence Score (CCF-FIS) Fecal Incontinence Quality of Life score (FIQoL) Fecal Incontinence Severity Index (FISI) St. Mark's Incontinence Score EORTEC SF-36 Other scoring instruments
Physical examination	Inspection: patulous anus, folds, perineal body, keyhole, skin irritation, perineal descent, prolapse, cloaca, rectovaginal fistula (stool in the vagina) Digital exam: sphincter integrity, tone (rest/squeeze), compensatory contraction/discoordination, rectocele, mass Sensation/anal reflex Instrumentation/visualization: rule out other pathologies (i.e., rectal tumor, proctitis)
Anorectal physiology testing	Manometry Anorectal sensation and volume tolerance Compliance measurement Nerve studies: PNTML, occasionally EMG Placement of SNS trial electrode (phase I)
Additional evaluation in select cases	Imaging: anorectal and endovaginal ultrasound and dynamic MRI Defecating proctogram Evaluation by other specialties (urogynecology, urology, gastroenterology, etc.)

Table 4.3 Incontinence scoring systems

Authors	Scoring systems
Kelly [5]	Points: 0–2 = poor; 2–4 = fair; 5–6 = good
Parks [6]	1 = Normal 2 = Difficult control of flatus and diarrhea 3 = No control of diarrhea 4 = No solid stool control
Lane [7]	True incontinence = loss of feces without perception or control Partial incontinence = passage of flatus or mucus under the same conditions Overflow incontinence = result of rectal distention without sphincter relaxation
Rudd [8]	1 = Continence 2 = Minor leak 3 = Acceptable leak 4 = Unsatisfactory major leak 5 = Total failure
Holschneider [9]	Continence [resting tone at manometry (rt) > 16 mmHg] Partial continence (rt = 9–15) Incontinence (rt < 8)
Keighley and Fielding [10]	Minor = fecal leakage once a month or less, to diarrhea Moderate = incontinence once a week to solid stool Severe = incontinence in most days, perineal pad
Corman [11]	Excellent = continent all the time Good= continent, but may require enemas Fair= incontinent for liquid stool Poor= incontinent for solid stool
Hiltunen [12]	Continent, partially continent, totally incontinent
Broden [13]	1 = None 2 = Medium 3 = Severe incontinence
Womack [14]	A = Continence B = Incontinence for liquid stool C = Incontinence to flatus and diarrhea D = Totally incontinent
Rainey [15]	A = Continence B = Incontinence to liquid stool C = Incontinence to solid stool
Miller [16]	Grade I = incontinence less frequent than once a month Grade II = between once a month and once a week Grade III = more than once a week Score= flatus 1–3, fluid 4–6, solid 7–9
Pescatori [17]	Incontinence for A = flatus/mucus; B = diarrhea; C = solid stool 1 = Occasionally 2 = Weekly 3 = Daily Score = from 0 (continent) to 6 (severe totally incontinent)

parison and generally did not take into account the importance of altered quality of life. From the initial proposal of Pescatori et al. in 1992 [17], Jorge and Wexner [18] developed the first incontinence score that takes into account the impact of incontinence in quality of life (Table 4.4). This scoring system was later validated by Rothbarth et al. [19] and became known as the Cleveland Clinic Florida Fecal Incontinence Score (CCF-FIS), adopted worldwide, facilitating the comparisons between centers. Subsequently, Vaizey et al. [20] introduced a new scoring system, including the evaluation of urgency and oral medications (Table 4.5).

Table 4.4 Cleveland Clinic Florida Fecal Incontinence Score (CCF-FIS) [18]

	Frequency				
Type of incontinence	Never	Rarely	Sometimes	Usually	Always
Solid	0	1	2	3	4
Liquid	0	1	2	3	4
Gas	0	1	2	3	4
Wears pad	0	1	2	3	4
Lifestyle alteration	0	1	2	3	4

Never, 0 (never); rarely (<1/month); sometimes (<1/week >1/month); usually (<1/day >1/week); always (>1/day). 0 = perfect continence; 20 = complete incontinence

Table 4.5 Vaizey Continence Index [20]

	Never	Rarely	Sometimes	Weekly	Daily
Solid stool	0	1	2	3	4
Liquid stool	0	1	2	3	4
Gas	0	1	2	3	4
Lifestyle alteration	0	1	2	3	4
				No	Yes
Need to use anal plug				0	2
Taking constipating medications				0	2
Lack of ability to defer defecation for 15 min				0	4

Never, no episodes in the past 4 weeks; Rarely, one episode in the past 4 weeks; Sometimes, >1 episode in the past 4 weeks but <1 a week; Weekly, one or more episodes a week but <1 a day; Daily, one or more episodes a day. Add one score from each row: minimum score = 0 = perfect continence; maximum score = 24 = totally incontinent

Another important aspect of clinical evaluation is the evaluation of quality of life. The American Society of Colon and Rectal Surgeons (ASCRS) developed an instrument based on the quality of life questionnaire SF-36, where questions are separated into four domains: (1) lifestyle, (2) habits, (3) self-esteem and possible depression symptoms, and (4) shame due to the problem (Table 4.6) [21]. The use of validated and reproducible scoring systems and quality of life instruments constitutes an important aspect of the clinical evaluation process [22].

More recently, with the integration of specialties and development of a pelvic floor subspecialty, we understood the importance of assessing the pelvic floor as a unit. In 2014, Altomare [23] et al. proposed another scale to assess the organs of the pelvic floor in a more holistic way – the Three Axial Perineal Evaluation (TAPE), which includes evaluation of incontinence, constipation, urinary incontinence, sexual dysfunction, pelvic organ prolapse, and urinary retention (Table 4.7). In this three-axial scale, the data is transformed into a visual geometric graphic (Fig. 4.1). Finally, because one of the consequences of the treatment of rectal cancer is sphincter dysfunction, a specific scale was proposed by Laurberg and colleagues [24] to assess patients with incontinence and defecation disorders – low anterior resection score (LARS), which will be discussed in Chaps. 32 and 33 (Table 4.8).

As presented by Ruiz and Kaiser, our personal clinical workup for incontinent patients in the Department of Anorectal Physiology of Rio de Janeiro [25] also includes the following steps: (1) detailed clinical history to assess type and frequency of incontinence episodes with evaluation of relevant aspects such as previous surgeries, concomitant systemic diseases, congenital malformations, sexual abuse, and obstetrical history (Table 4.9); (2) application of the CCF-FIS and Fecal Incontinence Quality of Life (FIQL) scale to assess severity and impact of symptoms; (3) detailed physical exam looking for scars (Fig. 4.2), soiling, anorectal pathologies with wet anus (Fig. 4.3), pelvic floor prolapses, including rectoceles (Fig. 4.4), perianal dermatitis, patulous anus

Table 4.6 Fecal Incontinence Quality of Life (FIQoL) Scale [21]

Question 1: In general would you say that your health is
1. ( ) excellent; 2. ( ) very good; 3. ( ) fair; 4. ( ) regular; 5. ( ) poor

Question 2: For each of the items, please indicate how often is this issue a concern for you due to accidental bowel leakage. [If it is a concern for you for reasons other than accidental bowel leakage, then check the box under not applicable (N/A).]

Due to accidental bowel leakage:	**Most of the time**	**Some of the time**	**A little of the time**	**None of the time**	**Not applicable (N/A)**
A. I am afraid to go out	1	2	3	4	
B. I avoid visiting friends	1	2	3	4	
C. I avoid staying overnight away from home	1	2	3	4	
D. It is difficult for me to get out and do things like going to a movie or to church	1	2	3	4	
E. I cut down on how much I eat before I go out	1	2	3	4	
F. Whenever I am away from home, I try to stay near a restroom as much as possible	1	2	3	4	
G. It is important to plan my schedule (daily activities) around my bowel pattern	1	2	3	4	
H. I avoid traveling	1	2	3	4	
I. I worry about not being able to get to the toilet in time	1	2	3	4	
J. I feel I have no control over my bowels	1	2	3	4	
K. I can't hold my bowel movement long enough to get to the bathroom	1	2	3	4	
L. I leak stool without even knowing it	1	2	3	4	
M. I try to prevent bowel accidents by staying very near a bathroom	1	2	3	4	

Question 3:Due to accidental bowel leakage, indicate the extent to which you agree or disagree with each of the following items. [If it is a concern for you for reasons other than accidental bowel leakage, then check the box under not applicable (N/A).]

Due to accidental bowel leakage:	**Strongly agree**	**Somewhat agree**	**Somewhat disagree**	**Strongly disagree**	**Not applicable (N/A)**
A. Feel ashamed	1	2	3	4	
B. I cannot do many of things I want to do	1	2	3	4	
C. I worry about bowel accidents	1	2	3	4	
D. I feel depressed	1	2	3	4	
E. I worry about others smelling stool on me	1	2	3	4	
F. I feel like I am not a healthy person	1	2	3	4	
G. I enjoy life less	1	2	3	4	
H. I have sex less often than I would like to	1	2	3	4	
I. I feel different from other people	1	2	3	4	
J. The possibility of bowel accidents is always on my mind	1	2	3	4	
K. I am afraid to have sex	1	2	3	4	
L. I avoid traveling by place or train	1	2	3	4	
M. I avoid going out to eat	1	2	3	4	
N. Whenever I go someplace new, I especially locate where the bathrooms are	1	2	3	4	

Question 4: During the past month, have you felt so sad, discouraged, hopeless, or had so many problems that you wondered if anything was worthwhile?
1. ( ) extremely so – to the point that I have just about given up
2. ( ) Very much so
3. ( ) Quite a bit
4. ( ) Some– enough to bother me
5. ( ) A little bit
6. ( ) Not at all

or a cloacal anus (Figs. 4.5 and 4.6), anorectal and vaginal fistulas, cutaneous pinprick reflex, digital exam to assess sphincter tonus and presence of anismus and retained stool; rectosigmoidoscopy to exclude rectal tumors, solitary rectal ulcer, and internal intussusception; additional endoscopic examinations should be indicated on a case-by-case basis; and (4) anorectal studies, specifically anal manometry and endoanal ultrasound; other imaging modalities such as cinedefecography, tridimensional echodefecogaphy, or MRI defecography could be performed when organ prolapse or internal intussusception is suspected. In Fig. 4.7 an algorithm for evaluation of patients with incontinence is proposed.

Table 4.7 TAPE score

Parameter	Scales used
Fecal incontinence	St. Mark's score (Vaisey) = 0–24 points
Constipation	Altomare = 0–31 points
Urinary incontinence	ICIQ-SF = 0–21 points
Pelvic organs prolapse	Baden–Walker Halfway System= 0–3 points
Sexual dysfunction	PISQ-IR = gross values
Urinary retention	Graduation according to retained urinary volume = 0–3 points

Clinical Evaluation of Defecation

Individuals who present a clinical condition of evacuation difficulty are frequently regarded as constipated. In fact, the definition of bowel constipation has been subject of much discussion, as it is a complex multifactorial condition that sets off a variety of symptoms, which are described in various ways, such as "small fecal quantity," "decrease in the number of evacuations," "hard stools," "sensation of incomplete evacuation," "difficulty during evacuation," or "retained feces." These conditions may occur in isolation or not. Based on a study performed in constipated patients, it was observed that 31% of individuals who were considered "normal" reported they had strived to evacuate from one to four times a month [26]. In contrast, in 34% of the same "normal" individuals who complained about incomplete evacuation at least once a month, only 5% frequently reported such complaint. According to Yang et al. [27], about 10% of the normal population consider they are constipated. It is believed that up to 95% of the adult Western population shows an evacuation rate ranging from once every 2 days to twice a day [28]. Many patients who are at times classified as chronically constipated notice a brief alteration in their bowel habits, while other patients show symptoms related to the irritable bowel syndrome, which hinders data interpretation.

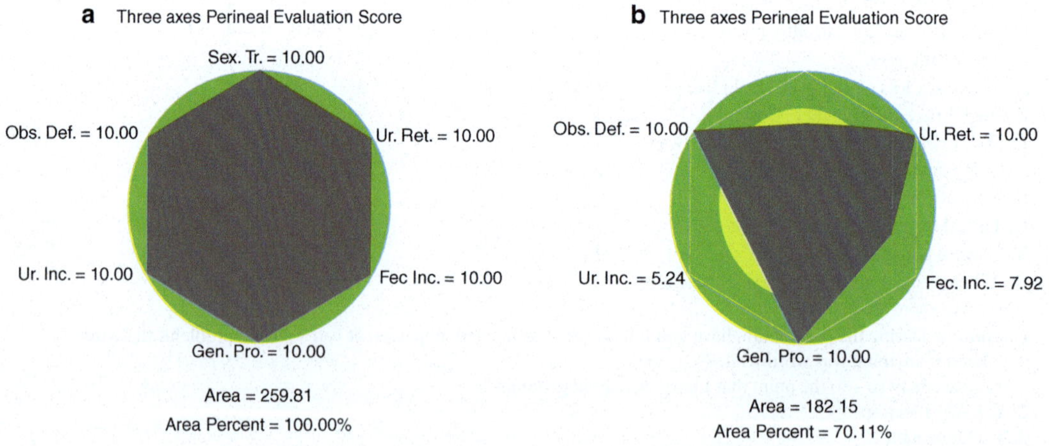

Fig. 4.1 Tape score. Pelvic floor function (PFF) polygon defining anorectal, urinary, and gynaecological functions in a normal subject (**a**) and a patient with urinary incontinence, sexual problems and minor fecal incontinence (**b**). (Reused with permission © John Wiley and Sons [23])

Table 4.8 Low anterior resection score (LARS)

1. Do you ever have occasions when you cannot control your flatus (wind)?	
( ) No, never	0
( ) Yes, less than once per week	4
( ) Yes, at least once per week	7
2. Do you ever have any accidental leakage of liquid stool?	
( ) No, never	0
( ) Yes, less than once per week	3
( ) Yes, at least once per week	3
3. How often do you open your bowels?	
( ) More than 7 times per day (24 hours)	4
( ) 4–7 Times per day (24 hours)	2
( ) 1–3 Times per day (24 hours)	0
( ) Less than once per day (24 hours)	5
4. Do you ever have to open your bowels again within 1 hour of the last bowel opening?	
( ) No, never	0
( ) Yes, less than once per week	9
( ) Yes, at least once per week	11
5. Do you ever have such a strong urge to open your bowels that you have to rush to the toilet?	
( ) No, never	0
( ) Yes, less than once per week	11
( ) Yes, at least once per week	16
Total score	
0–20, no LARS; 21–29, minor LARS; 30–42 major LARS	

Table 4.9 Physical examination of incontinent patient

Abdominal evaluation	Palpation: Masses, distension Auscultation: meteorism
Neurologic evaluation	Perineal sensitivity Anal reflex (*pinprick reflex*) Mental state
Perianal evaluation	Inspection: Abrasions Signs of infection Soiling Scars Mucous ectropion Mucosal and rectal prolapse Patulous anus Perineal body Muscle deficit Excessive descent of the perineum with the Valsalva maneuver Fistulas, abscesses, fissures
Palpation (DRE)	Anal reflex Resting tone to DRE Voluntary contraction Puborectalis sensitivity and tone Muscle injuries Fecaloma Tumors Inflammatory lesions Internal prolapse Rectocele, enterocele
Anoscopy	Hemorrhoids Rectal solitary ulcer Inflammatory proctitis Fistulas Tumors

In general, constipation can be defined as the evacuation of scarce, infrequent, and excessively hard stools, with an evacuation frequency inferior to once every 72 hours or a fecal mass below 35 g per day [29]. According to the Rome Criteria [30, 31], a constipated individual must show two or more of the following criteria:

- Difficulty to evacuate in 25% or more of evacuations
- The sensation of incomplete evacuation in 25% or of evacuations
- Hardened or petrous feces in 25% or more of evacuations

Therefore, when a patient complains about constipation, a detailed history about the dietary habits must be investigated. Patients with a low-fiber diet and little water ingestion are in an increased risk for developing constipation. The use of a food frequency questionnaire and the integrated work with a nutrition team can help to understand the exact quantity of fibers and liquid utilized by the patient, enabling proper adjustments. In addition, evaluation of the patient's physical activity and the defecation attempts are investigated. The voluntary inhibition of the evacuation urge can interfere with the defecation reflex: the retained stool dilates the rectum and the individual stops feeling the defecation urge, starting to retain the feces for a longer time. This mechanism is common in children and women and it is related to the psychogenic megacolon condition. The use of a fecal form scale (Bristol Stool Scale) [32] is a simple way to identify individuals that present with very hard or lumpy stools, commonly associated with straining on defecation and constipation.

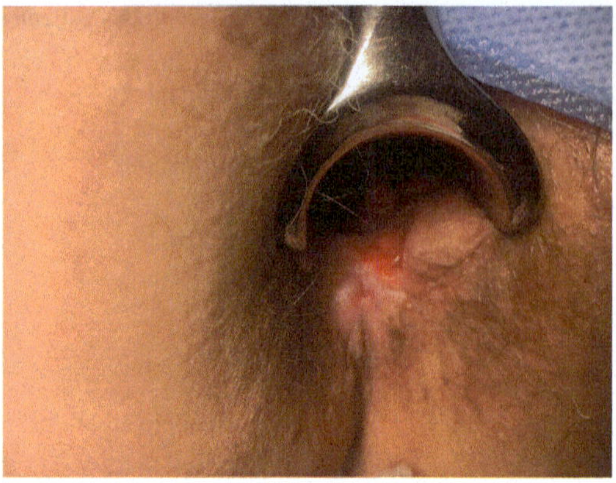

Fig. 4.2 Detailed physical examination looking for scars

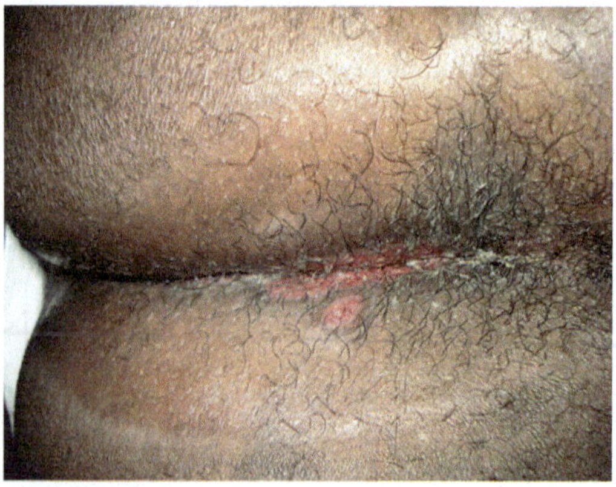

Fig. 4.3 Wet anus

The clinical history must also investigate the presence of extracolonic causes of constipation, such as neurologic disorders, previous pelvic surgeries, hypothyroidism, medular trauma, sexual abuse, psychiatric and eating disorders, use of constipating agents, and other secondary causes of constipation (Table 4.10) [33, 34].

Patients that complain of anal pain during defecation can have an associated hypertonic sphincter or anal fissure (Fig. 4.8). Bleeding hemorrhoids and mucosal prolapse can also occur due to straining or in the presence of anal fissure. The chronic consumption of irritant laxatives, such as bisacodyl and phenolphthalein, can be associated with melanosis coli (Fig. 4.9) [35].

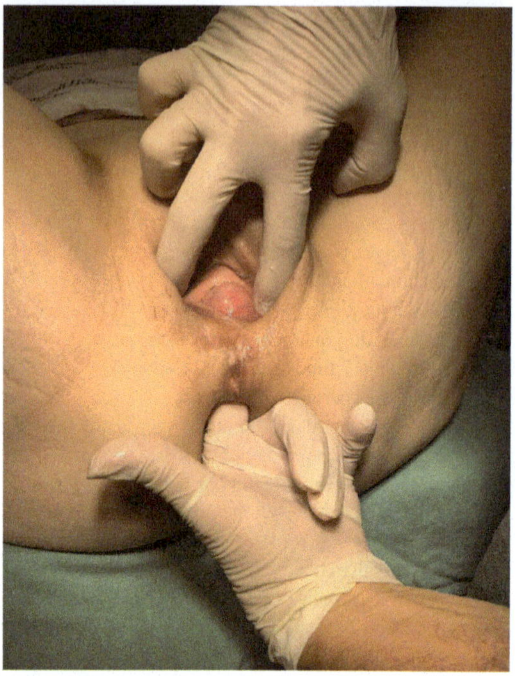

Fig. 4.4 Rectocele

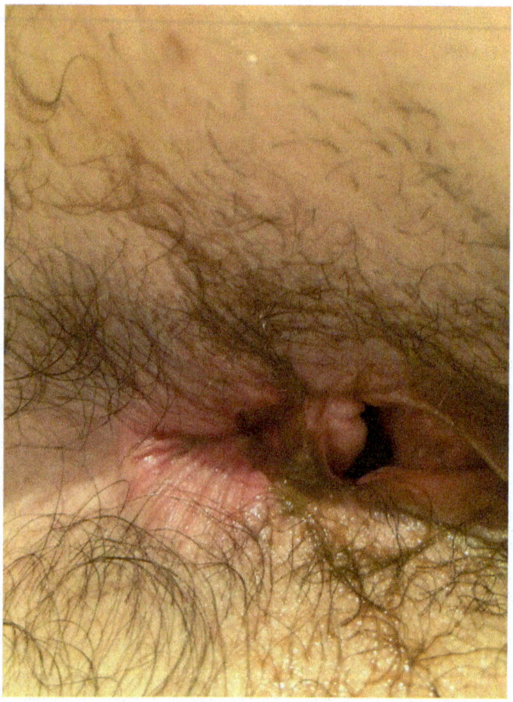

Fig. 4.5 Patulous anus

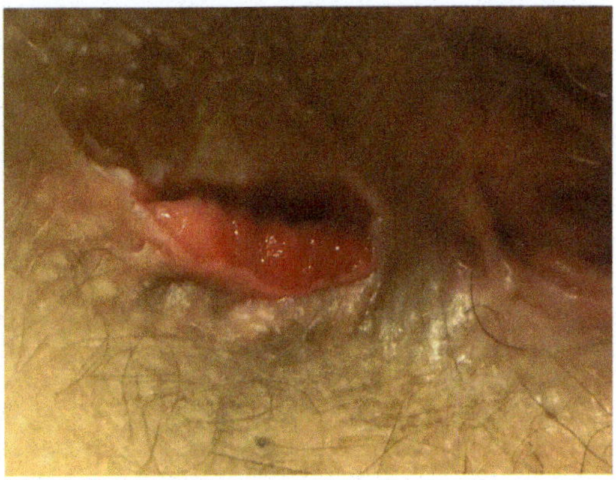

Fig. 4.6 Cloacal anus

In an attempt to establish a classification system for constipation, Agachan et al. [36] developed a questionnaire to evaluate this condition (Table 4.11). Based on this questionnaire, the authors established graduation of the severity of a symptom, similar to the severity scales for fecal incontinence. Although it has not been statistically validated, this study is the pioneer in graduating constipation severity, showing an intimate relation between the symptoms and the objective findings of the physical exam and the data resulting from the diagnostic methods. It is fundamental to assess the number of evacua-

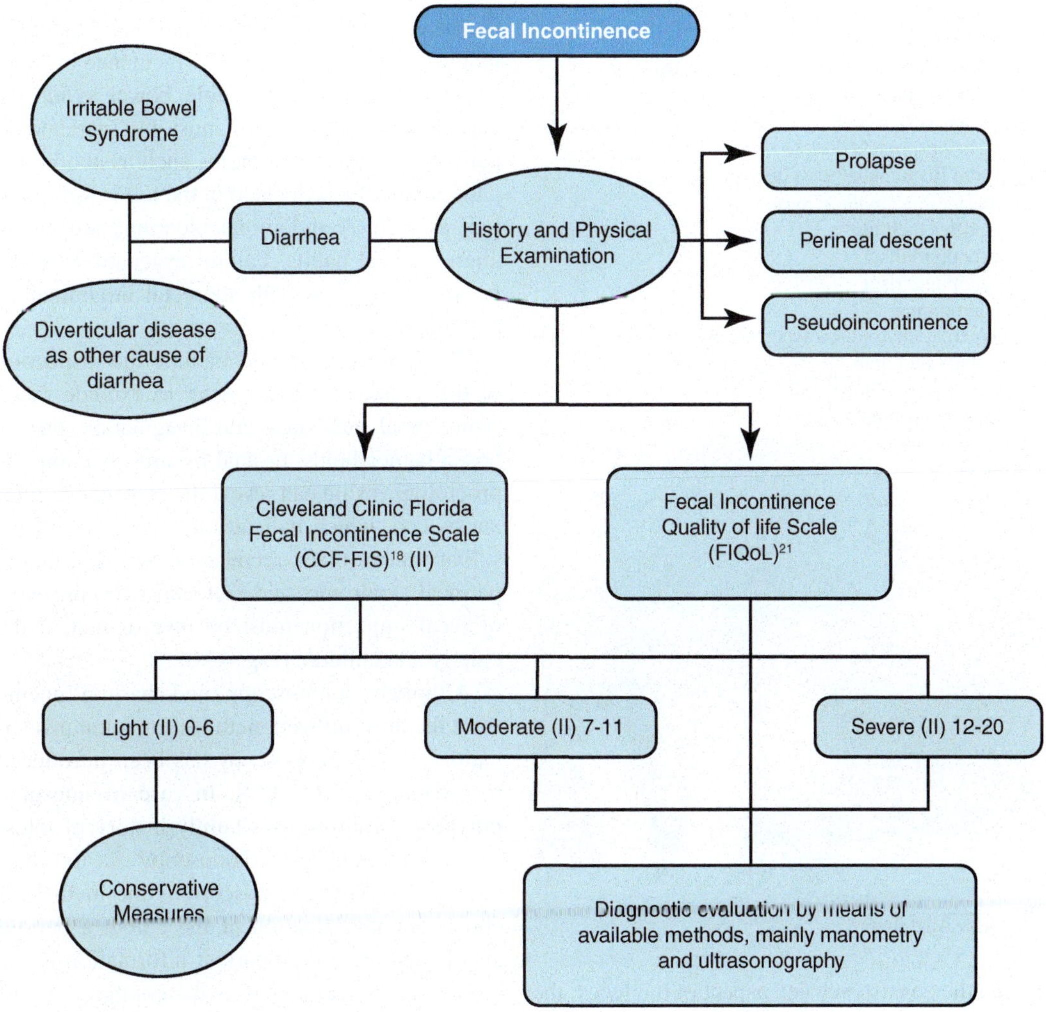

Fig. 4.7 Algorithm for the treatment of fecal incontinence

Table 4.10 Extracolonic cause of constipation

Medications
Analgesics
Anticonvulsants, antidepressants, anticholinergics
Antiparkinsonians, opioids, diuretics, ferrous sulfate.
Antacids, anesthetics (paralytics agents)
Antihypertensives (monoamine oxidase inhibitors [MAO])
Neurological diseases
Spinal cord injuries
Iatrogenics (erection nerve injury)
Tumors in cauda equina
Meningocele *tabes dorsalis*
Trauma
Paraplegia
Brain injuries
Parkinson's disease
Stroke
Tumors
Peripheral lesions
Multiple endocrine neoplasia type IIB
Autonomic neuropathy
Chagas's disease
Hirschsprung's disease
von Recklinghausen's disease
Endocrines/metabolic disease
Diabetes mellitus
Hypothyroidism
Hypopituitarism
Scleroderma
Hyperparathyroidism/hypercalcemia
Porphyria/uremia
Pheochromocytoma
Hypopotassemia

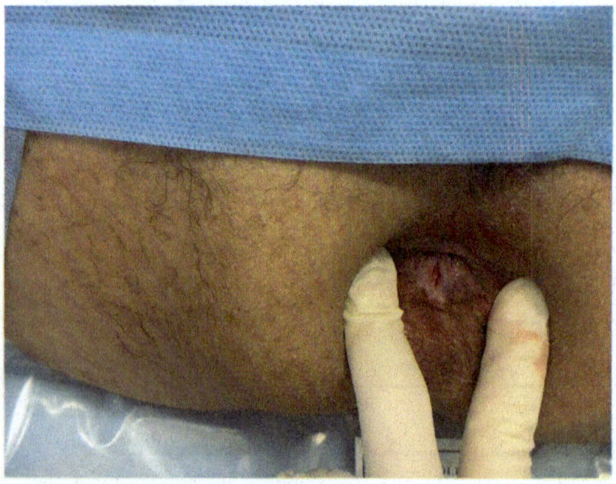

Fig. 4.8 Anal fissure

tions; the consistency and aspect of the feces; the presence of mucus or bleeding, evacuation pain, and sensation of incomplete evacuation; and the necessity of digital maneuvers to aid defecation.

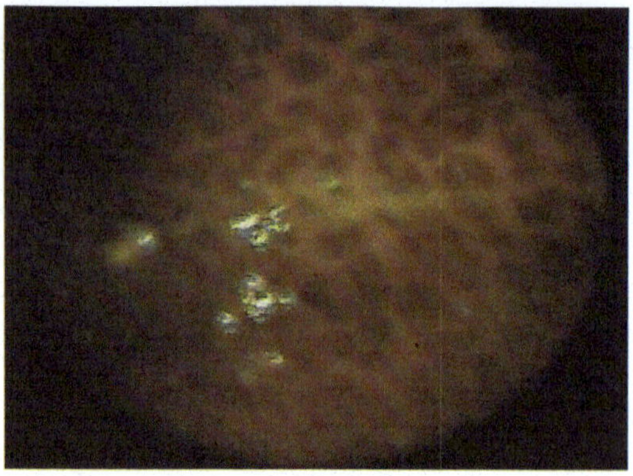

Fig. 4.9 Melanosis coli

Detailed information about laxative and diet habits must also be obtained. The incidence of functional diseases (e.g., irritable bowel syndrome and diverticular disease) must be searched in patients that show symptoms such as abdominal pain, mainly those located in the left lower quadrant as well as abdominal bloating, and recent altered bowel habits. Painful evacuation can be present in patients with anorectal inflammatory pathologies.

Physical exam should include the inspection of the anorectal area in order to exclude rectal tumors anal stenosis, rectal invagination, thrombosed hemorrhoids, or anal fissures. A complete proctologic exam can reveal the presence of melanosis coli, which indicates the use of laxatives, solitary rectal ulcer, rectal prolapse, descending perineal syndrome, and rectocele. The presence of fecal impaction must be investigated in the elderly and children (Fig. 4.10).

Although colonoscopy and barium enema were the most utilized methods for colon investigation, virtual colonoscopy has been introduced recently (Fig. 4.11) [37]. In case of intestinal polyposis syndromes or family history of intestinal neoplasia, the colonoscopy is the best choice. Although colonoscopy is the method of choice for investigation of the colon, a barium enema can provide important information to the colorectal surgeon, such as the configuration of the colon, the loss of haustrations, the presence of a hypertonic colon associated with diverticular disease, and a redundant colon (Fig. 4.12).

Table 4.11 Constipation Scoring System [36]

Frequency of bowel movements	*Index*
One to two times per 1–2 days	0
Two times per week	1
Once per week	2
Less than once per week	3
Less than once per month	4
Difficulty: painful evacuation effort	
Never	0
Rarely	1
Sometimes	2
Usually	3
Always	4
Completeness: feeling incomplete	
Never	0
Rarely	1
Sometimes	2
Usually	3
Always	4
Pain: abdominal pain	
Never	0
Rarely	1
Sometimes	2
Usually	3
Always	4
Time: minutes in lavatory per attempt	
Less than 5	0
5–10	1
10–20	2
20–30	3
More than 30	4
Assistance: type of assistance	
Without assistance	0
Stimulative laxatives	1
Digital assistance or enema	2
Failure: unsuccessful attempts for evacuation per 24 hours	
Never	0
1–3	1
3–6	2
6–9	3
>9	4
History: duration of constipation (yr)	
0	0
1–5	1
5–10	2
10–20	3
>20	4

Minimum score, 0; Maximum score, 30

When a suspected aganglionic colon is the present case, a barium enema with the lateral views can demonstrate the transitional zones between aganglionic segments and the dilated proximal colon. In children, the barium enema has represented a safe and reliable method for

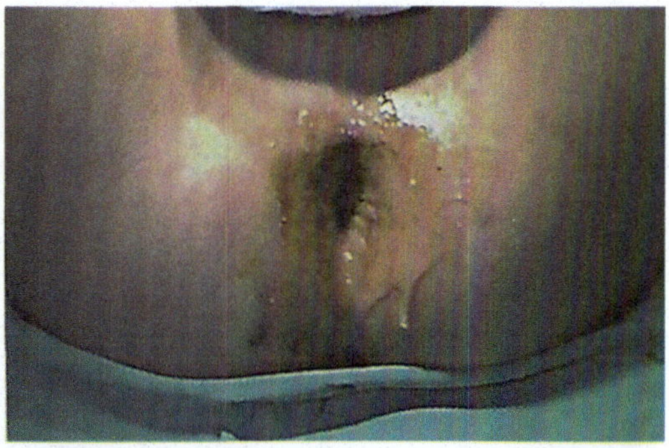

Fig. 4.10 Fecal impaction

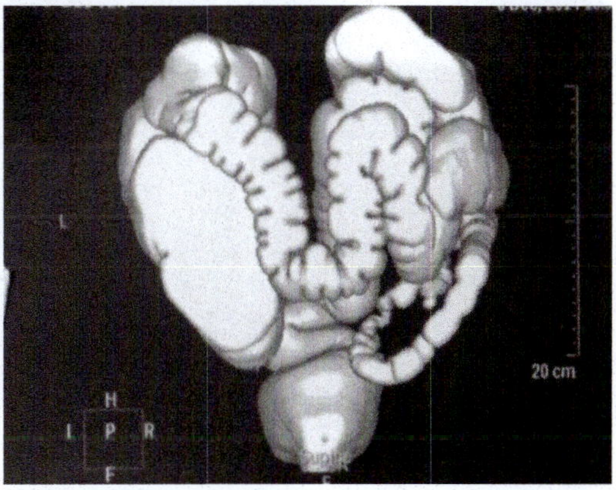

Fig. 4.11 Vitual colonoscopy

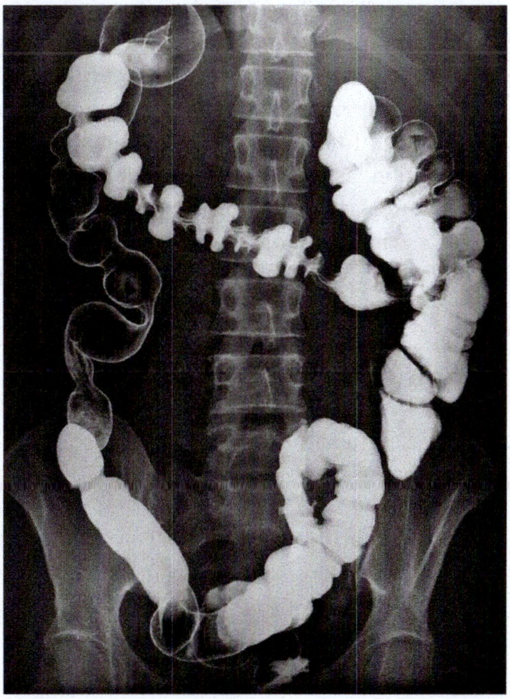

Fig. 4.12 Redundant colon

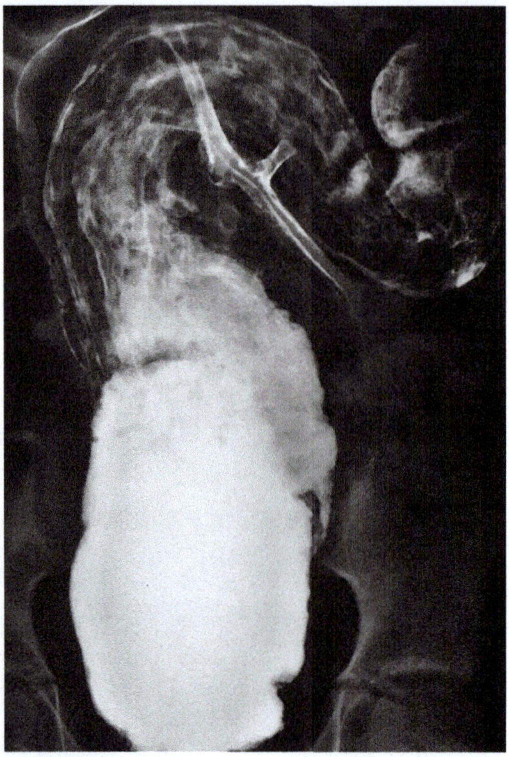

Fig. 4.13 Volumnous fecaloma

Table 4.12 Methods of physiology for the investigation of intestinal constipation

Tests	
Anorectal manometry	Rectoanal inhibitory reflex RAIR present or absent Anal hypertonia Paradoxical puborectalis contraction Short or ultrashort waves
Colonic transit time	Colonic inertia Obstructed defecation syndrome
Balloon expulsion test	Paradoxical puborectalis contraction
Videodefecogram	Paradoxical puborectalis contraction Sigmoidocele Rectocele Perineal descent syndrome Anal intussusception Rectal prolapse
Anal electromyography	Paradoxical puborectalis contraction
Echodefecography and Dynamic MRI defecography	Paradoxical puborectalis contraction Sigmoidocele Rectocele Perineal descent syndrome Anal intussusception Enterocele
Expired hydrogen test	Small bowel slow transit

the investigation of psychogenic megacolon or encopresis, and, sometimes, it can demonstrate fecal impact and voluminous fecalomas (Fig. 4.13).

In patients with constipation, it is important to exclude all the organic causes and specially constipation associated with inadequate diet. Once those factors are ruled out, especially in patients with evacuation frequency lower than three times a week, a functional evaluation of the colon is indicated in order to diagnose conditions such as colonic inertia or obstructed defecation syndrome or anismus [38] (Table 4.12). The algorithm in Fig. 4.14 lists the most commonly used functional methods for the evaluation of the functional constipation.

Finally, the terms and definitions of the most used exams in the physiological evaluation of anorectal disorders are described based on the American Society of Colon and Rectal Surgeons guidelines [2, 39, 40].

Terms and Definitions of the Anorectal Physiology Guidelines

Anal sensation: It is the sensation in the anal canal measured by anal mucosal electrosensitivity. It reflects the somatic sensory component of the pudendal nerve.

Anal pressures: *High-pressure zone* is the length of the anal canal with resting pressures at least 30% higher than rectal pressure. *Resting pressure* is the pressure in the high-pressure zone at rest after a period of stabilization. M*aximum resting pressure* is the highest resting pressure recorded. *Mean resting pressure* is the mean of the resting pressures recorded within the high-pressure zone. *Maximum voluntary pressure* is the highest pressure recorded above the baseline (zero) at any level of the anal canal during maximum squeeze effort by the patient. *Squeeze*

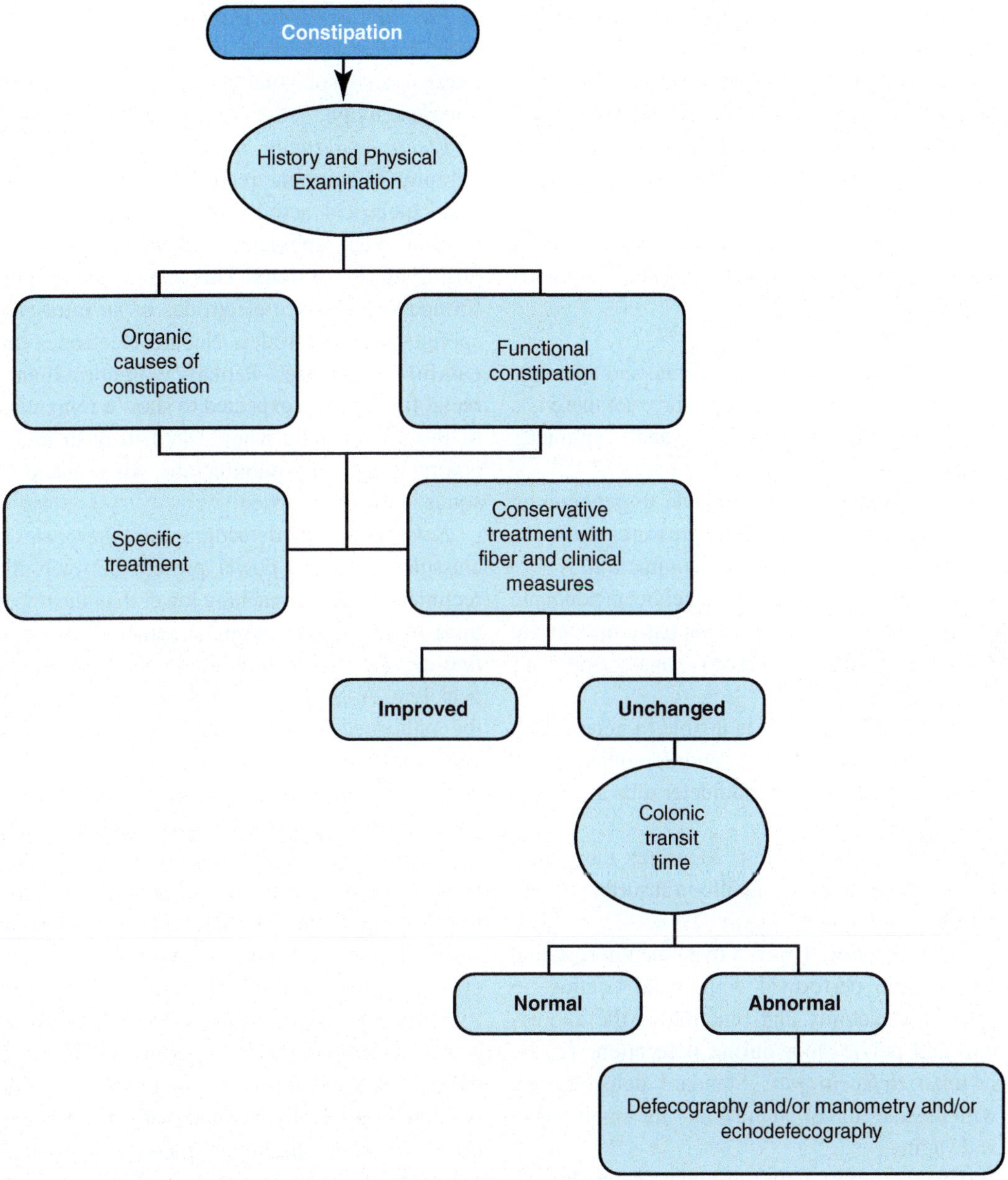

Fig. 4.14 Algorithm for treatment of functional constipation

pressure is the pressure increment above resting pressure following voluntary squeeze contraction and is a calculated value that is the difference between the maximum voluntary pressure and the resting pressure at the same level of the anal canal. *Cough pressure* is the pressure increment above resting pressure following a cough and is a calculated value that is the difference between the maximum pressure recorded during cough and the resting pressure at the same level in the anal canal. *Sphincter endurance* is the length of time the patient can maintain a squeeze pressure above the resting pressure.

Anorectal angle: It is the proctographic angle between the mid-axial longitudinal axis of the rectum and the anal canal. Normal values range from 90 to 110 degrees at rest.

Anismus (nonrelaxation of the puborectalis and paradoxical contraction of the puborectalis): It is defined by either the observation of a nonrelaxation of the puborectalis or actual paradoxical contraction of the puborectalis at the anorectal junction during evacuation. Anismus can be additionally categorized into four types on anorectal manometry based on patient ability to generate adequate pushing force and the type of sphincter contraction.

- *Type I*: The patient can generate and adequate rise in intra-abdominal pressure, yet there is a paradoxical increase in anal sphincter pressures.
- *Type II*: The patient is unable to generate an adequate rise in intrarectal pressures and has a paradoxical anal sphincter contraction.
- *Type III*: The patient can generate an adequate intra-abdominal pressure but has either absent or incomplete (<20%) anal sphincter relaxation.
- *Type IV*: The patient is unable to generate an adequate pushing force and demonstrates an absent or incomplete sphincter relaxation.

Balloon expulsion test: Measures the ability of the patient to expel a balloon insufflated with 50 ± 60 cc of water.

Defecography: This is a dynamic fluoroscopic examination performed with rectal contrast to study the anatomy and function of the anorectum and pelvic floor during defecation. *Triple-contrast defecography*: Standard defecography with the addition of contrast into the small bowel and vagina.

Dynamic MRI defecography: The rectum is opacified with ultrasound gel. The patient is then asked to evacuate gel (supine or sitting), while the MRI captures dynamic evacuation images.

Dynamic ultrasound defecography: The rectum is opacified with ultrasound gel. The patient is then asked to evacuate gel (in lithotomy or in a left lateral decubitus position), while the ultrasonographer captures dynamic evacuation images with an intrarectal or a perineal ultrasound probe.

Electromyography: It registers the change of base electric activity in the motor units of the internal anal sphincter and levator ani during muscle activity.

Electromyography recruitment: Electromyography recruitment records the change from basal electrical activity of motor units of the external anal sphincter and levator muscles during muscle activity. This test is usually performed with surface electrodes or an intra-anal sponge with electrodes. Needle electrodes are painful and not used. Patients with normal anorectal function are expected to show a relaxation of the puborectalis when asked to push and a contraction of the puborectalis when asked to squeeze the anus closed.

Enterocele: On defecography, *enterocele* is classified as small bowel present between the rectum and vagina, reaching lower than the upper third of the vagina during evacuation effort. A *first-degree enterocele* is above the pubococcygeal line. A *second-degree enterocele* is below the pubococcygeal line but above the ischiococcygeal line, and a *third-degree enterocele* is below the ischiococcygeal line. Alternatively, herniations of the peritoneal sac with contained peritoneoceles, omentoceles, sigmoidoceles, and enteroceles can be graded as small (<3 cm), moderate (3–6 cm), and large (>6 cm) by measuring the largest distance between the pubococcygeal line and the most inferior point of the sac.

Measurements for rectal sensation: Evaluation of rectal sensation is performed by placing a balloon catheter above the anorectal ring. The balloon is gradually inflated with air. *Sensory threshold* ± the minimum rectal volume perceived by the patient. *Urge sensation* ± the volume associated with the initial urge to defecate. *Maximum tolerated volume* ± the volume at which the patient experiences discomfort and an intense desire to defecate.

Mucosal prolapse: Protrusion of rectal mucosa into or beyond the anal canal.

Peritoneocele/omentocele: On defecography, *peritoneocele* is a protrusion of the peritoneum between the rectum and the vagina that does not contain any abdominal viscera. A *first-degree peritoneocele/omentocele* is above the pubococcygeal line, a *second-degree peritoneocele/omentocele* is below the pubococcygeal line but is above the ischiococcygeal line, and a *third-degree peritoneocele/omentocele* is below the ischiococcygeal line. Alternatively, herniations of the peritoneal sac with contained peritoneoceles, omentoceles, sigmoidoceles, and enteroceles can be graded as small (<3 cm), moderate (3–6 cm), and large (>6 cm) by measuring the largest distance between the pubococcygeal line and the most inferior point of the sac.

Perineal descent: Perineal descent is the caudad movement of the pelvic floor with straining. It is measured clinically by the position of the anal verge in relationship to the plane of the ischial tuberosities at rest and during maximal straining. It is measured on defecography as the difference between the position of the anorectal junction at rest and during maximal straining. Normal values are <2 cm. Additional measurements of the degree of descent can be generated with two additional reference lines.

A. *H line* measures hiatal width. This measurement is taken from the inferior aspect of the symphysis pubis to the posterior wall of the rectum at the level of the anorectal junction.
B. *M line* measures the movement of the pelvic floor away from pubococcygeal line. This measurement is taken by extending a perpendicular line from the pubococcygeal line to the posterior end of the H line.
C. *Abnormal perineal descent* is present when the H line exceeds 6 cm and when the M line exceeds 2 cm in length.

Pudendal nerve latency: Pudendal nerve latency is the measurement of the time from stimulation of the pudendal nerve at the ischial spine to the response of the external anal sphincter. Normal PNTML is <2.2 milliseconds.

Rectal anal inhibitory reflex (RAIR): Transient decrease in resting anal pressure by >25% of basal pressure in response to rapid insufflation of a rectal balloon with subsequent return to baseline.

Rectal compliance: Rectal compliance is the measured change in pressure in response to change in volume within a water or air-filled balloon within the rectum.

Rectal prolapse: Rectal prolapse is a circumferential full-thickness intussusception of the rectal wall with protrusion beyond the anal canal.

Rectocele: On defecography, a *rectocele* is a bulging of the rectal wall toward the vagina. A comment regarding its size and whether the rectocele empties with defecation with or without digitation can help to guide treatment. Radiographically, rectoceles are graded as small (<2 cm), moderate (2–4 cm), and large (>4 cm). Additional comment on the ability to empty with and without digitation can help guide therapy.

Sigmoidocele: It is a protrusion of the peritoneum between the rectum and vagina that contains sigmoid colon. Sigmoidoceles are classified by the position of the lowest loop of sigmoid during evacuatory effort on defecography. A first-degree sigmoidocele is above the pubococcygeal line, a second-degree sigmoidocele is below the pubococcygeal line but is above the ischiococcygeal line, and a third-degree sigmoidocele is below the ischiococcygeal line.

Slow-transit constipation: Subset of functional constipation in which patients report <1 bowel movement every 3 days. The presence and severity of STC are suspected and measured by querying patients on the degree of the decrease in their bowel movements, the extent of the loss of urge to defecate, the degree of need to use laxatives to assist a bowel movement, and the degree of bother caused by these symptoms. STC can coexist with other functional bowel disorders, such as obstructive defecation syndrome or IBS. Additional characterization and confirmation of STC require documentation of delayed colonic transit.

Rectoanal inhibitory reflex: This is the transient decrease in resting anal pressure by >25% of basal pressure in response to rapid inflation of a rectal balloon with subsequent return to baseline.

Summary

Patients with pelvic floor disorders including incontinence and constipation deserve a very detailed clinical history and physical exam. The severity of the condition as well as the impact in quality of life are important aspects that should be well addressed with the validated available quality-of-life instruments and specific scales.

References

1. Whitehead WE, Wald A, Norton NJ. Treatment options for fecal incontinence. Dis Colon Rectum. 2001;44:131–42.. discussion 142–4
2. Abrams P, Cardozo L, Wagg A, Wein A, editors. Incontinence 6th Edition. 2017. ICI-ICS. International Continence Society, Bristol UK, ISBN: 978–0956960733.
3. Saldana Ruiz N, Kaiser AM. Fecal incontinence – challenges and solutions. World J Gastroenterol. 2017;23(1):11–24.
4. Paquette IM, Varma MG, Kaiser AM, Steele SR, Rafferty JF. The American Society of Colon and Rectal Surgeons' clinical practice guideline for the treatment of fecal incontinence. Dis Colon Rectum. 2015;58(7):623–36.
5. Kelly JH. Cinedefecography in anorectal malformation. J Pediatr Surg. 1968;4:538.
6. Parks AG. Anorectal incontinence. J R Soc Med. 1975;68:21–30.
7. Lane RN. Clinical application of anorectal physiology. Proc Soc Med. 1975;68:28–30.
8. Rudd WW. The transanal anastomosis: a sphincter-saving operation with improved continence. Dis Colon Rectum. 1979;22:102–5.
9. Holschneider AM. Treatment and functional results of anorectal continence in children with imperforate anus. Acta Chir Belg. 1983;3:191–204.
10. Keighley MR, Fielding WL. Management of faecal incontinence and results of surgical treatment. Br J Surg. 1983;70:463–8.
11. Corman M. Gracilis muscle transposition for anal incontinence. Late results. Br J Surg. 1985;72:S21–2.
12. Hiltunen KNM, Matikainen M, Auvinen O, Hietanen P. Clinical and manometric evaluation of anal sphincter function in incontinent patients. Am J Surg. 1986;151:489–92.
13. Broden G, Dolk A, Holmstroem B. Recovery of the internal anal sphincter following rectopexy: a possible explanation for continence improvement. Int J Color Dis. 1988;3:23–8.
14. Womak NR, Morrison JF, Williams NS. Prospective study of the effects of postanal repair in neurogenic fecal incontinence. Br J Surg. 1988;75:48–52.
15. Rainey JB, Donaldson DN, Thomson JP. Postanal repair: which patients derive most benefit? J R Coll Surg Edinb. 1990;35:101–5.
16. Miller R, Bartolo DC, Lock-Edminds JC, Mortensen NJ. Prospective study of conservative and operative treatment for faecal incontinence. Br J Surg. 1988;75:101–5.
17. Pescatori M, Anastasio G, Bottini C, Mentasti A. New grading and scoring for anal incontinence. Evaluation of 335 patients. Dis Colon Rectum. 1992;35:482–7.
18. Jorge JMN, Wexner SD. Etiology and management of fecal incontinence. Dis Colon Rectum. 1993;36:77–97.
19. Rothbarth J, Bemelman WA, Meijerink WJHJ, Stiggnelbout AM, Zwinderman AH, Buyze-Westerweel ME, Delemarre JBVM. What is the impact of fecal incontinence on quality of life? Dis Colon Rectum. 2001;44:67–71.
20. Vaizey CJ, Carapeti E, Cahill JA, Kamm MA. Prospective comparison of faecal incontinence garding systems. Gut. 1999;44:77–80.
21. Rockwood TH, Church J, Fleshman JW, Kane RL, Mavrantonis C, Thorson AG, Wexner SD, Bliss D, Lowry AC. Fecal incontinence quality of life scale. Quality of life instrument for patients with fecal incontinence. Dis Colon Rectum. 2000;43:9–17.
22. Oliveira L. Quality of life issues. In: Wexner SD, Zbar A, Pescatori M, editors. Complex Ano-rectal disorders. London: Springer; 2005. p. 595–605.
23. Altomare DF, Di Lena M, Giuratrabocchetta S, Giannini I, Falagariot M, Zbar AP, Rockwood T. The Axial Perineal Evaluation (TAPE) score: a new scoring for comprehensive evaluation of pelvic floor function. Color Dis. 2014;16:459–70.
24. Emmertsen KL, Laurberg S. Low anterior resection syndrome score. Ann Surg. 2012;255(5):922–8.
25. Oliveira L, Wexner SD. Anal Incontinence. In: Beck, Wexner, editors. Fundamentals of anorectal surgery. 2nd ed. London: W.B. Saunders; 1999. p. 115–52.
26. Dent OF, Goulston KJ, Zubrzychi J, et al. Bowelsymptoms in an apparently well population. Dis Colon Rectum. 1986;29:243–7.
27. Yang P, Banwell JG. Dietary fiber. Pract Gastroenterol. 1986;10:28–32.
28. Lennard Jones JE. Constipation. In: Feldman M, Scharsschmidt BF, Sleisenger MH, editors. Sleisenger and Fordtran's gastrintestinal and liver disease. 6th ed. Philadelphia: WB Saunders; 1988. p. 174–97.
29. Wexner SD, Dailey TH. Diagnosis and surgical treatment of chronic constipation. Contemp Surg. 1988;32:58–60.
30. Talley NJ, Spiller R. Irritable bowel syndrome: a little understood organic bowel disease? Lancet. 2002;360:555–64.
31. Badia X, Mearin F, Balboa A, et al. Burden of illness in irritable syndrome comparing Rome I and Rome II criteria. PharmacoEconomics. 2002;20:749–58.
32. Lane MM, Czyzewski DI, Chumpitazi BP, Shulman RJ. Reliability and validity of a modified Bristol Stool

Form Scale for children. J Pediatr. 2011;159(3):437–441.e1.
33. Gemlo B, Wong WD. Etiology of acquired colorectal disease. In: Wexner SD, Bartolo DCC, editors. Constipation. Oxford: Butterworth-Heinemann; 1995. p. 17–22.
34. Poison J, Devroede G. Severe chronic constipation as a surgical problem. Surg Clin North Am. 1983;63:193.
35. Freeman HJ. "Melanosis" in the small and large intestine. World J Gastroenterol. 2008;14(27):4296–9.
36. Agachan F, Chen T, Pfeifer J, Reissman P, Wernex SD. A constipation scoring system to simplify evaluation and management of constipated patients. Dis Colon Rectum. 1996;39:681–5.
37. Bosna J, Morrin MM, Copel L, Raptopoulos V, Kruskal JB. Computed tomography colonography (virtual colonoscopy): update on technique, applications, and future developments. Surg Technol Int. 2003;11:102–10.
38. Rao SS, Mudipalli RS, Stessman M, Zimmerman B. Investigation of the utility of colorectal function tests and Rome II criteria in dyssynergic defecation (Anismus). Neurogastroenterol Motil. 2004;16(5):589–96.
39. Lowry AC, Simmang CL, Boulos P, Farmer KC, Finan PJ, Hyman N, Killingback M, Lubowski DZ, Moore R, Penfold C, Savoca P, Stitz R, Tjandra JJ. Consensus statement of definitions for anorectal physiology and rectal cancer. ANZ J Surg. 2001;71(10):603–5.
40. Bordeianou LG, Carmichael JC, Paquette IM, Wexner S, Hull TL, Bernstein M, Keller DS, Zutshi M, Varma MG, Gurland BH, Steele SR. Consensus statement of definitions for anorectal physiology testing and pelvic floor terminology (revised). Dis Colon Rectum. 2018;61(4):421–7.

Introduction to Methods of Anorectal Physiology Evaluation

5

Jason Lei, Colleen Nofi, and David E. Rivadeneira

Introduction

Pelvic floor disorders are a common group of heterogeneous pathologies that range from being inconvenient to extremely debilitating. Although it is essential to complete a thorough history and physical examination in patients with functional pelvic disorders, specialized tests can greatly assist in the evaluation, diagnosis, and management of these complex patients. In recent years, major advances in anorectal physiology testing and imaging have provided a better understanding of pelvic floor disorders. The methods used in evaluating anorectal function are evolving, becoming more sophisticated, and providing more clinically relevant data.

This chapter will focus on a brief introduction to the methods of anorectal physiology testing including

- Endoanal ultrasound
- Cinedefecography
- Magnetic resonance imaging (MRI)
- Anorectal manometry
- Electromyography
- Pudendal nerve assessment
- Colonic transit study

Some of these exams will be detailed in a separate chapter of this book.

Endoanal Ultrasound

Anatomic evaluation of a patient with a functional pelvic floor disorder often begins with a endoanal ultrasound. Ultrasound uses sound waves that are transmitted to and reabsorbed by an ultrasound probe. Different tissues will appear differently based on its echogenicity. Tissues with high water content such as muscle will reflect the ultrasound waves less and appear hypoechoic. Tissues with less water content such as connective tissue and fat will reflect ultrasound waves more and appear hyperechoic.

The main utility of endoanal ultrasound plays an essential role in evaluating the structural integrity of the anal sphincter complex and determining surgical candidacy for sphincteroplasty, artificial anal sphincters, sacral nerve stimulation, and injectable biomaterials [1–5].

To perform the procedure, patients are instructed to self-administer enemas to evaluate stool from the rectal vault the day of the proce-

J. Lei
Jefferson Health, Philadelphia, PA, USA

C. Nofi
New York Institute of Technology College of Osteopathic Medicine, Huntington, NY, USA

D. E. Rivadeneira (✉)
Northwell Health, Huntington Hospital, Hofstra School of Medicine, Huntington, NY, USA
e-mail: DRivadeneira@northwell.edu

L. C. C. Oliveira (ed.), *Anorectal Physiology*, https://doi.org/10.1007/978-3-030-43811-1_5

dure. While the patient is in left lateral decubitus, a lubricated ultrasound probe is placed into the upper anal canal. The anal canal can be divided into three levels. In the upper anal canal, the horseshoe-shaped, hyperechoic puborectalis and the hypoechoic internal anal sphincter (IAS) will be visualized (Fig. 5.1). In the mid anal canal, the concentric rings of the hyperechoic external anal sphincter (EAS) and the hypoechoic IAS will be visualized (Fig. 5.2). In the lower anal canal, the IAS terminates, and the EAS appears thick (Fig. 5.3).

Upon reviewing the images of endoanal ultrasound, abnormalities in the structural integrity of the sphincter complex can be seen. Scarring, thinning, or disruption of the sphincters can be visualized (Fig. 5.4).

Although endoanal ultrasound has the advances of low cost, convenience, and accessibility, the main drawback of endoanal ultrasound is that its accuracy depends on operator experience [6–9]. However, diagnostic accuracy and resolution of the ultrasound can be improved with the use of 3D ultrasound and dynamic ultrasound, which are beyond the scope of this chapter.

Cinedefecography

Cinedefecography also known as evacuation proctography is a fluoroscopic study that is used to investigate the physiology of defecation. Using this modality, morphological changes in the rec-

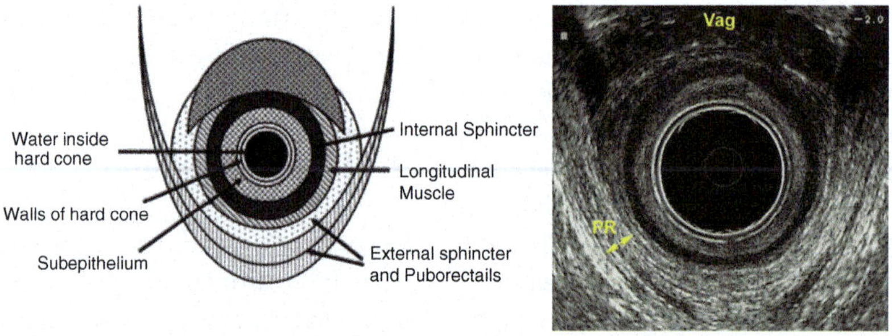

Fig. 5.1 Endoanal ultrasound of the upper anal canal showing the puborectalis muscle. Schematic representation and axial image. (Reused with permission © Springer Nature)

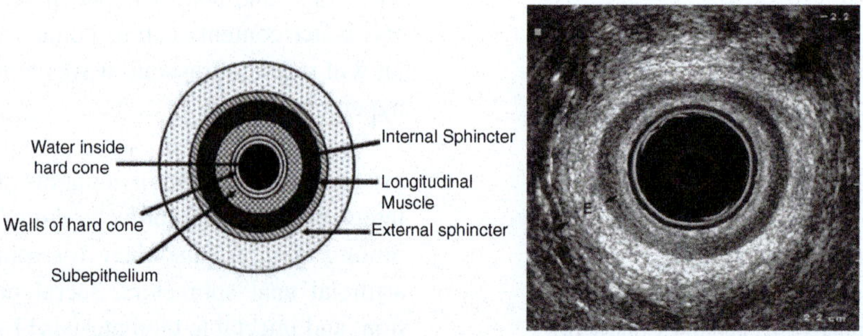

Fig. 5.2 Endoanal ultrasound of the mid anal canal showing the two concentric sphincter muscle rings. Schematic representation and axial image. (Reused with permission © American Gastroenterological Association)

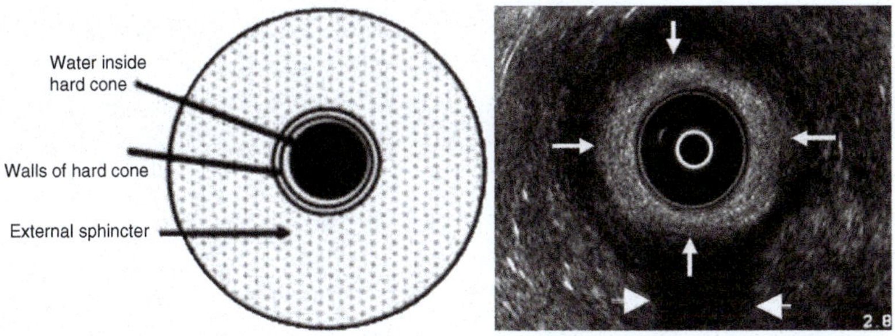

Fig. 5.3 Endoanal ultrasound of the lower anal canal showing the only the EAS. Schematic representation and axial image. (Reused with permission © John Wiley and Sons)

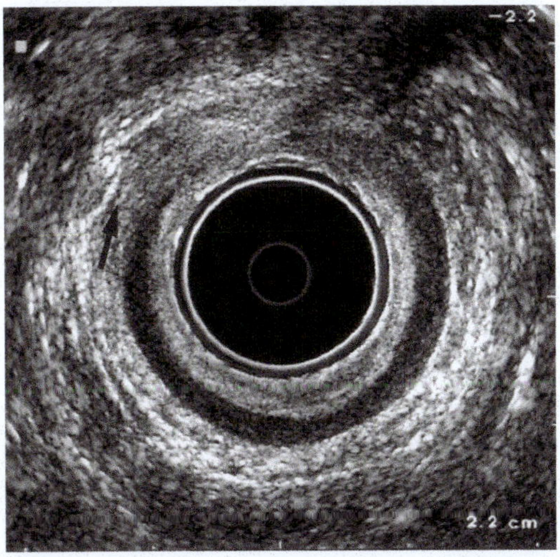

Fig. 5.4 Endoanal ultrasound image of an anterior EAS damage due to obstetrical injury. (Reused with permission © American Gastroenterological Association)

tal wall, anal canal, vagina, and pelvic floor are simultaneously visualized [10–12].

The patient is prepared for the study by administering phosphate enemas prior to the study. Then, the patient is placed in left lateral decubitus, and 50 mL of liquid barium is instilled into the rectum with a catheter followed by insufflation of a small amount of air. Also, 100–200 mL of barium paste is injected into the rectum until the patient feels rectal fullness. This study can be modified by simultaneous instillation of vaginal contrast/radiopaque tampons and bladder contrast for dynamic pelvicography [13]. The patient then sits on a radiolucent commode (Fig. 5.5), and standard fluoroscopic equipment is used to take images and video of the patient in rest, squeeze, push, evacuation, and post-evacuation phases (Fig. 5.6). Three important data points obtained from this study are measured including the anorectal angle, perineal descent, and puborectalis length.

The anorectal angle can be defined as the angle between the axis of the anal canal and the tangential line of the posterior rectal wall. As expected, the anorectal angle changes during the different phases of defecation. It approximates 90 degrees at rest. During the squeezing phase, puborectalis contraction makes the angle more acute at 75 degrees resulting in elevation of the anorectal junction. Normally, the puborectalis relaxes during straining thereby changing the angle to 110–180 degrees allowing the anorectal junction to descend to a maximum of 3.5 cm. Lack of these dynamic changes during relaxation can be interpreted as paradoxical contraction of the pelvic floor during defecation [14–16].

The pubococcygeal length or line is the distance from the tip of the coccyx to the pubis. It can be used as a surrogate to delineate the degree of puborectalis relaxation, and failure of the pubococcygeal length to change during resting and pushing phases suggests non-relaxation of the puborectalis muscle [15].

Perineal descent is in reference to the degree of the rectum that is below the pubo-

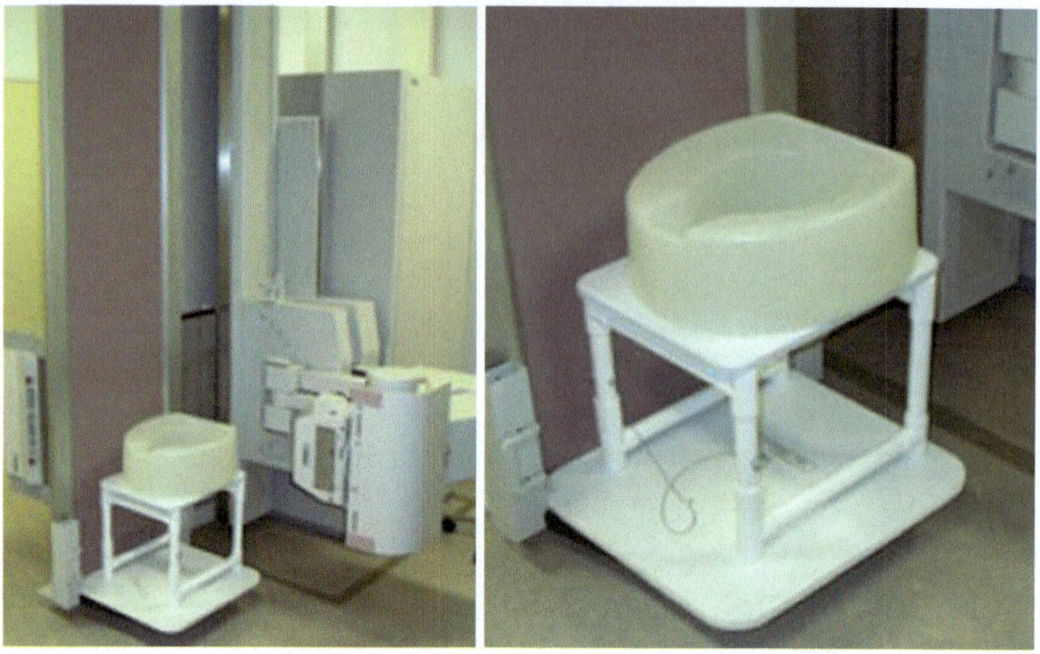

Fig. 5.5 Radiolucent commode. (Wiersma, T. 2006. Rectum-Dynamic Evaluation. Retrieved from http://www.radiologyassistant.nl/en/p4412ca5e2c21a/rectum-dynamic-examination.html. Accessed Nov 2017)

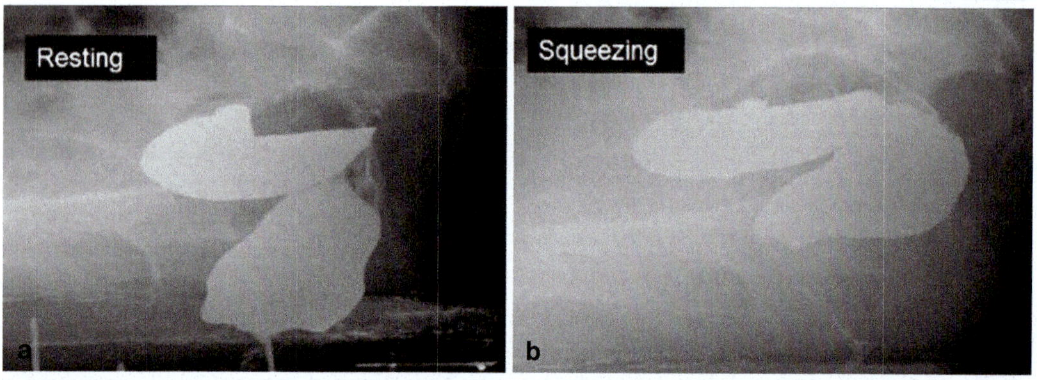

Fig. 5.6 Appearance of normal defecography phases. (a) Resting. (b) Squeeze. (c) Straightening of anorectal angle. (d) Post defecation. From: Kumar et al. [14]

coccygeal line during the study. If the rectum is more than one-third below the pubococcygeal line, radiologists interpret this as pelvic organ prolapse.

The main advantage of cinedefecography is that it gives dynamic imaging of a patient's defecatory function under normal physiologic conditions [12]. For instance, if the patient required

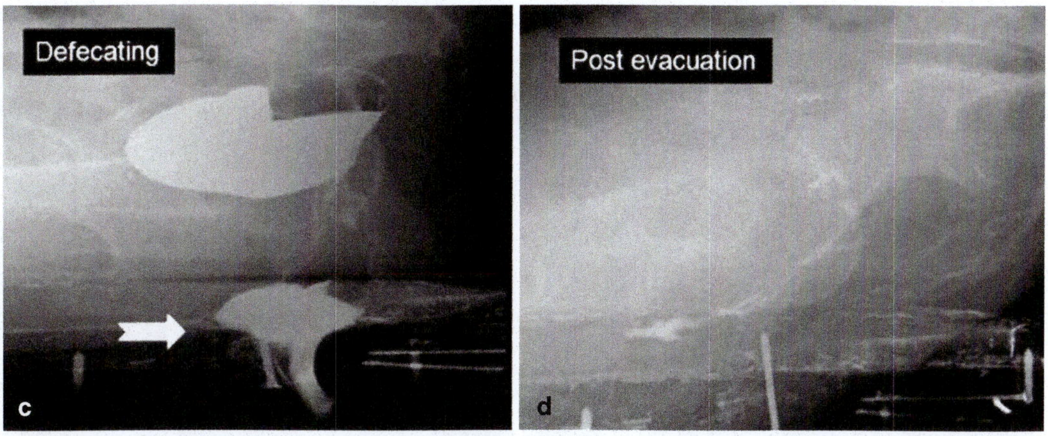

Fig. 5.6 (continued)

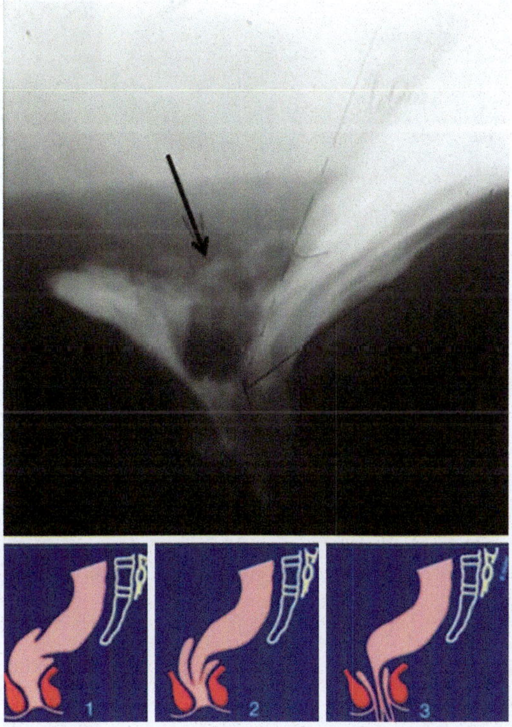

Fig. 5.7 Rectal Intussusception. (Wiersma, T. 2006. Rectum-Dynamic Evaluation. Retrieved from http://www.radiologyassistant.nl/en/p4412ca5e2c21a/rectum-dynamic examination.html. Accessed Nov 2017)

digitation to evacuate stool, this can be imaged as well. It also provides anatomic detail of mucosal prolapse, intussusception (Fig. 5.7), rectocele, and enterocele.

However, the findings from this study must be interpreted in conjunction with symptomatology because it can demonstrate findings of questionable clinical significance [17, 18].

Magnetic Resonance Imaging

Another modality in anatomic evaluation of the pelvic floor is magnetic resonance imaging (MRI). High-resolution cross-sectional images are obtained either by an external-phased array coil or and endoanal coil. MRI has been used extensively in the evaluation of fecal incontinence [10]. Dynamic evaluation of the pelvic floor can be accomplished with magnetic resonance defecography. While the patient is in a sitting position, rectal contrast is instilled into the rectum. Parameters measured are similar to cinedefecography described above. This modality has the advantage of providing great spatial resolution and soft tissue details (Fig. 5.8). However, this exam is not easily accessible.

Anorectal Manometry

Anorectal manometry enables objective evaluation of the sphincter complex of pelvic floor (Fig. 5.9) [19]. It provides a comprehensive

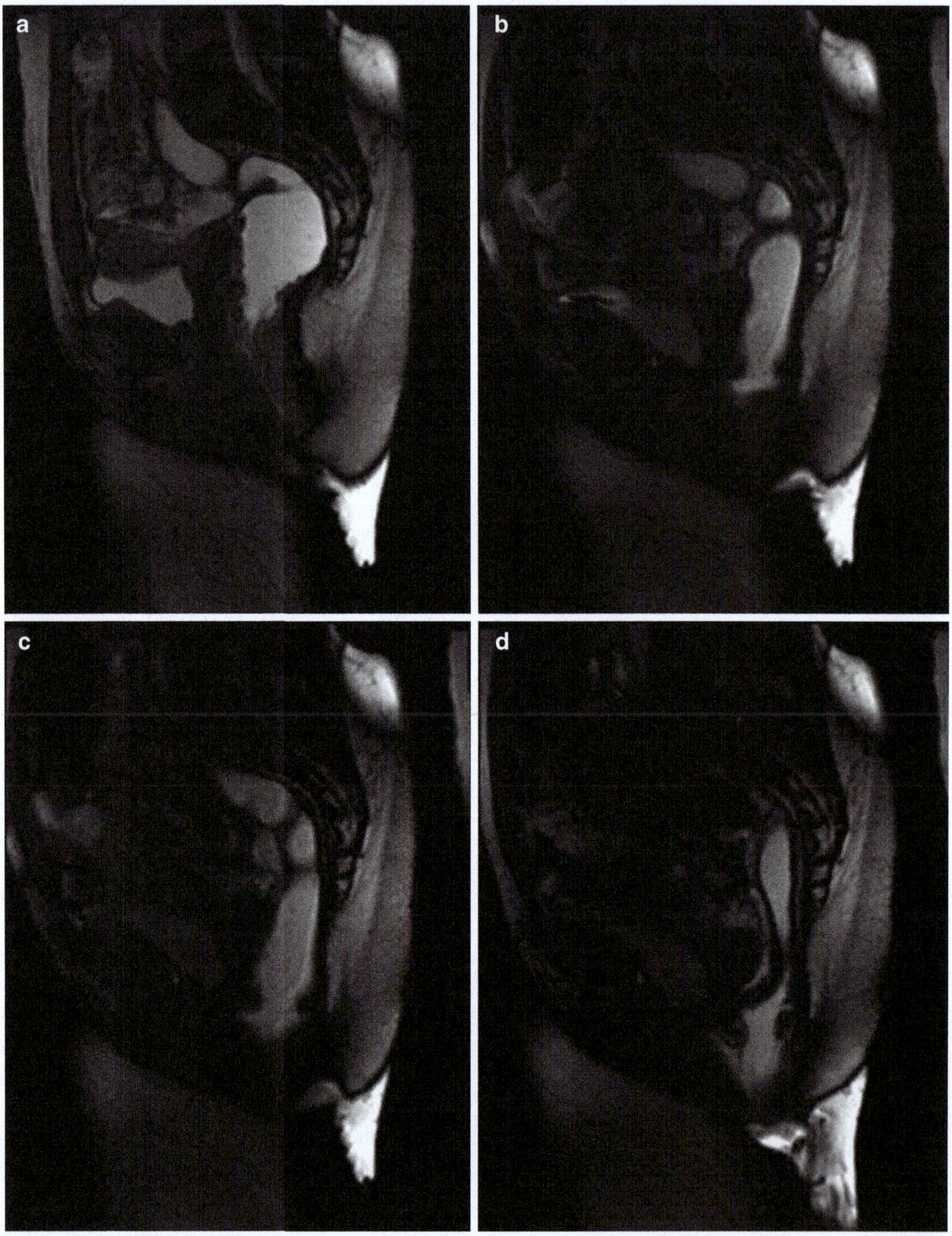

Fig. 5.8 Normal MR defecogram. Normal position of the anorectal junction at rest (arrow in **a**) with mild pelvic floor lift on squeeze (**b**). On straining (**c**) and defecation (**d**), there is mild descent of the anorectal junction, with the rectum and anal canal aligned in almost a straight line. The broken white line in (**d**) is the pubococcygeal line. The broken black line is the "H line" corresponding to the anteroposterior dimension of the hiatus. The solid black line is the "M line" which is the perpendicular distance between the pubococcygeal line and the posterior anorectal junction. (Reused with permission © Thieme)

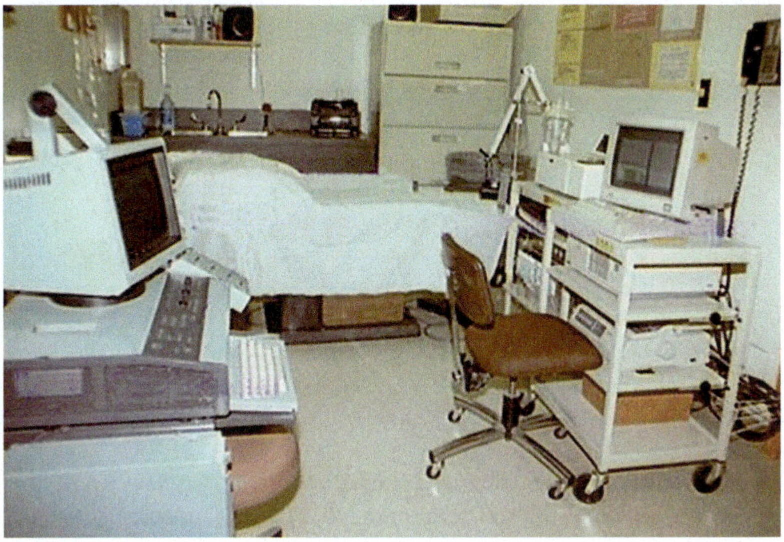

Fig. 5.9 Anorectal physiology lab set up and equipment

picture of the anorectal sphincter function by measuring parameters such as resting pressure, squeeze pressure, cough reflex pressure, Valsalva pressure changes, rectoanal inhibitory reflex, and rectal capacity. Clinically, this information can help in the diagnosis and management of disorders such as fecal incontinence, constipation, anal pain, and Hirschsprung's disease.

In order to be useful, data gathered from anal manometry must be compared to a standard of normal values that is specific to the institution performing the study. These normal values must undergo routine calibration and standardization to minimize inter-examiner variability allowing for accurate interpretation of the data.

Several systems and probes exist to perform anorectal manometry, each with its advantages and disadvantages; however, these systems require two basic components: a catheter and a transducer. The pressure measured by the system is generated by the resistance to the flow of perfusion through the catheter channels. Measured pressures are transmitted to specialized software which produces a polygraph. A large amount of data can be acquired from multiple channels to further enhance the study. Much of the software currently available for these studies allows for easy interpretation of the data.

Knowledge of normal pelvic floor anatomy as well as possible dysfunction is essential for evaluation of this complex patient population. The anal canal measures 2–5 cm and is surrounded by the internal and external anal sphincters. Importantly, there are gender differences in canal length, and it is well known that the anal canal is longer in men than it is in women [18]. The IAS is about 0.15–0.5 cm thick and consists of involuntary muscle that is innervated by the autonomic nervous system. The IAS is contracted at rest and is responsible for 50–85% of resting pressure. The external anal sphincter is made of striated muscle that is under voluntary control via the somatic innervation, specifically, the inferior branch of the pudendal nerve (S2–S3) and perineal branch of S4. When contracted, it is responsible for about 20% of the resting anal canal pressure.

Anal Manometry Systems and Techniques

Perfusion Systems

This system was originally developed by Arndorfer et al. [11] and uses a combination of flexible or rigid small catheters with ranging in diameter of 2.5–7 mm. These catheters have multiple channels or lumens that are arranged radially around the main axis of the catheter. Distilled water is infused through these catheters

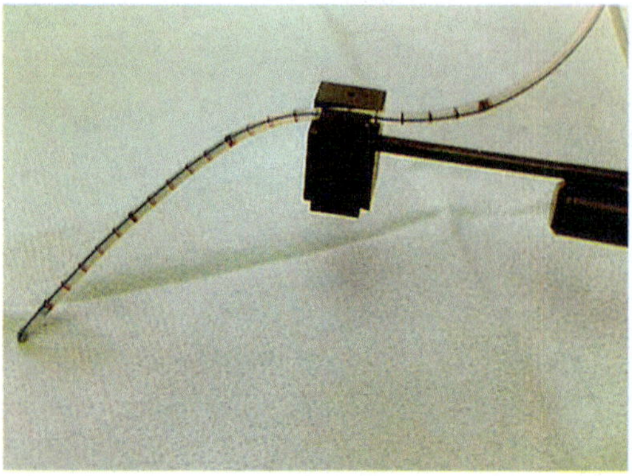

Fig. 5.10 1A de Coller® manometry catheter

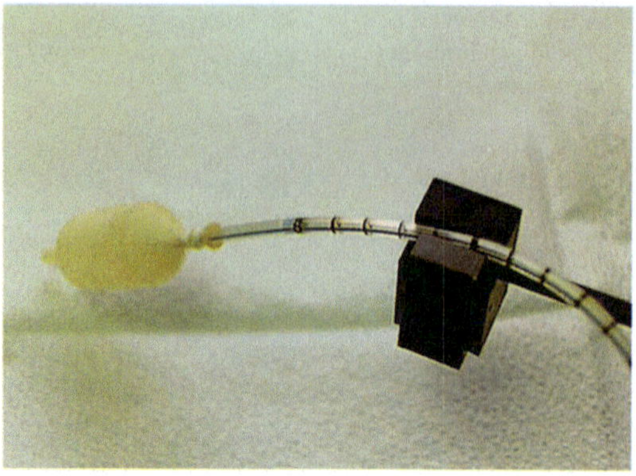

Fig. 5.11 B3 de Coller® manometry catheter

by a pneumohydraulic pump used to attain a consistent perfusion rate of 0.2–0.4 mL/min [20, 21].

Occlusion of the channels increases intraluminal pressure and produces resistance to the flow of water. This resistance is measured by transducers and interpreted as compliance of the tissue. Different types of catheters are available and allow the measurement of distinct parameters. Although many anal physiology labs use four channel systems, the author's lab utilizes an eight-channel system (Narco Biosystem, Austin, Texas).

In fact, the authors use two different catheters for the collection of data. Our first catheter is a Coller type 1A catheter (Fig. 5.10), and it has eight channels radially aligned into a single catheter. This catheter is used to measure the resting and squeeze pressure with a technique of continuous withdrawal. It is withdrawn from the anal canal at a velocity of 1 mm/s. The circumferential pressure differentiates are recorded eight times. Also, anal canal pressures can be measured at stationary intervals of 0.5 cm resulting in 64 recordings. At intervals of 1 cm, only 32 recordings are measured. In contrast, a 4-channel catheter measures only 32 and 12 pressure recordings at 0.5 and 1 cm intervals, respectively.

The second catheter we use is a Coller type B3 (Fig. 5.11) which has channels aligned in a linear orientation. Moreover, it is equipped with a balloon whose capacity is about 60 mL. It features a central channel that measures intrarectal pressure. Also, it measures the rectoanal inhibitory reflex (RAIR) and other parameters of rectal volumetrics (first sensation, desire to evacuate, and maximal tolerable volume).

Perfusion manometry systems have the advantage of wide availability and low cost. However, they are limited by artifacts during acquisition of data related to the number of channels utilized, the velocity of perfusion, and the left lateral decubitus position.

As inferred by the above discussion, there are two commonly used techniques for the evaluation of anorectal pressure by the perfusion manometry systems: continuous withdrawal and withdrawal at intervals.

The continuous withdrawal (pull through) method allows pressure readings to be made of the length of the anal canal, from the rectum to the anal margin. To perform this study, the catheter is gently introduced into the rectum with the most distal channel positioned above 6 cm proximal to the anal margin, and it is withdrawn at a speed of 1 mm/s via a motorized or computer-controlled arm. As the channels in the catheter pass over the high-pressure zone (HPZ), a pressure spike should be recorded. This method allows one to evaluate the resting and squeeze pressure, determination of the HPZ, and sphincter dysfunction/asymmetry. To minimize artifacts from this method, adequate lubrication and maintenance of constant withdrawal velocity are required.

The interval withdrawal method is performed by advancing the catheter 6 cm into the rectum so that the most distal channel is adjacent to the HPZ. Next, it is withdrawn at intervals of 0.5–1 cm every 30 seconds between withdrawals to allow stabilization of pressures. Static measurements at rest and squeeze are taken at these intervals. Although this method provides reliable pressure recordings at these intervals, it is hard to get a complete picture of sphincter function.

Microtransducer Systems

This system uses catheters with fine and flexible pressure sensors and microtransducers. These microtransducers consist pressure-sensitive diaphragms with semiconductors. This type of catheter allows the pressure sensor to come in direct contact with the area being investigated thereby providing direct measurements of pressures in the anal canal. A unique feature of this system is that it can be used in any patient and even while the patient is ambulating. Currently, this is the only system that allows for continuous ambulatory evaluation; however, its use is limited by greater fragility and high cost.

Balloon Systems

The largest evidence for use of balloon systems in manometry is from Shuster et al. [20] who used a system with two to three balloons filled with air. The device consisted of a small hollow cylinder surrounded by a latex balloon to create two compartments. The two parts of the balloon are connected to pressure transducers. The catheter is positioned in the anal canal and the balloon inflated until it is fixated in the rectum. The internal balloon measures the internal pressure of the anal canal, while the outer balloon measures the pressures of the anal sphincter. A modification can be made where a third balloon is placed in the proximal portion of the cylinder to measure pressures in the rectum.

High-Resolution Anorectal Manometry (HR-ARM)

Since its initial introduction in 2007, high-resolution anometry catheters are increasingly used in clinical practice. There are two types of catheters utilized: high-resolution anorectal manometry (HR-ARM) and high-definition anorectal manometry (HD-RAM). Unlike earlier catheters which have up to six unidirectional sensors, HR-ARM and HD-ARM catheters contain several closely spaced circumferential sensor elements along the longitudinal axis with the pressure-sensing element varying among different systems.

High-resolution anometry catheters provide several advantages over traditional catheters. These systems provide a continuous and dynamic spatiotemporal mapping of anorectal pressures allowing for easier and more detailed data interpretation [22]. Patient comfort with the procedure is also improved with HR-ARM because the time needed for the exam is much shorter because the catheters do not require pull-through and a topographic display enables rapid positioning of the probe [23].

The main disadvantage of this system is that the catheters are expensive, fragile, and are less durable [23]. Also, these catheters are temperature-sensitive, which require a thermal compensation algorithm built into the software [24–26]. While pressures recorded with HR-ARM and traditional manometry correlate well, anal sphincter pressures at rest and squeeze are often higher with HR-ARM. It is thought that this is due to improved sensitivity of measurements with sensors in the high-resolution probe [27].

Like traditional anorectal manometry, HR-ARM are either water-perfused or solid state. A brief review of these catheters will be given here. Given Imaging HR-ARM systems are solid-state catheters (ManoScan AR catheters) with an outer diameter of 4.2 mm. There are two different types of probes. The regular probe (AAN) has 12 circumferential sensors, including tensors at 5 mm intervals along the anal canal and 2 sensors in the rectal balloon. The balloon is 3.3-cm-long and has a maximum capacity of 400 mL. In contrast, the small probe (APN) has 8 circumferential sensors and 1 balloon sensor. The balloon is again 3.3-cm-long and has a maximum capacity of 300 mL. Either of these catheters have 36 circumferentially oriented, pressure-sensing elements that acquire data at 35 Hz. These 36 sector pressures are averaged to yield a single value. The data acquired are displayed using ManoScan AR analysis software.

The eSleeve option in the software produces a single value derived from all the recorded pressure across the anal value. This eSleeve value is used to calculate the average and maximum anal resting pressure and the maximum squeeze pressure over 20 seconds during these maneuvers. During simulated evacuation, the eSleeve identifies the rectoanal gradient between over a 20-second interval [25].

Sandhill HR-AM (Denver, CO, USA) system uses a 4-mm-diameter probe that has eight directional solid-state sensors. The sensors are spaced at 1 cm intervals in the following locations: rectal balloon (1), rectum (1), anus (5), and external to anal verge (1). The pressures recorded are averaged to provide a mean sphincter pressure and analyzed by the InSight system (Sandhill Scientific) [25].

Medical Measurement Systems uses a 12G catheter probe that incorporates eight directional sensors along its axis. Six of these eight sensors are equidistant from each other and span 5 cm. The proximal sensor is located within the rectal balloon and is spaced 2.5 cm proximal to the other sensors. The most distal sensor is 2 cm below the most distal anal sensor and is used as an external reference. This catheter requires submersion in water for about 3 minutes to pre-wet the sensors and then zeroed to atmospheric pressure. Data is analyzed using the Solar GI HRM software (MMS, Enschede, the Netherlands) [25].

Given Imaging's HD-ARM (Yokne'am Illit, Israel) catheter is 6.4 cm in length with an outer diameter of 10.75 mm. The sensing segment is composed of 256 sensing elements that are arranged in 16 rows which are circumferentially oriented. The spacing between the sensors is 3 mm axially and 2 mm radially. Unique to this probe is that it displays pressures recorded by individual sensors around the circumference. Manometry and topographic images are displayed using the Motility Acquisition AR System (Given Imaging). The probe is calibrated immediately before the procedure by placing it in a calibration chamber, where it is zeroed to atmospheric pressure and set to a range of pressures up to 300 mmHg [25].

The sophisticated software provides the clinician with an intuitive color topographic analysis of anal canal pressures [26]. Cool colors (blue and green) correlate with low pressure. Warm colors (red and yellow) correlate to higher pressures (Fig. 5.12). This can lead to improved understanding of anal canal function.

Parameters Measured with Anal Manometry

Anorectal manometry measures the following parameters listed in no particular order:

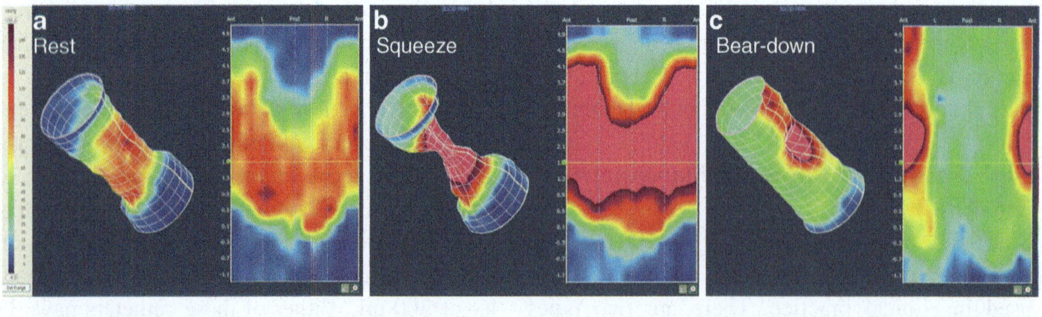

Fig. 5.12 High-definition anorectal manometry in a healthy individual. (**a**) Rest frame shows that the high-pressure band is seen in the middle of the image. (**b**) Squeeze frame shows an "λ" shape on 2-D mapping indicating normal functioning of the EAS muscle. (**c**) Bear-down frame shows a green low-pressure zone appearing in the end (i.e., a low-pressure area in the distal posterior wall of the anorectum). (Reused with permission © The Korean Society of Neurogastroenterology and Motility)

- Anal sphincter function including resting and squeeze pressures
- Anal canal length
- Anal motility
- Rectal sensation and compliance
- Pressures during forced evacuation
- Anorectal reflexes

Resting Anal Pressure

The normal sphincter rest pressure reflects the sum of the tone of the internal and external anal sphincters. Resting anal pressure is defined as the difference between intrarectal pressure and anal canal pressure. It is important to remember that measures of pressures vary according to gender, age, and technique used. In general, these pressures are higher in men and younger individuals. In our institution, normal resting pressure usually varies between 65 and 85 mmHg above rectal pressure at 1 cm intervals. At typical longitudinal profile of normal anal sphincter, resting pressure is shown in Fig. 5.13.

The proximal portion of the sphincter is determined by an increased pressure of at least 5 mmHg compared to intrarectal pressure, and it is considered the zero point. The HPZ is defined as the length of the anal canal over which the pressure is greater than half the maximum resting pressure [28]. In men, it measures about 3–3.5 cm in length, and in women it is about 2–2.5 cm in length. The distal end of the anal canal is determined when the pressure abruptly reduces to zero. Defects in the anal sphincter musculature can be clearly observed in a longitudinal profile as seen in Fig. 5.14.

It is important to recognize that the sphincter muscles have longitudinal and radial asymmetry. The pressure is typically higher in the posterior portion of the anal sphincter and then increases distally. Cross-sectional measurements of the sphincter can create a 3-D image such as Fig. 5.15. An example of a right anterior sphincter defect in an obstetric patient can be seen in Fig. 5.16.

Squeeze (Contraction) Pressure

The evaluation of the external sphincter may be better demonstrated as the pressure during vol-

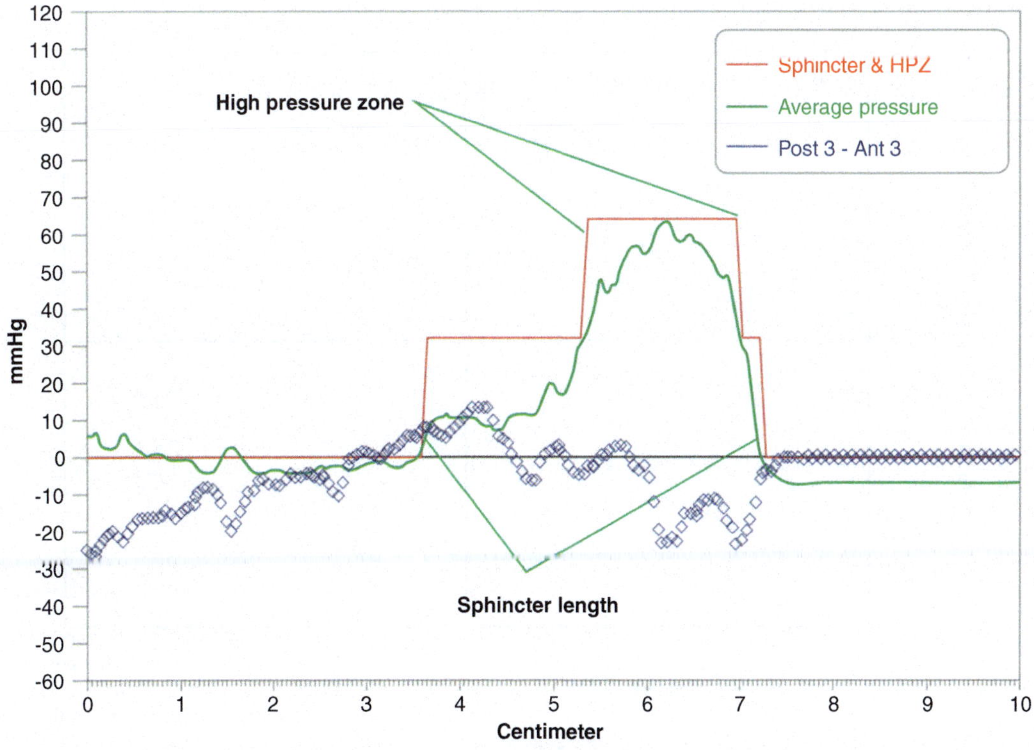

Fig. 5.13 Manometric graph showing a longitudinal profile typical of the pressure at rest

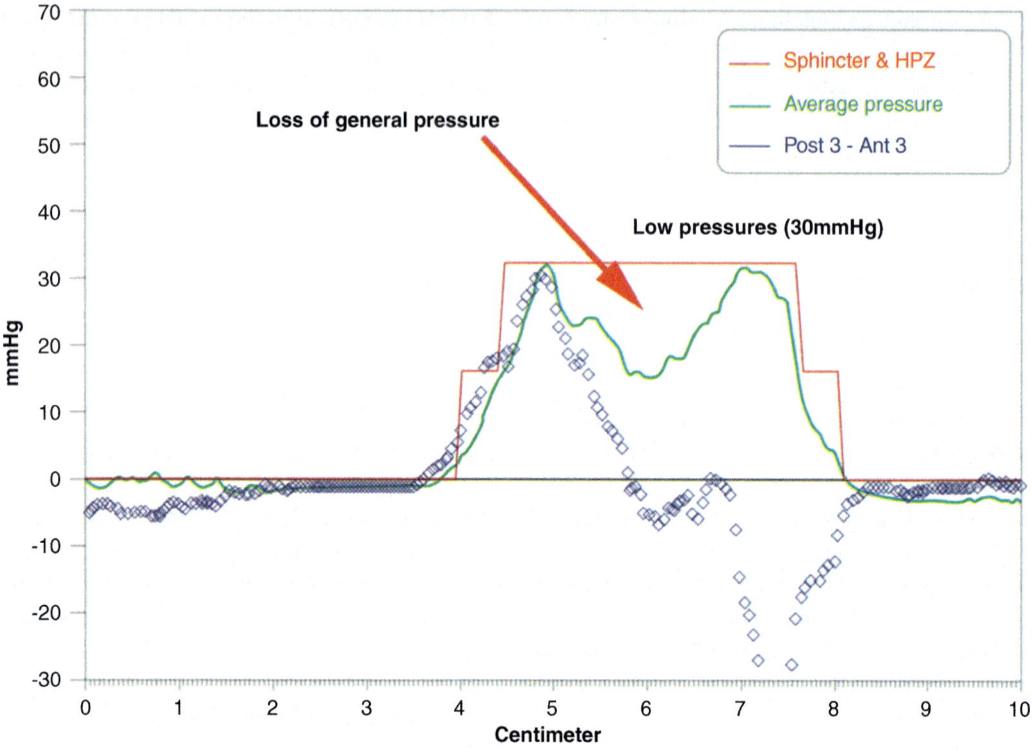

Fig. 5.14 Manometric graph showing the high-pressure zone

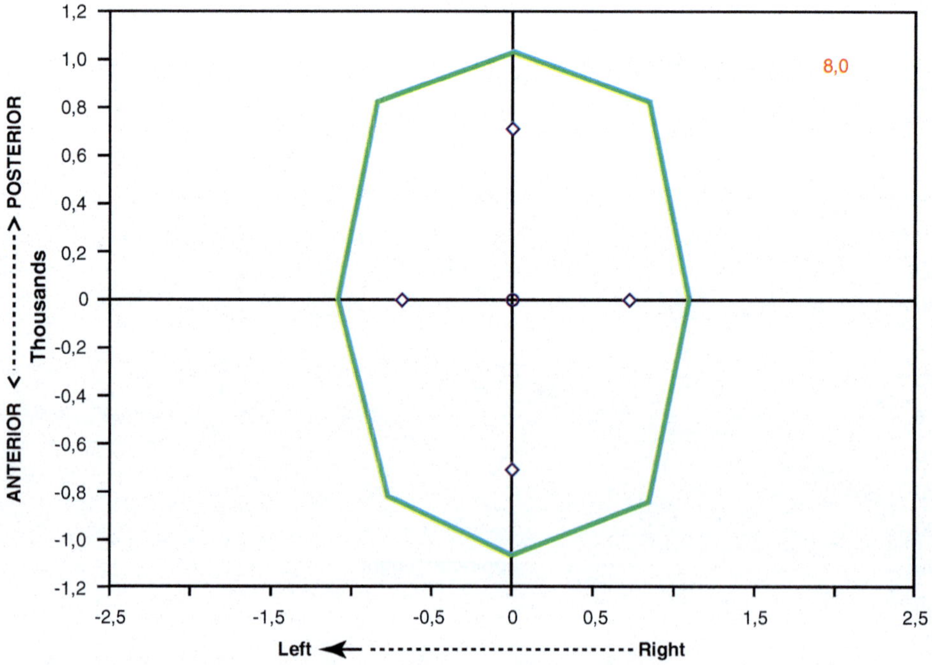

Fig. 5.15 Three-dimensional graph showing sphincter symmetry

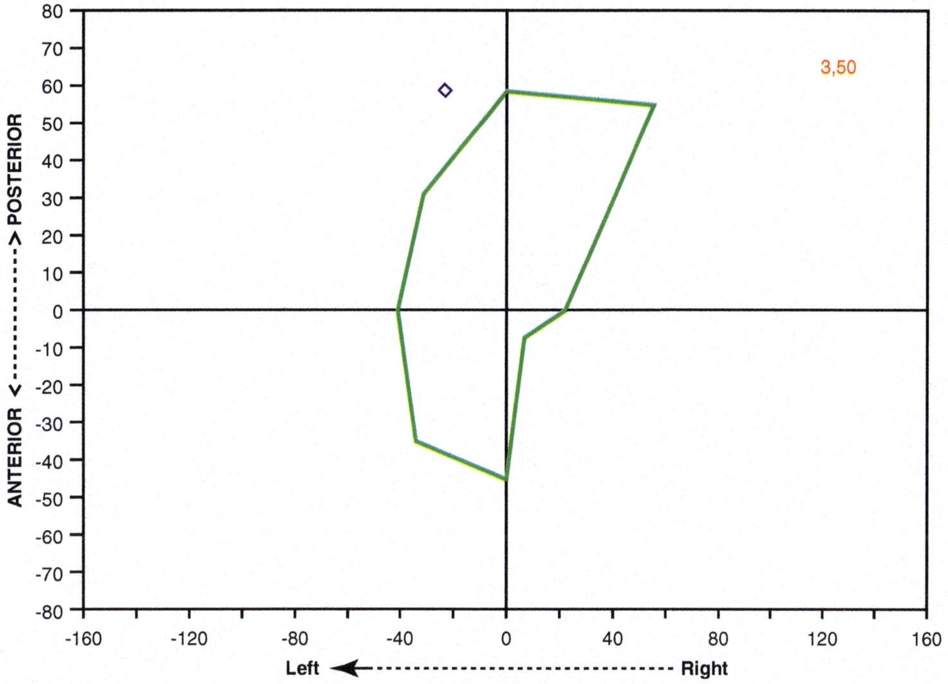

Fig. 5.16 Sphincter asymmetry in a patient with defect in the right anterior quadrant

untary contraction. As the perfusion catheter is continuously pulled through the anal canal, the patient instructed to contract their external anal sphincter which is measured as an increased pressure. The contraction pressure maximum is up to 155 mmHg in normal controls; however, incontinent patients have averages that are well below these values [29]. Importantly, there is a large variation in the average squeeze pressure due to a variety of factors. Figures 5.17 and 5.18 demonstrate normal voluntary squeeze pressures during continuous or stationary techniques, respectively. A diminished contraction pressure can be an important parameter in incontinent patients and could indicate sphincter deficiency, neurologic dysfunction, or low rectal compliance.

Normally, patients can maintain maximum contraction tone for about 45–50 seconds which is followed by a refractory period. The muscle fiber complement in the external anal sphincter determines the duration of maximum contraction until fatigue. Types I and II muscle fiber complement in the external anal sphincter changes with age. The duration of squeeze and fatigue index has been shown to be much lower in patients with incontinence [30–34].

Anal Motility

Internal anal sphincter resting tone is fundamentally different from other muscles because there is evidence that there are cyclical variations in the electrical activity. Analysis of the configuration of the IAS tone demonstrates short, ultrashort, and intermediate waves.

The frequency of the short wave ranges from 9 to 15 cycles per minute (CPM), with an amplitude of 0 to 40 mmHg (Fig. 5.19). The shortwave frequency in the IAS is greater than any other gastrointestinal muscle and is highest in the most distal portion of the sphincter. The tonal frequency depends on factors such as prandial vs. postprandial state and awake vs. asleep states. Although the shortwave pattern is present in most patients and is the most frequently present wave, its clinical significance is not none.

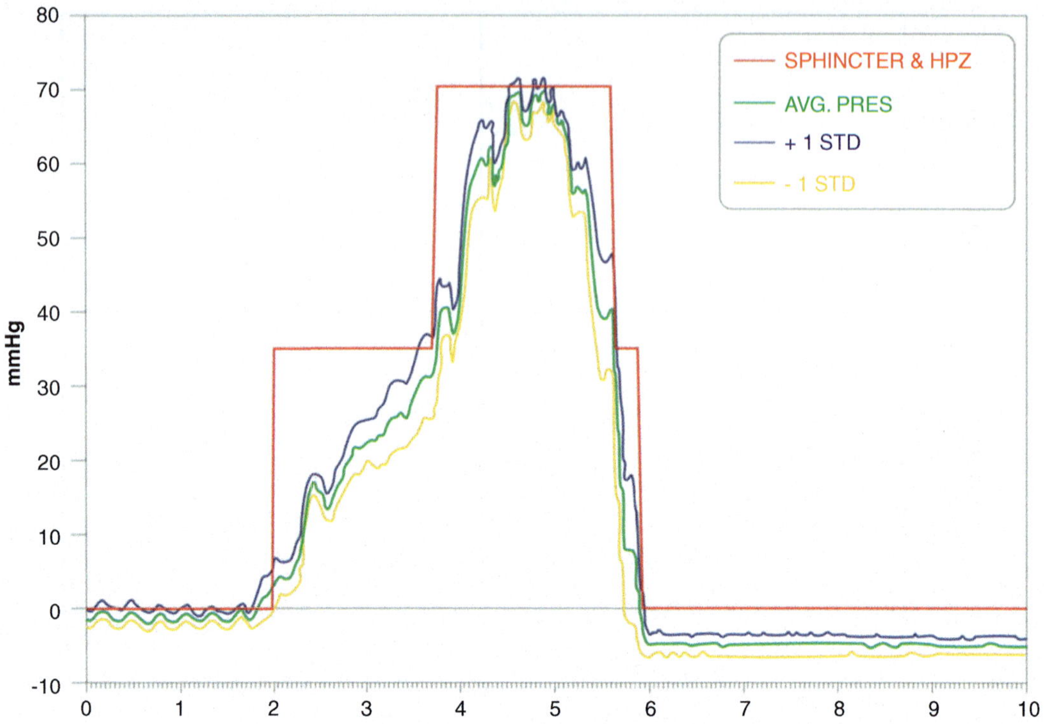

Fig. 5.17 Pressure profile in manometry performed with continuous withdrawal method

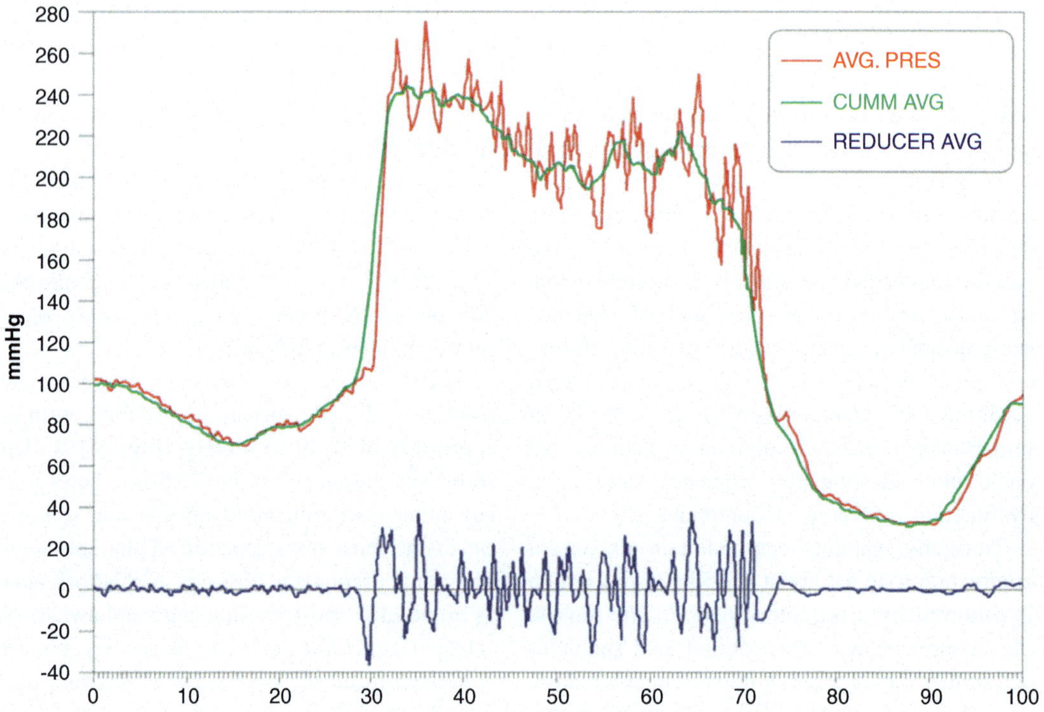

Fig. 5.18 Manometric tracing with stationary withdrawal technique

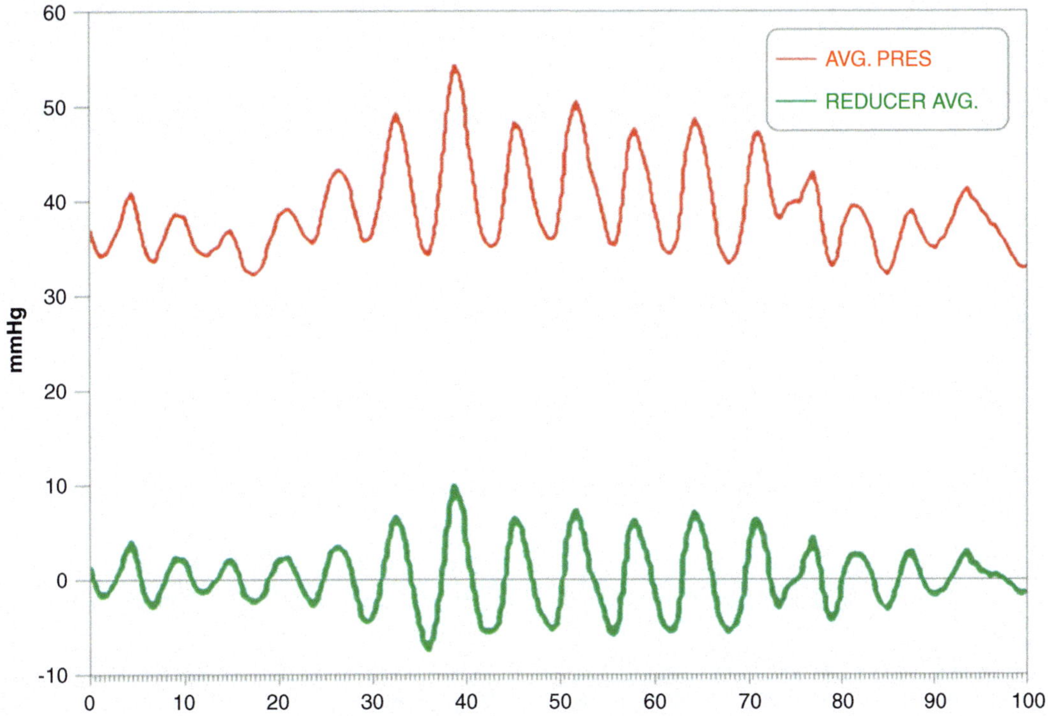

Fig. 5.19 Manometric tracing showing resting pressure with short and ultrashort waves

Ultrashort waves vary in frequency from 0.5 to 1.5 CPM and present a wider range reaching about 100 mmHg (Fig. 5.20). It is believed that these waves originate from the IAS and can be seen in patients with anal fissures and hemorrhoids. Although shortwaves and ultrashort waves can be attributed to the IAS, a correlation with patients with continence has yet to be demonstrated.

The intermediate waves are slightly faster than short waves and range from 4 to 8 CPM and present with pressures reaching about 3 to 70 mmHg (Fig. 5.21). We believe that they relate to intrinsic fasciculations in the IAS and can be observed in patients with neurogenic fecal incontinence and those with ileoanal anastomosis [29].

Volumetric Measurements

Rectal Sensation

The ability of the rectum to distend in response to increasing rectal volume enables it to store substantial quantity of stool before defecation. Gradual intrarectal balloon distension enables accurate measurement of rectal sensation. The first sensation measured refers to the threshold of rectal sensitivity, and it can be described to the patient as the first sensation of cold or nuisance. The second sensation recorded is the urge to defecate. Finally, the third sensation recorded is the maximum tolerable volume.

Mucosal receptors in the rectum are innervated by nerves in the pelvic fascia and pelvic floor musculature through S2 to S4 nerve roots [35]. Patients can have a hypo- or hypersensitive rectum which correlates to increased or decreased rectal capacity.

Rectal Compliance

The ability of the rectum to distend and accommodate stool is rectal compliance. This is calculated as the change in rectal pressure with the change in volume. Rectal compliance can be altered in conditions such as proctitis or external beam radiation.

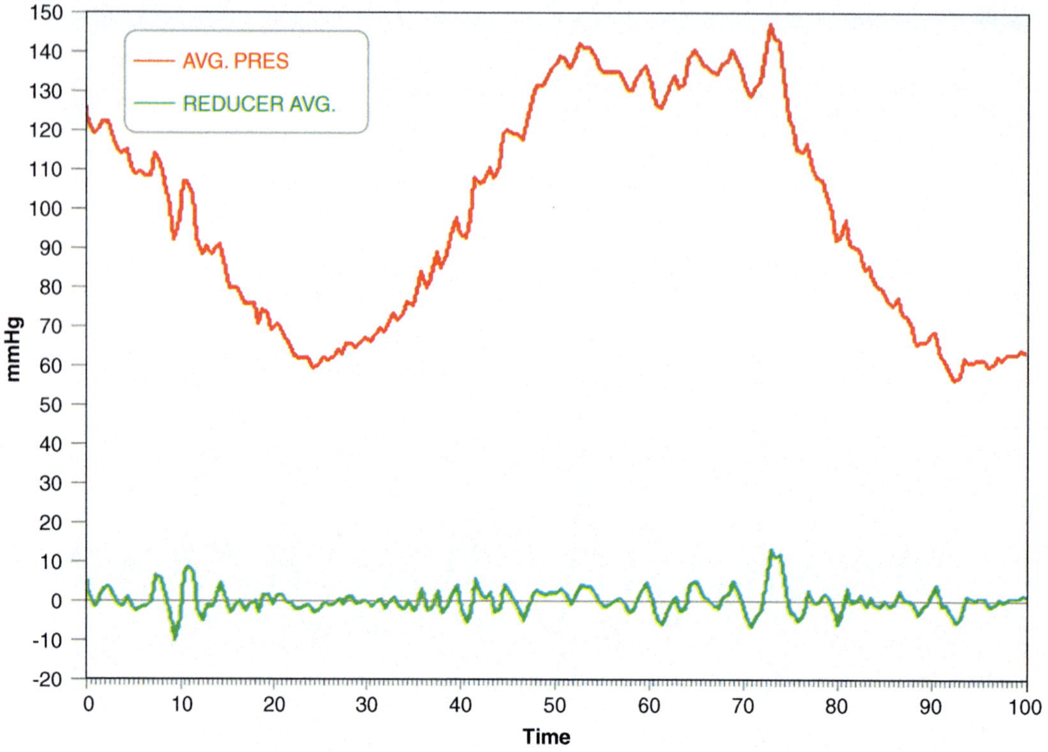

Fig. 5.20 Graphic showing ultrashort waves

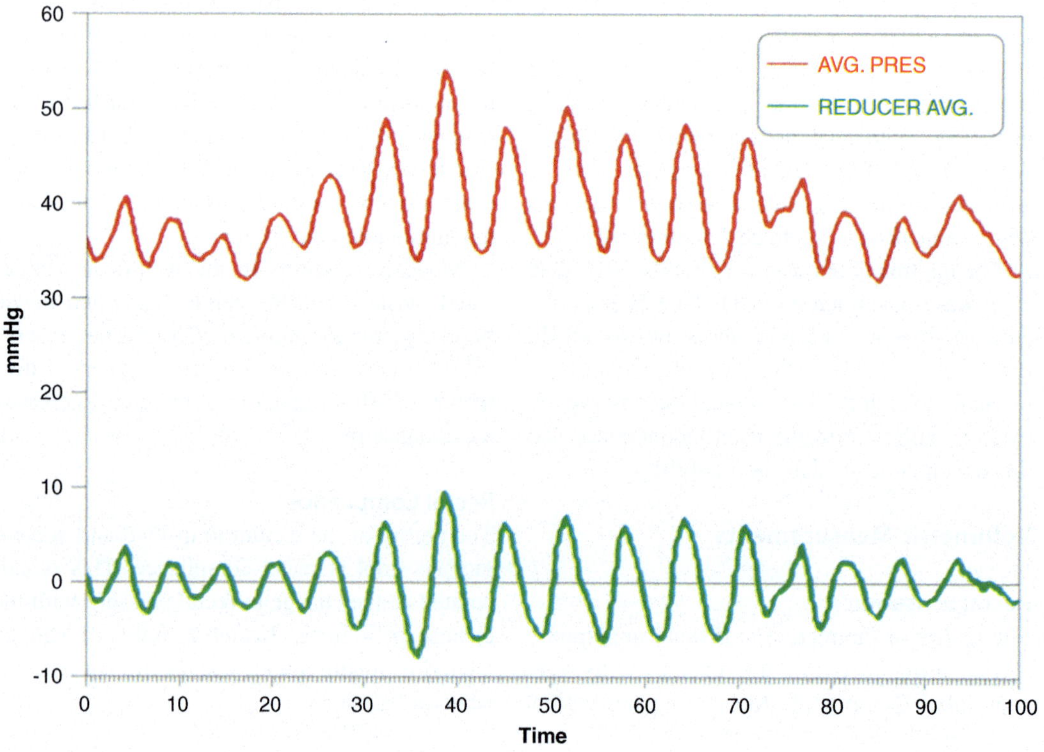

Fig. 5.21 Manometric tracing showing giant short waves

Balloon Expulsion Test

The balloon expulsion test is used to evaluate forced evacuation. This test is performed by placing a 4 cm rectal balloon into the rectum. Next, the patient is asked to expel the balloon while time is recording. The ability to expel the balloon and time taken to expel the balloon are recorded. This is clinically significant in patients with normal-transit constipation and megarectum as they are often unable to expel the inflated balloon.

Rectoanal Inhibitory Reflex (RAIR)

The rectoanal inhibitory reflex is defined as the relaxation of the IAS during rapid rectal distension [32]. This reflex allows the anal epithelium to sample rectal contents enabling one to discern between liquid and solid stool [33].

On manometry, the RAIR is demonstrated as a steep drop in IAS tone when rectal balloon is inflated to 50–100 mL of air (Fig. 5.22). The degree of RAIR stimulation and inhibition as well as time needed for the curve to return to baseline are analyzed. Volumes larger than 100 mL are required in condition such as megarectum and hyposensitivity. The magnitude of the pressure reduction depends on the volume in the balloon used to distend the rectum. In fact, the IAS tone can be totally inhibited at a certain rectal volume. With time, the IAS pressure returns to normal because it adapts to the increased rectal volume.

The RAIR is mediated by the enteric nervous system composed of myenteric plexi with input from the autonomic nervous system. Its absence is pathognomonic of patients diagnosed with Hirschsprung's or rectal Chagas disease. Of note, the RAIR is still present after head, spinal cord, and hypogastric nerve trauma. However, it is affected by low colorectal or coloanal anastomoses. In addition, fecal constipation and incontinence and have been associated with altered patterns of the RAIR [34].

Valsalva

Normally, the external anal sphincter relaxes during the Valsalva maneuver to allow stool

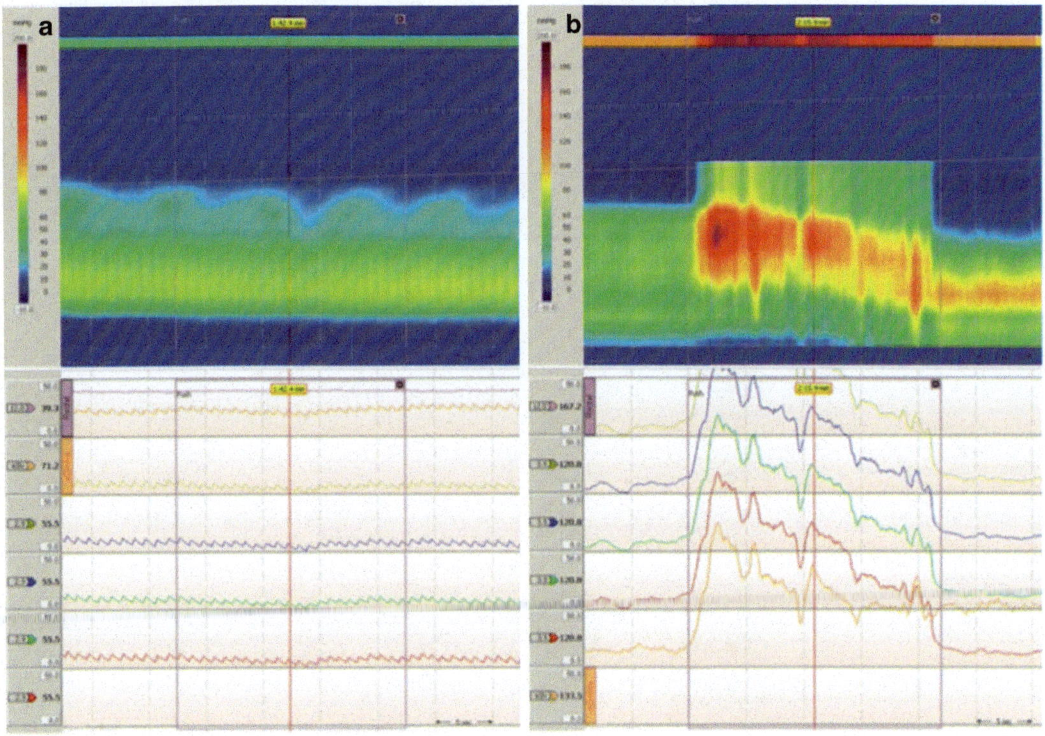

Fig. 5.22 Appearance of Valsalva on high-resolution manometry. (**a**) Normal relaxation. (**b**) Paradoxical contraction. (Reused with permission © Springer Nature)

evacuation. On manometry, this is demonstrated by an increase in intrarectal and a decrease in anal canal pressures (Fig. 5.23). However, puborectalis and/or EAS contraction can lead to outlet obstruction–induced constipation [31]. Anal manometry can detect paradoxical contraction of these muscles and diagnose dyssynergic defecation (Fig. 5.23). Like other manometry studies, these findings must be interpreted in the context of the symptomatology and examination of the patient because paradoxical sphincter contraction can be a finding in healthy patients as well.

Neurophysiologic Examination

Electromyography

Anal electromyography quantifies the electrical activity of the external anal sphincter and puborectalis muscles by providing measures of amplitude as well as duration of action potentials. The study provides a global evaluation of each motor unit and can be used to assess the functional activity of pelvic floor muscles during rest, squeeze, and attempted defecation. Before widespread use of endoanal ultrasound,

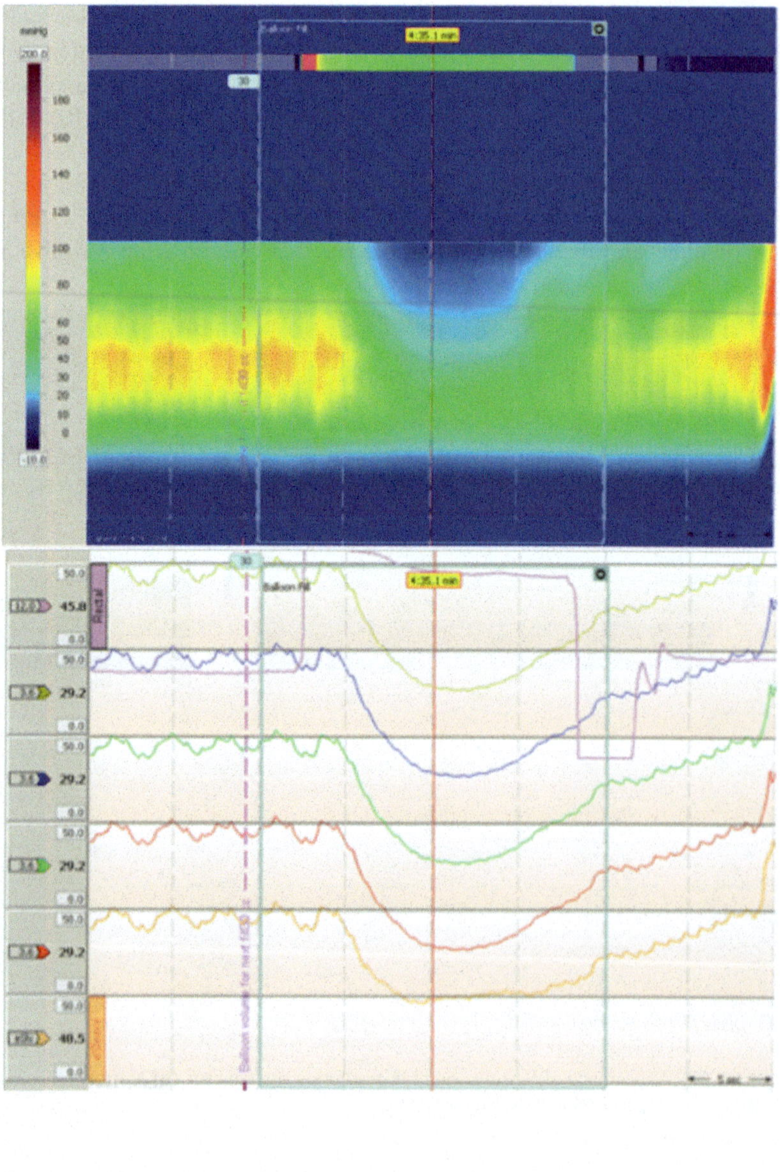

Fig. 5.23 Appearance of RAIR (rectoanal inhibitory reflex) on high-resolution manometry. (Reused with permission © Springer Nature)

EMG was the used primarily for evaluating anal sphincter defects [36]. Currently, EMG can still be helpful in mapping sphincter defects when there is dense scarring that can cause artifact on ultrasound.

The study can be performed via three methods: a needle electrode, a surface electrode on the perianal skin, or a cone-shaped plug in the anal canal. Needle electrodes come in two varieties: a concentric needle and a monopolar wire. Using this device, the clinician can precisely assess motor unit function by each quadrant. The number of motor units recruited during voluntary contraction correlates with sphincter pressures. For example, incontinent patients may have areas of sphincter damage that will display prolonged or absent action potentials. Also, disorganized polyphasic responses can be seen in motor units that have undergone reorganization and reinnervation [37]. Unfortunately, this study is not well tolerated due to the pain associated with this exam.

On the other hand, surface and anal plug electrodes are painless and well tolerated; however, it only provides a global assessment of motor function instead of each quadrant of the external anal sphincter. The anal canal responses during voluntary contraction are recorded. In constipated patients, failure to decrease or increase motor unit recruitment during attempted defecation can signal anismus or paradoxical contraction of the puborectalis respectively [38]. This information can be used therapeutically during biofeedback sessions.

Pudendal Nerve Terminal Motor Latency

Pudendal nerve fibers originate from the nerve sacral nerve roots of S2, S3, and S4; then, they traverse Alcock's canal and terminate in the fibers of the levator muscle and external anal sphincter. Due to its anatomic location, the pudendal nerve is vulnerable to traumatic injuries to the pelvic floor, particularly those related to forceps-assisted vaginal childbirth. Other causes of pudendal nerve damage include chronic rectal prolapse, perineal descent, and other conditions associated with intestinal constipation,

First described in 1984 at St. Mark's Hospital [39], pudendal nerve terminal motor latency (PNTML) is a test that assesses the integrity of the pelvic floor innervation via conduction through the pudendal nerve [40]. A neurogenic sphincter injury can be demonstrated as prolonged PNTML. This test has effectively replaced concentric needle EMG in the evaluation of patients with incontinence [41].

To perform this test, an electrode mounted to a gloved finger is introduced into the rectum (Fig. 5.24). After palpating the coccyx, the finger is slid laterally and positioned over the ischial spine which serves as a landmark for the pudendal nerve. Next, 5–15 mA impulses are delivered by the fingertip electrode, and responses are captured by another electrode at the base of the finger. The interval between nerve stimulation and muscle contraction at the level of the external anal sphincter corresponds to the PNTML, and normally it is less than 2.2 milliseconds. Values above 2.2 milliseconds are abnormal and can indicate neuropathy (Fig. 5.25a, b).

Of note, the reliability of this test is operator-dependent because proper placement of the fingertip electrode over the pudendal nerve is critical [42]. However, the test is reproducible with low intra- and inter-observe variability [43].

Colonic Transit Study

Colonic transit studies are used to assist in the evaluation of patients with intestinal constipation due to slow transit or colonic inertia [44].

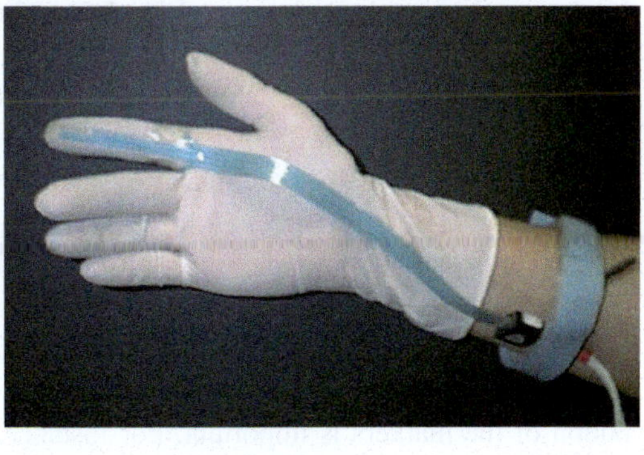

Fig. 5.24 St. Mark's electrode

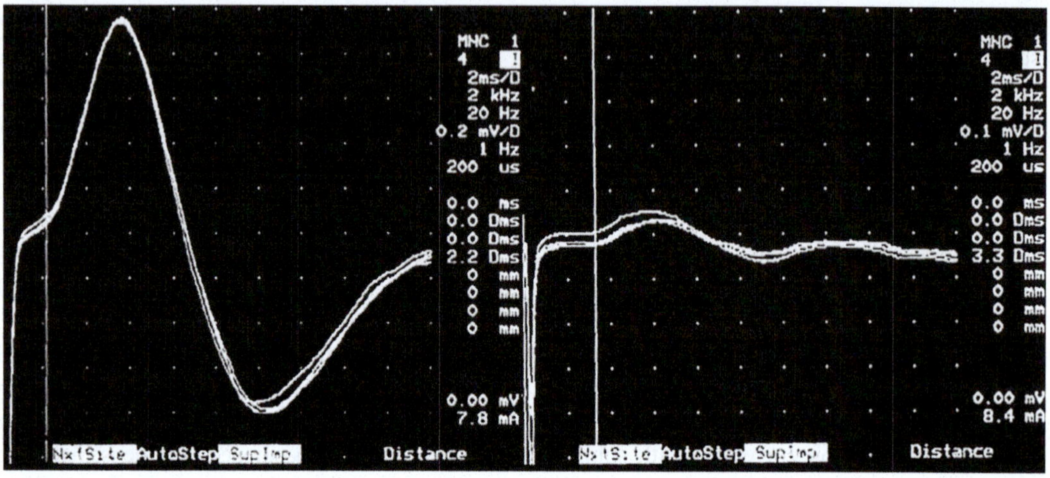

Fig. 5.25 Graph showing normal latency time of the pudendal nerve (**a**). And prolonged latency time of the pudendal nerve (**b**)

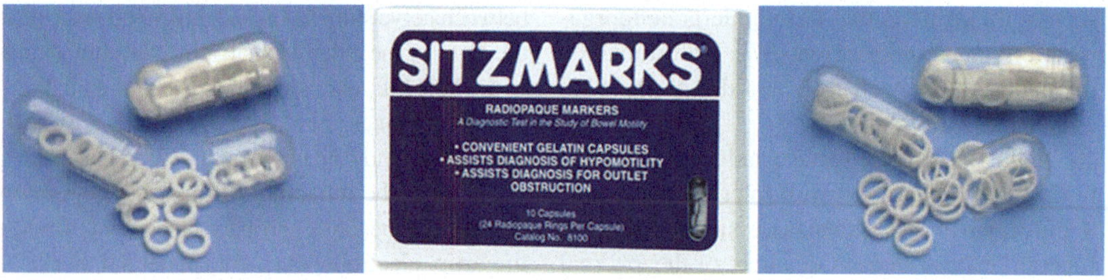

Fig. 5.26 Sitzmarks® Radiopaque Markers

Although there are many methods to assess colonic transit, this chapter will review only the main studies currently used in clinical practice.

Single-Capsule Technique

In the single-capsule technique (Fig. 5.26), 24 radiopaque markers are administered to the patient as a single capsule (Sitzmarks® Konsyl Pharmaceuticals, Fort Worth, TX, USA). Plain anterior–posterior (AP) abdominal films are on day 5 (120 hours) post-ingestion to visualize the location and distribution of the markers. A normal study should have less than 20% or less than five of the markers remaining in the colon. When a patient retains more than five markers, the distribution of the markers is important. For instance, patients with colonic inertia will have the markers spread out through the colon. On the other hand, patients with outlet obstruction will have most of the retained markers in the rectosigmoid [45–47].

The main advantage of this technique is its ease of administration, tolerability, and reduced radiation exposure; however, it does not permit exact quantification of transit time nor does it assess segmental transit time. Also, it is important to counsel the patient to stop all laxatives 48 hours prior to the study.

Multiple-Capsule Technique

In the multiple-capsule technique, a patient ingests a Sitzmarks capsule daily for 3 days consecutively. A plain AP radiograph is obtained on day 4 and day 7. Total and segmental colonic transit times are calculated using a specialized formula. The average total colonic transit time is about 35 hours (95% eliminate by 68 hours), and average segmental transit time is about 12 hours (95% have segmental transit by 26 hours). The main advantages of this technique is quantification of total transit time and segmental transit time at the cost of increased radiation exposure.

Summary

Pelvic floor disorders are a common group of heterogeneous pathologies that range from being inconvenient to extremely debilitating. Although it is essential to complete a thorough history and physical examination in patients with functional pelvic disorders, specialized tests can greatly assist in the evaluation, diagnosis, and management of these complex patients.

References

1. Law PJ, Kamm MA, Bartram CI. Anal endosonography in the investigation of faecal incontinence. Br J Surg. 1991;78:312–4.
2. Dobben AC, Terra MP, Deutekom M, Slors JFM, Janssen LWM, Bossuyt PMM, Stoker J. The role of endoluminal imaging in clinical outcome of overlapping anterior anal sphincter repair in patients with fecal incontinence. AJR Am J Roentgenol. 2007;189:W70–7.
3. Parker SC, Spencer MP, Madoff RD, Jensen LL, Wong WD, Rothenberger DA. Artificial bowel sphincter: long-term experience at a single institution. Dis Colon Rectum. 2003;46:722–9.
4. Matzel KE, Kamm MA, Stosser M, et al. Sacral spinal nerve stimulation for faecal incontinence: multicentre study. Lancet. 2004;363:1270–6.
5. Maeda Y, Vaizey CJ, Kamm MA. Pilot study of two new injectable bulking agents for the treatment of faecal incontinence. Color Dis. 2008;10:268–72.
6. Stroker K, et al. Endoanal ultrasound of the upper anal canal showing the puborectalis muscle. From: Endoanal ultrasound. Imaging pelvic floor disorders. 2nd ed. Berlin: Springer; 2008.
7. AGA review on anorectal testing techniques. Gastroenterology 1999;116(3):735–60.
8. Engin G. Endosonographic imaging of anorectal diseases. J Ultrasound Med. 2006;25:57–73.
9. Emblem R, Dhaenens G, Stien R, Morkrid L, Aasen AO, Bergan A. The importance of anal endosonography in the evaluation of idiopathic fecal incontinence. Dis Colon Rectum. 1994;37:42–8.
10. Williams AB, Malouf AJ, Bartram CI, Halligan S, Kamm MA, Kmiot WA. Assessment of external anal sphincter morphology in idiopathic fecal incontinence with endocoil magnetic resonance imaging. Dig Dis Sci. 2001;46:1466–71.
11. Thapar RB, Patankar RV, Kamat RD, Thapar RR, Chemburkar V. MR defecography for obstructed defecation syndrome. Indian J Radiol Imaging. 2015;25:25–30.
12. Ekberg O, Nylander G, Fork F. Defecography. Radiology. 1985;155:45–8.
13. Kelvin FM, Maglinte DD, Benson T, Brubaker L, Smith C. Dynamic cystoproctography: a technique for assessing disorders of the pelvic floor in women. Am J Roentgenol. 1994;163:368–70.
14. Kumar S, Sharma P, Andreisek G. Does conventional defecography has a role to play in evaluation of evacuatory disorders in Indian population? Indian J Radiol Imaging. 2013;23:92–6.
15. Jorge JMN, et al. How reliable are currently available methods of measuring the anorectal angle? Dis Colon Rectum. 1992;25:332–8.
16. Halligan S, Bartram CI, Park H, Kamm M. Proctographic features of anismus. Radiology. 1995;197:679–82.
17. Shorvon PJ, et al. Defecography in normal volunteers: results and implications. Gut. 1989;30:1737–49.
18. Sboarina A, et al. Shape and volume of internal anal sphincter showed by three-dimensional anorectal ultrasonography. Eur J Radiol. 2012;81(7):1479–82.
19. Arndorfer RC, et al. Improved infusion system of intra luminal esophageal manometry. Gastroenterology. 1977;73(1):23–7.
20. Schuster MM, et al. Simultaneous manometric recording of internal and external anal sphincter reflexes. Bull Johns Hopkins Hosp. 1965;166:79–88.
21. Rao SS, Azpiroz F, Diamant N, et al. Minimum standards of anorectal manometry. Neurogastroenterol Motil. 2002;14:553–9.
22. Kang HR, Lee JE, Lee JS, et al. Comparison of high-resolution anorectal manometry with water-perfused anorectal manometry. J Neurogastroenterol Motil. 2015;21:126–32.
23. Sauter M, Heinrich H, Fox M. Toward more accurate measurements of anorectal motor and sensory function in routine clinical practice: validation of high-resolution anorectal manometry and Rapid Barostat Bag measurements of rectal function. Neurogastroenterol Motil. 2014;26:685–95.
24. Babaei A, Lin EC, Szabo A, Massey BT. Determinants of pressure drift in ManoscanTM esophageal high-resolution manometry system. Neurogastroenterol Motil. 2015;27:277–84.
25. Lee TH, Bharucha AE. How to perform and interpret a high-resolution anorectal manometry test. J Neurogastroenterol Motil. 2016;22(1):46–59.

26. Clouse RE, Staiano A, Alrakawi A, Haroian L. Application of topographical methods to clinical esophageal manometry. Am J Gastroenterol. 2000;95:2720–30.
27. Vitton V, Ben Hadj Amor W, Baumstarck K, et al. Water perfused manometry vs three-dimensional high-resolution manometry: a comparative study on a large patient population with anorectal disorders. Color Dis. 2013;15:726–31.
28. Jorge JM, Wexner SD. Anorectal manometry: techniques and clinical applications. South Med J. 1993;86:924–31.
29. Coller JA. Clinical application of anorectal manometry. Gastro Clin N Amer. 1987;16:17–34.
30. Kaur G. Rectoanal reflex parameters in incontinence and constipation. Dis Colon Rectum. 2002;45:928–33.
31. Parkman HP, McCallum RW, Rao SSC. GI motility testing: laboratory and office handbook. Thorofare: SLACK; 2011.
32. Gowers WR. The automatic action of the sphincter ani. Proc R Soc Lond B Biol Sci. 1877;26:77–84.
33. Miller R, Bartolo DC, Cervero F, et al. Anorectal sampling: a comparison of normal and incontinent patients. Br J Surg. 1988;75:44–7.
34. Kaur G, Gardiner A, Duthie GS. Rectoanal reflex parameters in incontinence and constipation. Dis Colon Rectum. 2002;45:928–33.
35. Rogers J. Testing for and the role of anal and rectal sensation. Baillieres Clin Gastroenterol. 1992;6:179–91.
36. Enck P, von Giesen H, Schafer A, et al. Comparison of anal sonography with conventional needle electromyography in the evaluation of anal sphincter defects. Am J Gastroenterol. 1996;91:2539–43.
37. Podnar S, Zalewska E, Hausmanowa-Petrusewicz I. Evaluation of the complexity of motor unit potentials in anal sphincter electromyography. Clin Neurophysiol. 2005;116:948–56.
38. Axelson HW, Edebol Eeg-Olofsson K. Simplified evaluation of the paradoxical puborectalis contraction with surface electrodes. Dis Colon Rectum. 2010;53:928–31.
39. Kiff ES, Swash M. Slowed conduction in the pudendal nerves in idiopathic (neurogenic) faecal incontinence. Br J Surg. 1984;71:614–6.
40. Ger GC, Wexner SD. Electromyography and pudendal nerve terminal motor latency. In: Wexner SD, Bartolo DC, editors. Constipation: etiology, evaluation, and management. Oxford: Butterworth-Heinemann; 1995. p. 91–102.
41. Thomas C, et al. Respective value of pudendal nerve terminal motor latency and anal sphincter electromyography in neurogenic fecal incontinence. Neurophysiol Clin. 2002;32:85–90.
42. Yip B, Barrett RC, Coller JA, et al. Pudendal nerve terminal motor latency testing: assessing the educational learning curve: can we teach our own? Dis Colon Rectum. 2002;45:184–7.
43. Tetzschner T, Sørensen M, Rasmussen OO, et al. Reliability of pudendal nerve terminal motor latency. Int J Color Dis. 1997;12:280–4.
44. Hutchinson R, Kumar D. Colonic and small bowel transit studies. In: Wexner SD, Bartolo DC, editors. Constipation: etiology, evaluation, and management. Oxford: Butterworth-Heinemann; 1995. p. 52–62.
45. Arhan P, Devroede G, Jehannu B, et al. Segmental colonic transit time. Dis Colon Rectum. 1981;24:625–9.
46. Paquette I, Varma, MG, Kaiser AM, Steele SR, Rafferty, JF. The American Society of Colon and Rectal Surgeons, Clinical Practice Guideline for the Treatment of Fecal Incontinence. 2015.
47. Paquette I, Varma, MG, Ternet C, Melton-Meaux G, Rafferty JF, Feingold D, Steele SR. The American Society of Colon and Rectal Surgeons, Clinical Practice Guideline for the Evaluation and Management of Constipation. 2015.

Specifics of Anal Manometry

6

Claudio Saddy Rodrigues Coy

Introduction

Anorectal manometry is an important tool in the evaluation of bowel disorders. It is considered an adequate method to quantify resting pressures and anal sphincter contraction. In anal incontinence, it is possible to use anorectal manometry and add additional information in 53–98% of the cases. In addition, it influences therapeutic options in 75–85% of those cases [1].

The most utilized system is the pneumohydraulic perfusion system, mainly because it is less expensive and more available. The pressure values result from the resistance to fluid flow through the catheter, and a graphic register is generated by a specific software, allowing several parameters to be evaluated. The catheter is connected to an infusion pump, with capillaries and pressure sensors that will transmit the data to a polygraph. It must be considered that the system is subject to variations or losses, which will impair the trace and may be due to inadequate calibration, poor quality catheters with different diameters between channels, poorly sensitive membranes, or perfusion pumps that do not maintain a constant pressure.

Manometry made possible a better understanding of the anorectal physiology and established the basis for other diagnostic methods such as electroneuromyography or ultrasound. It is accessible, has low morbidity, and allows therapeutic orientation. The association with endoanal ultrasonography completes the evaluation, particularly in anal incontinence, since it is possible to correlate the parameters of functional reserve of the anal canal with the evaluation of the structural integrity of the anal sphincters.

However, not uncommonly, anorectal manometry is the only exam available, and the examiner must then have the knowledge and experience to correctly extract and interpret all the information available. It is also necessary to provide the data in a manner accessible to the often-unfamiliar requesting physician with particularities of anorectal physiology. Thus, besides the quantitative evaluation with pressure values, it is necessary to make a qualitative analysis of the manometric tracing.

Anal continence is maintained by a complex mechanism, involving several factors: the pressure gradient between the rectum and the anal canal, the structural integrity of the pelvic floor musculature and the anal sphincters, adequate supply by sensory and motor neurogenic pathways, receptors in the anal canal and pelvic floor, hemorrhoidal plexus, rectal complacency, and stool consistency. What is evaluated by this method is the graphic representation of the pressure variations in the anal canal, but it is not possible to identify the etiopathogenic factors related to the dysfunction. Thus, anorectal manometry is

C. S. R. Coy (✉)
Group of Coloproctology, Discipline of Diseases of the Digestive System, Department of Surgery – Campinas State University (UNICAMP), Campinas, SP, Brazil

L. C. C. Oliveira (ed.), *Anorectal Physiology*, https://doi.org/10.1007/978-3-030-43811-1_6

used to evaluate the functional reserve of the anal muscles, and this important aspect is what should be considered for the correct reading of the manometric tracing. A descriptive report that is based only on reference values of resting pressure, voluntary contraction, and presence or absence of rectoanal inhibitory reflex will be misinterpreted and will not provide the necessary information for therapeutic orientation. Therefore, the examiner should consider other aspects of the layout.

There are several publications demonstrating that anorectal manometry is non-discriminatory between healthy individuals and patients with anal incontinence [2–6], but in most studies the assertions are based exclusively on numerical analysis. The need for improvement in the interpretation of the tracings has led to the development of new measures and qualitative analysis aimed at a more elaborate evaluation and thus improves the data so that they correlate adequately with the clinical symptoms [7–11].

Some aspects of anorectal manometry are discussed below.

Technical Aspects

The manometric investigation of the anal canal can be performed in several ways, which is interesting because it allows distinct interpretations of normal motility as well as different "readings" of anorectal dysfunctions. There are two conventional methods to perform anal manometry: a dynamic and a static method. The catheters for both methods can have channels arranged axially or radially. The axial catheter is always used with the static method, while the radial catheter allows measurements at various levels to be used by means of manual or mechanical traction in a continuous manner.

This method is ideal for the study of the motility pattern at rest and with voluntary squeeze or contraction of the anal muscles, evaluation of the capacity, and quality of the contraction and interpretation of fatigue during and after voluntary squeeze or contraction. Particularities of the rectoanal inhibitory reflex (RAIR) such as the relaxation pattern (maximum amplitude and duration) as well as the analysis of relaxation latency and post-RAIR pressure recovery velocity at different anal canal levels can be performed.

Dynamic Method

Using a radial eight-channel catheter arranged at 45°, the dynamic method consists in introducing the catheter into the rectum with the radial orifices positioned just above the anal canal and mobilizing it using an automatic mechanical arm at a constant velocity of 10 mm/s. This method allows obtaining a simultaneous radial pressure values at each 5 mm length. Resting and squeeze pressures can be measured. According to the software utilized, a vector composition analysis can generate a three-dimensional graphic figure, allowing a correlation with the vector volume pressures, functional length of the anal canal, and the asymmetry index at different levels of the location of the high pressure zone.

Radial or Axial Catheter

The radial catheter is the most commonly utilized and enables simultaneous measurement in the various quadrants and at various levels of the anal canal. The channels are located in the rectum and traction is performed until the elevation of the pressure values at rest occurs, and this location corresponds to the more cranial portion of the anal canal. The manual traction is then performed every 10 mm and the corresponding measurements are recorded until the anal margin is reached. The procedure is repeated with voluntary anal contraction effort. The high pressure zone is identified and the RAIR is investigated. The radial catheter also allows two important items for the interpretation of the results: the functional length of the anal canal and the position of the high pressure zone.

As previously mentioned, there are systems where the equipment can make a drawing of the catheter continuously from a predetermined point, and with specific softwares, it is possible to correlate the position of the channels with the corresponding pressure values. In this way the figure of the volume vector is obtained and in a more practical way the identification of the values in the various quadrants, the high pressure zone, the length of the anal canal, and the asymmetry evaluation. The vector volume is the graphic representation of the forces that compose the anal canal, but it has been difficult to correlate with anatomical defects. A defect in the external anal muscle does not necessarily present a decrease in pressure values in the anterior quadrant in the anal canal.

In the axial catheter, the channel outlet holes are arranged along the longitudinal axis, 5 mm apart, and no manual or mechanical traction is required. Therefore, it is best to utilize the eight-channel systems. It makes the examination faster and allows the simultaneous identification and comparison of curve patterns such as the presence of slow and ultraslow waves and the different patterns of the voluntary squeeze pressures.

In practice, the type of catheter utilized does not matter for the obtained results, as interpretation is similar. The choice of the catheter depends on the preference of the examiner. In our service when the mechanical traction system is available, all exams are performed with both types of catheters.

Technical Standardization

The standardization of the anorectal manometric examination in the Anorectal Physiology Department of Gastrocentro – UNICAMP, is the following:

The water reservoir of the infusion pump must only be filled with distilled water to avoid obstruction of the capillaries of the system. The infusion pump is connected to allow a continuous flow of 0.56 ml/min/in all channels. The examination is performed with the patient in left lateral decubitus position. For incontinent patients, no bowel preparation is necessary. Before introducing the axial catheter into the anal canal, the system is calibrated to establish the baseline. The software that accompanies the manometry equipment usually allows this function and should be performed to all patients. The catheter then is elevated by 10–15 cm and it is verified if all the channels are registering the pressure variation resulting from this variation. The catheter is placed in the anal canal and the resting pressure is recorded for 60 s and the voluntary squeeze pressure for 40 s. RAIR investigation is performed by rapid inflation of the latex balloon at the end of the catheter, positioned in the rectum, with volumes varying from 15 to 50 ml. Volumetric parameters such as first rectal sensation, urgency, and rectal capacity are performed by the inflation of different volumes into the balloon.

Evaluation of Manometric Parameters

Quantitative Analysis

Pressure Measurements

The anal resting pressure corresponds to the action of the internal anal sphincter, responsible for the resting anal canal tone. Thus, the measure that best represents this physiological mechanism is the average anal pressure (mean anal resting pressure). In the radial catheter, the mean values of all the channels are obtained at the level of the anal canal where the catheter is positioned.

The squeeze pressure corresponds to the action of the external anal sphincter and corresponds to the maximum pressure or the mean anal pressure in each channel after voluntary contraction of a predetermined time can be considered. It should be considered that the external anal sphincter is composed of specialized striated fibers being able to maintain a basal tone up to 40 s. In our Department, for the evaluation of muscle fatigue, the patients are asked to perform a voluntary contraction or squeeze during 40 s, allowing the comparison of initial and final pressure in incontinent patients (Table 6.1, Fig. 6.1).

Rectoanal Inhibitory Reflex (RAIR)

Described by Gowers, it demonstrates the integrity of a local reflex between the rectum and the anal canal, manifested as a relaxation of the resting pressure of the anal canal in response to distention of the rectum. It is considered positive when the distention of the rectum is accompanied by a fall of 20% in relation to the resting pressure, a measure that is feasible in most devices. In normal conditions, it occurs several times a day and allows, consciously or not, the perception of rectal contents. Rectal distension promotes relaxation of the upper portion of the anal canal, allowing the rectal wall receptors to discriminate the rectal contents. Its role in the maintenance of continence is controversial, since individuals submitted to coloanal anastomosis with preserved anal continence do not present this reflex. Variations in RAIR relaxation values (intensity and duration) may be associated with incontinence, and the evaluation of this reflex can be utilized to discriminate between functional constipation and Hirschsprung's disease in children. It is also not detected in megacolon secondary to Chagas disease as well as in severe fecal incontinence, when the values of resting pressure are very low.

Table 6.1 Reference values [12]

Resting pressure	40–70 mmHg
Squeeze pressure	100–180 mmHg
RAIR	Positive
First rectal sensation	10–30 ml
Maximal rectal capacity	100–250 ml

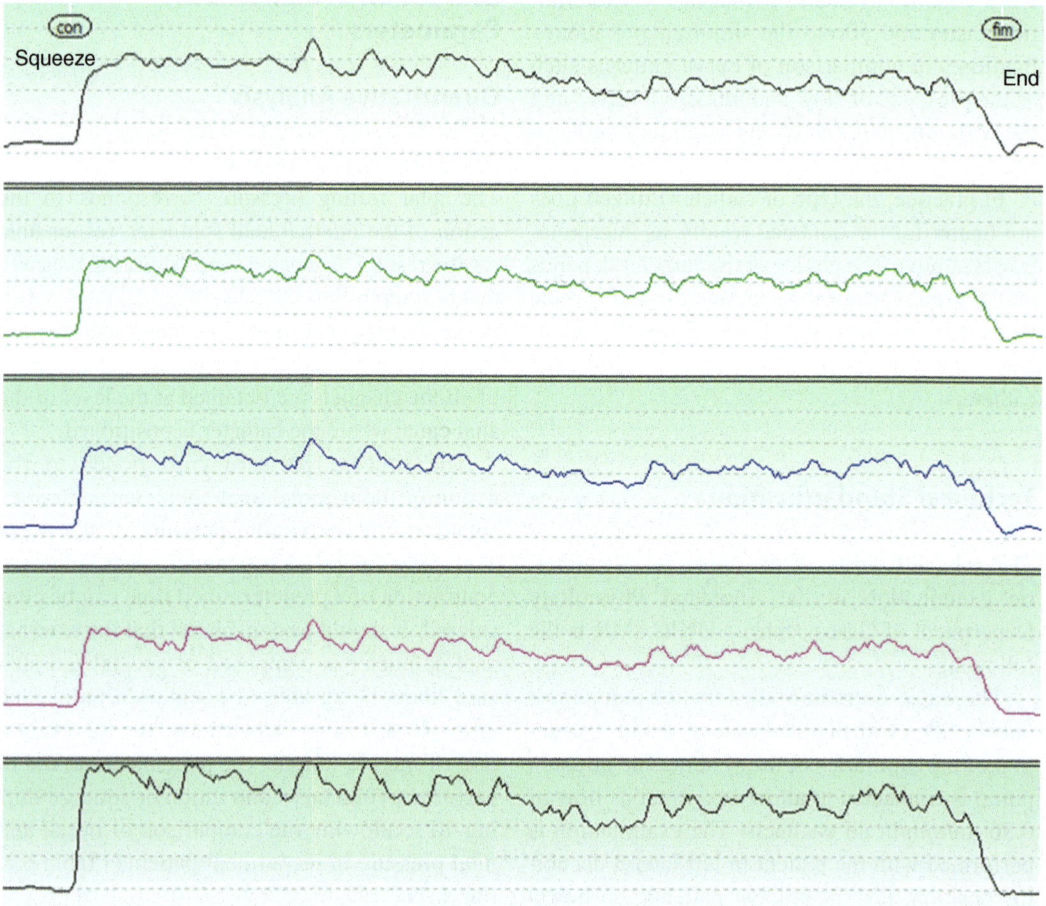

Fig. 6.1 Axial catheter, rest and voluntary contraction. (Coloproctology Group, Laboratory of Anorectal Physiology, GASTROCENTRO—UNICAMP)

The relaxation amplitude is dependent on the volume of rectal insufflation, with larger volumes inducing more intense relaxation and longer duration in the proximal portion of the anal canal (Fig. 6.2).

The role of the RAIR in the evacuation mechanism needs further investigation, specifically the meaning in the different phases in patients with constipation and incontinence. In this Department, RIRA has been divided into phases

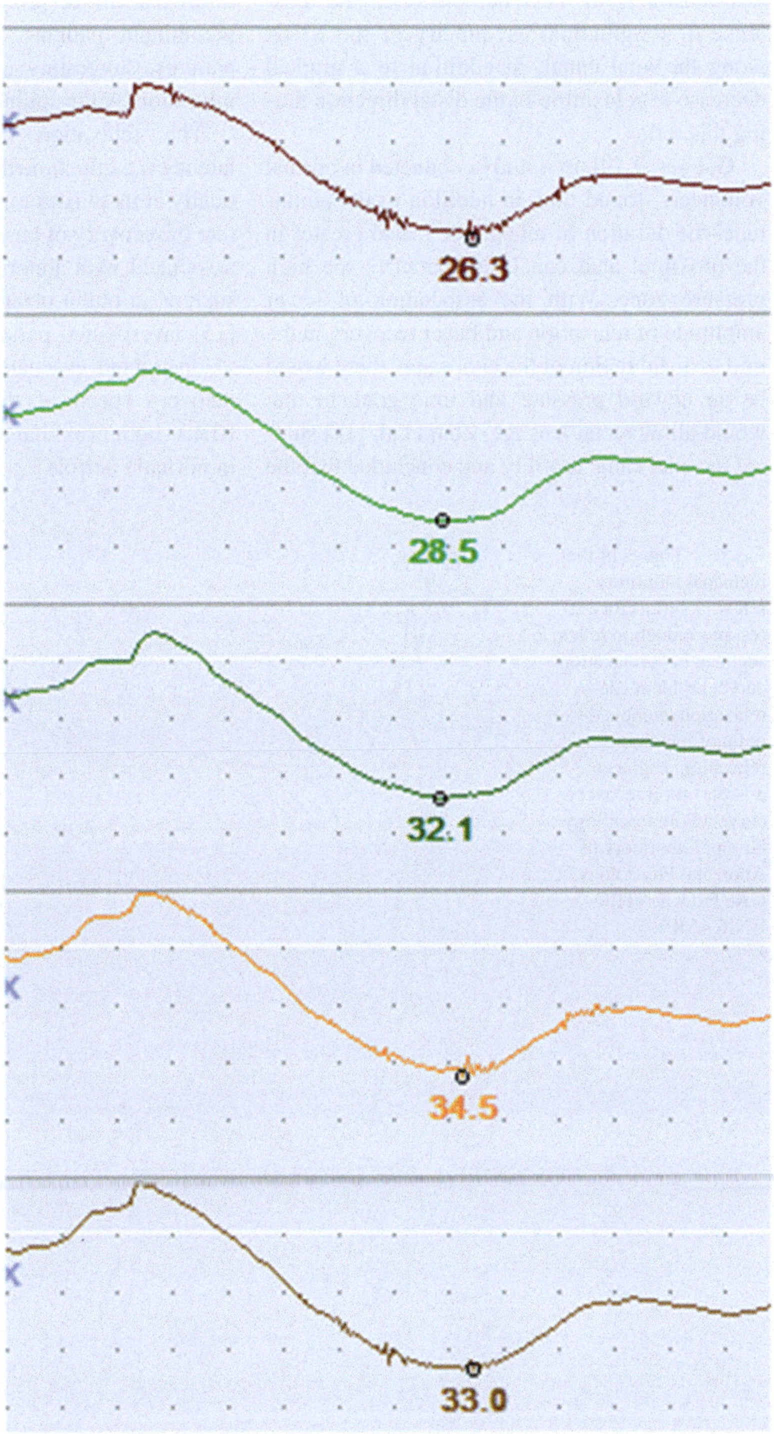

Fig. 6.2 Rectoanal inhibitory reflex. Greater amplitude and duration in the proximal portion of the anal canal. The relaxation depends on the inflation volume of the balloon. (Group of Coloproctology, Laboratory of Anorectal Physiology, GASTROCENTRO—UNICAMP)

in order to better correlate its variations with possible clinical events (Fig. 6.3).

The intensity and duration of RIRA are the main components of this reflex, but further investigations need to be done to understand its role in the functional behavior of the anal canal. Williamson et al. [13] demonstrated the existence of longitudinal asymmetry of the RIRA along the anal canal, in addition to a gradual decrease of relaxation in the distal direction during this reflex.

Góes et al. [9], in a study conducted in normal volunteers, found that, in addition to the amplitude, the duration of relaxation is also greater in the proximal anal canal, compared to the high pressure zone. With the association of lower amplitude of relaxation and faster recovery in the mid to distal region of the anal canal, there would be an upward pressure and time gradient that would allow rectal leakage. Zbar et al. [14] studied the anal canal motility and concluded that the relaxation amplitude did not differ significantly between normal individuals with anal incontinence and constipation. However, when comparing the duration of relaxation between patients with constipation and incontinence, these authors showed that recovery time in the distal region is faster than in the proximal region, as well as in incontinent patients, compared to constipated patients, suggesting a coordinated response of inhibition by the sphincter.

The relaxation duration patterns, called latency (i.e., the time the relaxation persists practically at its maximum amplitude), as well as the rate of recovery of basal pressure at rest could be associated with abnormal functional conditions, such as in outlet obstructed defecation. Netinho [15] investigated patients with constipation due to obstructed evacuation and observed that the recovery speed of the relaxation induced by RIRA, both proximal and distal, was higher than in normal controls.

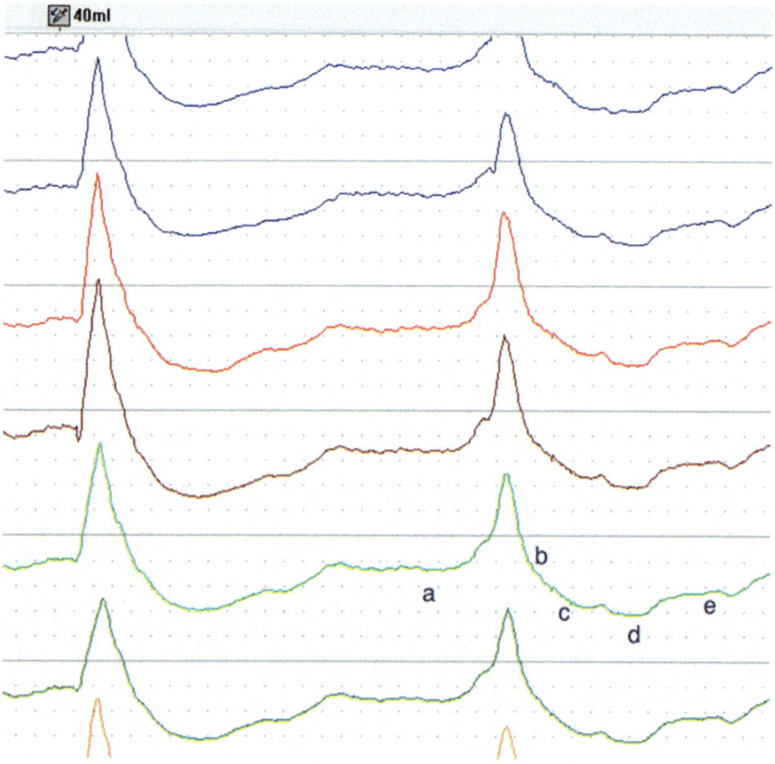

Fig. 6.3 Phases of the rectoanal inhibitory reflex: a—trajectory at rest pre-induction reflex; segment bc—relaxation curve; segment cde—relaxation latency; d—point of maximum relaxation; segment e—relaxation recovery curve. (Coloproctology Group, Laboratory of Anorectal Physiology, GASTROCENTRO—UNICAMP)

Qualitative Analysis

The analysis of particularities of the manometric tracing, referring to non-numerical data, is rarely employed. However, we must consider that there are limitations when the value pressures are the only aspects evaluated.

Motility of the Anal Canal

The manometric tracing translates anal canal motility. In normal individuals, the pattern most often found is short-wave and ultra-short waves, and patients with functional loss usually present a decrease in pressure activity, with a decrease or almost absence of these waves. It can be interpreted as a response to pressure changes in the rectum and due to the integrity of autonomic pathways between the rectum and the anal canal (author's note). The graphic pattern compatible with a visually straight, linear trajectory, presenting only small oscillations, is more associated with fecal incontinence (Figs. 6.4 and 6.5).

Evaluation of the Quality of the Voluntary Contraction

The ability to voluntarily raise the pressure gradient of the anal canal relative to the rectum with contraction represents, as previously expressed, the anal pressure of voluntary contraction by the action of the external anal sphincter muscle. This information can be obtained either with the radial or axial catheter, but in this Department it has been investigated for 40 s, mainly with the axial catheter, since it allows the simultaneous evaluation at all levels of the anal canal (Fig. 6.6).

The evaluation of the squeeze anal pressure with a radial catheter allows visualization of the pressures radially at all levels of the anal canal. Thus, eventually localized functional losses of the external anal sphincter and the puborectalis muscle can be identified and quantified. However, patients with fecal incontinence may have normal voluntary contraction pressure value. Thus, in these cases, it is relevant to study the capacity to sustain this contraction satisfactorily. This can be

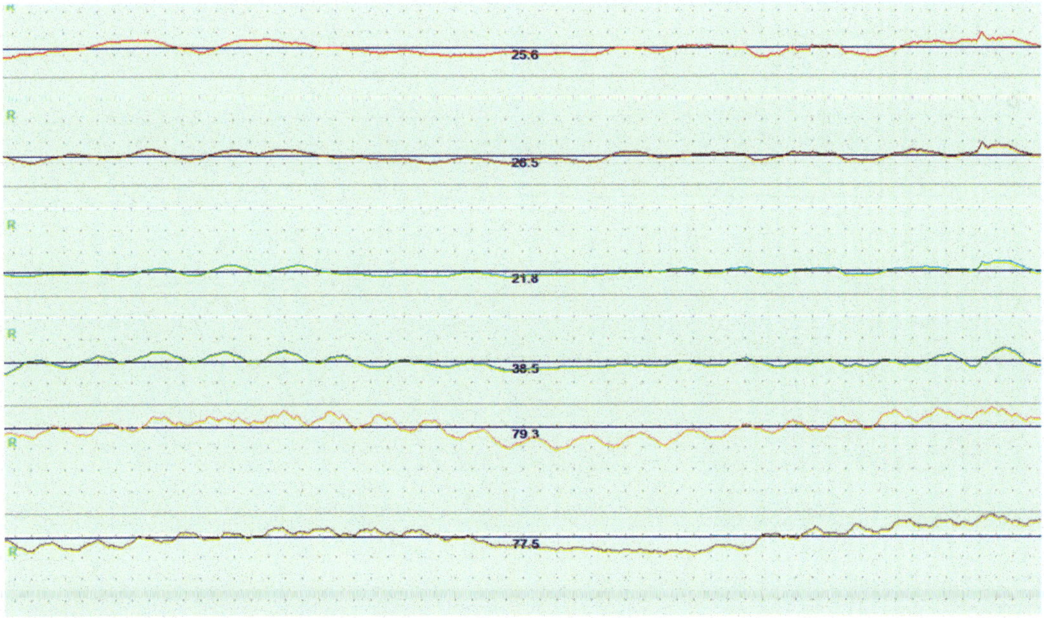

Fig. 6.4 Anal canal motility. Short and ultra-short waves. (Coloproctology Group, Laboratory of Anorectal Physiology, GASTROCENTRO—UNICAMP)

performed by identifying the curve pattern, with maintenance or drop of pressure values throughout the period considered or presence of contractions in "spikes" (Fig. 6.7). In the evaluation of the incontinent patient, the confirmation of their ability to voluntarily raise the pressure in the anal canal can be of great value, since it suggests a functional reserve that can be rescued with physiotherapy techniques of the pelvic floor.

Post-contraction Fatigue

Another important parameter is the demonstration of fatigue that can follow a period of voluntary contraction lasting 40–50 s. Post-contraction fatigue translates into relaxation of the anal canal from 20% to 30% of the resting pressure value before contraction. When well evidenced, this phase of pressure drop lasts about 8–10 s, maintaining the anal canal at maximum

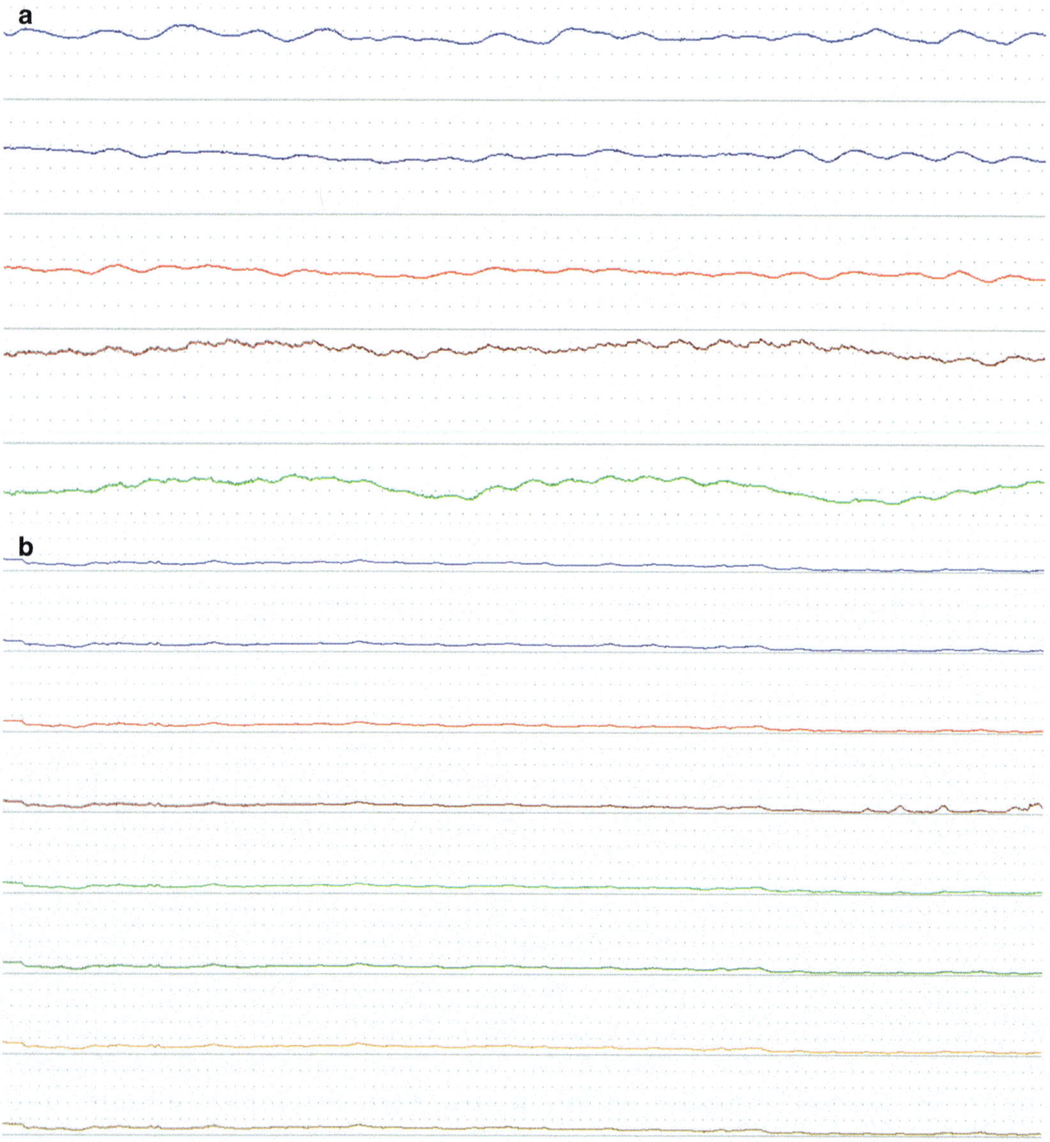

Fig. 6.5 Anal canal motility. (**a**) Normal appearance, (**b**) decreased motility, (**c**) increased motility in the patient with constipation. (Coloproctology Group, Laboratory of Anorectal Physiology, GASTROCENTRO—UNICAMP)

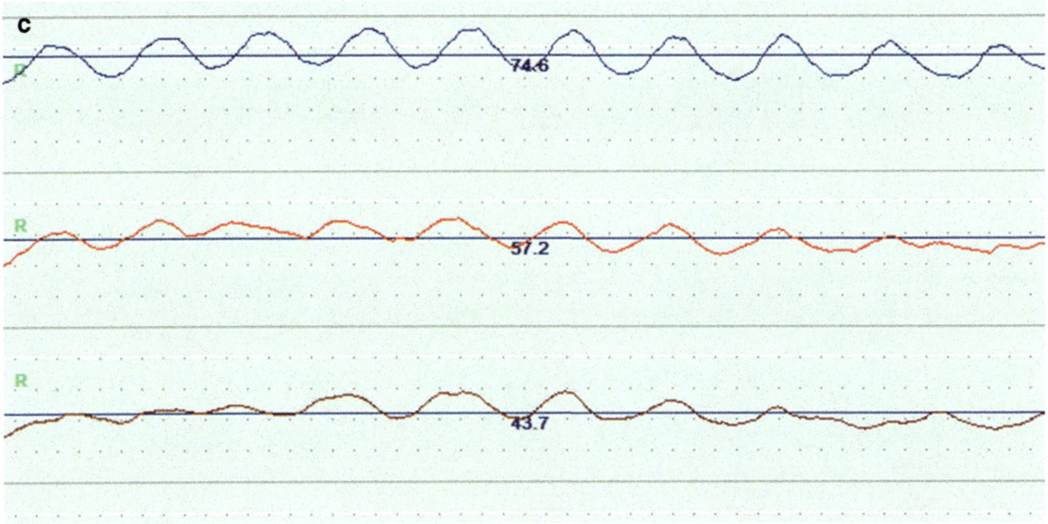

Fig. 6.5 (continued)

Fig. 6.6 Evaluation of the quality of the tracing in voluntary contraction. The pressure values were maintained for 40 s. (Coloproctology Group, Laboratory of Anorectal Physiology, GASTROCENTRO—UNICAMP)

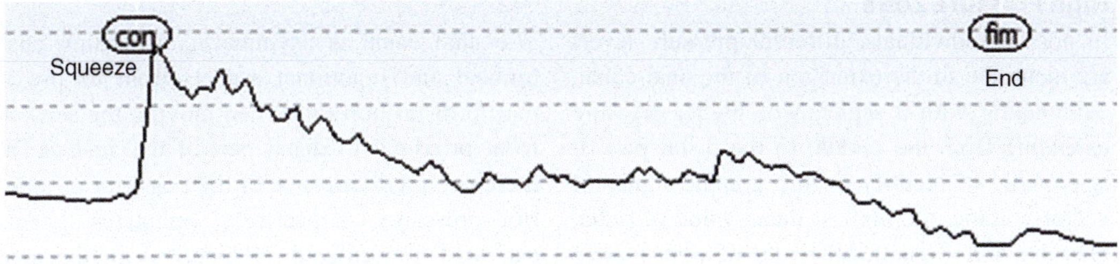

Fig. 6.7 Incontinence. Normal voluntary contraction pressure value at the beginning of the trajectory with drop after 40 s. (Coloproctology Group, Laboratory of Anorectal Physiology, GASTROCENTRO—UNICAMP)

relaxation for about 5–10 s – relaxation latency phase (Fig. 6.8).

Post-fatigue relaxation has been an important aid in the orientation of constipated patients due to obstructed evacuation, especially those with elevated blood pressure levels (usually 80 mmHg or more). After the effort of voluntary contraction (from 40 to 50 s), the duration of the relaxation until its maximum point occurs after 5–8 s. In the next phase of relaxation latency, the patient is asked to perform an evacuation effort for 10–12 s. What is observed is that with the effort, the patient keeps contraction of the anal canal, delaying the recovery of the basal pressure at rest. This aspect has been useful for the analysis of the possible efficacy in the treatment of constipation caused by obstructed evacuation using the post-effort biofeedback method. With

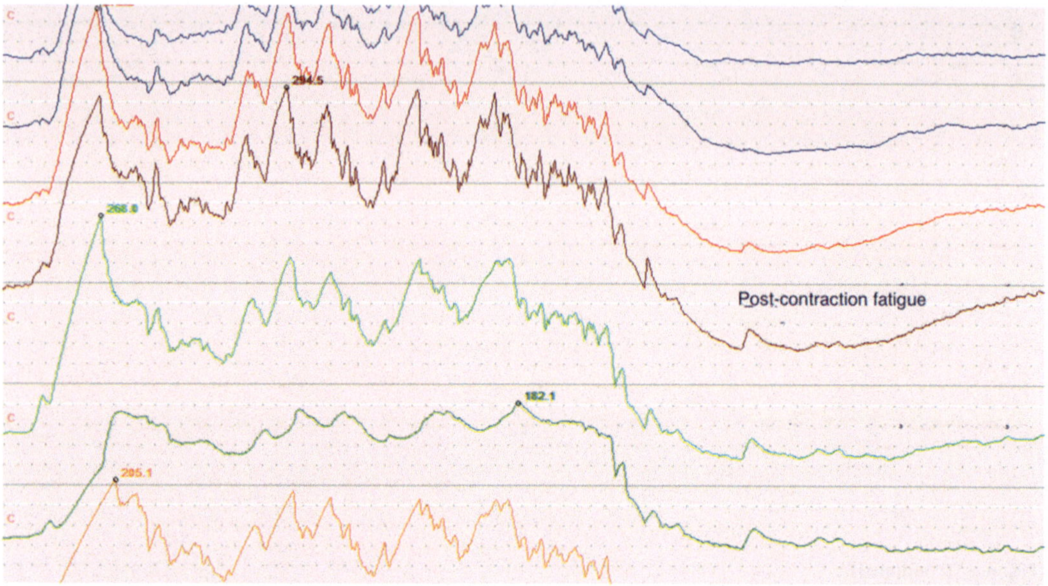

Fig. 6.8 Post-fatigue anal canal relaxation. (Coloproctology Group, Laboratory of Anorectal Physiology, GASTROCENTRO—UNICAMP)

this method, the patient with obstructed evacuation (anismus), after voluntary contraction with an intra-rectal balloon, is trained to relax the anal canal during the relaxation latency phase and so facilitating the balloon expulsion.

High Pressure Zone

In normal individuals, different pressure levels are identified in the extension of the anal canal, culminating with a segment of higher pressure extending from the middle to the distal part. It is possible to identify, in the graphic figure of vector volume, the highest mean value of radial pressure, and it is called the zone of highest pressure (ZHP) of the anal canal and it predominantly translates the greater sphincter action of the internal and external sphincter during the resting phase. The use of this parameter is important to demonstrate possible functional losses in the anal canal sectors, by observing the variation of the asymmetry indices to the segment of higher pressure. In cases of traumatic lesions or after sphincterotomy for the treatment of fistula or anal fissure, an association with distal loss of function can be seen, leading to a more proximal location of the segment of HPZ. Thus, in the vector volume figure, the proximal location of the HPZ is associated with a functional loss of the distal canal anal.

Symmetry/Asymmetry of Anal Canal

The anal canal is asymmetric, regarding anatomical and functional aspects from its proximal to distal portion. When moving the sensors from proximal to distal part of the anal canal, there is a predominance of the posterior or anterior pressures, respectively. Sphincter defects can lead to localized dysfunctions in the anal canal and changes in normal asymmetry may be useful for interpretation of localized lesions (Fig. 6.9). Oliveira et al. [16], in a study that evaluated manometric parameters related to clinical improvement after use of injectable agents in patients with internal sphincter structural damage, found that the alterations found were the increase in the length of the high pressure zone and the improvement of the asymmetry index. The authors then emphasized the importance of the asymmetry index for a better understanding of mechanisms related to continence.

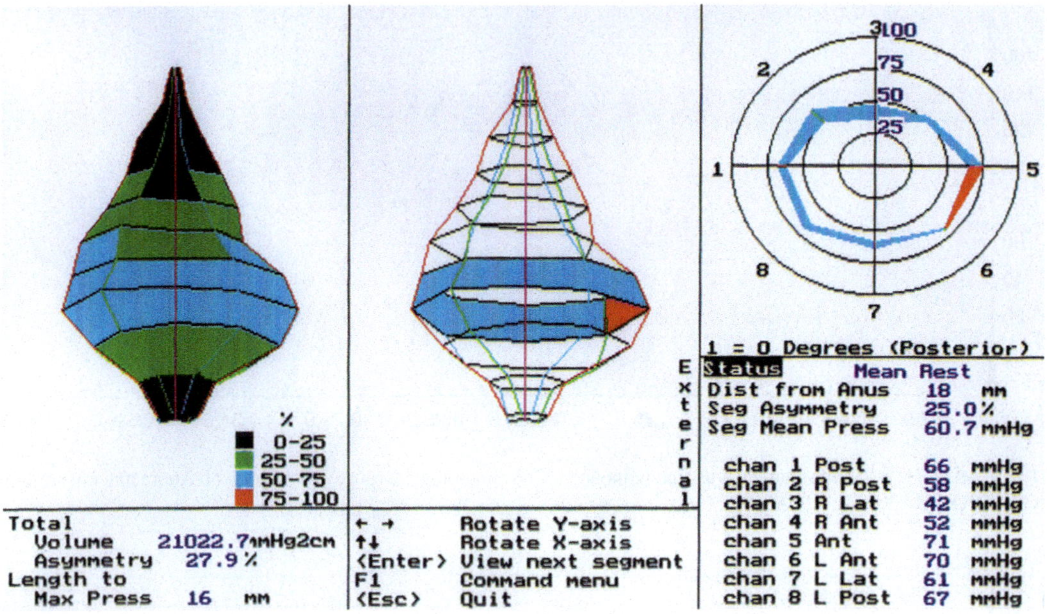

Fig. 6.9 Normal example of the symmetry in the anal canal obtained in the zone of higher pressure. (Coloproctology Group, Laboratory of Anorectal Physiology, GASTROCENTRO—UNICAMP)

Evaluation of the Quality of Voluntary Contraction

The difficulties for the identification of the muscular action by manometry are evaluated by numerical parameters that may be close to the values considered normal and by the analysis of manometric traces that often do not present a regular pattern. Thus, measures that estimate the quality of the voluntary contraction were described in the literature, in order to facilitate the interpretation of the tracing. Marcello et al. [7], employing a mathematical model, estimated the theoretical time necessary for the external sphincter to become completely fatigued. It was called fatigue rate index (IFR) and showed that with this measure it is possible the discrimination between normal and incontinent individuals, as well as the evaluation of the response to biofeedback (Fig. 6.10). However, the IFR presents discrepant reports in the literature. Telford et al. [17] demonstrated that IRF discriminates and correlates with the severity of fecal incontinence, unlike the findings of Baliali and Pfeifer [18].

The squeeze capacity (SC) described by Saad et al. [9] is based on the fact that maintaining voluntary contraction steadily is more important than isolated contraction peaks even if they are within normal values. These same authors showed that maximum squeeze pressure in incontinent individuals was within normal values in 54% of cases; meanwhile in 78% of incontinent individuals, SC was below normal values (Fig. 6.11).

It can be considered that both measures improve the quality of interpretation of the findings as well as make it possible to better establish the criteria for indication of pelvic floor physiotherapy and evaluation of its results.

Clinical Cases and Models of Reports

Case 1 (Fig. 6.12)

MS, 67 years old, female, occasional incontinence for solid stools and flatus. Three prior vaginal deliveries.

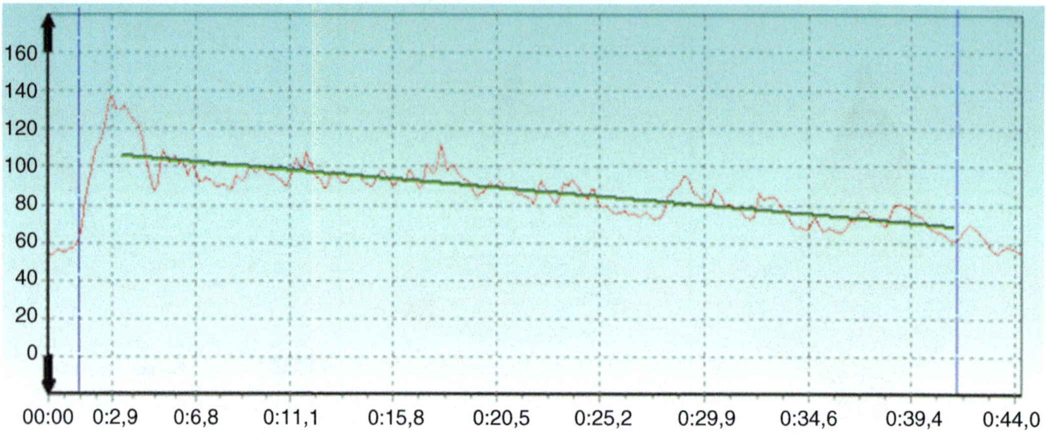

Fig. 6.10 FTI, 0.85 min patient with incontinence. (Coloproctology Group, Laboratory of Anorectal Physiology, GASTROCENTRO—UNICAMP)

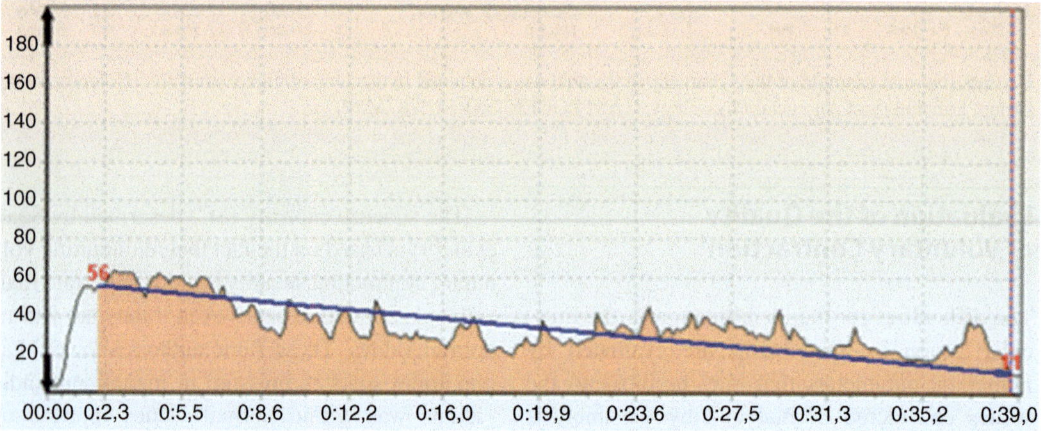

Fig. 6.11 Decreased ability to sustain voluntary contraction pressure in patients with anal incontinence. (Coloproctology Group, Laboratory of Anorectal Physiology, GASTROCENTRO—UNICAMP)

Mean anal pressure at rest:	20–50 mmHg
Maximum squeeze pressure with axial catheter:	80–120 mmHg

Report

Technical details: Exam performed with a perfusion system and an eight-channel axial catheter. Resting pressure evaluated for 60 s and voluntary contraction for 40 s, with the lowest and highest values in each channel being reported in mmHg. Resting pressure values expressed as mean values during the period considered and voluntary contraction expressed at maximum values. Rectoanal inhibitory reflex investigated by air insufflation in a latex balloon located in the distal portion of the rectum, being considered positive with fall of the values of resting pressure by 20%.

Findings

Numerical values show that the pressure at rest and voluntary contraction are within the parameters of normality, but it is considered that the lower values in both measures are lower than the normal values. In this way, it is possible to consider the presence of functionally short anal canal. The qualitative analysis of resting traces

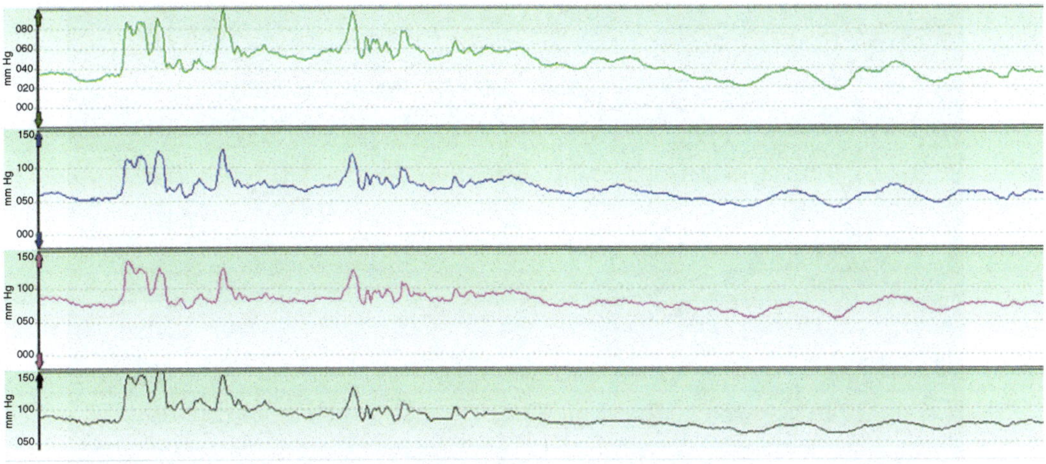

Fig. 6.12 Case 1

shows the presence of slow and ultra-slow waves evidencing neuromuscular functional integrity. The tracing of voluntary contraction shows adequate pressure values at the beginning of the tracing, with a marked decrease in the intermediate period, in which the values are similar to those of rest with pressure increase at the end of the period considered.

Summary

The data analysis shows absolute values of rest and squeeze pressures within the normal values, but functionally short anal canal and low ability to sustain the voluntary contraction. Partially preserved neuromuscular integrity can be assumed as evidenced by preserved anal canal motility, as well as functional reserve of the external sphincter. Thus, clinical improvement with pelvic floor physiotherapy can be obtained.

Case 2 (Fig. 6.13)

ARO, 65 years old, female, incontinence to liquid stools and flatus. Hemorrhoidectomy 30 years ago and diabetes for 15 years, body mass index 38 kg/m^2.

Mean anal pressure at rest	42–61 mmHg
Maximum squeeze pressure	86–115 mmHg

Findings

The values of resting pressure and voluntary contraction are within the normal parameters in all channels. The qualitative analysis of the traces at rest evidences the absence of slow and ultra-slow waves. The voluntary contraction tract shows adequate pressure values at the beginning of the tracing, but with a sharp decrease over the period considered and with values similar to those at rest at the end of the period considered.

Summary

The data analysis shows absolute values of rest and contraction within the normal parameters in all channels, but with low capacity to sustain the voluntary contraction. Thus, functional reserve is present and clinical improvement can be obtained with pelvic floor physiotherapy.

The patient was submitted to pelvic floor physiotherapy. Clinical improvement correlated with improvement in the quality of voluntary contraction (Fig. 6.14).

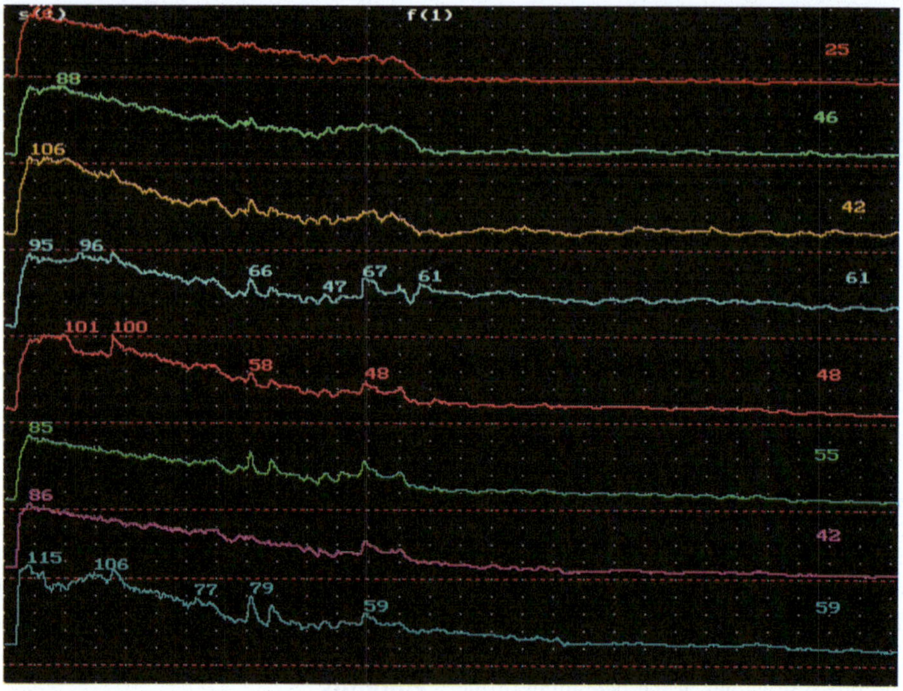

Fig. 6.13 Aspect of the tracing before treatment with pelvic floor physiotherapy, evidencing a decrease in pressure values of voluntary contraction. (Coloproctology Group, Laboratory of Anorectal Physiology, GASTROCENTRO—UNICAMP)

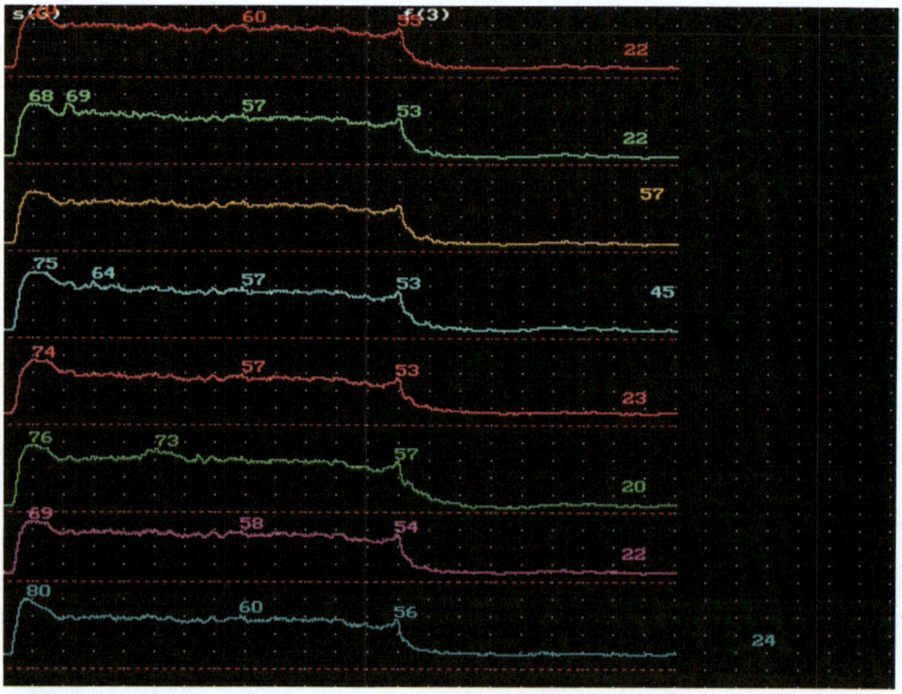

Fig. 6.14 Aspect of the tracing after treatment with pelvic floor physiotherapy, evidencing maintenance of pressure values of voluntary contraction. (Group of Coloproctology, Laboratory of Anorectal Physiology, GASTROCENTRO—UNICAMP)

Summary

Manometry is the most available anorectal physiology test. In order to obtain results that better express the patient's clinical condition, the examiner must interpret aspects of the tracing that are beyond the pressure values which are also relevant. The association of the functional data obtained with manometry with the clinical history and other evaluation modalities, can improve the overall assessment [19].

References

1. Wexner SD, Jorge JM. Colorectal physiological tests: use or abuse of technology? Eur J Surg. 1994;160(3):167–74.
2. Rao SS, Patel RS. How useful are manometric tests of anorectal function in the management of defecation disorders? Am J Gastroenterol. 1997;92(3):469–75.
3. Otto SD, Clewing JM, Gröne J, Buhr HJ, Kroesen AJ. Repeatability of anorectal manometry in healthy volunteers and patients. J Surg Res. 2013;185(2):e85–92.
4. Raza N, Bielefeldt K. Discriminative value of anorectal manometry in clinical practice. Dig Dis Sci. 2009;54(11):2503–11.
5. Read NW, Bartolo DC, Read MG. Differences in anal function in patients with incontinence to solids and in patients with incontinence to liquids. Br J Surg. 1984;71(1):39–42.
6. Lee BH, Kim N, Kang SB, Kim SY, Lee KH, Im BY, Jee JH, Oh JC, Park YS, Lee DH. The long-term clinical efficacy of biofeedback therapy for patients with constipation or fecal incontinence. J Neurogastroenterol Motil. 2010;16(2):177–85.
7. Marcello PW, Barrett RC, Coller JA, Schoetz DJ Jr, Roberts PL, Murray JJ, Rusin LC. Fatigue rate index as a new measurement of external sphincter function. Dis Colon Rectum. 1998;41(3):336–43.
8. Goes RN, Simons AJ, Beart RW Jr. Level of highest mean resting pressure segment in the anal canal. A quantitative assessment of anal sphincter function. Dis Colon Rectum. 1996;39(3):289–93.
9. Goes RN, Simons AJ, Masri L, Beart RW Jr. Gradient of pressure and time between proximal anal canal and high-pressure zone during internal anal sphincter relaxation. Its role in the fecal continence mechanism. Dis Colon Rectum. 1995;38(10):1043–6.
10. Saad LHC, Coy CSR, Fagundes JJ, Ayrizono ML, Shoji N, Góes JRN. Sphincteric function quantification by measuring the capacity to sustain the squeeze pressure of the anal canal. Arq Gastroenterol. 2002;39:233–9.
11. Pucciani F. Faecal soiling: pathophysiology of postdefecatory incontinence. Color Dis. 2013;15(8):987–92.
12. Papaconstantinou HT. Evaluation of anal incontinence: minimal approach, maximal effectiveness. Clin Colon Rectal Surg. 2005;18(1):9–16.
13. Williamson JL, Nelson RL, Orsay C, Pearl RK, Abcarian H. A comparison of simultaneous longitudinal and radial recordings of anal canal pressures. Dis Colon Rectum. 1990;33(3):201–6.
14. Zbar AP, Aslam M, Gold DM, Gatzen C, Gosling A, Kmiot WA. Parameters of the rectoanal inhibitory reflex in patients with fecal incontinence and chronic constipation. Dis Colon Rectum. 1998;41(2):200–8.
15. Netinho JG. Intensity and speed of the recovery of the sphincterial relaxation induced by the retinal inhibitory reflex and its correlation with intestinal constipation by obstructed evacuation. Campinas, 2002. (Doctoral Thesis Faculty of Medical Sciences, State University of Campinas).
16. Oliveira LC, Neve Jorge JM, Yussu FS, Habr-Gama A, Kiss D, Cecconello I. Anal incontinence improvement after silicone injection maybe related to restoration of sphincter asymmetry. Surg Innov. 2009;16(2):155–61.
17. Telford KJ, Ali AS, Lymer K, Hosker GL, Kiff ES, Hill J. Fatigability of the external anal sphincter in anal incontinence. Dis Colon Rectum. 2004;47:746–52; discussion 752.
18. Baliali S, Pfeifer J. Anorectal manometry: are fatigue rate and fatigue rate index of any clinical importance? Tech Coloproctol. 2005;9:225–8.
19. Roos AM, Abdool Z, Thakar R, Sultan AH. Predicting anal sphincter defects: the value of clinical examination and manometry. Int Urogynecol J. 2012;23(6):755–63.

Endoanal, Endovaginal, and Transperineal Ultrasound

7

Lucia Camara Castro Oliveira

Introduction

Ultrasonography or ultrasound is one of the best imaging modalities for evaluation of the pelvic floor and the anal sphincter muscles. The development of high-resolution transducers, with higher frequencies and three-dimensional configuration, has enabled the evaluation of the pelvic floor in a static and dynamic way.

Endoanal ultrasound was first described in 1989 by Law et al. [1], who established a correlation between the anatomical aspects and the ultrasound images of the anal canal. Since then, numerous publications have shown the importance of this examination in the identification of sphincter lesions in cases of incontinence, when often occult lesions that affect the delicate internal anal sphincter muscle in both men and women can be identified. In addition, in the case of suppurative processes, complex fistulas, and benign or malignant tumors, ultrasound images facilitate the demonstration of inflammatory cavities, fistulous tracts, and the degree of penetration of malignant lesions in the wall of the rectum and anal canal.

The use of three-dimensional transducers via the vaginal route has allowed the evaluation of obstetric findings associated with injury to the puboviseral muscles with consequent widening of the genital hiatus.

Transperineal ultrasound (TPUS) was introduced a few years later and can be performed with conventional, linear transducers. This method allows demonstration of the anal region and may favor the diagnosis of most anorectal diseases, with images similar to those of endoanal ultrasound.

The development and improvement of the ultrasound images resulted in more detailed evaluation of the anal canal and pelvic floor in different planes, in both static and dynamic views.

Technical Aspects

The ultrasound waves produced by the transducers have frequencies inaudible to the human ear. When these waves penetrate into the different tissues they produce images of hyper- or hypoechogenicity. Therefore, the images are described as hyper-, hypoechogenic or mixed ones. The images of hyperechogenicity reflect the denser tissues, muscles, and ligaments and are visualized as whitish images. An example of a hyperechoic image is the imaging of the external anal sphincter muscle (Fig. 7.1).

Conversely, the more delicate tissues – subcutaneous, smooth muscles, and cystic lesions – are visualized as dark images of hypoechogenicity. The most easily visualized example in the anal canal is imaging of the internal anal sphincter muscle (Fig. 7.2).

L. C. C. Oliveira (✉)
Director Department of Anorectal Physiology of Rio de Janeiro and CEPEMED, Rio de Janeiro, Brazil

L. C. C. Oliveira (ed.), *Anorectal Physiology*, https://doi.org/10.1007/978-3-030-43811-1_7

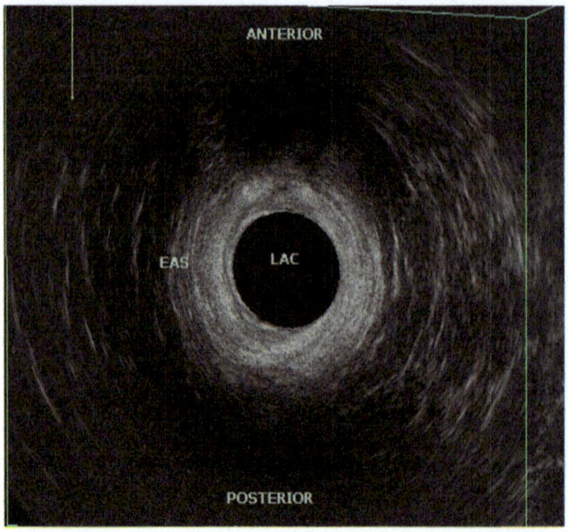

Fig. 7.1 External anal sphincter muscle (EAS) as a hyperechoic ultrasound image at the low anal canal (LAC)

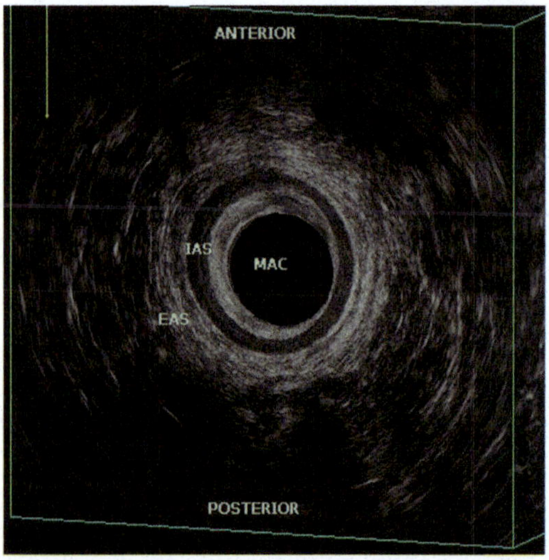

Fig. 7.2 Hypoechoic imaging of the internal anal sphincter muscle (IAS) at the middle anal canal (MAC)

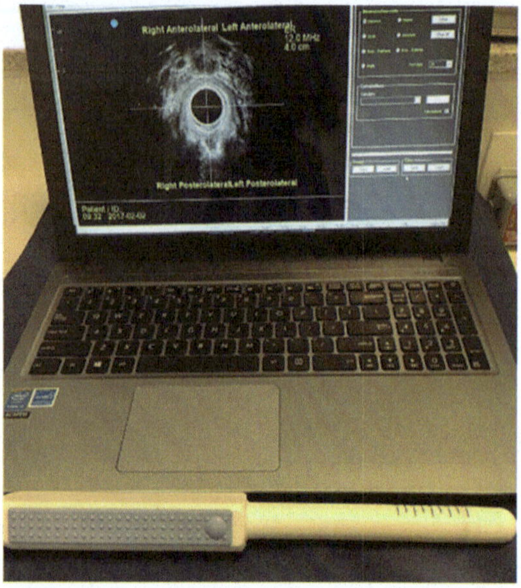

Fig. 7.3 Bidimensional transducers

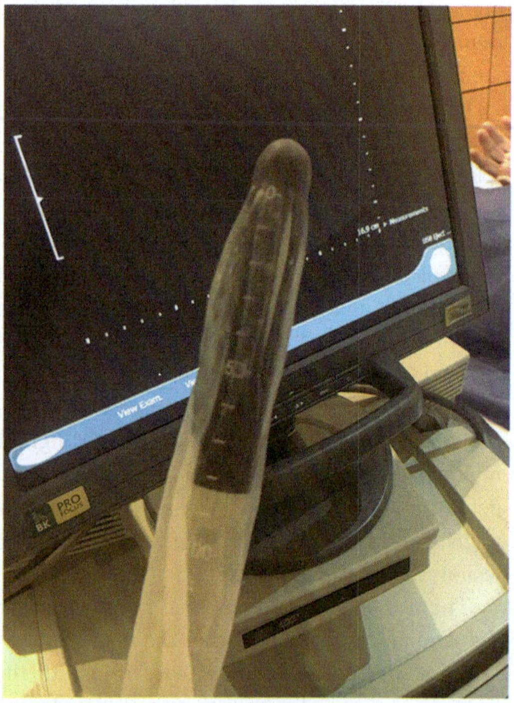

Fig. 7.4 Tridimensional B-K transducer

Equipment

In order to perform endoanal and endovaginal ultrasound, bidimensional (Morpheus 360, Prometheus Group, Dover NH, USA) (Fig. 7.3) and three-dimensional transducers (Pro Focus 2050 transducer, B-K Medical, Herlev, Denmark) (Fig. 7.4) are available. In general, ultrasound equipment have circular transducers, with a viewing angle ranging from 180 to 360 degrees, thus obtaining an evaluation of the entire anal canal. The B-K Profocus equipment allows evaluation of the entire pelvic floor using the different transducer modalities. The circular transducers have rotating crystals, with up to four to six cycles with a focal length of 2–5 cm and high frequencies of 9–16 MHz (Pro Focus

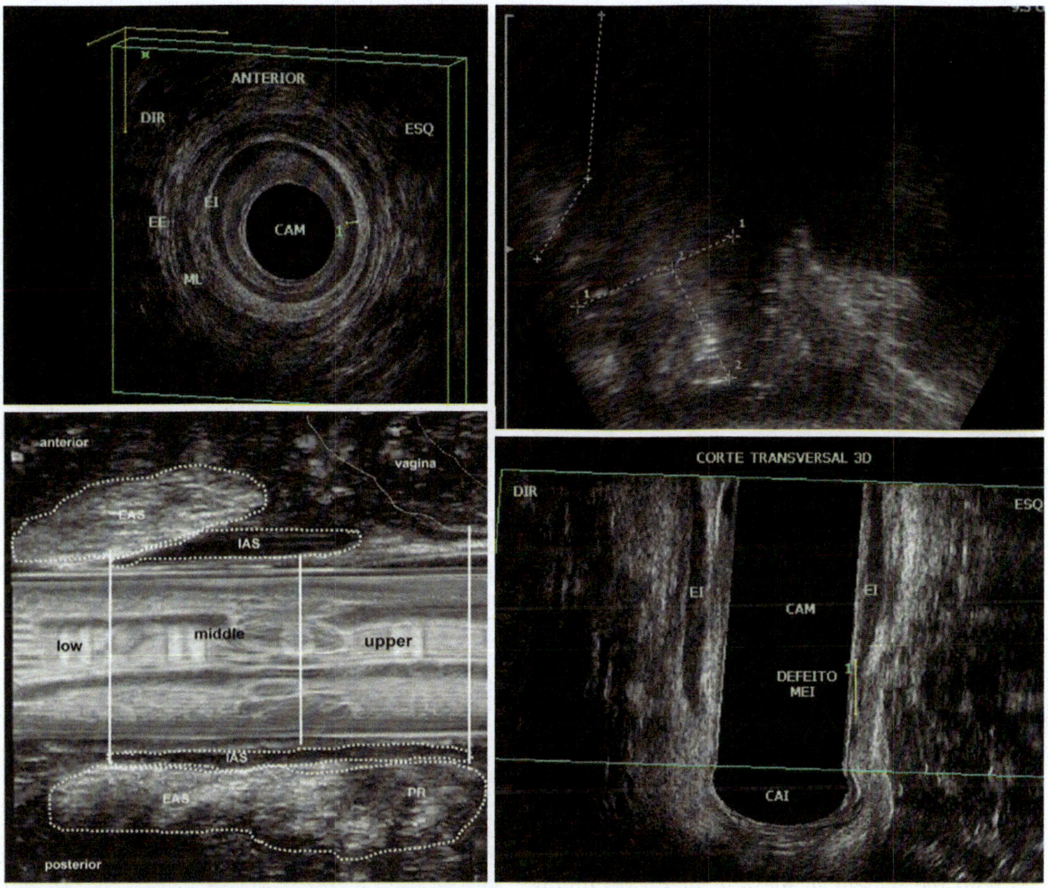

Fig. 7.5 Pro Focus B-K 2050 Equipment with the possibility of utilizing circular and convex transducer s that can provide images of the entire pelvic floor in different planes (axial, sagittal, coronal and transperineal)

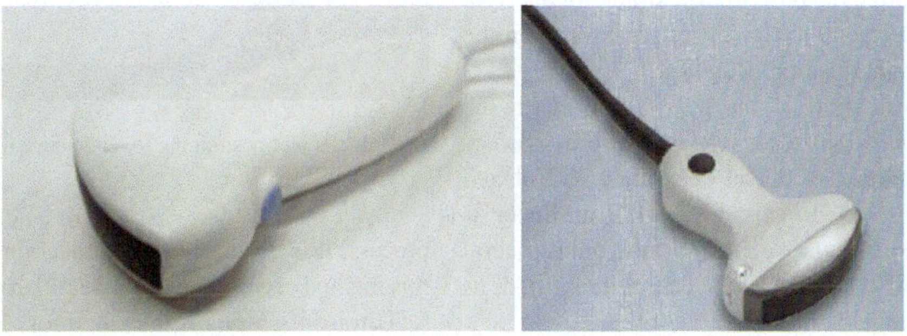

Fig. 7.6 Transperineal convex ultrasound transducer

2050 transducer, B-K Medical, Herlev, Denmark). Together with the convex transducer the entire pelvic floor structures can be evaluated through different planes (Fig. 7.5). For transperineal ultrasound, most of the transducers are similar to the ones utilized for abdominal evaluation – curvilinear or convex (Fig. 7.6). In general, they present frequencies ranging from 4 to 8 MHz and are positioned externally over the perineum, in the region of the great labia, allowing visualization of the structures of the anal canal as well as the bladder, urethra, and vagina.

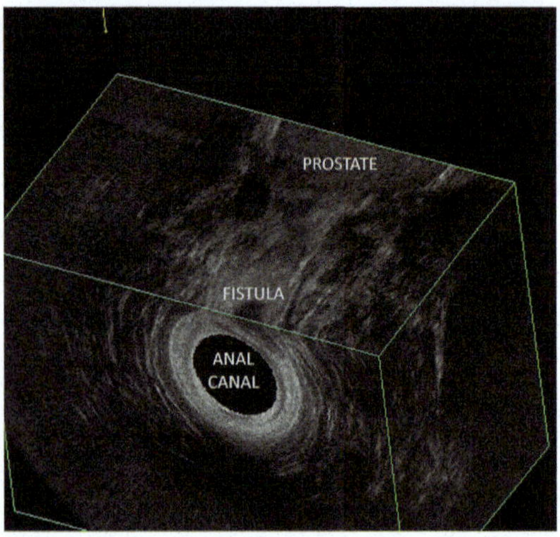

Fig. 7.7 Three-dimensional cube

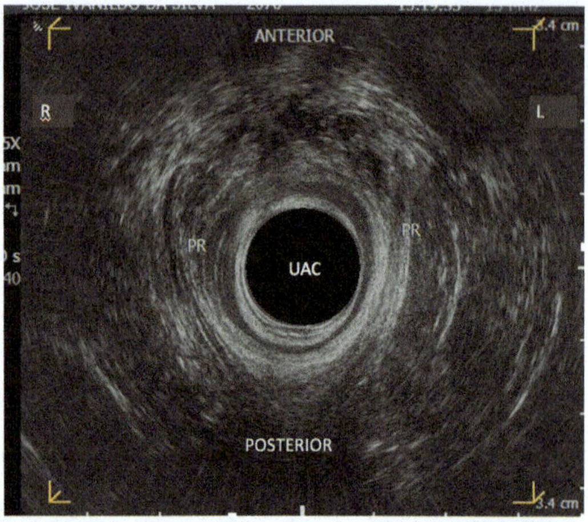

Fig. 7.8 Orientation of the structures on the monitor is as follows: in the upper portion of the screen, the anterior quadrant is observed and, in the lower portion, the posterior quadrant; on the right side of the monitor, the left quadrant and, on the left, the right quadrant

3D transducer images are captured in a three-dimensional cube, and the dynamic acquisition remains saved for further analysis (Fig. 7.7). Axial, sagittal, and coronal planes of 3D transducers allow observation from different angles. These new transducers have allowed a more detailed study of the anal sphincter and diagnosis of diseases that affect the pelvic floor.

With the patient in the left lateral decubitus position, orientation of structures on the monitor is as follows: in the upper portion of the screen, the anterior quadrant is observed and, in the lower portion, the posterior quadrant; on the right side of the monitor, the left quadrant and on the left, the right quadrant (Fig. 7.8).

For evaluation of the pelvic floor with the endovaginal transducer, we utilize the gynecological position of dorsal decubitus, with flexed legs (Fig. 7.9). In this case, the transducer is introduced into the vagina, and the pubovisceral muscles of the pelvic floor can be identified, as shown in Fig. 7.10.

For those beginning to use these imaging techniques, it is crucial to understand the anatomy of the anal canal and the pelvic floor and its correlation with ultrasound images. It is interesting to correlate the physical examination with the ultrasound images, for example, when evaluating an anterior sphincter defect in a patient with a thin perineal body, especially at the beginning of the learning process. It is estimated that the learning curve for this method corresponds to about 50 exams.

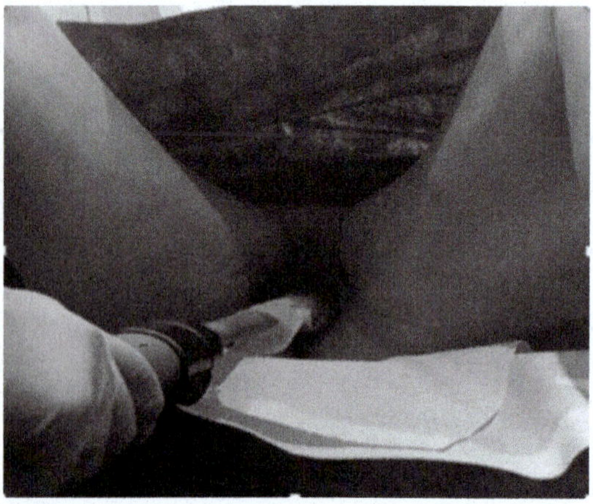

Fig. 7.9 Patient positioning for endovaginal evaluation

Ultrasonography is indicated for various situations involving the anal canal structures and is especially for fecal incontinence (Table 7.1). Disorders of the pelvic floor have been increasingly observed in postmenopausal women. Among the factors that lead to the involvement and weakening of the pelvic floor muscles is the effects of aging and lesions of the fascia, ligaments, and muscles, which are mainly

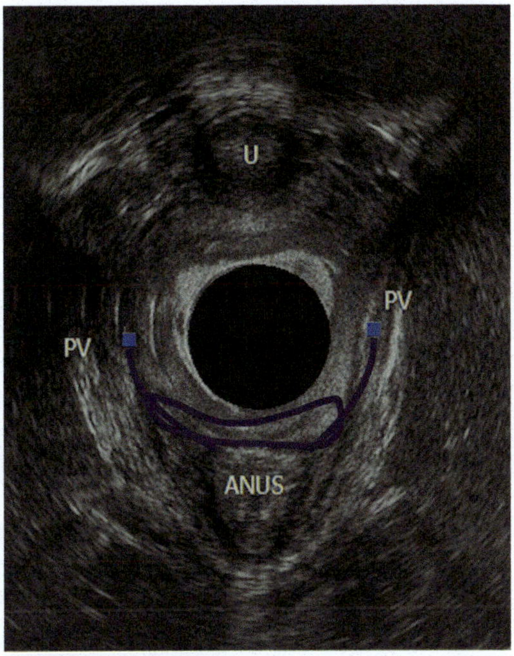

Fig. 7.10 Endovaginal ultrasound with the vagina and the pubovisceral muscles of the pelvic floor

Table 7.1 Indications for endoanal ultrasound in the evaluation of incontinence

Traumatic, obstetric, or postsurgical incontinence
Complex anal fistulae and abscess
Hypertonic sphincter in patients with anal fissures before surgery
Rectal prolapse
Anorectal tumors
Anal pain
Endometriosis and cysts

Table 7.2 Indications for pelvic floor ultrasound

Double incontinence
Enterocele, rectocele, cystocele
Follow-up after urinary slings
Complex obstetrical trauma
Rectovaginal fistulas
Cysts, anorectal, and vaginal tumors
Genital hiatal evaluation
Pubovisceral muscles evaluation

associated with obstetric history. The primary indications for endoanal, endovaginal, and transperineal ultrasonography for the evaluation of pelvic floor disorders are shown in Table 7.2.

Anorectal Ultrasound and Anatomy

The anorectal region is composed of different muscles, including the internal anal sphincter (IAS) and the external anal sphincter (EAS) muscles, the puborectalis (PR), and the longitudinal muscle (Fig. 7.11).

We divide the anal canal, which usually measures 2–4 cm, into three levels: upper, middle, and low (Fig. 7.11). Depending on the level of the anal canal, one expects to find the three main muscles involved in the mechanism of anal continence:

Upper anal canal: puborectalis (PR) muscle and internal anal sphincter (Fig. 7.12).

Middle anal canal: external sphincter muscles (EAS) and internal sphincter (IAS) (Fig. 7.13). In male patients, it is possible to see the longitudinal muscle (Fig. 7.14).

Low anal canal: EAS in its outer or superficial portion (Fig. 7.15).

In fact, when we correlate anatomy and ultrasound images of the anal canal, we observe that there are five layers, as described below (Fig. 7.16).

First layer: hyperechoic, corresponds to the contact between the transducer and its balloon and the surface of the anal mucosa.

Second layer: low ecogenicity, represents the subepithelial tissues and part of the submucosa.

Third layer: hypoechoic, corresponds to the internal sphincter itself.

Fourth layer: not always well visualized, corresponds to the longitudinal muscle, which is visualized as mixed echogenicity image.

Fifth layer: hyperechoic, corresponds to the external sphincter muscle.

Puborectalis Muscle

The puborectalis (PR) muscle consists of two branches that originate in the posterior portion of the pubis and form a V-shaped loop around the upper anal canal, immediately superior to the deep portion of the external sphincter of the anus (Fig. 7.12).

Because it is a striated muscle, it is seen as an image of mixed echogenicity (clear images) in

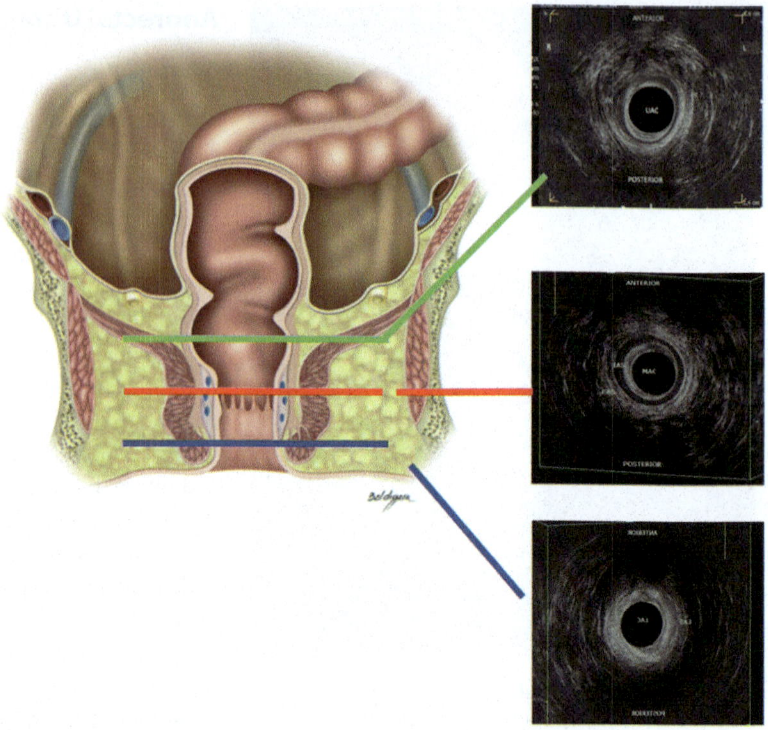

Fig. 7.11 The green line is the level of the upper anal canal with the puborectalis muscle with its fibers forming a loop or a "U" and the internal anal sphincter muscle. The red line corresponds to the middle anal canal with the internal and external anal sphincter muscles. The blue line corresponds to the lower anal canal, where we observe the superficial portion of the external anal sphincter

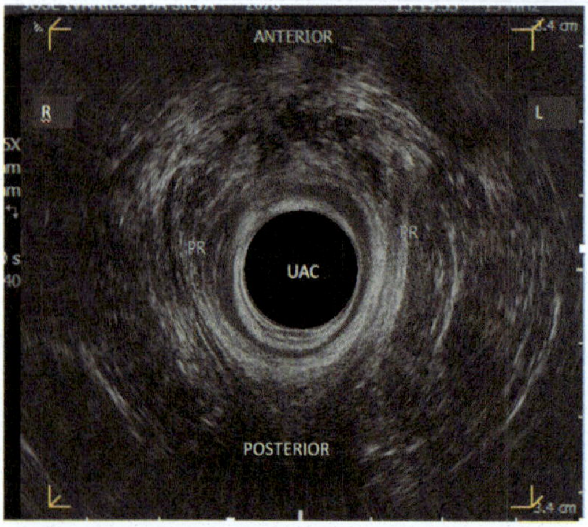

Fig. 7.12 Upper anal canal: Puborectalis muscle (PR) and internal anal sphincter

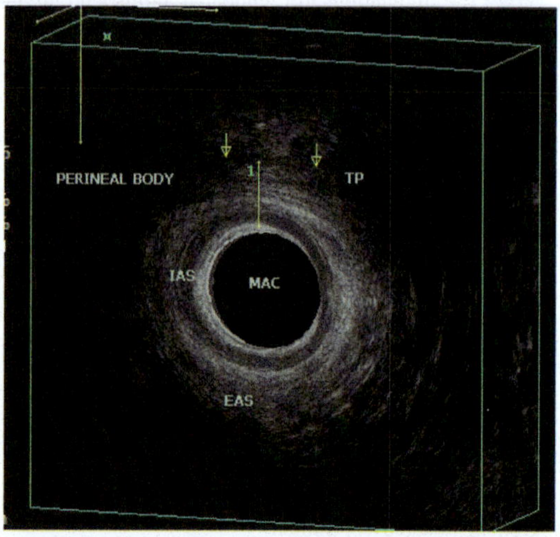

Fig. 7.13 Middle anal canal: external anal sphincter (EAS) muscle and internal anal sphincter (IAS) The perineal body with the transverse perineal muscle (TP)

the form of an arch or "horseshoe," at the height of the superior anal canal. Traumatic injuries of this muscle can occur in one of the branches or "cables," right or left (Fig. 7.17) or in both branches. More commonly, different thicknesses are found between the muscular cables, an aspect called asymmetry (Fig. 7.18).

It is important to remember that the PR is attached to the pubis and encircles the rectum in its posterior portion in the form of an arch; thus, it is not visualized in the anterior quadrant of the anal canal. It is necessary to differentiate the

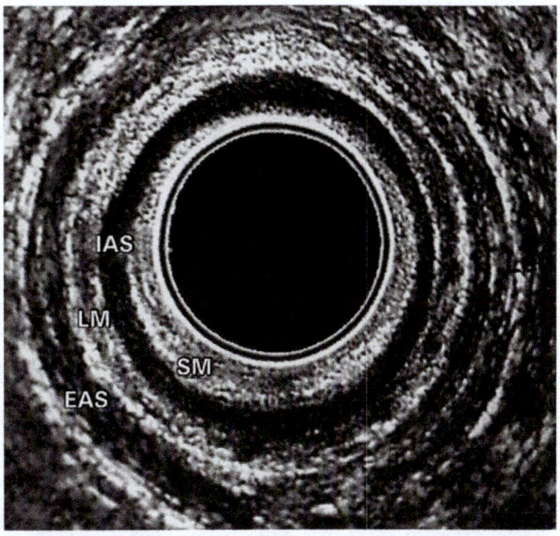

Fig. 7.14 Longitudinal muscle in the - Middle anal canal; IAS = internal anal sphincter; LM = longitudinal muscle; SM = submucosal; EAS = external anal sphincter

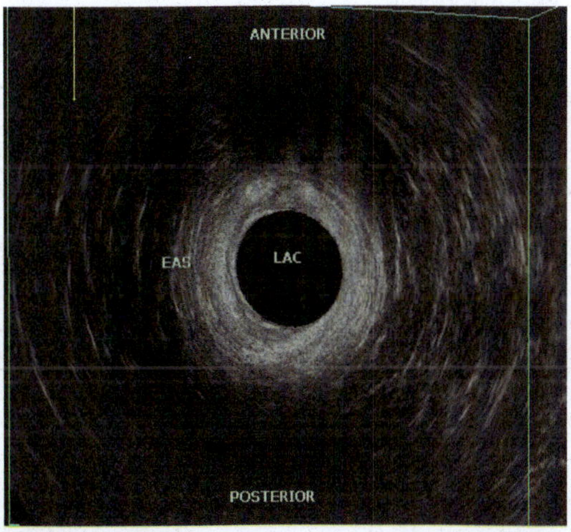

Fig. 7.15 External anal sphincter muscle (EAS) in the low anal canal (LAC)

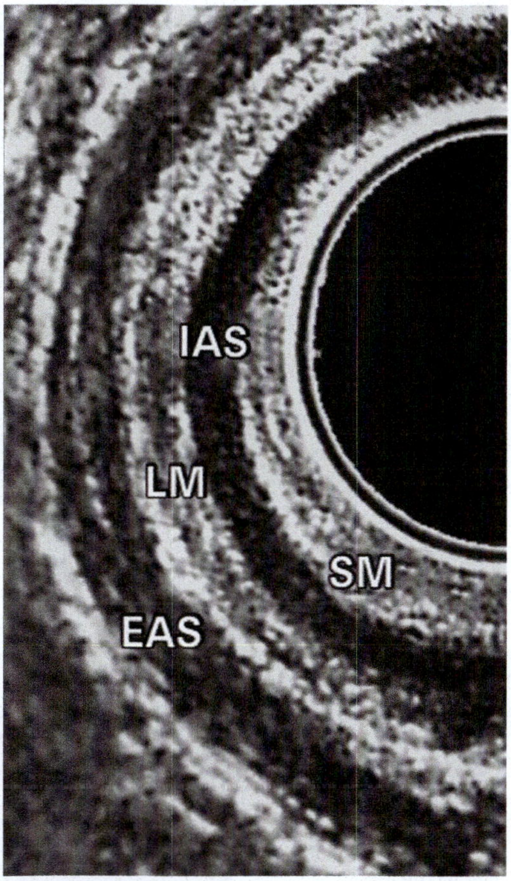

Fig. 7.16 Layers of the anal canal by ultrasound; IAS = internal anal sphincter; LM = longitudinal muscle; SM = submucosal; EAS = external anal sphincter

images of the superior anal canal in the anterior quadrant, in which the presence of striated musculature is not observed, with previous defects of the EAS. For this differentiation, the PR fibers form a "V"-shaped lateral arch, unlike the EAS fibers, which run in a circular fashion along the anal canal.

The images acquired with the three-dimensional transducer allow evaluation of these muscular structures from other angles. In longitudinal or sagittal sections, the PR is visualized in continuity with the external sphincter as a mixed image (Fig. 7.19).

External Anal Sphincter

The EAS is visualized in the middle anal canal (Fig. 7.13), and its most superficial portion is seen in the lower anal canal (Fig. 7.14). It is also a striated muscle, producing an image of hyper-echogenicity or mixed echogenicity that surrounds the anal canal circumferentially. Its external margins, however, are poorly defined which sometimes makes it difficult to establish its thickness. However, studies in the literature estimate that this muscle has an average thickness of about 5 to 8 mm [2–4], measuring about

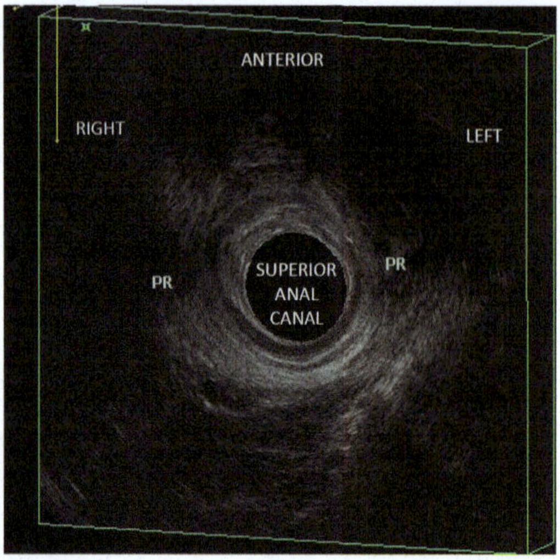

Fig. 7.17 Traumatic injuries of the puborectalis muscle can occur in one of the branches or "cables," right or left; PR = puborectalis

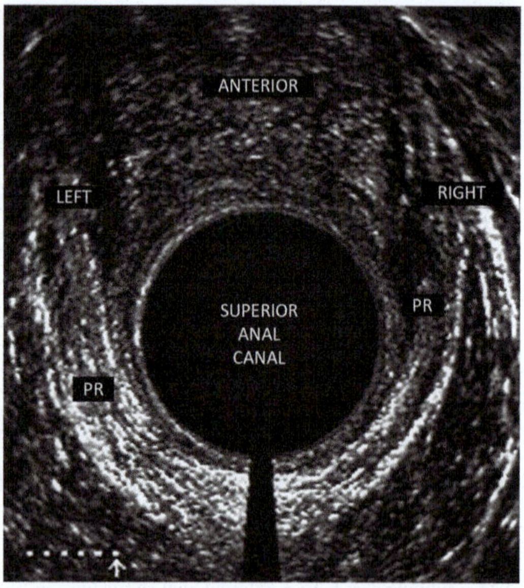

Fig. 7.18 Asymmetry of the puborectalis muscle; PR = puborectalis

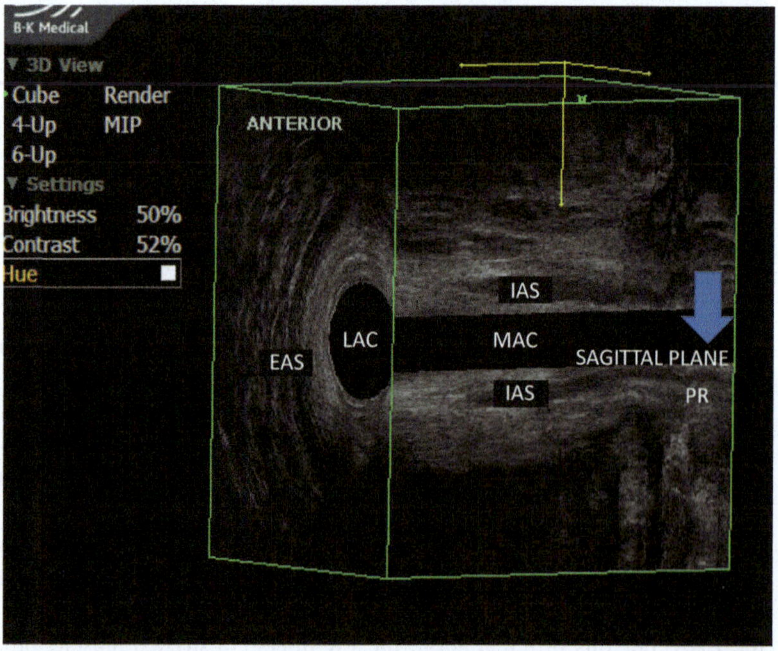

Fig. 7.19 Sagittal plane on the 3D endoanal ultrasound with external anal sphincter (EAS), internal anal sphincter (IAS) and puborectalis muscle (PR); MAC = middle anal canal

4 cm in extension. The thickness decreases with age, due to the process of degeneration and atrophy of the muscles [5]. Adequate recognition of lesions involving EAS is of great importance, since obstetrical injury is a major cause of incontinence among women [6, 7]. Obstetrical muscle injury may involve the EAS and/or the IAS, usually in the anterior quadrant of the anal canal.

Endoanal ultrasonography is the ideal method for recognition of these lesions, but up to 25% of false-positives have been reported [9]. Since it is well tolerated by patients, it can be performed before and after surgical repair allowing very clear and detailed images of the affected muscles. Figure 7.20 shows two examples of typical obstetric lesions involving the EAS in the anterior anal canal. For patients who have undergone anterior sphincteroplasty, the overlapped muscle can be well demonstrated (Fig. 7.21).

Although easy to visualize, the anterior lesions of the EAS cannot be misinterpreted as the anatomical "defect" normally found in women in the region of the perineal body. This concept was proposed by Bollard et al. [10] who evaluated 57 nulliparous women and verified that there is actually a reduction in the fibers of the EAS in the anterior portion of the anal canal (Fig. 7.22).

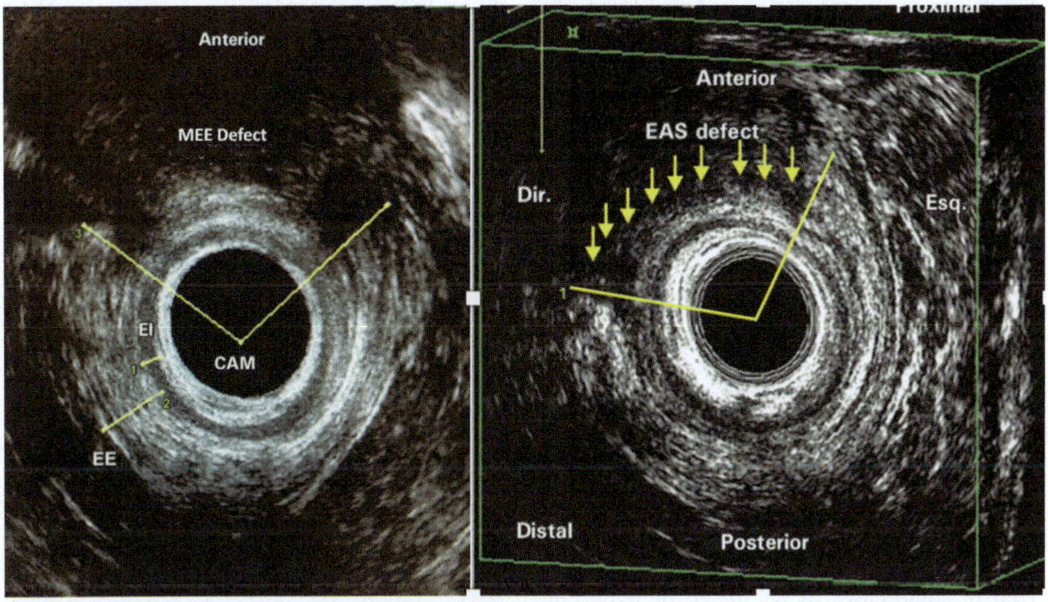

Fig. 7.20 Typical obstetric lesion involving the external anal sphincter (EE) in the anterior anal canal; Esq = left; Dir = right; EI = internal anal sphincter; CAM = middle anal canal

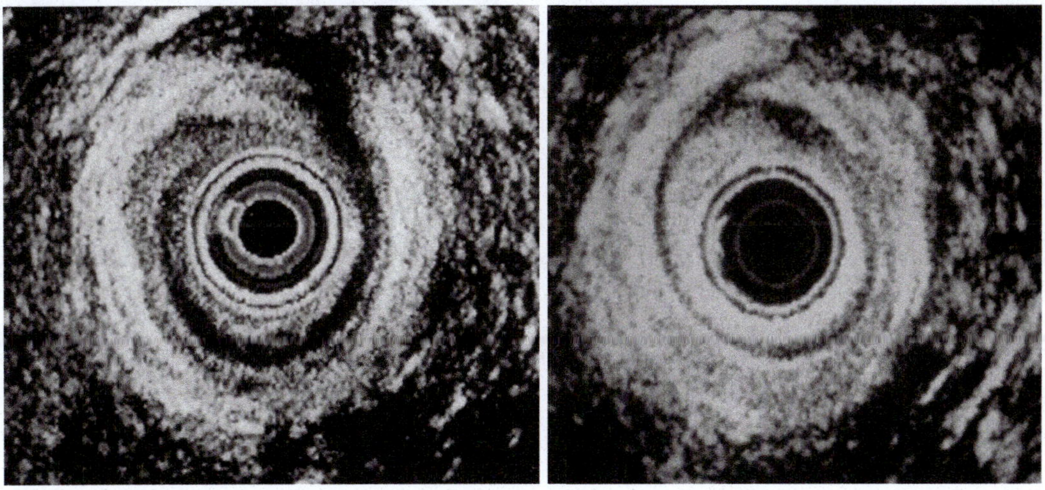

Fig. 7.21 Anterior overlap of the external anal sphincter

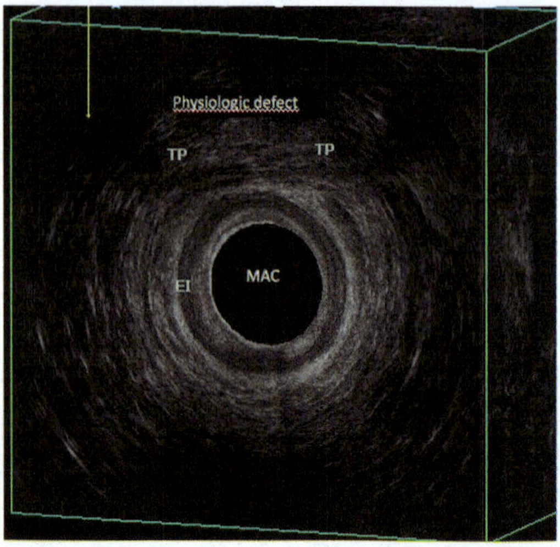

Fig. 7.22 Anatomical physiological anterior defect of the external anal sphincter; the transverse perineal (TP) muscle in the perineal body; EI = internal anal sphincter; MAC = middle anal canal

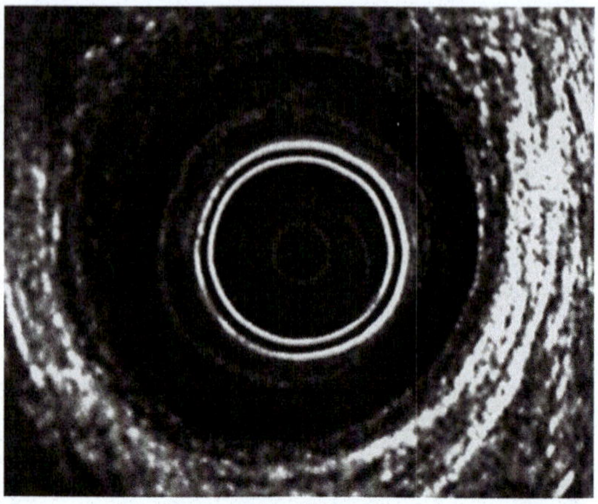

Fig. 7.23 Internal anal sphincter muscle hypertrophy

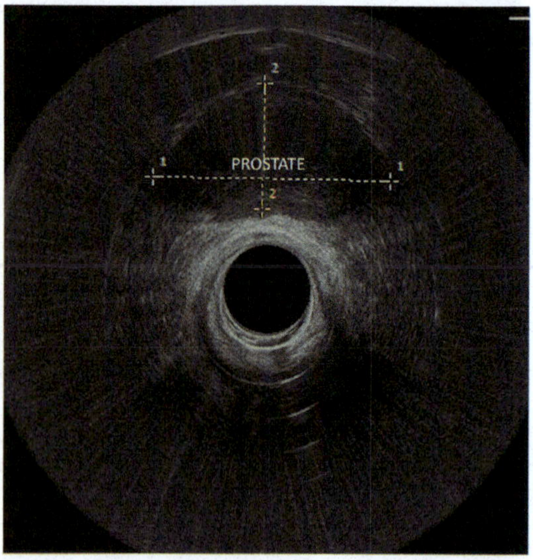

Fig. 7.24 Prostate

Longitudinal Muscle

The longitudinal muscle is presented as a continuation of the longitudinal fibers of the rectum in the intersphincteric space and may be visualized in some cases as an image of hypoechogenicity (Fig. 7.14).

Internal Anal Sphincter

This smooth and thin muscle measures about 1–3 mm and appears as a hypoechogenic image (dark circle) surrounding the anal circumference in the middle anal canal (see Fig. 7.13). Unlike the EAS, the IAS suffers mild hypertrophy with advancing age [11]. An IAS thickness of >3.5 mm is associated with dystrophic diseases that present with symptoms of obstructive defecation resulting from muscular hypertrophy and consequential hypertonia at rest (Fig. 7.23) [12].

Other Structures

The combination of endoanal, transperineal, and endovaginal ultrasonography allows visualization of other structures such as the prostate (Fig. 7.24), the seminal vesicles in males (Fig. 7.25) and the vagina and urethra in females (Fig. 7.26), and the muscles of the pelvic hiatus and the pelvic floor (Fig. 7.27).

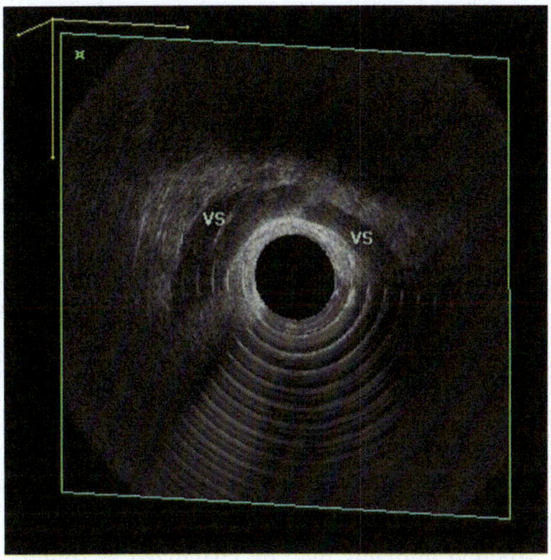

Fig. 7.25 Seminal vesicles (VS) and the urethra in males

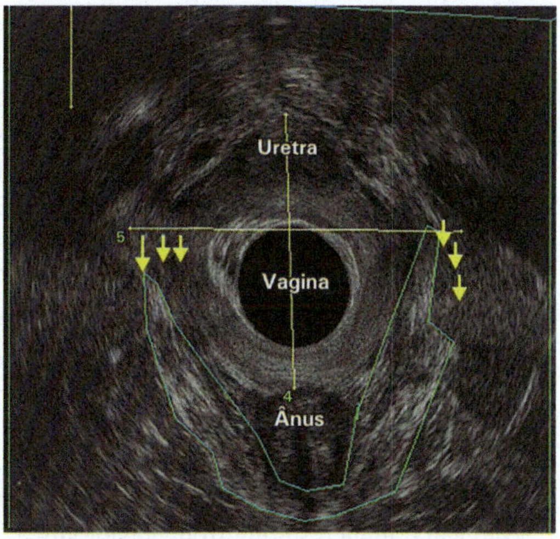

Fig. 7.27 Muscles of the pelvic hiatus and the pelvic floor

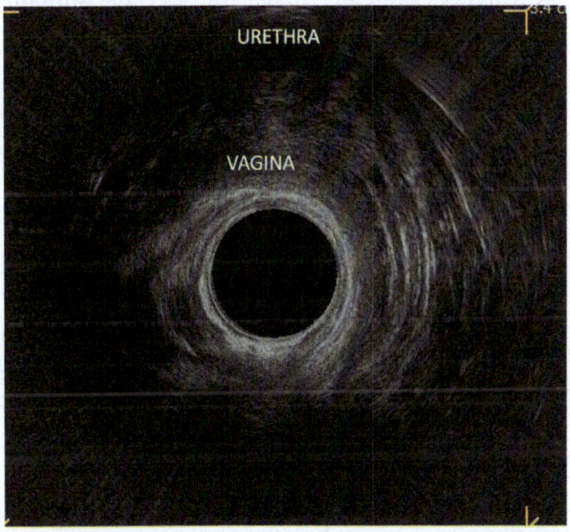

Fig. 7.26 Vagina and urethra in females

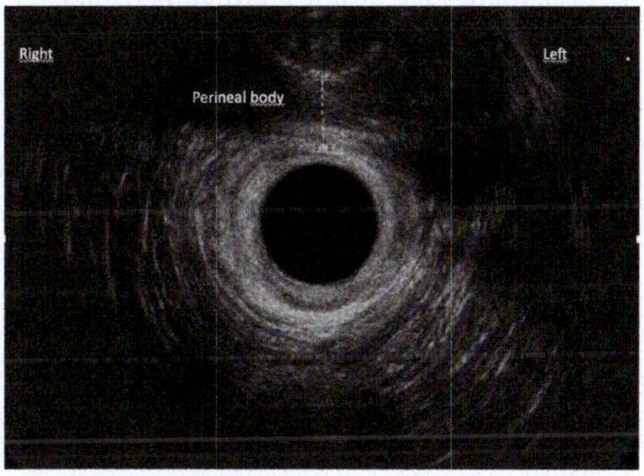

Fig. 7.28 The perineal body

The perineal body can be assessed, and its thickness can be measured by performing a digital exam at the time the mid anal canal is examined (Fig. 7.28). It is estimated that the normal thickness of the perineal body is at least 10 mm [13].

There are different ways of evaluating the location and size of defects. Defects can be described based the quadrants of the anal canal, in a clockwise fashion, and by the level in the anal canal. The angle of the defects can also be measured. In general, EAS defects are found more frequently in the mid anal canal in the anterior quadrant [14]. With regard to the thickness of the sphincter muscles, the results of different studies are controversial [15].

Evaluation of Anal Incontinence

Endoanal Ultrasound

Endoanal ultrasound has been considered the most informative and utilized exam to evaluate anal incontinence because it is a painless, simple, easy-to-perform method that provides excellent imaging of the entire anal canal, including the IAS (Fig. 7.13) [16, 17]. In addition, it provides sensitivity and specificity for the detection of sphincter defects ranging from 83% to 100% [18]. It is well tolerated by patients and can be repeated postoperatively with no harm to the previous repair. Among women, obstetric history is one of the main causes of traumatic incontinence [8]. Obstetric injury can be presented as two different mechanisms:

1. Direct injury of the musculature by stress or surgical trauma resulting from episiotomy
2. Stretch lesion of the developing nerve, leading to sphincter denervation

Recent studies have shown a high percentage of obstetric injury associated with anal incontinence. Although severe obstetric lesions account for only 1% of natural births, endoanal ultrasonography can show hidden defects in 3–60% of women after the first birth [19].

In the investigation of obstetric lesions, the thickness of the perineal body can also be evaluated by vaginal digital examination in addition to ultrasound of the anal canal. This method proved to be useful in the evaluation of 42 incontinent patients, allowing better visualization in 74% of the cases in one study [13]. The perineal body thickness is considered to be at least 10 mm.

Anorectal surgeries involving the sphincter muscles can cause muscle injury and subsequent fecal incontinence. In a study published by Stamatiadis et al. [20], of the 123 patients who underwent different anorectal surgeries, lesions of the EAS and IAS were found in 21% of cases. For patients who underwent Milligan–Morgan hemorrhoidectomy, 5.5% had internal sphincter lesions. The incidence of IAS and EAS defects corresponded to 57% and 29% of fistulotomies and 76% and 24% of anal dilatations, respectively.

Endoanal ultrasound has also proven useful in assessing post-sphincteroplasty results, anterior or posterior (postanal repair) [21–25], because it allows visualization of the repair and correlates with optimal surgical outcomes [26].

Endoanal ultrasound has replaced electromyography (EMG) mapping due to its simplicity and tolerability to patients. In addition, EMG does not provide information about the IAS and is more invasive, more painful, and less widely available. The role of EMG is restricted to cases where an assessment of neurofunctional integrity is required or confirmation of inaccurate ultrasonographic findings (i.e., when a posterior EAS defect of the anus is not well defined as a result of echographic changes of adjacent layers) [27, 28].

Follow-up of incontinent patients who were treated by the injection of a bulking agent can also be easily performed by endoanal ultrasound. Silicone implants may be visualized as rounded hyperechogenic images in the quadrants corresponding to the injection sites (Fig. 7.29a, b) or as a hypoechogenic image for acrylate implants (Fig. 7.30).

The development of scores for ultrasound evaluation in incontinent patients was initially advocated by Starck et al. [29] in a study that evaluated women in the early postpartum period. In this study, the authors evaluated women with third- and fourth-degree sphincter lesions who underwent a sphincter repair 2–7 days prior. Evaluating the appearance of the sphincter muscles, they established a classification and defined a sphincter defect as a discontinuity in the endosonographic image of the IAS (hypoechoic ring) or EAS (mixed echo ring).

Defects involving less than half of the EAS and/or IAS thickness were not classified as defects; those involving more than half but not the entire sphincter thickness were classified as partial defects. Finally, those involving the entire sphincter thickness were classified as total defects. The location and size of any defect of the IAS or EAS were described using the positions of a clock: a defect at 12 hours would be earlier, defective 6 hours later. In addition, longitudinal location and extent of the sphincter defect have been described (e.g., proximal, distal, or full-length defect, less than half of the length of the sphincter, more than half the length of the sphincter, or the length of the entire sphincter). Thus, they described a scoring system (Table 7.3)

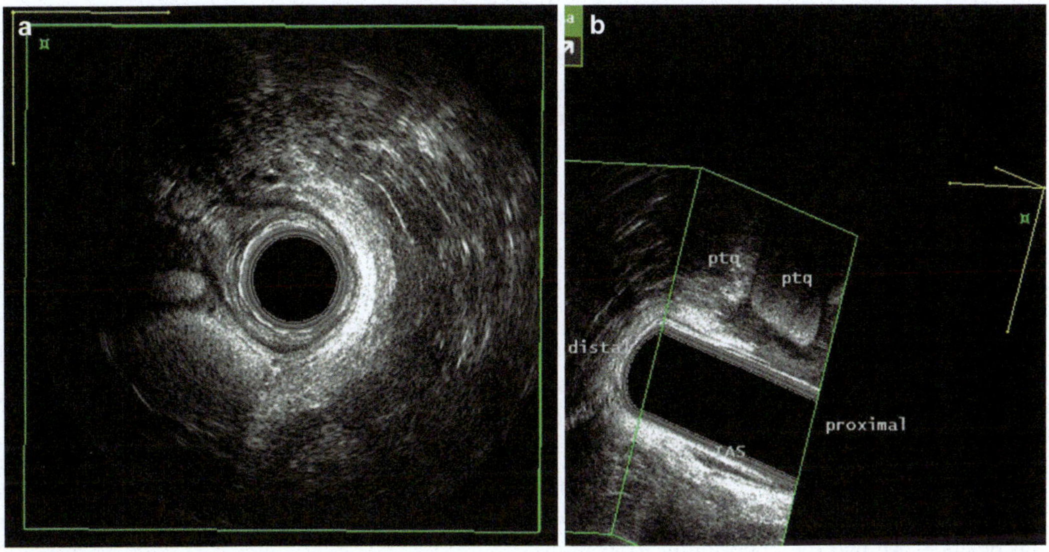

Fig. 7.29 (**a**, **b**) Silicone-PTQ- bulking agent

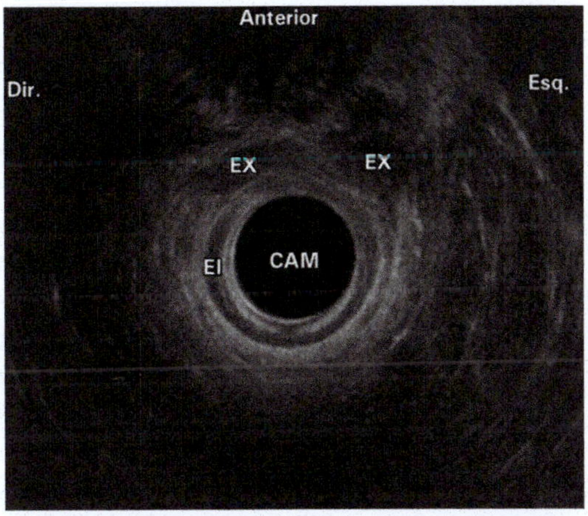

Fig. 7.30 Acrylate bulking agent (Ex = Exantia bulking agent); Dir = right; Esq = left; EI = internal anal sphincter; CAM = middle anal canal)

Table 7.3 Starck score

	Score			
Parameter	0	1	2	3
External sphincter				
Length of defect	None	≤1/2	>1/2	Whole
Depth of defect	None	Partial	Total	
Size of defect	None	≤90 degrees	91–180 degrees	>180 degrees
Internal sphincter				
Length of defect	None	≤1/2	>1/2	Whole
Depth of defect	None	Partial	Total	
Size of defect	None	≤90 degrees	91–180 degrees	>180 degrees

where the defect receives a score of 0–3 with a maximum score of 16.

This work by Starck et al. [29] highlighted one important aspect of obstetric injury – many patients have occult defects with a difference in outcomes depending on the experience of the surgeon at the time of repair. The results showed that 71% of the women had a partial third-degree rupture, 23% had a total third-degree rupture, and 6% had a fourth-degree rupture. Thirty-four (71%) defects were sutured by practicing OB/GYNs and 14 (29%) by physicians in training. This high incidence of sphincter defects was also reported by Sultan et al. [30–32] who, concerned with the quality of repairs, developed training courses for gynecologists. Thus, ultrasonography is a very valuable imaging modality for monitoring and diagnosing the incontinent patient.

In 2008, Nordeval et al. [33] compared the results of patients who were evaluated using a score they proposed based on the Starck score and observed a good correlation. In a retrospective evaluation of 55 patients, they established a scoring system that takes into account the presence of defects and the echogenicity of the muscle, as either normal or with a pattern of atrophy. Defects were found in 51 (93%) patients, 22 patients had partial defects of the EAS, 15 had complete EAS defects, and 14 had defects of both the IAS and EAS muscles. The proposed score ranges from 0 to 7, being simpler and correlating with the score proposed by Starck.

Our experience with endoanal ultrasound started in 1995. During these 25 years of colorectal practice, we had the opportunity to examine more than 3000 patients with anal incontinence. Complete evaluation of those patients, including a good physical exam followed by endoanal ultrasound, was important for a precise diagnosis and suitable treatment options.

With the introduction of three-dimensional ultrasonography, additional information can be obtained, such as the presence of internal mucosal prolapse, cystoceles and enteroceles, which may be associated with pelvic floor disorders in incontinent patients. Three-dimensional transducers allow evaluation of the anal canal in different planes, and it is possible to enlarge these images and obtain additional features such as measurement of angles and the distance and thickness of the sphincters. In Fig. 7.31 and Fig. 7.32, combined axial and coronal views at the US 3D cube can facilitate visualization of sphincter defects and thickness of the EAS and IAS. The echodefecography technique will be discussed in Chap. 8 in more detail.

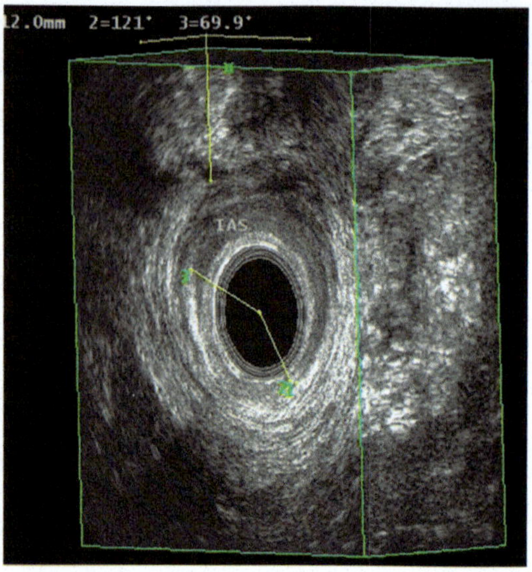

Fig. 7.31 Internal anal sphincter (IAS) defect as observed in the 3D cube

Evaluation of the Internal Anal Sphincter

As already mentioned, the IAS is a smooth, thin muscle that surrounds the anal canal. It appears as a hypoechogenic image, with a mean thickness of 1–3 mm (Fig. 7.13). In the coronal section of the 3D cube, the IAS can resemble a baseball bat, similar to that seen on MRI (Fig. 7.33). Isolated or multiple lesions of this muscle are easily visualized as areas of discontinuity of the hypoechogenic circle. IAS thickness increases with age in asymptomatic individuals [5], and hypertrophy can be observed in some children and adults with intestinal constipation, and endoanal ultrasonography may be useful in this evaluation [34]. The IAS is considered to be hypertrophied with a thickness of more than 3.5–4 mm in patients under 50 years of age, whereas in patients over 50 years of age, a thickness of 5 mm or more is considered abnormal (Fig. 7.23) [35]. Some rare diseases of the IAS such as familial myopathy are manifested by constipation and proctalgia, and MRI and endoanal ultrasound allow visualization of the hypertrophied muscle [12]. In those cases, histological evaluation of the IAS muscle reveals muscular fibers with several vacuoles in addition to hypertrophy [36].

Neuronal intestinal dysplasia is a rare congenital disease that resembles Hirschsprung's disease clinically and is manifested by constipation. In these cases, hyperplasia of the submucosal and myenteric plexuses occurs, in addition to an increase in the activity of ace-

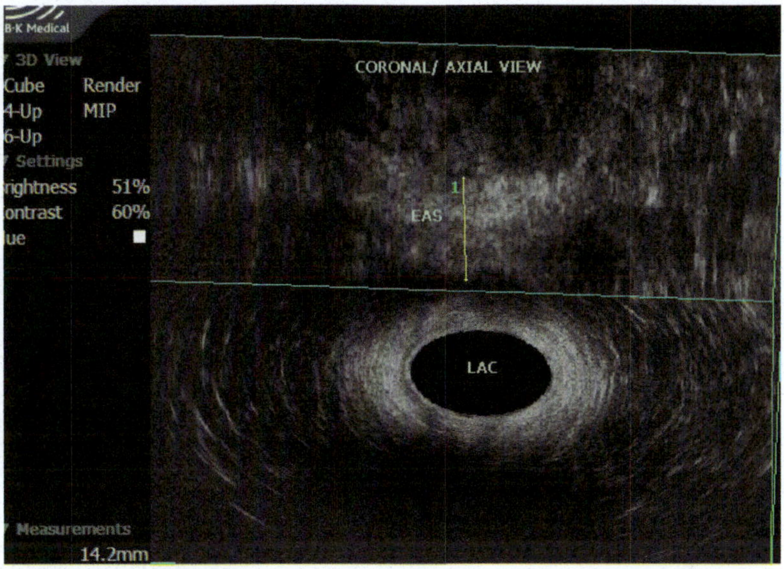

Fig. 7.32 External anal sphincter (EAS) thickness in combined coronal and axial planes; LAC = low anal canal

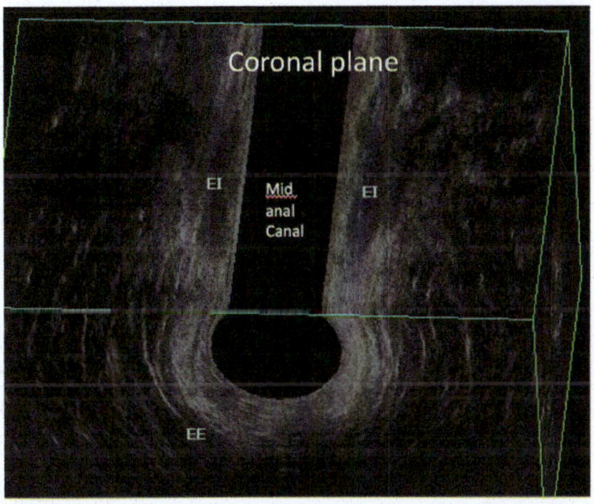

Fig. 7.33 Coronal plane of the 3D cube with bilateral internal anal sphincters; EI = internal anal sphincter; EE = external anal sphincter

tylcholinesterase [37]. Although diagnosis is based on immunohistochemical analysis of the IAS, ultrasound can also demonstrate muscular hypertrophy and guide surgical treatment, when indicated. In patients with solitary rectal ulcer and constipation, a thick IAS on ultrasound is suggestive of an associated internal intussusception. This was demonstrated by Marshall [38] et al. in a study evaluating 20 patients with solitary rectal ulcers.

In systemic sclerosis, anal incontinence may occur as a result of replacement of the internal fibers of the sphincter muscles by fibrosis or myopathy due to vasculitis. A recent study based on endoanal ultrasonography showed that, in fact, the IAS can be thickened or not [39]. Another interesting study evaluated the correlation of postoperative anal pain after anal surgeries with thickening of the IAS [40]. In 95 patients evaluated by endoanal ultrasound, more than 80% of the causes of anal pain were related to IAS hypertrophy.

Senile degeneration of IAS has been well documented in different studies [41, 42].

Evaluation of the IAS by endoanal ultrasound is a simple method and provides important information. In partial ruptures, the finding of a sphincter injury in a patient with incontinence may indicate the cause of incontinence (Fig. 7.34). In males, this is a common cause of fecal leakage and can be treated by injection of a bulking agent [43].

Evaluation of the Pelvic Floor

Disorders of the pelvic floor occur in one in four adult women in the USA [44]. With an aging population, a growing number of women with

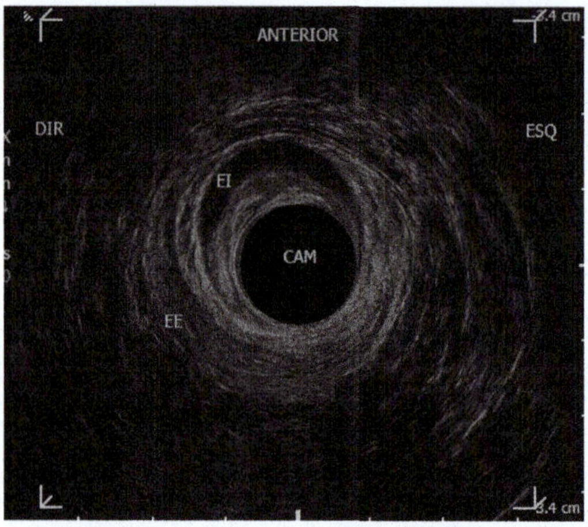

Fig. 7.34 Lesion of the internal anal sphincter (EI) in the middle anal canal (CAM); Dir = right; Esq = left; EE = external anal sphincter

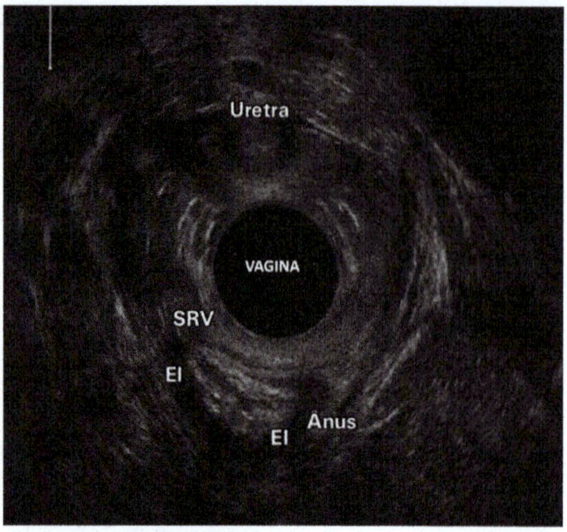

Fig. 7.35 Lesion of the levators and pubovisceral, which leads to widening of the genital hiatus; EI = internal anal sphincter; SRV = perineal body

pelvic floor dysfunction, represented by fecal and urinary incontinence and pelvic organ prolapse, can be expected.

Evaluation of the pelvic floor by endovaginal and transperineal ultrasound with three-dimensional circular transducers can detect defects in the pubovisceral muscles and ballooning of the genital hiatus as well as the more superficial layers that compose the pelvic floor, represented by the transverse perineal, bulbospongiosus and EAS muscles.

The use of high-resolution three-dimensional circular transducer allows images to be acquired in a virtual cube for 30–60 seconds, with visualization of all layers and anatomy of the pelvic floor; this method has great reproducibility regardless of the examiner.

The patient is positioned in dorsal decubitus, with flexed legs; after lubrication, the transducer is inserted and maintained centrally for image acquisition (Fig. 7.9). The transducer should be inserted to about 6 cm until the pubic symphysis and urethra can be seen in the anterior quadrant. It is possible to visualize the transverse perineum, puboperineal, puboanal, pubovaginal, iliococcygeal, and puborectalis muscles. Shobieri [45] et al. proposed evaluation by endovaginal ultrasonography from three levels:

- *Level 1:* Contains the muscles of the perineal body, represented by the transverse perineum, puborectalis and puboanal
- *Level 2*: Contains the pubovaginal, puborectalis, puboanal, and ileococcygeus muscles
- *Level 3*: Contains the subdivisions of the pubococcygeus and ileococcygeus

Considering that up to 55% of women with pelvic organ prolapse have obstetric complications, the identification of these lesions by means of ultrasonography has an important clinical implication. Measurement of the pelvic gap in 80 nulliparous women revealed a mean extension of 13–14 cm. Widening of the hiatus has been associated with injury to the levators, with avulsion more frequently causing obstetric injury.

Studies by Santoro et al. [46] and Regadas et al. [47] in the ultrasound arena have provided new perspectives for evaluation of the pelvic floor.

The use of endovaginal three-dimensional transducers has allowed the perineal body and genital hiatus to be detailed. Observation of images in nulliparous women and those with incontinence and obstetric history has allowed the detection of levator lesions, which leads to widening of the genital hiatus (Fig. 7.35).

Regadas et al. [47] performed endovaginal ultrasound in 20 asymptomatic nulliparous women with the objective of providing anatomic and functional measurements of the pelvic floor at rest and during the Valsalva maneuver. During the Valsalva maneuver, the hiatal area appeared significantly larger, the urethra appeared significantly shorter, and the anorectal angle appeared larger. Measurements at rest and during the Valsalva maneuver were significantly different in relation to the position of the anorectal junction and the neck of the bladder. The mean decrease in the perineum and bladder was 0.6 cm and 0.5 cm above the pubic symphysis, respectively (Fig. 7.36).

Dynamic transperineal ultrasound was developed as an alternative to fluoroscopic defecography as it can demonstrate pelvic floor hernias and distinguish organ deformation from dyssynergic defecation related to paradoxical contraction of the puborectalis muscle [48]. This ultrasound modality can be performed in the left lateral position or prone position with the legs flexed see (Fig. 7.9). The compartments of the pelvic floor can be well observed (Fig. 7.37).

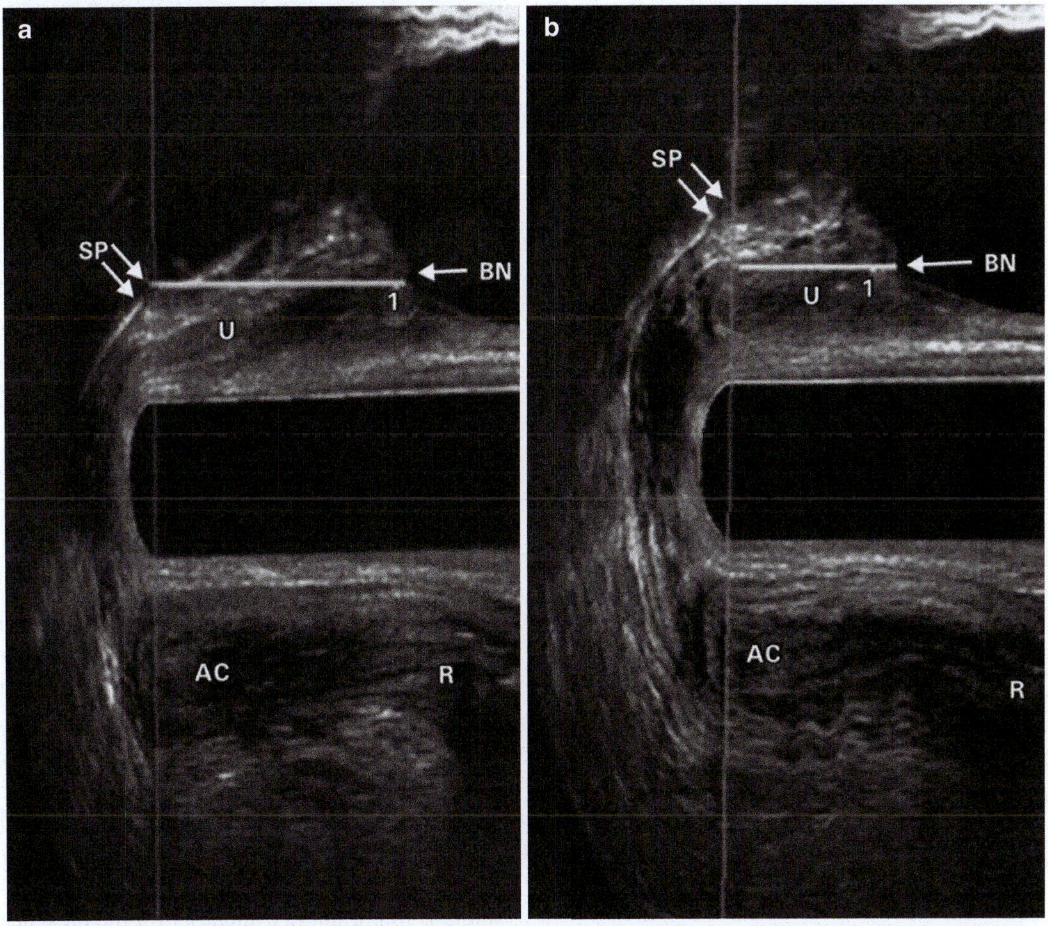

Fig. 7.36 Mean decrease in the perineum and decrease of the bladder was 0.6 cm and 0.5 cm above the pubic symphysis, respectively. (**a**) evaluation of the descending of the perineum by endovaginal ultrasound. (**b**) decrease in the perineum and bladder above the pubic symphysis. SP = pubic symphysis; BN = bladder neck; U = urethra; R=right

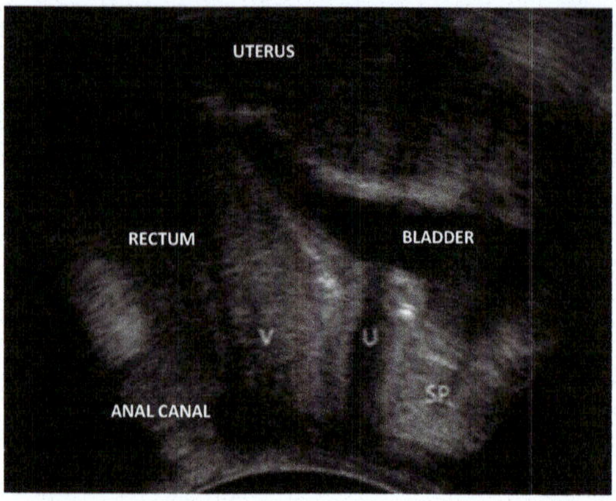

Fig. 7.37 Transperineal ultrasound with three compartments and structures: bladder, urethra, vagina, rectum, puborectalis, and synphase pubic

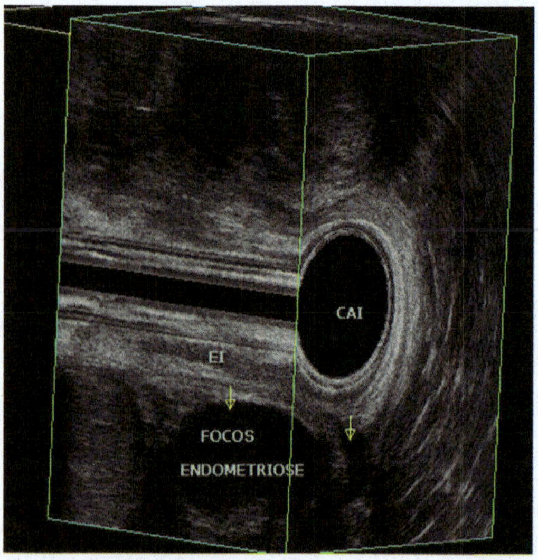

Fig. 7.38 Endometriosis with hypoechoic cystic lesions in the sagittal plane; CAI = low anal canal; FOCOS ENDOMETRIOSIS = endometriosis; EI = internal anal sphincter

Evaluation of Endometriosis and Other Cysts

Endoanal ultrasound can be used to assess endometrial lesions in the rectal and anal canal, which appear as mixed hypoechogenic images (Fig. 7.38) [49]. A history of exacerbation of pain symptoms and perianal discomfort in the

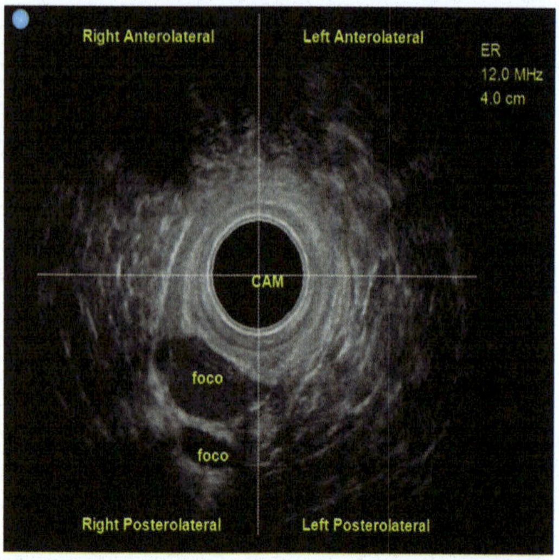

Fig. 7.39 Endometriosis cysts (foco) ; CAM = middle anal canal

menstrual period is indicative of the diagnosis, which can then be confirmed by the ultrasound evaluation. The use of a three-dimensional transducer allows the examiner to evaluate the anatomical relationship of the lesion with the perianal muscular structures and to identify infiltration into the vaginal septum. Hypoechoic or mixed echogenic images are visualized by infiltrating the sphincter muscles and the perineal body and vagina.

With the use of the most modern three-dimensional transducers that reach focal distances of up to 6 cm, the presence of retrorectal cysts and tumors localized in the retrorectal space can be detected. The most common lesions are congenital, representing up to 50% of all lesions, most of which are cystic in nature (Fig. 7.39).

Retrorectal or presacral tumors are rare and more common in the females, with the majority being congenital. Endoanal or endorectal ultrasound is important in evaluating for the presence of cystic masses and to help distinguish between mixed and heterogenic images, which are malignant in origin. In Fig. 7.40 a significant retrorectal cyst can be observed in the axial and sagittal plane.

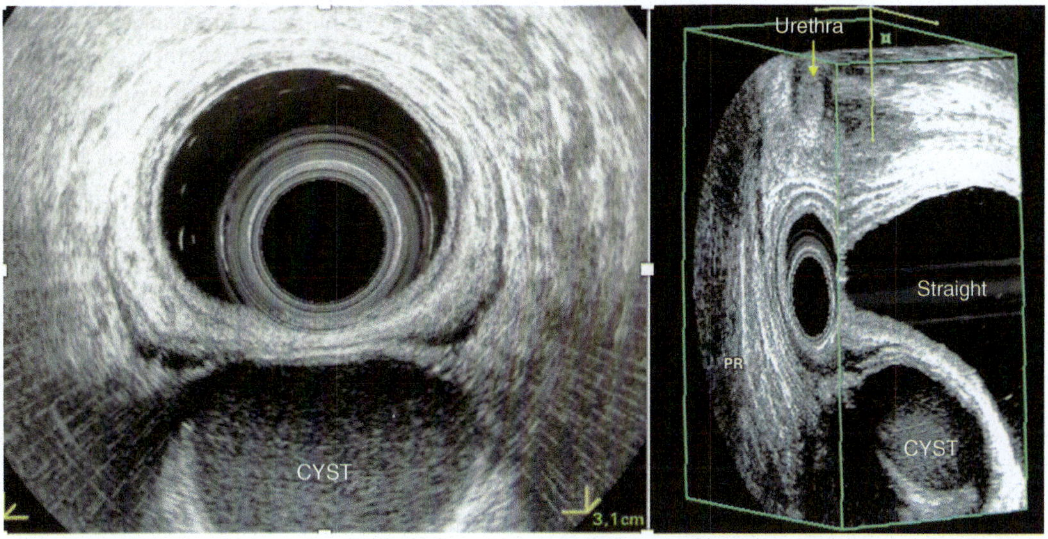

Fig. 7.40 Retrorectal cysts

Fig. 7.41 Anal fistula tract in the sagittal plane; EAS = external anal sphincter

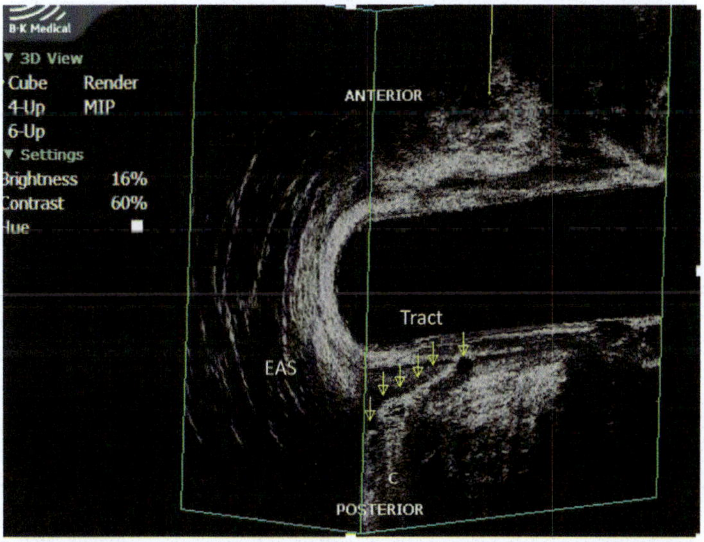

Evaluation of Anal Fistula

Ultrasound evaluation of patients with complex anal fistulas allows identification of primary and secondary tracts. In these cases, evaluation of the sphincter muscles also yields important information to distinguish between intersphincteric or transsphincteric fistulas. The information provided is valuable for proper planning of the surgical approach. The use of three-dimensional transducers has allowed identification of the fistula tract in all planes (Fig. 7.41). In addition, when an external opening is present, it is possible to introduce 1–2 ml of hydrogen peroxide to improve visualization of the tracts and the internal openings (Fig. 7.42a, b).

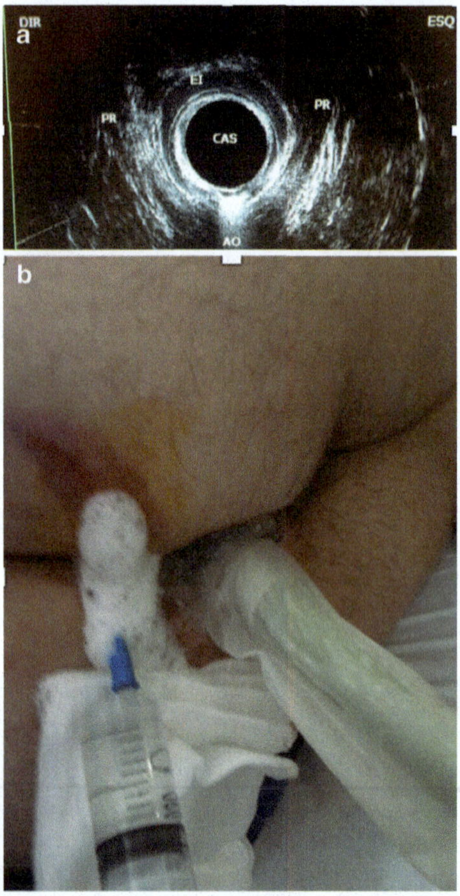

Fig. 7.42 (**a**, **b**) Hydrogen peroxide to improve visualization of the tracts and the internal opening CAS = Upper anal canal; Esq = right; Dir = left; PR = puborectalis; EI = internal anal sphincter; AO = hydrogen peroxide

Summary

Endoanal ultrasound and other ultrasound modalities are important tools for the evaluation of pelvic floor and anorectal disorders. In a review by Albuquerque [50], the sensitivity of ultrasonography for the detection of sphincter defects was approximately 100% in most studies. In comparison with MRI, ultrasound has superior sensitivity for the detection of IAS lesion, an equivalent sensitivity for the diagnosis of EAS and IAS defects, and is less sensitive only to the evaluation of atrophy of this muscle. As for other anorectal pathologies including cysts, tumors, fistulas, and abscesses, endoanal and endovaginal ultrasound enhances evaluation of the relationship between these pathologies and the sphincter complex by a simple and well-tolerated procedure. Finally, utilization of the linear and convex transducers can also help in evaluation of the three compartments of the pelvic floor, adding dynamic images to the evaluation.

References

1. Law PJ, Bartram CI. Anal endosonography – technique and normal anatomy. Gastrointest Radiol. 1989;14:349–53.
2. Rottenberg GT, Williams AB. Endoanal ultrasound. Br J Radiol. 2002;75:482–8.
3. Williams AB, Bartram CI, Halligan S, et al. Endosonographic anatomy of the normal anal canal compared with endocoil magnetic resonance imaging. Dis Colon Rectum. 2002;45(2):176–83.
4. Tjandra JJ, Milsom JW, Stolfi VM, et al. Endoluminal ultrasound defines anatomy of the anal canal and pelvic floor. Dis Colon Rectum. 1992;35:465–70.
5. Frudinger A, Halligan S, Bartram CI, et al. Female anal sphincter: age-related differences in asymptomatic volunteers with high-frequency endoanal US. Radiology. 2002;224(2):417–23.
6. Oliveira L, Wexner SD. Anal incontinence. In: Beck D, Wexner SD, editors. Fundamentals of anorectal surgery. 2nd ed. London: W.B. Saunders; 1999. p. 115–52..
7. Faridi A, Willis S, Schelzig P, et al. Anal sphincter injury during vaginal delivery – an argument for cesarean section on request? J Perinat Med. 2002;30(5):379–87.
8. Oliveira L, Pfeifer J, Wexner SD. Clinical and physiological outcome of anterior sphincteroplasty. Br J Surg. 1996;83:502–5.
9. Sentovich SM, Wong WD, Blatchford GJ. Accuracy and reliability of transanal ultrasound for anterior anal sphincter injury. Dis Colon Rectum. 1998;41:1000–4.
10. Bollard RC, Gardiner A, Lindow S, et al. Normal female anal sphincter: difficulties in interpretation explained. Dis Colon Rectum. 2002;45:171–5.
11. Ramirez Rodriguez JM, Palacios Fanlo MJ, Artigas Marco C et al. Anal endosonography for the study of anal canal anatomy: normal images and sonographic variants. Ver Esp Enferm Dig 2001; 93(6):364–371.
12. Guy RJ, Kamm MA, Martin JE. Internal anal sphincter myopathy causing proctalgia fugax and constipation: further clinical and radiological characterization in a patient. Eur J Gastroenterol Hepatol. 1997;9(2):221–4.
13. Zetterstrom JP, Mellgren A, Madoff RD, et al. Perineal body measurement improves evaluation of anterior sphincter lesions during endoanal ultrasonography. Dis Colon Rectum. 1998;41:705–13.

14. Deen KI, Kumar D, Williams JG, et al. The prevalence of anal sphincter defects in faecal incontinence: a prospective endosonic study. Gut. 1993;34:685–8.
15. Enck P, Heyer T, Gantke B, et al. How reproducible are measures of the anal sphincter muscle diameter by endoanal ultrasound? Am J Gastroenterol. 1997;92(2):293–6.
16. Buhr HJ, Kroesen AJ. The importance of diagnostics in faecal incontinence. Endosonography. Chirurg. 2003;74(1):4–14.
17. Rieger NA, Sweeney JL, Hoffmann DC, et al. Investigation of fecal incontinence with endoanal ultrasound. Dis Colon Rectum. 1996;39(8):860–4.
18. Farouk R, Bartolo DC. The use of endoluminal ultrasound in the assessment of patients with faecal incontinence. J R Coll Surg Edinb. 1994;39(5):312–8.
19. Karoui S, Savoye-Collet C, Koning E, et al. Prevalence of sphincter defects revealed by sonography in 335 incontinent patients and 115 continent patients. Am J Roentgenol. 1999;173:389–92.
20. Stamatiadis A, Konstantinou E, Theodosopoulou E, et al. Frequency of operative trauma to anal sphincters: evaluation with endoanal ultrasound. Gastroenterol Nurs. 2002;25(2):55–9.
21. Engel AF, Kamm MA, Sultan AH, et al. Anterior anal sphincter repair in patients with obstetric trauma. Br J Surg. 1994;81(8):1231–4.
22. Nielsen MB, Dammegaard L, Pedersen JF. Endosonographic assessment of the anal sphincter after surgical reconstruction. Dis Colon Rectum. 1994;37(5):434–8.
23. Fleshman JW, Peters WR, Shemesh EI, et al. Anal sphincter reconstruction: anterior overlapping muscle repair. Dis Colon Rectum. 1991;34(9):739–43.
24. Ternent CA, Shashidharan M, Blatchford GJ, et al. Transanal ultrasound and anorectal physiology findings affecting continence after sphinctero- plasty. Dis Colon Rectum. 1997;40(4):462–7.
25. Matsuoka H, Mavrantonis C, Wexner SD, et al. Postanal repair for fecal incontinence–is it worthwhile? Dis Colon Rectum. 2000;43(11):1561–7.
26. Savoye-Collet C, Savoye G, Koning E, Thoumas D, Michot F, Denis P, et al. Anal endosonography after sphincter repair: specific patterns related to clinical outcome. Abdom Imaging. 1999;24(6):569–73.
27. Tjandra JJ, Milsom JW, Schroeder T, et al. Endoluminal ultrasound is preferable to electromyography in mapping anal sphincteric defects. Dis Colon Rectum. 1993;36(7):689–92.
28. Enck P, von Giesen HJ, Schäfer A, et al. Comparison of anal sonography with conventional needle electromyography in the evaluation of anal sphincter defects. Am J Gastroenterol. 1996;91(12):2539–43.
29. Starck M, Bohe M, Valentin L. Results of endosonographic imaging of the anal sphincter 2–7 days after primary repair of third- or fourth-degree obstetric sphincter tears. Ultrasound Obstet Gynecol. 2003;22:609–15.
30. Sultan AH. Anal incontinence after childbirth. Curr Opin Obstet Gynecol. 1997;9:320–4.
31. Sultan AH, Kamm MA, Talbot IC, Nicholls RJ, Bartram CI. Anal endosonography for identifying external sphincter defects confirmed histologically. Br J Surg. 1994;81:463–5.
32. Sultan AH, Johanson RB, Carter JE. Occult anal sphincter trauma following randomized forceps and vacuum delivery. Int J Gynecol Obstet. 1998;61:113–9.
33. Norderval S, Markskog A, Røssaak K, Vonen B. Correlation between anal sphincter defects and anal incontinence following obstetric sphincter tears: assessment using scoring systems for sonographic classification of defects. Ultrasound Obstet Gynecol. 2007;31(1):78–84.
34. Suarez L, Olivares F, Lobo E, et al. Anal endosonography in children with chronic constipation. Pediatr Radiol. 1999;29(1):28–30.
35. Nielsen MB, Pedersen JF. Changes in the anal sphincter with age. An endosonographic study. Acta Radiol. 1996;37(3 Pt 1):357–61.
36. Martin JE, Swash M, Kamm MA, et al. Myopathy of internal anal sphincter with polyglucosan inclusions. J Pathol. 1990;161(3):221–6.
37. Gillick J, Tazawa H, Puri P. Intestinal neuronal dysplasia: results of treatment in 33 patients. J Pediatr Surg. 2001;36(5):777–9.
38. Marshall M, Halligan S, Fotheringham T, et al. Predictive value of internal anal sphincter thickness for diagnosis of rectal intussusception in patients with solitary rectal ulcer. Br J Surg. 2002;89:1281–5.
39. Koh CE, Young CJ, Wright CM, et al. The internal anal sphincter in systemic sclerosis. Dis Colon Rectum. 2009;52(2):315–8.
40. Pascual I, García-Olmo D, Martínez-Puente C, et al. Ultrasound findings in spontaneous and postoperative anal pain. Rev Esp Enferm Dig. 2008;100(12):764–7.
41. Knowles AM, Knowles CH, Scott SM, et al. Effects of age and gender on three-dimensional endoanal ultrasonography measurements: development of normal ranges. Tech Coloproctol. 2008;12(4):323–9.
42. Lewicky-Gaupp C, Hamilton Q, Ashton-Miller J, et al. Anal sphincter structure and function relationships in aging and fecal incontinence. Am J Obstet Gynecol. 2009;200(5):559.
43. Kenefick NJ, Vaizey CJ, Malouf AJ, et al. Injectable silicone biomaterial for faecal incontinence due to internal anal sphincter dysfunction. Gut. 2002;51(2):225–8.
44. Nygaard IE, Shaw JM. Physical activity and the pelvic floor. Am J Obstet Gynecol. 2016;214(2):164–71.
45. Shobeiri SA, Leclaire E, Nihira MA, et al. Appearance of the levator ani muscle subdivisions in endovaginal three-dimensional ultrasonography. Obstet Gynecol. 2009;114(1):66–72.
46. Santoro GA, Wieczorek AP, Stankiewicz A, et al. High-resolution three-dimensional endovaginal ultrasonography in the assessment of pelvic floor anatomy: a preliminary study. Int Urogynecol J Pelvic Floor Dysfunct. 2009;20(10):1213–22.

47. Murad-Regadas SM, Bezerra LR, Silveira CR, et al. Anatomical and functional characteristics of the pelvic floor in nulliparous women submitted to three-dimensional endovaginal ultrasonography: case control study and evaluation of interobserver agreement. Rev Bras Ginecol Obstet. 2013;35(3):123–9.
48. Beer-Gabel M, Teshler M, Barzilai N, Lurie Y, Malnick S, Bass D, Zbar A. Dynamic transperineal ultrasound in the diagnosis of pelvic floor disorders: pilot study. Dis Colon Rectum. 2002;45(2):239–45; discussion 245–8.
49. Kołodziejczak M, Sudoł-Szopińska I, Santoro GA, Bielecki K, Wiączek A. Ultrasonographic evaluation of anal endometriosis: report of four cases. Tech Coloproctol. 2014;18(11):1099–104.
50. Albuquerque A. Endoanal ultrasonography in fecal incontinence: current and future perspectives. World J Gastrointest Endosc. 2015;7(6):575–81.

8 Echodefecography: Technique and Clinical Application

Sthela M. Murad-Regadas
and Francisco Sérgio Regadas

Introduction

In recent years, alternatives to defecography, such as dynamic magnetic resonance imaging (MRI) and dynamic ultrasonography, have been developed for the assessment of pelvic floor dysfunction, and the results were similar [1–8]. In addition, evaluation of the images has the advantage of showing the anal sphincter, the levator ani muscles, and the pelvic organs [6–10]. Studies utilizing dynamic ultrasonography to assess pelvic floor dysfunction related to obstructed defecation syndrome have been ongoing since the year 2000 [3]. Various techniques have been described using different approaches, and the results are similar when compared with defecography [3, 4, 6–9, 11]. Barthet et al. [3] tested a rigid linear endoanal probe with the patient in the lateral position and reported a good correlation with defecography. In 2004, Beer-Gabel et al. [4] published the technique of using dynamic transperineal ultrasound to evaluate pelvic floor dysfunction and compared it with defecography, demonstrating a high degree of concordance, followed by a series of published studies in the literature. Dietz et al. [5] described a technique for the assessment of pelvic organ descent with two-dimensional translabial ultrasound, measuring the distance between the base of the rectal ampulla and the lower margin of the symphysis pubis during the Valsalva maneuver. The technique was shown to correlate well with clinical measurements.

Recent advances in imaging technologies and development of ultrasound equipment with 3D acquisition have opened new possibilities for research on anal canal, rectum, and pelvic floor disorders [6, 8, 9–11]. The advantage of 3D with automatic scanning is the measurement of the length and thickness of the sphincter muscles without manual movement of the transducer, which may potentially be more comfortable.

Murad-Regadas et al. developed the echodefecography technique, a 3D dynamic anorectal ultrasonography technique using a 360° transducer, automatic scanning, and high frequencies for high-resolution images, to evaluate the anal canal and pelvic floor anatomy in patients with evacuation disorders affecting the posterior compartment (rectocele, intussusception, anismus), the middle-apical compartment (grade II or III sigmoidocele/enterocele), and the anterior compartment (cystocele). The authors established the technique including parameters and reproducible values [6, 9, 11].

S. M. Murad-Regadas (✉)
University Hospital, School of Medicine, Federal University of Ceara, Anorectal Physiology and Pelvic Floor Unit, Sao Carlos Hospital, Ceara, Brazil
e-mail: smregadas@hospitalsaocarlos.com.br

F. S. Regadas
School of Medicine, Federal University of Ceara, Ceara, Brazil

L. C. C. Oliveira (ed.), *Anorectal Physiology*, https://doi.org/10.1007/978-3-030-43811-1_8

Echodefecography Technique

Echodefecography is performed with a 3D ultrasound device (Pro Focus, endoprobe model 2052, B-K Medical®, Herlev, Denmark) with proximal-to-distal 6.0-cm automatic scans. By moving two crystals on the extremity of the transducer, axial and longitudinal images are merged into a single cube image, recorded and analyzed in multiple planes. Following a rectal enema, the patient is examined in the left lateral position.

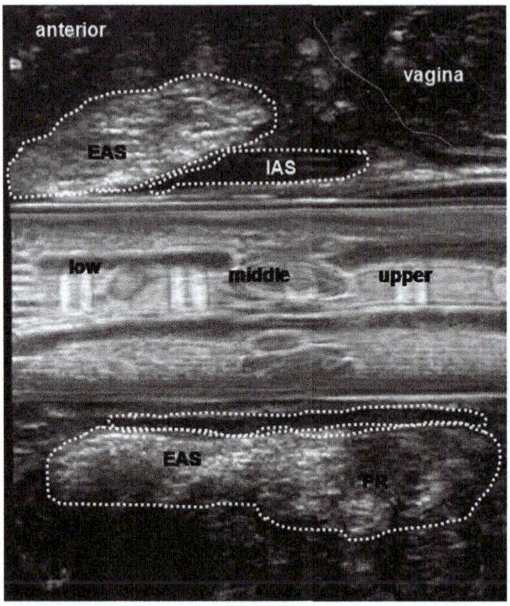

Fig. 8.1 Female anal canal anatomic configuration (sagittal plane). EAS, external anal sphincter; IAS, internal anal sphincter; PR, puborectalis muscle

The dynamic assessment consists of four automatic scans analyzed in the axial, sagittal, and, if necessary, in the oblique plane. The result of the exam depends on the degree of patient cooperation.

Scans 1, 2, and 4 utilize a slice width of 0.25 mm and lasts 50 seconds each.

Scan 3 lasts 30 seconds with a slice width of 0.35 mm.

Scan 1 (At Rest Position Without Gel)

The transducer is positioned at 4.0–5.0 cm from the anal margin to visualize the anatomic integrity of the anal sphincter muscles (Fig. 8.1) and to identify any occult defect (Fig. 8.2a, b). The anorectal angle is measured at rest. The angle is formed between a line traced along the internal border of the posterior external anal sphincter–puborectalis (EAS-PR) muscles

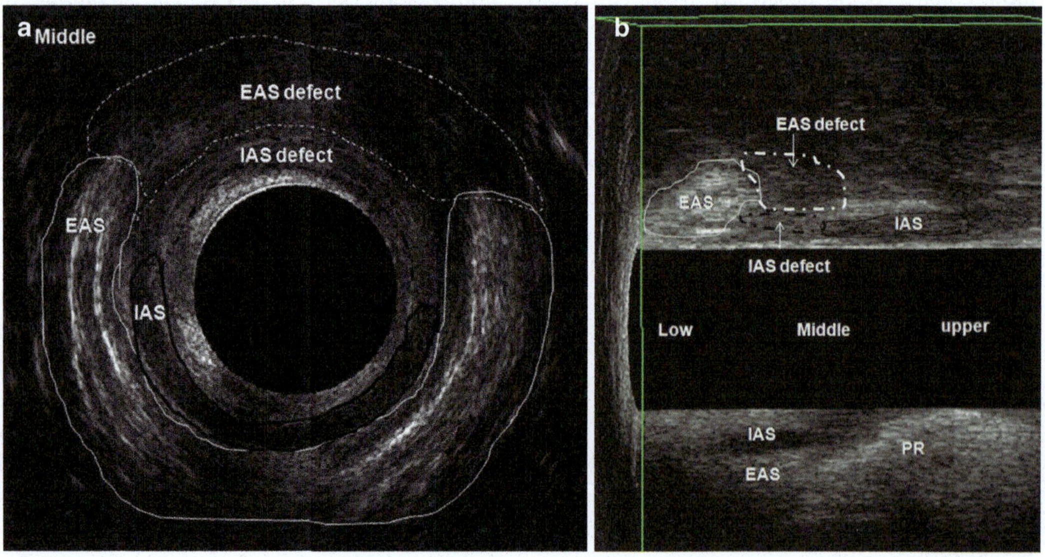

Fig. 8.2 Combined EAS and IAS defects after vaginal delivery. (**a**) Mid-anal canal: external and internal anal sphincter defects (dotted line) (axial plane). (**b**) The lesion compromised only the middle anal canal. Measurements of the length of the residual EAS and IAS (sagittal plane). EAS, external anal sphincter; IAS, internal anal sphincter EAS, external anal sphincter; IAS, internal anal sphincter; PR, puborectalis muscle

(1.5 cm) and a line traced perpendicular to the axis of the anal canal. The following scans are dynamic evaluations.

Scan 2 (At Rest – Straining – At Rest Without Gel)

The transducer is positioned at 6.0 cm from the anal verge. The patient is requested to keep at rest during the first 15 seconds, and then to maximally strain for 20 seconds, then to relax again, with the transducer following the movement. The purpose of this scan is to evaluate the movement of the PR-EAS muscles during straining in order to identify normal relaxation, non-relaxation, or paradoxical contraction (anismus). The result of the posterior EAS-PR muscles' position (represented by the angle size) is compared between scans 1 and 2. Normal relaxation is recorded if the angle increases by a minimum of 1 degree (Fig. 8.3), whereas paradoxical contraction (anismus) is recorded if the angle decreases by a minimum of 1 degree (Fig. 8.4). Non-relaxation is recorded if the angle changes less than 1 degree.

Scan 3: The Transducer Is Positioned Proximally to the PR (Anorectal Junction)

The scan starts with the patient at rest (3.0 seconds), followed by maximum straining with the transducer in fixed position (the transducer does not follow the descending muscles of the pelvic floor). Scanning continues distally until the PR muscle becomes visible.

Perineal descent is quantified by measuring the distance between the position of the proximal border of the PR at rest and the point to which it has been displaced by maximum straining (PR descent). Straining time is directly proportional to the distance of perineal descent.

Even with patients in the lateral position, the displacement of the PR muscle is easily visualized and quantified. Normal perineal descent during straining is defined as a difference in PR muscle position of ≤2.5 cm and perineal descent >2.5 cm (Fig. 8.5a, b). The normal range values were established by comparing EDF findings with DF.

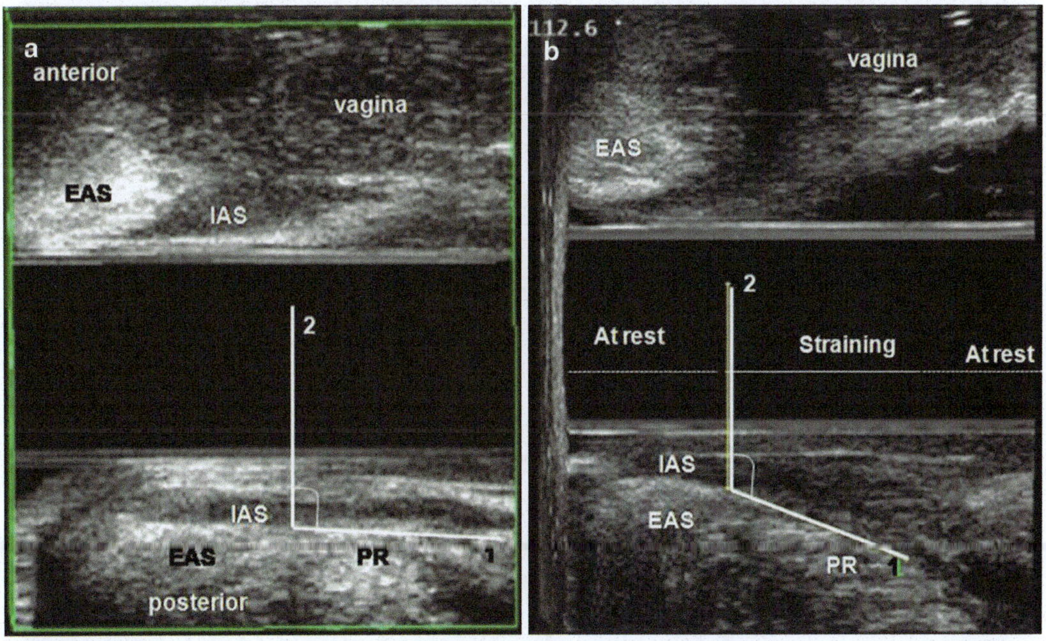

Fig. 8.3 Patient with normal relaxation. (sagittal plane). Line 1: a line traced along the internal border of the posterior EAS-PR muscles (1.5 cm). Line 2: a line traced perpendicular to the axis of the anal canal. (**a**) Angle measured at rest position (lines). (**b**) Increased angle during straining (lines). EAS , external anal sphincter; IAS, internal anal sphincter; PR , puborectalis muscle

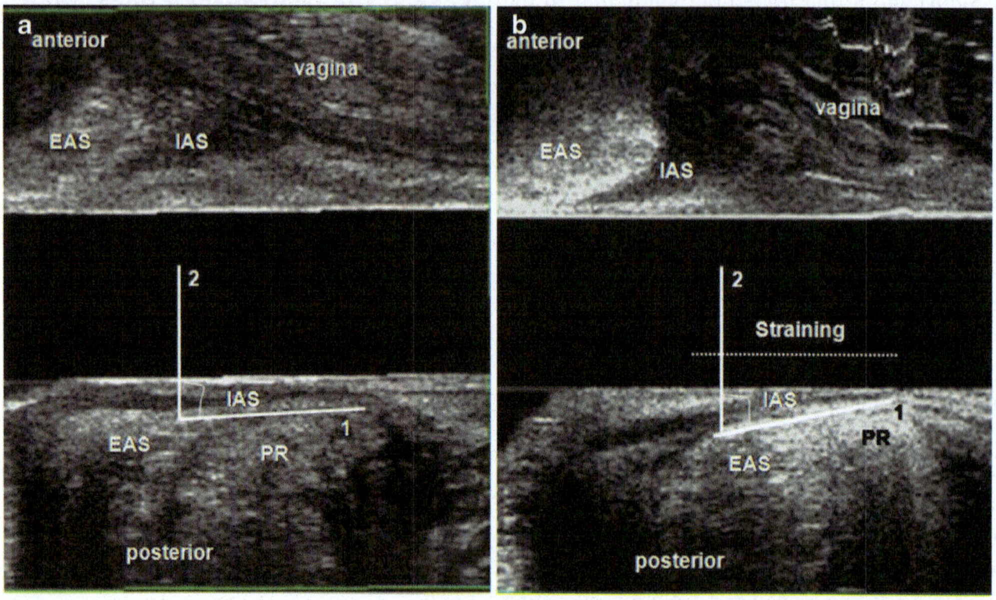

Fig. 8.4 Patient with *anismus* (sagittal plane). Line 1: a line traced along the internal border of the posterior EAS-PR muscles (1.5 cm). Line 2: a line traced perpendicular to the axis of the anal canal. (**a**) Angle measured at rest position (lines). (**b**) Decreased angle (*anismus*) during straining (lines). EAS , external anal sphincter; IAS, internal anal sphincter; PR , puborectalis muscle

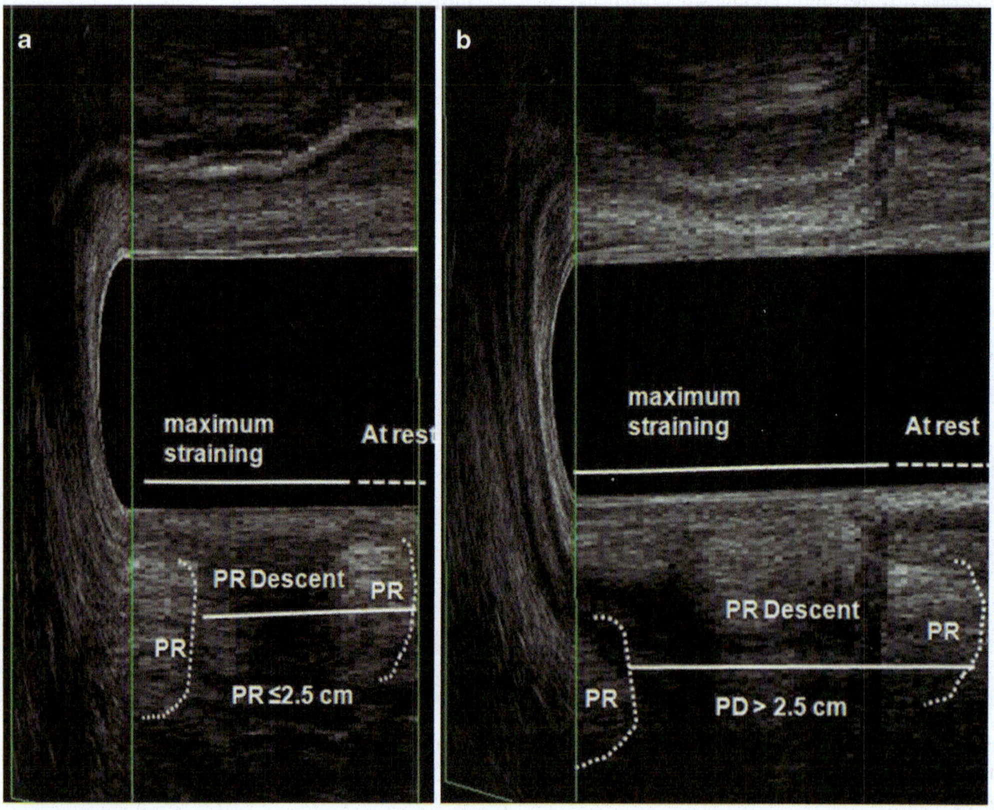

Fig. 8.5 Perineal descent measurement (sagittal plane). (**a**) PD ≤2.5 cm, normal perineal descent. (**b**) PD >2.5 cm. PD, puborectalis descent

Scan 4

Following injection of 120 mL ultrasound gel into the rectal ampulla, the transducer is positioned at 7.0 cm from the anal verge. The scanning sequence is the same as Scan 2 (at rest for 15 seconds, strain maximally for 20 seconds, then relax again, with the transducer following the movement). The purpose of the scan is to visualize and quantify all anatomical structures and functional disorders (rectocele, intussusception, grade II or III sigmoidocele/enterocele, and cystocele).

In normal patients, the posterior vaginal wall displaces the lower rectum and upper anal canal inferiorly and posteriorly but maintains a straight horizontal position during defecatory effort (Fig. 8.6). If rectocele is identified, it is classified as grade I (<6.0 mm), grade II (6.0–13.0 mm), or grade III (>13.0 mm). Measurements are calculated by first drawing two parallel horizontal lines along the posterior vaginal wall, with one line placed in the initial straining position and the other line drawn at the point of maximal straining. The distance between the two lines (vaginal wall positions) determines the size of the rectocele (Fig. 8.7a, b). Intussusception is clearly identified by observing the rectal wall layers protruding through the anorectal lumen. No classification is used to quantify intussusception (Fig. 8.8a, b). Grade II or III sigmoidocele/enterocele is recognized when the bowel bulges downward to the pelvis, between the posterior vagina and the anterior lower rectal wall (on the projection of the lower rectum and upper anal canal) (Fig. 8.9). Cystocele is identified using a reference line drawn perpendicular to the proximal margin of the puborectalis muscles and measured by a displacement of the bladder or bladder neck below the proximal margin of the PR ≥ 0.5 cm (Figs. 8.10, 8.11 and 8.12).

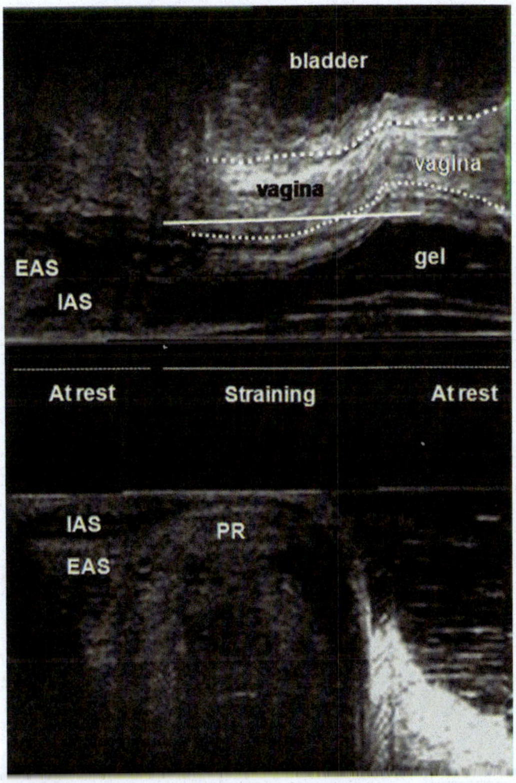

Fig. 8.6 Patient without rectocele using intrarectal gel (sagittal plane). The vagina maintains a straight horizontal position during defecatory effort. EAS, external anal sphincter; IAS, internal anal sphincter; PR, puborectalis muscle

In patients in whom a sling or mesh has been placed (such as a urethral sling), it is possible to identify its position in relation to the organ.

The main advantage of using ultrasound is the possibility to simultaneously evaluate the anatomy of the anal canal as well as any dynamic dysfunction. In patients with pelvic floor dysfunction with a previous history of vaginal delivery or anorectal surgery resulting in sphincter division, such as fistulotomy or sphincterotomy, ultrasonographic assessment may reveal occult defects. Similarly, in patients who underwent vaginal delivery, it is possible to evaluate levator ani defects with the same 360° rotating transducer when utilizing the endovaginal approach (see Chap. 7).

Echodefecography has been validated in a multicenter study of 86 women at 6 colorectal surgery centers in the United States, Brazil, and Venezuela [9]. A high degree of agreement was observed between echodefecography and conventional defecography in the diagnosis of anorectal disorders. In a recent study, the EDF was compared with dynamic translabial ultra-

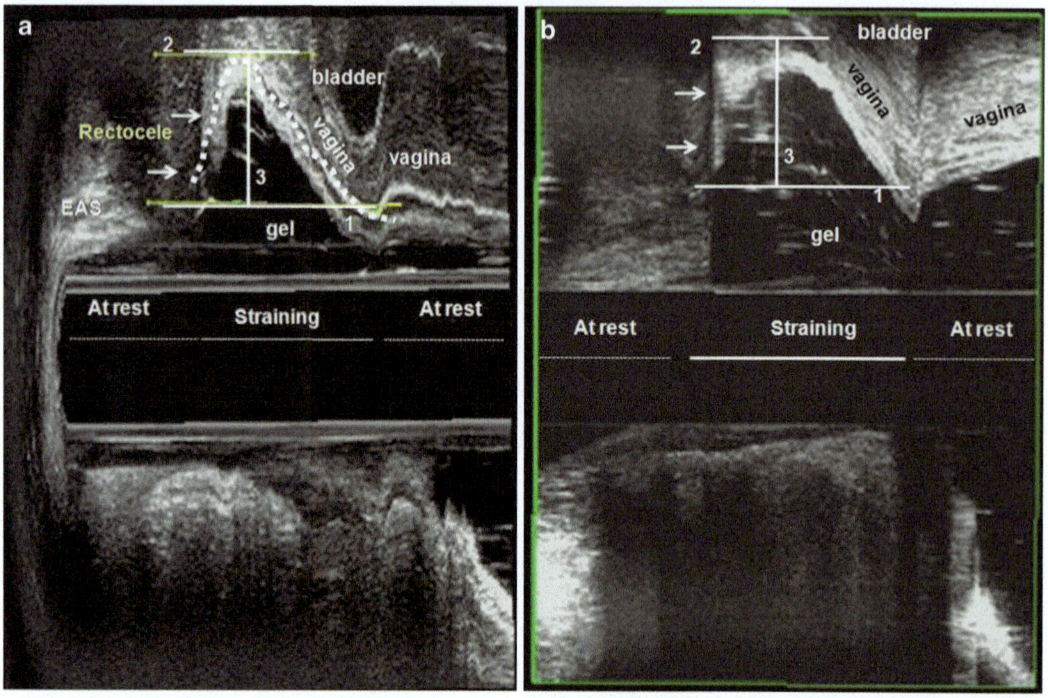

Fig. 8.7 Patient with rectocele (arrows) using intrarectal gel. Rectocele grade (sagittal plane) is measured by one line placed in the initial straining position (1) and the other line drawn at the point of maximal straining (2). The distance between the two lines (vaginal wall positions) determines the size of the rectocele (3). (**a**) Case 1. (**b**) Case 2, rectocele grade III. EAS , external anal sphincter; IAS, internal anal sphincter; PR, puborectalis muscle

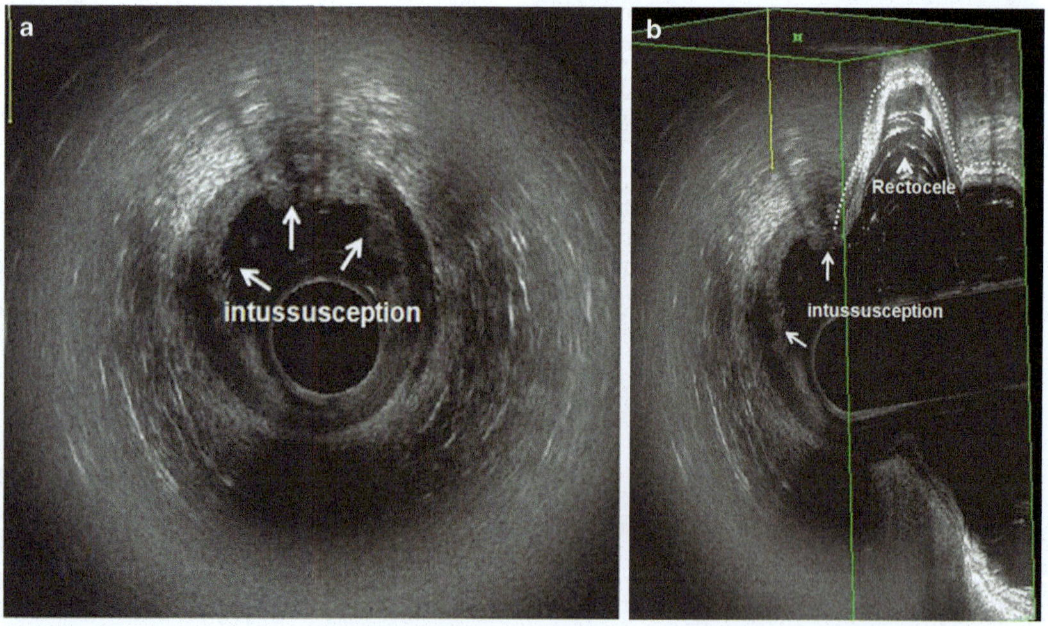

Fig. 8.8 Patient with grade III rectocele (arrow head) and intussusception (arrows) using intrarectal gel. (**a**) Anterior, right, and left lateral intussusception(arrows) (axial plane). (**b**) Grade III rectocele (arrow head) and intussusception (arrows) (sagittal with coronal plane). EAS, external anal sphincter; IAS, internal anal sphincter; PR , puborectalis muscle

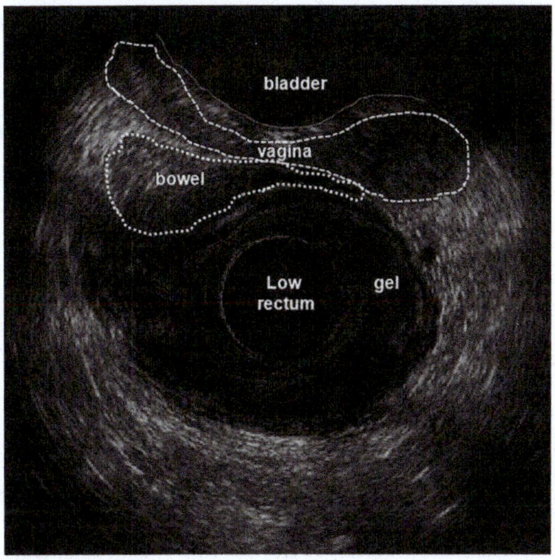

Fig. 8.9 Patient with III sigmoidocele/enterocele. The bowel bulges downward to the pelvis (axial plane)

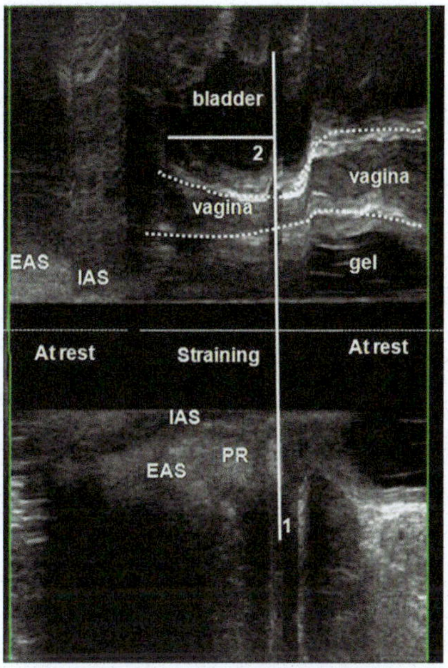

Fig. 8.10 Patient with cystocele and without rectocele. Cystocele measured by the displacement of the bladder below (line 2) the proximal margin of the PR (lines 1). EAS, external anal sphincter; IAS, internal anal sphincter; PR, puborectalis muscle

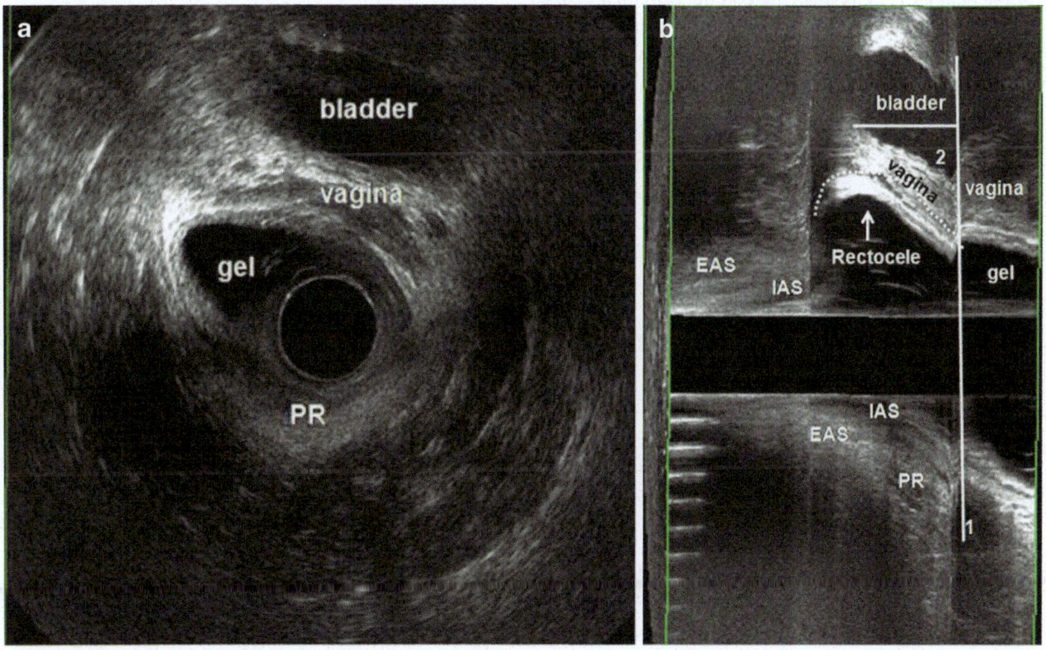

Fig. 8.11 Patient with cystocele and rectocele grade II (sagittal plane). (**a**) Axial plane – bladder below the PR muscles. (**b**) Cystocele measured by the displacement of the bladder neck below (lines 2) the proximal margin of the PR (line 1) and rectocele (arrow head). EAS, external anal sphincter; IAS, internal anal sphincter; PR , puborectalis muscle

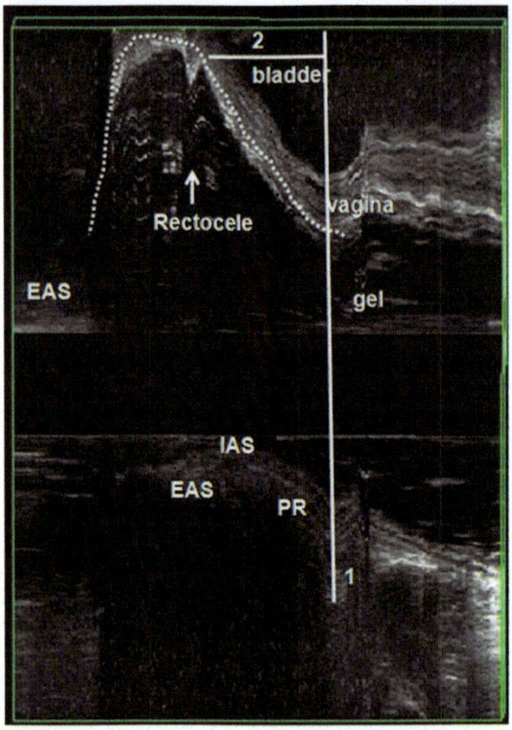

Fig. 8.12 Patient with cystocele and rectocele grade III (sagittal plane). Cystocele measured by the displacement of the bladder neck below (lines 2) the proximal margin of the PR (line 1) and rectocele (arrow head). EAS, external anal sphincter; IAS, internal anal sphincter; PR, puborectalis muscle

sound, and the results demonstrated good correlation for diagnosis of anismus, rectocele, and cystocele.

Summary

In conclusion, dynamic ultrasound scanning using echodefecography technique is a helpful tool in the evaluation of patients with symptoms of obstructed defecation syndrome. It has the advantage of concomitant evaluation of fecal incontinence and genital prolapse. This technique clearly shows the anatomical structures with high spatial and optimal resolution. Finally, this is a valuable option for the assessment of constipated patients as well as for patients with pelvic floor dysfunction. The procedure is simple, quick, can be performed in the office setting with the advantage of being inexpensive, and well tolerated by patients without exposure to radiation.

References

1. Mahieu P, Pringot J, Bodart P. Defecography: I. Description of a new procedure and results in normal patients. Gastrointest Radiol. 1984;9:247–51.
2. Felt-Bersna RJ, Luth WJ, Janssen JJ, Meuwissen SG. Defecography in patients with anorectal disorders. Which findings are clinically relevant? Dis Colon Rectum. 1990;33:277–84.
3. Barthet M, Portier F, Heyries L. Dynamic anal endosonography may challenge defecography for assessing dynamic anorectal disorders: results of a prospective pilot study. Endoscopy. 2000;32:300–5.
4. Beer-Gabel M, Teshler M, Schechtman E, Zbar AP. Dynamic transperineal ultrasound vs. defecography in patients with evacuatory difficulty: a pilot study. Int J Color Dis. 2004;19:60–7.
5. Dietz HP, Steensma AB. Posterior compartment prolapse on two-dimensional and three-dimensional pelvic floor ultrasound: the distinction between true rectocele, perineal hypermobility and enterocele. Ultrasound Obstet Gynecol. 2005;26:73–7.
6. Murad-Regadas SM, Regadas FSP, Rodrigues LV, et al. A novel three-dimensional dynamic anorectal ultrasonography technique (echodefecography) to assess obstructed defecation, a comparison with defecography. Surg Endosc. 2008;22:974–9.
7. Perniola G, Shek C, Chong CC, et al. Defecation proctography and translabial ultrasound in the investigation of defecatory disorders. Ultrasound Obstet Gynecol. 2008;31:567–71.
8. Steensma AB, Oom DMJ, Burger CW, Schouten WR. Assessment of posterior compartment prolapse: a comparison of evacuation proctography and 3D transperineal ultrasound. Color Dis. 2010;12:533–9.
9. Regadas FSP, Haas EM, Jorge JM, et al. Prospective multicenter trial comparing echodefecography with defecography in the assessment of anorectal dysfunctions in patients with obstructed defecation. Dis Colon Rectum. 2011;54:686–92.
10. Santoro GA, Wieczorek AP, Dietz HP, et al. State of the art: an integrated approach to pelvic floor ultrasonography. Ultrasound Obstet Gynecol. 2011;37:381–96.
11. Murad-Regadas SM, Soares GS, Regadas FSP, et al. A novel three-dimensional dynamic anorectal ultrasonography technique for the assessment of perineal descent, compared with defaecography. Color Dis. 2012;14:740–7.

Cinedefecography

9

Carlos Walter Sobrado Jr,
Isaac José Felippe Corrêa Neto,
and Lucas Faraco Sobrado

Introduction

The mechanisms that control normal defecation and anal continence are complex and were discussed in previous chapters. Disorders involving the pelvic floor may be manifested by a variety of symptoms such as anal incontinence, constipation, obstructed defecation, pelvic pain, and/or the sensation of incomplete evacuation [1]. It is estimated that pelvic floor dysfunctions affect more than 15% of multiparous women [2]. Within the 17% of patients that present chronic constipation, 50% present with symptoms of obstructed defecation [3, 4].

Pelvic floor dysfunction (PFD) is a significant cause of morbidity, with a negative impact in the patient's quality of life, and appear to have increased in frequency in recent years [5]. Although a medical history and physical examination are required to provide an appropriate and detailed assessment, it is essential to include an imaging technique when preparing a more accurate diagnosis [6, 7]. Anorectal manometry, static and dynamic endoanal ultrasound, cinedefecography, electromyography, and pelvic MRI may be used for this purpose.

The first reports of radiological studies of pelvic dynamics during defecation were provided by Walldén [8] in 1952. However, it was only after the study of Mahieu et al. [9] in 1984 that the examination gained the interest of the global medical community. Defecography is a dynamic radiological method for studying defecation. It is performed through the expulsion of a radio-opaque material, which provides real-time images of the morphological and functional changes of the pelvis and anorectal segment [10–13] based on anatomical changes and relationships with neighboring organs and bone components of the pelvis [14]. Therefore, it is a valuable method for the physiological study of pelvic dynamics and colorectal disorders such as dyssynergic defecation, constipation, fecal incontinence, anal pain, and tenesmus.

Cinedefecography is primarily used in cases of chronic constipation and is particularly relevant in situations where there are obstructed defecation symptoms, including straining, sensation of incomplete evacuation, and necessity of digitation. It is therefore an adjunctive examination indicated in defecation disorders that result in functional and anatomical changes in the pelvic compartments, producing rectocele, intussusception, enterocele, perineal descent,

C. W. Sobrado Jr (✉) · I. J. F. C. Neto · L. F. Sobrado
Department of Gastroenterology, Division of Colorectal Surgery, University of São Paulo, São Paulo, Brazil
e-mail: sobrado@iconet.com.br

L. C. C. Oliveira (ed.), *Anorectal Physiology*, https://doi.org/10.1007/978-3-030-43811-1_9

absence of sphincter relaxation, or paradoxical contraction of the voluntary sphincter muscles (anismus) [15]. However, while pelvic dysfunctions are quite common, the exact identification of the pathophysiological mechanism of patients who will actually benefit from surgical treatment is quite difficult, even when using imaging examinations [16]. In this regard, Palit et al. [17] demonstrated the presence of rectocele in 93% of their asymptomatic female volunteers. Moreover, intussusception was present in 20% of their volunteers, all of whom were asymptomatic.

In order to confirm that those abnormalities can be present in patients without symptoms, Sobrado et al performed videodefecography on 20 healthy volunteers and identified morphological changes in the pelvis (intussusception, sigmoidocele, paradoxical puborectalis muscle contraction, rectocele) during straining in 10 volunteers (50%) [18].

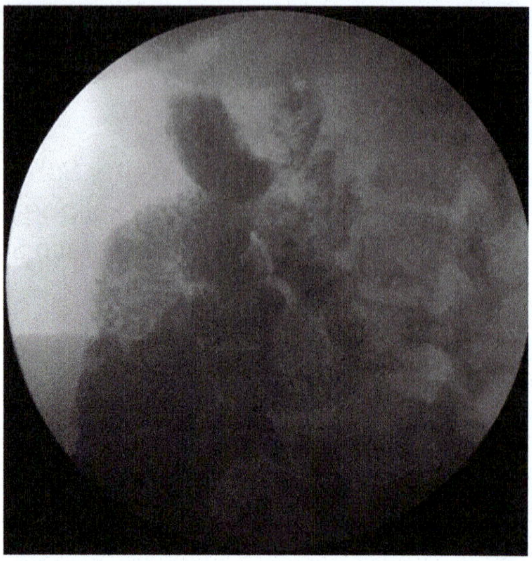

Fig. 9.1 Contrast phase in small bowel loops

Technique

Conventional

Although different techniques can be employed within each Radiology Department, the majority utilize the defecography method standardized by Mahieu et al. in 1984 [9]. In all of these techniques, care must be taken to make the examination as physiological as possible, avoiding embarrassment and reducing the patient's exposure to radiation.

Prior to the examination, a rectal preparation (minimum of 30 minutes before) should be performed with an enema. In addition, oral barium contrast (150–200 ml) must be administered 1 hour before the defecation simulation to contrast small intestine loops on the pelvis [6] (Fig. 9.1). The patient is placed in the left lateral decubitus position, and 50 ml of liquid barium followed by 200 ml of barium paste are introduced into the rectal ampulla [10, 11, 19]. The use of a vaginal tampon with iodinated contrast as well as any intravesical contrast are optional. However, the latter is seldom used due to the risk of urinary tract infection [20].

The radiological examination table is then raised to 90° so that the patient adopts a sitting position on the radiolucent seat (Fig. 9.2).

Once the patient is in this position, the analysis proceeds, in four examination phases:

- Resting phase (Fig. 9.3)
- Voluntary contraction of the pelvic floor muscles: anal canal closure and puborectalis muscle contraction (Fig. 9.4)
- Defecation phase (Fig. 9.5)
- Post-defecation phase: recorded 1 minute after total evacuation of the rectal contents

Computed Videodefecography

Computed videodefecography is performed in the same manner as the conventional technique, except that it dispenses static radiographs. First, a panoramic evaluation is performed of the pelvis at rest, with emphasis on the identification of bone repairs, i.e., the pubis and the coccyx. It is known that obtaining good-quality images of these regions is of great importance in facilitating the tracing of reference lines.

To compare the two described techniques, Sobrado et al. [21] analyzed images obtained through dynamic radiographs and videos in ten asymptomatic volunteers and demonstrated that there was no significant difference between the two techniques in terms of the analyzed parameters. However, Sobrado et al. realized that exposure to radiation, calculated by thermoluminescent dosimeters, was significantly lower in the group undergoing computed videodefecography [22]. Therefore,

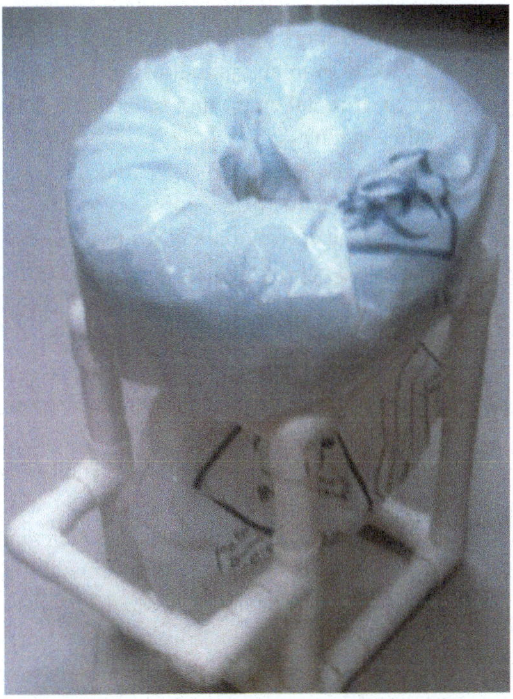

Fig. 9.2 Radiolucent seat

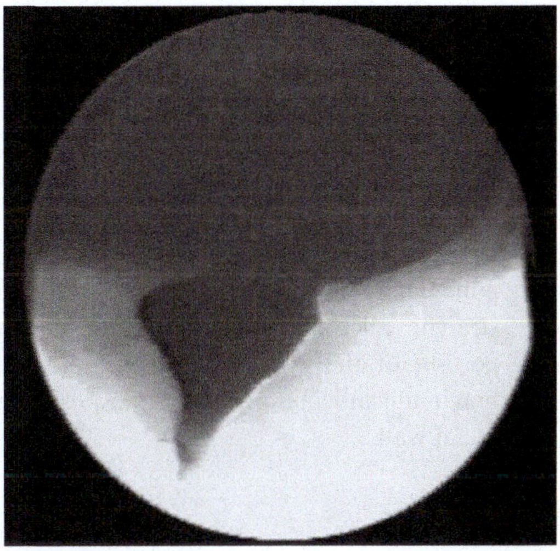

Fig. 9.4 Cinedefecography squeeze phase

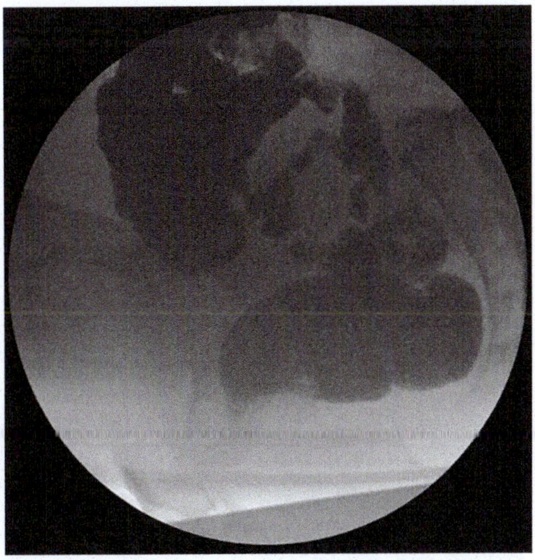

Fig. 9.3 Cinedefecography: resting phase

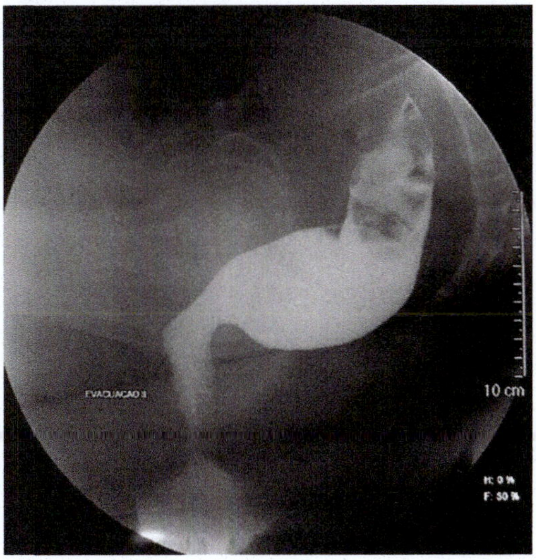

Fig. 9.5 Cinedefecography defecation phase

they concluded that priority should be given to fluoroscopy rather than static radiographs.

Examination

After completion of the defecography, whether by static radiographs or filming technique, the main parameters must be analyzed in the examination (Fig. 9.6):

1. Rectal morphology and position: rectal contour, positioning, and dimensions.
2. Anorectal angle (ARA): angle formed by a straight line passing through the anal canal axis and another that passes through the posterior rectal wall.
3. Puborectalis muscle length (PRL): the puborectalis muscle can be measured by a straight line extending from the lower portion of the pubic symphysis and the maximum inflection point in the posterior rectal wall.
4. Puborectalis muscle relaxation: evaluated based on the presence or absence of its impression on the posterior rectal face during the act of defecation. Indirectly, non-relaxation of the puborectalis muscle can also be verified by excessive post-defecation residue.
5. Anal canal length: distance between the anal orifice and anorectal junction.
6. Degree of rectal emptying: reflects the defecation capacity.
7. Perineal descent: variation in the anorectal junction relative to the pubococcygeal line (PCL) in the defecation and resting phases.
8. Pubococcygeal line: distance between the bottom edge of the pubic symphysis and coccyx and represents the pelvic floor.
9. Presence of anismus (Fig. 9.7), rectocele (Fig. 9.8), sigmoidocele and intussusception (Fig. 9.9).
10. Time elapsed for elimination of rectal contrast: there is little consensus, but defecation occurs in full in approximately 1 minute and may last up to 2 minutes to eliminate any minor or more compact stools [20]. However, some consider anismus to be present when the time to eliminate the rectal contrast exceeds 30 seconds [23].
11. Number of contractions necessary for complete evacuation [18].

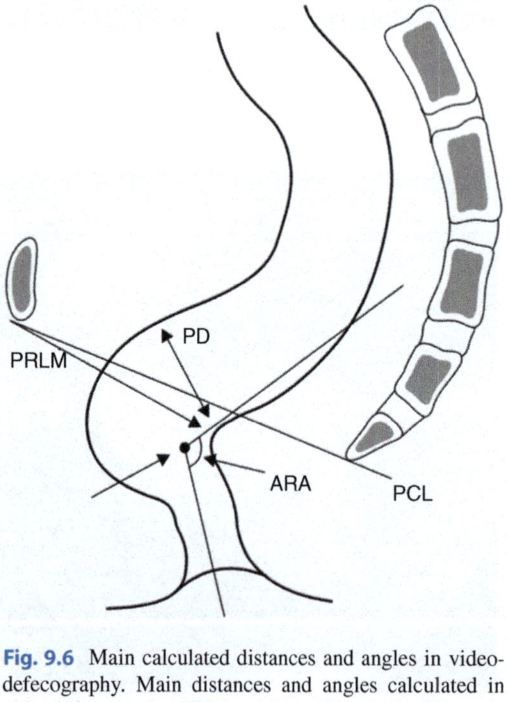

Fig. 9.6 Main calculated distances and angles in videodefecography. Main distances and angles calculated in videodefecography. PRM, puborectalis muscle; PD, perineal descent; ARJ, anorectal junction; PCL, pubococcygeal line; ARA, anorectal angle

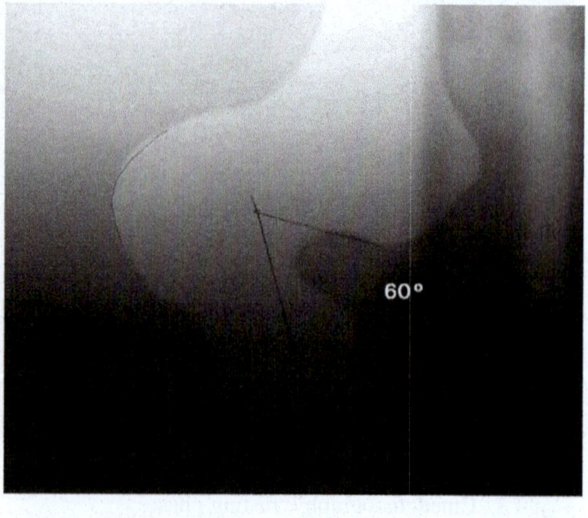

Fig. 9.7 Anismus

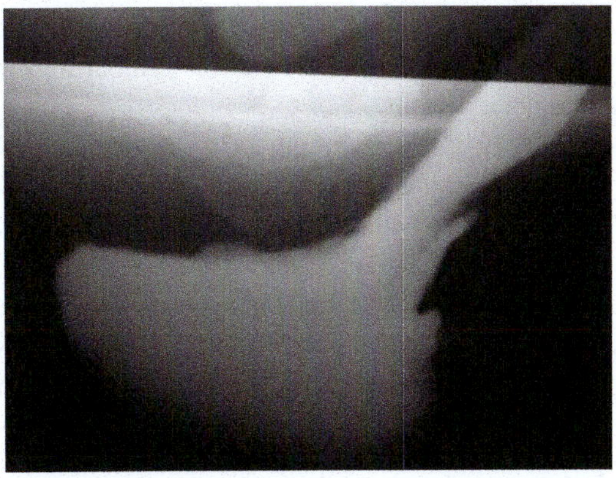

Fig. 9.8 Anterior rectocele

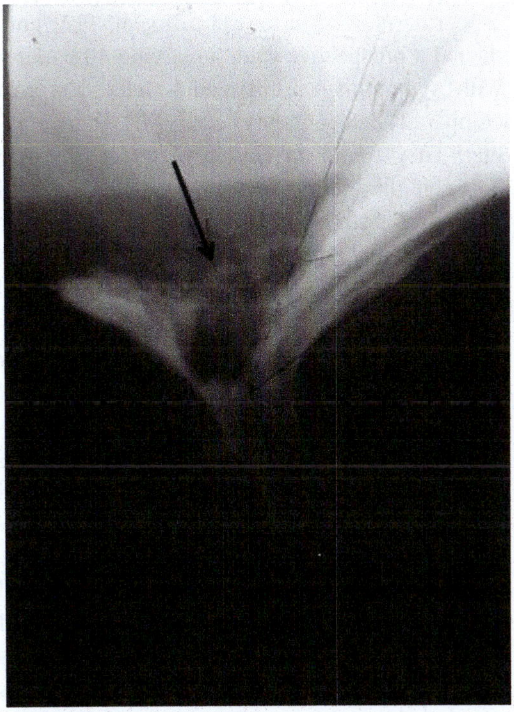

Fig. 9.9 Rectal intussusception

Literature Review

The study of dynamic defecation has fundamental importance in the evaluation and treatment of the constipated patient, especially those with obstructed defecation symptoms, i.e., those with symptoms of straining, sensation of incomplete defecation, digital maneuvers, vaginal bulging, and tenesmus. In addition, an in-depth and adjunctive study is also required for the population of patients with chronic constipation that fails to improve with hygiene, including dietary and behavioral measures and the use of laxatives. In this patient group, conducting an adjunctive investigation that evaluates the defecation mechanism plays a key role. Cinedefecography, echodefecography, and MR defecography are available methods utilized for this purpose.

Comparison among these methods reveals that cinedefecography is a minimally invasive procedure that is safe and technically simple and, in addition, is performed in the physiological defecation position. However, it has the disadvantage of not clearly showing the anatomical structures involved in pelvic floor disorders; it is uncomfortable and exposes the patient to radiation [15], which should not exceed 5 minutes [19].

In contrast, MR defecography evaluates multiple pelvic compartments and relationships between the organs, without radiation exposure [24]. However, it is a quite costly method with few specialized centers and, in addition, is performed in the supine position, which is not the physiological position adopted in defecography [6, 25]. Echodefecography, as demonstrated by Murad-Regadas et al. [15], enables static and dynamic evaluations of the anorectal and pelvic floor anatomy, identifying the different positions of the anatomical anorectal structures involved in defecation. It also does not expose the patient to radiation. However, it has disadvantages, such as cost, being examiner-dependent, the presence of the probe inside the anal canal, and the fact that the examination position is in the left lateral decubitus.

Comparing the examinations, Regadas et al. [26] analyzed 86 female patients suffering from obstructive defecation and demonstrated agreement between echodefecography and cinedefecography in the diagnosis of rectocele (Kappa = 0.61; 95% CI), rectal intussusception (Kappa = 0.79, 95% CI), anismus (Kappa = 0.61, 95% CI), and grade III enterocele (Kappa = 0.87, 95% CI).

Based on a literature survey, Reginelli et al. [6] demonstrated the superiority of cinedefecography over MR defecography in detecting enteroceles, sigmoidoceles, and omentoceles. Similar to Bertschinger et al. [27], they believed that this finding can be attributed to the supine position adopted in MR defecography, as verification of these pelvic floor hernias is more prominent during the increase in intra-abdominal pressure in the sitting position. Likewise, Faccioli et al. [13] found that defecography has the highest accuracy among adjunctive pelvic floor examinations in the diagnosis of rectal prolapse, rectal intussusception, and enterocele.

Kumar et al. [11] note that dynamic magnetic resonance imaging of the pelvic floor is especially well indicated in patients with multicompartmental pelvic disorder involvement and in those who have already undergone surgery or previous pelvic repair.

In a recent study, Pilkington et al. [28] compared videodefecography with MR defecography in the same individual to establish whether there were differences between the examinations in terms of clinically relevant findings. Their study included 42 patients (38 women), with a mean age of 59 (37–76) years. Rectocele was an extremely common finding in both examinations, and there was substantial agreement between the methods (Kappa = 0.69).

Anismus was found in 12 patients (29%) using cinedefecography and in 18 (43%) with the use of MR defecography, demonstrating a moderate agreement between the examinations (Kappa = 0.493) and a high correlation in terms of detecting enterocele (Kappa = 0.69). In turn, rectal intussusception was diagnosed in 35 patients (83.3%) using defecography and in 26 (61.9%) with the aid of MR defecography, showing low agreement between the methods (Kappa = 0.209). Furthermore, there was no significant difference between the two examinations regarding anorectal angle measurement.

In the same study, 52% of patients reported that they performed the usual defecation movement during the two examinations, 16% only during defecography and 12% only during MR defecography, and 20% reported that they did not experience their usual excretion movement in either examination. When comparing patient preference, 62% stated preferring MR defecography. The main reported reason for this response was that MR defecography is less embarrassing than defecography.

Similarly, Foti et al. [29] evaluated 19 obstructed defecation patients using videodefecography and MR defecography and concluded that for the evaluation of an isolated compartment, defecography is still the gold standard. However, when evaluating the various pelvic compartments and the interactions between organs, especially in females and multiparous patients, MR defecography is an important adjunct examination in terms of appropriate patient selection and treatment and reduces postoperative recurrences of pelvic floor disorders [30], as argued by Pisano et al. [31].

Furthermore, Martin-Martin et al. [32] conducted a prospective study involving 40 patients with symptoms of obstructed defecation that compared defecography with MR defecography. They found an almost perfect correlation between the examinations in the diagnosis of anismus, substantial agreement in cases of intussusception and grade III rectocele, and low agreement in cases of grade I rectocele and perineal descent. Rectocele was the most common diagnosis in both examinations. Video defecography is a dynamic investigation which can influence surgical decision-making in constipated patients. In order to assess the inter and intraobserver variability in video defecography, Pfeifer et al [33] conducted a study with four independent observers with the same training, guidelines, and standards. Two of these four observers were blinded to the patient's history and they all reviewed 100 randomly sequenced videodefecographies performed in constipated patients.The authors concluded that the overall accuracy of the issues discussed in this study was 83.3%, confirming that videodefecography is a reliable tool in clinical decision-making in constipated patients.

Summary

Cinedefecography, especially its computerized version, plays an important role in the investigation and treatment of constipated patients with symptoms of obstructed defecation. It has been recognized as one of the best dynamic modalities of evaluation of the defecation process over the past many years. Despite being less expensive, it has been replaced by MRI defecography because of the availability of MRI in Radiology Departments. Nevertheless, regardless of the method utilized, complete and adequate clinical evaluation should be performed before referring patients to any of the dynamic methods of rectal evacuation.

References

1. Koughnett JAMV, Silva G. Anorectal physiology and testing. Gastroenterol Clin N Am. 2013;42:713–28.
2. Roos JE, Weishaupt D, Wildermuth S, et al. Experience of 4 years with open MR defecography: pictorial review of anorectal anatomy and disease. Radiographics. 2002;22:817–32.
3. Andomanakos N, Skandalakis P, Troupis T, Filippou D. Constipation of anorectal outlet obstruction: pathophysiology, evaluation and management. J Gastroenterol Hepatol. 2006;21:638–46.
4. Choung R, Locke G III, Schleck C, Zinsmeister A, Talley N. Cumulative incidence of chronic constipation: a population-based study 1988–2003. Aliment Pharmacol Ther. 2007;26:1521–8.
5. McNevin MS. Overview of pelvic floor disorders. Surg Clin N Am. 2010;90:195–205.
6. Reginelli A, Grezia GD, Gatta G, Iacobellis F, Rossi C, Giganti M, Coppolino F, Brunese L. Role of conventional radiology and MRi defecography of pelvic floor hernias. BMC Surg. 2013;13(Suppl 2):S53.
7. Reginelli A, Pezzullo MG, Scaglione M, Scialpi M, Brunese L, Grassi R. Gastrointestinal disorders in elderly patients. Radiol Clin N Am. 2008;46:755–71.
8. Walldén L. Defecation block in cases of deep rectogenital pouch. Acta Chir Scand. 1952;165:1–121.
9. Mahieu P, Pringot J, Bodart P. Defecography: I. Description of a new procedure and results in normal patients. Gastrointest Radiol. 1984;9:247–51.
10. Sobrado CW, Pires CEF, Amaro E, Cerri GG, Habr-Gama A, Kiss DR. Videodefecografia: aspectos técnicos atuais. Radiol Bras. 2004;37(4):283–5.
11. Kumar S, Sharma P, Andreisek G. Does conventional defecography has a role to play in evaluation of evacuatory disorders in Indian population? Indian J Radiol Imaging. 2013;23(1):92–6.
12. Ganeshan A, Anderson EM, Upponi S, et al. Imaging of obstructed defecation. Clin Radiol. 2008;63:18–26.
13. Faccioli N, Comai A, Mainardi P, Perandini S, Moore F, Pozzi R. Defecography: a practical approach. Diagn Interv Radiol. 2010;16:209–16.
14. Seong MK, Kim TW. Significance of defecographic parameters in diagnosing pelvic floor dyssynergia. J Korean Surg Soc. 2013;84:225–30.
15. Murad-Regadas SM, Regadas FSP, Rodrigues LV, Escalante RD, Silva FRS, Lima DMR, et al. Ecodefecografia Tridimensional Dinâmica. Nova Técnica para Avaliação da Síndrome da Defecação Obstruída (SDO). Rev bras Coloproct. 2006;26(2):168–77.
16. Kenton K, Mueller ER. The global burden of female pelvic floor disorders. BJU Int. 2006;98(Suppl 1):1–5.
17. Palit S, Bhan C, Lunniss PJ, Boyle DJ, Gladman MA, Knowles CH, Scott SM. Evacuation proctography: a reappraisal of normal variability. Color Dis. 2014;16:538–46.
18. Sobrado CW. Contribuição da videodefecografia dinâmica computadorizada no estudo de doentes submetidos a graciloplastia. Tese (Doutorado). Faculdade de Medicina da Universidade de São Paulo, 1999.
19. Jorge JMN, Habr-Gama A, Wexner SD. Clinical applications and techniques of cinedefecography. Am J Surg. 2001;182:93–101.
20. Sobrado CW, Nadal SR, Sousa AFS Jr. Manual de Fisiologia Anorretal e Assoalho Pélvico. Office Editora. 2012;1:49–55.
21. Sobrado CW, Pires CEF, Araújo SEA, Amaro E, Habr-Gama A, Kiss DR. Computerized videodefecography versus defecography: do we need radiographs? Sao Paulo Med J. 2005;123(3):105–7.
22. Sobrado Júnior CW, Pires CEF, Araújo SEA, Lopes RMG, Habr-Gama A, Kiss DR. Dose de radiação na defecografia e na videodefecografia computadorizada. Rev bras Coloproct. 2003;23(1):20–4.
23. Hompes R, Harmston C, Wijffels N, Jones OM, Cunningham C, Lindsey I. Excellent response rate of anismus to botulinum toxin if rectal prolapse misdiagnosed as anismus ('pseudoanismus') is excluded. Color Dis. 2012;14:224–30.
24. O'Brien AS. Defecografia por Resonancia Magnética. Revista Chilena de Radiología. 2010;16(1):11–6.
25. Noelting J, Bharucha AE, Lake DS, Manduca A, Fletcher JG, Riederer SJ, et al. Semi-automated vectorial analysis of anorectal motion by magnetic resonance defecography in healthy subjects and fecal incontinence. Neurogastroenterol Motil. 2012;24:e467–75.
26. Regadas FS, Hass EM, Abbas MA, Marcio Jorge J, Habr-Gama A, Sands D, et al. Prospective multicenter trial comparing echodefecography with defecography in the assessment of anorectal dysfunction in patients with obstructed defecation. Dis Colon Rectum. 2011;54(6):686–92.
27. Bertschinger KM, Hetzer FH, Roos JE, Treiber K, Marincek B, Hilfiker PR. Dynamic MR imaging of the pelvic floor performed with patient sitting in an open magnet unit versus with patient supine in a closed-magnet unit. Radiology. 2002;223:501–8.
28. Pilkington SA, Nugent KP, Brenner J, Harris S, Clarke A, Lamparelli M, et al. Barium proctography vs magnetic resonance proctography for pelvic floor disorders: a comparative study. Color Dis. 2012;14:1224–30.
29. Foti PV, Farina R, Riva G, Coronella M, Fisichella E, Palmucci S, et al. Pelvic floor imaging: comparison

between magnetic resonance imaging and conventional defecography in studying outlet obstruction syndrome. Radiol Med. 2013;118:23–39.

30. Li M, Jiang T, Peng P, Yang XQ, Wang WC. Association of compartment defects in anorectal and pelvic floor dysfunction with female outlet obstruction constipation (OOC) by dynamic MR defecography. Eur Rev Med Pharmacol Sci. 2015;19:1407–15.
31. Pisano U, Irvine L, Szczachor J, Jawad A, MacLeods A, Lima M. Anismus, physiology, radiology: is it time for some pragmatism? A comparative study of radiological and anorectal physiology findings in patients with Anismus. Ann Coloproctol. 2016;32(5):170–4.
32. Martin-Martin GP, Garcia-Armengol J, Roig-Vila JV, Matinez-Sanjua V, Minguez-Perez M, et al. Magnetic resonance defecography versus videodefecography in the study of obstructed defecation syndrome: is videodefecography still the test of choice after 50 years? Tech Coloproctol. 2017;21(10):795–802.
33. Pfeifer J, Oliveira L, Park UC, Gonzalez A, Agachan F, Wexner SD. Are interpretations of video defecographies reliable and reproducible?. Int J Colorectal Dis. 1997;12(2):67–72.

MRI Defecography

10

Alice Brandão, Anelise Oliveira,
and Lucia Camara Castro Oliveira

The anatomy of the pelvic floor is complex and is generally divided into three compartments:

- Anterior: bladder, urethra, and prostate in men
- Middle: vagina, uterus and cervix in women
- Posterior: rectum and anus

Patients with pelvic floor dysfunction may present different symptomatology: urinary symptoms such as urinary urgency and incontinence, dyspareunia, and bowel dysfunction such as fecal incontinence, obstructed defecation, and rectal prolapse [1].

The use of magnetic resonance imaging (MRI) for the evaluation of pelvic floor disorders has allowed a better understanding of different pathologies associated with pelvic floor dysfunction [2]. Imaging correlation with anatomical findings can help in the treatment decision process.

MRI defecography, three-dimensional (3D) endoanal ultrasound, and cinedefecography are imaging methods that complement each other and, when well indicated, are essential tools for demonstrating that the pelvic organs function as a unit: the absence of an organ may have consequences on neighboring structures.

Dynamic assessment of defecation is clearly demonstrated in these three modalities, even when performed in the decubitus position. The results are comparable, and in this chapter, we discuss the most important findings of MRI defecography in the assessment of pelvic floor disorders.

The great advantage of this method over cinedefecography is the absence of radiation and its long-term consequences, as well as the direct and simultaneous identification of the pelvic floor and sphincter muscles. With respect to echodefecography, it also allows the evaluation of the dynamics of the anterior compartment, thus providing more complete information in a single examination.

MRI Defecography

MRI defecography is a new imaging modality of the pelvis that provides excellent evaluation of the entire pelvic floor and pelvic compartments, both at rest and dynamically [3]. This imaging method can evaluate the opening of the anal canal and the anorectal angle during contraction of the anorectal muscles and during evacuation movement, quantifying the elimination of the gel injected into the rectum as a contrast medium. The rectal wall can also be assessed, and changes

A. Brandão
Fonte Imagem Medicina Diagnóstica,
Rio de Janeiro, Brazil

A. Oliveira
Diimagem Grupo Alliar, Campo Grande, MS, Brazil

L. C. C. Oliveira (✉)
Director Department of Anorectal Physiology of Rio de Janeiro and CEPEMED, Rio de Janeiro, Brazil

L. C. C. Oliveira (ed.), *Anorectal Physiology*, https://doi.org/10.1007/978-3-030-43811-1_10

such as rectoceles and intussusception can be clearly identified. This global evaluation of the pelvic floor is crucial in the assessment of many complex pathologies [4, 5].

Technical Aspects

MRI defecography is usually available in closed equipment systems with configuration for the acquisition of images with the patient lying down and with the legs flexed. However, it can also be performed in open systems with configuration for acquisition of images with the patient sitting or in a vertical position. Unfortunately, open systems are less available.

Patients are prepared with a rectal enema at least 2 hours prior to evaluation. Approximately 250 ml of gel is injected into the rectal ampulla to simulate the presence of feces and enable evaluation of herniations and perineal descent.

The three compartments of the pelvis can be identified before and after administration of the gel into the rectal ampulla, with no need for further opacification of the vagina or bladder, since soft tissues provide excellent contrast during MRI acquisition (Fig. 10.1).

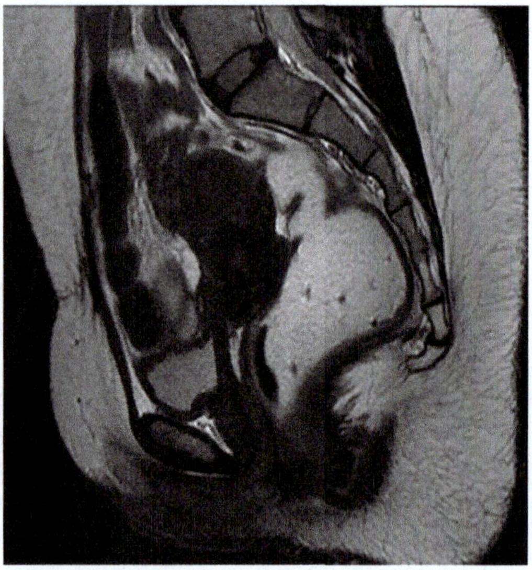

Fig. 10.1 Sagittal T2 MRI defecography at rest

The exam is divided into stages (Table 10.1):

1. Analysis of the pelvic floor at rest, in T2 with high spatial resolution, in the axial, sagittal, and coronal planes: in these images, it is possible to analyze the components of the levator ani and pelvic fascia, as well as the position of the pelvic organs (e.g., vagina, urethra, and uterus), as well as the presence of associated lesions, such as leiomyoma and solitary ulcer of the rectum, which may delay evacuation and be associated with intussusception (Fig. 10.2a, b).
2. Dynamic evaluation in the sagittal and coronal planes: the dynamic study uses the FIESTA sequence, fast (less than 1 s) and with excellent spatial resolution, repeated for 170 s, having a filmlike presentation. It allows the identification of subtle lesions and in time of enhancement, such as defecography.

The dynamic evaluation is performed in the sagittal plane, in three phases:

1. During the Valsalva maneuver (Fig. 10.3a, b): detects decrease or loss of the rectal gel.
2. During sphincter contraction: the degree of muscle contraction of the puborectalis.
3. Evacuation of the gel: this dynamic phase allows the demonstration and graduation of excessive descent of the perineum, intussusception or rectal prolapse, and enteroceles (Fig. 10.4). In the absence of this step, some findings may not be identified, and there is a risk that the severity of the pathology will be underestimated [1].

Table 10.1 Defecography MRI protocol

DEFECOGRAPHY PROTOCOL MRI	
Patient position	supine
Equipment	At least 1,5T closed
Contrast	Oral or venous is not used, the rectum should be distended with ultrasound gel
Static MRI sequences	High resolution T2w in axial, sagital and coronal planes
Dynamic MRI sequences	Balance-FFE or FIESTA in sagital plane during rest, contraction, Valsalva maneuver and defecation

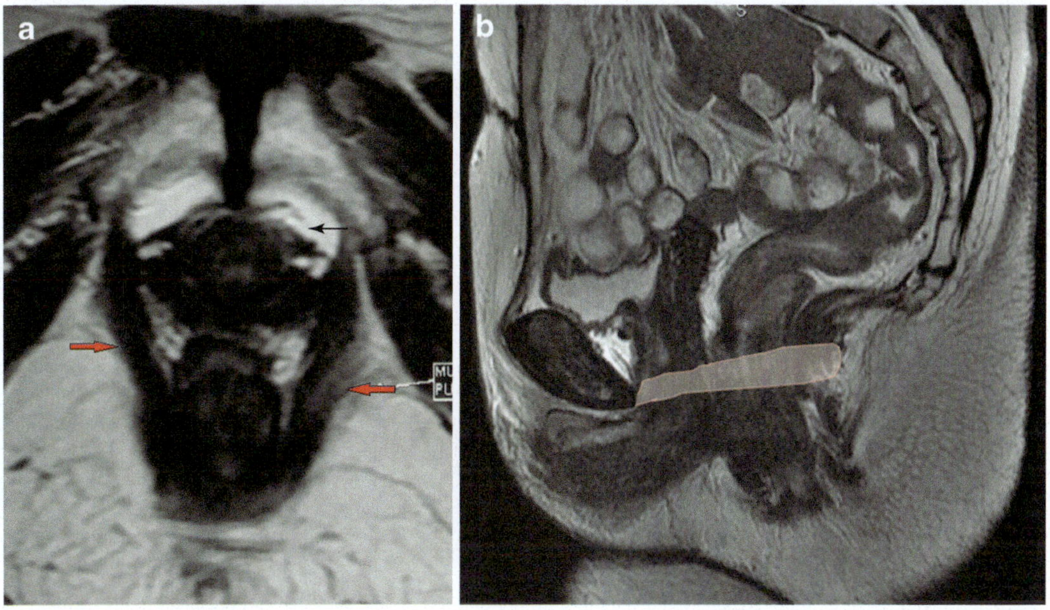

Fig. 10.2 (**a**) Axial plane. Analysis of the pelvic floor at rest, and levator ani. (**b**) Levator ani position at the sagittal plane

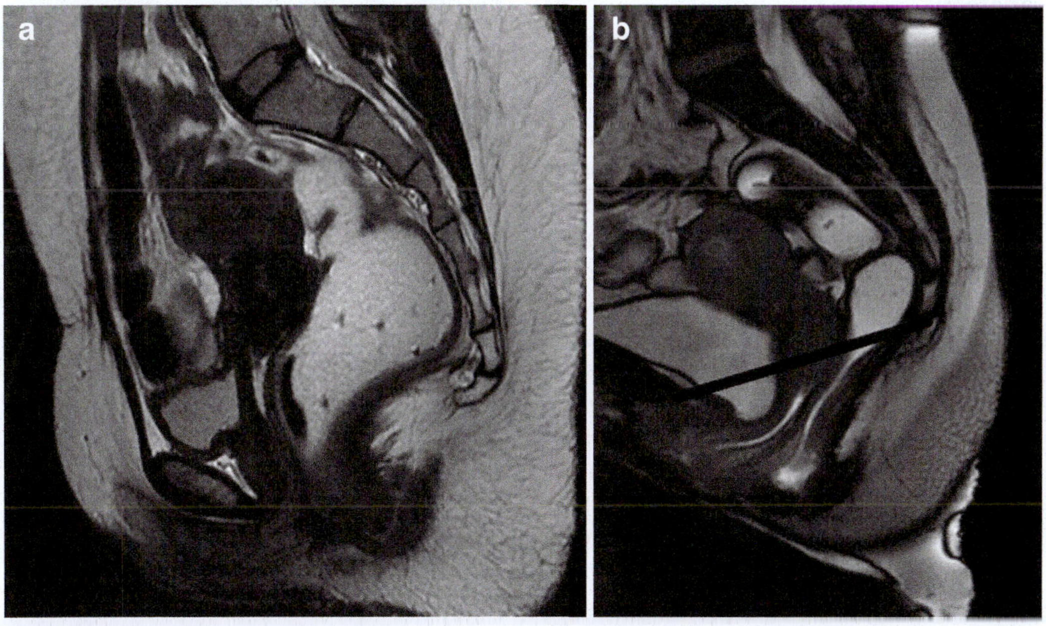

Fig. 10.3 Defecography at rest and Valsalva maneuver. (**a**) High-resolution – sagittal T2 – rest. (**b**) Dynamic study – Valsalva maneuver: Descent of the bladder, urethra, vagina, and anorectal junction after evacuation of ultrasound gel

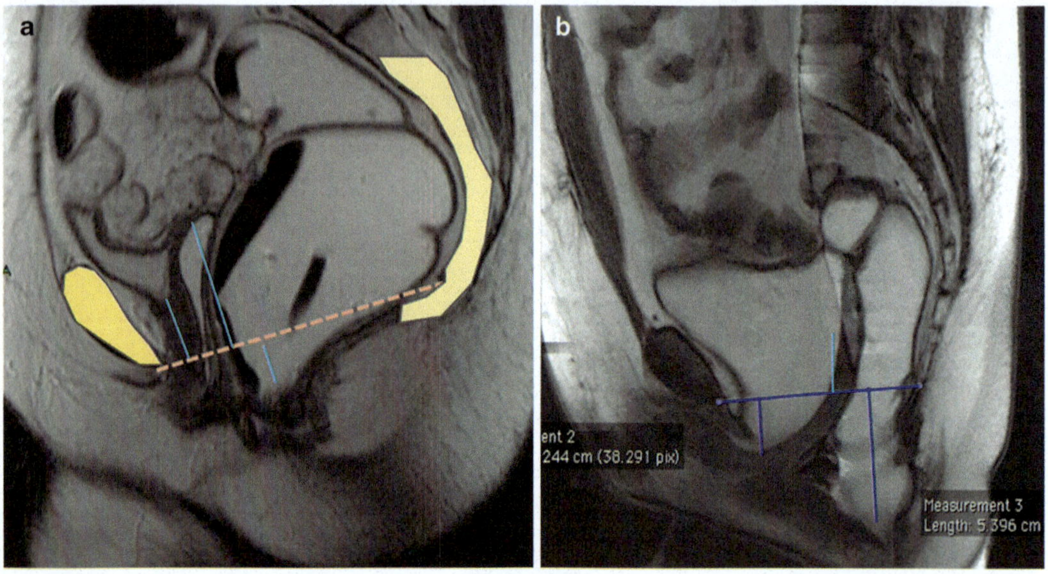

Fig. 10.4 MRI defecography: Evacuation – patient with perineum descent, anterior, and posterior compartment. (**a**) Sagital plane at rest. (**b**) MRI defecography with ´pelvic floor descent

Interpretation

Anatomy

A basic knowledge about pelvic floor anatomy is fundamental for the interpretation of images and understanding the alterations that can generate dysfunction. The pelvic floor is divided into three compartments: anterior, medial, and posterior. The anterior compartment contains the bladder and urethra; the middle contains the uterus, cervix, and vagina; and the posterior one contains the rectum and the anal canal. The support of these structures is obtained by a complex of muscles, fascia, and ligaments that are inserted in the pelvic bones. The pelvic floor consists of a complex and integrated, multilayer system that provides active and passive support. From top to bottom, the pelvic floor has three layers: the pelvic fascia, the pelvic diaphragm, and the urogenital diaphragm. These structures are integrated and interact and compensate one another [3]. The pelvic fascia and ligaments provide passive support, while the floor muscles, especially the anus lift, provide active support (Fig. 10.5) [3].

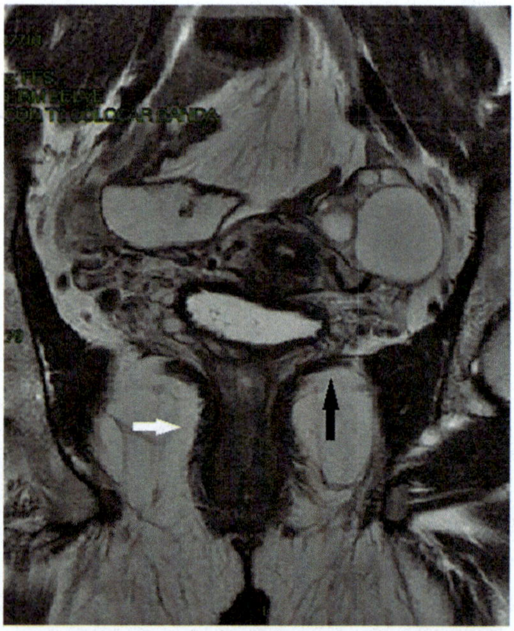

Fig. 10.5 Levator ani evaluation. T2 coronal – levator ani (black arrow) and external anal sphincter muscle (white arrow)

Endopelvic Fascia

The endopelvic fascia is a structure that recovers the pelvic organs and maintains the support of the bladder, urethra, vagina, and rectum [6]. It is a continuous layer that extends from the peritoneum to the perineum, inserting laterally into the pelvic bones.

The endopelvic fascia consists of layers:

- *Pubocervical*: between the bladder and the vagina
- *Rectovaginal*: between the vagina and the rectum
- *Parametrium*: at the height of the uterus
- *Paracolpos*: at the height of the vagina
- *Arch tendinous*

The pubocervical fascia is an anterior transverse layer extending from the pubis, anteriorly, to the uterine cervix posteriorly, between the bladder, urethra, and vagina, from the pericervical ring to the perineal membrane of the urogenital triangle. It gives origin to ligaments of the urethral support system and the bladder neck, which connect the urethra to the vagina, pubis and levator ani muscle [6]. Injury to the pubocervical fascia may result in prolapse of the anterior wall of the vagina, including cystocele and urethral hypermobility.

The rectovaginal fascia is a layer of connective tissue adhered to the posterior wall of the vagina and anterior to the rectum, suspended superiorly by the insertion of the uterosacral ligaments, distally in the perineal body, and laterally in the fascia of the arch tendinous. Lesions in this fascia result in posterior compartment prolapse such as rectocele and enterocele [6].

The parametrium and the paracolpos are components of the fascia at the height of the uterus and vagina and inserts into the lateral wall of the pelvis, being related to the uterosacral and cardinal ligaments. The uterosacral ligaments form the pericervical ring that surrounds the cervix and the upper portion of the vagina and, together with the cardinal ligaments, form the support of the cervix and the upper third of the vagina (Fig. 10.6) [6]. Lesions in the pericervical ring lead to prolapse of the vagina and uterus.

Pelvic Diaphragm

The pelvic diaphragm comprises the levator ani and coccygeal muscles.

The levator ani, the most important muscle in the pelvic diaphragm, maintains a constant basal tone and closes the urogenital gap, preventing incontinence and prolapse.

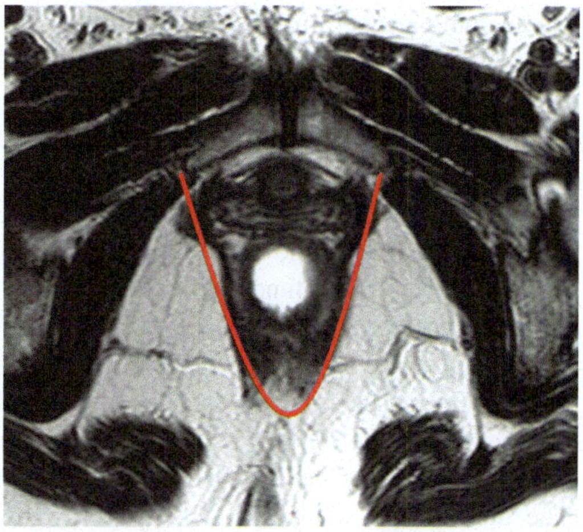

Fig. 10.6 T2 axial plane. In the axial plane, the levator ani presents a hypointense and homogeneous signal in the T2-weighted sequence and can be seen as the red line

It is inserted anteriorly to the pubis and laterally to the tendon arch of the levator ani. This muscle is formed by inseparable parts of a unit. However, for teaching purposes, various components of the levator ani are described, in accordance with their insertions, such as the iliococcygeus and pubococcygeus muscles.

The ventromedial part of the levator ani, the so-called pubovisceral or pubococcygeal component, corresponds to a bundle of U-shaped medial fibers, which are inserted into the inner face of the pubis and involve the urethra, vagina, lower rectum, and anal canal. This configuration makes this muscle an important component of the pelvic floor and in the genesis of prolapse and urinary incontinence. The constant tonus closes the urogenital and anorectal hiatus and provides support during normal and standing activity, as well as being responsible for the contraction reflex during increased intra-abdominal pressure.

The pubovisceral muscle is composed by the pubovaginal, puborectalis, and pubococcygeal muscles. The pubovaginal muscle inserts into the lateral and posterior vaginal wall and helps to support the vagina. The puborectalis muscle suspends the rectum and anorectal junction, with a U-shaped configuration around the anal canal and rectum. This muscle controls stool descent and is considered part of the external anal sphincter. Its lateral components form a space (the raising hiatus) that contains the urethra, vagina, rectum, and anal canal.

The iliococcygeal muscle is the least dynamic component of the levator ani. It is located above the pubococcygeus muscle and originates along the tendinous arch of the anus elevator and later extends to the rectum, functioning as a musculofascial layer (Fig. 10.5).

These muscles are easily visible on MRI. The puborectalis muscle is better evaluated in the axial plane, whereas the iliococcygeus presents better visualization in the coronal and sagittal planes. The morphology, thickness, and type of signal should be evaluated. In the axial plane, the elevators present a hypo-intense and homogeneous signal in the T2-weighted sequence (Fig. 10.6).

The radiologist should focus on the muscle's morphology, thickness, and signal intensity. Muscle injury is characterized by reduced asymmetrical or diffuse thickness and may be accompanied by fat infiltration into the muscle fibers. Fat infiltration appears as a high signal intensity in muscle fibers in T1 and T2 (Fig. 10.7) [7, 8].

In more severe lesions, the distance between the puborectalis fibers and the levator gap increases in the axial plane. In patients with intact pelvic floor, the length should be at most 4.5 cm. There may be a rupture in the pubic insertion of the puborectalis muscle, with sharpening or retraction, with the proximal end retracted and irregular, causing a paravaginal defect: the vagina retracts toward the lesion, and the urethra rotates (Fig. 10.7).

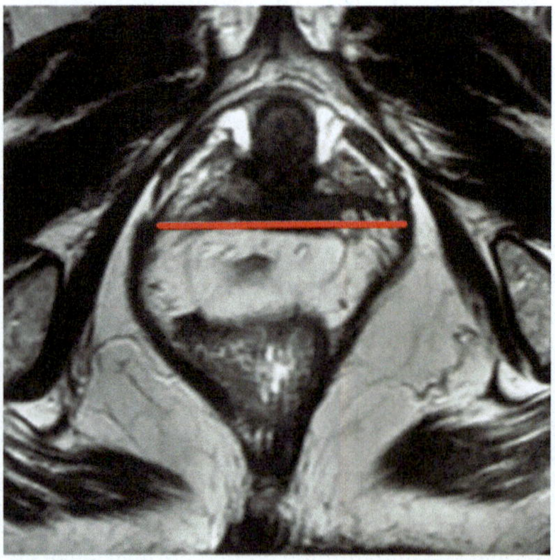

Fig. 10.7 Levator ani lesion – hiatus levator enlargement – transverse diameter

Urogenital Diaphragm

Also called the perineal membrane, it is composed of the transverse muscle of the perineum and the muscles that make up the urethral and urethrovaginal sphincter, and enters the vagina, the perineal body, external anal sphincter, and the bulbocavernosus muscle [6].

Urethra

The female urethra is about 4 cm long and posterior to the pubic bone. In continent patients, the urethra should be retropubic, above or up to the level of the lower pubis. When there is deficiency of the supporting structures, the urethra may undergo a lower translation and be positioned below the pubis [6]. In urethral hypermobility, there is excessive rotation of the urethra (over 30 °) from its axial axis, from vertical to horizontal (descending urethral rotation), and may be related to urinary stress incontinence (Fig. 10.8) [6]. The urethral sphincter comprises the internal involuntary smooth muscle, which is continuous to the bladder, extending through the proximal two-thirds of the urethra, and the voluntary external muscle.

Vagina

The supporting structures of the vagina include the pericervical ring, pelvic fascia, urogenital diaphragm, and arch tendinous of the levator ani. Three levels of vaginal support have been described:

- *Level I*: cervix and upper third of the vagina, by the complex formed by uterosacral and cardinal ligaments
- *Level II*: middle third of the vagina, through the vaginal muscles and tendinous arch
- *Level III*: lower third of the vagina, maintained by the rectovaginal fascia and the perineal membrane [3]

When there is integrity of the supporting structures, the vagina is H-shaped and when there is a

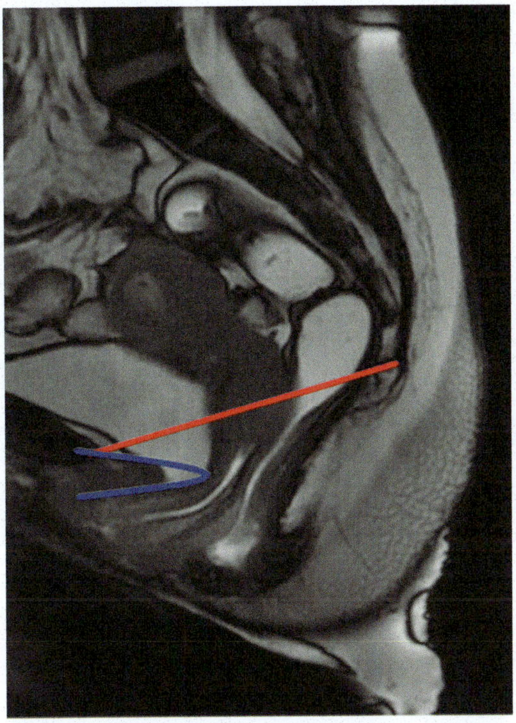

Fig. 10.8 Bladder neck and distal urethra underneath the symphysis pubic level (red line). Blue line shows urethra rotation

lesion of these structures presents asymmetrical and elongated.

Anal Sphincter

The anal sphincter is composed of three muscles and is surrounded by the ischioanal space.

The internal anal sphincter extends from the anorectal junction about 1.0 to 1.5 cm below the dentate line. The external anal sphincter is the most external muscle of the distal anal canal; it comprises several parallel bands and extends about 1 cm beyond the internal sphincter.

MRI Findings

Pelvic Floor Reference Lines

Pubococcygeal Line

The pubococcygeal line is an important parameter for the assessment of the level of the pelvic floor [9, 10]. It is a landmark also utilized in cinedefecography and allows the evaluation of perineum descent and sigmoidocele classification. It is traced electronically, joining the lower margin of the pubis to the last coccygeal joint. In continent patients, the bladder, vagina, and anorectal junction are located above this line (Fig. 10.9a, b) [9]. Some studies suggest that it should not be more than 2 cm below this line [6]. The pubococcygeal line serves as a reference for determining the graduation of cystocele, enterocele, rectal descending, and vaginal dome. The prolapse is graded as light (<3 cm), moderate (3–6 cm), and severe (> 6 cm) [3].

The Medial Pubic Line

The medial pubic line has been used as an auxiliary marker for pelvic organ prolapse quantification system (POP-Q) by magnetic resonance imaging. There is a difficulty in obtaining a reference point that is equivalent to resonance and physical examination. Therefore, some studies have proposed the use of this line as a reference point for prolapsed pelvic organs visualized by means of the MRI defecography. It is drawn in the midline of the pubis, passing along the long axis of the vagina, and is equivalent to the hymen on the physical examination (Fig. 10.10) [11].

In the MRI sequences, the anatomic milestones used are the median pubic line, the hymenal reference, and the distances to the anterior and posterior uterine cervix or vaginal dome.

For evaluation of prolapse, we measure (in centimeters) the distance between the maximum limit of prolapse and the remanescent hymen [11].

Pathological Findings

Perineal Descent Syndrome

Perineal descent syndrome is a complex syndrome characterized by a weak pelvic floor resulting in an excessive descend of the pelvic organs which fall, resulting in symptoms of incontinence, bowel movement, and various prolapses [4]. This imaging modality has shown good correlation with the clinical evaluation.

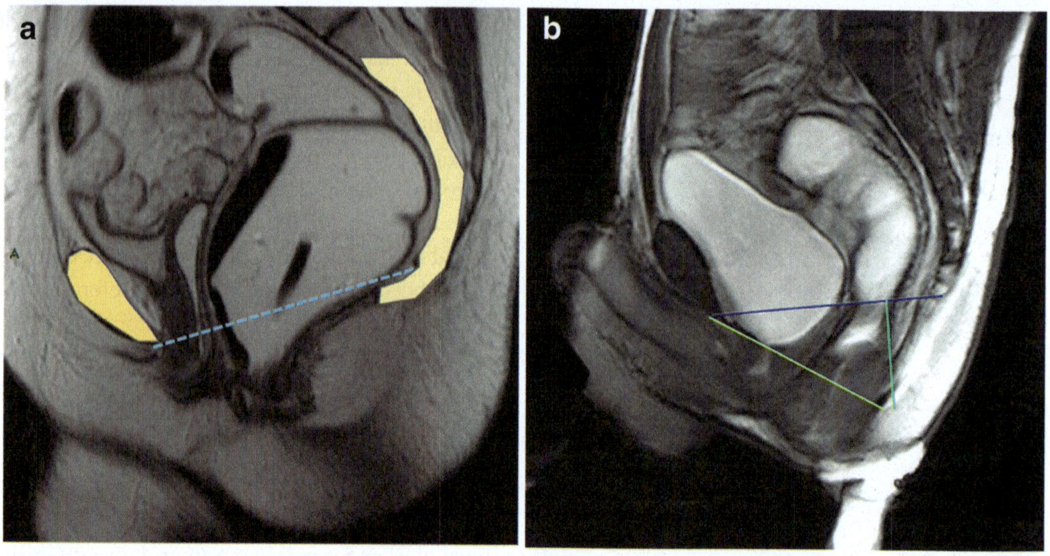

Fig. 10.9 The pubococcygeal line (blue line). (**a**) It is traced electronically, joining the lower margin of the pubis to the last coccygeal articulation. (**b**) The H line (yellow line) (normal inferior to 5 cm) corresponds to the AP distance of levator hiatus, and M line (green line) (normal inferior to 2 cm) is the perpendicular distance between the pubococcygeal line and H line. It represents the longitudinal extent of the levator ani

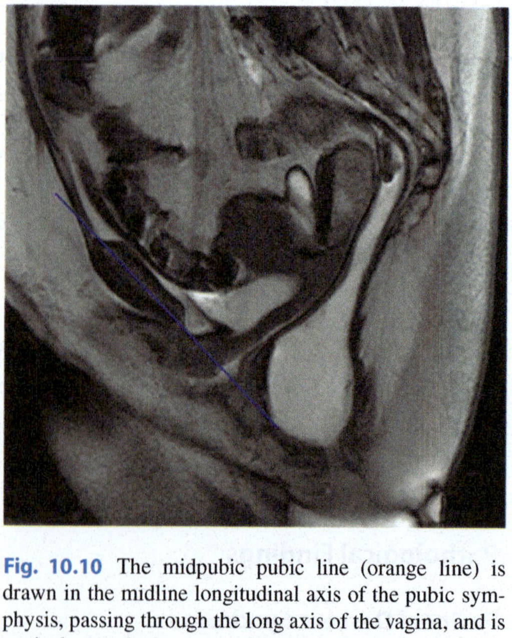

Fig. 10.10 The midpubic pubic line (orange line) is drawn in the midline longitudinal axis of the pubic symphysis, passing through the long axis of the vagina, and is equivalent to the hymen on the physical examination

Some studies have shown that patients with clinical complaints of a single compartment in fact had a multi-compartment syndrome.

The evaluation is performed in the resting position and during evacuation effort. The reference lines used in the evaluation of prolapses in general are the pubococcygeal line and the median pubic line, already mentioned; line H, extending from the lower pubis to the posterior wall of the rectum at the anorectal junction (Fig. 10.11).

Anterior Compartment

The bladder is a component of the anterior compartment, and, in normal individuals, the urethra is anterior to the bladder in the sagittal plane. The evaluation of cystocele can be performed during the Valsalva maneuver during evacuation, comparing the position of the bladder with respect to the pubococcygeal line (Fig. 10.12) [10]. During contraction, the bladder neck should be less than 1cm from the pubococcygeal line [3].

Medial Compartment

Many authors consider that the descent of the base of the vagina beyond the pubococcygeal line is abnormal, but some continent patients may also

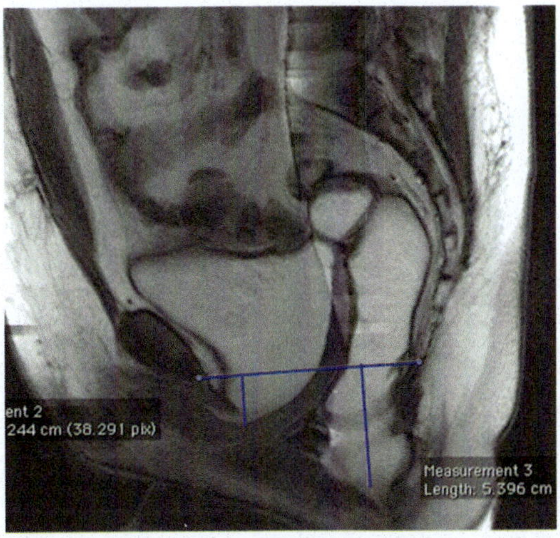

Fig. 10.11 Perineal descent syndrome. Bladder and posterior descend below the pubococcygeal line

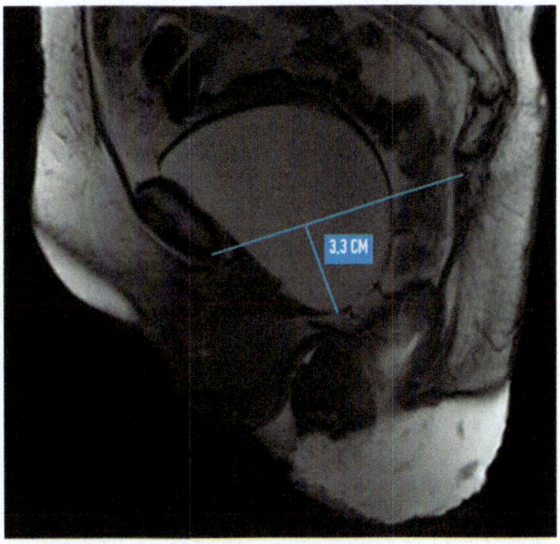

Fig. 10.12 Cystocele during evacuation phase. Bladder descent inferior to pubococcygeal line

show a slight descent beyond this limit. Thus, it is suggested that this prolapse be graded as small (<3 cm), medium (3–6 cm), or large (>6 cm) [6].

Many authors consider that the descent of the vaginal dome or the cervicovaginal junction below the pubococcygeal line is abnormal, but certain asymptomatic patients may present down to 3 cm below this line [4].

When patients present the uterus more than 3 cm below the pubococcygeal line, there is a greater chance of being prevented from evacuating. Uterus prolapse should also be classified as small (<3 cm), medium (3–6 cm), or large (> 6 cm) [4].

The peritoneocele is another disturbance of this compartment that occurs when there is herniation of the peritoneal fat between the rectum and the vagina, enlarging the rectovaginal septum. For this diagnosis, the lower point of the hernia anterior to the rectal wall and an increase in the distance between the vagina and the rectum should be identified [4]. Peritoneocele may be accompanied by herniation of intestinal loops (enterocele) [4].

Enterocele

Enteroceles are herniations of the thin or sigmoid loops that occur in the middle compartment, between the bladder and vagina and the rectum, due to rupture of the endopelvic fascia and the rectovaginal septum [13]. They are more frequent in women undergoing hysterectomy. The protrusion of the herniation is maximal in the dynamic phase of the evacuation effort, seen in the blood plane (Fig. 10.13). Enteroceles can be classified into three types, as shown in Table 10.2.

Posterior Compartment

Anorectal Angle/Anal Canal/Perineal Descent

The anorectal angle is formed by a line parallel to the anal canal and tangent or parallel to the posterior wall of the rectum (Fig. 10.14a, b). At rest, it varies about 95° (between 70° and 134°) 0.11 As in cinedefecography, this angle becomes more acute at the time of voluntary contraction representing contraction of the puborectalis muscle, and more obtuse to exertion normal evacuation, which indicates their relaxation. The anal canal has an average length of 16 mm in women, being longer in men (22 mm). During the contraction and the evacuation effort, small changes in these measurements can be observed in MRI defecography. The posterior perineal descent is evaluated

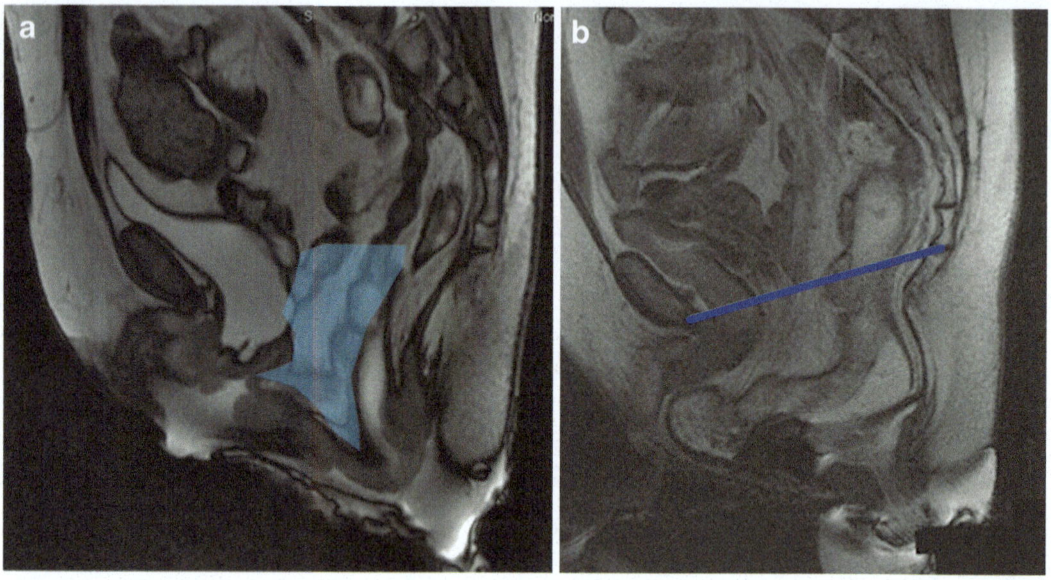

Fig. 10.13 (**a**) Enterocele (blue image) – containing small bowel herniation in the middle compartment, between the bladder and vagina and rectum, due to rupture of the rectovaginal septum. (**b**) enterocele (white arrow) descending bellow the pubococcygeal line (blue line)

Table 10.2 Enterocele classification

Grade I	Herniation to the distal half of the vagina
Grade II	Herniation to the perineum
Grade III	Herniation with protrusion out of the anal canal

in the expulsion phase, observing the position of the perineum in relation to the pubococcygeal line. A "descending" perineum is positioned 2 cm below the pubococcygeal line (Fig. 10.15) [12]. This descent should also be graded as small (<3 cm), medium (3–6 cm), or large (>6 cm) [4].

Rectocele

Rectoceles are herniations of the rectal wall on the posterior wall of the vagina [14]. Although they represent a common finding in asymptomatic women in cinedefecography and in MRI, they may be the cause of defecation disorders, mainly when they are bulky and do not empty (Fig. 10.16). The posterior rectoceles can be classified according to the its size [15, 16] as shown in Table 10.3.

Paradoxical Contraction of the Puborectalis or Anismus

Anismus is a clinical syndrome characterized by difficulty initiating evacuation and incomplete rectal emptying. Paradoxical contraction of the puborectalis, or anismus, presents in the MRI defecography as a non-relaxation of the puborectalis during evacuation attempts, just as in the cinedefecography. The anorectal angle opening is not observed, and there is no relaxation of the puborectalis, resulting in spastic pelvic floor syndrome or anismus (Fig. 10.17).

Intussusception

Intussusception is the invagination of the wall and mucosa of the rectum toward the rectum and anal canal itself. It can be internal (intrarectal and intra-anal) or external (rectal prolapse) [4].

The internal invagination of the rectal wall, or internal intussusception, consists of a fold of the rectal wall on itself that can, through the evacuation effort, descend to the anorectal junction. It is one of the causes of obstructive

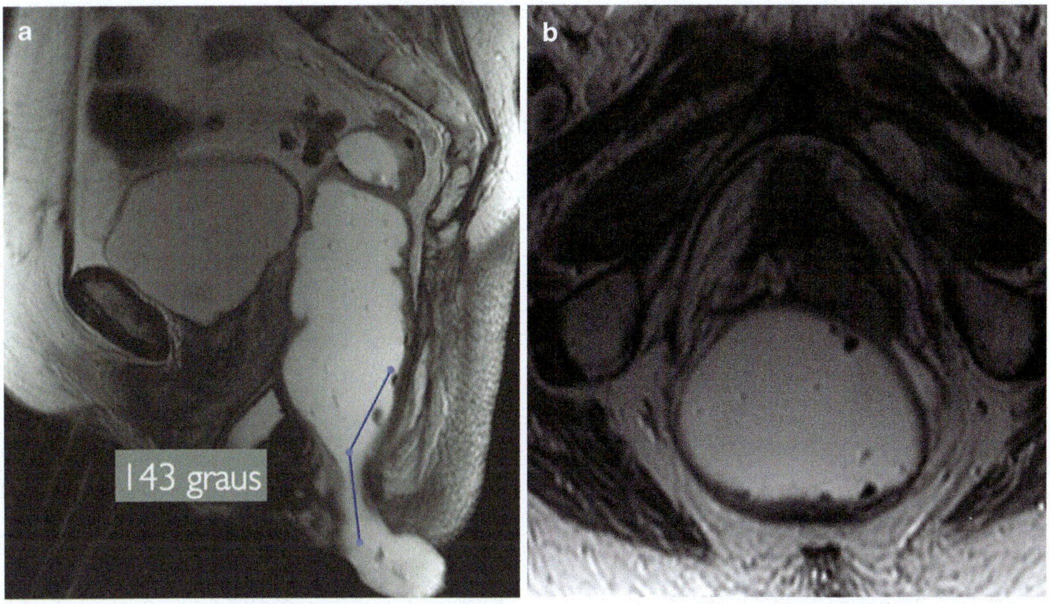

Fig. 10.14 (**a**) Sagittal plane. The angle is increased (143) related to levator ani rupture (axial plane) (**b**)

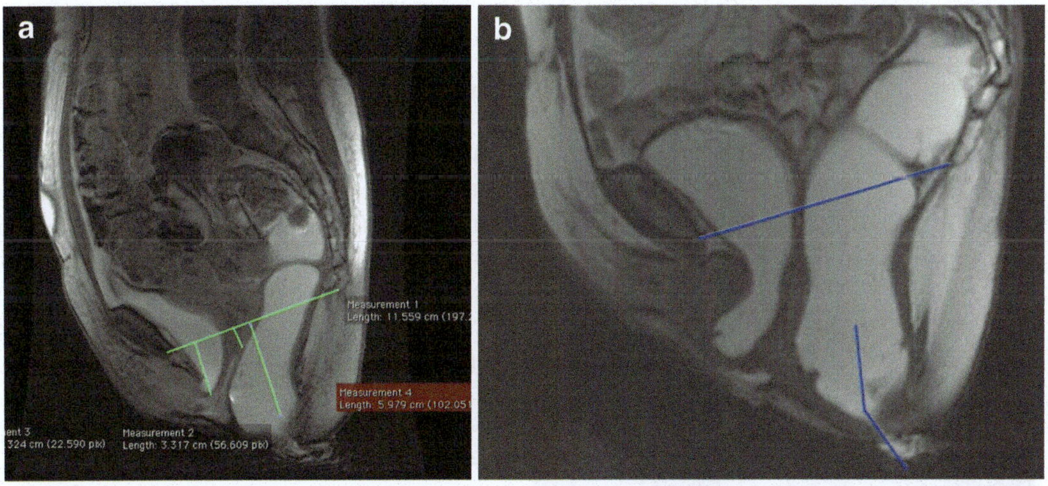

Fig. 10.15 Two cases of descending multicompartmental pelvic floor syndrome (**a** and **b**). Bladder and rectum are 3 cm bellow the puboccocygeal line

obstruction syndrome, leading to the sensation of incomplete evacuation (Fig. 10.18).

Intussusception can be classified according to the degree of rectal exteriorization at the end of the evacuation phase, such as intrarectal (minimal involvement of the rectum wall or circumferential restricted to the rectum), intussusception restricted to the anal canal, and, finally, which progresses toward the anus [4, 17].

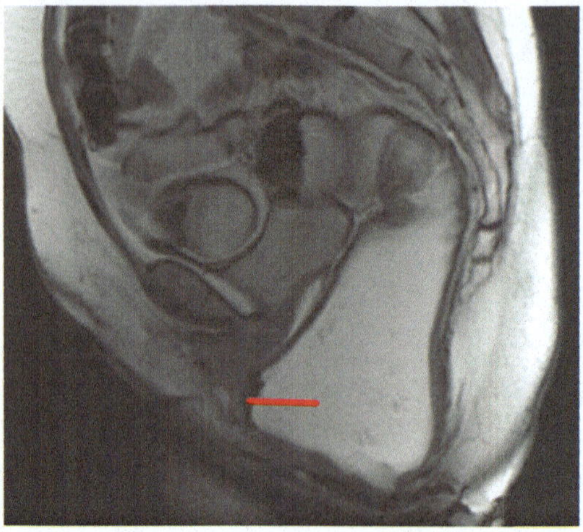

Fig. 10.16 Rectocele – anterior protrusion or herniation of the rectal wall on the posterior wall of the vagina

Table 10.3 Rectocele classification by size

Grade I	Small rectocele <2 m
Grade II	Rectocele between 2 cm and 4 cm
Grade III	Rectocele >4 cm

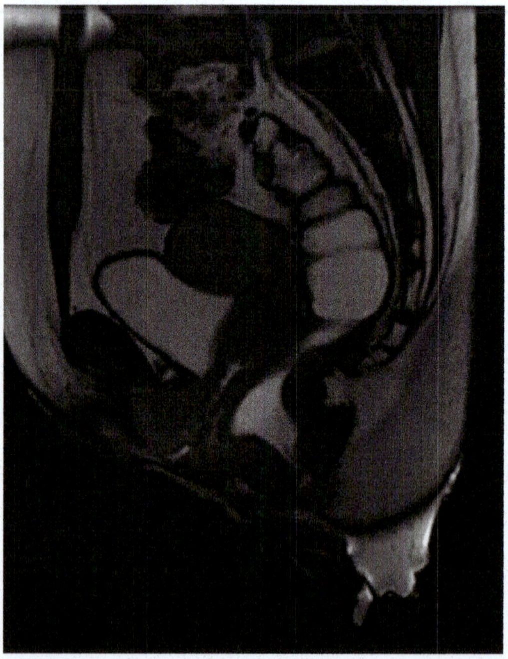

Fig. 10.17 Anismus – paradoxical contraction of the puborectalis during evacuation (white arrow)

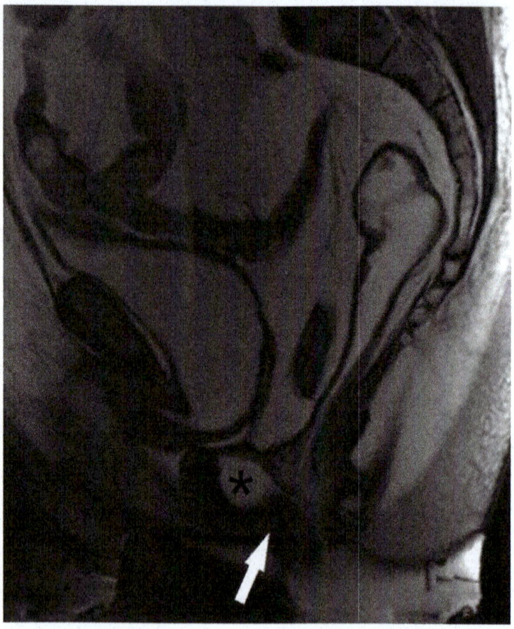

Fig. 10.18 Intussusception (white arrow): invagination of the rectal wall and mucous membrane toward the rectal or the anal lumen

Summary

MRI defecography is an imaging modality of the pelvis that provides excellent evaluation of the entire pelvic floor and pelvic compartments, both at rest and dynamically. One of the advantages over the other methods is the possibility of evaluating the three compartments of the pelvis, providing important information on pelvic floor dysfunction.

References

1. Pierce H, Perry L, Chiarelli P, et al. A systematic review of prevalence and impact of symptoms of pelvic floor dysfunction in identified workforce groups. J Adv Nurs. 2016;72(8):1718–34.
2. Attenberger UI, Morelli JN, Budjan J, et al. The value of dynamic magnetic resonance imaging in interdisciplinary treatment of pelvic floor dysfunction. Abdom Imaging. 2015;40(7):2242–7.
3. Brandâo A. Dynamic Magnetic Resonance Defecography. In: Pescatori M, Regadas FSP, Murad Regadas, SM, Zbar AP. Imaging Atlas of the Pelvic Floor and Anorectal Diseases. Milan: Springer; 2008. p.219–235.

4. Rociu E, Stoker J, Eijkermans MJ, et al. Fecal incontinence: endoanal US versus endoanal MR imaging. Radiology. 1999;212(2):453–8.
5. Malouf AJ, Halligan S, Williams AB, et al. Prospective assessment of interobserver agreement for endoanal MRI in fecal incontinence. Abdom Imaging. 2001;26(1):76–8.
6. Brandão MR, Ianez MR. Imaging of the pelvic floor defecography. Magn Reson Imaging Clin N Am. 2013;21:427–45.
7. Fielding JR, Dumanli H, Schreyer AG, et al. MR- based three-dimensional modeling of the normal pelvic floor in women: quantification of muscle mass. AJR Am J Roentgenol. 2000;174(3): 657–60.
8. Jaap S, Halligan S, Bartram CI. Pelvic floor imaging. Radiology. 2001;218:621–41.
9. Fielding JR. Practical MR imaging of female pelvic floor weakness. Radiographics. 2002;22(2): 295–304.
10. Unterweger M, Marincek B, Gottstein-Aalame N, et al. Ochsenbein-Imhof N, Perucchini D, Kubik-Huch RA. Ultrafast MR imaging of the pelvic floor. AJR Am J Roentgenol. 2001;176(4):959–63.
11. Nosseir S, Schachar J, Naidich J, et al. Midpubic line to stage pelvic organ prolapse by MRI defecography as compared to POP-Q. Obstet Gynecol Int J. 2014;1(2):00011. https://doi.org/10.15406/ogij.2014.01.00011.
12. Mariëlle ME, Lakeman FM, Zijta J, et al. Dynamic magnetic resonance imaging to quantify pelvic organ prolapse: reliability of assessment and correlation with clinical findings and pelvic floor symptoms. Int Urogynecol J. 2012;23:1547–54.
13. Roos JE, Weishaupt D, Wildermuth S, et al. Experience of 4 years with open MR defecography: pictorial review of anorectal anatomy and disease. Radiographics. 2002;22(4):817–32.
14. Yoshioka K, Matsui Y, Yamada O, et al. Physiologic and anatomic assessment of patients with rectocele. Dis Colon Rectum. 1991;34(8):704–8.
15. Hetzer F, Andreisek G, Tsagari C, et al. MR defecography in patients with fecal incontinence: imaging findings and their effect on surgical management. Radiology. 2006;240(2):449–57.
16. Harewood GC, Coulie B, Camilleri M, et al. Descending perineum syndrome: audit of clinical and laboratory features and outcome of pelvic floor retraining. Am J Gastroenterol. 1999;94(1):126–30.
17. Martín G, García Armengol J, Roig Vila JV, et al. Analysis of experience in the use of dynamic pelvic magnetic resonance imaging in the assessment of obstructive defaecation syndrome. Cir Esp. 2012;90(5):292–7.

Neurophysiology of the Pelvic Floor

11

Nucelio L. B. M. Lemos, Corey Sermer, Laura Cancelliere, Adrienne Li, Kinshuk Kumar, and Virginia C. Roncatti

Introduction

The pelvic floor functions to provide support and suspension necessary to the pelvic viscerae in order to allow for proper function – urethral and anal continence, bladder and rectal emptying, and sexual function – and maintain anatomical integrity in one of the most dynamic regions of the human body. This complex function results from the interaction between muscles, fascias, ligaments, and autonomic and somatic nerves.

In this chapter, we will discuss the functional anatomy of the pelvic floor, with emphasis on neurophysiology, since an adequate understanding of pelvic neurophysiology requires a deep understanding of the anatomy of the muscles and fascias that execute the commands carried by the nerves. Given the greater prevalence of pelvic floor dysfunction (PFD) in women, the availability of studies and literature is much higher in this group of individuals. Thus, this chapter will be presented as it pertains to the anatomy of the female pelvis, keeping in mind the parallels with male pelvic anatomy.

Pelvic Floor Muscles

The Pelvic Diaphragm Muscles

Much disagreement can be found in the literature regarding the nomenclature of the levator ani muscle and the muscular bundles that compose it. In this chapter, we will reference the *Terminologia Anatomica*, an update of the *Nomina Anatomica*, published by the International Federation of Associations of Anatomists (IFAA) [1] in 1998 and a review on the anatomy of the levator ani muscle, published in 2004 by Kearney, Sawhney, and DeLancey [2], in order to unify the published nomenclature based on *Terminologia Anatomica*.

According to *Terminologia Anatomica*, the pelvic diaphragm is formed by the levator ani and ischiococcigeus muscles. The levator ani is

N. L. B. M. Lemos (✉)
Faculty of Medical Sciences, Santa Casa de São Paulo, Department of Gynecology, UNIFESP-EPM, Department of Neurodisorders in Gynecology, Sao Paulo, Brazil

C. Sermer
Faculty of Medicine, University of Toronto, Department of Obstetrics and Gynecology, Mount Sinai Hospital and Women's College Hospital, Toronto, ON, Canada

L. Cancelliere · K. Kumar
Department of Obstetrics and Gynecology, Faculty of Medicine, University of Toronto, Toronto, ON, Canada

A. Li
Division of Urogynecology, Department of Obstetrics and Gynecology, Mount Sinai Hospital and Women's College Hospital, University of Toronto, Toronto, ON, Canada

V. C. Roncatti
Department of Urogynecology and Pelvic Reconstruction, Heliópolis Hospital, Sao Paulo, Brazil

L. C. C. Oliveira (ed.), *Anorectal Physiology*, https://doi.org/10.1007/978-3-030-43811-1_11

formed by three muscle groups and three more subgroups. Thus, they constitute the pelvic diaphragm:

- Pubococcygeus or pubovisceral muscle: originates from the pubic bone and is anchored in the coccyx (hence the name of the muscle) by means of ligaments that function as pulleys; it is divided into three muscles, according to their insertions (Fig. 11.1).
- Pubovaginal muscle (puboprostatic musce in males): inserts into the medial and lateral

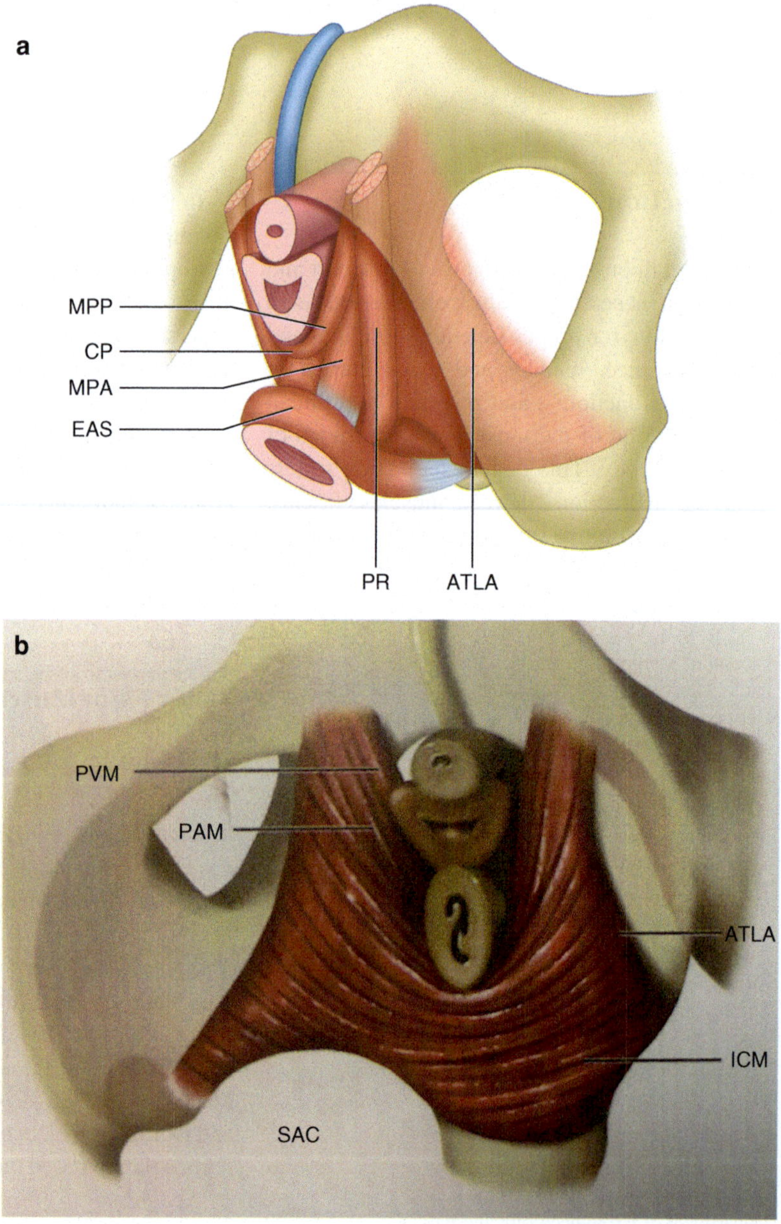

Fig. 11.1 (**a**, **b**) Levator ani muscle and subdivisions of the puboviseral muscle. (Adapted from Kearney et al. [2]). PPM, puboperineal muscle; PB, perineal body, PAM, puboanal muscle; PVM, pubovaginal muscle; PRM, puborectal muscle; ICM, iliococcygeus muscle; ATLA, tendinous arch of levator ani muscle; EAS, external anal sphincter; SAC, sacrum

pubovesical (puboprostatic)[1] and urethropelvic ligaments (see pubocervical fascia).
- Puboperineal muscle: inserts proximally to the deep transverse perineal muscle, in the lateral portion of the perineal body, tensioning it anteriorly.
- Puboanal muscle: inserts in the anus, between the external and internal sphincter muscles.
- Puborectalis muscle: originates on the posterior surface of one pubic bone and inserts on the contralateral pubic bone, forming a loop around the distal rectum; its contraction forms the anorectal angle.
- Iliococcygeus muscle: originates from the arcus tendineus levator ani muscle and extends posteromedially, inserting itself in the anterior aspect of the sacrum the levator ani muscle plate (levator plate).
- Ischiococcygeus muscle: originates on the ischial spine and inserts on the levator plate, covering the anterior surface of the sacrospinous ligament.

Later, we will study the result of the coordinated contraction of the pelvic diaphragm and perineum muscle groups on the dynamics of urinary and fecal continence and emptying.

As for the shape of the levator ani muscle, it is interesting to note that it is classically described as a muscle in the form of a basin or funnel. However, this description stems from a biased view of cadaver studies, whose pelvic floor is distended by increased postmortem abdominal pressure. Currently, studies with three-dimensional reconstruction of images generated by dynamic magnetic resonance demonstrate that such conformation would correspond to the muscular relaxation situation during the Valsalva maneuver [3, 4].

Endopelvic Fascia

The connective tissue that connects the viscera to the pelvic wall is called endopelvic fascia (Fig. 11.2). In spite of the name, this structure is not formed solely by parallel collagen fibers that normally characterize the fasciae (such as the fascia of the rectus abdominis muscle)–it is comprised of blood vessels, lymph tissue, smooth muscle, and large amounts of proteoglycans and collagen. The regions of the endopelvic fascia are named according to their location and function.

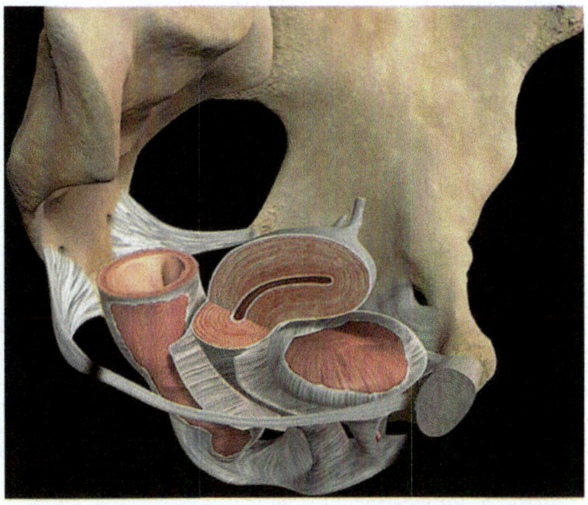

Fig. 11.2 Endopelvic fascia is the name given to tissue that connects the viscera to the pelvic wall, represented in gray in this figure. Morphofunctionally, the following regions are shown: uterosacrocardinal ligament complex, pubocervical fascia, and rectovaginal fascia. In men, these two fascias are fused and are called rectovesical fascia

Cardinal-uterossacral Complex (Posterior and Lateral Prostatic Fascia)

The cardinal-uterosacral complex is formed by the uterosacral (posterior prostatic fascia ligament), the cardinal (parametrial) ligament, and the paracolpos. This forms a single fanlike structure that originates from the anterior surface of the sacral bone, just medially to the S4, S3 and S2 foramina and inserts in the posterolateral anterior portion of the pericervical ring [3]. Therefore, this ligament complex provides suspension for the uterus and the vaginal apex [5].

Pubocervical Fascia

The pubocervical fascia (Fig. 11.3) originates in the retropubic region and inserts laterally into the

[1] The medial pubovesical (puboprostatic) ligament is commonly called “pubourethral ligament” in urogynecologic literature; this is the ligament that is reconstituted by retropubic slings.

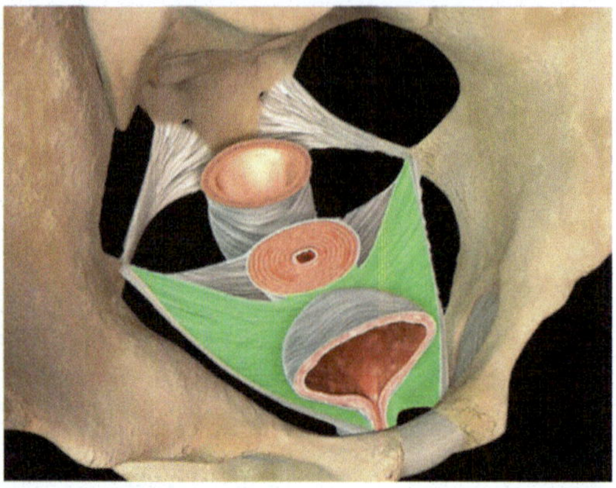

Fig. 11.3 Pubocervical fascia

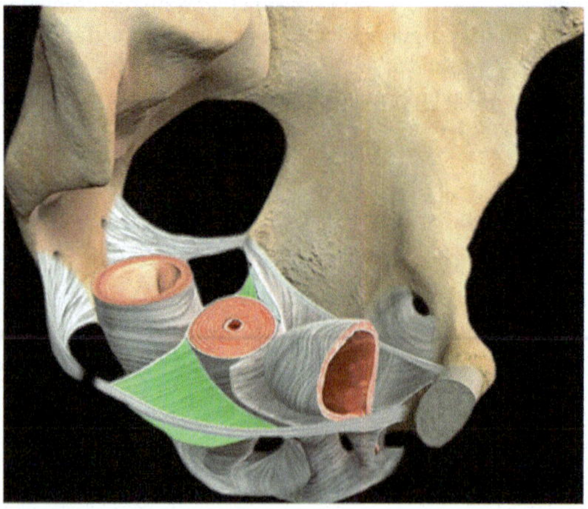

Fig. 11.4 Rectovaginal fascia

the arci tendinei fascia pelvis (which are thickenings of the obturator internus muscles fasciae) and proximally into the pericervical ring.[2]

Rectovaginal Fascia

The rectovaginal fascia, as the name implies, runs between the rectum and the vagina and connects proximally to the cardinal-uterosacral complexes and the levator plate, laterally to the arci tendinei of the rectovaginal fascia and distally to the perineal body[3] [6] (Fig. 11.4).

Mechanics of Urinary Continence and Bladder Emptying

The extrinsic mechanisms of urination and continence are due to the action of two muscle groups [7–9]:

1. Anterior group: the pubovaginal muscle (which tenses the urethro-pelvic ligament) and the ischiocavernosus muscle (which tenses the perineal membrane and, through this, the pubourethral ligament).
2. Posterior group: the iliococcygeus and ischiococcygeus muscles (the levator plate) apply tension posteriorly to the cardinal-uterosacral complex, which in turn pulls the pericervical ring and transmits the tension to the pubocervical fascia.

For urinary continence, the two muscle groups contract simultaneously, causing the proximal urethra to be tensioned posteriorly and the mid-urethra to be angled, resisting the increased pressure from the bladder. In order to promote urination, the muscles of the posterior group contract, generating a force vector that opens the bladder neck, while the muscles of the anterior group relax, reducing the angulation of the and allowing the flow to occur (Fig. 11.5a, b).

Mechanics of Anal Continence and Rectal Emptying

Anal continence and rectal emptying occur in a very similar way to urinary continence and bladder emptying, also by a balance of forces, as follows:

1. Posterior group: the iliococcygeus and ischiococcygeus muscles tense the rectovaginal fascia and the posterior wall of the rectum.

[2]The cardinal ligament and the paracolpos are equivalent to the lateral prostatic fascia.

[3]In men, the rectovaginal and pubocervical fascias are fused, since there is no vagina, forming the rectoprostatic fascia.

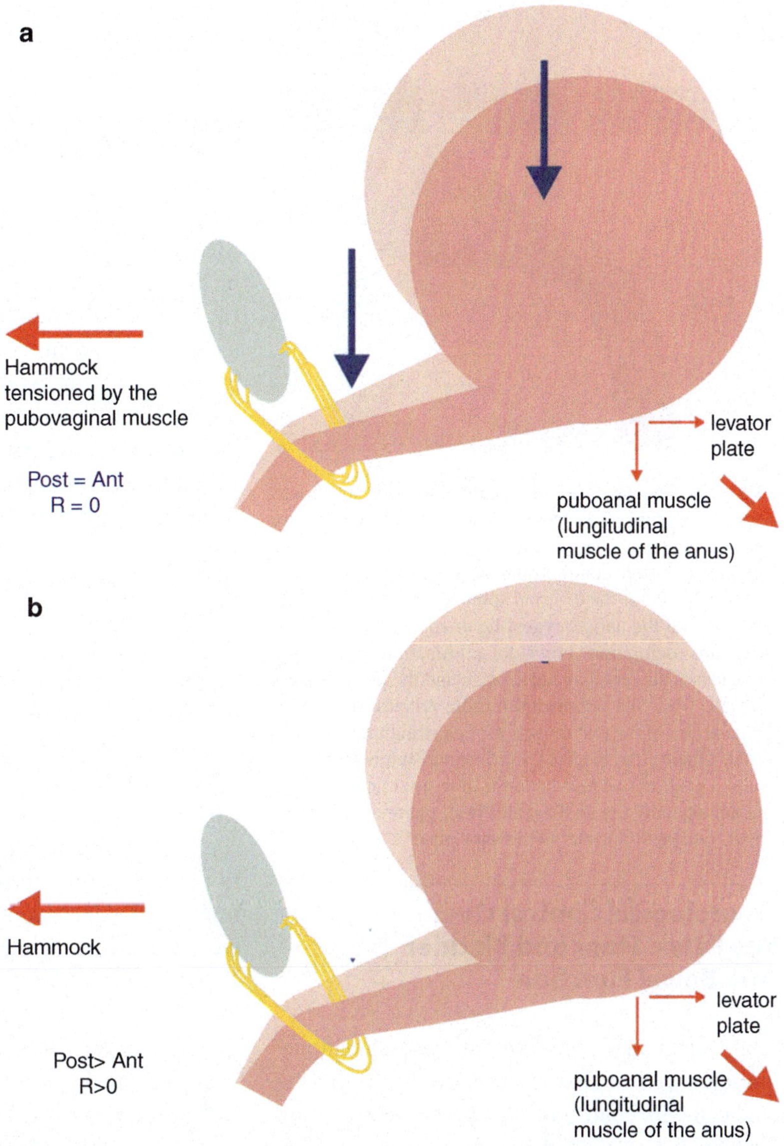

Fig. 11.5 Continence dynamics and micturition. (**a**) Continence – the anterior and posterior muscle groups contract simultaneously, causing the proximal urethra to be tensioned posteriorly and the mid-urethra to anguish, resisting the increase of the bladder pressure. (**b**) Urination – the muscles of the posterior group contract, generating a force vector that opens the bladder neck, while the muscles of the anterior group relax, reducing the angulation of the average urethra and allowing flow (Adapted from Palma & Netto, Uroginecologia Ilustrada, Ed. Roca, São Paulo, Brazil, 2005)

2. Anterior group: puborectalis muscle – contraction determines the anorectal angle.

For anal continence, the posterior group muscles generate a postero-inferior vector. In contrast, the puborectalis muscle contracts, angling the anal canal, while contraction of the external sphincter pulls the perineal body toward the anococcygeal ligament, further accentuating this angulation.

For elimination of feces, the posterior group muscles stress the rectovaginal septum and the posterior rectal wall, while the puborectalis muscle relaxes; this synergistic action of the two muscle groups straightens the anal canal. Finally, the external and internal sphincters relax, and the puboanal muscle (lungitudinal muscle of the anus) contracts, shortening and opening the anal canal (Fig. 11.6).

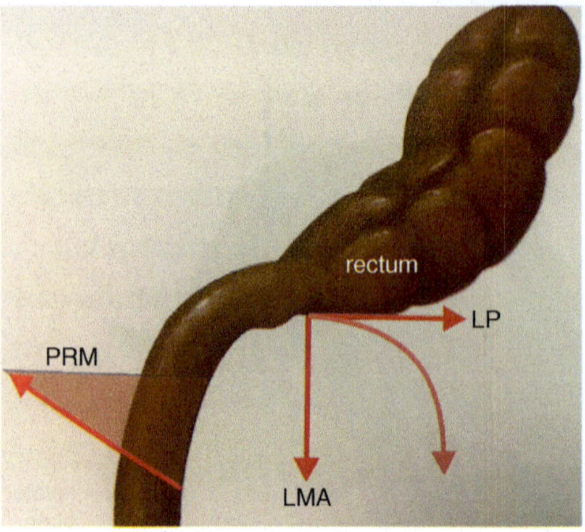

Fig. 11.6 Dynamics of anorectal emptying and continence emptying. To promote continence, the levator plate generates a postero-inferior vector. In contrast, the puborectal muscle contracts, angling the anal canal, while the contraction of the external sphincter pulls the perineal body toward the anococcygeal ligament, further accentuating this angulation. For stool elimination, the levator plate tenses the rectovaginal septum and the posterior rectal wall, while the puborectal muscle relaxes; this synergistic action of the two muscle groups straightens the anal canal. Finally, the external and internal sphincters relax, and the longitudinal muscle of the anus contracts, shortening and opening the anal canal. PRM, puborectal muscle; LMA, puboanal muscle; LP, levator plate

Neurological Control Over the Pelvic Floor and Bladder and Bowel Function

The interaction between the aforementioned muscular and soft tissue structures is due to a fine and complex control of the autonomous (sympathetic and parasympathetic) and somatic nervous systems. The mechanisms involved in this control are not yet completely elucidated, but we will seek, in the following section, to summarize the current knowledge about neuroanatomy and neurophysiology of the pelvic floor.

Efferent Innervation (Motor)

Three groups of nerves control the activity of the pelvic floor and viscera: the nerves of the autonomic sympathetic and parasympathetic nervous system and the nerves of the somatic nervous system.

Nerves of the Sympathetic Nervous System

Sympathetic innervation of the bladder originates in the periaortic sympathetic ganglia of T10 to L2, from which the fibers that form the superior hypogastric plexus (presacral plexus) emerge gives origin to the hypogastric nerves, which run distally and laterally over the uterosacral ligaments to meet the pelvic splanchnic nerves and form the inferior hypogastric plexus. This plexus contains sympathetic, parasympathetic, and sensory fibers, which form the vesical bundle and enter the bladder wall together with the ureters before branching to the urethra, trigone, and detrusor [11, 12].

The innervation of the descending colon, the sigmoid colon, and the proximal rectal colon comes from the lumbar splanchnic nerves (L1 to L3), which synapse in the inferior mesenteric ganglion and follow the arterial supply up to the wall of the loop. The fibers that innervate the middle and distal rectum, anal canal, and internal anal sphincter have the same origin (L1 to L3) but follow from the mesenteric ganglion to the superior hypogastric plexus. Here, they form the hypogastric nerves, making up the lower rectal beam of the hypogastric plexus. Hence, they follow the fascia of the pubococcygeus muscle and enter the anus in the intersphincteric space, integrating into the myoenteric plexus (Auerbach's plexus). The portions proximal to the splenic flexure of the colon are innervated by the vagus nerve [13–15].

The release of noradrenaline by the sympathetic fibers activates type 1 ($\alpha 1$) alpha adrenergic receptors on the inner sphincter of the urethra and beta 3-adrenergic receptors on the detrusor, promoting urethral contraction and bladder relaxation – thus, continence [16, 17]. Similarly, it promotes contraction of the internal sphincter of the anus via $\alpha 1$ receptors [17, 18].

Nerves of the Parasympathetic Nervous System

The pelvic splanchnic nerves originate from S2, S3 and S4 nerve roots and cross a short distance in the pararectal fossa to form the inferior hypogastric plexus. The nerves then travel laterally and distally toward the urethra and bladder (in the vesical branch of the inferior hypogastric plexus), or medially, in the direction of the medial and proximal rectum, through the rectal branch of the inferior hypogastric plexus (Fig. 11.7) [12, 17, 19, 20].

The release of acetylcholine from the vesical branch fibers activates muscarinic type 3 (M3) receptors, promoting detrusor contraction and urethral relaxation – thus, micturition [16]. The activation of the fibers of the rectal branch stimulates the Auerbach (myenteric) plexus [15].

Nerves of the Somatic Nervous System

The pudendal nerves originate from nerve roots S2, S3 and S4, leaving the pelvis through the pudendal canal (Alcock's canal) and traveling toward the perineum where it branches into the inferior rectal nerves, perineal nerve, and dorsal nerve of the clitoris (or the penis, in men) [11]. The inferior rectal nerve provides the main motor innervation of the external anal sphincter. The perineal nerve innervates the transverse perineal muscles, bulbospongiosus muscle, the ischiocavernosus muscle, the striated sphincter of the urethra, and the anterior pubococcygeus muscle fibers, specifically the pubovaginal (puboprostatic) muscles [21, 22]. The levator ani nerves originate from S2, S3 and S4 and provide motoric and sensory innervation to the iliococcigeus and coccigeus (ischiococcigeus) muscles [22].

Afferent Innervation

Bladder and rectal proprioception is performed by Aδ myelin fibers that ascend to the pontine and hypothalamic centers [17] primarily by the hypogastric nerves [11]. Bladder nociception is carried mainly by unmyelinated C-fibers to the central nuclei, predominantly through the pelvic splanchnic nerves. In addition to responding to nociceptive stimuli, these fibers trigger with

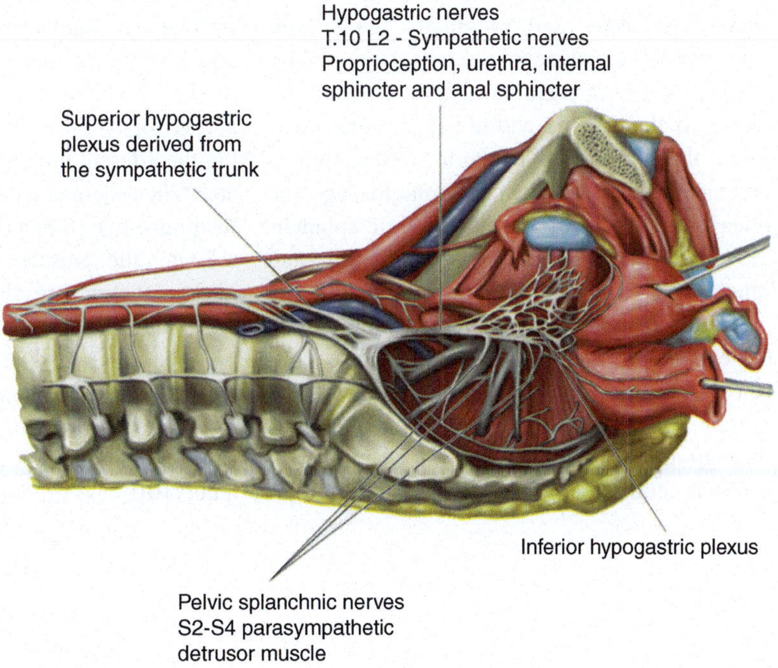

Fig. 11.7 The nerves of the autonomic nervous system. Fibers originating in the sympathetic ganglia of T10 to L3 form the superior hypogastric plexus, which in turn gives rise to the hypogastric nerves. The pelvic splanchnic nerves originate from S2 to S4 and join the hypogastric nerves to form the inferior hypogastric plexus, which has sympathetic, parasympathetic, and sensory fibers

higher pressures and are associated with intense voiding urge [11]. In the descending colon, sigmoid, and rectum, the hypogastric and lumbar nerves are responsible for nociception, while the pelvic splanchnic nerves transmit the proprioceptive stimuli [23]. The muscle spindles of the pelvic floor give signals for the Aδ and Aβ fibers of the pudendal nerves (muscles of the anterior group) and nerves of the levator ani (muscles of the posterior group).

Neurophysiology of Continence and Emptying

During the bladder-filling stage, the low-pressure trigone receptors trigger low-frequency stimuli via hypogastric nerves to the pontine micturition center. In response to the low-frequency stimuli, the pons maintains a sympathetic predominance, with release of norepinephrine by the hypogastric nerves, promoting detrusor relaxation and contraction of the internal urethral sphincter. The pons also sends stimulatory signals to Onuf's nucleus in the *conus medullaris*, which maintains the tonic contraction of the external urethral sphincter, via pudendal nerves [8, 11, 17].

As the pressure rises, the frequency of the signals increases, and eventually the stimuli from the pelvic splanchnic nerves, triggered by the high-pressure receptors of the trigone, are added to them. This additional stimulus activates the pontine micturition reflex, which increases the parasympathetic tone through the release of acetylcholine by the pelvic splanchnic nerves, causing detrusor contraction and relaxation of the internal urethral sphincter. The pons also emits sensitive signals to the frontal cortex, which makes the voiding desire conscious. The brain analyzes the situation and, if it is not suitable for emptying, sends signals through the pudendal and levator ani nerves ordering contraction of the external urethral sphincter and pelvic floor muscles. The concomitant contraction of the anterior and posterior compartment muscles stresses the pubocervical fascia, supporting the trigone and decreasing the tension on its receptors. In addition, the brain sends modulating signals to the pontine micturition center, interrupting the parasympathetic reflex and restoring the pattern of continence and inhibiting urination. With progressive bladder filling, there is a new activation of the pontine reflex of urination, and this cycle repeats itself, until the point where bladder capacity is reached. From there, the elevation of the trigone is no longer sufficient to decrease the tension on the receptors, and the voiding desire becomes intense. If the situation is adequate for emptying, the brain controls the relaxation of the muscles of the anterior compartment and the contraction of the muscles of the posterior compartment, promoting urethral opening and emptying (Fig. 11.8) [7, 8, 11, 17].

The role of extrinsic innervation in the dynamics of rectal emptying is less important than in the bladder, since sympathetic and parasympathetic stimuli exert a modulating or excitatory influence on the myoenteric plexus [23]. The role of the pelvic floor musculature, however, is important in anorectal function. Similar to what occurs in the bladder, simultaneous contraction of the muscles of the anterior and posterior compartments of the pelvic floor promotes anorectal angling and retropulsion of the rectal contents, decreasing the afferent impulses and the desire to evacuate. When there is a desire to trigger evacuation, the brain slows down the pudendal nerve stimuli, relaxing the muscles of the anterior compartment, and stimulates the nerves of the levator ani, promoting the contraction of the posterior compartment muscles, which proximally tract the rectovaginal (rectoprostatic) fascia and the posterior wall of the rectum. This straightens the anal canal and facilitates the elimination of rectal contents [10].

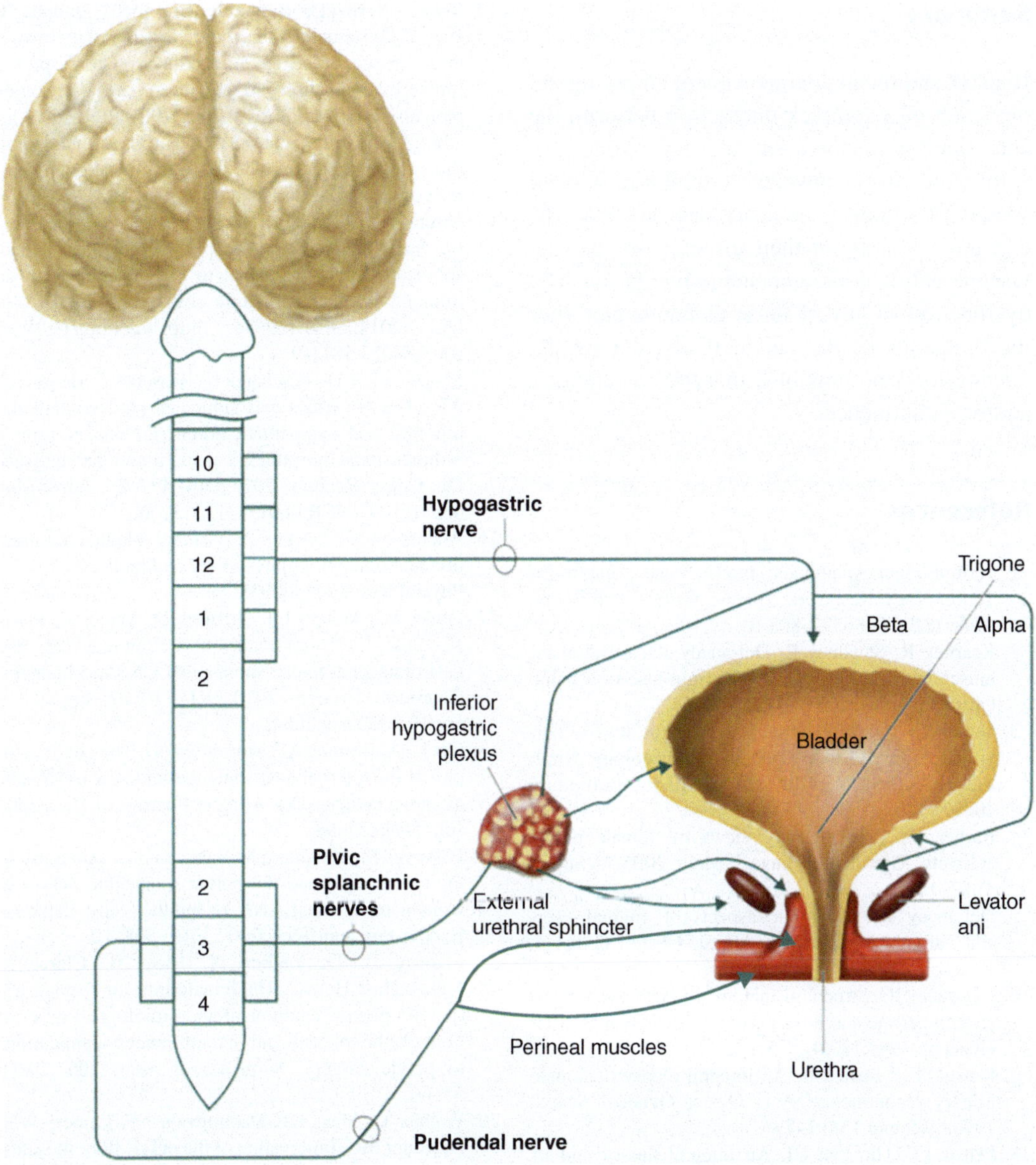

Fig. 11.8 Neurophysiology of micturition: the hypogastric nerves provide the sympathetic fibers to the inferior plexus, while the pelvic splanchnic nerves provide the parasympathetic innervation. Mixed bundles from the inferior hypogastric plexus reach the bladder together with the ureters. They control continence by beta-adrenergic fibers in the detrusor and alpha-adrenergic in the urethra and emptying by means of muscarinic fibers in the detrusor and urethra. Somatic control of the external anal sphincter and pelvic floor muscles is done by the pudendal nerves and direct fibers to the levator ani muscles

Summary

Bladder and rectal continence and emptying are the result of a complex interaction between the autonomic sympathetic (promoting continence), autonomic parasympathetic (emptying), and somatic (voluntary control) nervous systems and their actions on their effector organs: the viscera, pelvic floor, and endopelvic fascia. The dysfunction of any of these elements can alter the dynamics of the pelvic floor as a whole, leading to symptoms of urinary/fecal retention and/or incontinence.

References

1. Terminologia Anatomica: International Anatomical Terminology. New York: Thieme Medical Publishers. 1998. ISBN 0–86577–808-6.
2. Kearney R, Sawhney R, DeLancey JO. Levator ani muscle anatomy evaluated by origin-insertion pairs. Obstet Gynecol. 2004;104(1):168–73.
3. DeLancey JO. Surgical anatomy for the female pelvic floor. In: TeLinde's operative gynecology. Ninth Edition. Lippincot, Williams & Wilkins, Baltimore, 2003.
4. Barber M. Contemporary views on female pelvic anatomy. Cleveland Clinic J Med. 2005;72(Suppl. 4):S3–11.
5. DeLancey JO. Anatomic aspects of vaginal eversion after hysterectomy. Am J Obstet Gynecol. 1992;166:1717–28.
6. DeLancey JO. Structural anatomy of the posterior pelvic compartment as it relates to rectocele. Am J Obstet Gynecol. 1999;180(4):815–23.
7. Petros PE, Ulmsten U. An integral theory of female urinary incontinence. Acta Obstet Gynecol Scand. 1990;69(Suppl 153):1–79.
8. Petros PE, Ulmsten UI. An integral theory and its method for the diagnosis and management of female urinary incontinence. Scand J Urol Nephrol Suppl. 1993;153:1–93.
9. DeLancey JO. Structural aspects of the extrinsic continence mechanism. Obstet Gynecol. 1988;72(3 Pt 1): 296–301.
10. Petros PE, Swash M. The Musculo-elastic theory of anorectal function and dysfunction. Pelviperineology. 2008;27:89–93. http://www.pelviperineology.org.
11. DeGroat WC, Yoshimura N. Anatomy and physiology of the lower urinary tract. In: Handbook of clinical neurology. 3rd Series. Ed. Elsevier. Oxford, Reino Unido, 2015.
12. Mauroy B, Bizet B, Bonnal JL, Crombet T, Duburcq T, Hurt C. Systematization of the vesical and uterovaginal efferences of the female inferior hypogastric plexus (pelvic): applications to pelvic surgery on women patients. Surg Radiol Anat. 2007;29(3):209–17.
13. Chung EA, Emmanuel AV. Gastrointestinal symptoms related to autonomic dysfunction following spinal cord injury. Prog Brain Res. 2006;152:317–33.
14. Kinugasa Y, Arakawa T, Murakami G, Fujimiya M, Sugihara K. Nerve supply to the internal anal sphincter differs from that to the distal rectum: an immunohistochemical study of cadavers. Int J Color Dis. 2014;29(4):429–36. https://doi.org/10.1007/s00384-013-1811-9.
15. Moszkowicz D, Peschaud F, Bessede T, Benoit G, Alsaid B. Internal anal sphincter parasympathetic-nitrergic and sympathetic-adrenergic innervation: a 3-dimensional morphological and functional analysis. Dis Colon Rectum. 2012;55(4):473–81. https://doi.org/10.1097/DCR.0b013e318245190e.
16. Andersson KE, Arner A. Urinary bladder contraction and relaxation: physiology and pathophysiology. Physiol Rev. 2004;84:935–86.
17. Drake MJ, Fowler CJ, Griffiths D, Mayer E, Paton JF, Birder L. Neural control of the lower urinary and gastrointestinal tracts: supraspinal CNS mechanisms. Neurourol Urodyn. 2010;29(1):119–27. https://doi.org/10.1002/nau.20841.
18. Cook TA, Brading AF, Mortensen NJ. The pharmacology of the internal anal sphincter and new treatments of anorectal disorders. Aliment Pharmacol Ther. 2001 Jul;15(7):887–98.
19. Possover M, Chiantera V, Baekelandt J. Anatomy of the sacral roots and the pelvic splanchnic nerves in women using the LANN technique. Surg Laparosc Endosc Percutan Tech. 2007;17(6):508–10.
21. Mauroy B, Demondion X, Bizet B, Claret A, Mestdagh P, Hurt C. The female inferior hypogastric (pelvic) plexus: anatomical and radiological description of the plexus and its afferences--applications to pelvic surgery. Surg Radiol Anat. 2007;29(1): 55–66.
20. Wallner C, Maas CP, Dabhoiwala NF, Lamers WH, DeRuiter MC. Innervation of the pelvic floor muscles: a reappraisal for the levator ani nerve. Obstet Gynecol. 2006;108(3 Pt 1):529–34.
22. Grigorescu BA, Lazarou G, Olson TR, Downie SA, Powers K, Greston WM, Mikhail MS. Innervation of the levator ani muscles: description of the nerve branches to the pubococcygeus, iliococcygeus, and puborectalis muscles. Int Urogynecol J Pelvic Floor Dysfunct. 2008;19(1):107–16.
23. Feng B, Brumovsky PR, Gebhart GF. Differential roles of stretch-sensitive pelvic nerve afferents innervating mouse distal colon and rectum. Am J Physiol Gastrointest Liver Physiol. 2010;298(3):G402–9. https://doi.org/10.1152/ajpgi.00487.2009.

12 Electromyography and Pudendal Nerve Terminal Motor Latency

Guillermo O. Rosato
and Lucia Camara Castro Oliveira

Introduction

Electromyography (EMG) is an assessment method used in the treatment of neuromuscular diseases, including pelvic floor [1] disorders. This method, initially developed by Piper [2] in 1908, facilitates the evaluation of the motor unit components—the second neuron and the muscle fibers it innervates. Thus, EMG records the electrical activity generated by muscle fibers during voluntary contractions and at rest [3, 4]. EMG is a neurophysiological assessment technique that can help explain the pathophysiology of symptoms in patients with a clear diagnosis [5]. EMG may be valuable in the assessment of patients with anorectal dysfunction and is complementary to imaging and manometry. The neurophysiology of the pelvic floor demonstrates muscle denervation and quantitatively estimates muscle reinnervation and the level of motor neuron excitability [6].

The most frequently used neurophysiological tests include pudendal nerve terminal motor latency, sympathetic skin response, perineal electromyography, and evoked potentials.

G. O. Rosato (✉)
Hospital Universitario Austral, Department of Colorectal Surgery, Buenos Aires, Argentina

L. C. C. Oliveira
Director Department of Anorectal Physiology of Rio de Janeiro and CEPEMED, Rio de Janeiro, Brazil

Based on current neuroanatomical and neurophysiological knowledge of the pelvic floor, we can choose the most accurate method to assess the central and peripheral nervous system and help diagnose different clinical disorders, such as neurogenic bladder, fecal incontinence, constipation, paradoxical contraction of puborectalis muscle, urinary retention, male sexual impotence, and chronic pelvic pain. There are at present various studies to quantify electrical activity; however, not all of them can be used for the electromyographic evaluation of pelvic floor disorders. Thus, EMG can provide important information relative to the external anal sphincter and the puborectalis muscle, as well as facilitate the assessment of the electrical activity during voluntary contraction and straining.

Injury of the pelvic floor contributes to the loss of pelvic structure and organ support, accounts for fecal and urinary incontinence, genital and rectal prolapse, and constitutes the pathophysiological background for the failure of pelvic floor syndrome.

Neurophysiological testing of the pelvic floor involves two basic steps: (1) electroneurography to assess neuroconduction of the pudendal nerve and (2) electromyography to record the muscular electrical activity by concentric needle EMG or single-fiber EMG [7]. In general terms, anal EMG is better to evaluate the functional activity of pelvic floor muscles during voluntary contractions, cough reflex, or straining at stool, as well as to detect denervation of the anal sphincter; however, it is not an adequate method to

L. C. C. Oliveira (ed.), *Anorectal Physiology*, https://doi.org/10.1007/978-3-030-43811-1_12

differentiate the abnormal electric activity of myopathic or neurogenic diseases.

Anal EMG was introduced in 1929 by Adrian and Bronk [4] through the development of concentric electrodes. In recent years, this method has become one of the ways to test pelvic floor muscles functional activity and thus a series of functional disorders, particularly anal incontinence and paradoxical puborectalis contraction syndrome, among others (Table 12.1).

Table 12.1 Indication of anal EMG and pudendal nerve terminal motor latency

Fecal Incontinence
Sphincteroplasty—preoperative evaluation
Neurogenic or idiopathic incontinence
Increased perineal descent
Patients over 60 years of age
Constipation
Anismus
Solitary rectal ulcer syndrome
Increased perineal descent
Adjunct to botulinum toxin injection
Miscellaneous
Neurological diseases
Parkinson's disease
Multiple system atrophy
Myelomeningocele
Arachnoiditis
Cauda equina tumors
Sacral agenesis
Vulvodynia
Urinary incontinence

Electromyography

Electromyography records the electrical activity that is generated by muscle fibers and helps evaluate the integrity of their innervation [3, 4]. Muscle activity can be recorded using surface electrodes, concentric needle electrodes, and wire electrodes (Fig. 12.1a–d) [7].

The purpose of EMG in the assessment of pelvic floor disorders is to determine the following:

- Muscle fiber denervation
- Muscle fiber reinnervation (single-fiber EMG)
- Sphincter integrity
- Adequate contraction/relaxation during muscle activity (surface electrodes) [2, 3]

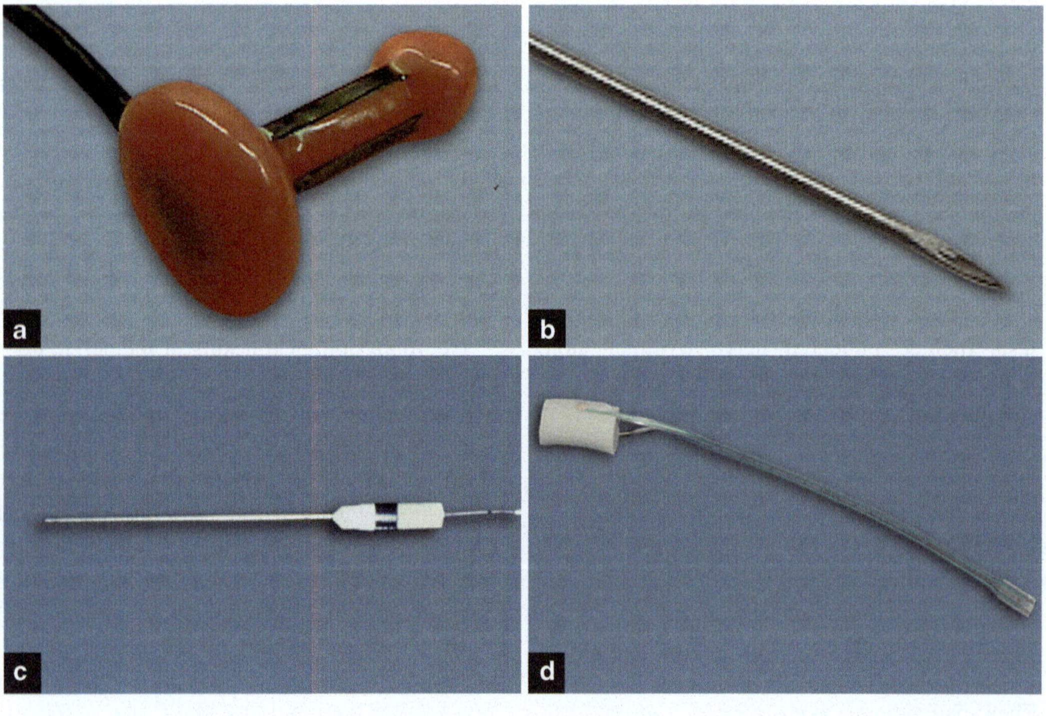

Fig. 12.1 (**a**) Anal plug surface electrode. (**b**) Wire electrode. (**c**) Concentric needle electrode. (**d**) Anal plug sponge electrode

Concentric EMG

Patients should be examined in a quiet room and receive a full detailed explanation about the proceeding. Rectal cleansing with enema 2 hours before the examination is recommended. Patients are placed in the left lateral decubitus position, with a ground surface electrode fixed on one of the lower limbs. The perianal skin is cleansed with a swab soaked in an antiseptic solution. A 75-mm concentric needle electrode is introduced through the skin in both the left and the right sides of the perianal region, with no anesthetic, at 1–1.5 cm from the anal verge.

The introduction of this electrode is followed by a reactive discharge of motor unit potentials (MUPs), which is more evident in anxious patients. It is necessary to distinguish this activity from that known as "insertion activity," which results from the mechanical stimulation of the muscle fibers. This reaction disappears rapidly as the patient relaxes. In order to avoid conceptual confusion, it is preferable to use the term "reactive activity" to discriminate from insertion activity. It is very difficult to see or register this insertion activity at the anal sphincter because of the reactive activity of the MUPs.

The insertion activity in skeletal muscles is a parameter which allows for the identification of the state of excitability and contractile capacity of the muscle fibers. In the puborectalis muscle and the external anal sphincter, MUP discharges also provide information concerning the contractile capacity of muscle fibers. In cases of fibrosis, there is a loss of contractile capacity of these muscle fibers, and subsequently no MUP or fibrillation is recognized. In the case of denervation, MUP activity is replaced by fibrillation. The response to these different maneuvers (squeeze or cough) in healthy individuals shows an increase in activity (recruitment of MUPs) and a significant decrease or eventual electrical silencing during straining [8–11]. The needle electrode is then advanced four to five centimeters parallel to the anal canal, where electrical activity is detected. The electrode reaches the puborectal level, and electrical activity is recorded at rest. Then, patients are asked to squeeze, cough, and strain, and each of these events is registered on a print-out EMG paper (Fig. 12.2). The electrode is then withdrawn, passing across a zone where no electrical activity is registered, until the examination reaches another area with myoelectrical potentials, assumed to be at the external anal sphincter level. Here, again, recordings are made of the reactive activity, and as for the puborectalis muscle, recordings are made at rest and during squeezing, coughing, and straining (Fig. 12.3).

Single-Fiber EMG

This is a technique complementary to concentric needle EMG which provides additional information.

In the early 1970s, Stålberg and Trontelj [12] described a method to record individual muscle fiber action potentials. Patients are examined in the same position as used in concentric needle EMG. A special 25 mm-needle electrode records muscle fibers electrical activity over a recording surface of 250 mm. A reference electrode is placed in a zone of electrical quiescence. Single fiber potentials, unlike those obtained with a concentric needle, are of shorter duration, higher amplitude, and shorter rise time [13–15].

The main information provided by single-fiber EMG is fiber density. This is the mean number resulting from the analysis of single muscle potentials in 20 different positions of the recording electrode at the same muscle. Potentials accepted for analysis should be greater than 100 microvolts. Normal fiber density is 1.5 ± 0.16 but tends to increase after the age of 60 years [16].

EMG Abnormalities

The following describe some of the abnormalities that can be found with EMG.

- *Muscle injury:* Regardless of the cause, the EMG assessment of muscle injury shows a significant reduction of MUPs in all the situations, that is, at rest, during squeezing, coughing, or straining (Fig. 12.4).

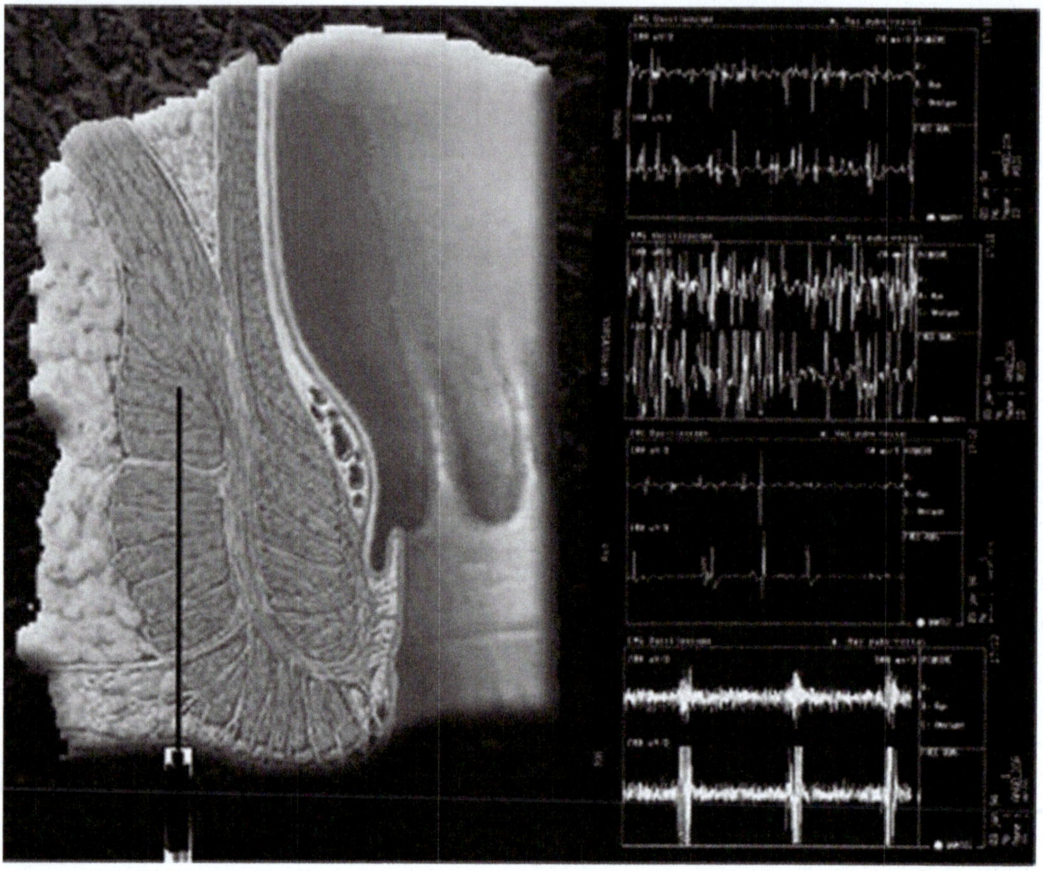

Fig. 12.2 EMG recordings at the puborectalis muscle during squeezing, coughing, and straining

- *Neurogenic injury:* Muscle denervation leads to the progressive atrophy of muscle fibers. In the case of partial nerve injury, reinnervation of the damaged axons and redistribution to the muscle fibers in the motor units can occur. These alterations result in changes in MUP amplitude and duration. In severe cases, the electrical activity can be extremely reduced or absent. Thus, these alterations involve an increase in waveform amplitude and action potentials duration, the presence of polyphasic potentials and increased density of muscle fibers (Fig. 12.5).
- *Functional diseases:* The increase in the pelvic floor muscle activity during straining, resulting from failure to relax by the puborectalis muscle, is the best example of functional disorders and justifies the use of EMG to assess patients with intestinal constipation (Fig. 12.6).

Electroneuromyography or Nerve Conduction Studies

Nerve stimulation techniques or conduction testing can be used in combination with conventional or single-fiber EMG. These tests include pudendal nerve terminal motor latency and spinal latency. This chapter will specifically deal with pudendal nerve terminal motor latency.

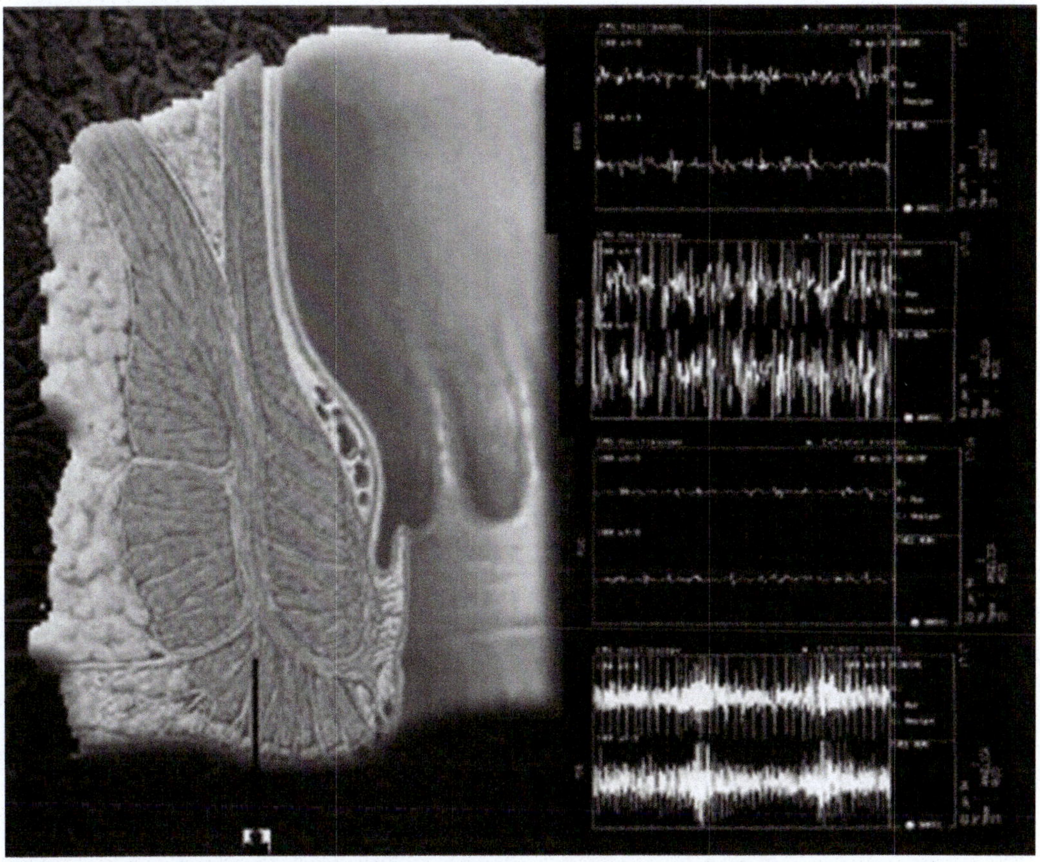

Fig. 12.3 EMG recordings in the external anal sphincter at rest and during squeezing, coughing, and straining

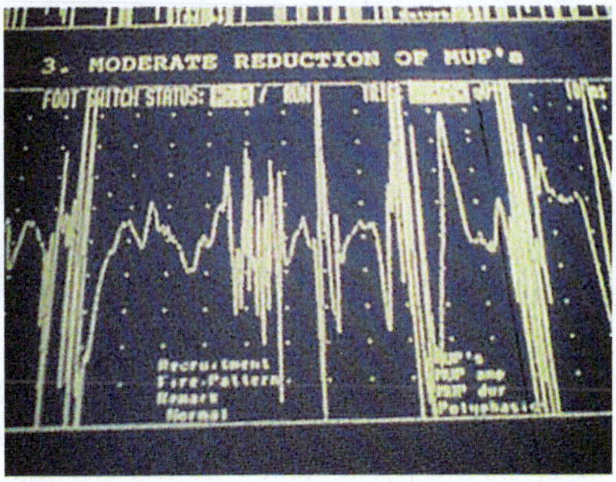

Fig. 12.4 EMG assessment of muscle injury: significant reduction of MUPs

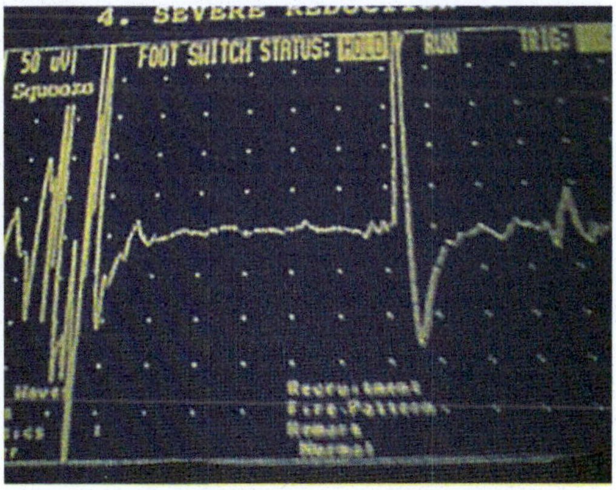

Fig. 12.5 EMG showing a neurogenic injury with increased waveform amplitude and action potential duration

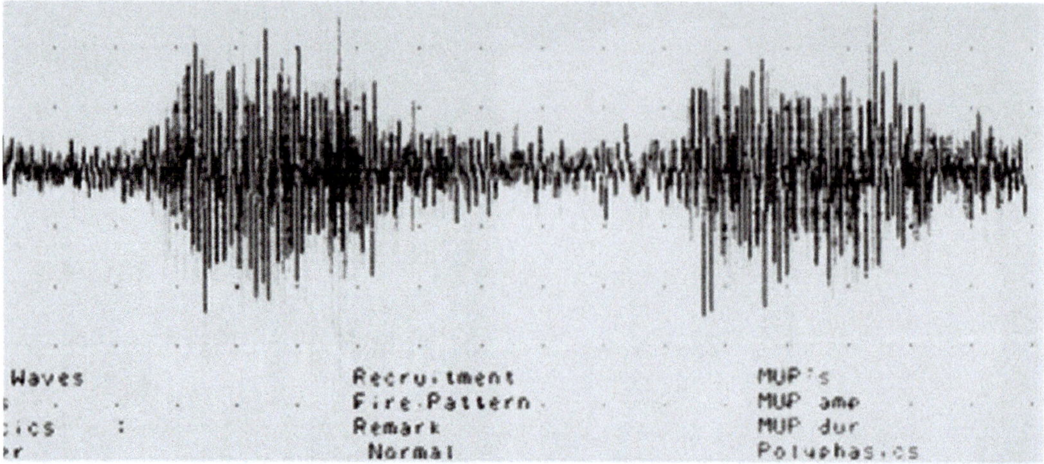

Fig. 12.6 EMG showing an increase in the pelvic floor muscle activity during straining, which characterizes paradoxical puborectalis contraction or anismus

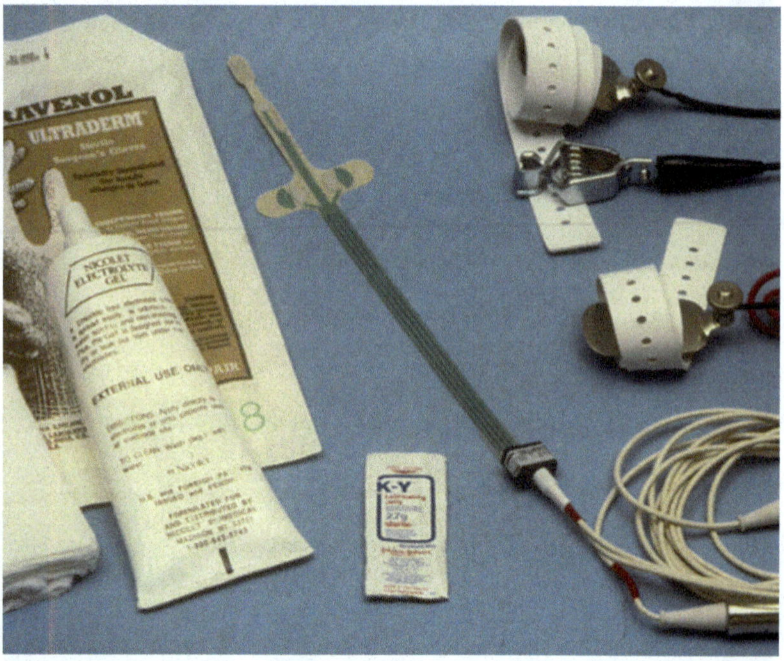

Fig. 12.7 Electrodes and wires used in the assessment of the pudendal nerve terminal motor latency

Pudendal Nerve Terminal Motor Latency

The pudendal nerve terminal motor latency (PNTML) technique was initially used by Kiff and Swash in 1984 [16]. PNTML consists in the transrectal stimulation of bilateral pudendal nerves with the patient in the left lateral decubitus position. The electrical stimulation of the pudendal nerves allows for the measurement of conduction speed along the terminal portion of the nerve, which is the most vulnerable and likely to suffer traumatic injuries. A disposable stimulating electrode (St. Mark's®) is used, attached to a surgical glove and an EMG equipment. The examining finger is directed towards the ischial spine (Figs. 12.7 and 12.8). A stimulus of 0.1 milliseconds in duration is delivered at 1-s intervals.

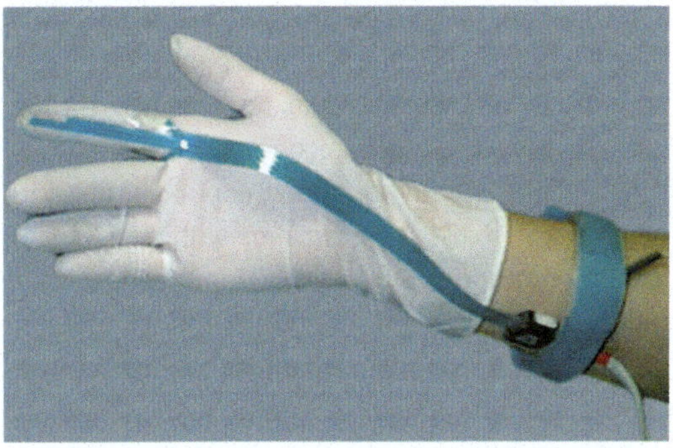

Fig. 12.8 St. Mark's® stimulating electrode attached to a surgical glove and the EMG cable

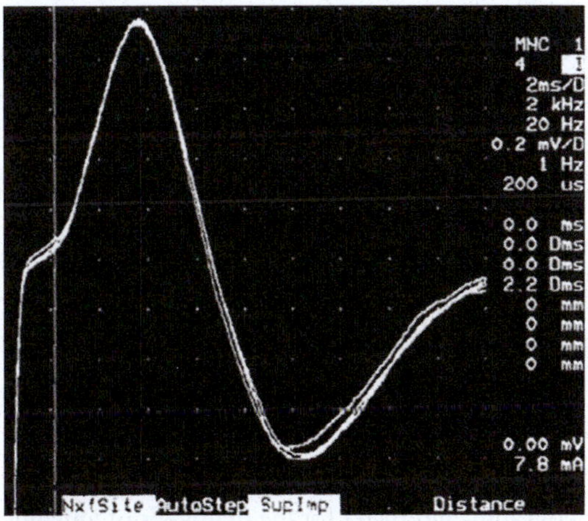

Fig. 12.9 Normal latency of the pudendal nerve

Pudendal nerve injury may occur as result of two mechanisms:

- Stretching during labor or chronic straining at stool.
- Direct injury of pelvic nerves causing excessive perineal descent and stretching of pudendal nerves [17].

The normal latency reference value is up to 2.2 milliseconds (Fig. 12.9). It should be evaluated on each side, whenever possible, since the injury may be asymmetrical. Latency increases with age [17], and in healthy subjects it is normally higher in men than in women [16].

The use of PNTML is particularly important in the assessment of patients with incontinence. Vaccaro et al. [18] analyzed 395 consecutive

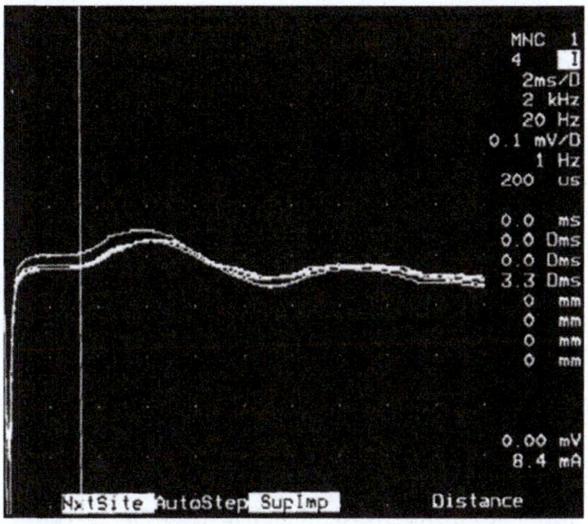

Fig. 12.10 Pudendal nerve latency recording with normal response

patients, 223 of whom suffered incontinence, and 172 constipation. The global frequency of latency prolongation found was 31.4%, with 37.2% of incontinent patients presenting prolonged latency. Pudendal nerve latency measurement is an important test to assess patients with incontinence, particularly those with surgical indication. Presence of neuropathy represents a negative prognosis for anal sphincter repair [19–21].

Gilliland et al. [22] studied a group of female patients who underwent anterior overlapping sphincteroplasty due to obstetric defects. Complete physiological investigations were performed, including manometry, PNTML, and ultrasound. The authors concluded pudendal neuropathy, particularly bilateral lesions, represents a negative predictive factor for surgical treatment.

Pudendal latency can also be prolonged in patients with excessive perineal descent and rectal prolapse (Fig. 12.10).

In patients with double incontinence, fecal and urinary, nerve conduction in the pudendal nerve branches innervating the periurethral sphincter and the external anal sphincter is also prolonged. The degree of the lesion in the perineal nerve is greater in patients with double incontinence than in those with fecal incontinence alone [23].

Transcutaneous lumbar stimulation at the L1 and L4 level has been used in patients with myelographic evidence of lumbar canal stenosis, cauda equina tumors, arachnoiditis, and sacral agenesis.

EMG in Anal Incontinence

Anal incontinence is a condition of complex mechanisms and with various associated factors [24]. In addition to musculature integrity, stool consistency, and intestinal transit time, pelvic floor innervation represents one of the factors in normal continence. Denervation of the pelvic floor resulting from obstetric trauma, advanced age, pudendal neuropathy, collagen diseases, diabetes, or a combination of these can lead to various degrees of incontinence.

In these cases, EMG testing facilitates muscle injury mapping in the presence of a series of abnormalities, such as a true electrical silence or total absence of electrical signals in old and severe injuries, denervation patterns or polyphasic potentials, and in cases of reinnervation, increased amplitude and duration of muscle activity, presence of polyphasic potentials, and higher density of fibers [25]. In patients with neurogenic incontinence, EMG abnormalities are characteristic—in the most severe cases muscle activity is reduced or absent [17].

Although EMG can provide these data, it has been superseded by the anal canal sonography, a method that causes no discomfort and is tolerated by patients, in addition to facilitating internal anal sphincter evaluation. Likewise, sonography yields sensitivity and specificity values of almost 100% to detect sphincter abnormalities as compared to EMG [26, 27].

EMG in Anismus

Although anismus, or paradoxical puborectalis contraction, remains a controversial diagnosis in the literature [28], it can be understood as a distal functional defecation disorder, whose most frequent symptoms include defecation difficulty, feeling of incomplete evacuation, tenesmus, and the need for enemas or suppositories [29, 30].

Diagnosis should be initially based on patient history, considering the symptoms above described. Additional information can be obtained through anal manometry, cinedefecography, colonic transit time, anal EMG, and balloon expulsion test [30]. The criteria to establish anismus with cinedefecography are puborectalis contraction or puborectalis relaxation during the expulsion phase, and prolonged and incomplete evacuation of the contrast media in absence of other factors, such as large rectoceles or intussusception.

The external anal sphincter and puborectalis muscles have basal tone at rest, even during sleep. Their activity increases as abdominal pressure increases and is inhibited during defecation. The increase in EMG activity during straining has been established for a series of disorders which share the symptom of increased straining, such as solitary rectal ulcer syndrome, hemorrhoids, and symptomatic uterovaginal prolapse. EMG findings in anismus show failure of muscle relaxation during the expulsion phase, that is, prolongation of the puborectalis and external anal sphincter electrical activity with the patient on attempted evacuation (Fig. 12.6).

In a prospective study comparing cinedefecography and EMG for the diagnosis of anismus, Jorge et al. [31] observed sensitivity and specificity of 70% and 80%, and 67% and 83%, respectively. Therefore, a combination of these two tests is suggested for the diagnosis of anismus. In these cases, EMG should be performed in a quiet room with the patient in the left lateral decubitus position, and preferably with bipolar concentric electrodes.

Other EMG Indications

EMG has been used to assess patients with multiple system atrophy (MSA), a degenerative disease manifesting a combination of parkinsonism, cerebellar, pyramidal, and autonomic (including urinary, sexual, and anorectal) dysfunction [32]. EMG has also been proposed for the early diagnosis of MSA since it can detect various sphincter abnormalities, such as motor unit potential changes, abnormal spontaneous activity of the anal sphincter, and alteration of the neurogenic pattern.

In addition, EMG has been useful for distinguishing MSA from Parkinson's disease [33]. However, the method shows low sensitivity, and further studies are required.

Surface electrode EMG is also helpful in the treatment of anismus by means of biofeed-back (Fig. 12.1a). More recently, the introduction of botulinum toxin for the treatment of anismus has become an indication for EMG and for monitoring the most suitable injection site, that is, the most active portion of the puborectal musculature [31].

Summary

Pelvic floor EMG is one of a group of available tests to assess continence and defecation disorders. They can be used in combination with clinical data to obtain the best possible results. These methods have been superseded by other pelvic floor studies such as high resolution manometry, ultrasonography, and magnetic resonance imaging of the pelvis, since they cause little discomfort to the patient.

According to the guidelines of the American Society of Colon and Rectal Surgeons (ASCRS), those examinations may be performed but have limited impact on the diagnosis and management of patients with fecal incontinence. Therefore, it is currently not routinely recommended, although it has a strong recommendation and level 1B evidence [34].

References

1. Rosato GO, Lumi CM, Miguel AM. Anal sphincter electromyography and pudendal nerve terminal motor latency assessment. Semin Colon Rectal Surg. 1992;3(2):68–74.
2. Piper H. Uber die leitungsgeschwindigkeit in den markhaltigen, menslichen nerven. Pflug Arch Gesamte Physiol Mensch Tiere. 1908;124:591–600.
3. Snooks SJ, Swash M. Nerve stimulation techniques. In: Henry MM, Swash M, editors. Coloproctology and the pelvic floor: pathophysiology and management. London: Butterworths; 1985. p. 184–206.
4. Adrian ED, Bronk DW. The discharge of impulses in motor nerve fibers: II. The frequency of discharge in reflex and voluntary contractions. J Physiol. 1929;67:119–51.
5. Marangone TNFP, Squintani G, Crovato M, TNFP, Donato F, Leopardi F, D'Amico A, Romito S, Moretto G. Predictive value of neurophysiological testing and the importance of multidisciplinary approach in the pelvic floor disorders. Clin Neurophysiol. 2005;126(2015):e1–e28.
6. Podnar S. Electrodiagnosis of the anorectum: a review of techniques and clinical applications. Tech Coloproctol. 2003;7(2):71–6.
7. Swash M. Electrophysiological investigation of the posterior pelvic floor musculature. In: Pemberton JH, Swash M, Henry M, editors. The pelvic floor. Its function and disorders. London: W.B. Saunders; 2002. p. 213–36.
8. Buchtal F. The electromyogram: its value in the diagnosis of neuromuscular disorders. World Neurol. 1962;3:16–34.
9. Buchthal F, Pinelli P, Rosenfalck P. Action potentials parameters in normal human muscle and their physiological determinants. Acta Physiol Scand. 1954;32:219–29.
10. Buchthal F, Guld C, Rosenfalck P. Multielectrode study of the territory of a motor unit. Acta Physiol Scand. 1957;39:83–104.
11. Buchthal F. Electromyography in evaluation of muscle disease. Method Clin Neurophysiol. 1991;2:28.
12. Stalberg E, Trontelj V. Single fiber electromyography. Surrey: Mirvalle Press Old Working; 1979.
13. Goodgold J, Everstein A. Electrodiagnosis of neuromuscular disease. 2nd ed. Baltimore: Williams & Wilkins; 1977. p. 210.
14. Swash M, Snooks SJ. Electromyography in pelvic floor disorders. In: Henry MM, Swash M, editors. Coloproctology and the pelvic floor: pathophysiology and management. London: Butterworths; 1985. p. 189–91.
15. Laurberg S, Swash M. Effects of ageing on the anorectal sphincters and their innervation. Dis Colon Rectum. 1989;32:734–42.
16. Kiff ES, Swash M. Slowed conduction in the pudendal nerves in idiopathic (neurogenic) faecal incontinence. Br J Surg. 1984;71:614–6.
17. Henry MM, Parks AG, Swach M. The pelvic floor musculature in the descending perineum syndrome. Br J Surg. 1982;69:470–2.
18. Vaccaro CA, Cheong DMO, Wexner SD, Nogueras JJ, Salanga VD, Hanson MR, et al. Pudendal neuropathy in evacuatory disorders. Dis Colon Rectum. 1995;38:166–71.
19. Wexner SD, Marchetti F, Jagelman D. The role of sphincteroplasty for fecal incontinence reevaluated: a prospective physiologic and functional review. Dis Colon Rectum. 1991;34:22–30.

20. Jacobs PP, Scheuer M, Kuijpers JHC. Obstretic fecal incontinence: role of pelvic floor denervation and results of delayed sphincter repair. Dis Colon Rectum. 1990;33:494–7.
21. Sangwan YP, Coller JA, Barret RC, et al. Unilateral pudendal neuropathy: impact on outcome of anal sphincter repair. Dis Colon Rectum. 1996;39:689.
22. Gilliland R, Altomare DF, Moreira H Jr, Oliveira L, Gilliland JE, Wexner SD. Pudendal neuropathy is predictive of failure following anterior overlapping sphincteroplasty. Dis Colon Rectum. 1998;41:1516–22.
23. Snooks SJ, Barnes PRH, Swash M. Abnormalities of the innervation of the voluntary anal and urethral sphincters in incontinence: an electrophysiological study. J Neurol Neurosurg Psychiatry. 1984;47:1269–73.
24. Oliveira L, Wexner SD. Anal incontinence. In: Beck D, Wexner SD, editors. Fundamentals of anorectal surgery. 2nd ed. New York: Saunders; 1998. p. 115–52.
25. Wexner SD, Marchetti F, Salanga VD, Corredor C, Jagelman D. Neurophysiologic assessment of the anal sphincters. Dis Colon Rectum. 1991;34:606–12.
26. Enck P, von Giesen HJ, Schafer A, Heyer T, Gantke B, Flesch S, et al. Comparison of anal sonography in the evaluation of anal sphincter defects. Am J Gastroenterol. 1996;91:2539–43.
27. Law PJ, Kamm MA, Bartram CI. A comparison between electromyography and anal endosonography in mapping external anal sphincter defects. Dis Colon Rectum. 1990;33:370–3.
28. Herbaut AG, Van de Stadt J, Panzer JM, Lalmand B, Crick DH. Paradoxical contraction of pelvic floor muscles: clinical significance. Acta Gastroenterol Belg. 1994;57:13–8.
29. Kuijpers HC, Bleijenberg G. The spastic pelvic floor syndrome: a cause of constipation. Dis Colon Rectum. 1985;28:669–72.
30. Preston DM, Lennard-Jones JE. Anismus in chronic constipation. Dig Dis Sci. 1985;30:413–8.
31. Jorge JMN, Wexner SD, Ger GC, Salanga VD, Nogueras JJ, Jagelman DG. Cinedefecography and electromyography in the diagnosis of nonrelaxing puborectalis syndrome. Dis Colon Rectum. 1993;36:668–76.
32. Vodusek DB. Sphincter EMG and differential diagnosis of multiple system atrophy. Mov Disord. 2001;16(4):600–7.
33. Libelius R, Johansson F. Quantitative electromyography of the external anal sphincter in Parkinson's disease and multiple system atrophy. Muscle Nerve. 2000;23(8):1250–6.
34. Paquette IM, Varma MG, Kaiser AM, Steele SR, Rafferty JF. The American Society of Colon and Rectal Surgeons' clinical practice guideline for the treatment of fecal incontinence. Dis Colon Rectum. 2015;58(7):623–36.

13 Anal Incontinence: Etiology and Clinical Treatment

Anuradha R. Bhama and Scott R. Steele

Introduction

Fecal incontinence is a potentially debilitating condition that can have a grave impact on quality of life. Patients may be fearful to participate in their normal social activities, and it is not uncommon for patients to report that they avoid leaving their house due to fear of embarrassment due to accidental bowel leakage. Unfortunately, the epidemiology of the disease is challenging to pinpoint given variations in definitions of incontinence and the reliance on self-reporting of the condition. Moreover, there is a wide range of etiologic factors that may result in fecal incontinence. Treatment of the disease starts with medical therapy and physical therapy but may also involve surgical management (Table 13.1).

Table 13.1 Clinical treatment of anal incontinence

Lifestyle modifications
  Cessation of smoking
  Weight loss

Dietary strategies
  Sodium and protein reduction
  Caffeine restriction
  Dietary timing manipulation
  Elimination of aggravating foods (spicy foodstuffs, cabbage, onions)
  Selective lactose restriction
  Fiber supplementation insolubles (whole grain breads, cereals, nuts, beans, fruits and vegetables with skin and sweet corn), psyllium
  Adequate fluid intake
  Exercise regime

Avoidance of drugs that exacerbate diarrhea
Mechanical barriers   The anal plug

Medications
  Antidiarrheal treatments
  Phenylephrine gel

Physical therapies
  Pelvic floor muscle training
  Biofeedback
  Electrical stimulation

Epidemiology

The prevalence of fecal incontinence is difficult to assess given that estimates of this condition rely on self-reporting and vary based upon patient population and definitions of incontinence. The definition of fecal incontinence varies based upon the type and frequency of incontinence. The generally accepted definition of fecal incontinence is any uncontrolled passage of feces or gas over at least 1 month's duration in a person over the age of 4 who previous had achieved control [1–4]. Frequently, fecal incontinence may be associated with fecal urgency—which is a sudden urge to defecate with difficulty in postponing defecation. Reports of the prevalence of FI is variable and

A. R. Bhama
Rush University, Division of Colorectal Surgery, Chicago, IL, USA

S. R. Steele (✉)
Department of Colorectal Surgery, Digestive Disease and Surgery Institute, Cleveland Clinic Foundation, Cleveland, OH, USA
e-mail: steeles3@ccf.org

L. C. C. Oliveira (ed.), *Anorectal Physiology*, https://doi.org/10.1007/978-3-030-43811-1_13

has been stated to be anywhere from 2.0% to 20.7%; this broad range reflects the inherent difficulty with defining the true prevalence of this condition [5, 6]. In patients residing in nursing homes, the reported incidence is between 50% and 70%, 18% and 33% in those who are hospitalized, and 7–15% in the community [7, 8]. Not surprisingly, fecal incontinence is the primary reason for admission to nursing homes [9].

Nearly 20% of women report at least one episode of fecal incontinence per year and 9.5% report one episode per month [10]. The incidence of fecal incontinence is supposedly equivalent in men and women, though a majority of patients seen in clinical practice are women. Risk factors for fecal incontinence in men include prostate cancer, symptomatic hemorrhoids, perianal sepsis, rectal cancer, and a history of restorative rectal resection [11]. The quality of life impact of fecal incontinence is also difficult to assess, as available scales are not comparable due to a wide range of methodology in their development and in psychometric rigor [12].

Risk Factors

Main risk factors for fecal incontinence include female gender, multiparity, complex vaginal deliveries, and a history of anorectal surgery. Complex vaginal deliveries include spontaneous tearing of varying degrees, use of vacuum assistance, and use of forceps assistance. There are several medical risk factors as well, including diabetes and several neurologic conditions, described below.

Etiology

Obstetric Trauma

Obstetric trauma is the most common cause of fecal incontinence in women. The rate of sphincter injury following vaginal deliver has been studied in several large population-based reviews. A study of over 100,000 vaginal deliveries identified a sphincter injury rate of 2.1%; forceps and vacuum delivery were significantly associated with sphincter injury [13]. Another large population based study of over 200,000 women identified a 6.5% rate of sphincter injury; vacuum extraction and higher birth weight were identified as risk factors for sphincter injury while episiotomy was found to be protective against sphincter injury [14]. Another study confirmed forceps and vacuum were risk factors for sphincter injury and also found that a previous fourth degree tear, higher birth weight, and maternal age over 35 to be risk factors for sphincter injury as well [15].

A large retrospective review over a 12-year period of nearly 1500 patients with obstetric 3rd and 4th degree tears evaluated sphincter anatomy and function at 4 months postpartum. Fifty two percent of patients has no evidence of sphincter injury, 44% of patients had normal sonographic sphincter anatomy without any evidence of prior repair, and 3% had evidence of successful repair. Of the patients with a sphincter defect, 63% had defects of both the internal and external sphincter, 36% had isolated defects of the external sphincter, and 1% had isolated defects of the internal sphincter. Anal manometry was also performed and demonstrated that patients with a combined sphincter defect had significantly lower resting pressures than those with an intact sphincter. Patients with isolated external anal sphincter defects also had significantly lower resting pressures, whereas those with isolated internal anal sphincter defects had no significant difference in resting pressures. Incontinence questionnaires were returned by 59% of patients, and no significant difference in CCF incontinence score was identified, suggesting that the effect of obstetric injury on continence likely does not present until later in life [16]. Another large Danish study of 1490 women evaluated those who had experienced a second vaginal delivery after a sphincter injury associated with the first delivery. A second sphincter injury was experienced by 7% of women who were also found to have significantly worse incontinence compared to those that did not experience a second injury. Subsequent elective cesarean was not protective in the development of fecal incontinence [17]. Obstetric trauma remains the most common cause of sphincter injury in women and has a substantial influence on continence.

Congenital Disease

There are several congenital diseases that may result in fecal incontinence. These are typically categorized as anorectal malformations/anomalies and spinal anomalies. Though the working definition of incontinence is that in one who has previously gained control, it is important to be cognizant of these congenital etiologies, as those born with anorectal malformations may present to a colorectal surgery clinics as an adult seeking management for their ongoing incontinence symptoms. In fact, 16.7–76.7% of patients with anorectal malformations report long-term fecal incontinence [18]. Anorectal malformations account for 1 in 5000 live births; several congenital conditions (syndromic and nonsyndromic) result in various types of anorectal malformations that may result in fecal incontinence. These include, but are not limited to: Trisomy 13, 18, 21, VACTERL (vertebral anomalies, anal atresia, cardiac malformations, tracheoesophageal fistula, renal anomalies, and limb abnormalities), MURCS (Mullerian duct aplasia, renal aplasia, and cervicothoracic somite dysplasia), rectoperineal fistulas, cloacal defects, and several others. The most frequently encountered diagnoses include imperforate anus and Hirschsprung's disease. Many chromosomal syndromes have associated anorectal malformations, most commonly the VACTERL syndrome. Some patients with anorectal malformations never achieve continence and effectively have a perineal colostomy. 75% of patients are able to have voluntary bowel movements, but 50% of patients have occasional soiling. The remaining 25% have full fecal incontinence and rely on a bowel management program [19]. Treatment options for children with incontinence is beyond the scope of this discussion. However, most children are managed with an aggressive bowel program and/or various types of antegrade/retrograde enemas or a colostomy.

Neurologic Conditions

There are several neurologic conditions, structural and functional, that can cause fecal incontinence. Systemic diseases that can result in fecal incontinence include multiple sclerosis, diabetes mellitus, tabes dorsalis, and dementia. Structural causes may include traumatic brain injury, spinal cord injury, stroke, or iatrogenic injury due to spinal cord or brain surgery.

Patients with multiple sclerosis (MS) can experience fecal incontinence, often alternating with constipation, which correlate with MS involving the spinal cord. Incontinence is thought to be due to anal sphincter weakness and rectal hyposensitivity [20, 21]. The average Cleveland Clinic Florida Fecal Incontinence Score (CCF-FIS) in patients with MS is 12 (range: 3–15), which has been shown to significantly improve after treatment with biofeedback up to an average score of 8 (range: 4–14) [22]. Patients with MS have been found to have impaired function of the external anal sphincter and decreased volumes of rectal distension to inhibit the internal anal sphincter [23].

Diabetes mellitus is a common disease affecting 9.4% of the population in the United States, with 25.5% of people 65 and older having the condition (CDC). Patients with diabetes have demonstrate lower resting and maximal anal sphincter pressures [23]. Diabetes has been found to have an independent risk factor for fecal incontinence in elderly women [24]. Additionally, diabetes has been found to be associated with increased fecal incontinence symptom severity with symptoms worsening with increased duration of disease [25, 26].

The incidence of spinal cord injury in the United States is around 12,000 people per year [27]. The pathophysiology of fecal incontinence in patients with spinal cord injury remains unclear [28]. The incidence of fecal incontinence in patients with traumatic brain injury is upwards of 68% in patients admitted to inpatient rehabilitation after experience an injury, with a decline to 12.4% on discharge from rehabilitation and 5.2% at 1 year follow up [29]. Fecal incontinence in these patients is associated with lower GCS scores, length of coma prior to admission to nursing home, and frontal lobe contusion. Similarly, in patients who have experienced a stroke, the incidence of fecal incontinence is 30% in the first 7–10 days, 11% at 3 months, 11% at 1 year, and 15% at 3 years post-stroke [30]. Poor mobility, depression, and anticholinergic medications were found to be significantly associated with fecal incontinence. Dementia is a risk factor for fecal incontinence, with a prevalence of 11.1% in men and 10.1% in women with dementia [31].

Iatrogenic

There are several iatrogenic causes of fecal incontinence. Anorectal surgery may cause incontinence. Common anorectal procedures include hemorrhoidectomy, lateral internal sphincterotomy, botulinum toxin chemodenervation of the internal anal sphincter, and fistulotomy.

Hemorrhoidectomy, stapled hemorrhoidectomy and Ferguson technique excisional hemorrhoidectomy, carries a 2% incidence of incontinence [32]. Reports of incontinence with hemorrhoidectomy are low, but patients should always be counselled as to the risks. Sphincterotomy also carries a risk of incontinence and the length of sphincterotomy is directly related to the development of incontinence [33]. It has also been suggested that to mitigate risks of incontinence, no greater than 1 cm (25%) of sphincter muscle should be cut in sphincterotomy in women [34]. A meta-analysis of randomized control trials comparing botulinum toxin and lateral internal sphincterotomy showed a significantly lower rate of incontinence with botulinum toxin [35]. Given the transient nature of the effect of botulinum toxin, there is typically no long-term incontinence after this procedure.

Fistulotomy should only be performed when there is a limited amount of muscle involved and the risk to continence is low. A study of 624 patients who underwent surgery for anal fistula reported that 45% of patients complained of some degree of postoperative incontinence, which was associated with female sex, high anal fistulas, type of surgery, and a history of anal fistula surgery [36]. Other studies have reported minor incontinence rates of 5–20% with fistulotomy [37–40].

Incontinence has also been reported in conjunction with and after surgery for other colorectal diseases. This includes: rectal prolapse, rectal cancer, Crohn's disease, radiation proctitis, low anterior resection, or restorative proctocolectomy.

Clinical Treatment

Conservative measures are always the first line treatment in the management of fecal incontinence—which includes dietary changes, medication management, stool bulking agents/antidiarrheal medications, bowel management programs, and biofeedback. Dietary and medical management are strongly recommended by the American Society of Colon and Rectal Surgeons (ASCRS), based on level 1C evidence. Bowel management programs carry a weak recommendation by the ASCRS based on level 2C evidence. Biofeedback is discussed in detail elsewhere in this text.

Dietary Changes

Frequently, many foods and drinks may contribute to incontinence. It is important for patients to keep a food and stool diary in order to pinpoint the exact offending agents. Caffeine, alcohol, fatty foods, dairy, and artificial sweeteners are commonly known to worsen diarrhea, which can then worsen incontinence. Patients should be advised to avoid these foods if they are contributing.

Medication Management

The first encounter with a patient with fecal incontinence should include a thorough review of past medical, surgical, and social history. This should include a detailed review of the patient's current medications, both prescription and over the counter. Because stool quality is a key component of continence, it is important to examine any medications that may affect stool quality causing loose stools. There are several mechanisms of drug-induced diarrhea, including osmotic, secretory, inflammatory, and motility diarrhea. Table 13.2 lists the various drugs that cause diarrhea by category [41]. Any suspected

Table 13.2 Medications that may cause diarrhea; NSAID = nonsteroidal anti-inflammatory drugs

Osmotic Diarrhea	
	Artificial sweeteners (mannitol/sorbitol/xylitol)
	Alpha glucosidase inhibitors
	Ampicillin, clindamycin
	Enteral feeding
	Magnesium laxatives and antacids
	Phosphates
	Polyethylene glycol
	Prebiotics
	Sugars/poorly absorbable carbohydrates
	Methyldopa
	Quinidine
	Propranolol
	Hydralazine
	ACE inhibitors
	Procainamide
Secretory Diarrhea	
	Antiarrhythmics
	Amoxicillin–clavulanate
	Auranofin
	Caffeine
	Calcitonin
	Carbamazepine
	Chemotherapeutic agents
	Chenodeoxycholic acid
	Cimetidine
	Colchicine
	Diaccrcin
	Flavonoid veinotonics
	Laxatives
	Misoprostol
	Metformin
	NSAIDS
	Olsalazine
	Theophylline
	Levodopa–benserazide
	Cholinesterase inhibitors
Motility drugs	
	Acetylcholine esterase inhibitors
	Cholinergics
	Cisapride
	Metoclopramide
	Tegaserod
	Irinotecan
	Macrolides
	Ticlopidine
	Thyroid hormones
	Colchicine
	Anticholinergics
Inflammatory diarrhea	
	Antibiotics
	Auranofin
	Carbamazepine
	Etanercept
	Flutamide
	Lovastatin/pravastatin/simvastatin
	Ipilimumab
	Isotretinoin
	Mercaptopurine
	NSAIDs
	Olmesartan
	Oral contraceptives
	Oral cyclosporine
	Penicillamine
	Proton pump inhibitors
	Rituximab
	Sodium phosphate
	Ticlopidine
	Tyrosine kinase inhibitors
	Laxatives

agents should be reduced in dose or discontinued, particularly laxatives and stool softeners. If diarrhea persists, endoscopy and histologic evaluation may be necessary to assess for microscopic colitis. It is important to also review over-the-counter medications and supplements, as these may be contributing to loose stools.

Stool Bulking Agents and Antidiarrheal Medications

After appropriate medication adjustments are made, the next step is the addition of stool bulking agents as bulkier stools are easier to control than loose stools. The first step is adding fiber supplementation. Daily fiber intake should be a total of 25–30 grams per day, including diet and supplements. There are several over-the-counter fiber supplement formulas available; patients should be instructed to examine the fiber content to ensure the product contains an adequate amount of fiber. Fiber pills typically have an extremely low fiber content and should be discouraged. There are two types of fiber: soluble and insoluble. Soluble fiber (psyllium, guar gum) forms a jelly-like matrix when mixed with fluids. Examples of insoluble fiber include psyllium husk, methyl cellulose, and calcium polycarbophil. Side effects of fiber may include bloating and flatulence. These side effects may abate after continued use. Several studies demonstrate that fiber is an effective treatment for incontinence. When fiber is added to enteral feedings, the frequency of loose stool in hospitalized patients improves [42, 43]. A randomized control trial comparing psyllium, gum arabic, and placebo to treat fecal incontinence found that treatment with fiber supplementation resulted in a 50% reduction of incontinence episodes [44].

If fiber is minimally effective or ineffective, antidiarrheal medications may be added. It is necessary to ensure there are no concerns for infectious diarrhea (especially *C. Difficile*) prior to initiating treatment with antidiarrheals. Table 13.3 outlines the dosages and side effects for common antidiarrheal medications [45]. Stool-bulking agents/antidiarrheal medications should be used in conjunction with biofeedback. A randomized control trial compared stool bulking agents (loperamide plus fiber) to biofeedback and to the combination of the two treatment modalities. This study demonstrated that both groups had improved continence with combination therapy; urgency, and rectal sensory thresholds were also improved [46]. Cholestyramine has also been shown to be a useful adjunct in the treatment of incontinence [47]. Specifically, this is useful in patients with malabsorptive states, or extensive resections of the terminal ileum where bile acid enterohepatic circulation is compromised resulting in diarrhea. The bile acids will form insoluble complexes with the medication, which will be excreted in the stool and help diarrhea associated with bile acid malabsorption.

Topical Treatments

Phenylephrine has been investigated as a possible treatment for anal incontinence, with the underlying theory that internal sphincter dysfunction is not amenable to simple surgical repair. Topically applied phenylephrine has been shown to increase the resting pressure of the internal sphincter [48, 49]. A small randomized control trial did not demonstrate any objective difference in incontinencies scores, resting anal pressures, or anodermal blood flow; but there were some subjective reports by patients of improvement in symptoms [50]. The data for the use of phenylephrine to treat fecal incontinence is sparse, and at this time it is not included as a recommended treatment by the American Society of Colon and Rectal Surgeons.

Bowel Management Programs

Select patients with lifestyle limiting incontinence may benefit from scheduled enemas or suppositories to aid in rectal evacuation.

Table 13.3 Antidiarrheal medications, doses, and side effects

Medication	Dose	Side effects
Loperamide	2 mg twice daily. Maximum 4 mg 4 times a day	Constipation, nausea, vomiting, cramping
Diphenoxylate/atropine	2 tablets (2.5 mg/0.025 mg) once a day. Maximum: 2 tablets, 4 times per day	Toxic megacolon, neurologic effects
Cholestyramine	4 g daily. Maximum: 24 g per day.	Nausea, vomiting, flatulence, dyspepsia, dysgeusia, rash
Colestipol	Begin 2 g PO daily. Titrate to a maximum of 16 g/day.	Nausea, vomiting, flatulence, dyspepsia, abdominal pain, rash
Clonidine	0.1 mg PO twice daily. May increase to 0.3 mg twice daily.	Rebound hypertension, dry mouth, neurologic effects, constipation, headache, rash, nausea
Tincture of opium	1–2 drops PO twice daily. Slowly titrate up to a maximum dose of 12 drops twice daily.	Sedation, nausea, anorexia, urinary retention, hypotension, bradycardia, respiratory depression, neurologic effects
Alosetron	0.5 mg PO daily—twice daily Maximum dose is 1 mg twice daily.	Constipation, ischemic colitis

Emptying the rectum on a scheduled basis may minimize incontinence episodes. This may be particularly helpful in children or adults with neurologic conditions such as spina bifida [51]. Antegrade colonic enema (ACE) is a surgical procedure that is discussed elsewhere.

Summary

Fecal incontinence is a common but underreported condition that is seen frequently by colorectal surgeons. Identifying the underlying etiology of the disease is helpful in selecting the ideal treatment, but all patients who present with fecal incontinence may benefit from clinical treatment. Dietary alterations, medication management, fiber supplementation, antidiarrheal medication, and biofeedback are frequently successful in reducing incontinence episodes. Patients who do not respond to clinical treatment can be considered for more aggressive treatment strategies.

References

1. Rao SSC, Bharucha AE, Chiarioni G, Felt-Bersma R, Knowles C, Malcolm A, et al. Anorectal disorders. Gastroenterology. 2016;150(6):1430–1442e4.
2. Tjandra JJ, Dykes SL, Kumar RR, Ellis CN, Gregorcyk SG, Hyman NH, et al. Practice parameters for the treatment of fecal incontinence. Dis Colon Rectum. 2007;50(10):1497–507.
3. Wald A. Clinical practice. Fecal incontinence in adults. N Engl J Med [Internet]. 2007;356(16):1648–55. Available from: papers3://publication/doi/https://doi.org/10.1056/NEJMcp067041.
4. Madoff RD, Parker SC, Varma MG, Lowry AC. Faecal incontinence in adults. Lancet. 2004;364:621–32.
5. Wu JM, Vaughan CP, Goode PS, Redden DT, Burgio KL, Richter HE, et al. Prevalence and trends of symptomatic pelvic floor disorders in U.S. women. Obstet Gynecol. 2014;123(1):141–8.
6. Ng K-S, Sivakumaran Y, Nassar N, Gladman MA. Fecal incontinence: community prevalence and associated factors--a systematic review. Dis Colon Rectum [Internet]. 2015;58(12):1194–209. Available from: http://content.wkhealth.com/linkback/openurl?sid=WKPTLP:landingpage&an=00003453-201512000-00012.
7. Nelson R, Furner S, Jesudason V. Fecal incontinence in Wisconsin nursing homes prevalence and associations. Dis Colon Rectum. 1998;41(10):1226–9.
8. Bliss DZ, Harms S, Garrard JM, Cunanan K, Savik K, Gurvich O, et al. Prevalence of incontinence by race and ethnicity of older people admitted to nursing homes. J Am Med Dir Assoc. 2013;14(6):451.e1–7.
9. Landefeld CS, Bowers BJ, Feld AD, Hartmann KE, Hoffman E, Ingber MJ, et al. National Institutes of Health State-of-the-Science conference statement: prevention of fecal and urinary incontinence in adults. Ann Intern Med. 2008;148:449–58.
10. Brown HW, Wexner SD, Segall MM, Brezoczky KL, Lukacz ES. Accidental bowel leakage in the mature women's health study: prevalence and predictors. Int J Clin Pract. 2012;66(11):1101–8.
11. Kim T, Chae G, Chung SS, Sands DR, Speranza JR, Weiss EG, et al. Faecal incontinence in male patients. Color Dis. 2008;10(2):124–30.
12. Lee JT, Madoff RD, Rockwood TH. Quality-of-life measures in fecal incontinence: is validation valid? Dis Colon Rectum. 2015;58:352–7.
13. Hehir MP, O'Connor HD, Higgins S, Robson MS, McAuliffe FM, Boylan PC, et al. Obstetric anal sphincter injury, risk factors and method of delivery – an 8-year analysis across two tertiary referral centers. J Matern Neonatal Med [Internet]. 2013;26(15):1514–6. Available from: http://www.tandfonline.com/doi/full/10.3109/14767058.2013.791268.
14. Jangö H, Langhoff-Roos J, Rosthøj S, Sakse A. Modifiable risk factors of obstetric anal sphincter injury in primiparous women: a population-based cohort study. Am J Obstet Gynecol. 2014;210(1):59.e1–6.
15. Jha S, Parker V. Risk factors for recurrent obstetric anal sphincter injury (rOASI): a systematic review and meta-analysis. Int Urogynecol J Pelvic Floor Dysfunct [Internet]. 2015;1–9. Available from: https://doi.org/10.1007/s00192-015-2893-4
16. Thomas GP, Gould LE, Casunuran F, Kumar DA. A retrospective review of 1495 patients with obstetric anal sphincter injuries referred for assessment of function and endoanal ultrasonography. Int J Color Dis. 2017;32(9):1321–5.
17. Jangö H, Langhoff-Roos J, Rosthøj S, Saske A. Long-term anal incontinence after obstetric anal sphincter injury—does grade of tear matter? Am J Obstet Gynecol. 2018;218(2):232.e1–232.e10.
18. Rigueros Springford L, Connor MJ, Jones K, Kapetanakis VV, Giuliani S. Prevalence of active long-term problems in patients with anorectal malformations: a systematic review. Dis Colon Rectum. 2016;59:570–80.
19. Levitt MA, Peña A. Anorectal malformations. In: Fundamentals of pediatric surgery. New York: Springer; 2011. p. 499–512.
20. Nordenbo AM, Andersen JR, Andersen JT. Disturbances of ano-rectal function in multiple sclerosis. J Neurol. 1996;243(6):445–51.

21. Wiesel PH, Norton C, Glickman S, Kamm MA. Pathophysiology and management of bowel dysfunction in multiple sclerosis. Eur J Gastroenterol Hepatol. 2001;13(4):441–8.
22. Preziosi G, Raptis DA, Storrie J, Raeburn A, Fowler CJ, Emmanuel A. Bowel biofeedback treatment in patients with multiple sclerosis and bowel symptoms. Dis Colon Rectum. 2011;54(9):1114–21.
23. Caruana BJ, Wald A, Hinds JP, Eidelman BH. Anorectal sensory and motor function in neurogenic fecal incontinence. Comparison between multiple sclerosis and diabetes mellitus. Gastroenterology [Internet]. 1991;100(2):465–70. Available from: http://www.ncbi.nlm.nih.gov/pubmed/1985043.
24. Halland M, Koloski NA, Jones M, Byles J, Chiarelli P, Forder P, et al. Prevalence correlates and impact of fecal incontinence among older women. Dis Colon Rectum. 2013;56(9):1080–6.
25. Menees SB, Smith TM, Xu X, Chey WD, Saad RJ, Fenner DE. Factors associated with symptom severity in women presenting with fecal incontinence. Dis Colon Rectum. 2013;56(1):97–102.
26. Epanomeritakis E, Koutsoumbi P, Tsiaoussis I, Ganotakis E, Vlata M, Vassilakis JS, et al. Impairment of anorectal function in diabetes mellitus parallels duration of disease. Dis Colon Rectum. 1999;42(11):1394–400.
27. Selassie A, Snipe L, Focht KL, Welldaregay W. Baseline prevalence of heart diseases, hypertension, diabetes, and obesity in persons with acute traumatic spinal cord injury: potential threats in the recovery trajectory. Top Spinal Cord Inj Rehabil [Internet]. 2013;19(3):172–82. Available from: http://www.pubmedcentral.nih.gov/articlerender.fcgi?artid=3743967&tool=pmcentrez&rendertype=abstract.
28. Pan Y, Liu B, Li R, Zhang Z, Lu L. Bowel dysfunction in spinal cord injury: current perspectives. Cell Biochem Biophys. 2014;69:385–8.
29. Foxx-Orenstein A, Kolakowsky-Hayner S, Marwitz JH, Cifu DX, Dunbar A, Englander J, et al. Incidence, risk factors, and outcomes of fecal incontinence after acute brain injury: findings from the Traumatic Brain Injury Model Systems national database. Arch Phys Med Rehabil. 2003;84(2):231–7.
30. Harari D, Coshall C, Rudd AG, Wolfe CDA. New-onset fecal incontinence after stroke: prevalence, natural history, risk factors, and impact. Stroke [Internet]. 2003;34(1):144–50. Available from: http://stroke.ahajournals.org/content/strokeaha/34/1/144.full.pdf.
31. Grant RL, Drennan VM, Rait G, Petersen I, Iliffe S. First diagnosis and management of incontinence in older people with and without dementia in primary care: a Cohort Study using the health improvement network primary care database. PLoS Med. 2013;10(8):e1001505.
32. Aytac E, Gorgun E, Erem HH, Abbas MA, Hull TL, Remzi FH. Long-term outcomes after circular stapled hemorrhoidopexy versus Ferguson hemorrhoidectomy. Tech Coloproctol. 2015;19(10):653–8.
33. García-Aguilar J, Montes CB, Perez JJ, Jensen L, Madoff RD, Wong WD. Incontinence after lateral internal sphincterotomy: anatomic and functional evaluation. Dis Colon Rectum. 1998;41(4):423–7.
34. Murad-Regadas SM, Da Silva Fernandes GO, Regadas FSP, Rodrigues LV, Pereira JDJR, Filho FSPR, et al. How much of the internal sphincter may be divided during lateral sphincterotomy for chronic anal fissure in women? Morphologic and functional evaluation after sphincterotomy. Dis Colon Rectum. 2013;56(5):645–51.
35. Chen H-L, Woo X-B, Wang H-S, Lin Y-J, Luo H-X, Chen Y-H, et al. Botulinum toxin injection versus lateral internal sphincterotomy for chronic anal fissure: a meta-analysis of randomized control trials. Tech Coloproctol [Internet]. 2014;18(8):693–8. Available from: http://www.ncbi.nlm.nih.gov/pubmed/24500725.
36. Garcia-Aguilar J, Belmonte C, Douglas Wong W, Goldberg SM, Madoff RD. Anal fistula surgery: factors associated with recurrence and incontinence. Dis Colon Rectum. 1996;39:723–9.
37. Ho YH, Tan M, Leong AFPK, Seow-Choen F. Marsupialization of fistulotomy wounds improves healing: a randomized controlled trial. Br J Surg. 1998;85(1):105–7.
38. García-Aguilar J, Belmonte C, Wong DW, Goldberg SM, Madoff RD. Cutting seton versus two-stage seton fistulotomy in the surgical management of high anal fistula. Br J Surg. 1998;85(2):243–5.
39. Pescatori M, Ayabaca SM, Cafaro D, Iannello A, Magrini S. Marsupialization of fistulotomy and fistulectomy wounds improves healing and decreases bleeding: a randomized controlled trial. Color Dis. 2006;8(1):11–4.
40. Siddiqui MRS, Ashrafian H, Tozer P, Daulatzai N, Burling D, Hart A, et al. A diagnostic accuracy meta-analysis of endoanal ultrasound and mri for perianal fistula assessment. Dis Colon Rectum. 2012;55:576–85.
41. Philip NA, Ahmed N, Pitchumoni CS. Spectrum of drug-induced chronic diarrhea. J Clin Gastroenterol. 2017;51:111–7.
42. Vandewoude MFJ, Paridaens KMJ, Suy RAL, Boone MAA, Strobbe H. Fibre-supplemented tube feeding in the hospitalised elderly. Age Ageing. 2005;34(2):120–4.
43. Shankardass K, Chuchmach S, Chelswick K, Stefanovich C, Spurr S, Brooks J, et al. Bowel function of long-term tube-fed patients consuming formulae with and without dietary fiber. J Parenter Enter Nutr. 1990;14(5):508–12.
44. Bliss DZ, Jung HJ, Savik K, Lowry A, LeMoine M, Jensen L, et al. Supplementation with dietary fiber improves fecal incontinence. Nurs Res. 2001;50(4):203–13.
45. Costilla VC, Foxx-Orenstein AE, Mayer AP, Crowell MD. Office-based management of fecal incontinence. Gastroenterol Hepatol. 2013;9(7):423–33.

46. Sjödahl J, Walter SA, Johansson E, Ingemansson A, Ryn AK, Hallböök O. Combination therapy with biofeedback, loperamide, and stool-bulking agents is effective for the treatment of fecal incontinence in women - a randomized controlled trial. Scand J Gastroenterol. 2015;50(8):965–74.
47. Remes-Troche JM, Ozturk R, Philips C, Stessman M, Rao SSC. Cholestyramine---a useful adjunct for the treatment of patients with fecal incontinence. Int J Colorectal Dis [Internet]. 2008;23(2):189–94. Available from: https://doi.org/10.1007/s00384-007-0391-y.
48. Cheetham MJ, Kamm MA, Phillips RKS. Topical phenylephrine increases anal canal resting pressure in patients with faecal incontinence. Gut. 2001;48(3):356–9.
49. Carapeti EA, Kamm MA, Evans BK, Phillips RKS. Topical phenylephrine increases anal sphincter resting pressure. Br J Surg [Internet]. 1999 [cited 2018 Jan 10];86(2):267–70. Available from: http://www.blackwell-synergy.com/links/doi/10.1046/j.1365-2168.1999.01021.x.
50. Carapeti EA, Kamm MA, Phillips RKS. Randomized controlled trial of topical phenylephrine in the treatment of faecal incontinence. Br J Surg. 2000;87(1):38–42.
51. Vande Velde S, Van Biervliet S, Van Laecke E, De Bruyne R, Verhelst H, Hoebeke P, et al. Colon enemas for fecal incontinence in patients with spina bifida. J Urol. 2013;189(1):300–4.

Anal Incontinence: Minimally Invasive Options

14

Lucia Camara Castro Oliveira

Introduction

Anal incontinence is a condition involving complex and often multifactorial mechanisms, already discussed in previous chapters. Despite etiology, patients are usually managed initially with medical and conservative treatment. For patients with more severe incontinence, a number of minimally invasive options are available (Table 14.1). According to the latest fecal incontinence guidelines by the American Society of Colon and Rectal Surgeons (ASCRS) [1] evidence for the use of minimally invasive options is moderate (recommendations 2B and 2C). In this chapter, we will discuss the indications of bulking agents and other minimally invasive options.

Bulking Agents

Injection of bulking agents or injectables were initially described by urologists for the treatment of urinary incontinence. For anal incontinence, the first utilized agent was polytetrafluoroethylene as described by Shafik [2] in 1993. Eleven patients had an injection of polytetrafluoroethylene with satisfactory results. The authors subsequently utilized autologous fat as a new agent in four other patients. However, this agent was discontinued after the occurrence of fat embolism in a patient who underwent the procedure for urinary incontinence [3].

Currently, different agents or substances are being utilized with a variety of techniques, making comparisons among results more difficult. Bulking agents utilized are listed in Table 14.2.

One of the great advantages of this technique is the possibility to perform the procedure on an outpatient basis, without the need for general anesthe-

Table 14.1 Minimally invasive options for the treatment of anal incontinence

Bulking agents
GateKeeper and SphinKeeper
Radiofrequency
Minislings
Magnetic sphincter
Vaginal/analplugs
Neuromodulation
Stem cells

Table 14.2 Bulking agents for anal incontinence

Autologous fat
Polytetrafluoroethylene (Teflon®)
Porcine dermal collagen (Permacol™)
Synthetic bovine collagen (Contigen®)
Hydrogel cross-linked with polyacrylamide (Bulkamid™)
Polydimethylsiloxane (Silicone biomaterial -PTQ™)
Carbon-coated microbeads (Durasphere®/ACYST)
Calcium hydroxylapatite (Coaptite®)
Stabilized nonanimal hyaluronic acid with dextranomer (NASHA™Dx)
Polyacrylate polyalcohol copolymers (Exantia®)

L. C. C. Oliveira (✉)
Director Department of Anorectal Physiology of Rio de Janeiro and CEPEMED, Rio de Janeiro, Brazil

L. C. C. Oliveira (ed.), *Anorectal Physiology*, https://doi.org/10.1007/978-3-030-43811-1_14

sia or sedation. In addition, it is easy to apply with low morbidity. Perianal injection of bulking agents was described for the treatment of incontinence associated with isolated lesions of the internal sphincter muscle (MEI), such as after sphincterotomy (Fig. 14.1) [4]. It was believed that injecting the agent at the site of the muscular defect would correct the incontinence by increasing the resting anal pressure. Although the increase in resting pressure was utilized as an explanation for the mechanism of action of these agents, different studies using anorectal manometry before and after injection of agents were not able to demonstrate any statistically significant results.

Our experience with 35 patients that underwent injection of silicone agent for anal incontinence revealed an important manometric parameter for the understanding of its mechanism: correction of the sphincter asymmetry index was the main parameter related to clinical improvement in these patients [5, 6]. The mean incontinence score among the 35 patients presented a significant change before and after treatment, with a significant improvement in quality of life. There was also a significant correlation between improvement in the incontinence score and the domains of constraint and behavior of the quality of life instrument. Clinical follow-up of the first patients submitted to the treatment has demonstrated maintenance of the initial good results; a follow-up of more than 10 years is currently in place and the use of silicone in new patients is encouraged. Maeda et al. [7] demonstrated long-term effects of silicone injection in six incontinent patients. Although at a follow-up period of 61 months one of the patients had undergone a colostomy due to the persistence of incontinence, the others presented with significant improvement in quality of life, evaluated through the SF-36 instrument. Improvements in incontinence and quality of life rates were also observed by Tjandra et al. [8] in the largest published series that included 82 patients who received silicone injection.

Different studies have shown that silicone presents a lower migration potential when compared to other bulking agents [9, 10]. One of the reasons is the size of its particles, with an average diameter of approximately 200 μ. In animal studies it was demonstrated that the carrier gel is eliminated by the urinary tract 3 days later, leaving the nonabsorbable part of the silicone in the injected site, thereby forming a foreign body reaction, wherein the macrophages of the tissue's neighbors end up encapsulating the particles and forming a granuloma, allowing for adequate ultrasonographic visualization of the silicone 2–3 months after the injection [9, 10].

In our study, anal ultrasound was performed 3 months after injection in the 35 patients studied, when the injected agent was confirmed and easily demonstrated (Fig. 14.2).

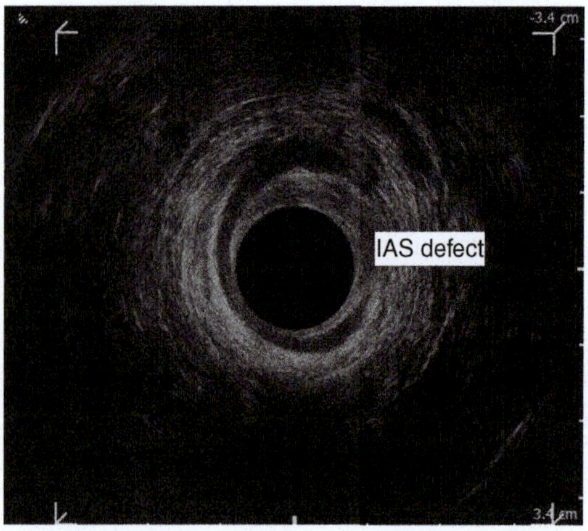

Fig. 14.1 Internal anal sphincter defect after lateral sphincterotomy

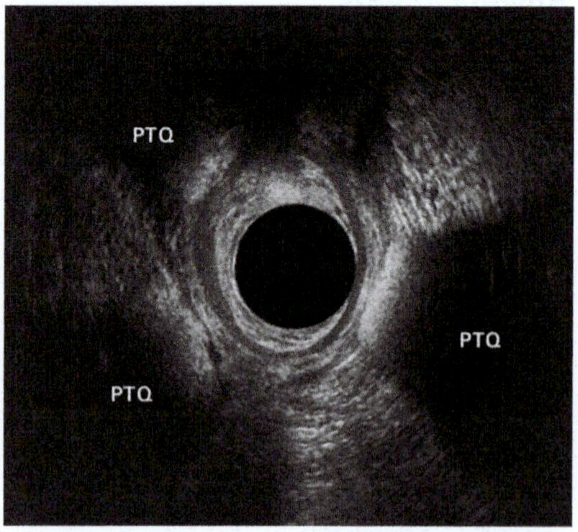

Fig. 14.2 PTQ or silicone injections after 3 months

The use of collagen as a bulking agent in a recent series of 73 incontinent patients with mean follow-up of 1 year showed satisfactory clinical results, mainly in patients over 60 years of age and with idiopathic incontinence [11]. Comparing silicone to collagen, the latter has two major disadvantages: the need for a skin test prior to application and the repetition of injections after a period of 12–30 months since the substance appears to degrade after this time interval.

Particles of pyrolytic carbon have also been used as a bulking agent [12, 13]. Among the disadvantages are the particle size and the higher resistance found for its application, which is generally performed in the submucosa, along the anal canal. The first series in the literature was reported by Davis et al. [12] in 2003. Eighteen incontinent patients underwent submucosal pyrolytic carbon injection and were followed up for a period of 28 months. The authors observed clinical improvement in 15 patients, mainly translated into changes in incontinence scores and quality of life. In another study, a comparison between silicone and pyrolytic carbon beads for the treatment of anal incontinence in 40 patients, Tjandra et al. [14] showed better results with the use of silicone. In incontinent post-hemorrhoidectomy patients, Chan and Tjandra [15] reported significant clinical improvement after silicone injection. In seven patients, the authors observed a significant increase in the resting pressures after treatment. Although most of the studies report improvement in the anal resting pressures, not all published series have been able to demonstrate such a correlation by anal manometry [15–17]. Correction of sphincter asymmetry and increase in functional anal canal length were considered the main mechanisms associated with clinical improvement in our patients [6]. The complexity of mechanism that maintain anal continence is likely contributing to the difficulties encountered in demonstrating the mechanism of action of filling agents.

When choosing a bulking agent, it is important to consider the size of the particles. It is well known that particles over 80 μ can prevent migration. Substances must also be biocompatible and nonallergenic. Silicone is a paste in 2.5 mL samples, requiring a special injecting gun for transsphincteric application (Figs. 14.3 and 14.4). Pyrolytic carbon coated graphite beads is marketed in 3 mL syringes (Fig. 14.5). The pyrolytic carbon is injected into the submucosa with the aid of an anoscope above the dentate line. Cost and availability of the bulking agents have been fac-

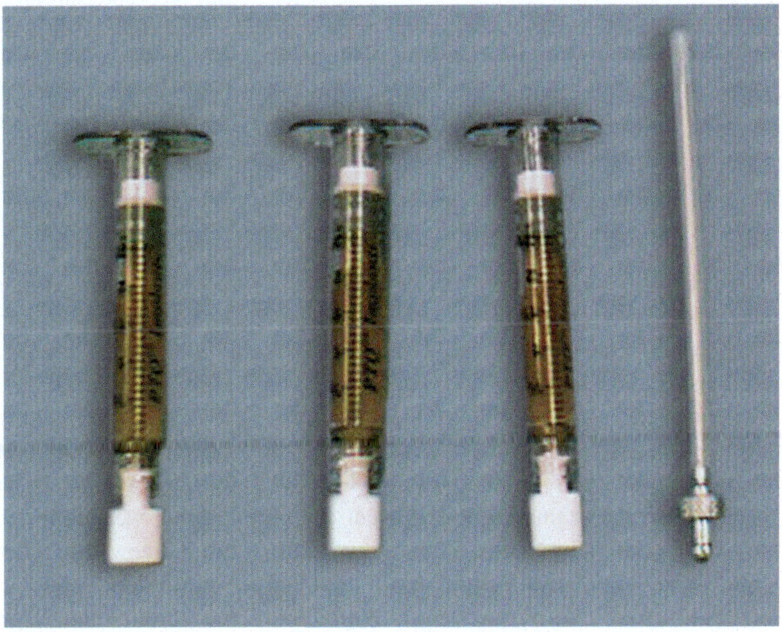

Fig. 14.3 PTQ bulking agents in 2.5 mL syringes

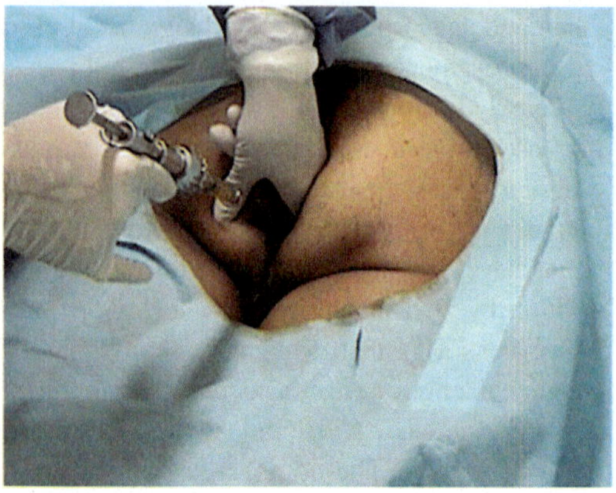

Fig. 14.4 Injection of PTQ with special gum guided by digital examination

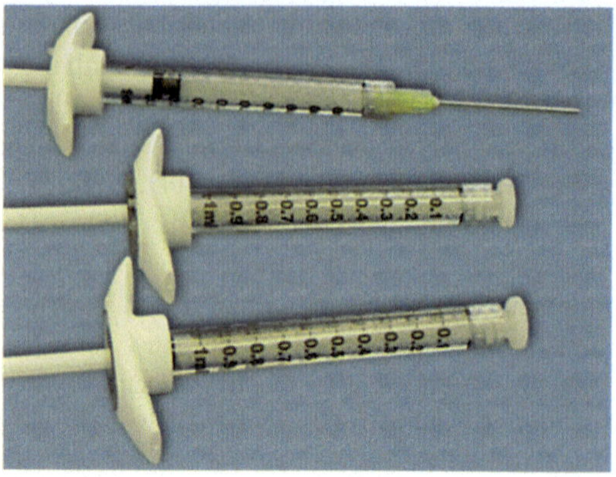

Fig. 14.6 Polyvinyl alcohol polymer of polyvinyl acetate in syringes of 1 mL

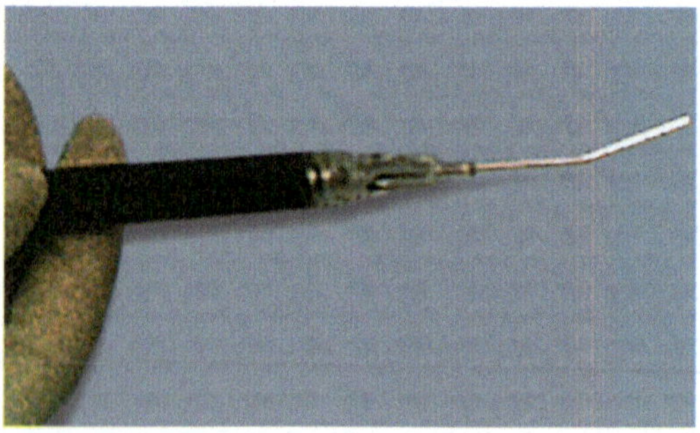

Fig. 14.5 Pyrolytic carbon coated graphite beads syringes containing 3 mL of the substance

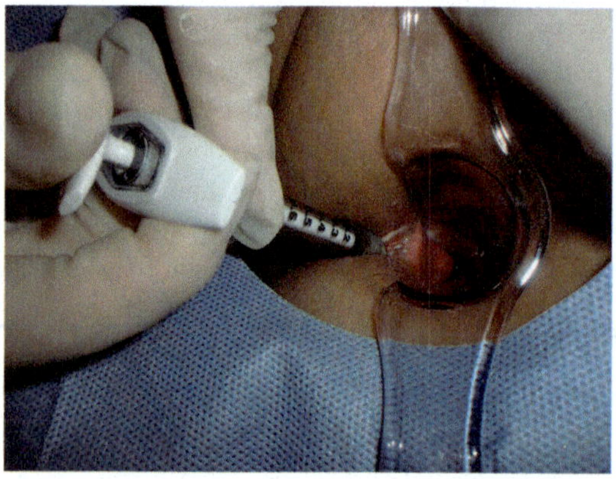

Fig. 14.7 Injection of polyvinyl acetate agent

tors contributing to the limitation of their widespread use. Currently in Brazil, the substance with regulated registration is polyacrylate. We conducted a prospective multicenter study in Latin America, in which we utilized this agent in 58 incontinent patients. After a follow-up of up to 3 years, results showed an improvement in the Cleveland Clinic Florida Fecal Incontinence Score (CCF-FIS) quality of life scale in 60.4% of cases [18]. The agent consists of a polyvinyl alcohol polymer of polyvinyl acetate. The particles are amorphous, flexible and 800 μ in size, which promote good adhesion to tissues and are available in 1 mL syringes (Fig. 14.6); submucosal injection is recommended under anoscopy (Fig. 14.7) or guided by rectal examination. The agent can be easily recognized by endoanal ultrasonography (Fig. 14.8a, b). Another agent recently used and approved by the FDA in the USA is NASHA, a dextranomer of nonanimal origin of hyaluronic acid. This agent was initially used as part of a multicenter study involving 15 European and Canadian centers [19]. The authors included 115 patients with mild-to-moderate incontinence with symptoms for more than 12 months. Twenty-four patients were excluded from the analysis and results were based 91 patients treated. There was a reduction in the number of incontinent episodes in all patients and the CCF-FIS decreased from 13.5 to 9.2 ($p < 0.001$). Subsequently, Mellgren et al. [20] published the results of a multicenter randomized trial involving 206 patients, 136 received the agent and 70 received placebo. Patients were followed up for 36 months. The results showed a reduction in the mean number of accidents from 15 to 7 after 36 months and a reduction of incontinence in 52% of the cases.

This agent has particles of 80–120 μ of size, which contributes to its absorption and fixation in the tissues. The treatment is simple and can be performed on an outpatient basis, once the substance is introduced into the submucosa above the dentate line. Approximately 82% of patients underwent a new injection during follow-up. In fact, the ideal amount of the substance to be injected has not yet been established. We believe that a minimum of 4–6 mL is required. The results of patients treated with silicone injection were higher, probably due to the amount of agent injected, which was generally 7.5 mL.

Differences between bulking agents can be seen in Table 14.3.

Complications described with the use of these agents are of minimal severity—pain or mild discomfort at the site of the injections, bleeding when the agent is used on the submucosa and when there is associated hemorrhoidal disease. There are also reports of hyperthermia and chills in the first 24 hours after the injection. The most serious complication is infection translated by anal abscess; therefore, broad-spectrum prophylactic antibiotic therapy is recommended. Our experience with the use of bulking agents includes over 100 patients. All patients presented with clinical improvement of incontinence, especially those whose incontinence was mild-to-moderate, with minimal complications. Only

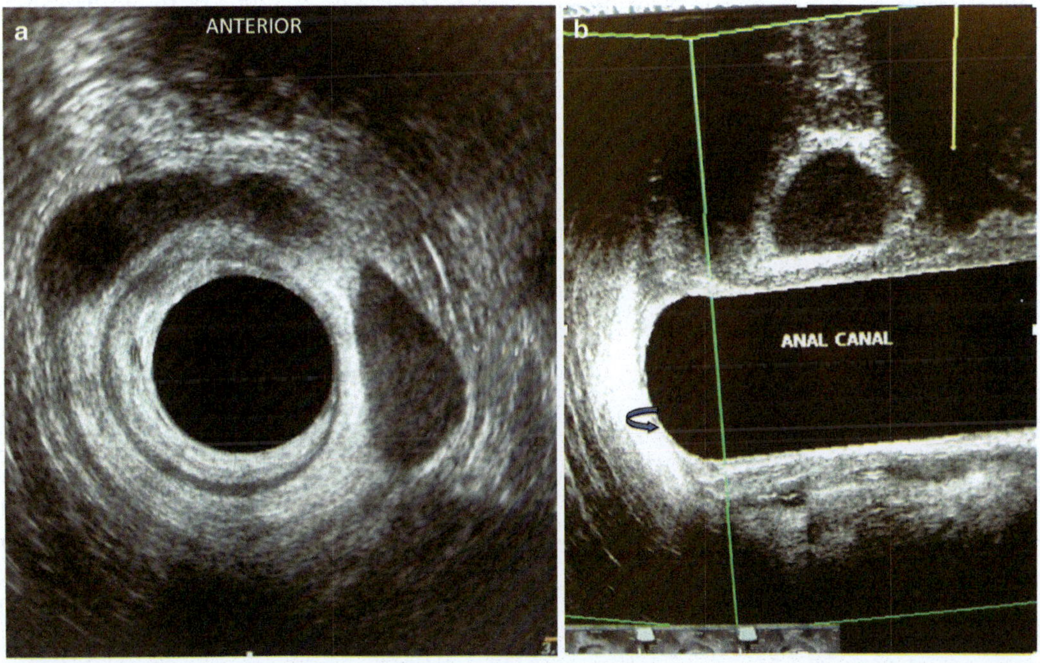

Fig. 14.8 (**a**, **b**) Endoanal Ultrasound images of polyvinyl acetate agent

Table 14.3 Differences between bulking agents for incontinence

PTQ	Pyrolytic carbon	Polyvinyl acetate	NASHA
Transcutaneous	Submucosal	Submucosal	Submucosal
Local anesthesia	No anesthesia	Local anesthesia	No anesthesia
Prophylactic antibiotics	Prophylactic antibiotics	Prophylactic antibiotics	Prophylactic antibiotics
2.5 mL syringes	1 and 3 mL	1 mL	1 mL
Guided by digital examination or US	Through anoscope	Through anoscope Above dentate line	Through anoscope above dentate line
Three injections	1–4 injections	1–4 injections	1–4 injections

one patient presented with an anal abscess, which was drained on an outpatient basis and without any further intervention (Fig. 14.9).

The greatest challenges with respect to injectables is identifying the ideal agent and the adequate amount to be used. All meta-analyses and systematic reviews have shown that there are still few well-designed studies with weak evidence. Despite the low quality of published studies, all have demonstrated an improvement in quality of life [21, 22].

In any case, the possibility of offering a minimally invasive modality for incontinent patients has modified the treatment protocol. In the new treatment algorithm, injectables are considered a first option together with neuromodulation and sphincteroplasty. These agents have demonstrated safety and efficacy, and significant improvement in quality of life. Unfortunately, availability of these agents has significantly decreased.

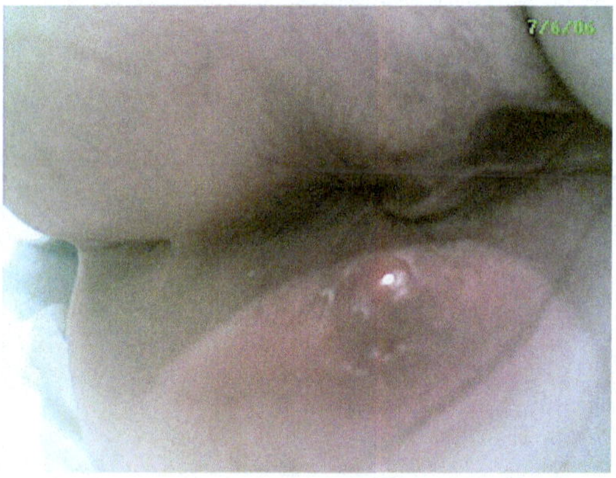

Fig. 14.9 Perianal abscess following bulking agent injection

GateKeeper and SphinKeeper

GateKeeper was launched in 2011 in Italy as a new bulking agent. The material consists of solid cylinders, self-expanding when in contact with the fluids of the human body. It functions as a bulking agent because it is injected through needles and a gun specially designed to release the substance [23].

GateKeeper (Hyspan) is a polyacrylonitrile with an initial diameter of 1–2 mm that expands to about 7 mm and is positioned in the four quadrants of the anal canal (Fig. 14.10a, b). The initial study included 14 incontinent patients with a CCF-FIS of 12.7. After 33 months of follow-up, there was a significant reduction in the number of incontinence episodes from 7.1 to 1.4 per week, and improvement in CCF-FIS from 12.7 to 5.1 at 3 months ($p < 0.001$) [23]. Subsequently, the same authors conducted a multicenter study that included 54 patients who were followed up for 1 year [24]. The improvement in incontinence after 12 months was 75% in 56% of patients, and seven patients were considered totally continent. They described migration of the agent in three patients with no need for replacement.

Finally, an adaptation of the initial material was developed, and the agent was renamed SphinKeeper, with the advantage of having an initial thickness of 3 mm, expanding to 7 mm;

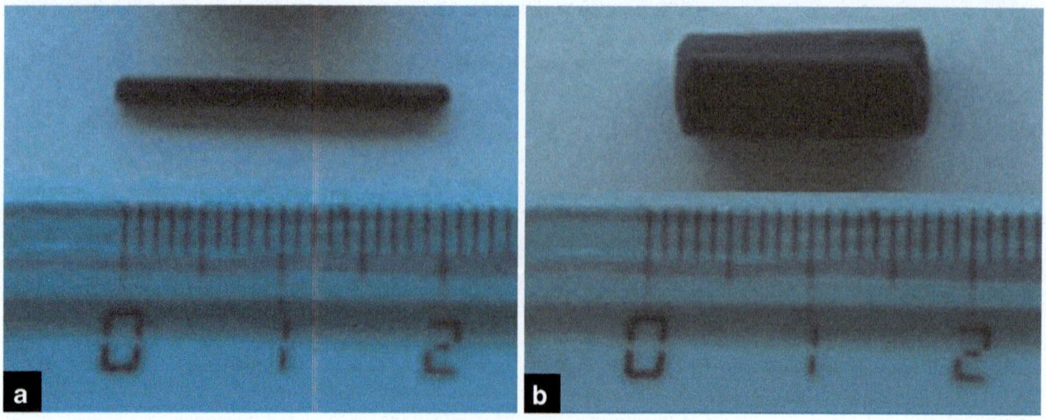

Fig. 14.10 (**a**, **b**) GateKeeper initial diameter and after expansion

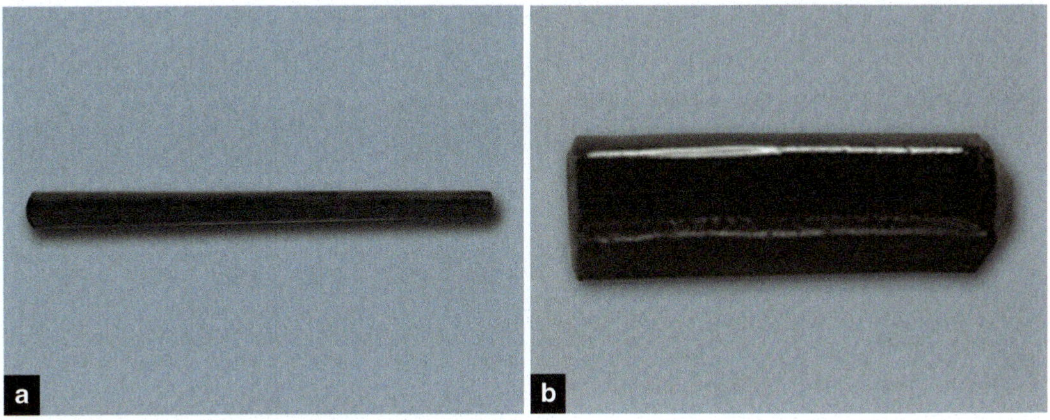

Fig. 14.11 (**a**, **b**) SphinKeeper before and after expansion

through a special pistol, the agent is now positioned around the entire anal canal with a greater number of implant sites—usually 8–12 (Fig. 14.11a, b) [25]. The result is complete encirclement of the anal canal by the implants, forming a kind of "prosthesis" or new sphincter after the expansion of the material.

The procedure is performed with mechanical and oral preparation of the colon, with the patient in the lithotomy position. Small incisions are made in the skin to release the product into the intersphincteric space, guided by ultrasonography. In this initial study, ten patients were treated, and presented with improved continence without migration or other complications.

In a recent multicenter retrospective and longitudinal study of patients with FI who were treated with GateKeeper and followed up for a 5-year period that included 49 consecutive patients [26], no postoperative and long-term complications were observed. Prosthesis migration was observed in 51% of patients. Twenty-three patients (48%) were classified as responders and 25 (52%) as nonresponders. The authors concluded that the GateKeeper is a safe and effective procedure in more than 50% of patients for at least 1 year after implantation.

Radiofrequency

The first authors who described radiofrequency for the treatment of mild-to-moderate incontinence believed that this therapy would act through a process of fibrotic formation in the anal canal [27, 28]. Radiofrequency energy is applied through an apparatus introduced into the anal canal (Secca procedure) (Fig. 14.12). In this technique, the special anoscope releases thermal energy for 90s in the four quadrants at 16 application points, 2.5 cm above the dentate line (Fig. 14.13). The patients who most benefited from the Secca procedure were those in whom conservative treatment with biofeedback and dietary modifications had failed. Patients with CCF-FIS between 9 and 17 who have normal anorectal anatomy with no significant scarring, tissue loss, or rupture of the external sphincter >30% confirmed by anal ultrasound or clinical examination are the most ideal indications. Another multicenter study involved five centers

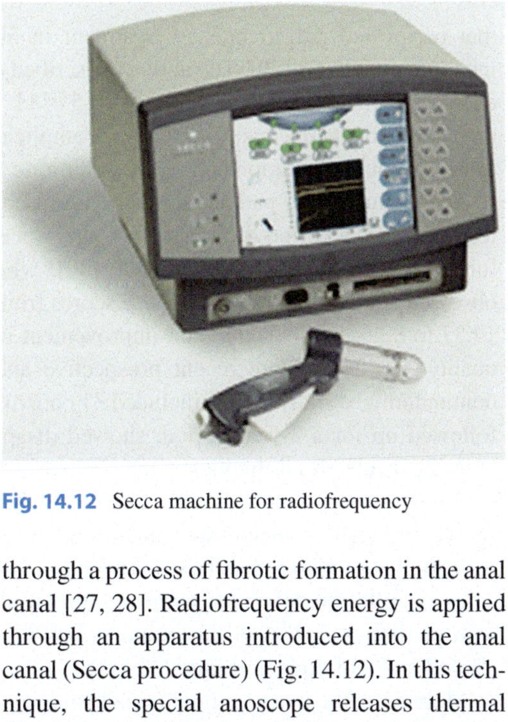

Fig. 14.12 Secca machine for radiofrequency

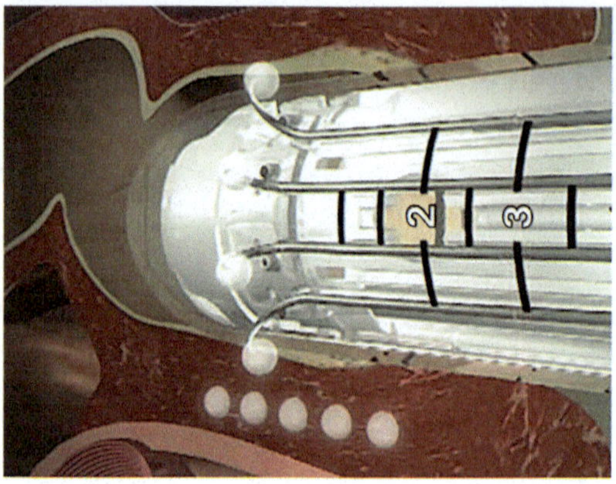

Fig. 14.13 Radiofrequency anoscope with energy delivery points

that performed radiofrequency treatment in 50 incontinent patients [29]. The authors described a decrease in incontinence scores from 14.5 to 11.1 after 6 months and a small number of complications including ulceration and anal bleeding.

The evaluation of patients treated with radiofrequency for a period of 5 years was again performed by Takahashi et al. in 2008 [30], who observed a reduction in incontinence scores from 14.37 to 8.26, with a significant improvement in quality of life. A more recent prospective and nonrandomized study, which included 31 patients followed up for a 3-year period, showed disappointing results. Radiofrequency may be a useful option in combination with other treatments, but subsequent studies, including a small number of patients, have shown that this is an option for cases of mild incontinence, with no correlation with predictive factors in the examinations performed and with low morbidity [31–33].

More recently, a new animal model study has tried to elucidate the mechanisms of action of radiofrequency, perhaps explaining the clinical improvement, not by the barrier effect and formation of fibrosis, but by stimulating growth of the smooth muscle of the sphincter collagen deposition and the reduction of Cajal cells [34]. In a review of 220 patients, the authors showed that radiofrequency is a good option for well-selected patients with mild-to-moderate incontinence [35].

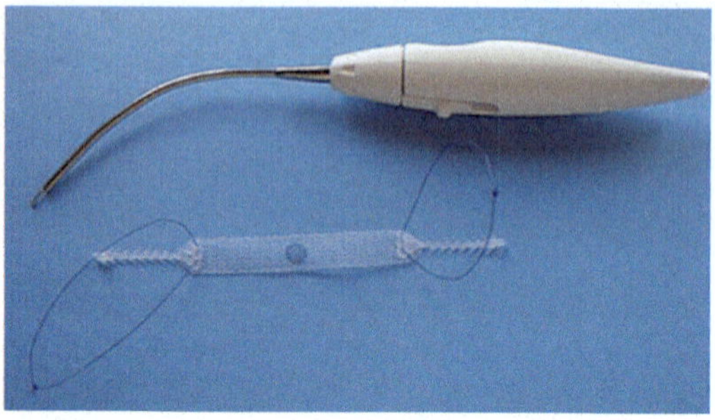

Fig. 14.14 Anorectal Polipropilene Minisling

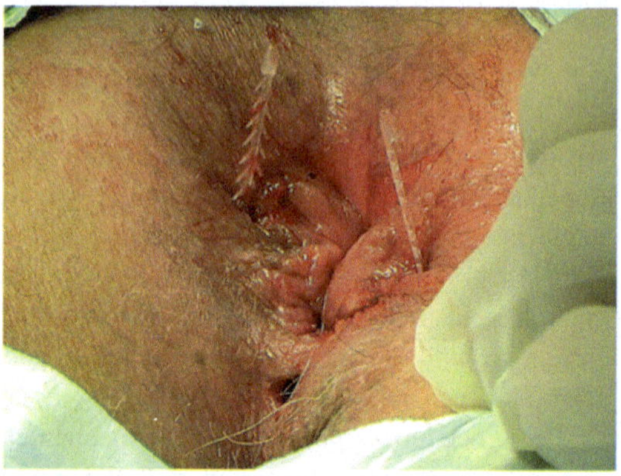

Fig. 14.15 Anorectal minisling inserted before being anchored

Minislings

The use of mesh or slings for the treatment of different genital dystopias and urinary incontinence has been performed by urogynecologists with good results [36, 37]. In the field of coloproctology, the use of small monofilament polypropylene slings (Ophira Minisling System, Promedon, Córdoba, Argentina) has been proposed for the treatment of anal incontinence, mainly in patients with mild-to-moderate absence of structural muscular defect, or as a complement to sphincteroplasty. The procedure is simple and performed with the patient in the jackknife position. The sling is inserted through a trocar specifically developed for this procedure (Fig. 14.14). The sling is attached to the trocar so that it is positioned around the anal canal. Two lateral incisions in one posterior quadrant are made for insertion of the trocar (Fig. 14.15). The sling has spicules that adhere to the tissues and hold the fixation around the anal canal.

This new minimally invasive technique was recently proposed, thus there are still no published results in the literature. The results of the minisling for urinary incontinence have motivated the development of a similar technique for anal incontinence. It is believed that, with adequate patient selection, this technique may be included in the range of options for patients with anal incontinence.

Magnetic Ring

The search for effective surgical alternatives for the treatment of anal incontinence has motivated the development of new devices such as the magnetic ring. The Magnetic Sphincter Augmentation Continence Restoration System (FENIX™ MSA), consists of a ring of 14–20 titanium beads with a magnetic core that are linked together to form an annular structure to be surgically placed around the anal canal (Fig. 14.16). To defecate, the patient strains in a normal way and the force generated separates the beads to open the anal canal. This device provides flexibility to the anus, which is important for reproducing the physiological function of the anal canal and allowing its opening during evacuation. This device was presented by Lehur et al. [38] in 2010 and has since been under review and by the FDA. A multicenter study has reported on the long-term effectiveness and safety of this new treatment modality [39]. The study was performed at four clinical sites in Europe and the United States. A total of 35 patients (34 women) underwent magnetic anal sphincter augmentation. The median length of follow-up was 5.0 years (range, 0–5.6 years), with 23 patients completing assessment at 5 years. Eight patients underwent a subsequent operation (seven device explantations) because of device failure or complications [defecatory dysfunction (20%), pain (14%), erosion (11%), and infection (11%)] In patients who retained their device, the number of incontinent episodes per week and CCF-FIS significantly decreased from baseline, and there were significant improvements in all 4 scales of the Fecal Incontinence Quality of Life instrument. This device was recently discontinued by the company. In fact, the long-term results were disappointing.

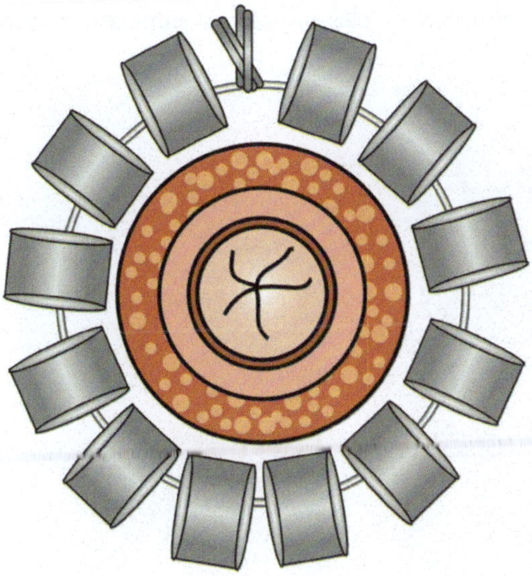

Fig. 14.16 Magnetic anal sphincter

Currently a randomized comparison of the FENIX™ MSA to sacral neuromodulation in terms of safety, efficacy, quality of life, and cost-effectiveness in being conducted [40].

Vaginal/Analplugs

The idea of using a mechanical barrier for the treatment and prevention of fecal incontinence episodes is an old concept. Regardless of whether vaginal or anal inserts are used, these devices could be used as a stand-alone therapy or in conjunction with other conservative therapies [41–43]. The first in-human study utilizing a vaginal bowel control system in 13 incontinent patients demonstrated that the system was comfortable, well tolerated, and showed objective evidence of occluding the rectum [44]. A multicenter prospective trial was then conducted including 110 incontinent patients. Participant eligibility was determined using 2-week bowel diaries (baseline) that recorded the severity, consistency, and associated urgency of incontinent episodes. Sixty-one women entered the treatment period with 93% (n = 56) of subjects completing the 1-month period per protocol (three withdrew from the trial, two did not fully complete treatment diaries). Eighty-six percent of the per-protocol population had treatment success (≥50% reduction in episodes) with a significant decrease in ABL frequency from 11.66 ± 9.65 per 2 weeks (baseline) to 2.16 ± 2.9 episodes per 2 weeks with the VBC system use for 1 month [45]. Subsequently, a secondary analysis of the vaginal device was conducted to assess the effect on bowel movement frequency, urgency, stool consistency, and evacuation completeness (Fig. 14.17). Accordingly, patients experienced a significant decrease in bowel movements and fewer women reported liquid stools, urgency with bowel movements, or impaired evacuation [46]. A new anal device was released in United States in 2015 (Fig. 14.18). Despite having methodological limitations, the studies demonstrated that 78% of those who completed treatment were very or extremely satisfied with the anal insert [47–49].

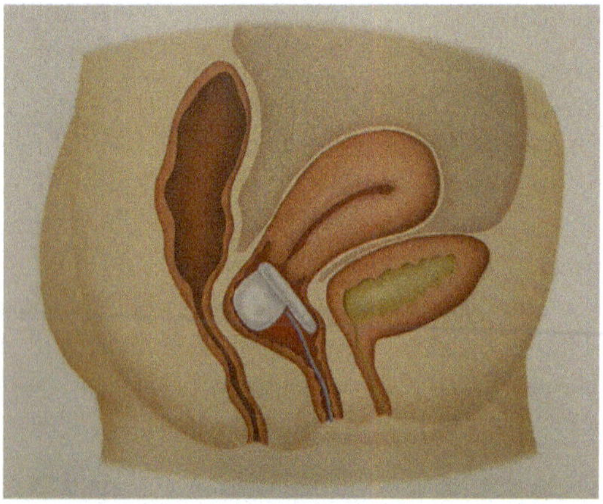

Fig. 14.17 The vaginal insert device (Eclipse System (Pelvalon) continence restoration system, Ethicon)

Neuromodulation

Since the introduction of this therapy in 1995, patients with pelvic floor dysfunction, specifically fecal incontinence, have experienced an improvement in their quality of life [50]. Sacral neuromodulation has a complex mechanism of action that is discussed in further detailed in other chapters. Nevertheless, the results of most series have demonstrated that this is one of the best available options for the treatment of fecal incontinence, regardless of the origin [51–53]. In our experience in South America, it can be very successful when patients are well selected and managed by trained and experienced surgeons [54].

Stem Cells

The concept of injecting stem cells into the anal sphincter muscles in order to stimulate the pro-

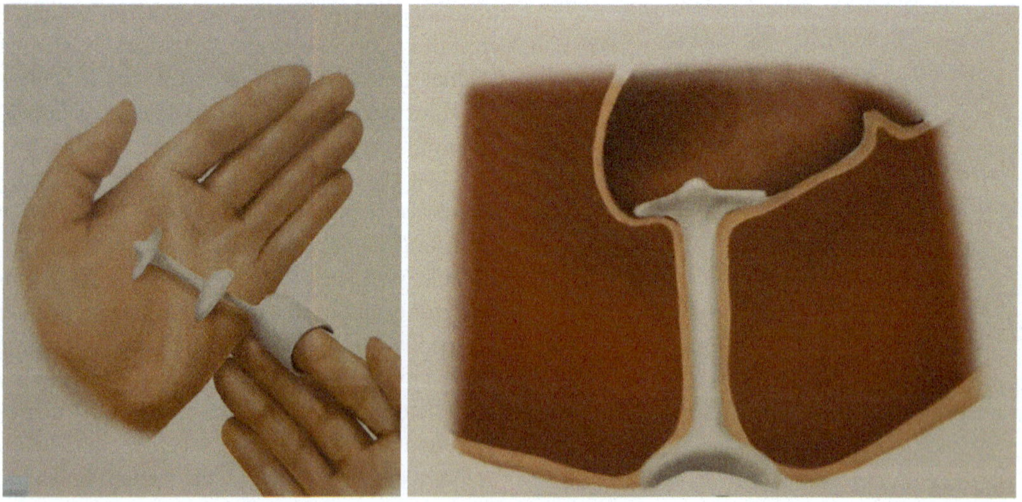

Fig. 14.18 Renew Anal Device (Renew Medical)

duction of new muscle cells is very attractive for patients with fecal incontinence, especially for those with a sphincter defect. This topic will be fully addressed in Chap. 15.

Summary

There were several minimally invasive options for the treatment of incontinent patients, however, some have been discontinued. Once continence control is lost, colorectal surgeons face a great challenge and a variety of situations that should be tailored according to the individual case scenario. The severity of incontinence and the impact on quality of life, evaluated by the use of valid instruments, together with other factors may help in the selection of patients who may be candidates for these options. The choice of treatment should be individualized and based on the available options.

References

1. Paquette IM, Varma MG, Kaiser AM, et al. The American Society of Colon and Rectal Surgeons' clinical practice guideline for the treatment of Fecal incontinence. Dis Colon Rectum. 2015;58(7):623–36.
2. Shafik A. Polytetrafluoroethylene injection for the treatment of partial fecal incontinence. Int Surg. 1993;78:159–61.
3. Shafik A. Perianal injection of autologous fat for the treatment of sphincteric incontinence. Dis Colon Rectum. 1995;38:583–7.
4. Rotholtz NA, Bun M, Mauri MV, et al. Long-term assessment of fecal incontinence after lateral internal sphincterotomy. Tech Coloproctol. 2005;9(2):115–8.
5. Oliveira L, Jorge JMN, Yusuf S, et al. Novos tratamentos para a incontinência anal: injeção de silicone melhora a qualidade de vida em 35 pacientes incontinentes. Rev Bras Coloproct. 2007;27(2):167–73.
6. Oliveira LC, Neves Jorge JM, Yussuf S, et al. Anal incontinence improvement after silicone injection may be related to restoration of sphincter asymmetry. Surg Innov. 2009;16(2):155–61.
7. Maeda Y, Vaizey CJ, Kamm MA. Long-term results of perianal silicone injection for faecal incontinence. Color Dis. 2007;9(4):357–61.
8. Tjandra JJ, Lim JF, Hiscock R. Injectable silicone biomaterial for fecal incontinence caused by internal anal sphincter dysfunction is effective. Dis Colon Rectum. 2004;47:2138–46.
9. Nijhuis PH, van den Bogaard TE, Daemen MJ, et al. Perianal injection of polydimethylsiloxane (Bioplastique implants) paste in the treatment of soiling: pilot study in rats to determine migratory tendency and locoregional reaction. Dis Colon Rectum. 1998;41:624–9.
10. Dewan PA, Byard RW. Histological response to injected Polytef and Bioplastique in a rat model. Br J Urol. 1994;73:370–6.
11. Stojkovic SG, Lim M, Burke D, et al. Intra-anal collagen injection for the treatment of faecal incontinence. Br J Surg. 2006;93(12):1514–8.
12. Davis K, Kumar D, Poloniecki J. Preliminary evaluation of an injectable anal sphincter bulking agent (Durasphere) in the management of faecal incontinence. Aliment Pharmacol Ther. 2003;18:237–43.
13. Weiss EG, Efron JE, Nogueras JJ, et al. Submucosal injection of carbon-coated beads is a successful and safe office-based treatment for fecal incontinence. Chicago: Am Soc of Colon and Rectal Surgeons; 2002.
14. Tjandra JJ, Chan MK, Yeh HC. Injectable silicone biomaterial (PTQ) is more effective than carbon-coated beads (Durasphere) in treating passive faecal incontinence – a randomized trial. Colorectal Dis. 2009;11(4):382–9.
15. Chan MK, Tjandra JJ. Injectable silicone biomaterial (PTQ) to treat fecal incontinence after hemorrhoidectomy. Dis Colon Rectum. 2006;49(4):433–9.
16. Malouf AJ, Vaizey CJ, Norton CS, et al. Internal anal sphincter augmentation for faecal incontinence using injectable silicone biomaterial. Dis Colon Rectum. 2001;44:595–600.
17. Kenefick NJ, Vaizey CJ, et al. Injectable silicone biomaterial for faecal incontinence due to internal anal sphincter dysfunction. Gut. 2002;51(2):225–8.
18. Rosato G, Piccinini P, Oliveira L, et al. Initial results of a new bulking agent for fecal incontinence: a multicenter study. Dis Colon Rectum. 2015;58(2):241–6.
19. Dodi G, Jongen J, de la Portilla F, et al. An open-label, noncomparative, multicenter study to evaluate efficacy and safety of NASHA/Dx Gel as a bulking agent for the treatment of fecal incontinence. Gastroenterol Res Pract. 2010;2010:467136.
20. Mellgren A, Matzel KE, Pollack J, et al. Study Group. Long-term efficacy of NASHA Dx injection therapy for treatment of fecal incontinence. Neurogastroenterol Motil. 2014;26(8):1087–94.
21. Maeda Y, Laurberg S, Norton C. Perianal injectable bulking agents as treatment for faecal incontinence in adults. Cochrane Database Syst Rev. 2013;2:CD007959.
22. Luo C, Samaranayake CB, et al. Systematic review on the efficacy and safety of injectable bulking agents for passive faecal incontinence. Color Dis. 2010;12:296–303.
23. Ratto A, Parello L, Donisi F, Litta V, De Simone L, et al. Novel bulking agent for faecal incontinence. Br J Surg. 2011;98:1644–52.
24. Ratto C, Buntzen S, Aigner F, Altomare DF, Heydari A, Donisi L, Lundby L, Parello A. Multicentre observational study of the Gatekeeper for faecal incontinence. Br J Surg. 2016;103(3):290–9.

25. Ratto C, Donisi L, Litta F, et al. Implantation of SphinKeeper™: a new artificial anal sphincter. Tech Coloproctol. 2016;20(1):59–66.
26. Trenti L, Biondo S, Noguerales F, Nomdedeu J, Coret A, Scherer R, Fraccalvieri D, Frago R, Kreisler E. Outcomes of Gatekeeper™ prosthesis implantation for the treatment of fecal incontinence: a multicenter observational study. Tech Coloproctol. 2017;21(12):963–70.
27. Takahashi T, Garcia-Osogobio S, Valdovinos MA, et al. Extended two-year results of radiofrequency energy delivery for the treatment of fecal incontinence (the Secca procedure). Dis Colon Rectum. 2003;46(6):711–5.
28. Efron JE. The Secca procedure: a new therapy for treatment of fecal incontinence. Surg Technol Int. 2004;13:107–10. Review
29. Efron JE, Corman ML, Fleshman J, et al. Safety and effectiveness of temperature-controlled radiofrequency energy delivery to the anal canal (Secca procedure) for the treatment of fecal incontinence. Dis Colon Rectum. 2003;46(12):1606–16.
30. Takahashi-Monroy T, Morales M, Garcia-Osogobio S, et al. Secca procedure for the treatment of fecal incontinence: results of five-year follow-up. Dis Colon Rectum. 2008;51(3):355–9.
31. Felt-Bersma RJ, Szojda MM, Mulder CJ. Temperature-controlled radiofrequency energy (Secca) to the anal canal for the treatment of faecal incontinence offers moderate improvement. Eur J Gastroenterol Hepatol. 2007;19(7):575–80.
32. Lefebure B, Tuech JJ, Bridoux V, et al. Temperature-controlled radiofrequency energy delivery (Secca procedure) for the treatment of fecal incontinence: results of a prospective study. Int J Color Dis. 2008;23(10):993–7.
33. Ruiz D, Pinto RA, Hull TL, et al. Does the radiofrequency procedure for fecal incontinence improve quality of life and incontinence at 1-year followup? Dis Colon Rectum. 2010;53(7):1041–6.
34. Herman RM, Berho M, Murawski M, et al. Defining the histopathological changes induced by nonablative radiofrequency treatment of faecal incontinence – a blinded assessment in an animal model. Color Dis. 2015;17(5):433–40.
35. Frascio M, Mandolfino F, Imperatore M, et al. The Secca procedure for faecal incontinence: a review. Color Dis. 2014;16(3):167–72.
36. Netto OF, Figueirêdo O, Garcia E, Figueirêdo PG. Minisling transobturatório para tratamento cirúrgico da incontinência urinária de esforço. Femina. 2004;32(4):293–7.
37. Alinsod R. Recent advances in tape slings for female urinary stress incontinence. Rev Obstet Gynecol. 2009;2(1):46–50.
38. Lehur PA, McNevin S, Buntzen S, et al. Magnetic anal sphincter augmentation for the treatment of fecal incontinence: a preliminary report from a feasibility study. Dis Colon Rectum. 2010;53(12):1604–10.
39. Sugrue J, Lehur PA, Madoff RD, McNevin S, Buntzen S, Laurberg S, Mellgren A. Long-term experience of magnetic anal sphincter augmentation in patients with fecal incontinence. Dis Colon Rectum. 2017;60(1):87–95.
40. Williams AE, Croft J, Napp V, Corrigan N, Brown JM, Hulme C, Brown SR, Lodge J, Protheroe D, Jayne DG. SaFaRI: sacral nerve stimulation versus the FENIX magnetic sphincter augmentation for adult faecal incontinence: a randomized investigation. Int J Color Dis. 2016;31(2):465–72.
41. Christiansen J, Roed-Petersen K. Clinical assessment of the anal continence plug. Dis Colon Rectum. 1993;36:740–42.
42. Norton C, Kamm MA. Anal plug for faecal incontinence. Colorectal Dis. 2001;3:323–27.
43. Giamundo P, Welber A, Weiss EG, et al. The Procon incontinence device: a new nonsurgical approach to preventing episodes of fecal incontinence. Am J Gastroenterol. 2002;97:2328–32.
44. Sokol ER. Management of fecal incontinence - focus on a vaginal insert for bowel control. Med Devices (Auckl). 2016.
45. Richter HE, Matthews CA, Muir T, et al. A vaginal bowel-control system for the treatment of fecal incontinence. Obstet Gynecol. 2015;125:540–47.
46. Varma MG, Matthews CA, Muir T, et al. Impact of a novel vagi- nal bowel control system on bowel function. Dis Colon Rectum. 2016;59(2):127–31.
47. Lukacz ES, Segall MM, Wexner SD. Evaluation of an anal insert device for the conservative management of fecal incontinence. Dis Colon Rectum. 2015;58:892–98.
48. Leo CA, Thomas GP, Hodgkinson JD, Segal JP, Maeda Y, Murphy J, Vaizey CJ. The Renew® anal insert for passive faecal incontinence: a retrospective audit of our use of a novel device. Colorectal Dis. 2019;21(6):684–88.
49. Segal JP, Leo CA, Hodgkinson JD, Cavazzoni E, Bradshaw E, Lung PFC, Ilangovan R, Vaizey CJ, Faiz OD, Hart AL, Clark SK. Acceptability, effectiveness and safety of a Renew(®) anal insert in patients who have undergone restorative proctocolectomy with ileal pouch-anal anastomosis. Colorectal Dis. 2019;21(1):73–78.
50. Matzel KE, Stadelmaier U, Hohenfellner M, Gall FP. Electrical stimulation of sacral spinal nerves for treatment of faecal incontinence. Lancet. 1995;346(8983):1124–27.
51. Wexner SD, Coller JA, Devroede G, et al. Sacral nerve stimulation for fecal incontinence: results of a 120-patient prospective multicenter study. Ann Surg. 2010;251(3):441–49.
52. Koh H, McSorley S, Hunt S, Quinn M, MacKay G, Anderson J. Sacral neuromodulation for faecal incontinence - 10 years' experience at a Scottish tertiary centre. Surgeon. 2018;16(4):207–13.
53. Patton V, Abraham E, Lubowski DZ. Sacral nerve stimulation for faecal incontinence: medium-term follow-up from a single institution. ANZ J Surg. 2017;87(6):462–66.
54. Oliveira L, Hagerman G, Torres ML, et al. Sacral neuromodulation for fecal incontinence in Latin America: initial results of a multicenter study. Tech Coloproctol. 2019;23(6):545–50.

Stem Cells: The State of Cellular Therapy in Treatment of Fecal Incontinence

15

Massarat Zutshi

Introduction

Stem cells are specialized undifferentiated cells that reside in various parts of the body and have the ability to transform into different cells. The current theory is that stem cells are attracted to a site of injury by chemotaxis due to proteins called chemokines produced by the injured cells. Stem cells function via the paracrine action when they secrete various factors which influence cells in the vicinity of the damaged structures, to produce cells similar to those that are damaged. This current understanding seems due to the fact that no study has shown labeled cells in the regenerating areas which are treated with stem cells, yet beneficial effects are reported.

Skeletal muscle regeneration is facilitated by satellite cells [1]. These cells are present in the muscle tissue and become activated during injury with the release of cytokines from the injured tissue. Hence, not only are the local cells responsible for regeneration but signaling factors are important to initiate, continue, and determine when the process of self-renewal should begin and end.

Fecal incontinence is a symptom that is multifactorial in etiology [2]. For continence to occur there is a fine balance of colonic motility, rectal sensation, and structural integrity of the anal sphincter mechanism. Colonic motility is dependent on multiple factors including pathologies of the colonic mucosa, neuronal abnormalities, and functional diagnoses. Sensation is a function of local reflexes and may reflect pathology of the pudendal nerve [3]. Structural integrity is usually damaged by childbirth in women or can be iatrogenic after surgery in the perineal and anal area. Cellular and noncellular therapies that target muscle regeneration can correct the structural defect but may need other therapies to affect motility and sensation.

Maintaining optimal bowel function is a priority before any therapy can be successful. Currently available therapies include anal sphincteroplasty, bulking agents, radiofrequency stimulation (SECCA® Procedure), sacral neuromodulation, and muscle interposition. Mechanical devices include anal plugs, like the Renew® insert and Peristeen device, and a vaginal insert – the Eclipse® device (see also Chap. 14). All device-related therapies have a fair effect but are challenged by complications, device malfunctions, and battery replacements [4]. There is a need to find a treatment which is relatively cost-effective, easy to use, and can be globally utilized.

Mesenchymal Stem Cells

Mesenchymal stem cells (MSC) are pluripotent cells that give rise to tissue of mesodermal origin. They can be derived from bone marrow

M. Zutshi (✉)
Department of Colorectal Surgery, Cleveland Clinic Foundation, Cleveland, OH, USA
e-mail: ZUTSHIM@ccf.org

L. C. C. Oliveira (ed.), *Anorectal Physiology*, https://doi.org/10.1007/978-3-030-43811-1_15

(BM-MSC), adipose tissue (AD-MSC), muscle (MDSC) umbilical cord, Wharton's Jelly, and placenta and give rise to tissues like bone, cartilage, muscle, tendon/ligament, fat, dermis, and other connective tissues (Fig. 15.1). MSC can be delivered directly to the site of injury or via the bloodstream when they home to the site of injury. However, the numbers that ultimately reach the desired site after systemic delivery may be low due to entrapment in the lungs and spleen. In very few studies have MSC been shown to engraft and regenerate new muscle leading to the theory of paracrine effects of MSC. Paracrine effects imply that the stem cells affects the local tissues to aid in regeneration by secretion of certain cytokines that cause differentiation of satellite cells into muscle and aid in the development of blood vessels and nerve endings to make the new muscle functional (Fig. 15.2) [5–25].

Acute Injury and Stem Cell Treatments

Most animal studies have dealt with the concept of treating an injury in the environment of acute inflammation as after an injury or after injury followed by a repair. The majority of the studies have used a rodent model while some have used rabbits. Many models of injury have been developed; these include a sphincterotomy, sphincterotomy with a repair, sphincter excision of variable lengths, and cryoinjury to study the effects after an acute injury.

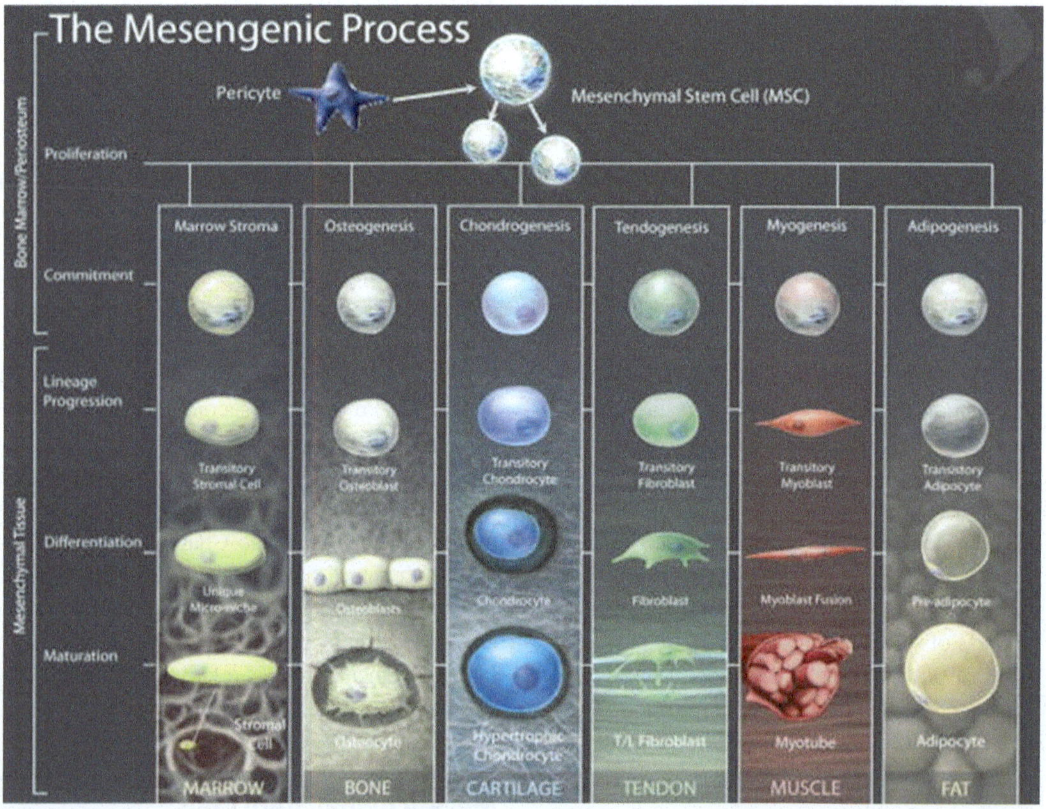

Fig. 15.1 The mesengenic process. The original version of this figure was generated in the late 1980s and has been modernized in this rendition. The figure proposes that an MSC exists in the bone marrow and that its progeny can be induced to enter one of several mesenchymal lineage pathways. The lineage format was constructed from what was known about the hematopoietic lineage pathway, and this figure depicts the predicted differentiation hierarchy of the most prominent candidate lineages. (Current image graphics produced by Michael Gilkey – National Center for Regenerative Medicine [26]. Reused with permission from © Elsevier)

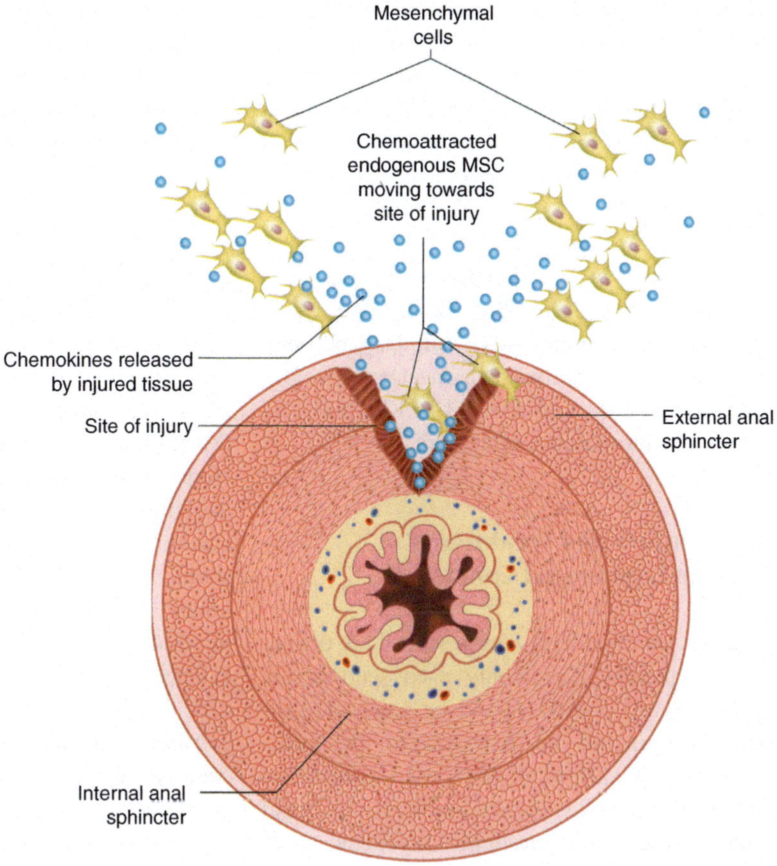

Fig. 15.2 Artists impression of endogenous MSC attracted to the site of injury via chemoattraction to cytokines released during the process of injury. MSC have the receptor to these cytokines

There have also been different cell types used in various studies. Most studies have used BM-MSC while others have used a myoblast preparation, satellite cells, MDSC, or commercially available cell lines. Most studies have evaluated function 2 weeks after treatment. To assess this type of treatment for incontinent patients, we have shown in a rodent model that most small injuries of 25% excision of the circumference or less are fully healed at 4 weeks after injury; hence timing of evaluation is critical after an acute injury. The main outcomes evaluated have been anal pressures, electromyography (EMG), histology, immunohistochemistry, response to electrical stimulation, muscle tensile strength and fatigue studies, and detection of various markers of muscle regeneration.

Lorenzi et al. [10] were the first to publish in 2007 and their model was a sphincterotomy and a repair. They used rat BM-MSC and had a group that used immune suppression as well. Their outcomes were histology which evaluated muscle area fractions at the site of repair and the contractile response to electrical and chemical stimulation. They reported increased muscle area fraction with MSC and greater contractility and no untoward effects of immune suppression. In comparison using a similar model of a repaired anal sphincter, Fritzwater et al. [12] in 2015 used a commercially available myogenic stem cell to evaluate if stem cell-treated sphincter repair had greater muscle mass compared to saline-treated group at 90 days. Their findings were an increase in the contractile force in the muscle treated with stem cells but could not find any increase in muscle characteristics when compared to the saline-treated groups.

Jacobs et al. [24] in 2013 used MDSC in a similar model of a sphincter repair. They reported EMG amplitude findings at 2 weeks which were significantly increased but were similar at 4 weeks. Anal resting pressures were reported as being significantly increased compared to the controls. However, they reported no increase in the internal anal sphincter muscle or external anal sphincter muscle in the transplanted group or an increase in the number of nuclei in this group.

In 2009 Aghaee-Afshar et al. [7] reported a rabbit model of a sphincterotomy and used umbilical cord cells and BM-MSC. Their results 2 weeks after injury were similar and they reported on an improvement in electromyography (EMG) study with the BM-MSC which also showed better muscle architecture.

Acute Injury and Treatments with an Engineered Construct

A few researchers have used a scaffold to deliver cellular therapy. Commonly used scaffold are hydrogels. A scaffold should be easily degradable and should be nontoxic to the cells. Monotoya et al. [16] in 2014 used a model of sphincter transection (7 mm) and used a hydrogel with commercially available myogenic stem cells. The control groups were no treatment, saline, and a collagen group. They concluded that although contractile forces were superior in the group with the scaffold and cells, they could not demonstrate a clear superiority of this group and alluded to the fact that healing mechanisms may differ in the groups. In our lab, we have used a hydrogel scaffold with BM-MSC and did not show any superiority over using only BM-MSC in a model of chronic injury with a plasmid encoding for a cytokine.

Chronic Injury and Treatment with Cell Therapy

Treating a chronic injury is the ultimate goal of all researchers in this field. This is because most of the anal sphincter injuries become chronic and only recognized clinically when they become symptomatic. Therefore, successful treatments will need a tissue environment that can mimic that of an acute injury, directing the stem cells to carry out their regenerative effects on local tissues. Oh et al. [17] used polycaprolaptone beads loaded with basic fibroblastic growth factor and autologous myoblass in a dog model of 25% transection of the posterior sphincter. Their premise was that the growth factor prolonged the effect of the stem cells causing greater differentiation of the muscle cells. They surmised that the beads with the growth factor and cells were successfully grafted and showed slightly increased pressures and contractile forces than cells alone. They did not have a group with the beads with factors alone. In the study by Sun et al. [4] from our lab we similarly used, a model of a chronic injury transecting 50% of the external and internal anal sphincter as our injury model. Three weeks after injury, a plasmid encoding for SDF-1 was injected alone or 3 days prior to injections of cells or cells in a scaffold. The controls were not used as treatment. We concluded that the SDF-1 plasmid alone did as well as the SDF 1 plasmid + cells or cells in a scaffold in terms of increased pressures, muscle quantification, decreased fibrosis, and more organized muscle fibers in the area of the defect (Fig. 15.3).

Neo-sphincters

Creating a neoanal sphincter to replace either the internal (IAS) or external anal sphincter (EAS) has its limitations in a clinical setting. Bitar et al. have created a neo-sphincter to replace the IAS. They used a construct to plate muscle cells that can be obtained from a colonic, small bowel, or appendicular biopsy. They implanted first in the back muscle of a rodent and moved to the anal sphincter in a rabbit. They have shown contractility and relaxation of these constructs and have demonstrated the development of both blood vessels and nerve terminals into the construct. Currently they are awaiting FDA clearance to embark on a clinical trial. It is unclear what patient population they would target and if it would require excision of the existing IAS.

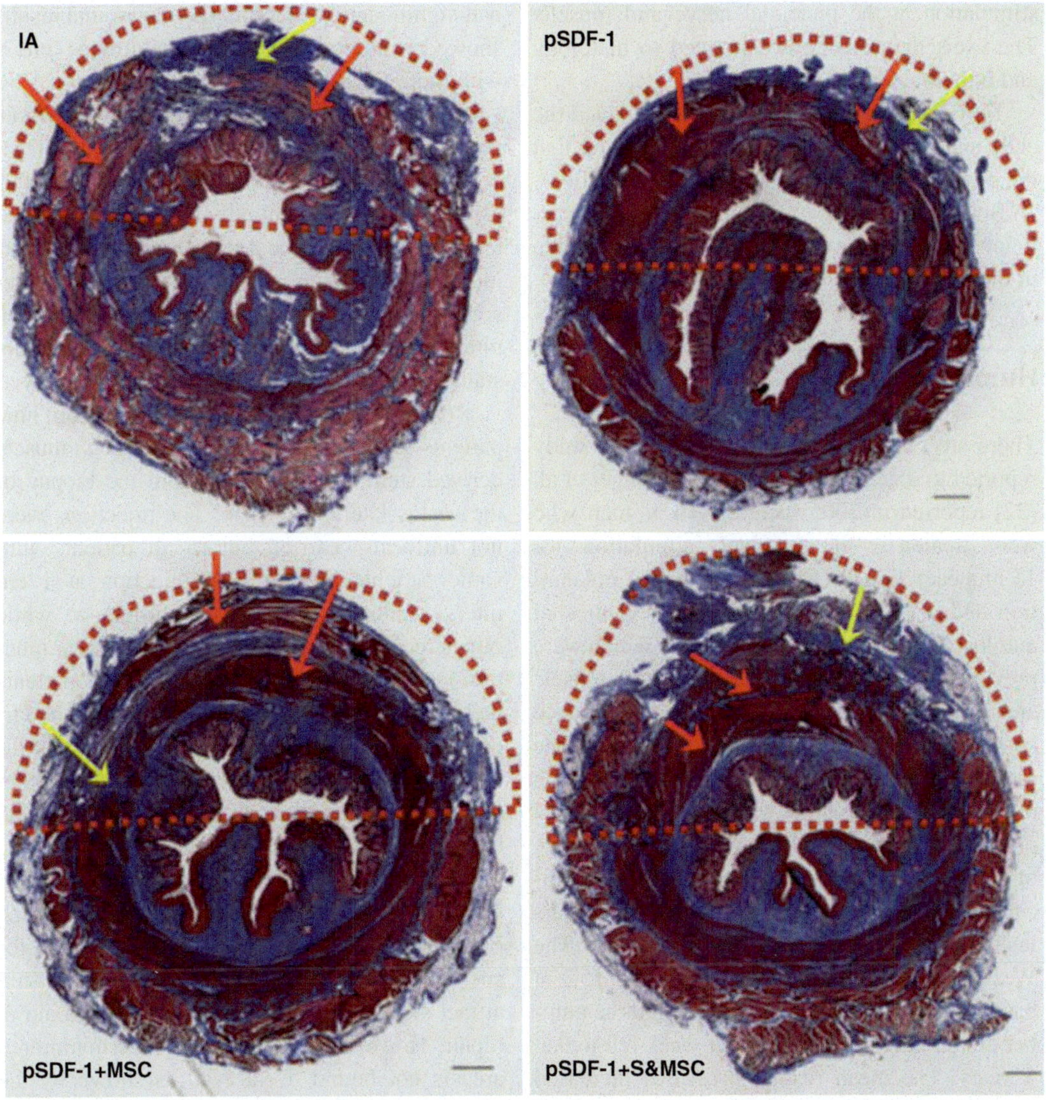

Fig. 15.3 Representative pictures of transverse anal sphincter sections stained by Masson trichrome and 4 weeks after various treatments after a partial anal sphincter excision treated 3 weeks after injury. In the area of the defect (circled by red oval), muscle is indicated by a red arrow and connective tissue is indicated by a yellow arrow. A higher percentage of muscle is seen at the area of injury compared with the uninjured normal muscle in the same section noted in all three groups with treatment (pSDF-1, pSDF-1 + MSC, pSDF-1 + S&MSC) versus the IA group. IA = injury without treatment; pSDF-1 = stromal derived factor 1-encoded plasmid local injection at the site of the defect; pSDF-1 + MSC = pSDF-1 and MSC injected at the site of the defect; pSDF-1 + S&MSC = pSDF-1 injected at the site of the defect with insertion of a gelatin scaffold mixed with MSC; MSC = mesenchymal stem cell. Scale bar = 500 [micro]m. (Reused with permission from © Wolters Kluwer [27])

Kajdafzadeh et al. [8] have replaced the EAS with a neo EAS in a rabbit model. They excised the EAS and decellularized it. Next it was seeded by satellite cells taken from a muscle biopsy and expanded. This grafted construct was then implanted 6 months later in the same rabbits and followed up for 2 years. Outcomes included electrical stimulation monthly till 6 months and every 3 months thereafter to evaluate contractility. At 2 years the sphincters were subjected to electrical stimulation at the pudendal nerve and muscle. The excised muscle was subjected to histology and immunohistochemistry.

They had two groups, one which received the cells and one that did not. They concluded that the group with cells had increased contractility.

Both the above may be a substitute for the artificial anal sphincter that is currently not available in the United States.

Human Trials

There are two human trials and one case study reported to date. The first one by Frudinger et al. [22] reported in 2005 included 10 women who were treated with electrical stimulation for 15 min each day for 10 weeks prior to implantation and 28 days after injection. They injected autologous muscle derived stem cells which were previously harvested from the pectoral muscle and expanded. Outcomes included incontinence based on the Wexner scale, fecal incontinence quality of life scale, and anal pressures on manometry. The Wexner score decreased by 14 points and the quality of life increased by a score of 30 points at 1 year while there was no significant increase on the anal pressures. Five years later the same patients were revaluated. The Wexner score decreased from a mean of 15.3 at baseline to a mean of 0.7 at 5 years, and the number of incontinence episodes per week fell to 0 at 5 years. The mean resting pressures increased from 32 to 41 and the squeeze pressures increased from 44 to 61. Quality of life was maintained at 5 years.

Sarveazad et al. [21] in 2017 reported on 18 patients in a randomized controlled trial who underwent a sphincteroplasty followed by either injection of adipose derived stem cells or saline. Adipose-derived stem cells were obtained from the abdomen via a liposuction and cultured and expanded. These were injected into the sphincteroplasty site. Outcomes included endoanal ultrasound, EMG, and incontinence scores using the Wexner score. The scores 2 months after surgery were similar. The EMG in the treated group was significantly different. The ultrasound results showed scattered muscle in the treated group with more fibrosis in the untreated group.

In 2015, Cesaro et al. reported on three patients treated with a lipoaspirate using a specific technique when an unknown quantity of this aspirate was injected blindly into the intersphincteric groove under local anesthesia. They report increase in resting pressures and the Wexner scores at 6 months. They report on an endoanal ultrasound as an assessment modality but do not state any results postoperatively.

Romaniszyn et al. [25] in 2015 subjected nine patients to implantation of cultured muscle derived stem cells obtained from the biopsy of the vastus lateralis muscle. The injections were not uniformly carried out in all patients with some patients receiving an injection at 1 cm intervals along the entire circumference while others received injections at the muscle scar junction and some into the scar tissue. The patients had a muscle defect of less than 25% of the circumference. Six out of nine patients did well for 6 months and two deteriorated at 1 year. Other outcome measures were anal manometry and ultrasound and EMG (surface).

When cells are injected after a sphincter repair or a stimulus like electrical study as in the Frudinger study [23], there may be a basis for success as there is some inflammation that could attract stem cells to remain at the site and aid in repair. However injecting cells in a noninjured area is not bound to succeed as shown by the study by Bisson et al. where cells injected in an area of no injury had no effect (Fig. 15.4).

Potential Adverse Effects

BM-MSC has been deemed safe in many disciplines. One study by Jacobs et al. [24] in 2013 reported no evidence of cell migration to liver or lung, but two transplanted rats developed abnormal foci of growth, i.e., tumors, from the external anal sphincter, raising further safety questions. This requires further research before they can be used in humans.

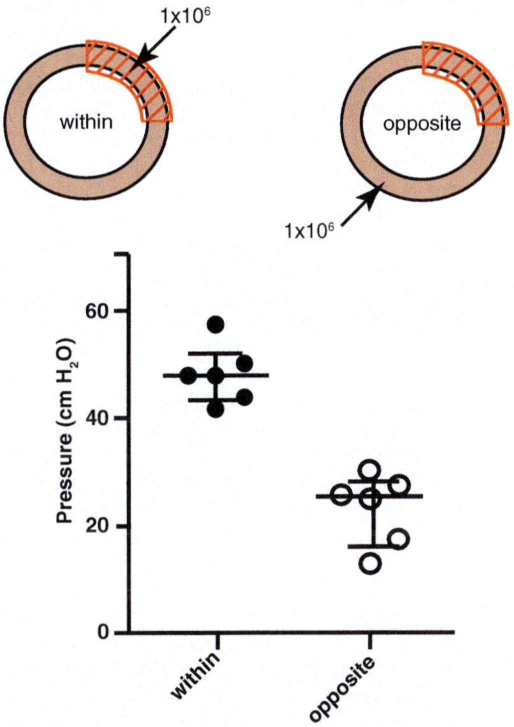

Fig. 15.4 Influence of site of injection on efficacy of myoblast cell therapy. Topography of the injection site within or opposite to the cryolesion (upper). 1 × 106 myoblasts were injected within (black circles, n = 6) or opposite to the lesion site (white circles, n = 6). Pressures under electrostimulation (6.35 V, 5 s) were measured at day 60. Data are expressed as median ± 25th percentile [18]

Summary

Advances in cell therapy are simplifying the use of cells, cytokines, and scaffolds. The future may see something available that is tailored to specific tissues. Researchers should be cognizant of the outcome measures they use to evaluate success. Clinical trials should be based on well performed bench research. The anal sphincter is easily accessible for an injection treatment of stem cells; however, it should not be the reason to inject cells in humans without good science to direct it. Current researchers have the onus of providing accurate scientific data so that cellular or noncellular therapy is successful in clinical trials in the future.

References

1. Relaix F, Zammit PS. Satellite cells are essential for skeletal muscle regeneration: the cell on the edge returns centre stage. Development. 2012;139:2845–56.
2. Bharucha AE, Zinsmeister AR, Locke GR, et al. Risk factors for fecal incontinence: a population-based study in women. Am J Gastroenterol. 2006;101:1305–12.
3. Haas S, Brock C, Krogh K, et al. Abnormal neuronal response to rectal and anal stimuli in patients with idiopathic fecal incontinence. Neurogastroenterol Motil. 2015;27:954–62.
4. Sun L, Yeh J, Xie Z, et al. Electrical stimulation followed by mesenchymal stem cells improves anal sphincter anatomy and function in a rat model at a time remote from injury. Dis Colon Rectum. 2016;59:434–42.
5. Salcedo L, Sopko N, Jiang HH, et al. Chemokine upregulation in response to anal sphincter and pudendal nerve injury: potential signals for stem cell homing. Int J Color Dis. 2011;26:1577–81.
6. Salcedo L, Penn M, Damaser M, et al. Functional outcome after anal sphincter injury and treatment with mesenchymal stem cells. Stem Cells Transl Med. 2014;3:760–7.
7. Aghaee-Afshar M, Rezazadehkermani M, Asadi A, et al. Potential of human umbilical cord matrix and rabbit bone marrow-derived mesenchymal stem cells in repair of surgically incised rabbit external anal sphincter. Dis Colon Rectum. 2009;52:1753–61.
8. Kajbafzadeh AM, Elmi A, Talab SS, et al. Functional external anal sphincter reconstruction for treatment of anal incontinence using muscle progenitor cell auto grafting. Dis Colon Rectum. 2010;53:1415–21.
9. Kang SB, Lee HN, Lee JY, Park JS, Lee HS, Lee JY. Sphincter contractility after muscle-derived stem cells autograft into the cryoinjured anal sphincters of rats. Dis Colon Rectum. 2008;51:1367–73.
10. Lorenzi B, Pessina F, Lorenzoni P, et al. Treatment of experimental injury of anal sphincters with primary surgical repair and injection of bone marrow-derived mesenchymal stem cells. Dis Colon Rectum. 2008;51:411–20.
11. Pathi SD, Acevedo JF, Keller PW, et al. Recovery of the injured external anal sphincter after injection of local or intravenous mesenchymal stem cells. Obstet Gynecol. 2012;119:134–44.
12. Fitzwater JL, Grande KB, Sailors JL, et al. Effect of myogenic stem cells on the integrity and histomorphology of repaired transected external anal sphincter. Int Urogynecol J. 2015;26:251–6.
13. Salcedo L, Mayorga M, Damaser M, et al. Mesenchymal stem cells can improve anal pressures after anal sphincter injury. Stem Cell Res. 2013;10:95–102.

14. Salcedo L, Penn M, Damaser M, Balog B, Zutshi M. Functional outcome after anal sphincter injury and treatment with mesenchymal stem cells. Stem Cells Trans Med. 2014;3:760–7.
15. Sun L, Yeh J, Xie Z, Kuang M, Damaser MS, Zutshi M. Electrical stimulation followed by mesenchymal stem cells improves anal sphincter anatomy and function in a rat model at a time remote from injury. Dis Colon Rectum. 2016;59:434–42.
16. Montoya TI, Acevedo JF, Smith B, et al. Myogenic stem cell-laden hydrogel scaffold in wound healing of the disrupted external anal sphincter. Int Urogynecol J. 2015;26:893–904.
17. Oh HK, Lee HS, Lee JH, et al. Functional and histological evidence for the targeted therapy using biocompatible polycaprolactone beads and autologous myoblasts in a dog model of fecal incontinence. Dis Colon Rectum. 2015;58:517–25.
18. Bisson A, Freret M, Drouot L, et al. Restoration of anal sphincter function after myoblast cell therapy in incontinent rats. Cell Transplant. 2015;24:277–86.
19. Boulaiz H, Marchal JA, Prados J, et al. Non-viral and viral vectors for gene therapy. Cell Mol Biol (Noisy-le-Grand). 2005;51:3–22.
20. Mazzanti B, Lorenzi B, Borghini A, et al. Local injection of bone marrow progenitor cells for the treatment of anal sphincter injury: in-vitro expanded versus minimally-manipulated cells. Stem Cell Res Ther. 2016;7:85.
21. Sarveazad A, Newstead GL, Mirzaei R, et al. A new method for treating fecal incontinence by implanting stem cells derived from human adipose tissue: preliminary findings of a randomized double-blind clinical trial. Stem Cell Res Ther. 2017;8:40.
22. Frudinger A, Kolle D, Schwaiger W, Pfeifer J, Paede J, Halligan S. Muscle-derived cell injection to treat anal incontinence due to obstetric trauma: pilot study with 1 year follow-up. Gut. 2010;59:55–61.
23. Frudinger A, Pfeifer J, Paede J, Kolovetsiou-Kreiner V, Marksteiner R, Halligan S. Autologous skeletal muscle-derived cell injection for anal incontinence due to obstetric trauma: a five-year follow-up of an initial study of ten patients. Color Dis. 2015;17:794–801.
24. Jacobs SA, Lane FL, et al. Safety assessment of myogenic stem cell transplantation and resulting tumor formation. Female Pelvic Med Reconstr Surg. 2013;19(6):362–8.
25. Romaniszyn M, Rozwadowska N, Malcher A, Kolanowski T, Walega P, Kurpisz M. Implantation of autologous musclederived stem cells in treatment of fecal incontinence: results of anexperimental pilot study. Tech Coloproctol. 2015;19:685–96.
26. Caplan AI, Correa D. The MSC: an injury drugstore. Cell Stem Cell. 2011;9(1):11–5.
27. Sun L, Xie Z, Kuang M, Penn M, Damaser MS, Zutshi M. Regenerating the anal sphincter: cytokines, stem cells, or both? Dis Colon Rectum. 2017;60(4):416–25.

Surgical Treatment of Fecal Incontinence

16

Anne Sirany, Steven D. Wexner,
and Giovanna Da Silva

Introduction

Fecal incontinence (FI) is the uncontrolled passage of feces or gas over at least 1 month's duration, in an individual who has previously achieved control [1]. The prevalence rates vary from 1.4% to 18% and up to 50% in institutionalized patients [2, 3]. More than 90% of patients with incontinence are female [3]. A 2013 survey of women with FI found that more than two thirds of women did not seek out care for their FI [4]. Factors predictive of patients who did seek care where those who had an established primary care physician, who had heard of FI and those that had suffered longer with the condition. The severity of FI varies widely but has been shown to have a negative impact on individuals with reduced quality of life, negative psychological effect, and associated social stigma [5, 6].

Normal continence is complex and involves coordination among the sphincter muscles, pelvic floor muscles and is affected by stool frequency and volume, sensory function, rectal compliance, and consistency [7]. Therefore, failure in any of these functions can alter continence. The etiology of FI may be multifactorial, and risk factors include old age, female sex, sphincter injury, obstetrical trauma, postsurgical complications, diarrhea, and constipation [8, 9]. Sphincter disruption from obstetrical trauma is clinically observed in up to 10% of all vaginal deliveries, but occult injury can be found in 21–35% of deliveries [10].

The goals of treatment are to decrease the frequency and severity of episodes and improve quality of life. The decision of which treatment to employ is based on the severity of symptoms and integrity of the anal sphincter. Several instruments have been designed to evaluate continence, including a bowel diary and continence scoring systems such as the Cleveland Clinic Florida Fecal Incontinence Score (CCF-FIS or Wexner score) [11], St. Mark's Incontinence Score (SMIS) [12], or the Fecal Incontinence Severity Index (FISI) [13]. CCF-FIS is the most common used scoring system [11]. The American Society of Colon and Rectal Surgeons (ASCRS) Clinical Practice Guidelines recommends that first-line therapy for FI should include dietary modifications, medical management with antidiarrheal and/or fiber supplement to bulk the stool, and biofeedback exercises, which can lead to improvement in a significant portion of patients with mild FI [14].

Patients with more severe disease and/or sphincter defects will require more invasive procedures. These interventions can be categorized into methods that repair (sphincteroplasty), augment (inject-

A. Sirany
University of Minnesota, Department of Surgery, Minneapolis, MN, USA

S. D. Wexner
Chair, Department of Colorectal Surgery, Cleveland Clinic Florida, Digestive Disease Center, Weston, FL, USA

G. Da Silva (✉)
Department of Colorectal Surgery, Cleveland Clinic Florida, Weston, FL, USA
e-mail: DASILVG@ccf.org

L. C. C. Oliveira (ed.), *Anorectal Physiology*, https://doi.org/10.1007/978-3-030-43811-1_16

ables, radiofrequency remodeling), and replace the sphincter (adynamic muscle transfers, dynamic graciloplasty, artificial bowel sphincter, magnetic anal sphincter), and those that stimulate (sacral neuromodulation, posterior tibial nerve stimulation) or divert fecal transit (stoma, antegrade stoma procedure) (Table 16.1) [15]. The 3rd edition of the ASCRS text outlines a treatment algorithm for the surgical management of FI (Fig. 16.1). This chapter will focus on the surgical management of FI.

Table 16.1 Surgical treatment options for fecal incontinence

	Surgical management	Indications	Other
Repair	Sphincteroplasty	Sphincter defect	
Augment	Injection of bulking agent	Minor leakage. Internal sphincter defect	
	Radiofrequency energy delivery	Mild-to-moderate FI. Try alternative treatments first	
	Anal sling		Not available in the USA
Sphincter replacement	Stimulated Graciloplasty	Severe FI, after failure of less invasive therapies	Not available in the USA
	Magnetic anal sphincter	Severe FI, after failure of less invasive therapies	Not available in the USA
	Artificial bowel sphincter	Severe FI	Not available in the USA
Neurostimulate/modulate bowel function	Sacral neuromodulation	Moderate-to-severe FI	
	Percutaneous tibial nerve stimulation	Moderate FI	Not approved in the USA for FI
Divert fecal transit	Antegrade continence enema	Severe FI. Failed alternative treatment modalities	
	Stoma	Severe FI. Failed alternative treatment modalities	

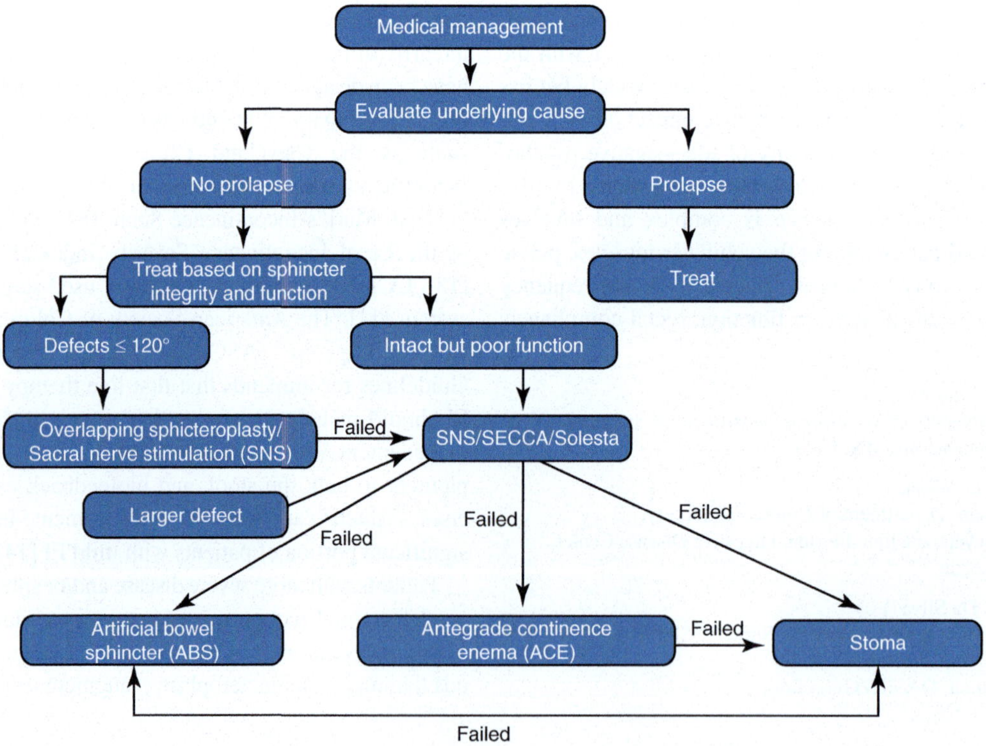

Fig. 16.1 Stepwise algorithm aimed at managing patients presenting with fecal incontinence

Procedures for Repair of Sphincter Injury

Sphincteroplasty

Traditionally, overlapping sphincteroplasty was the most common first-line surgical treatment approach to patients with FI due to a sphincter defect [16]. Parks [17] described this procedure in 1971 in 20 patients with anterior and lateral sphincter defects. He noted excellent functional results in 18 of the 20 patients, although the exact assessment of function was not described. Currently the procedure has been mainly reserved for young patients with a sphincter defect due to isolated, well-defined obstetrical injury but can also be offered to older patients with external defects.

The technique of separate internal and external sphincter repair was initially championed at Cleveland Clinic Florida [18]. Sphincteroplasty can be performed in the prone jackknife or lithotomy position. A curvilinear incision is made anteriorly overlying the perineal body (Fig. 16.2a). The injured external sphincter and associated scar are mobilized and separately

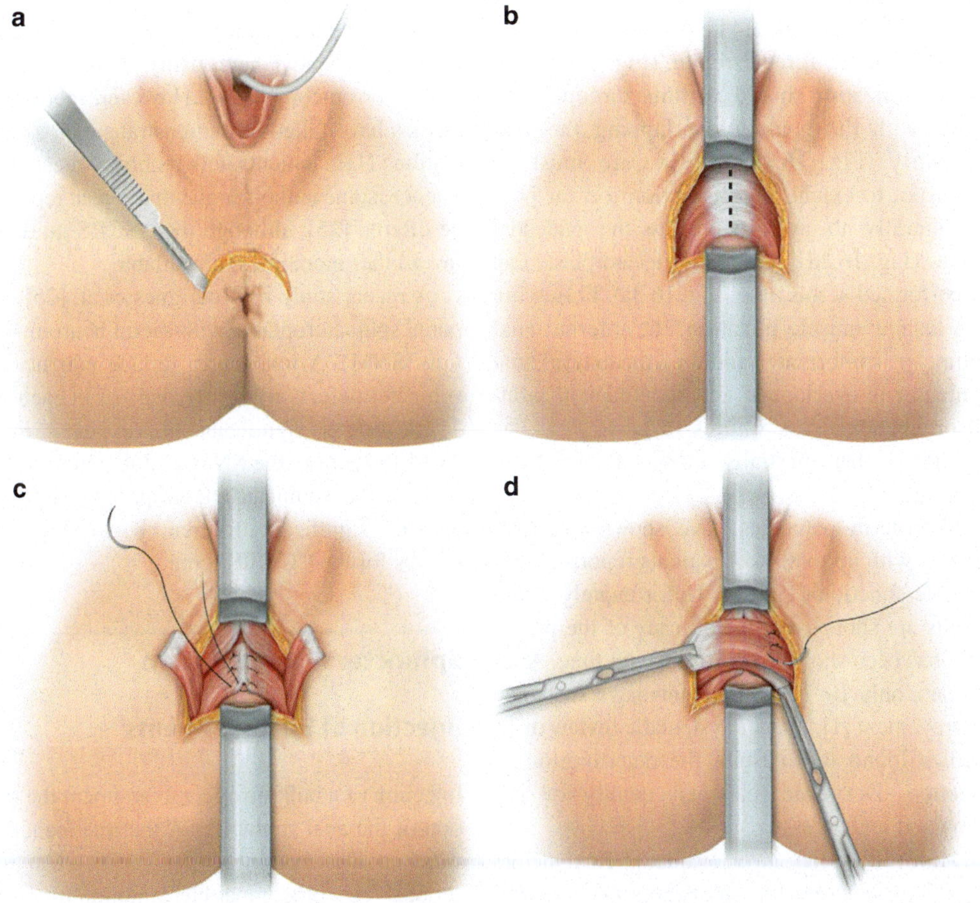

Fig. 16.2 Sphincteroplasty. (**a**) A curvilinear incision is made anteriorly overlying the perineal body. (**b**) Sphincter scar divided but not excised. (**c**) Internal anal sphincter imbricated when a layered repair is performed. (**d**) Overlapping repair of the anal sphincter with sutures. (**e**) External anal sphincter overlapped. (**f**) Edges of the wound are approximated in a V-shape or longitudinally with interrupted 3-0 absorbable mattresss sutures. The centerof the wound is left open for drainage. The perineal body is bulkier than it was preoperatively. (Reused with permission © Wolters Kluwer [90])

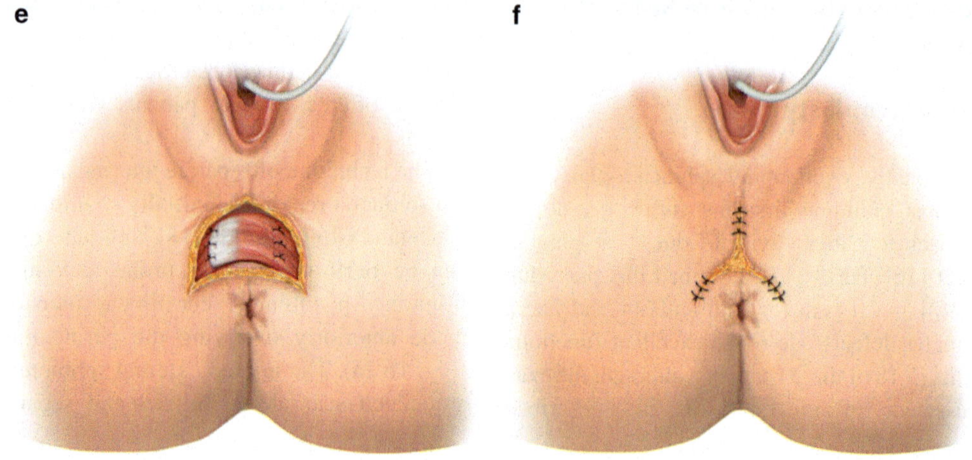

Fig. 16.2 (continued)

isolated. The scar should be preserved and used to hold sutures for the repair (Fig. 16.2b, c). The mobilization should continue circumferentially, with the goal of overlapping the scar anteriorly (Fig. 16.2d–f) [7]. Care must be taken not to extend the dissection too far posterolaterally to avoid injury to the pudendal nerves (Fig. 16.2d). An anterior levatorplasty can be performed at this time (Fig. 16.3a). Following this levator muscle plication, the internal anal sphincter is imbricated and then the externl anal sphincter is overlapped and secured with nonabsorbable mattress sutures (Fig. 16.3b). The wound is then partially closed to facilitate drainage.

Short-term outcomes have shown excellent results, with up to 80% of patients reporting reasonable function [19, 20]. Unfortunately, these results are short lived; the efficacy of the procedure has been shown to decrease with time. After 5 years, only 10–14% of patients have sustained improvement [1]. A 2012 systematic review of 16 studies found that despite the deterioration of function over time, patients' quality of life remained high [19].

Multiple studies have evaluated risk factors for failure or predictors of success of the repair with little consensus. Several studies have shown a correlation between short and long-term results, meaning patients who do well from the onset are more likely to have durable outcomes [21–23]. Some older studies have also shown unilateral or bilateral pudendal neuropathy to be associated with poor outcomes [24]. However, more recent studies have failed to demonstrate such as association [1]. In patients with recurrent fecal incontinence, some authors argue that repeat repair can be offered [25], although the ASCRS guidelines favor other modalities, if available.

A recent study by Rodrigues et al. [26] compared sphincteroplasty with sacral neuromodulation (SNM). A total of 26 female patients with sphincter defects were included, 13 in each arm. There was a significant improvement in the CCFFIS score in the SNM group, which was not seen in the sphincteroplasty group. There was also no significant different between the two groups for the CCFFIS score.

Sphincter Augmentation

Injection of Bulking Agents

Injection of a bulking agent to augment the function of the anal sphincter for the treatment of FI was first described in 1993 [27]. The ideal agent should be biocompatible, nonimmunogenic, small enough to inject, and with minimal migration potential [1]. Multiple agents have been proposed including autologous fat, synthetic bovine dermal collagen, Teflon™ (The Chemours

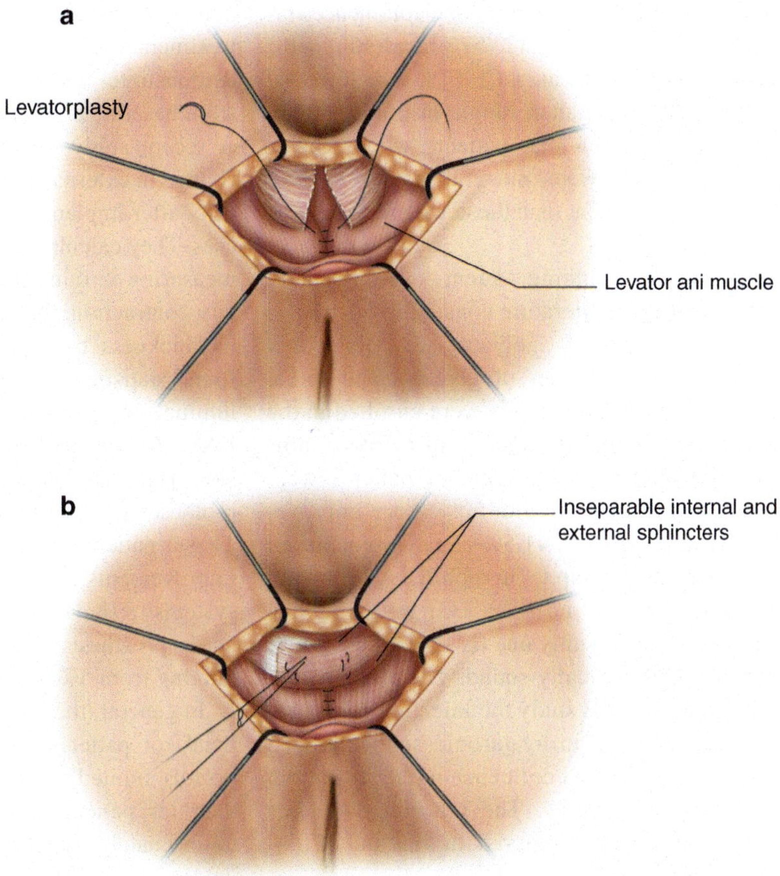

Fig. 16.3 (**a**) Levatorplasty is performed when appropriate. (**b**) Scar–sphincter complex is overlapped to allow the ends of the retracted muscles to be realigned as close as possible to each other. (Reused with permission © Springer Nature [90])

Company, Wilmington, DE) silicone, carbon beads, and stabilized hyaluronic acid. The method of injection is dependent on the agent of choice; final sites of implantation may include submucosal, intersphincteric, or intrasphincteric, and the route of injection may be transmucosal, transsphincteric, or intersphincteric. Local anesthetic and/or endorectal ultrasound may be used to assist in the injection of the agent. Complications are rare but include abscess and fistula. Injectable agents are contraindicated in patients with inflammatory bowel disease (IBD), rectocele, prolapse, previous pelvic irradiation, and anorectal malformations.

Results of studies evaluating this technique have been inconsistent and difficult to interpret due to the multiple materials used and variations in technique of injection. A Cochrane Review in 2010 [28] concluded that there was little evidence to support the use of bulking agents in the treatment of FI. In 2011, the US Food and Drug Administration (FDA) approved a non–animal stabilized hyaluronic acid–dextranomer gel for submucosal injection (NASHA Dx, Solesta®) for the treatment of FI. The largest series to date compared injection of NASHA Dx with sham injection in a randomized, sham controlled trial [29]. The primary end point was a >50% reduction in the number of incontinence episodes and an increase in incontinence-free days. Eighty percent of patients in the NASHA Dx group actually had a second injection within 1 month after no improvement with the first injection. Patients in the NASHA Dx group had a 52% reduction compared to 31% in the sham group ($P = 0.0089$). Despite this reduction, there was no significant difference in the FI scores between the groups at 2 months. In the published long-term follow-up [30], the reduction in incontinence episodes was sustained at 12 (57%) and 36 months (52%), and the mean CCF-FIS was lower at 36 months. A smaller, single-institution study [31] demonstrated that patients who received two injections were more likely to achieve >50% improvement than those who received one injection.

A newer bulking agent was recently described in the literature consisting of copolymer particles of polyacrylate–polyalcohol immersed in a carrier of glycerol and saline [32]. A total of 58 patients were enrolled in the study that followed patients up to 36 months post-injection. Sixty percent of patients were successfully treated (with 50% more improvement), although only 41.5% of patients achieved long-term follow-up to 36 months. A follow-up Cochrane review in 2013 [33] reviewed five studies, of which only one (NASHA DX Study) was methodologically sound. The authors concluded that in this study the injection of dextranomer stabilized in hyaluronic acid improved the symptoms of incontinence in the short term in >50% of patients. The other four studies were considered at high risk of bias and of limited value.

Other agents are currently being explored in order to improve long-term outcomes and include stem cells and the use of self-expandable agents. One such self-expandable agent being investigated is the Gatekeeper™ prosthesis (THD, Correggio, Italy). It is made of the inert polymer resin polyacrylonitrile and was originally intended for use to bulk the lower esophageal sphincter in the setting of gastric reflux. The material is implanted in six locations circumscribing the intersphincteric space using a specially designed delivery system. The resin material reshapes to its environment by water absorption over time and thus is purported as an ideal bulking agent. A multicenter observational study including 54 patients with moderate fecal incontinence was performed in Europe and demonstrated greater than 75% improvement in all FI parameters at 12 months, with 13% of patients reporting full continence during the same timeframe [32, 34, 35]. Another observational study noted that patients who initially responded to the treatment were likely to sustain a response and demonstrate >50% improvement in FI scores from baseline at least at the 1-year interval [36]. A recent study by Grossi et al. evaluated the change in external anal sphincter (EAS) contractility after implantation of the Gatekeeper protheses. They calculated the EAS muscle tension by measuring intraluminal pressures during voluntary contraction, the inner radius of the EAS and its thickness both pre- and postimplant. They found statistically significant increases in intraluminal pressures and the inner radius of the EAS, thereby increasing the EAS muscle tension. The EAS compression from the Gatekeeper protheses was felt to improve EAS contractility. These patients also demonstrated significant improvements in FI scores on multiple scoring systems [37]. The primary issue with this product is prosthesis migration, with reported rates ranging from 5% to more than 50%.

In general, the injectables may be more beneficial for patients with minor, passive FI with defects limited to the internal sphincter rather than for patients with more severe FI, which unfortunately is the case of most of the studies in the literature [34].

Anal Sling

Parks emphasized the importance of the anorectal angle in the maintenance of continence [38]. Based on this theory, the use of an artificial sling to support the posterior rectal wall has been developed for the treatment of fecal incontinence [39–41]. Most recently, Mellgren et al. [41] reported on the 1-year data of a prospective, multicenter study involving 152 patients at 14 centers in the USA using the Transobturator Posterior Anal Sling (TOPAS, American Medical Systems) (Fig. 16.4). The mesh is placed behind the anorectum via two small incisions in the buttocks, with each arm of the mesh exiting through the obturator foramen [42]. The mechanism of action is not exactly clear but is thought to support the puborectalis muscle and reinforce the anorectal angle. At 12 months, 69% of patients met the requirement for treatment success (≥50% reduction in the FI episodes from baseline to 12 month postoperatively), with 19% reporting complete continence. The posterior anal sling has not been compared to other treatment options for fecal incontinence, such as SNS and is not available in the USA.

Radiofrequency Energy Delivery

Radiofrequency energy delivery (SECCA ®) in the treatment of FI is a minimally invasive procedure and approved by the FDA for use in the USA in 2002. The mechanism of action of SECCA® (Mederi therapeutics) is through muscle regeneration and scar reduction [43]. This procedure can be performed in the operating room or the endoscopy suite under light sedation [44]. A custom-designed anoscope is placed in the anal canal after which four electrodes are deployed in four quadrants meant to enter the internal sphincter. The radiofrequency treatment is applied at four levels in the anal canal, above and below the dentate line (Fig. 16.5).

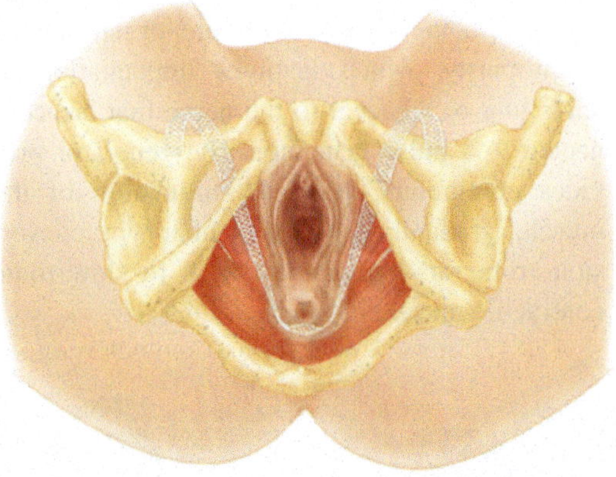

Fig. 16.4 Anal sling. (Reused with permission © Elsevier [10])

Complications include pain, infection, excessive scarring of the anus and rarely formation of a rectovaginal fistula [7]. The device is an option for patients with mild to moderate FI with intact or limited sphincter defect (less than 30 degrees).

Most studies have been small, single center series with short-term follow-up. The pilot study by Takahashi et al. [45] reported a success rate of 80% (>50% reduction of CCF-FIS score) at 1-year follow-up. In a follow-up study, Takahashi et al. [46] demonstrated that this response persisted at 5 years. Despite these promising results, other studies have not been able to replicate this level of success. Mandolfino et al. [47] reviewed the outcomes

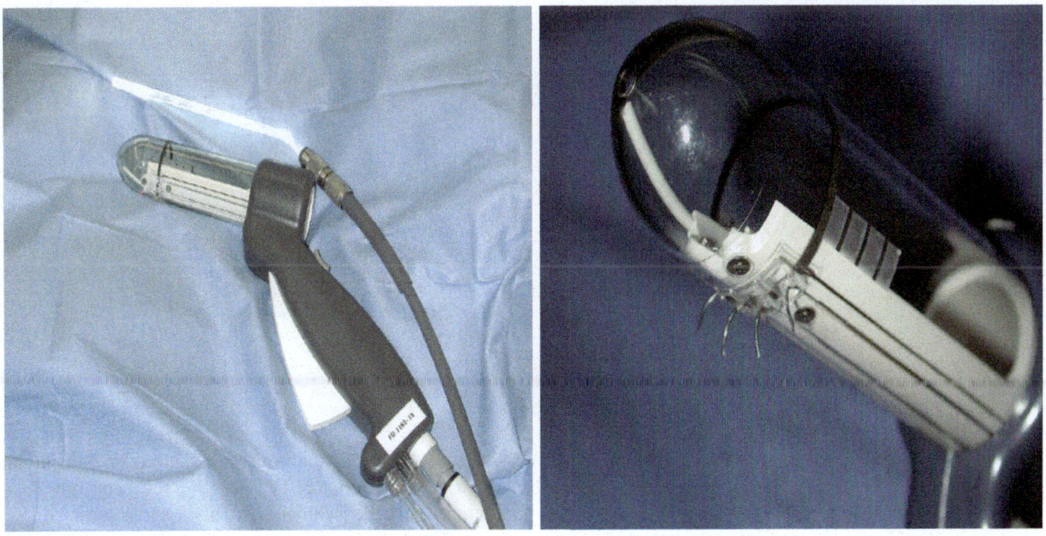

Fig. 16.5 Radiofrequency energy delivery (SECCA ®)

across several studies and found that 55–80% of patients are deemed responders at 12 months, showing some improvement in CCF-FIS scores. However, most series did not show a ≥50% improvement in CCF scores. A recent randomized sham-controlled trial compared the clinical response of SECCA® to a sham procedure in patients with FI [48]. Forty patients were randomized. They demonstrated a statistically significant decrease in the FI score, as evaluated by the Vaizey incontinence score, in patients undergoing the SECCA® procedure compared to sham, although only two patients met the criteria for clinically relevant improvement of ≥50% reduction in FI score or episodes. The authors concluded that there was some improvement in incontinence scores compared with placebo, but improvement was clinically negligible. The ASCRS clinical practice guidelines state that this treatment may be used to treat FI but recommend that alternative treatments be pursued before considering this therapy [14].

Sphincter Replacement

Graciloplasty

Muscle transposition has been described as a means to create a new anal sphincter giving patients the ability to control the passage of stool (Fig. 16.6a–c) [49]. The gluteus muscle was initially used but has been replaced by the gracilis muscle due to its superficial location and ease of mobilization [50]. The gracilis muscle is bilaterally or unilaterally mobilized and then used to encircle the anus. The drawbacks of the gracilis muscle transposition included its quick fatigability and low resting tone [49]. This problem was then overcome by the dynamic or stimulated graciloplasty, which involves implantation of leads into or adjacent to the nerve that are connected to a pulse generator. The generator provides continuous electrical stimulation of the muscle, gradually inducing a fiber-type conversion from rapid fatigable twitch to slow tonic contractile muscle [51].

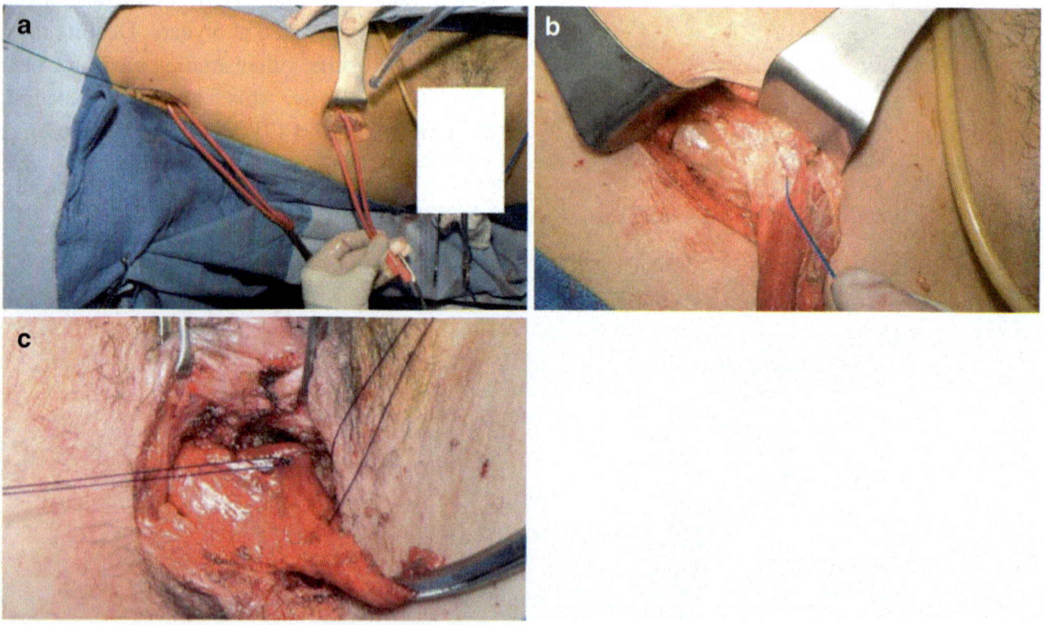

Fig. 16.6 Graciloplasty muscle transposition. (**a**) Gracilis muscle harvest: Separate incisions were made along the length of the muscle with rubber catheter retraction to ease the ligation of perforating vessels during muscle harvest. (**b**) The gracilis muscle is brought out through the most proximal incision at its insertion site at the ischiopubic ramus. (**c**) Defect coverage: The muscle is tunneled into the ischioanal fossa via a transperineal incision. The muscle is then secured to the apex once it is oriented circumferentially to cover the sphincter defect

While the procedure itself can be technically challenging in inexperienced hands, it has shown good results. In a prospective multicenter trial, Mander et al. [52] demonstrated a good functional result (continence to solids) in 56% of patients at a median of 10 months. Despite the promising functional results, the overall morbidity was quite high involving infectious complications, technical complications including electrode migration, fibrosis or lead fracture, and evacuatory dysfunction. In 2000, Baeten et al. [53] published results in 123 patients enrolled in a multicenter trial, again showing good functional results but at the expense of high morbidity. Complications occurred in 74% of patients, of which 49 patients (40%) required one or more operative procedures to treat the complications. Despite these complications, 63% and 57% of patients at 12 and 18 months achieved a 50% reduction in incontinence episodes. This procedure is not currently commercially available in the USA.

Magnetic Anal Sphincter (MAS)

The MAS is a newer therapeutic option approved by the US FDA in 2015 as a humanitarian use device for patients with FI who failed medical and other surgical managements. The MAS device is a ring of magnetic beads that is surgically implanted around the anal sphincter to reinforce weakened sphincter muscles (Fig. 16.7). The pressure generated during evacuation overcomes the magnetic attraction, allowing the beads to separate and the anal canal to open [7]. The device is placed as high as possible in the anal canal, with the preferred position just beneath the puborectalis muscle. The surgeon creates a tunnel encircling the entire external sphincter through an anterior or anterolateral incision [1]. A sizer is then used to determine the correct number of beads. Contraindications to the device include active infection, severe tissue rigidity, cancer, anoreceptive intercourse , and lack of sufficient tissue around the anus or rectovaginal septum.

Several studies have reported promising short-term outcomes [54, 55]. A recent publication by Sugrue et al. [56] reported on the long-term overall outcomes at a median follow-up of 5 years. A total of 35 patients underwent implantation of the device. Therapeutic success rates were reported as 63% at 1 year, 66% at 3 years, and 53% at 5 years. The device was explanted in seven patients due to major adverse events. There were a total of 30 adverse events reported in 20 patients ranging from pain to device erosion. The authors state that the majority (73%) of these were minor events and require little to no intervention and occurred during the first year post-implantation. There are two ongoing randomized controlled trials comparing MAS to SNM [57, 58]. Unfortunately, the device is not currently commercially available in the USA.

Artificial Bowel Sphincter (ABS)

The artificial bowel sphincter provides a patient-controlled replacement of a failed sphincter. Patients considered for ABS have severe FI but also must be motivated, relatively healthy and have sufficient amount of healthy tissue surrounding their anal canal. The device includes a silicone cuff that is implanted to encircle the anal canal. This is connected to a control pump located in the scrotum (male) (Fig. 16.8a) or labia (female) (Fig. 16.8b). A storage balloon is placed in the space of Retzius which acts as a reservoir. Continence is achieved when the cuff is inflated. Compressing the pump deflates the cuff and allows for stool passage. Fluid is then passively transferred back into the cuff and closes the anus.

This device was first described in the literature in 1987 [59] as a device for urinary incontinence. In 1998, Lehur and colleagues [60] reported their outcomes using with the Acticon Neosphincter (American Medical Systems, Minnetonka, MN), a device specifically designed for FI. Both papers described significant improvement in FI scores and quality of life. In 2002, Wong et al. [61] published their results from a multicenter trial evaluating the use of ABS for FI. They demonstrated

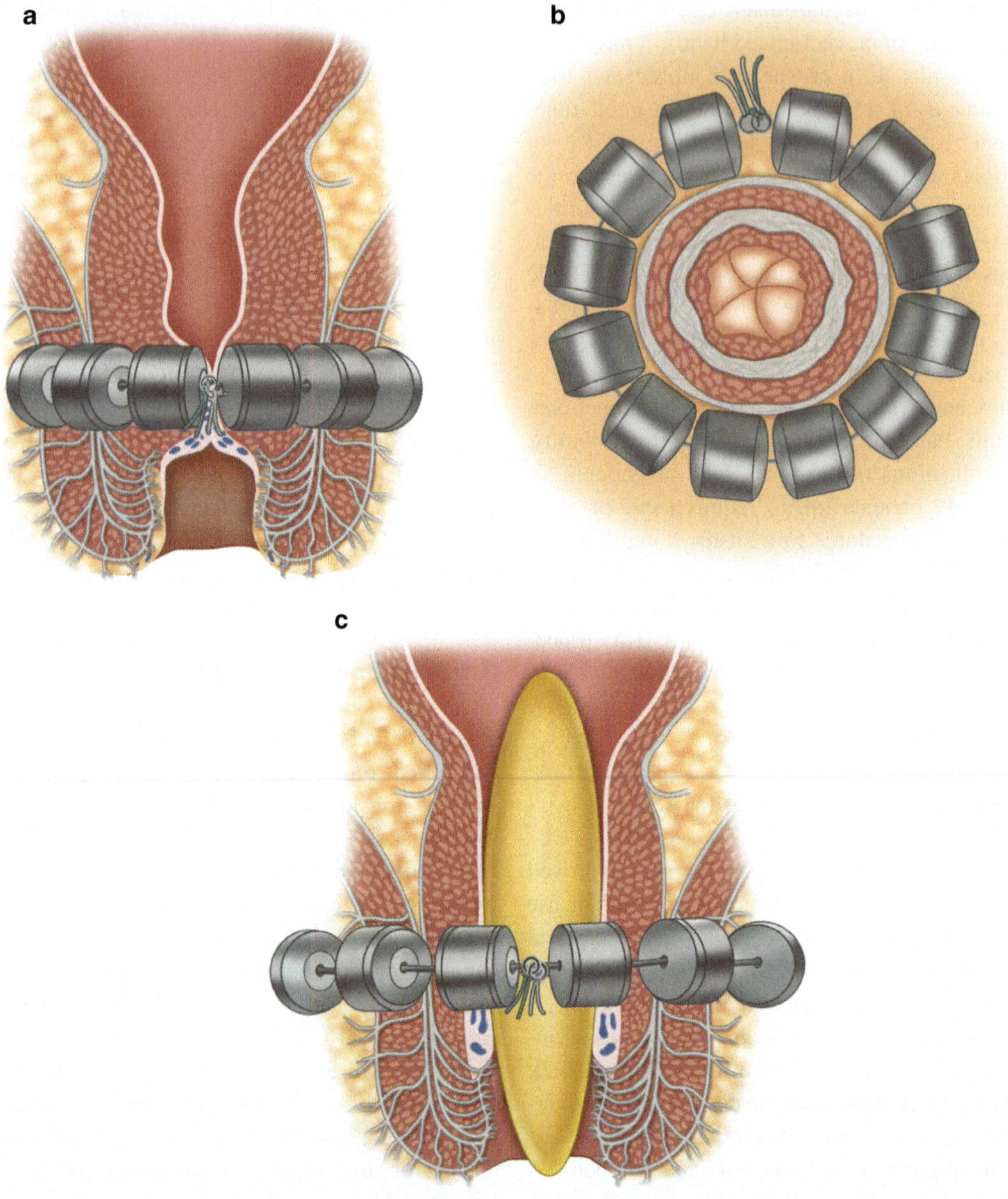

Fig. 16.7 Magnetic anal sphincter. (**a**) In the resting state the magnets keep the anal canal closed. (**b**) Axial view to demonstrate placement of the magnetic sphincter outside the sphincter complex. (**c**) With bowel movements and Valsalva, the magnets expand to allow for passage of stool. (Reused with permission © Springer [90])

an overall success rate of 53% and 86% in patients who had a functioning device. Despite success with the device, there was a high associated morbidity and device related complications. Device-related complications were reported in 86% of patients and 46% of these patients required revisional surgery to address the complications. A systematic review in 2013 [62] found that in functional devices continence decreases with time, but interestingly, quality of life remains high.

The most common complications include infection, erosions or ulcerations of the rectum, device malfunction from cuff rupture and bal-

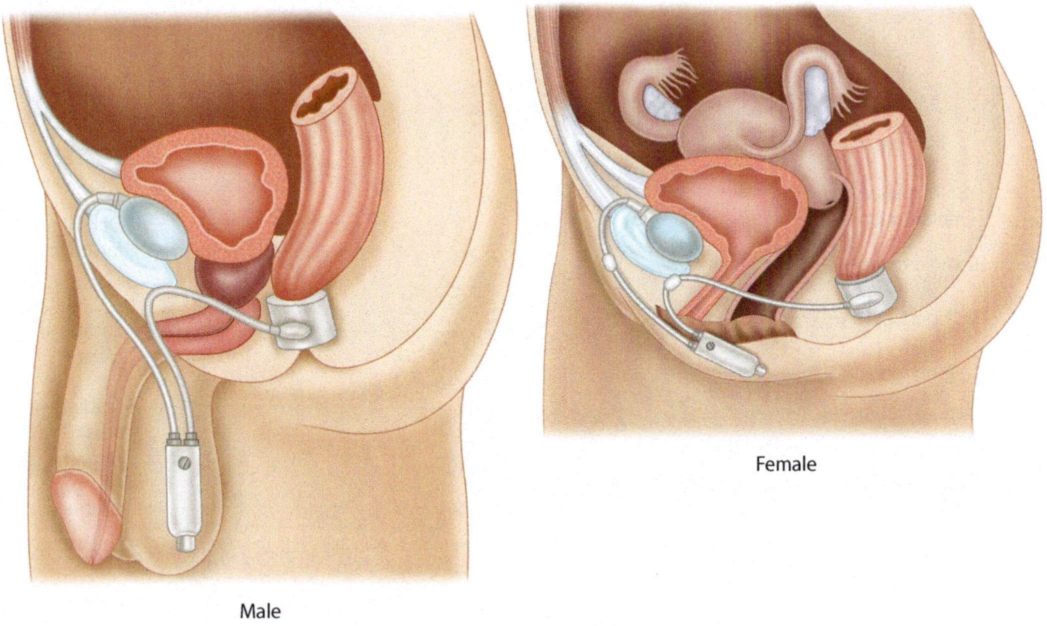

Fig. 16.8 Artificial bowel sphincter. The control pump is located in the scrotum in males (**a**) and labia in females (**b**). (Reused with permission © Springer [90])

loon and pump leaks, and device migration. Infection is the most common complication causing explantation, ranging from 4% to 38% [63]. In 2009, Wexner and colleagues [63] evaluated their outcomes in 51 patients implanted with the artificial bowel sphincter. Eighteen patients (35%) experienced early stage infections, requiring device explantation in all patients. A total of 25 patients had late-stage complications ranging from infection to device migration, with 13 devices requiring explantation due to late-stage complications. Multivariate analysis demonstrated that time to first bowel movement and history of perianal sepsis were risk factors associated with early stage infections. Late-stage failures appeared to be associated with device malfunction.

In 2013, Wong et al. [64] compared the MAS to the ABS at 30 days. Those implanted with the MAS had shorter operative times and length of hospitalization. There was no statistical difference in complications or explanation between the devices at 30 days. Both groups achieved statistically significant improvement in continence and quality of life. Because of the high rate of complications and high rates of success and better safety profiles of other treatments, ABS is reserved for patients in which all other options have failed, those with extensive sphincter defects, congenital malformations, neurogenic incontinence, postsurgical bowel dysfunction with intact anal canal anatomy [1]. Unfortunately, this device is no longer commercially available in the USA.

Nerve Stimulation

Sacral Neuromodulation (SNM)

Sacral neuromodulation (SNM) was approved for use in the management of FI by the US FDA in 2011. SNM works by electrical stimulation of the sacral nerve roots, through efferent modulation of spinal and cortical pathways [65]. The procedure involves an electrical stimulation via the S3 nerve root through the sacral foramen. There is an initial test phase of about 2 weeks to determine efficacy, followed by implantation of a permanent stimulator (Fig. 16.9).

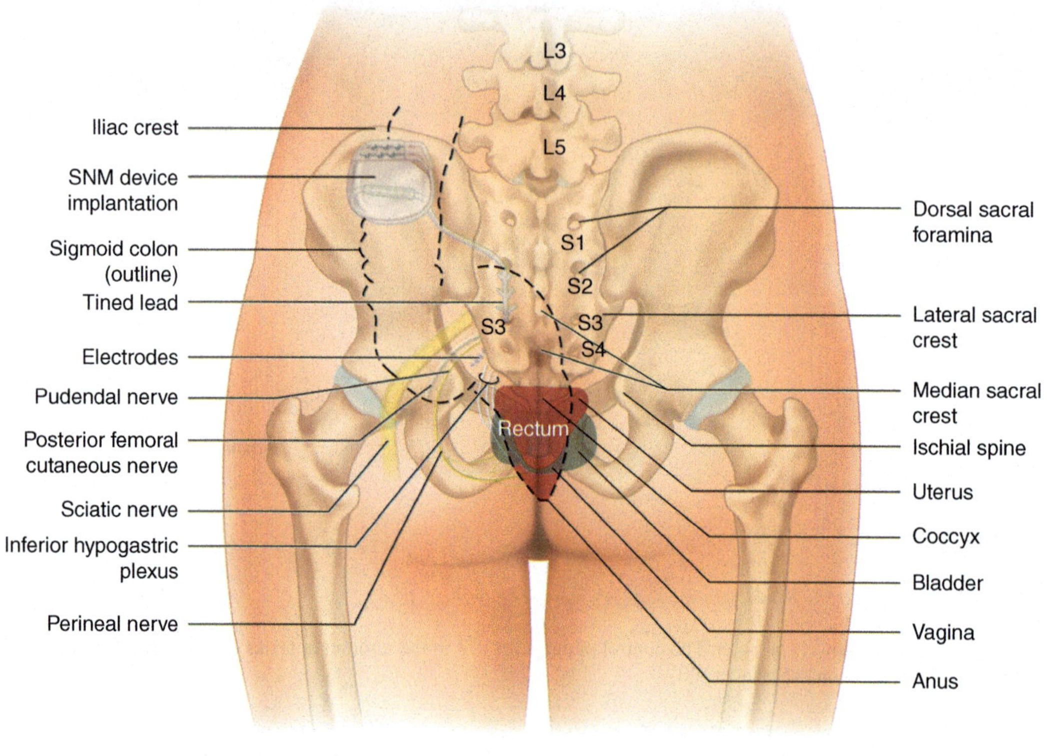

Fig. 16.9 Sacral neuromodulation. (Reused with permission © Springer [90])

Matzel et al. [66] initially described the procedure for use in fecal incontinence in 1995. Since this initial report, SNM has become an acceptable treatment option, and in some cases, the first line surgical treatment for patients with FI. Wexner et al. [67] demonstrated the efficacy of this device in the treatment of FI in a prospective, US multicenter trial. Success was defined as ≥50% reduction of incontinent episodes per week over 12 weeks in ≥50% of patients. They demonstrated an 83% therapeutic success at 12 months, and 41% of patients achieved 100% continence. In long-term follow-up at 5 years [68], 76 patients were evaluated and success was demonstrated in 89% of patients with 36% achieving complete continence. At 5 years, 35.5% of patients required a device revision, replacement, or explant. The safety profile of this device is fairly good, with an infection rate of about 11% [69]. All of the initial publications demonstrating the positive results of SNM were from North America, Europe, and Australia. Recently a Latin American collaborative series showed that these same salutatory results could be achieved. Multiple surgeons across several centers were involved in the study. All surgeons had training in a cadaver lab and underwent proctoring for their intial procedures. There was a statistically significant improvement in CCFFIS scores from the pre-operative baseline and 90% of patients reported their improvements as "significant". The infection rate was 3.8% and there was a low rate of device explantation at 2.2%. Moreover, the group offered useful guidelines to help ensure a cost effective approach to patient selection, device implantation and post-implantation programming [70].

Prior to SNM, the only option for patients with a sphincter defect up to approximately 120 degrees was sphincter repair. Soon after its advent, SNM was demonstrated as effective in patients with FI in the setting of sphincter injury. Brouwer and Duthie [71] found that SNM provided significant improvement of FI even in the presence of a sphincter defect or pudendal neuropathy. In the largest study to evaluate this, 91 patients with no sphincter defect were compared to 54 patients with imaging-documented external defect. There was no significant difference in the comparison of baseline CCFFIS scores and 12-month scores between the two groups. A pooled analysis of all studies to date found that 69–83% of patients achieve success [72].

Percutaneous Tibial Nerve Stimulation

Alternatives to SNM involving neuromodulation have recently been investigated. Percutaneous tibial nerve stimulation (PTNS) is thought to cause similar changes in anorectal neuromuscular function as SNM because of the shared sacral segmental innervation [73]. This treatment is not currently available in the USA. PTNS is a minimally invasive technique that can be performed in the outpatient setting. The initial data comes from several case series, but more recently, two randomized controlled trials have evaluated this therapy [73, 74]. In the CONFIDeNT trial [73], 227 patients were randomized to PTNS versus sham PTNS over a 12-week period. There was no significant clinical benefit of PTNS over sham stimulation in the treatment of FI as measured by a >50% reduction in the number of FI episodes, although the absolute number of FI episodes/week was significantly reduced in the PTNS group. However, a subsequent post hoc analysis revealed that patients with outlet obstructive symptoms fared significantly worse than did patients without such symptoms [73].

An additional randomized control trial by Van der Wilt et al. [74] compared PTNS to sham stimulation. A total of 59 patients were included in the trial. They demonstrated a statistically significant reduction in the median and mean number of FI episodes of severity over 9 weeks. However, this was not clinically significant when response to treatment was based on a >50% reduction in FI episodes per week. Both trials demonstrate that PTNS may be beneficial for some patients who have failed conservative management but have not established PTNS as a clinically meaningful treatment option, especially compared to SNM.

Diversion

Antegrade Continence Enema (ACE)

The antegrade continence enema was first described by Malone et al. [75] in 1990. This procedure can be used to control FI in both adults and children but is most commonly reported in the pediatric literature. The ACE does not alter the anorectal physiology or anatomy but allows the patient to empty their colon in a predictable fashion, thereby avoiding a fecal accident. The initial report of the ACE involved using the appendix as a conduit. The appendix and a cuff of cecum are amputated and the appendix reversed and reimplanted into the cecum through a submucosal tunnel. This was then brought out through the abdominal wall as a continent appendicostomy. Various techniques have been described for the creation of an ACE using the terminal ileum, cecum, left colon, or stomach as a conduit [76].

Overall, functional results in patients with ACE and FI are good, with about 75% of adults achieving continence [77]. In a meta-analysis including 17 studies involving over 400 adult patients [78], the overall pooled morbidity was 45%, with wound infection and stomal stenosis as the most common complications. Other reported complications include reflux through the conduit, fecal impaction and stomal prolapse [77, 79]. Stoma complications requiring stoma revision have been quoted as high as 30–45% in adult series [80, 81]. The operation is not widely used in the USA.

Stoma

Fecal diversion may be helpful in providing definitive control; the colostomy is the standard ostomy used for patients with FI. It is generally only offered to patients who have failed other treatment modalities and have severe FI. Patients may be resistant to the idea of a stoma due to the perceived stigma associated with having a stoma. In a survey of patients with a permanent stoma for FI, 83% reported improvement in lifestyle and 84% reported they would undergo the procedure again [82].

Future Developments

Stem Cell Therapy

Stem cell therapy has been evaluated for use in a variety of clinical settings such as cardiovascular, hematological, neurological, digestive, and trauma-associated conditions. The most commonly used stem cells are hematopoietic stem cells (HSCs), mesenchymal stem cells (MSCs), and adipose-derived stem cells (ASCs) [83]. The use of stem cell for the treatment of FI has been postulated as a means of providing recovery of sphincter function through regeneration of damaged striated sphincter muscle and allowing for reinnervation of newly formed myofibers [84]. The early literature on stem cells in FI was based on animal models demonstrating the safety and efficacy of stem cells for FI [84–86]. The first report of stem cell therapy for FI was by Lorenzi et al. in 2008 [86]. In this animal model of FI they demonstrated bone marrow–derived mesenchymal stem cell injection improved muscle regeneration and increased contractile function of anal sphincters after injury and repair.

The first study in humans was an observational study by Frudinger et al. [87]. They injected autologous myoblasts into the external anal sphincter in ten women with lesions that had not been operated on and were refractory to conservative management. The authors demonstrated improved scores on the Wexner Incontinence Score and quality of life at 12 months but did not demonstrate any significant physiological change to account for these improvements. These results were sustained at 5-year follow-up [88]. There were no significant side effects or adverse events noted in any patient. In 2015, Bisson et al. [84] published a proof-of-concept study evaluating the injection of myoblasts into the anal sphincter of rats. They demonstrated sustained increase in sphincter pressures after injection. The follow-up, phase II randomized, double-blind, placebo-controlled study in humans was just published in 2018 [89]. A total of 24 patients were enrolled, 12 in the treatment arm and 12 in the control arm. At 6 months, both groups demonstrated reduction in the CCF-FIS score, but at 12 months, the treatment group continued to show improvement in the CCF-FIS score while the control arm did not. No severe adverse events were reported. Injection of stem cells appears to be safe, without major adverse side effects and provides clinical benefit in the treatment of FI. There have been no phase III trials to date comparing stem cells therapy to the SNS and this option of treatment is not currently FDA approved in the USA. Refer to Chap. 15 for more in-depth discussion of stem cell therapy.

Summary

Fecal incontinence is a common problem noted in almost 20% of adult females. Several new promising therapies are unfortunately not currently commercially available in the USA. At present, sphincter repair, SNM, and stomas remain the only viable surgical options. Hopefully newer alternatives such as stem cells will offer additional possibilities for patients with FI.

References

1. Paquette IM, Madhulika G, Varma MD, et al. The American Society of Colon and Rectal Surgeons' clinical practice guidelines for the treatment of fecal incontinence. Dis Colon Rectum. 2015;58:623–36.
2. Ditah I, Devaki P, Luma NH, et al. Prevalence, trends, and risk factors for fecal incontinence in United States adults, 2005-2010. Clin Gastroenterol Hepatol. 2014;12:636–43.
3. Brown HW, Wexner SD, Segall MM, Brezoczky KL, Lukacz ES. Accidental bowel leakage in the mature women's health study: prevalence and predictors. Int J Clin Pract. 2012;66:1101–8.
4. Brown HW, Wexner SD, Lukacz ES. Factors associated with care seeking among women with accidental bowel leakage. Female Pelvic Med Reconstr Surg. 2013;19(2):66–71.

5. NIH state-of-the-science conference statement on prevention of fecal and urinary incontinence in adults. NIH Consensus State Sci Statements. 2007;24:1–37.
6. Brown HW, Wexner SD, Segall MM, Brezoczky KL, Lukacz ES. Quality of life impact in women with accidental bowel leakage. Int J Clin Pract. 2012;66(11):1109–16.
7. Steele SR, Hull TL, Read TE, et al. The ASCRS textbook of colon and rectal surgery. 3rd ed. New York: Springer Nature; 2016.
8. Whitehead WE, Borrud I, Goode PS, et al. Fecal incontinence in US adults: epidemiology and risk factors. Gastroenterology. 2009;137:512–517, e511-e512.
9. Shamiliyan T, Wyman J, Blis DZ, Kane RL, Wilt TJ. Prevention of urinary and fecal incontinence in adults. Evid Rep/Technol Assess. 2007;161:1–379.
10. Mellgren A. Fecal Incontinence. Surg Clin North Am. 2010;90:185–94.
11. Jorge JM, Wexner SD. Etiology and management of fecal incontinence. Dis Colon Rectum. 1993;36(1):77–97. Review.
12. Vaizey CJ, Carapeti E, Cahill JA, Kamm MA. Prospective comparison of faecal incontinence grading systems. Gut. 1999;44(1):77–80.
13. Rockwood TH, Church JM, Fleshman JW, Kane RL, Mavrantonis C, Thorson AG, Wexner SD, Bliss D, Lowry AC. Patient and surgeon ranking of the severity of symptoms associated with fecal incontinence: the fecal incontinence severity index. Dis Colon Rectum. 1999;42(12):1525–32.
14. Paquette IM, Varma MG, Kaiser AM, Steele SR, Rafferty JF. The American Society of Colon and Rectal Surgeons' clinical practice guideline for the treatment of fecal incontinence. Dis Colon Rectum. 2015;58(7):623–36.
15. Wexner SD. Percutaneous tibial nerve stimulation in faecal incontinence. Lancet. 2015;386(10004):1605–6.
16. Goetz LH, Lowry AC. Overlapping sphincteroplasty: is it the standard of care? Clin Colon Rectal Surg. 2005;18(1):22–31.
17. Parks AG, McPartlin JF. Late repair of injuries of the anal sphincter. Proc R Soc Med. 1971;64:1187–9.
18. Wexner SD, Marchetti F, Jagelman DG. The role of sphincteroplasty for fecal incontinence reevaluated: a prospective physiologic and functional review. Dis Colon Rectum. 1991;34(1):22–30.
19. Glasgow SC, Lowry AC. Long-term outcomes of anal sphincter repair for fecal incontinence: a systematic review. Dis Colon Rectum. 2012;55(4):482–90.
20. Oliveira L, Pfeifer J, Wexner SD. Physiological and clinical outcome of anterior sphincteroplasty. Br J Surg. 1996;83(4):502–5.
21. Vaizey CJ, Norton C, Thornton MJ, Nicholls RJ, Kamm MA. Long-term results of repeat anterior anal sphincter repair. Dis Colon Rectum. 2004;47:858–63.
22. Malouf AJ, Norton CS, Engel AF, Nicholls RJ, Kamm MA. Long-term results of overlapping anterior anal-sphincter repair for obstetric trauma. Lancet. 2000;355:260–5.
23. Bravo Gutierrez A, Madoff RD, Lowry AC, Parker SC, Buie WD, Baxter NN. Long-term results of anterior sphincteroplasty. Dis Colon Rectum. 2004;47:727–31.
24. Gilliland R, Altomare DF, Moreira H Jr, Oliveira L, Gilliland JE, Wexner SD. Pudendal neuropathy is predictive of failure following anterior overlapping sphincteroplasty. Dis Colon Rectum. 1998;41(12):1516–22. Review.
25. Giordano P, Renzi A, Efron J, et al. Previous sphincter repair does not affect the outcome of repeat repair. Dis Colon Rectum. 2002;45(5):635–40.
26. Rodrigues FG, Chadi SA, Cracco AJ, Sands DR, Zutshi M, Gurland B, Da Silva G, Wexner SD. Faecal incontinence in patients with a sphincter defect: comparison of sphinceroplasty and sacral nerve stimulation. Color Dis. 2017;19(5):456–61.
27. Shafik A. Polytetrafluoroethylene injection for the treatment of partial fecal incontinence. Int Surg. 1993;78:159–61.
28. Maeda Y, Laurberg S, Norton C. Perianal injectable bulking agents as treatment for faecal incontinence in adults. Cochrane Database Syst Rev. 2010:CD007959.
29. Graf W, Mellgren A, Matzel KE, Hul T, Johansson C, Bernstein M, NASHA Dx Study Group. Efficacy of dextranomer in stabilized hyaluronic acid for treatment of faecal incontinence: a randomised, sham controlled trial. Lancet. 2011;377:997–1003.
30. Mellgren A, Matzel KE, Pollack J, Hull T, Bernstein M, Graf W, NASHA Dx Study Group. Long-term efficacy of NASHA Dx injection therapy for treatment of fecal incontinence. Neurogastroenterol Motil. 2014;26:1087–94.
31. Danielson J, Karlbom U, Wester T, Graf W. Efficacy of quality of life 2 years after treatment for faecal incontinence with injectable bulking agents. Tech Coloproctol. 2013;17:389–95.
32. Rosato G, Piccinini P, Oliveira L, Habr-Gamma A, Chwat C. Initial results of a new bulking agent for fecal incontinence: a multicenter study. Dis Colon Rectum. 2015;58:241–6.
33. Madae Y, Laurberg S, Norton C. Perianal injectable bulking agents as treatment for faecal incontinence in adults. Cochrane Database Syst Rev. 2013;2:CD007959.
34. Kaiser AM, Orangio GR, Zutshi M, Alva S, Hull TL, Marcello PW, Margolin DA, Rafferty JF, Buie WD, Wexner SD. Current status: new technologies for the treatment of patients with fecal incontinence. Surg Endosc. 2014;28(8):2277–301.
35. Ratto C, Buntzen S, Aigner F, et al. Multicentre observational study of the Gatekeeper for faecal incontinence. Br J Surg. 2016;103(3):290–9.
36. Trenti L, Biondo S, Noguerales F, Nomdedeu J, Coret A, Scherer R, et al. Outcomes of Gatekeeper (TM) prosthesis implantation for the treatment of fecal incontinence: a multicenter observational study. Tech

Coloproctol. 2017;21(12):963–70. PubMed PMID: 29159781.
37. Grossi U, De Simone V, Parello A, et al. Gatekeeper improves Voluntary Contractility in Patients with Fecal Incontinence. Surg Innov. 2019;26(3):321–7.
38. Parks AG, Porter NH, Hardcastle J. The syndrome of the descending perineum. Proc R Soc Med. 1966;59:477–82.
39. Yamana T, Takahashi T, Iwadare J. Perineal Puborectalis sling operation for fecal incontinence: preliminary report. Dis Colon Rectum. 2004;47:1982–9.
40. Brochard C, Queralto M, Cabarrot P, et al. Technique of the transobturator puborectalis sling in fecal incontinence. Tech Coloproctol. 2017;21:315–8.
41. Mellgren A, Zutshi M, Lucente VR, et al. A posterior anal sling for fecal incontinence: results of a 152-patient prospective multicenter study. Am J Obstet Gynecol. 2016;214:349.e1–8.
42. Rosenblatt P. New developments in therapies for fecal incontinence. Curr Opin Obstet Gynecol. 2015;27:353–8.
43. Herman RM, Berho M, Murawski M, Nowakowski M, Ryś J, Schwarz T, Wojtysiak D, Wexner SD. Defining the histopathological changes induced by nonablative radiofrequency treatment of faecal incontinence--a blinded assessment in an animal model. Color Dis. 2015;17(5):433–40.
44. Efron JE. The SECCA procedure: a new therapy for treatment of fecal incontinence. Surg Technol Int. 2004;13:107–10.
45. Takahashi T, Garcia-Osogobio S, Valdovinos MA, et al. Radiofrequency energy delivery to the anal canal for the treatment of fecal incontinence. Dis Colon Rectum. 2002;45:915–22.
46. Takahashi-Monroy T, Morales M, Garcia-Osogobio S, et al. SECCA procedure for the treatment of fecal incontinence: results of five-year follow-up. Dis Colon Rectum. 2008;51:355–9.
47. Frascio M, Mandolfino F, Imperatore M, et al. The SECCA procedure for faecal incontinence: a review. Color Dis. 2014;16:167–72.
48. Visscher AP, Lam TJ, Meurs-Szojda MM, et al. Temperature-controlled delivery of radiofrequency energy in fecal incontinence: a randomized sham-controlled clinical trial. Dis Colon Rectum. 2017;60:860–5.
49. Alavi K, Chan S, Wise P, Kaiser AM, Sudan R, Bordeianou. Fecal incontinence: etiology, diagnosis, and management. J Gastrointest Surg. 2015;19:1910–21.
50. Cera SM, Wexner SD. Muscle transposition: does it still have a role? Clin Colon Rectal Surg. 2005;18(1):46–54.
51. Geerdes BP, Heineman E, Konsten J, Soeters PB, Baeten CG. Dynamic Graciloplasty. Complications and management. Dis Colon Rectum. 1996;39:912–7.
52. Mander BJ, Wexner SD, Williams NS, et al. Preliminary results of a multicenter trial of the electrically stimulated gracilis neoanal sphincter. Br J Surg. 1999;86:1543–8.
53. Baeten CG, Bailey HR, Bakka A, Belliveau P, Berg E, Buie WD, et al. Safety and efficacy of dynamic graciloplasty for fecal incontinence: report of a prospective, multicenter trial. Dynamic Graciloplasty Therapy Study Group. Dis Colon Rectum. 2000;43:743–51.
54. Wong MT, Meurette G, Stangherlin P, et al. The magnetic anal sphincter versus the artificial bowel sphincter: a comparison of 2 treatments for fecal incontinence. Dis Colon Rectum. 2011;54:773–9.
55. Lehur PA, McNevin S, Buntzen S, Mellgren AF, Laurberg S, Madoff RD. Magnetic anal sphincter augmentation for the treatment of fecal incontinence: a preliminary report from a feasibility study. Dis Colon Rectum. 2010;53:1604–10.
56. Sugrue J, Lehur PA, Madoff RD, et al. Long-term experience of magnetic anal Sphincter augmentation in patients with fecal incontinence. Dis Colon Rectum. 2017;60:87–95.
57. Williams AE, Croft J, Napp V, et al. SaFaRI: sacral nerve stimulation versus the FENIX magnetic sphincter augmentation for adult faecal incontinence: a randomised design. Int J Color Dis. 2016;31:465–72.
58. Lehur PA, Wyart V, Riche VP. SaFaRI: sacral nerve stimulation versus the Fernix magnetic sphincter augmentation for adult faecal incontinence: a randomised investigation. Int J Color Dis. 2016;31:1505.
59. Christiansen J, Lorentzen M. Implantation of artificial sphincter for incontinence. Lancet. 1987;2:244–5.
60. Lehur PA, Glemain P, Bruley des Varannes S, et al. Outcome of patients with an implanted artificial anal sphincter for severe faecal incontinence. A single institution report. Int J Color Dis. 1998;13:88–92.
61. Wong WD, Congliosi SM, Spencer MP, et al. The safety and effi cacy of the artificial bowel sphincter for fecal incontinence: results from a multicenter cohort study. Dis Colon Rectum. 2002;45(9):1139–53.
62. Hong KW, Dasilva G, Kalaskar SN, et al. Long-term outcomes of artificial bowel sphincter for fecal incontinence: a systematic review and meta-analysis. J Am Coll Surg. 2013;217(4):718–25.
63. Wexner SD, Jin HY, Weiss EG, et al. Factors associated with failure of the artificial bowel sphincter: a study of over 50 cases from Cleveland Clinic Florida. Dis Colon Rectum. 2009;52(9):1550–7.
64. Wong MT, Meurette G, Stangherlin P, et al. The magnetic anal sphincter versus the artificial bowel sphincter: a comparison of 2 treatments for fecal incontinence. Dis Colon Rectum. 2011;54(7):773–9.
65. Weledji EP. Electrophysiological basis of fecal incontinence and its implications for treatment. Ann Coloproctol. 2013;33:161–8.
66. Matzel KE, Stadelmaier U, Hohenfellner M, Gall FP. Electrical stimulation of sacral spinal nerves for treatment of faecal incontinence. Lancet. 1995;346:1124–7.
67. Wexner SD, Coller JA, Devroede G, et al. Sacral nerve stimulation for fecal incontinence: results of a

120-patient prospective multicenter study. Ann Surg. 2010;251:441–9.
68. Hull T, Giese C, Wexner SD, et al. SNS Study Group. Long-term durability of sacral nerve stimulation therapy for chronic fecal incontinence. Dis Colon Rectum. 2013;56:234–45.
69. Zbar AP. Sacral neuromodulation and peripheral nerve stimulation in patients with anal incontinence: an overview of techniques, complications and trouble-shooting. Gastroenterol Rep. 2014;2(2):112–20.
70. Oliveira L, Hagerman G, Torres ML, et al. Sacral neuromodulation for fecal incontinence in Latin America: initial results of amulticenter study. Tech Coloproctol. 2019;23(6):545–50.
71. Brouwer R, Duthie G. Sacral nerve neuromodulation is effective treatment for fecal incontinence in the presence of a sphincter defect, pudendal neuropathy, or previous sphincter repair. Dis Colon Rectum. 2010;53(3):273–8.
72. Thin NN, Horrocks EJ, Hotouras A, Palit S, Thaha MA, Chan CL, Matzel KE, Knowles CH. Systematic review of the clinical effectiveness of neuromodulation in the treatment of faecal incontinence. Br J Surg. 2013;100(11):1430–47.
73. Knowles CH, Horrocks EJ, Bremner SA, Stevens N, Norton C, O-Connell PR, Eldridge S. Percutaneous tibial nerve stimulation versus sham electrical stimulation for the treatment of faecal incontinence in adults (CONFIDeNT): a double-blind, multicenter, pragmatic, parallel-group, randomised controlled trial. Lancet. 2015;386:1640–8.
74. van der Wilt AA, Giuliani G, Kubis C, van Wunnik BPW, Ferreira I, Breukink SO, Lehur PA, La Torre F, Baeten CGMI. Randomized clinical trial of percutaneous tibial nerve stimulation versus sham electrical stimulation in patients with faecal incontinence. Br J Surg. 2017;104:1167–76.
75. Malone PS, Ransley PG, Kiely EM. Preliminary report: the Antegrade continence enema. Lancet. 1990;336:1217–8.
76. Patel AS, Saratzis A, Arasaradnam R, Harmston C. Use of antegrade continence enema for the treatment of fecal incontinence and functional constipation in aduluts: a systematic review. Dis Colon Rectum. 2015;58:999–1013.
77. Van Koughett JM, Wexner SD. Current management of fecal incontinence: choosing amongst treatment options to optimize outcomes. World J Gastroenterol. 2013;28:9216–30.
78. Chan DS, Delicata RJ. Meta-analysis of Antegrade continence enema in adults with faecal incontinence and constipation. Br J Surg. 2016;103:322–7.
79. Bleier JS, Kann BR. Surgical Management of Fecal Incontinence. Gastroenterol Clin N Am. 2013;42:815–36.
80. Hirst GR, Arumugam PJ, Watkins AJ, et al. Antegrade continence enema in the treatment of obstructed defaecation with or without faecal incontinence. Tech Coloproctol. 2005;9:217–21.
81. Hill J, Stott S, MacLennan I. Antegrade enemas for the treatment of severe idiopathic constipation. Br J Surg. 1994;81:1490–1.
82. Norton C, Burch J, Kamm MA. Patients' views of a colostomy for fecal incontinence. Dis Colon Rectum. 2005;18:42–5.
83. Trebol J, Carabias-Orgaz A, Gracia-Arranz M, Garcia-Olmo D. Stem cell therapy for faecal incontinence: current state and future perspectives. World J Stem Cells. 2018;10:82–105.
84. Bisson A, Freret M, Drouot L, et al. Restoration of anal sphincter function after myoblast cell therapy in incontinent rats. Cell Transplant. 2015;24:277–86.
85. Kang SB, Lee HN, Lee JY, et al. Sphincter contractility after muscle-derived stem cells autograft into the cryoinjured anal sphincters of rats. Dis Colon Rectum. 2008;51:1367–72.
86. Lorenzi B, Pessina F, Lorenzo P, et al. Treatment of experimental injury of anal sphincters with primary surgical repair and injection of bone marrow-derived mesenchymal stem cells. Dis Colon Rectum. 2008;51:411–20.
87. Frudinger A, Kölle D, Schwaiger W, et al. Muscle derived cell injection to treat anal incontinence due to obstetric trauma: pilot study with 1 year follow-up. Gut. 2010;59:55–61.
88. Frudinger A, Pfeifer J, Paede J, et al. Autologous skeletal-muscle-derived cell injection for anal incontinence due to obstetric trauma: a 5-year follow-up of an initial study of 10 patients. Color Dis. 2015;17:794–801.
89. Boyer O, Bridoux V, Giverne C, Bisson A, Koning E, Leroi AM, Chambon P, Déhayes J, Le Corre S, Jacquot S, Bastit D, Martinet J, Houivet E, Tuech JJ, Benichou J, Michot F, the Study Group of Myoblast Therapy for Faecal Incontinence. Autologous myoblasts for the treatment of fecal incontinence: results of a phase 2 randomized placebo-controlled study (MIAS). Ann Surg. 2018;267:443–50.
90. Paquette IM, Bordeianou L. Evaluation and treatment of FI. In: Steele SR, Hull TL, Read TE, editors. The ASCRS textbook of colon and rectal surgery. 3rd ed. New York: Springer Nature; 2016. p. 1091–105.

Complex Procedures for Fecal Incontinence

17

J. Manuel Devesa

Introduction

Normal continence is a very complex mechanism that includes the functional and anatomical integrity of a series of factors. From a practical viewpoint, causes of fecal incontinence (FI) can be grouped as those depending on the brain (depression, anxiety, senility, and other degenerative diseases with loss of mobility), the nerves (spinal cord injury, neuropathic, idiopathic), the rectum (inflammatory disease, partial or total absence), the sphincters (congenital absence, trauma, idiopathic), the pelvic floor function (damaged suspensory ligaments), and stool volume and consistency (diarrhea). An alteration in one or more of these structures can trigger soiling, urgency, or passive incontinence, whose severity will depend on the type and degree of dysfunction, injury, or absence of such structures. Surgical management with resolutive intent should be directed to repair or replace any of the defects or their function. Brain causes other than temporary alterations are not treatable and nerve dysfunction is better treated by neuromodulation. When the rectum has to be partially or totally removed the construction of a new rectum using a J-pouch is highly recommended.

J. M. Devesa (✉)
Colorectal Unit, Ruber International, General and Digestive Surgery, Madrid, Spain

FLIP (functional luminal imaging probe), a relatively recent test developed by Luft et al. [1] has shown that the puborectal muscle contributes to continence more by the rectoanal angle than by closing the anal canal, which must be investigated and taking into account as a part of the surgical approach if the rectoanal angle is abnormally open at rest.

Complementary medical treatment including perianal skin care, diet, scheduled toileting, and drugs (fecal mass formers, antidiarrheals, antidepressants) should always be considered [2]. When necessary, some instrumental help can be useful to improve a result or when there are no other alternatives: pelvic floor training (biofeedback, Kegel), transanal irrigation, local injection of various substances in the internal sphincter, plugs [3], and the novel Vaginal Bowel Control (VBC) system (Eclipse™ System, Pelvalon, Inc. Sunnyvale, CA, USA) consisting of a vaginal insert and pressure-regulated pump [4].

Although any surgery in the treatment of FI is complex, we exclude of this chapter the simple sphincteroplasty and other minimally invasive techniques, such as implanting biocompatible materials into the submucosa or in the intersphincteric space (SphinKeeper®), radiofrequency, electrostimulation techniques, or cell therapy. In relation to complex techniques, the reconstruction of cloacal malformations by anorecto-vagino-urethroplasty using a posterior sagittal approach or total urogenital mobilization

L. C. C. Oliveira (ed.), *Anorectal Physiology*, https://doi.org/10.1007/978-3-030-43811-1_17

is not included, since these techniques belong to pediatric surgeons and we have no experience own.

The greatest difficulty in the surgical treatment of FI is established when direct sphincter repair is not possible because it is absent or failed. These are the cases of congenital anal atresia, severe traumatic destruction, when it comes to restoring a continent perineal colostomy after the rectum amputation (total anal reconstruction: TAR), or if the sphincter is anatomically intact but functionally inactive and does not respond to electrostimulation techniques (sacral or the posterior tibial nerve) or, simply, if these last two options are neither available nor are they affordable.

In these cases of severe FI with significant impairment of quality of life, it is necessary to re-create a neosphincter either with an autologous muscle or resorting to implanting a device as a simple cerclage or using techniques more sophisticated such as an artificial anal sphincter.

In recent years surgical alternatives have been affected by the disappearance of certain devices, for example, the electrode for dynamic muscular plasties, the artificial sphincter Acticon® (AMS), and the even more recent, the magnetic sphincter MAS®. Therefore, attention will be focused on adynamic muscle plasties, the artificial sphincter currently available, levatorplasty, slings, tapes, and appendicostomy for antegrade colon enema (ACE, Malone procedure).

Muscular Neosphincter

The goal of a neosphincter is to keep the anus closed, usually a function of the internal anal sphincter, and maintain voluntary closure, a function of the external sphincter. Although smooth muscle and many striated muscles (semitendinosus, biceps femoris, long adductor, sartorius, palmaris longus) have been used, not stimulated or electrically stimulated, the two that have shown the most effectiveness as neosphincter have been the gluteus maximus and the gracilis.

Gluteus Maximus

The gluteus maximus is the most powerful muscle in the body. Its main actions are to extend and rotate the thigh laterally and stabilize the hip joint posteriorly, but it also functions as an auxiliary muscle of continence. The individual knows perfectly how to cause his voluntary contraction. Anatomically it is structured in two halves with their respective independent neurovascular pedicles, which allows for their division maintaining its viability. The inferior pedicle, formed by the inferior gluteal artery and the inferior gluteal nerve (L5, S1-2), is usually found about 6–8 cm from the medial border of the sacrum, entering at its deep part. The lower half of the gluteus can be detaching from its sacral insertion or from the gluteal tuberosity of the femur, respecting the innervation and vascularization, allowing for its transposition without affecting the mobility of the thigh or hip. The gluteus is a physically active muscle that contains at least 52% type I fibers (resistant to fatigue); less than the external sphincter (78%), but more than the sartorius (50%), rectus abdominis (46%), and gracilis (43%). Its anatomical and physiological characteristics, its natural synergism in the mechanism of continence and the technical possibility to carry out its transposition make it an appropriate muscle for the construction of a neosphincter.

Chetwood (1902) [5] successfully performed the first muscle transposition to re-create an anal neosphincter with the gluteus maximus. In the following decades different technical variants of bi- and unilateral transposition were described, some disinserting the gluteus of its distal insertion, the iliotibial tract, and others its proximal insertion, the sacrococcygeal fascia. Details of the surgical technique are well documented in the original papers of the different authors [6–15], but I would like to emphasize that identification of the neurovascular pedicle of the inferior half of the gluteus and the anterior encirclement of the anus constitutes the two more difficult and critical steps of the operation (Fig. 17.1) [7, 11]. A temporary colostomy is

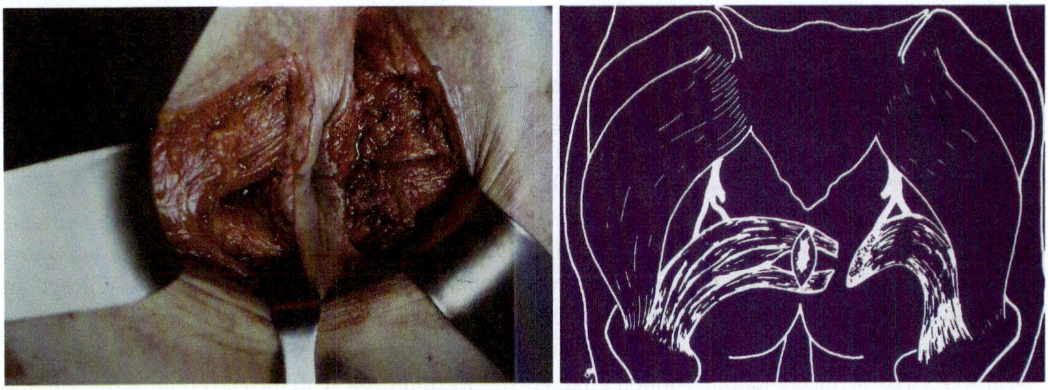

Fig. 17.1 Devesa technique of gluteoplasty. One end of the inferior half of the gluteus is divided and crossed anteriorly and posteriorly to the contralateral side, where they are sutured together or independently to the undivided contralateral end

indicated if there was perforation of the rectal wall during the creation of the perianal tunnel, if gross contamination of the surgical field occurs at a moment of the intervention, in the case of TAR [15] and, tactically, if there are doubts about the functional feasibility of the plasty due to ischemia, possible section of the nerve, or excessive tension. In turn, the irrigation of a colostomy can reeducate and regularize the intestinal habit before closing it.

Candidates for gluteoplasty are those patients with complete FI not amenable to sphincter repair or when it is absent, including TAR.

Denervation of the plasty and anterior disruption of the muscular sutures (caused by infection or excessive tension) may result in bad or worse-than-expected functional results. Diagnosis is ascertained by electromyography and magnetic resonance imaging (Fig. 17.2). Some other situations may negatively affect the result: multiple malformations, especially if there are associated deformations of the urinary tract [6]; pelvic irradiation; elderly, weak, and immobile patients; intestinal habit with a tendency to diarrhea and an attitude of the patient little collaborationist; postoperative complications and, obviously, an inadequate selection of the patient and a bad technical performance.

These techniques entail a significant morbidity, which in some series reaches up to 88%, mainly related to wound infection (35–43%), with special incidence and consequences in the perianal wounds [6–16].

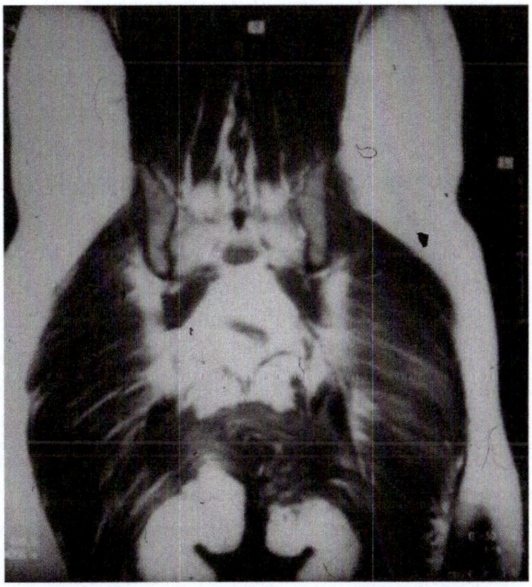

Fig. 17.2 Magnetic Resonance image showing the gluteoplasty encircling the anus

Complications are shared almost equally between those of the donor site of the plasty and the peri-rectum. Dehiscence of muscular plasty may be due to excessive tension or subcutaneous sepsis. Once the corresponding fibrosis is developed, local reconstruction may be useful, although the chances of functional success are lower than after the initial operation, as with sphincteroplasty.

Perirectal complications such as perirectal abscess, rectovaginal fistulae, or vaginal perforation are related with a difficult anterior tunneling mainly due to fibrous and scar tissue [16]. A high incidence of donor-site morbidity can be found explicitly after proximally based gluteus flap, the most common of which is neurologic: dysesthesias, posterior thigh numbness, and chronic pain or discomfort had been reported despite preservation of the posterior cutaneous nerve and coverage of the sciatic nerve with the advancement of the lower edge of the remaining donor gluteus muscle [10]. Hip dysfunction or altered gait are not an aftermath of unilateral or bilateral gluteoplasties [15].

No formal meta-analysis of the published functional results after gluteoplasty can be done. Different scoring systems used for fecal incontinence, techniques of gluteoplasty, scant number of patients, functional test measures, definitions of success and objectively documented results, length of follow-up, and so on preclude the objective measurement of the magnitude of gluteoplasty effect on end-stage fecal incontinence. In addition, most publications are based on the technique described by the author and not on the reproduction of others, so the subjective component may be important in the evaluation of the results. However, whatever the impact of these weaknesses may be, in most of the published series including different types of gluteoplasty it is shown that around two-thirds of patients reach a satisfactory continence status, although this varies from 0% to 100% [6–16].

In our series [11] of 20 patients undergoing bilateral unstimulated gluteoplasty, with a follow-up time between 2 and 11 years, the results were objectively estimated as excellent or good in 54% (Pescatori A-1, A-2, B-1), mild in 6% (Pescatori C-2) and bad in 41% (Pescatori C-3). Eight patients showed the ability to retain an enema of 200 ml of water instilled in the rectum between 5 minutes and 2 hours. In the patients with the best results there was a significant difference between the rest and voluntary contraction pressures before and after gluteoplasty, like those shown by Kong et al. (2012) [12] in their series of 25 patients followed between 1 and 9 years, accompanied by a significant improvement in the frequency of the number of bowel movements, the Jorge–Wexner scale and the quality of life (Rockwood).

In the case of TAR with gluteus, we would like to highlight the Puerta et al. (2013) [15] series. Although the number of patients is scarce, long-term follow-up between 1 and 10 years adds value. Four of the seven patients achieved an excellent function with scores on the Jorge–Wexner scale (range, 0–20) equal to or less than 5.

In summary, with the current information, it is not possible to establish the true value of gluteoplasty in severe FI, but according to the different published series [6–15] about two-thirds of patients significantly improve their status, although only a very low percentage may reach optimal levels of continence; nor can it be established which is the best of the techniques described. Restoration of the resting length of a muscle after transposition is difficult to obtain. Whatever the technique chosen, free-floating bilateral gluteoplasties may lose significant contraction force because reestablishment of resting length is compromised by the absence of a distal muscle insertion. In the author's experience, the stimulated gluteoplasty did not significantly improve the results of the unstimulated [16], and in any case this option is not currently available. The results of other techniques associated with sphincter reconstruction with muscular plasties, such as perineal colostomy with smooth muscle spiral graft [17] or neuromodulation of the transposition of the antro-pylorus with anastomosis of the pudendal nerve to the anterior vagus [18, 19] or using the inferior rectal nerve [20], both after TAR, cannot be analyzed because they are isolated publications.

Any of the techniques is difficult and all have a high morbidity. Selection of patients is a key factor for success. The rescue of some failures is possible in a number of patients [21].

Gracilis

Gracilis is the most superficial muscle on the medial side of the thigh. It is an adductor muscle of the lower limb, which also contributes to flexion of the hip and knee and internal rotation. Together with the tendon of the semitendinosus muscle and the sartorius muscle it forms the "goose's foot." Its transposition does not affect the original function because other muscles are able to replace it.

Unlike the muscular volume of the gluteus, the gracilis is a long and flat superficial muscle that allows for its unilateral mobilization and transposition to complete the wrapping of the anus with some ease.

Gracilis receives its vascular supply through a single dominant proximal artery from the deep femoral system and two or three distal pedicles, approaching the muscle between the long adductor and the short adductor about 8–10 cm distal to the pubic tubercle. The proximal blood supply plays a dominant role, since it provides the most blood to the muscle (75%). Its innervation comes from the obturator nerve (L2-4). It is a muscle predominantly formed by type II fibers, fast contraction, easily fatigable, unable to maintain a prolonged contraction.

The original description by Pickrell et al. (1952) [22] demonstrated 100% continence in pediatric patients. He attributes his success to appropriate patient selection including only young motivated patients without functional colonic dysmotility and disabling incontinence secondary to trauma or congenital anomaly. However, further reports by other authors [23–25] indicating inconsistent or poor results appeared, casting doubts about the suitability of the procedure In an attempt to improve outcome, the procedure was modified by Kumar et al. (1995) [26] to include bilateral gracilis transposition. They performed this procedure in ten patients with colostomy for diversion. All of the nine who underwent colostomy reversal were fully continent at 2 years. Another modification was described by Rosen et al. (1998) [27].

The configuration of the type of graciloplasty depends on the preference of the author and the facilities to develop one or the other, according to the local anatomical conditions (Fig. 17.3). Basically, the three configurations are the gamma (anterior delivery and contralateral ischial fixation), epsilon (posterior delivery and contralateral ischial fixation), or alpha loop (posterior delivery and ipsilateral ischial fixation). The decision between one or the other depends on the length of the muscle compared to the size of the tunnel, remembering that the cerclage should not be too tight to cause an outlet obstruction. By means of an adduction of the leg, an additional 2 cm of length can be gained. The important thing is that the plasty surrounds the anus with its muscular part, not the tendon, and covers the entire circumference. In the multicenter study of Madoff et al. (1999) [16] this was only possible in 65% of the patients; in the remaining 35% the muscle only surrounded 270°, completing the circumference with the distal tendon. The gracilis tendon was anchored to the ischial tuberosity in 65%, to the skin in 30%, and to the contralateral gracilis in 5%. The latter is performed by Cavina (1996) [28] in cases of TAR, in which the transposition of gracilis is bilateral, with one of the gracilis functioning as a puborectal ring. Interestingly, Madoff et al. (1999) [16] did not find any correlation between the type of plasty, the fact that the encirclement was complete, the anchoring point of the tendon and the functional results.

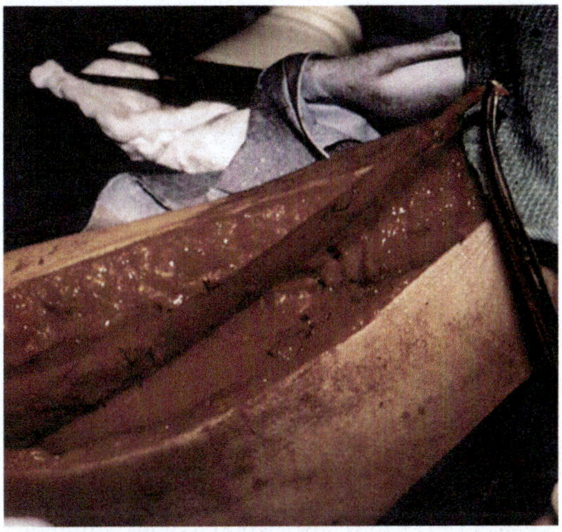

Fig. 17.3 The gracilis muscle transposition procedure

However, the distal end of the muscle is prone to ischemic injury due to the distribution of its irrigation, with which in many cases the distal segment that surrounds the anus becomes a fibrotic ring, acting more like an obstructive ring (Thiersch type) than as an active contractile muscle. Furthermore, the individual does not know how to voluntarily contract the transposed muscle, which lacks involuntary tone at rest, conditioning the poor result. So, the initial interest waned until it was revitalized by Corman [29] and Leguit et al. [30] in 1985. Baeten et al. [31], in 1988, successfully performed the first dynamic graciloplasty in a woman with a complete FI, using intramuscular perineural electrical stimulation by means of two electrodes connected to a programmable generator (Itrel II, Medtronic, Minneapolis, MN). Around that time, Williams et al. (1991) [33] also worked with stimulated graciloplasty but directly stimulated the nervous trunk (NICE, Neuromed Inc., Fort Lauderdale, FL). Seccia et al. (1994) [32] and Cavina et al. (1996) [28] also found that the previous unsatisfactory results were significantly improved with the stimulation and, in the following years, the dynamic graciloplasty became the most used technique for the treatment of severe FI and reports about adynamic graciloplasty disappeared from the literature.

Patients probably best suited for uni- or bilateral graciloplasty are those with extensive scarring of the rectovaginal septum, anterior sphincter defects or a diffuse deficient perineal body requiring an anterior approach, who retain some native sphincter function (squeeze pressure) and anal sensitivity is preserved [25]. The indications and selection of patients are the same as those exposed for gluteoplasty, including also those patients submitted to synchronous or metachronous TAR, after amputation of the rectum.

The Maastricht study [34] showed a success rate of 73% in 52 patients (mean follow-up, 2.1 years), with a good correlation between the evaluation of quality of life and functional outcome. Likewise, the evacuation frequency dropped from 5 uncontrolled stools per day to 2 controlled; the time to delay defecation increased from 9 seconds to 19 minutes, and the ability to hold the retention enema from 0 to 180 seconds. Regarding etiology, patients with traumatic incontinence responded much better (92% of 24 patients) than those with pudendal lesions (64% of 14 patients), caudal lesions (50% of 2 patients), or anal atresia (50% of 12 patients), probably due to lack of sensitivity. Williams et al. (1991) [33] had similar successful results. Unfortunately, despite these promising results, dynamic graciloplasty is not currently available.

Leaving apart those complications derived from stimulation, those of graciloplasty are mainly technical (loose ring, transitory edema of the leg, insufficient contraction and perforation of the anal canal during the intervention) and septic, which are the most frequent and serious ones, especially in the TAR group. Complications occur in more than 50 percent of patients and may be serious, including evacuation difficulties. Curiously, patients with stomas had a higher incidence of major wound complications and a lower percentage of success than patients without stomas [16].

Causes of failure and predictive factors of a poor outcome are the same as those already mentioned in gluteoplasty. In the Madoff et al. (1999) [16] multicenter study the results of the two most experienced centers were compared with those of the rest. The percentage of success was 80% for centers with experience vs. 47% for centers with no experience or less.

Given the nature of gracilis and the characteristics of the technique, the unstimulated gracilis wrap essentially serves as a static sling (Thiersch) with few contractile properties, yet this muscle is dependent on volition, and thus a sustained contraction is not possible, so its current use is very limited.

The bilateral graciloplasty, proposed by Kumar et al. (1995) [26], seems to offer better results, although larger series and prolonged follow-up are lacking thus far.

Although transpositions of skeletal muscle played an important role in the reconstruction of sphincters, today they are techniques of little use due to its complexity, high rate of complications needing frequent revision surgeries and lack of an accurate predictor of outcome [34–40]. With a better understanding of mechanisms of continence and technology development, newer and less invasive treatment modalities, including different types of nerve stimulation, injectable biomaterials, muscle cell transfer and anal slings are nowadays the preferred options. However, the transposition procedures can still be an alternative for a permanent stoma in cases of severe traumatic muscle loss, either for sphincter replacement or for sphincter augmentation prior to the placement of an anal or pelvic encirclement with a prosthetic device. These procedures are more likely to benefit young healthy patients with extensive tissue loss of the anal sphincter. Both inclusion and exclusion criteria must be clearly established. Even in well-selected candidates not suitable for other less invasive procedures, the Malone technique [41] can offer a better solution.The use of muscular plasties in the surgical treatment of anal incontinence has two main protagonists: the gluteus maximus and the gracilis. Adynamic gluteoplasty is effective in the medium and long term in more than 50% in a few series [6–15] with a small number of patients, so it is not possible to establish its real value. However, from the works with unilateral dynamic graciloplasty it can be deduced that the adynamic graciloplasty is inefficient [16, 23–25, 27–34, 40, 42]. The ideal muscle transposition has not yet elucidated. It is not clear which is the best type of muscle (gracilis vs. gluteus), the best configuration (alpha, gamma, or epsilon shape in gracilis transposition, or proximally vs. distally based gluteus flap), and whether the advantages of external stimulation outweigh the risk and cost. In any case, the unstimulated gluteoplasty [11] or (bilateral?) graciloplasty [26] remains an effective option in places where the stimulator is unavailable or the price of an artificial device is not affordable.

Artificial Anal Sphincter

The first successful implant of an artificial anal sphincter was published in 1987 by Christiansen et al. [43], who implanted a urinary sphincter (AMS 800, Minnetonka, MN, USA), in five patients with neurogenic incontinence. However, a specially designed intestinal anal sphincter, designated Acticon® Neosphincter (AMS 800, Minnetonka, MN, USA), was only available in 1996 (Fig. 17.4). In 2003, a new sphincter designed by A.M.I., the Soft Anal Band System® (Feldkirch, Austria) appeared in the market, proposing certain technical improvements and advantages in the implant with respect to Acticon®. The Circular Pre-Shaped form provides equal circular pressure on the anal canal resulting in improved continence of flatus and thin stool, and less wrinkle formation. The so called Fully Ti-Port makes easy and safe postoperative adjustments of the anal closing force. It is also claimed that the risk of erosion of the different components and the infection rate is lower than with Acticon®, which is associated with high infection and anal penetration rates of the cuff, leading to an explantation rate up to 60% of the implants. All these theoretical advantages need scientific demonstration in long series with enough follow-up time. The major problem with the soft anal band (A.M.I.) is a defunctioning valve which occasionally has to be replaced [44].

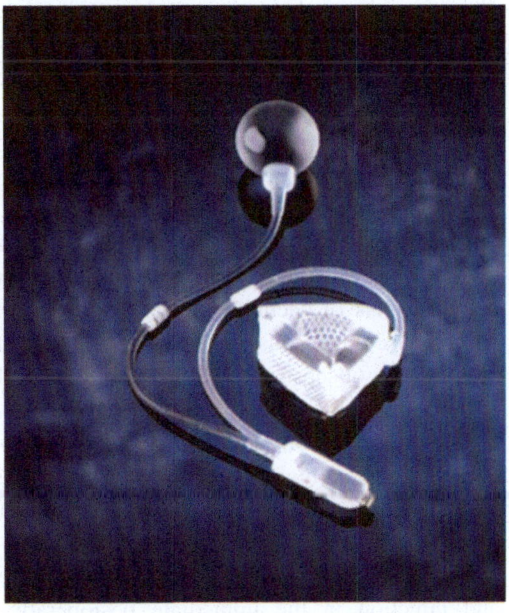

Fig. 17.4 The artificial bowel sphincter procedure

In any case, with the arrival of the artificial sphincters, the anal myoplasties fell into disuse. One study by Tan et al. (2008) [45] comparing dynamic graciloplasty, artificial bowel sphincter (AMS) and end stoma demonstrated that the implant of the artificial sphincter was most cost-effective after 10 years.

In our series [46] of 53 patients with Acticon®, of which 43 had the device in use in a follow-up period between 1 and 9 years, 65% reached normal continence and 98% were continent to solid stool. The score on the Jorge–Wexner scale changed from 17 ± 3 before to 4 ± 3 after, and the score on the scale of Quality of Life (Rockwood) changed significantly in all the subscales. In another series [47] of a multicenter study that includes 112 patients evaluated at 6 and 12 months, the incontinence score decreased by more than 50%, with a satisfactory result of 85%.

Whatever the reasons, Acticon® for anal incontinence has been removed from the market. However, the Anal Band System® (A.M.I.) is still available and the indications for its implant are the same as those previously explained for Acticon® and myoplasties, with the difference that there are no strict age or constitutional limits, as it is not a physical-dependent technique. Thus, the implant of an artificial sphincter is indicated in cases of congenital or traumatic absence of the sphincter, atonic sphincter without sphincter defect, and in cases of TAR, although there are only a few reports regarding this indication [48, 49].

In radiated patients, the implant of a prosthesis is a relative contraindication, for the almost inevitable risk of decubitus on damaged tissues and because it can cause severe perineal pain. In the presence of sepsis, Crohn's disease, and anal sex practices, the contraindication is absolute. Chronic constipation is a relative contraindication due to the high risk of fecal impaction. Chronic diarrhea is not a contraindication, but it is highly advisable to perform a temporary colostomy to protect the anal wounds from dehiscence and infection in the immediate postoperative period.

Certain factors such as the presence of extensive perianal fibrosis, an absent or very thin vaginal wall (e.g., congenital, traumatic, pronounced rectocele), or the absence of sensitivity (e.g., hereditary malformations, neurological diseases) may favor mechanical or functional complications, like cutaneous erosion by decubitus or difficulty for the evacuation and fecal impaction. Consider the implant in these circumstances. In our series [46] there was a greater tendency to develop fecalomas among patients with an abnormal sensitivity (31% vs. 18%).

Technical details of the procedures have been well documented by different authors. The operative difficulties are mainly focused on the sleeve implant. We favor the use of an anterior incision halfway between the anus and vagina or scrotum (80% in our series) [46] because it allows for a better control of the dissection in that more difficult area, with less risk of injury to the vagina and urethra. It also allows for filling the space with fat or muscle, if necessary, to reduce the risk of erosion. If lateral incisions are made, they should be at a minimum distance of 2 cm from the anus, to prevent them from partially staying inside the anus when the implant is placed. In any case they should never be done over a scar area because the risk of immediate dehiscence or future erosion is very high (Fig. 17.5). In women with a large anterior peri-

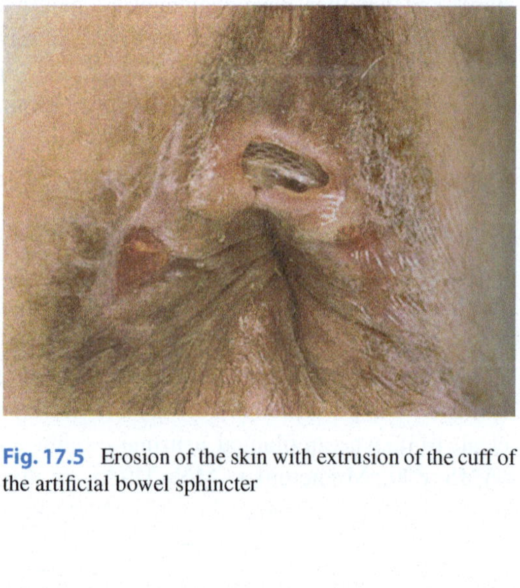

Fig. 17.5 Erosion of the skin with extrusion of the cuff of the artificial bowel sphincter

neal scar, the transvaginal route can be chosen, 2 cm above the introitus, as proposed by Michot et al. (2010) [50].

Our recommendation is that if there are doubts between two lengths, we prefer the shortest if the patient usually makes soft stool, and the longest if the stool is normal or hard. Care must be taken that the cuff is not too tight, because of the risk of rectal necrosis.

Indications for a temporary colostomy are exceptional. In our study [46] it was not shown that patients with a stoma had a lower risk of infection. In case of gross intraoperative contamination it is preferable to defer the implant.

Infection, erosion or ulceration, chronic pain, constipation or fecal impaction, malposition, malfunction, migration, recurrent incontinence, surgical revision, explantation, and re-explantation, all are complications common to any anal prosthesis. The key, apart from the type of device and the ease of implantation, is in the selection of patients and in the detailed execution of the technique.

In our series with Acticon® [46] partial dehiscence of the anal wounds were significantly associated with the presence of fibrosis and with tension in the closure, hematoma and infection. Late complications were associated more frequently to the episodes of fecal impaction and erosion of any of the elements of the device, almost equally the cuff and pump (Fig. 17.6). The risk of erosion seems to be much lower with the Soft Anal Band® (AMI) but there is still not enough experience with this device in number of patients and time tracking. In the Acticon® multicentric series [47] infection (34%), followed by erosion (21%, most of the cuff) were the most frequent. Altogether, the complications required one or more surgical revisions in 60% of the patients. The definitive explants rate oscillates between 19% [46] and 30% [47] at the end of the respective follow-ups, although it increases with time. Wexner et al. (2009) [51] found a cumulative risk of explant of 57% at 5 years, mainly related to problems arising from the device. Except for infection, which in almost 100% of cases involves the explant, many of the complications can be resolved by surgery, but ultimately, with the passage of time, for some or other causes less than 50% of patients retain the functioning sphincter [47].

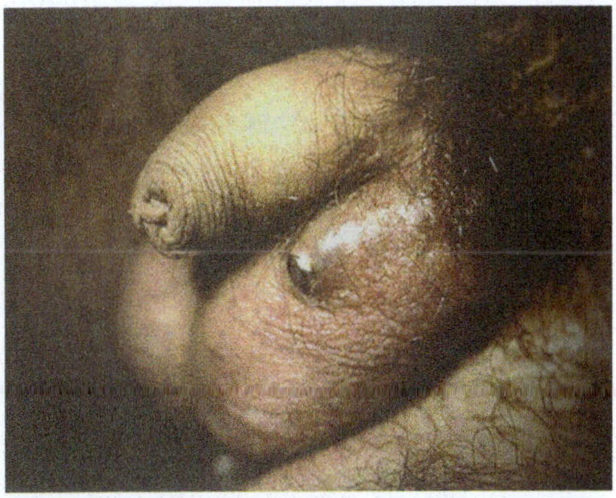

Fig. 17.6 Erosion and extrusion of the skin of the scrotum where the pump was placed

Although the morbidity of an artificial sphincter implant is very high, there is no associated mortality, and the percentage of successes maintained in the long term is higher than with myoplasties. After the withdrawal of Acticon® only A.M.I. Soft Anal Band® remains for clinical use. Larger series and longer follow-up are necessary for its scientific evaluation and criticism [52, 53].

There is little information [55–57] of other artificial sphincters like PAS® (Prosthetic Anal Sphincter, NPH Design Ltd., London, UK), designed by Hajivassiliou et al. (1997) [54], implanted intraperitoneally around the rectum in the supralevator plane, and GASS® (German Artificial Sphincter System), and the Magnetic Anal Sphincter (MAS®, (FENIX™, MN, USA) was withdrawn from the market in 2017.

In conclusion, the use of artificial sphincters for end-stage incontinence or following rectal excision for cancer is an acceptable management strategy to obtain continence and restore anal defecation in a considerable number of patients despite the high morbidity. Selection of patients and operator experience are key points in the successful outcome of the procedure. However, with the artificial sphincters available today, standardized data leading to significant evidence-based conclusions are still lacking.

Anal Slings and Tapes

Simple Silicone Band

Anal encirclement with different types of slings for treating all degrees of anal incontinence and rectal prolapse has proved successful over time in a significant proportion of patients in small series. Moreover, functional results are consistent along the time. Throughout last century, simple devices had only anecdotic evidence for their acceptance, due to the own nature of the device and the high incidence of complications mainly associated with fracture of the sling or episodes of fecal impaction. New materials and new concepts for re-creating or reinforcing weak or damaged sphincters have seen the light of day in recent years.

The simple cerclage was initially described by Thiersch [58], in 1891, to treat prolapse in patients with elevated surgical risk, but it was also evident that it was useful to reestablish continence in a percentage of patients. The arrival of the artificial sphincter made abandoning these techniques to treat incontinence, due to its poor long-term results derived from ring rupture and dysfunction. The dysfunction depends to a great extent on the elasticity of the ring and the ability to recover its basal tone.

There are very few studies related to the use of different types of cerclage rings in the treatment of incontinence, and their value is limited by the size of the samples, the absence of studies comparing with other groups and the time of follow-up.

At the end of 2004, Devesa et al. [59] began to perform the simple cerclage technique using the flat part of the Jackson-Pratt® drainage, to treat very old patients with prolapse and incontinence (Figs. 17.7 and 17.8). Subsequently, in view of the results, they extended the indication to patients with soiling and variable degrees of incontinence, in which other options would have been too aggressive, or were not possible due to technical or economic reasons or had failed. In a period of 5 years, they performed the operation on 33 patients, including patients with ileoanal pouch, ultralow colorectal anastomosis and

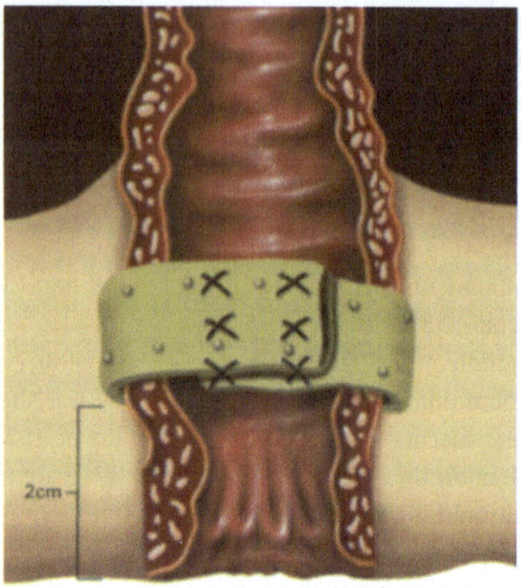

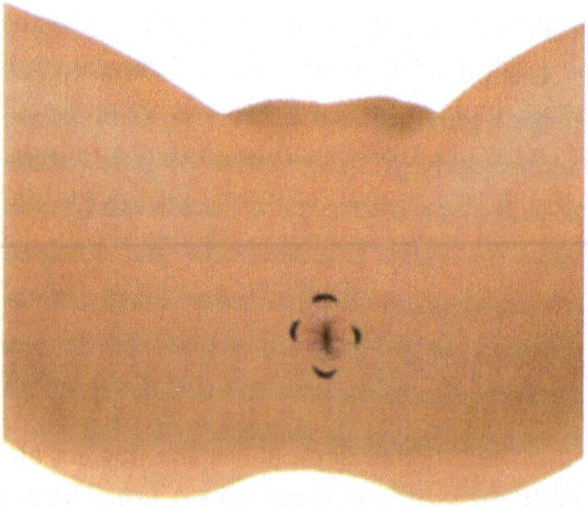

Fig. 17.7 Anal cerclage with an elastic band of silicone. The anal encirclement is made through a few perianal incisions

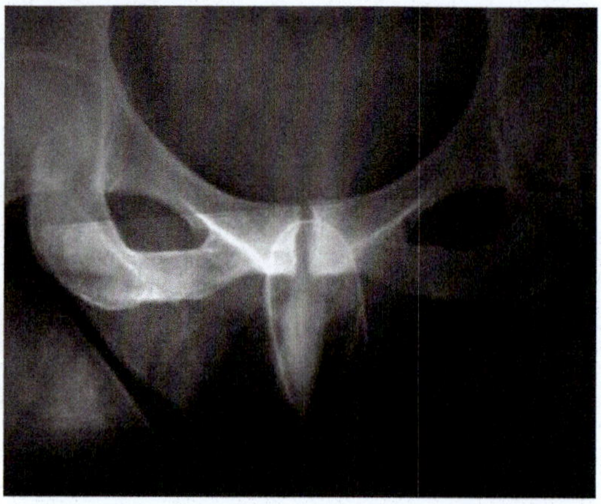

Fig. 17.8 Plain pelvis X-ray showing the anal cerclage

TAR. The contraindications were those already mentioned for the other procedures described. Early complications included the rupture of the ring in 2 patients and local infection, which forced its removal in both cases. The late complications were related to skin erosion in 2 patients, one at 3 years, and the rupture of the ring in 7, over a follow-up time between 2 and 60 months (mean, 37). The ring was reimplanted in 10 of the patients.

Except for one patient, all of them improved their functional status with a significant improvement in the Jorge–Wexner score and in the four subscales of quality of life (Rockwood). When comparing the results with the historical series of 53 patients with Acticon® implant, and similar demographics and follow-up time, we observed that the functional and objective results were very similar with the advantage that in the case of simple cerclage, except for the complication of the breakage of the device due to its artisanal nature, the morbidity was lower and more easily resolved. Other advantages are the simplicity of the technique and its cost. In any case, it is necessary to have the appropriate device to avoid the problems related to its breakage and the artisanal closure system, and to carry out the pertinent clinical studies for the recommendation of its use.

Anal Encirclement with Surgisis™

In 2012, Zutshi et al. [60] first reported the results of this procedure in a series of 13 patients with FI due to an anal sphincter defect combined with a weak anal sphincter. In this technique, a biological graft (Surgisis™; Indiana, USA) of 2 × 20 cm is inserted through a tunnel created under the damaged external sphincter and sutured to the muscle after being pulled firmly to close the patulous anus throughout its entire length. An overlapping repair is then carried out. Postoperative incontinence severity scores and quality of life scales showed improvement and incontinence episodes were markedly decreased, at a mean follow-up of 16.3 (range 6–24) months, despite continuing low resting pressure. The authors do not mention any change in the postoperative squeeze pressure. These facts are in agreement with the results of sphincteroplasty alone, which also demonstrated that good results are not associated with an improvement in the resting or squeeze pressures. Therefore, the mechanism of action of this procedure must be more to act as a barrier as to the restoration of a physiological action, and this might be due to the weak nature of the muscle, which was already present in all patients of this series. The authors report no early or late complications. It is understandable that infection or cutaneous dehiscence may not have occurred, probably related to technical performance and the nature of the procedure, but it is difficult to imagine that an orifice that only admits the tip of the little finger has not caused any episode of fecal impaction. This type of sling for reinforcing sphincteroplasty deserves attention. Further studies regarding long-term efficacy in comparison with sphincteroplasty alone are necessary.

Levatorplasty

Postanal repair was first reported by Sir Alan Parks [61], in 1975, who modified Nesselrod's [62] original operation. This procedure was designed to increase the length of the anal canal, restore the anorectal angle and re-create the flap valve mechanism, which at the time was thought essential for maintaining fecal continence. Success rates ranged from 15% to 83%, depending on the definition of the success, the length of follow-up, and possibly the cause of incontinence. Deen et al. (1993) [63] in a randomized controlled trial comparing three procedures in 36 women with neuropathic fecal incontinence found that complete continence was achieved in 42% of patients after postanal repair, 33% after anterior levatorplasty, and 67% after total pelvic floor repair. In contrast, van Tets et al. (1998) [64] conducted a randomized controlled trial comparing postanal repair and total pelvic floor repair in 20 women with neurogenic fecal incontinence. Complete continence to solid or liquid stool was achieved in 27% of patients after postanal repair and in 22% after total pelvic floor repair. In 2004, Yamana et al. [65] reported the perineal puborectalis sling operation on eight patients with idiopathic fecal incontinence. A rectal ulcer developed in one patient, necessitating sling removal. In the remaining seven patients, the Fecal Incontinence Severity Index improved from 27 to 9, and the Cleveland Clinic Score of Incontinence improved from 13 to 5 ($P < 0.05$). All parameters in the Fecal Incontinence Quality of Life Scale improved. No significant difference was found between preoperative and postoperative maximum resting pressure and maximum squeeze pressure. However, the median anorectal angle on defecography after the operation was significantly reduced ($P < 0.05$).

TOPAS (transobturator postanal sling, AMS, Minnetonka, MN, USA) was designed as a minimally invasive procedure to treat FI. It involves placing a polypropylene mesh under the rectum, to support weakened pelvic floor muscles, by restoring the function of the puborectalis recreating the normal anorectal angle.

In the study conducted by Mellgren et al. (2016) [66], 152 women were implanted at 14 centers in the USA. FI was assessed preoperatively and at the 12-month follow-up with a 14-day bowel diary, Cleveland Clinic Incontinence Scores, and FI Quality of Life questionnaires. Treatment success was defined as reduction in number of FI episodes of ≥50% compared to baseline. Average follow-up was 24.9 months. At 12 months, 69.1% of patients met the criteria for treatment success, and 19% of subjects reported complete continence. A total of 66 subjects experienced 104 procedure- and/or device-related adverse events (AEs). Most AEs were short in duration and 97% were managed without therapy or with nonsurgical interventions. No treatment-related deaths, erosions, extrusions, or device revisions were reported. However, for reasons I do not know, the device is not commercially available.

Tissue Fixation System (TFS)

TFS is a minimally invasive method for tape implantation in patients with vaginal prolapse, urinary and FI. In a study by Abendstein et al. (2008) [67] the contribution of arcus tendineus fascia pelvis (ATFP)/cardinal ligaments and their attached fascia, if any, to causation of FI was tested. All 33 patients with FI had intact external anal sphincters and were classified as having idiopathic FI. Symptomatic improvement >80% was noted in 88% of these patients for FI, 89% for stress incontinence ($n = 43$), and 80% for urgency and nicturia ($n = 50$). No erosions or dyspareunia have been reported to date. Both urinary and FI symptoms were simultaneously cured, indicating a causal link. Repair of ATFP and cardinal ligament defects (cystocele) did not produce a significantly higher cure rate for FI to that achieved by repairing just the anterior and/or posterior suspensory ligaments indicating perhaps, the primacy of anterior and/or posterior suspensory ligaments in FI control.

Although the longer-term efficacy for FI cure through TFS has not yet been evaluated in large series and controlled studies, this is an interesting logistic approach in those women complaining of multiple pelvic/perineal symptoms on whom FI plays a significant pathological role and the anorectal sphincters are otherwise intact.

Total Anorectal Reconstruction

Although total excision of the anorectum with construction of a permanent abdominal colostomy is accepted by many patients as a small price to pay for cancer cure, it greatly affects their quality of life and carries a high incidence over their lifetime. Around 10% of patients will require additional surgery for stoma-related complications. Furthermore, abdominoperineal resection (APR) has a high rate of postoperative morbidity of the perineal wound, reported to occur in between 40% and 60% of cases, particularly if preoperative radiotherapy has been administered to the perineum [68].

Chittenden (1930) [69] published the first attempt at total anorectal reconstruction (TAR) by performing a continent perineal colostomy using a flap of the gluteus maximus as a neosphincter. Perfect continence depends on the restoration of anatomical integrity and normal function of the rectum as an adaptable reservoir, the anorectal sphincters, and the anal mucosa, with its sensory receptors able to discriminate the quality of the rectal content and the rectal/pelvic receptors transmitting the sensation of rectal filling. Following APR, restoration of anal sensitivity is not possible, but re-creation of a neorectal reservoir with the descending colon and a neosphincter with an autologous muscle or an artificial sphincter has made TAR feasible in those patients. Furthermore, the performance of an appendicostomy or an ileal/colonic conduit for antegrade irrigation of the colon (either alone or in association with any of those procedures), contributes to achieve a pseudocontinent status in a significant number of patients [70].

Although scientific evidence of the benefit of a neorectum in TAR procedures and which type of pouch to use which would result in better function have yet to be demonstrated [71–73], the performance of a simple coloplasty or a myotomy proximal to the neosphincter does not seem to add significant morbidity and could improve the functional status.

In clinical practice, it is difficult to know the real functional benefit of re-creating a new internal sphincter in TAR techniques. Moreover, if the smooth-muscle cuff is created from a free graft it has neither intrinsic nor extrinsic innervation, therefore its mechanism of action is likely related to its action as a biological Thiersch graft exclusively.

The best results and the largest series of gluteoplasty in TAR have been published by Puerta et al. (2013) [15]. Regarding adynamic graciloplasty the optimal technique remains to be determined, although the alpha-loop has been the most commonly used configuration. Another unsolved question is whether bilateral transposition is better than unilateral deployment, although some data suggest that double graciloplasty is not the ideal neosphincter for anorectal reconstruction [74]. The most serious complication is significant necrosis of the anus or neoanus which occurred more frequently when the procedure was performed after an APR than for incontinence alone [34, 40]. Although stimulated graciloplasty showed excellent results according to Cavina et al. (1998) [75] who reported an 87% success rate in 98 patients with the longest follow-up of 55 months, morbidity was very high (37% of patients) and the procedure is no longer a therapeutic option. Although recovery of all lost structures with their corresponding functions should be necessary for achieving perfect continence, the re-creation of an efficient external neosphincter appears to be the most important aspect of TAR. Currently, it is difficult to imagine other suitable muscles better than the gluteus or the gracilis, but both options carry a high rate of complications.

Excellent functional results with Acticon® implant were reported by Romano et al. (2003) [76] for patients undergoing TAR, despite a high morbidity associated with the high rate of definitive explants, but this option is no longer available. Experience with the Soft Anal Band® (AMI) is still lacking. Regarding the modified Thiersch procedure proposed by Devesa et al. (2011) [5, 9] to the best of our knowledge there are no other reports concerning this simple approach and therefore no conclusions can be drawn.

Antegrade Colonic Enema (ACE)

It has been demonstrated that ACE is a useful and safe procedure to achieve a satisfactory pseudocontinent status in patients with end-stage incontinence or with a perineal colostomy, either resultant from a single procedure or as a complement to more complex reconstructions Furthermore, ACE provides better functional results than those achieved by retrograde enemas. Although it was introduced later than other options for TAR, it is likely that in the near future ACE is going to play a significant role.

Adequate selection of well-motivated patients and disclosure of full information about the expected benefits and risks are mandatory before such procedures are undertaken by those with a proven track record in this type of surgical expertise.

In summary, a perineal colostomy associated with the Malone's procedure [41] is the simplest, safest, and cheapest approach, providing, most likely, the best functional option currently available. Whether or not the functional results could be improved by adding the reconstruction of an internal neosphincter, and/or or a simple encirclement of the neoanus with the appropriate device, and/or a pouch, or a coloplasty, or a myotomy in the distal part of the descending colon, has yet to be determined in these specialized patient cohorts.

Summary

Surgical treatment of FI is subject to multiple factors that must be analyzed and commented to the patient before any decision. There is no doubt that the initial treatment of an injured sphincter is its surgical repair, including the internal sphincter reconstruction if it is viable. Continence to flatus is rarely restored. There is currently a general consensus that patients should not receive constipating agents in the postoperative period and that biofeedback does not help.

The problem arises when the repair is not possible, or if failed after a retry, or if the sphincters are absent or denervated, or when there are no resources to address the techniques of neuromodulation, including the tibial nerve, or if those options have failed.

In those cases, re-creation of a neosphincter is necessary. The mentioned muscular plasties may be the first approach; however, considering their technical complexity and high morbidity, they should be reserved when no other alternative is available. The artificial sphincter A.M.I® is another option which functional results are probably better than those achieved by muscle plasties although experience in a large number of cases with long follow-up is lacking. A simple cerclage with an elastic band can achieve results very similar to those of the artificial sphincter and is the most cost-effective, but the exposed procedure is artisanal and the number of breaks in the device means that the technique does not thrive until a secure closure and resistance system is available.

Levatorplasty alone seems to have some chances of improvement in the first 2 years, but this tendency is lost as time goes by. In any case it seems reasonable to try it as a complement to other procedures when that is possible.

New and simple approaches come from the use of a postanal sling (TOPAS) and tapes (TFS) that restore the anorectal angle and reinforce ligaments, but there is still not enough experience with those simple procedures applicable only to certain cases. In addition, TOPAS is not commercially available.

Personally, my unequivocal proposal is to always start with the simplest and most economical procedure, leaving sacred neuromodulation only for those specific cases in which it is an excellent indication and there are no other alternatives. Even in these cases, it must be considered that it is not free of complications and that its cost makes it unaffordable for the vast majority of the population.

Finally, the ACE technique is a very useful complement for other techniques if the results are imperfect, and indispensable in those of TAR, or as the last alternative before end colostomy.

References

1. Luft F, Fynne L, Gregersen H, Lundager F, Buntzen S, Lundby L, et al. Functional Luminal Imaging Probe (FLIP): a new technique for dynamic evaluation of mechanical properties of the anal canal. Tech Coloproctol. 2012;16(6):451–7.
2. Ehrenpreis ED, Chang D, Eichenwald E. Pharmacoterapy for fecal incontinence: a review. Dis Colon Rectum. 2007;50(5):641–9.
3. Margolin DA. New options for the treatment of fecal incontinence. Ochsner J. 2008;8(1):18–24.
4. Varma MG, Matthews CA, Muir T, Takase-Sanchez MM, Hale DS, Van Drie D, et al. Impact of a novel vaginal bowel control system on bowel function. Dis Colon Rectum. 2016;59(2):127–31.
5. Chetwood CH. Plastic operation for restoration of the sphincter ani with report of a case. Med Rec. 1902;61:529.
6. Prochiantz A, Gross P. Gluteal myoplasty for sphincter replacement: principles, results and prospect. J Pediatr Surg. 1982;17(1):25–30.

7. Devesa JM, Vicente E, Enríquez JM, Nuño J, Buchelli P, de Blas G. el al. Total fecal incontinence - a new method of gluteus maximus transposition: preliminary results and report of previous experience with similar procedures. Dis Colon Rectum. 1992;35(4):339–49.
8. Guelinckx PJ, Sinsel NK, Gruwez JA. Anal sphincter reconstruction with the gluteus maximus muscle: anatomic and physiologic considerations concerning conventional and dynamic gluteoplasty. Plast Reconstr Surg. 1996;98(2):293–302; discussion 303-4.
9. Farid M, Moneim HA, Mahdy T, Omar W. Augmented unilateral gluteoplasty with fascia lata graft in fecal incontinence. Tech Coloproctol. 2003;7(1):23–8; discussion 8.
10. Hultman CS, Zenn MR, Agarwal T, Baker CC. Restoration of fecal continence after functional gluteoplasty: long-term results, technical refinements, and donor-site morbidity. Ann Plast Surg. 2006;56(1):65–70.
11. Devesa JM, Madrid JM, Gallego BR, Vicente E, Nuño J, Enriquez JM. Bilateral gluteoplasty for fecal incontinence. Dis Colon Rectum. 1997;40(8):883–8.
12. Kong F, Li F, Liu J, Chen Y, Wu Y, Yang X, et al. Gluteus maximus transplantation for fecal incontinence after surgery of high anal atresia. Chin J Reparative Reconstr Surg. 2012;26(5):571–5.
13. Yoshioka K, Ogunbiyi OA, Keighley MR. A pilot study of total pelvic floor repair or gluteus maximus transposition for postobstetric neuropathic fecal incontinence. Dis Colon Rectum. 1999;42(2):252–7.
14. Meehan JJ, Hardin WD Jr, Georgeson KE. Gluteus maximus augmentation for the treatment of fecal incontinence. J Pediatr Surg. 1997;32(7):1045–7; discussion 1047-8.
15. Puerta JD, Castaño R, Lombana LJ, Restrepo JI, Gomez G. Use of the gluteus maximus muscle as the neosphincter for restoration of anal function after abdominoperineal resection. Tech Coloproctol. 2013;17(4):425–9.
16. Madoff R, Rosen HR, Baeten CG, LaFontaine LJ, Cavina E, Devesa M, et al. Safety and efficacy of dynamic muscle plasty for anal incontinence: lessons from a prospective, multicenter trial. Gastroenterology. 1999;116(3):549–56.
17. Hirche C, Mrak K, Kneif S, Mohr Z, Slisow W, Hünerbein M, et al. Perineal colostomy with spiral smooth muscle graft for neosphincter reconstruction following abdominoperineal resection for very low rectal cancer: long-term outcome. Dis Colon Rectum. 2010;53(9):1272–9.
18. Sato T, Konishi F, Endoh N, Uda H, Sugawara Y, Nagai H. Long-term outcomes of a neo-anus with a pudendal nerve anastomosis contemporaneously reconstructed with an abdominoperineal excision of the rectum. Surgery. 2005;137(1):8–15.
19. Chandra A, Malhotra HS, MN, Gupta V, Sing SK, Kumar N, et al. Neuromodulation of perineally transposed antropylorus with pudendal nerve anastomosis following total anorectal reconstruction in humans. Neurogastroenterol Motil. 2014;26(9):1342–8.
20. Chandra A, Kumar A, Noushif M, Gupta V, Kumar V, Srivastav PK, et al. Neurovascular antropylorus perineal transposition using inferior rectal nerve anastomosis for total anorectal reconstruction: preliminary results in humans. Tech Coloproctol. 2014;18(6):535–42.
21. Devesa JM, Rodriguez B, Fernandez JM, Lopez-Hervas P, Nuno J, Kamal S, et al. Anal incontinence: the fate of the initial failed gluteoplasty. Tech Coloproctol. 1998;2:67–72.
22. Pickrell KL, Broadbent TR, Masters FW, Metzger JT. Construction of a rectal sphincter and restoration of anal continence by transplanting the gracilis muscle: a report of four cases in children. Ann Surg. 1952;135(6):853–62.
23. Yoshioka K, Keighley MR. Clinical and manometric assessment of gracilis muscle transplant for fecal incontinence. Dis Colon Rectum. 1988;31(10):767–9.
24. Christiansen J, Sørensen M, Rasmussen OO. Gracilis muscle transposition for faecal incontinence. Br J Surg. 1990;77(9):1039–40.
25. Eccersley AJ, Lunniss PJ, Willimas NS. Unstimulated graciloplasty in traumatic fecal incontinence. Br J Surg. 1999;86(8):1071–2.
26. Kumar D, Hutchinson R, Grant E. Bilateral gracilis neosphincter construction for treatment of fecal incontinence. Br J Surg. 1995;82(12):1645–7.
27. Rosen HR, Novi G, Zoech G, Feil W, Urbarz C, Schiessel R. Restoration of anal sphincter function by single-stage dynamic graciloplasty with a modified (split sling) technique. Am J Surg. 1998;175(3):187–93.
28. Cavina E. Outcome of restorative perineal graciloplasty with simultaneous excision of the anus and rectum for cancer: a ten-year experience with 81 patients. Dis Colon Rectum. 1996;39(2):182–90.
29. Corman ML. Gracilis muscle transposition for anal incontinence: late results. Br J Surg. 1985;72(suppl):S21–2.
30. Leguit P Jr, van Baal JG, Brummelkamp WH. Gracilis muscle transposition in the treatment of fecal incontinence: long-term follow-up and evaluation of anal pressure recordings. Dis Colon Rectum. 1985;28(1):1–4.
31. Baeten C, Spaans F, Fluks A. An implanted neuromuscular stimulator for fecal continence following previously implanted gracilis muscle: report of a case. Dis Colon Rectum. 1988;31(2):134–7.
32. Seccia M, Menconi C, Balestri R, Cavina E. Study protocols and functional results in 86 stimulated graciloplasties. Dis Colon Rectum. 1994;37(9):897–904.
33. Williams NS, Patel J, George BD, Hallan RI, Watkins ES. Development of an electrically stimulated neo-anal sphincter. Lancet. 1991;338(8776):1166–9.
34. Baeten C, Geerdes BP, Adang EM, Heineman E, Konsten J, Engel GL, et al. Anal dynamic graciloplasty in the treatment of intractable fecal incontinence. N Engl J Med. 1995;332(24):1600–5.
35. Enriquez Navascues JM, Devesa Mugica JM. Traumatic anal incontinence. Role of unilateral

gluteus maximus transposition supplementing and supporting direct anal sphincteroplasty. Dis Colon Rectum. 1994;37(8):766–9.
36. Madoff RD, Baeten CG, Christiansen J, Rosen HR, Williams NS, Heine JA, et al. Standards for anal sphincter replacement. Dis Colon Rectum. 2000;43(2):135–41.
37. Niriella DA, Deen KI. Neosphincters in the management of faecal incontinence. Br J Surg. 2000;87(12):1617–28.
38. Cera SM, Wexner SD. Muscle transposition: does it still have a role? Clin Colon Rectal Surg. 2005;18(1):46–54.
39. Abou Zeid AA, Marzouk DM. Gluteus maximus neosphincter is a viable option for patients with end-stage fecal incontinence. Dis Colon Rectum. 2000;43(11):1635–6.
40. Mander BJ, Wexner SD, Williams NS, Bartolo DC, Lubowski DZ, Oresland T, et al. Preliminary results of a multicenter trial of the electrically stimulated gracilis neoanal sphincter. Br J Surg. 1999;86(12):1543–8.
41. Malone PS, Ransley PG, Kiely EM. Preliminary report: the antegrade continence enema. Lancet. 1990;336(8725):1217–8.
42. Wexner SD, Gonzalez-Padron A, Rius J, Teoh TA, Cheong DM, Nogueras JJ, et al. Stimulated gracilis neosphincter operation. Initial experience, pitfalls, and complications. Dis Colon Rectum. 1996;39(9):957–64.
43. Christiansen J, Lorentzen M. Implantation of artificial sphincter for anal incontinence. Lancet. 1987;2(8553):244–5.
44. Baumgartner U. The artificial sphincter: therapy for faecal incontinence. Zentralbl Chir. 2012;137(4):340–4.
45. Tan EK, Vaizey C, Cornish J, Darzi A, Tekkis PP. Surgical strategies for fecal incontinence-a decision analysis between graciloplasty, artificial bowel sphincter and end stoma. Color Dis. 2008;10(6):577–86.
46. Devesa JM, Rey A, Hervas PL, Halawa KS, Larrañaga I, Svidler L, et al. Artificial anal sphincter: complications and functional results of a large personal series. Dis Colon Rectum. 2002;45(9):1154–63.
47. Wong WD, Congliosi SM, Spencer MP, Corman ML, Tan P, Opelka FG, et al. The safety and efficacy of the artificial bowel sphincter for fecal incontinence: results from a multicenter cohort study. Dis Colon Rectum. 2002;45(9):1139–53.
48. Devesa JM, Lopez-Hervas P, Vicente R, Rey A, Die J, Fraile A. Total anorectal reconstruction: a novel technique. Tech Coloproctol. 2005;9(2):149–52.
49. Herman R, Walega P, Nowakowski M, Salowka J, Smeder K. First in Poland and second in Europe implantation of an artificial anal sphincter – A.M.I. Soft Anal Band to an end colostomy in a patient after abdominal-perineal amputation of rectum. PRO. 2005;6(4):256–63.
50. Michot F, Lefebure B, Bridoux V, Gourcerol G, Kianifard B, Leroi AM, et al. Artificial anal sphincter for severe fecal incontinence implanted by a transvaginal approach: experience with 32 patients treated at one institution. Dis Colon Rectum. 2010;53(8):1155–60.
51. Wexner SD, Jin HY, Weiss EG, Nogueras JJ, Li VK. Factors associated with failure of the artificial bowel sphincter: a study over 50 cases from Cleveland Clinic Florida. Dis Colon Rectum. 2009;52(9):1550–7.
52. Chatoor D, Emmanuel A, Cohen R, Abdel-Halim M, Taylor SA, Baumgartner U. The anal band for severe faecal incontinence. Clin Radiol. 2009;64(4):386–94.
53. Goos M, Baumgartner U, Löhnert M, Thomusch O, Ruf G. Experience with a new prosthetic anal sphincter in three coloproctological centres. BMC Surg. 2013;13:45.
54. Hajivassiliou CA, Carter KB, Finlay IG. Assessment of a novel implantable artificial anal sphincter. Dis Colon Rectum. 1997;40(6):711–7.
55. Schrag H, Padilla F, Goldschmidtböing F, Doll A, Woias P, Hopt U. German artificial Sphincter system: first report of a novel and highly integrated sphincter prosthesis for therapy of major fecal incontinence. Dis Colon Rectum. 2004;47(12):2215–7.
56. Finlay IG, Richardson W, Hajivassiliou CA. Outcome after implantation of a novel prosthetic anal sphincter in humans. Br J Surg. 2004;91(11):1485–92.
57. Zubaidi AM. Artificial bowel sphincters in severe fecal incontinence: are they a solution? Saudi Med J. 2010;31(9):965–73.
58. Thiersch K. Quoted in Carrasco AB Contribution a l'etude du Prolapsus du Rectum. Paris: Masson; 1943.
59. Devesa JM, Hervas P, Vicente R, Rey A, Die J, Moreno I, et al. Anal encirclement with a simple rectal sling for faecal incontinence. Tech Coloproctol. 2011;15(1):17–22.
60. Zutshi M, Hull T, Gurland B. Anal encirclement with sphincter repair (AESR procedure) using a biological graft for anal sphincter damage involving the entire circumference. Color Dis. 2012;14(5):592–5.
61. Parks AG. Anorectal incontinence. Proc R Soc Med. 1975;68(4):681–90.
62. Nesselrod JP. Clinical proctology. 3rd ed. Philadelphia: WB: Saunders; 1950. p. 76–9.
63. Deen KI, Oya M, Ortiz J, Keighley MR. Randomized trial comparing three forms of pelvic floor repair for neuropathic faecal incontinence. Br J Surg. 1993;80(6):794–8.
64. van Tets WF, Kuijpers JH. Pelvic floor procedures produce no consistent changes in anatomy or physiology. Dis Colon Rectum. 1998;41(3):365–9.
65. Yamana T, Takahashi T, Iwadare J. Perineal puborectalis sling operation for fecal incontinence: preliminary report. Dis Colon Rectum. 2004;47(11):1982–9.
66. Mellgren A, Zutshi M, Lucente VR, Culligan P, Fenner DE, et al. A posterior anal sling for fecal incontinence: results of a 152-patient prospective multicenter study. Am J Obstet Gynecol. 2016;214(3):349.e1–8. Available from: https://www.ajog.org/article/S0002-9378(15)01276-4/abstract.

67. Abendstein B, Petros P, Richardson P. Ligamentous repair using the tissue fixation system confirms a causal link between damaged suspensory ligaments and urinary and fecal incontinence. Pelviperineology. 2008;27:114–7.
68. Bullard KM, Trudel JL, Baxter NN, Rothenberger DA. Primary perineal wound closure after preoperative radiotherapy and abdominoperineal resection has a high incidence of wound failure. Dis Colon Rectum. 2005;48(3):438–43.
69. Chittenden AS. Reconstruction of anal sphincter by muscle slips from glutei. Ann Surg. 1930;92(1):152–4.
70. Penninckx F, D'Hoore A, Vanden BA. Perineal colostomy with antegrade continence enemas as an alternative after abdominoperineal resection for low rectal cancer. Ann Chir. 2005;130(5):327–30.
71. Mander BJ, Abercrombie JF, George BD, Williams NS. The electrically stimulated gracilis neosphincter incorporated as part of total anorectal reconstruction after abdominoperineal excision of the rectum. Ann Surg. 1996;224(6):702–9.
72. Rouanet P, Senesse P, Bouamrirene D, Toureille E, Veyrac M, Astre, et al. Anal sphincter reconstruction by dynamic graciloplasty after abdominoperineal resection for cancer. Dis Colon Rectum. 1999;42(4):451–6.
73. Geerdes BP, Zoetmulder FA, Heineman E, Vos EJ, Rongen MJ, Baeten CG. Total anorectal reconstruction with a double dynamic graciloplasty after abdominoperineal reconstruction for low rectal cancer. Dis Colon Rectum. 1997;40(6):698–705.
74. Rullier E, Zerbib F, Laurent C, Caudry M, Saric J. Morbidity and functional outcome after double dynamic graciloplasty for anorectal reconstruction. Br J Surg. 2000;87(7):909–13.
75. Cavina E, Seccia M, Chiarugi M. Total anorectal reconstruction supported by electrostimulated gracilis neosphincter. Recent Results Cancer Res. 1998;146:104–13.
76. Romano G, La Torre F, Cutini G, Bianco F, Esposito P, Montori A. Total anorectal reconstruction with the artificial bowel sphincter: report of eight cases. A quality of life assessment. Dis Colon Rectum. 2003;46(6):730–4.

Constipation

18

James Elvis Waha and Johann Pfeifer

Introduction

Chronic constipation can be a symptom or a disease per se. Unlike common belief, it is not a disease of modern civilization. The first available description of constipation and its therapy is found in the "Ebers Papyrus" and dates to 1600 BC. As life expectancy increases, we can witness a rising prevalence of constipation in the population, especially among the elderly [1] and with considerable costs [2]. Epidemiological data describe differences according to region, gender (female:male = 2:1), age, and definition (Rome Criteria vs. self-reported). The reported prevalence of patients suffering from constipation ranges between 2% and 27% in North America and Europe, respectively, and the pooled prevalence averages approximately 15% [3, 4]. Generally, constipation is a result of pelvic floor dysfunction (evacuation disorder), intestinal motility disorders (slow transit disorder), or a combination of both. Unfortunately, the pathophysiology of either one is not fully understood. The high number of discontent patients and their doctors makes constipation an ongoing challenge for modern medicine [5, 6].

J. E. Waha · J. Pfeifer (✉)
Department of General Surgery, University Surgical Clinic, Medical University of Graz, Graz, Austria
e-mail: johann.pfeifer@medunigraz.at

Definition

Defining chronic constipation can be difficult as physicians and patients often have different opinions on this condition. One approach could be to allow patients to decide on the level of their satisfaction with the frequency of defecation or to what extent their quality of life is affected, which is often influenced by psychological factors [7]. Many attempts to find a widely recognized definition, which is clinically applicable as well as useful for clinical trials dealing with chronic constipation, have been made in the past. The most internationally accepted effort was undertaken by the Rome Foundation. According to the Rome IV criteria, a patient is suffering from chronic constipation or functional constipation (FC), when he is complaining about unsatisfactory stools within the last 6 months over a period of 3 months in combination with at least two of the following symptoms:

- Straining in at least 25% of defecations
- Passing of lumpy or hard stools in at least 25% of defecations
- Sensation of incomplete evacuation in more than 25% of defecations
- Manual maneuvers to facilitate evacuation in more than 25% of defecations
- Less than three bowel movements per week

If accompanying symptoms of irritable bowel syndrome such as recurrent abdominal pain

L. C. C. Oliveira (ed.), *Anorectal Physiology*, https://doi.org/10.1007/978-3-030-43811-1_18

within the last 3 months are diagnosed in association with two or more of the symptoms shown in Table 18.1, then the patient should be classified into the group of irritable bowel syndrome with predominant constipation (IBS-C) [7–9]. Additionally, since the two groups, FC and IBS-C, often overlap, the Rome IV criteria describe them as a continuum rather than as two separate entities [8, 9].

Table 18.1 Rome IV criteria for IBS-C

1. Recurrent abdominal pain related to defecation
2. Recurrent abdominal pain and variation in stool frequency
3. Recurrent abdominal pain and variation in stool form

Classifications of the different causes of constipation are key to finding the right treatment for each group (Fig. 18.1). In this chapter, we focus

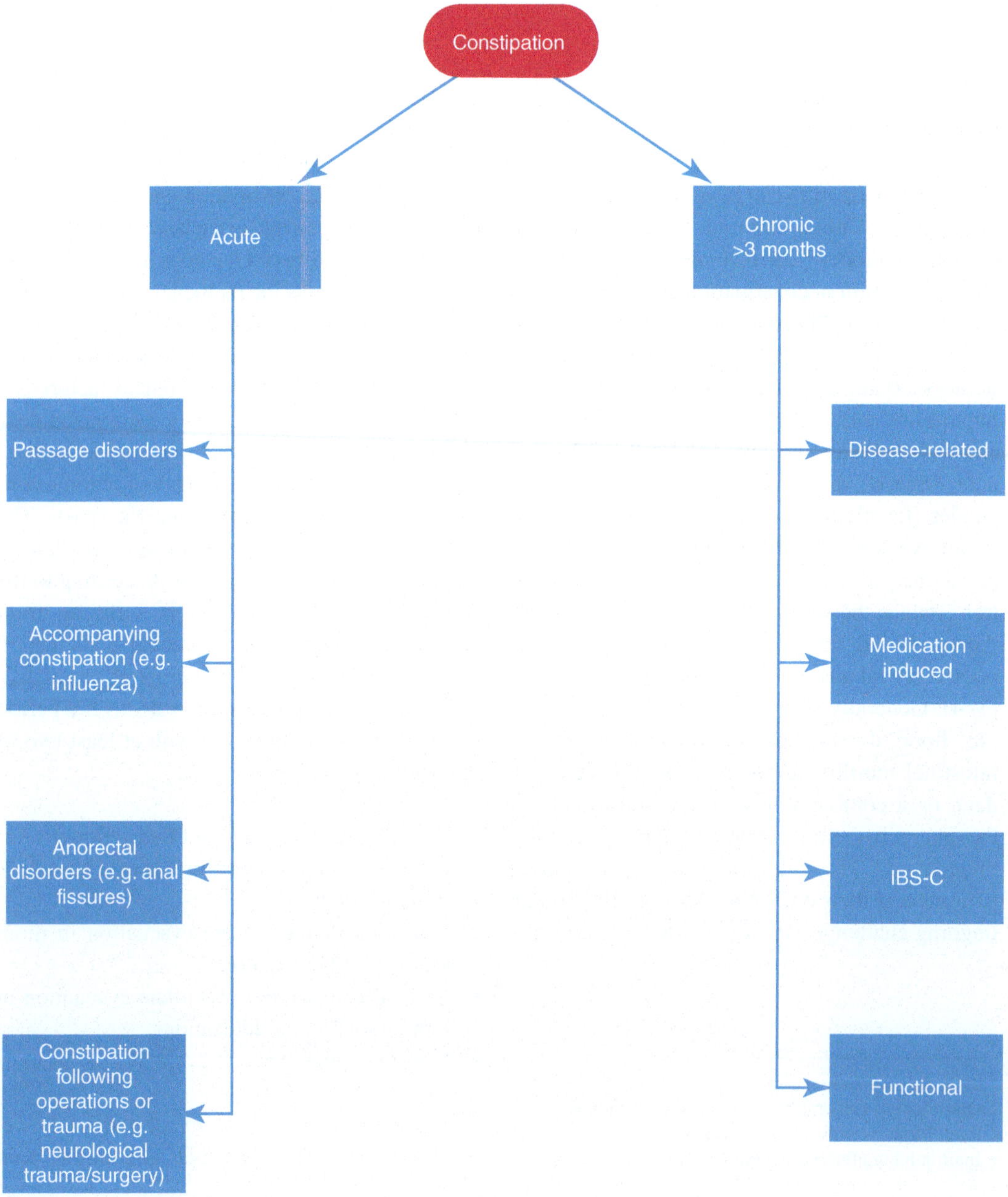

Fig. 18.1 Classification of constipation. (© Gernot Benko. Reused with permission)

on FC as these patients are potentially treatable by surgical interventions, if all conservative management has failed. Patients who suffer from extracolonic causes never benefit from surgery (Table 18.2).

Table 18.2 Classification of evacuation disorders (modified according to A. Herold)

A. Functional evacuation disorders (a) Pelvic floor: Faulty coordination Anism Spasticity Psychogenic causes (b) Colon and rectum: Idiopathic inertia recti Impaired autonomic innervation
B. Structural evacuation disorders (a) Pelvic floor: Myopathy of the internal sphincter muscle Hypertrophy of the internal sphincter muscle Dorsal sphincter dysplasia Anorectal stenosis Anal fissure (b) Colon and rectum: Postoperative inertia recti Dysgangliosis Deficient internal sphincter muscle Hirschsprung's disease Enterocoele Rectocoele Rectal prolapse Reduced rectal compliance Obstructive tumur
C. Combination of A and B

Etiology and Pathophysiology

Functional constipation can either be caused by a functional disorder within the colon (slow transit constipation) or a functional or structural disorder within the pelvis, (evacuation disorder) (Fig. 18.2 and Table 18.2).

Due to methodologically poor studies concerning the pathophysiology of chronic constipation, it is difficult to make a sophisticated statement of high evidence, not least because of the diversely used definitions and terms. The International Anorectal Physiology Working Party Group [IAPWG] is attempting to set standards by establishing a standard terminology and classification. Thus, reading through the literature can be quite confusing regarding the terminology.

Generally, constipation has multiple causes. The intestinal motility can be affected by nutrition and diet and water uptake. Also, many agents and drugs affect the intestinal passage, some through neurotransmitters. It is known that several endocrine disorders and neurological conditions influence bowel activity (Table 18.3). The colon follows a circadian rhythm with periodical alteration in the transport of the feces towards the rectum. Colonic motility reacts to movement, such as increased motility after walking. Furthermore, ageing

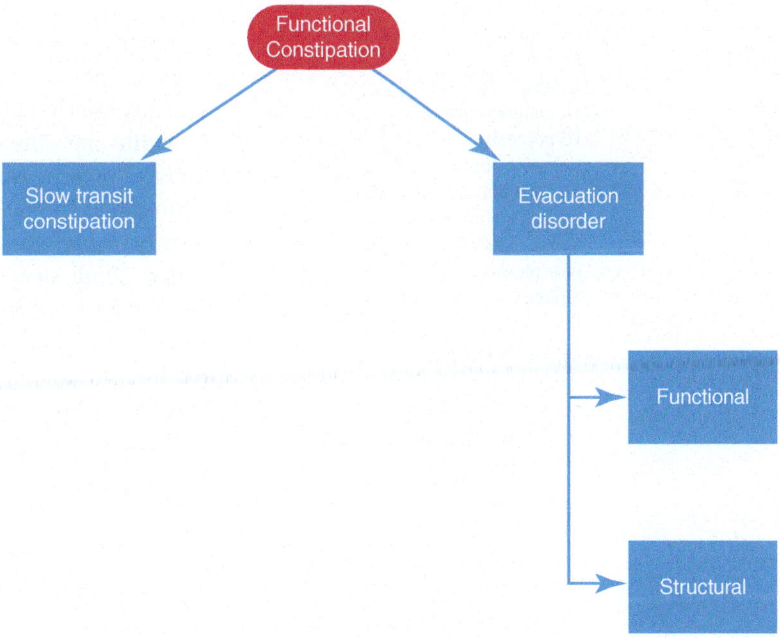

Fig. 18.2 Subtypes of functional constipation. (© Gernot Benko. Reused with permission)

Table 18.3 Extra-colonic causes of constipation

Endocrine and metabolic:	Diabetes mellitus Glycagonoma Hypercalcemia Hyperparathyroidism Hypokalaemia Hypopituitarism Hypothyroidism Milk–alkali syndrome Pheochromocytoma Porphyria Pregnancy Uremia
Neurologic/ Cerebral	Parkinson's disease Stroke Tumurs
Spinal	Cauda equina tumour Ischemia Iatrogenic Meningocoele Multiple sclerosis Paraplegia Shy–Drager syndrome Tabes dorsalis Trauma
Peripheral	Autonomic neuropathy Chagas disease Multiple endocrine neoplasia, Type 2B Von Recklinghausen's disease
Drugs	Anesthetic Analgetic Antacids (calcium and aluminium compounds) Anticholinergic Anticonvulsant Antidepressant Anti-Parkinsonian Barium sulphate Calcium channel blocker Diuretics Ganglion blockers Hematinics (iron) Hypotensives Laxative abuse Monoamine oxidase (MAO) inhibitor Metals (arsenic, lead, mercury, phosphorus) Opiates Paralytic agents Psychotherapeutics
Myopathy	Amyloidosis Dermatomyositis Myotonic dystrophy Scleroderma

affects intestinal motility, thus the elderly are at great risk for developing chronic constipation. It is hypothesized that oxidative stress and epigenetic alterations cause manipulations in biochemical pathways and lead to loss of myenteric cells [10]. Neuronal loss generally leads to disturbances in gut function. The underlying mechanisms go far beyond this chapter and are subject to current research. This chapter concentrates on the surgical options of treating FC. Therefore, we discuss only the pathophysiology of these types of constipation that are treatable by surgery.

Primary causes for constipation are very seldom (e.g. sporadic and familial genetic defects, which lead to enteric neuro- and/or myopathy, Hirschsprung's disease). More common are disorders that secondarily disturb intestinal motility (e.g. diabetes, Parkinson's disease, paraneoplastic syndrome) (Table 18.3). Some of these conditions may lead to degeneration of interstitial cells of Cajal (ICC), the intestinal pacemaker cells, which may result in slow transit constipation (see below).

Slow Transit Constipation

Almost half of the patients suffering from constipation who are refractory to a fiber diet have delayed colonic transit time [11]. These patients have fewer high-amplitude propagated contractions caused by an impaired motor activity in the colon due to either abnormalities of myenteric cells or reduced volume of ICC within the colon [12, 13]. Current basic research is investigating pharmacological pathways to influence the activity of ICC and, consequently, gut motility [14]. Evacuation disorders, both structural and functional, can subsequently lead to slow transit constipation (STC). In the case of functional evacuation disorder, the subsequent secondary STC might be successfully cured by behavioural or biofeedback therapy [15].

Functional Evacuation Disorder

Functional evacuation disorder is characterized by paradoxical contraction or inadequate relaxation of the pelvic floor muscles and/or inadequate propulsive forces during attempted defaecation. This group is estimated to be 40% of all constipated patients [16]. Patient complaints are primarily a feeling of incomplete evacuation and excessive straining. Physiological testing in these patients show pathological patterns in inappropriate contraction of the pelvic floor muscles, or incomplete relaxation of the anal sphincter, or a combination of both [17].

Structural Evacuation Disorder

The most common structural disorders in patients suffering from constipation are rectal prolapse, rectocoele, and descending perineum syndrome. Women are disproportionally affected by these conditions and a multidisciplinary approach should be taken in the treatment strategies. Dedicated multidisciplinary pelvic floor teams can ensure high quality therapy by providing a platform in which to exchange ideas and opinions among the different specialities and should include gynecology, urology, neurology, physiotherapy, and nutritional sciences.

Descending Perineum Syndrome

Due to chronic straining over time, and a variety of lesions of ligament, fascial elements and muscles, patients develop a weakness of the pelvic floor leading to bulging and descent of the perineum during defecation.

Solitary Rectal Ulcer Syndrome

Solitary rectal ulcer syndrome occurs when repeatedly forceful straining (due to either functional or structural evacuation disorder) causes irritation to the rectal mucosa (and histologically can appear similar to rectal cancer), often found in patients suffering from rectal intussusception. Patients present with rectal bleeding, mucus discharge, and/or a feeling of incomplete evacuation. This subject will be discussed in Chap. 24.

Rectal Prolapse

Constipated patients often use excessive straining to evacuate. Over time, the fixation of the mesorectum at the sacrum becomes worn out, pelvic floor muscles weaken, and parts of the rectum prolapse through a widened pelvic hiatus. Additionally, laxity, stretching, paralysis, or rupture of the muscular components and ligaments of the pelvic floor by surgery or obstetric interventions can increase the risk for a weakened pelvis. Prior to developing rectal prolapse, intussusception is often present. The mucosal prolapse is the beginning of a vicious cycle. Without treatment, evacuation can become increasing more difficult and straining increases until a full protrusion of all layers traumatizes the rectum. Neuromuscular defects, shear stress, and rectal ischemia are often the result and aggravate the inability to evacuate [18].

Rectocele

A rectocoele is a herniation of the rectal wall. In most cases, through the weakened anterior rectovaginal septum and, less commonly, posterior through the divergent muscles of the pelvic floor. Risk factors for developing rectocele are age, obesity, obstetric injury and/or multiple vaginal deliveries. Functional evacuation disorder, or pelvic floor dyssynergia, is often left untreated and results in a rectocele. Rectocele is classified by its position (low, middle or high) or by size (small <2 cm, medium 2–4 cm, large >4 cm). Small rectoceles often present without symptoms, but with increased size manifest as symptoms of constipation. If a rectocele is large enough to show contrast medium pooling during defecography (considered a significant rectocele), a surgical intervention should be considered.

Diagnosis

History

A detailed patient history is of paramount importance to rule out the presence of any warning signs and (red flags) (Table 18.4). If none of these symptoms are present, conservative and clinical treatment options can be safely initiated over a period of 4–8 weeks. Tailored treatment requires a detailed history consisting of the current medication profile and specific bowel activities. Extracolonic causes must also be identified (Table 18.2) and excluded. A detailed description of stool frequency and consistency, duration and completeness of each bowel movement is key to evaluate whether the constipation is due to slow colonic transit or if the patient is suffering from an evacuation disorder. Pressure and flatulence in the upper abdomen suggest slow transit constipation, while a sensation of incomplete evacuation after a bowel movement points more in the direction of an evacuation disorder. However, it should be taken into consideration that symptoms are not unique in this regard and subtypes cannot be distinguished with a medical history alone [19]. In addition, patients can present both situations, translating the complexity we face when treating constipation. Further investigation should consider the duration of the constipation, affected family members, onset (as early as childhood), comorbidities, and previous surgery.

Table 18.4 "Red flags"

Red flags:
Bleeding
Unexplained weight loss of more than 10%
Anemia
Malnutrition
Paradoxical diarrhea
Age over 50 years
Gastrointestinal tumour history (family and personal history)
Palpable resistance
Progressive course
Brief history of disease onset

Constipation Scores

Several constipation scores have been designed to estimate the degree of symptoms and also serve as a means of measuring the success of an ongoing treatment. The most widely utilized, although it was never fully validated, is the Cleveland Clinic Constipation Score [20], mainly because of its simplicity (Table 18.5). The Constipation Severity Instrument (CSI) [21] is a tool consisting of 78 items which aims at identifying and quantifying different subtypes of constipation (Table 18.6). Another scoring system worth mentioning is the obstructed defaecation syndrome score (ODS) [22], which has been prospectively validated (Table 18.7). To assess the quality of life of constipated patients, the Constipation-Related Quality of Life (CRQOL)

Table 18.5 Cleveland Clinic Constipation Score [20]

	0	1	2	3	4	Score
Frequency of bowel movements	1–2 times per 1–2 days	2 times per week	Once per week	Less than once per week	Less than once per month	
Difficulty: painful evacuation effort	Never	Rarely	Sometimes	Usually	Always	
Completeness: feeling incomplete evacuation	Never	Rarely	Sometimes	Usually	Always	
Pain: abdominal pain	Never	Rarely	Sometimes	Usually	Always	
Time: minutes in lavatory per attempt	Less than 5	5–10	10–20	20–30	More than 30	
Assistance: type of assistance	Without assistance	Stimulative laxatives	Digital assistance or enema			
Failure: unsuccessful attempts for evacuation per 24 hours	Never	1–3	3–6	6–9	More than 9	
History: duration of constipation (yr)	0	1–5	5–10	10–20	More than 20	
Total score:						

Table 18.6 Constipation Severity Instrument [21]

Obstructive Defaecation Subscale:					
CSI 1. Incomplete Bowel movements					
A) How often do you experience incomplete bowel movements?					
(0) Never *experience this* (Skip to #2)	(1) Occasionally *experience this*	(2) Sometimes *experience this*	(3) Usually *experience this*	(4) Always *experience this*	
B) How severe are these symptoms for you?					
(1) Not at all severe *(most of my bowel movement comes out)*	(2) Mild	(3) Sometimes severe *(There is still a lot of stool in me after I have a bowel movement)*	(4) Severe	(5) Extremely severe *(I feel constant pressure in my rectum from the stool or keep going back to the bathroom)*	
C) How much does this bother you?					
(1) Not at all *bothersome*	(2) A little *bothersome*	(3) Somewhat *bothersome*	(4) Very *bothersome*	(5) Extremely *bothersome*	
CSI2. Straining/ Difficulty in Having a Bowel Movement					
A) How often do you experience this?					
(0) Never *experience this* (Skip to #3)	(1) Occasionally *experience this*	(2) Sometimes *experience this*	(3) Usually *experience this*	(4) Always *experience this*	
B) How severe is this for you?					
(1) Not at all severe *(I push a little)*	(2) Mild	(3) Sometimes severe *(I bear down hard)*	(4) Severe	(5) Extremely severe *(I push on my belly, grunt and bear down very hard)*	
C) How much does this bother you?					
(1) Not at all *bothersome*	(2) A little *bothersome*	(3) Somewhat *bothersome*	(4) Very *bothersome*	(5) Extremely *bothersome*	
Colonic Inertia Subscale:					
CSI 3. Think about when you are having difficulty with your bowel habits:					
During a typical month, how many times do you usually have a bowel movement?					
(0) N/A—I never have difficulty with my bowel habits	(1) Daily	(2) A few times per week	(3) Once per week	(4) Once every 2 weeks	(5) Once a month
CSI 4. Infrequent Bowel Movement (Less than 1 bowel movement every 3 days)					
A) How often do you experience infrequent bowel movements?					
(0) Never *experience this* (Skip to #5)	(1) Occasionally *experience this*	(2) Sometimes *experience this*	(3) Usually *experience this*	(4) Always *experience this*	
B) How severe is this symptom for you?					
(1) Not at all severe *(I go almost every day)*	(2) Mild	(3) Somewhat severe *(I go 1–2 times per week)*	(4) Severe	(5) Extremely severe *(I can go up to 4 weeks without going)*	
C) How much does this symptom bother you?					
(1) Not at all *bothersome*	(2) A little *bothersome*	(3) Somewhat *bothersome*	(4) Very *bothersome*	(5) Extremely *bothersome*	
CSI 5. Lack of Urge to Have a Bowel Movement					
A) When you lack the urge to have a bowel movement, how severe is this for you?					
(0) Never experience this	(1) Not at all severe *(I have a pretty good sense when I have to go)*	(2) Mild	(3) Somewhat severe *(I only have a vague sense that I might have to go)*	(4) Severe	(5) Extremely severe *(I don't have any sensation in the pelvic area)*

(continued)

Table 18.6 (continued)

B) When you lack the urge to have a bowel movement, how much does this bother you?					
(0) Never experience this	(1) Not at all severe *bothersome*	(2) A little *bothersome*	(3) Somewhat *bothersome*	(4) Very *bothersome*	(5) Extremely *bothersome*
Pain subscale—Rectal/Anal Pain due to Your Bowel Problems:					
CSI 6. **During the last month, on average, how severe was the pain in your rectum/anus?**					
(0) I haven't experienced this	(1) Mild	(2) Somewhat severe	(3) Severe	(4) Extremely severe	
CSI 7. **Rate the level of your rectal/anal pain at the present moment**					
(0) No pain	(1) Mild	(2) Somewhat severe	(3) Severe	(4) Extremely severe	
CSI 8. **How much suffering do you experience because of rectal/anal pain?**					
(0) None	(1) Mild suffering	(2) Somewhat severe suffering	(3) Severe suffering	(4) Extremely severe	
CSI 9. **During the past month, due to your bowel habits, how often have you had bleeding during/after a bowel movement?**					
(0) Never	(1) Rarely	(2) Occasionally	(3) Usually	(4) Always	

Table 18.7 Obstructed defecation syndrome score questionnaire [22]

	Score				
Variables	0	1	2	3	4
Mean time spent at the toilet	≤5 min	6–10 min	11–20 min	21–30 min	>30 min
N attempts to defaecate per day	One	Two	Three-four	Five-six	>six
Anal/vaginal digitation	Never	>1/month, <1/week	Once a week	Two to three per week	Every defaecation
Use of laxatives	Never	>1/month, <1/week	Once a week	Two to three per week	Every day
Use of enemas	Never	>1/month, <1/week	Once a week	Two to three per week	Every day
Incomplete/fragmented defaecation	Never	>1/month, <1/week	Once a week	Two to three per week	Every defaecation
Straining at defaecation	Never	<25% of the time	<50% of the time	<75% of the time	Every defaecation
Stool consistency	Soft	Hard	Hard and few	Faecaloma formation	

(Table 18.8) is a statistically validated questionnaire [23]. There are several more scoring systems available. The ones mentioned here are recommended for clinical trials by the Italian Association of Hospital Gastroenterologists and the Italian Society of Colo-Rectal Surgery [19].

Physical Examination

Physical examination is mandatory in the assessment of constipation, including the shape of the abdomen, scars, bowel sounds, and localization of pain [24]. Proctological examination is performed in left lateral position and includes inspection of the anus and the perianal area. During rectal digitation, the patient is asked to squeeze, push down as during defecation, and to relax. With this simple test, functional evacuation disorders caused by dyssynergia of the pelvic floor muscles can easily be recognized.

Laboratory Examination

Blood testing in functional constipation is not mandatory but can be useful. Electrolytes, differential blood count, creatinine clearance, glomerular

Table 18.8 Constipation-Related Quality of Life [23]

I. Distress Subscale					
Over the past 12 months, how often have you experienced the following feelings regarding the symptoms of your bowel problems?					
	Never (1)	Rarely (2)	Occasionally (3)	Usually (4)	Always (5)
1. Discouraged that I am not getting better					
2. Helpless with my ability to solve my bowel problems					
3. Frustrated that the treatments I have tried do not work					
4. Worried that this problem will not go away					
5. Depressed that my bowel problems are controlling my life					
6. Nervous that this means something more serious is happening to my body					
II. Social Impairment Subscale					
Over the past 12 months, have your bowel problems had a negative impact on:					
	Not at all (1)	Slightly (2)	Moderately (3)	Quite a bit (4)	Extremely (5)
1. Your relationship with friends?					
2. Your relationship with coworkers?					
3. Your relationship with acquaintances?					
4. Your encounters with people you do not know?					
5. Your ability to make new friends?					
III. Eating subscale					
Over the past 12 months, because of your bowel problems how often have you:					
	Never (1)	Rarely (2)	Occasionally (3)	Usually (4)	Always (5)
1. Avoided foods you like?					
2. Restricted the amount of food you eat?					
3. Restricted the kinds of food you eat?					
IV. Bathroom Subscale:					
Over the past 12 months, to what extent have you been:					
	Not at all (1)	Slightly (2)	Moderately (3)	Quite a bit (4)	Extremely (5)
1. Embarrassed about having to go to the bathroom when you are away from home					
2. Anxious about being far from a bathroom					
3. Anxious about having to use a public restroom					
4. Embarrassed about the amount of time you spend in the bathroom					

filtration rate, as well as thyroxine levels help to exclude secondary conditions related to constipation.

Endoscopy

Proctoscopy without bowel preparation should complete every physical examination. If there is suspicion of a colon transit disorder, a colonoscopy should be performed after bowel preparation and under sedation.

Physiological Testing

If empirical conservative treatment does not provide the patient symptomatic relief, further physiological work up should be considered (Table 18.9). In the last decade, significant progress has been made in the understanding of the pathophysiology and development of chronic constipation. Physiological testing provides essential information on the given pathology of each case, narrowing down the treatment options. Under optimal conditions, several tests are available in order to make a the correct diagnosis.

Interpretation of Results

If structural causes for constipation are ruled out, a colonic transit study should be performed. If the colonic transit time is normal, pelvic floor disorder (functional or structural) should be considered and further investigated. After completing physiological testing, a diagnosis can be made from one of the following categories:

1. Normal colonic transit + normal evacuation: IBS-C
2. Slow colonic transit + normal evacuation: STC
3. Normal colonic transit + paradoxical contraction/dyssynergia: functional evacuation disorder
4. Normal colonic transit + rectocoele/descending perineum/rectal prolapse: structural evacuation disorder
5. Combination of STC and functional evacuation disorder
6. Combination of STC and structural evacuation disorder
7. Secondary constipation (metabolic, adverse drug effects)

The interpretation of the results is shown in Fig. 18.3.

Treatment

Before initiating therapy, the following red flag symptoms must be excluded: bleeding, weight loss, anemia, malnutrition, paradoxical diarrhea, age >50 years, gastrointestinal tumours, palpable resistance, fast progression, and a short clinical history. After these red flags are excluded, a non-surgical treatment should be initiated. Treatment should consist of a step-by-step approach, and surgery is never the first line therapy in this context. Depending on the cause of constipation, there are different strategies. In the case of STC, the approach is very different to that of an outlet

Table 18.9 Physiological testing

Test	Purpose	Message
Anorectal manometry	Measurement of pressures in the anal canal Assessment of the rectoanal inhibitory reflex (RAIR)	Resting pressure = function of the internal anal sphincter Squeeze pressure = function of the external anal sphincter Loss of the reflex typically in Hirschsprung's disease
Endoanal ultrasonography	Fecal incontinence, Constipation	Judgment of the integrity of sphincter muscles (defect?) Hypertrophy of the internal anal sphincter?
Colonic transit time	Kind of constipation	Diffuse spread of radio-opaque markers typically for STC Collection of markers in the small pelvis as sign of ODS
(MRI)—Defecography	Functional assessment of the pelvic floor and the internal organs and their mobility	Structural substrate (e.g. rectocoele) or only functional disorders (e.g. anismus)
EMG of the pelvic floor	Assessment of the motor unit potentials (MUP) interference pattern	Loss or alteration or signs of denervation or reinnervation; malfunction of muscle groups (e.g. anismus)
Pudendal Nerve Terminal Motor Latency (PNTML)	Function of the nerve supplying the pelvic floor	Useful for prognosis, if surgery is planned

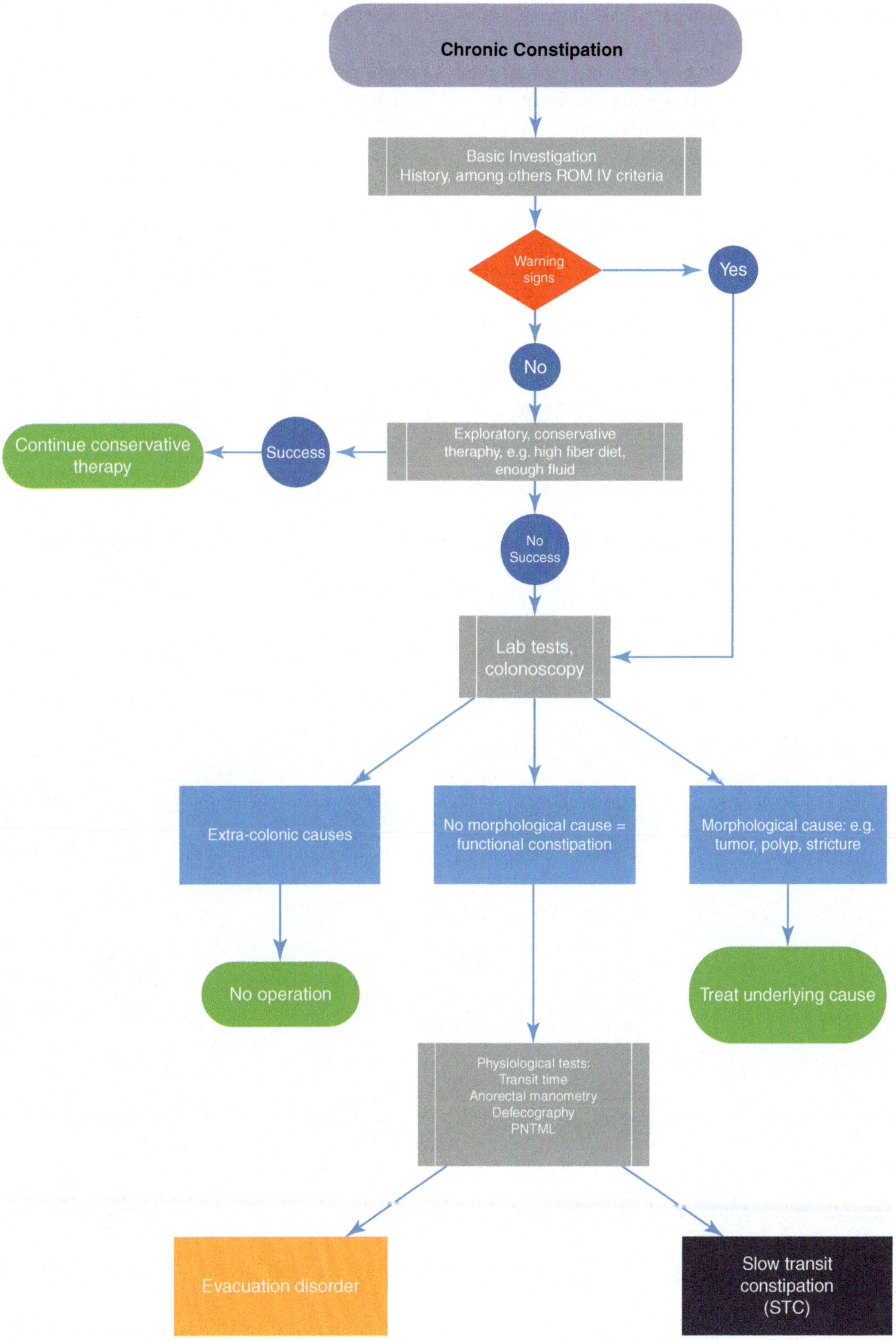

Fig. 18.3 Basic diagnostic algorithm. (© Gernot Benko. Reused with permission)

obstruction (Figs. 18.4 and 18.5). In general, surgery is only suitable for a very small group of patients. Less invasive approaches should first be considered before performing a colectomy or creating an ostomy. The correct approach is to start with conservative therapy for as long as possible.

Conservative Treatment

Behaviour and Diet

Initially, patients should be asked about their dietary and behaviural habits. Any drug which potentially influences stool consistency or has an

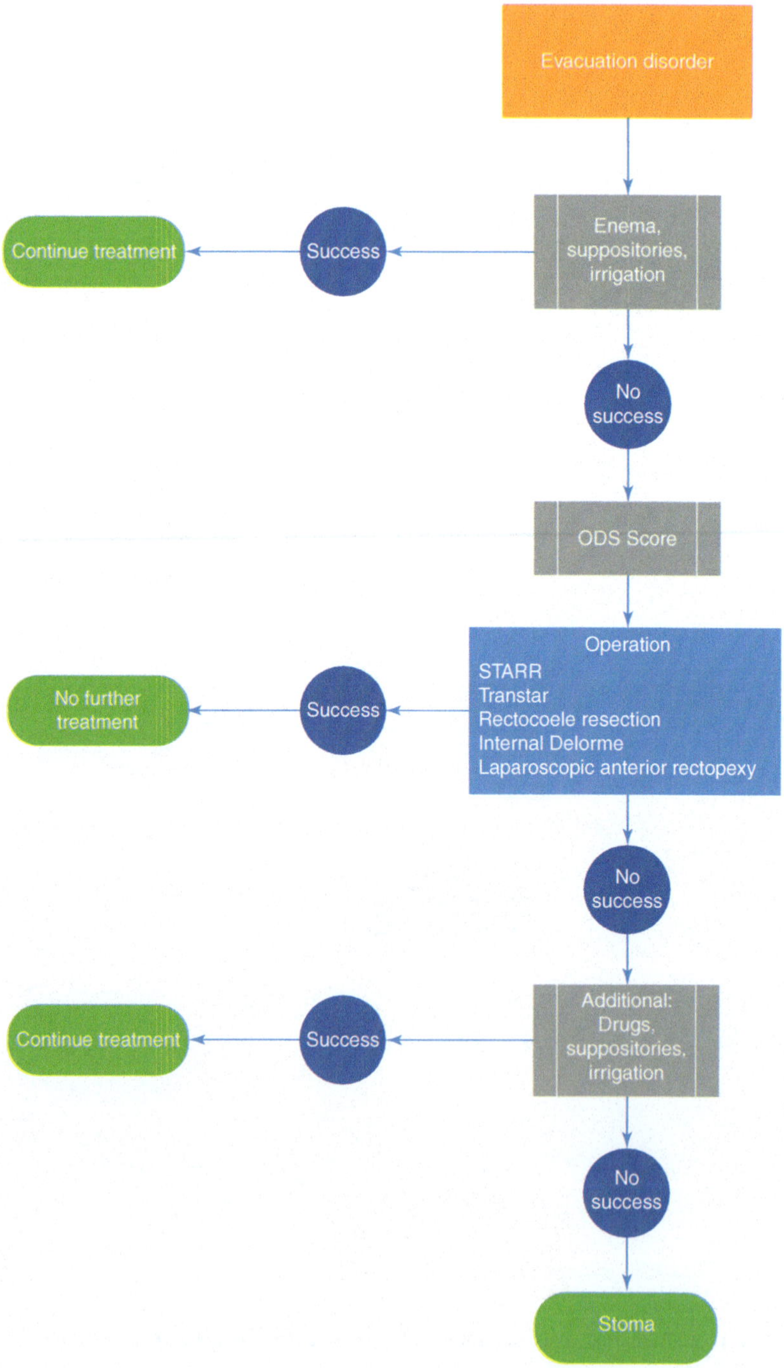

Fig. 18.4 Treatment algorithm of evacuation disorders. (© Gernot Benko. Reused with permission)

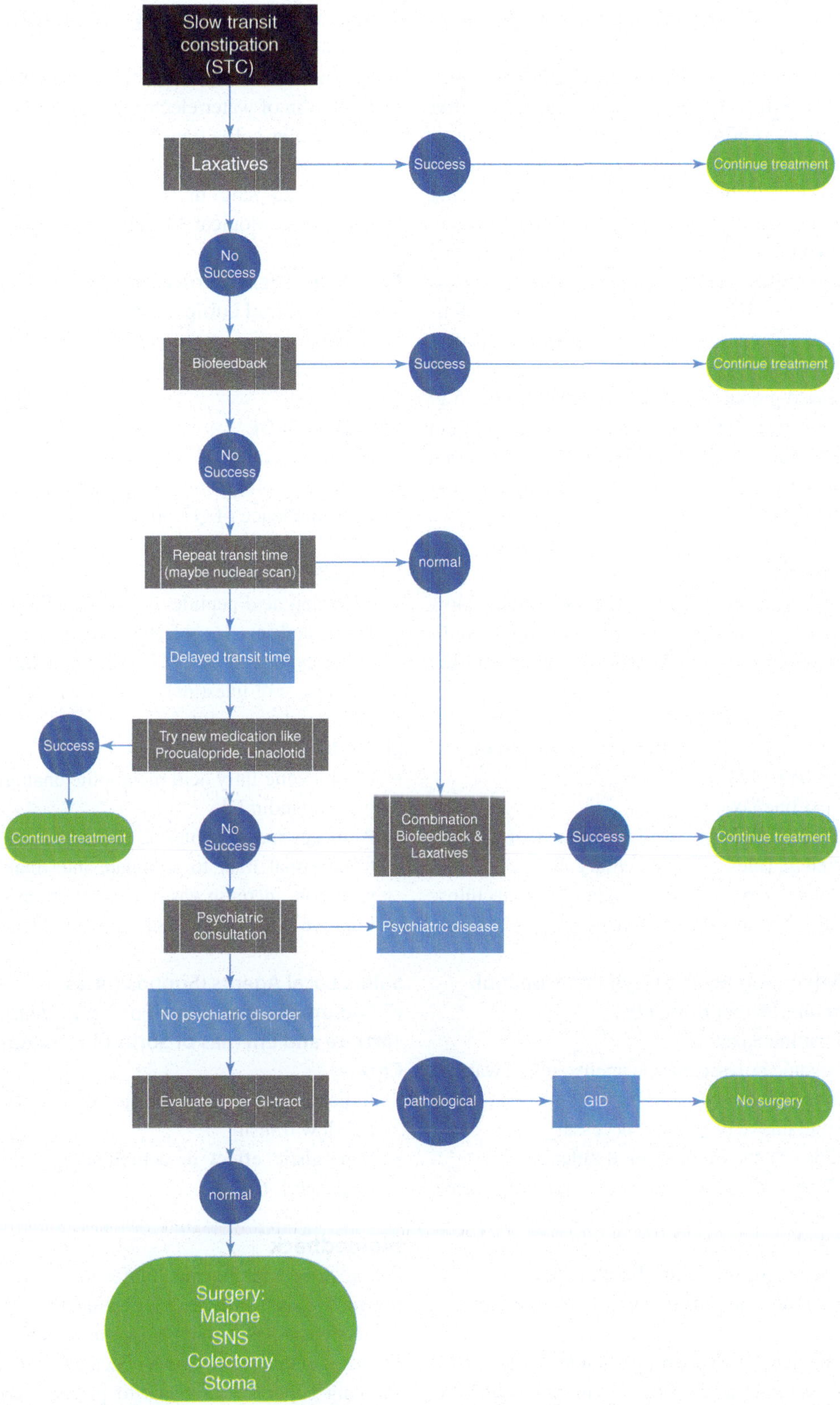

Fig. 18.5 Treatment algorithm of slow transit constipation; GID = Gastrointestinal dysmotility. (© Gernot Benko. Reused with permission)

impact on colonic motility should be discontinued. Patients should be instructed to take advantage of the gastro-colic response in the mornings after breakfast. Colonic motility increases after food intake. Furthermore, patients should arrange for a peaceful atmosphere while sitting on the toilet and give themselves enough time. The seating position on the toilet should be adjusted so that the anorectal angulation is open to reduce mechanical resistance (squatting position or footstool to lift the feet). Dietary recommendations often have regional and country specific differences [7]. In general, fiber intake should be increased gradually (to reduce bloating) to 25 g/day over a period of two weeks. In addition, fluid consumption should be 1,5 L to 2 L/day and rich in magnesium. Regular physical activity is highly recommended [25–29].

Laxatives

If these measures of lifestyle modification show no improvement, the use of laxatives is indicated. There are many different agents on the market.

Osmotics (Macrogol, Polyethylene Glycol, Lactulose, etc.)

- First line laxative
- Good efficacy, well tolerated regarding dietetic rules, complementary use
- Polyethylene glycol is superior to lactulose (less abdominal pain, bloating)

Bulk-Forming Agents (Psyllium, Ispaghul, Sterculia, Wheat Bran, etc.)

- First line laxative
- Organic polysaccharide agents, retain water in intestinal lumen
- High intake of water is necessary
- Side effects: Meteorism, flatulence
- Contraindications: Intestinal stenosis, faecal impaction, inflammatory colitis

Lubricant Agents (Paraffin Oil, etc.)

- Second line laxative, if first line laxative fails
- Mechanically softening colonic stool
- Hypovitaminosis (Loss of vitamin A/D/E/K)

Stimulant Agents (Bisacodyl, Docusate Sodium, Sennosides, etc.)

- Second line laxative, if first line laxative fails
- Inhibition of water/electrolytes absorption
- Accelerates colon transit [30]
- Intermittent use recommended
- Considered safe in regular use (American Gastroenterological Association) [29]

Prokinetic Agents (Prucalopride, Linaclotide, and Lubiprostone)

- Relatively new agents with high level of scientific evidence

PRUCALOPRIDE

- Highly selective 5-HT_4 agonist
- Increases colonic motility and transit
- Low incidence of QT prolongation [31–33]

LINACLOTIDE

- 14-amino acid peptide of the guanylin peptide family and acts as a selective agonist at the guanylate cyclase–C (GC–C) receptor on the luminal surface of intestinal enterocytes [34]

LUBIPROSTONE

- Is a bicyclic fatty acid metabolite analogue of prostaglandin E1
- Activates specific chloride channels in the gastrointestinal tract to stimulate intestinal fluid secretion, increase gastrointestinal transit, and improve symptoms of constipation [35]

Saline Local Agents (Suppositories of Sodium Bicarbonate + Potassium Acid Tartrate and Enemas of Sorbitol + Sodium Citrate)

- Indicated for evacuation disorders as well as for slow transit
- Synergistic effect in combination with biofeedback [36]

Biofeedback

Pelvic floor training and biofeedback are safe, useful, and effective treatment options. Biofeedback has gained importance since randomized controlled trials became available, showing superior outcomes to placebo, dietary measures, or laxatives [37–41]. Patients suffering from dyssynergic defecation have especially shown to respond well and should be introduced to biofeedback treatment. Secondary STC resulting from dyssynergia is also an indication for biofeedback treatment, whereas patients suffering from isolated STC will likely not benefit [39].

Surgical Treatment

If all of the above treatment options fail, it is time to talk to the patient about surgical interventions. Evidence level for surgery is low. But there is some consensus. The most important principle of any surgical procedure in chronic constipation is, "Do not harm the patient".

In the case of STC, it is important to repeat colonic transit studies before finally taking the patient to the operating room. The surgical interventions for STC can be divided into resecting and non-resecting techniques (Fig. 18.6).

The surgical approach for structural evacuation disorders focusses on restoring the anatomy as much as possible and consists of the STARR (Stapled Trans-anal Rectal Resection), TRANSTAR, Delorme operation, rectocele resection, and laparoscopic ventral rectopexy (Fig. 18.7).

Sacral Nerve Stimulation/Sacral Nerve Modulation (SNS/SNM)

The theory behind SNS is the increase of stool frequency, decrease of transit time, and amelioration of rectal sensitivity by continuously sending low amplitude electrical stimulus to the root of the sacral nerves (preferably S3). Furthermore, this stimulation modulates afferent nerves. SNS is a two-step procedure. In the first phase, the electrodes of an external mobile test stimulator are placed. If the patient is responding to the stimulus after 2–3 weeks, the pulse generator is then implanted in a second step. The procedure can be done under local anesthesia. In a recent review, the success rates were reportedly 57–87% in patients with permanent implanted pulse generators [42]. The removal rate due to complications ranges between 8% and 23%.

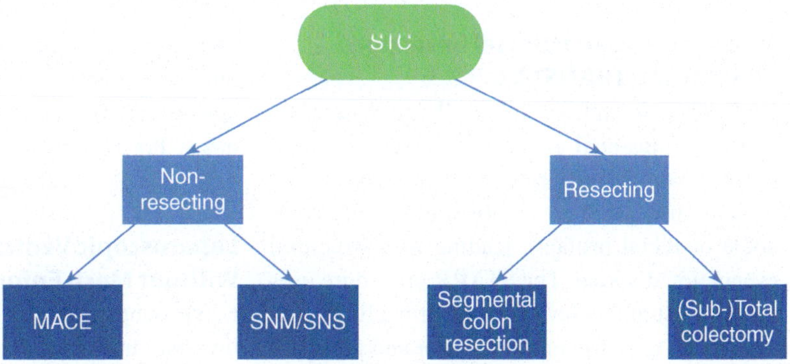

Fig. 18.6 Surgical treatment for STC. (© Gernot Benko. Reused with permission)

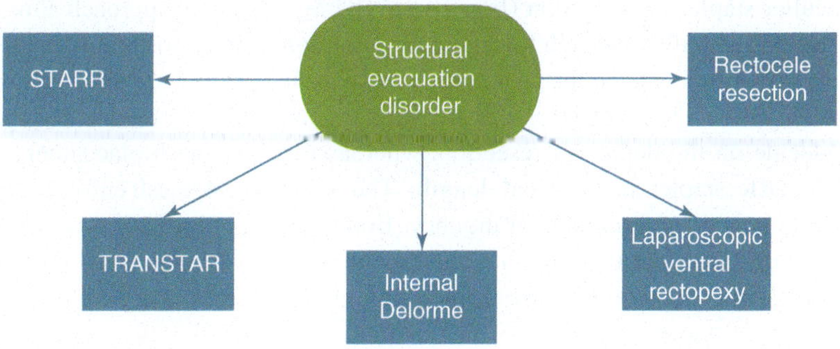

Fig. 18.7 Surgical treatment for structural evacuation disorders. (© Gernot Benko. Reused with permission)

Malone Antegrade Colonic Enema (MACE)

Appendiceal stoma is a procedure mostly conducted in pediatric patients. The idea is to flush the colon antegrade. Therefore, the appendix is sutured to the abdominal wall. Patients then perform antegrade enema using a catheter. High complication rates due to infection have been reported [43].

Segmental Colonic Resection and (Sub-)total Colectomy

In segmental colonic resection, a targeted open or laparoscopic resection of the ineffective bowel segment is performed to improve transit time. Patients with an isolated megasigmoid benefit the most from segmental colonic resection. Total colectomy (open or laparoscopic) can be undertaken by resecting or preserving the Bauhin's valve [ileorectal anastomosis (IRA) vs. cecorectal anastomosis (CRA)]. Complications occur in approximately 24% of cases, with the most common being small bowel obstruction. Conversely, the patient satisfaction rate is high. Significant psychological disorders seem to have a negative effect on colectomy [44].

Stapled Trans-anal Rectal Resection (STARR) and TRANSTAR

Both STARR and TRANSTAR are minimally invasive trans-anal procedures restoring anatomy in patients suffering from intussusception or intussusception in combination with rectocoele or rectal prolapse leading to a structural evacuation disorder. The STARR procedure uses a circular stapling device to perform a full thickness rectal excision of the anterior and posterior wall (Fig. 18.8). The TRANSTAR uses a curved cutter stapler for resection (Fig. 18.9). The rectal wall is pulled through the anus and resection is performed outside the anal canal. The advantage of the TRANSTAR is that the surgeon can decide on the degree of resection, whereas the STARR stapler has a fixed length. The most common side effect is fecal urgency. Both procedures are reported to be safe and the majority of patients show a decrease in the Longo-ODS scoring system [45]. The definition of safety can be challenged knowing that up to 23% of patients have complications and 10% needed another operation in the long run [46].

Intra-anal Delorme Procedure

The intra-anal Delorme procedure is a transanal resection of the redundant mucosa while gathering the underlying muscle for structural evacuation disorders due to intussusception or intussusception in combination with rectocoele (Fig. 18.10).

Rectocele Resection

Resection of the rectocele can be performed trans-anally, vaginally, transperineally, with or without levatorplasty. Large rectocoele is most commonly resected by vaginal access using the vaginal introitus, dissection of the thin rectocoele wall by means of an endostapling device and running suture over the staple line (Fig. 18.11). Levatorplasty causes dyspareunia and should only be done in sexually inactive or elderly women. There is very little evidence in the literature and most are observational studies [47]. In a small unpublished series of the author (JP), 11/12 patients did well after a stapled resection and postoperative behave workout; the one failure due to an additional enterocoele not seen on initital work up and diagnosis. For resection of rectocele, patient selection is the key to success.

Laparoscopic Ventral Rectopexy With/Without Mesh Enforcement

Laparoscopic ventral rectopexy is a minimally invasive transabdominal approach for treating high grade rectal intussusception. This laparoscopic approach aims at straightening the rectum or rectosigmoid with or without resection of a prolonged sigmoid, if necessary. The rectum is then sutured to the sacrum, if possible, enforced by mesh placement. In the case of bowel resection, mesh enforcement should be avoided. There is low quality evidence reporting a high (83%) satisfactory rate and a low (2–7%) recurrence rate [48].

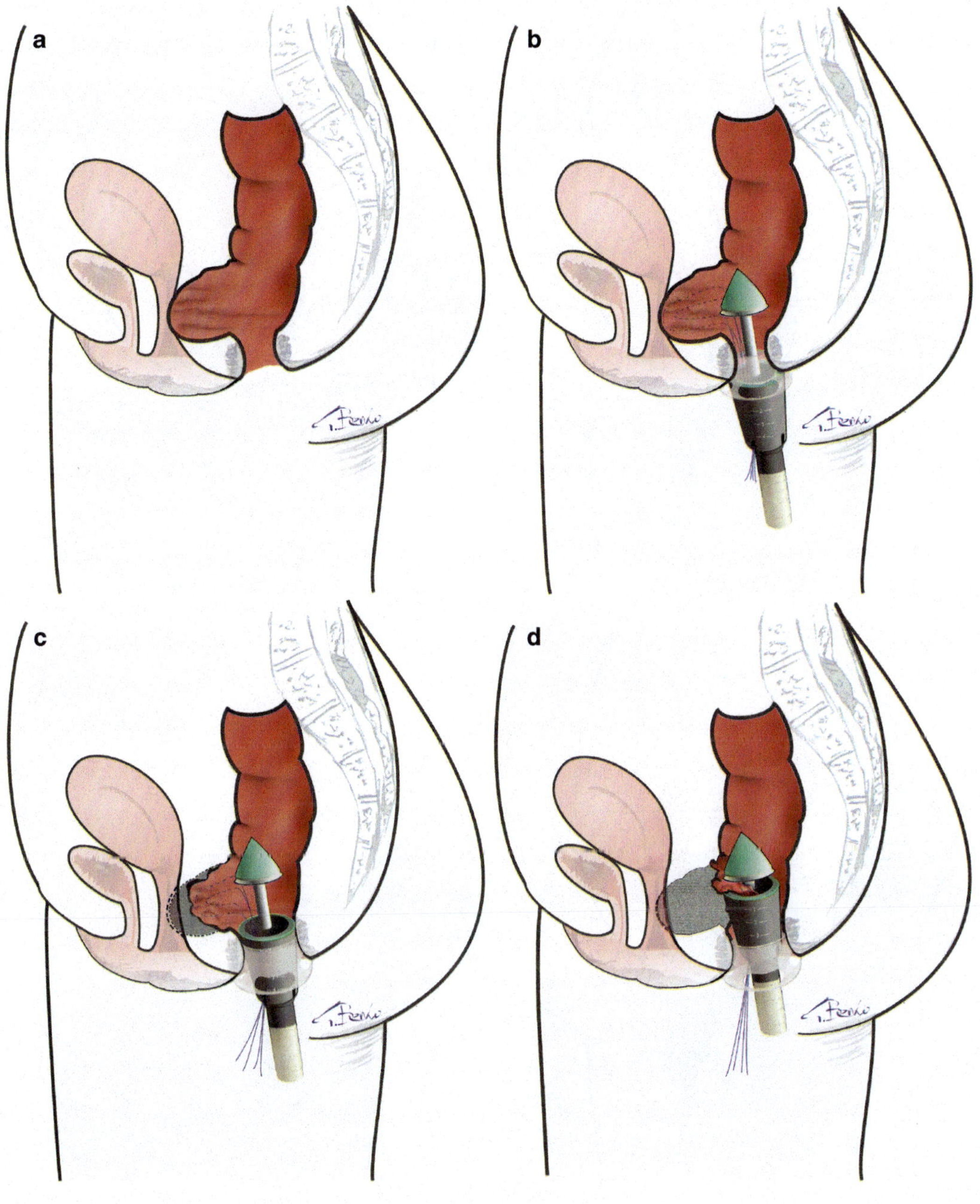

Fig. 18.8 (**a–g**) STARR procedure. (© Gernot Benko. Reused with permission)

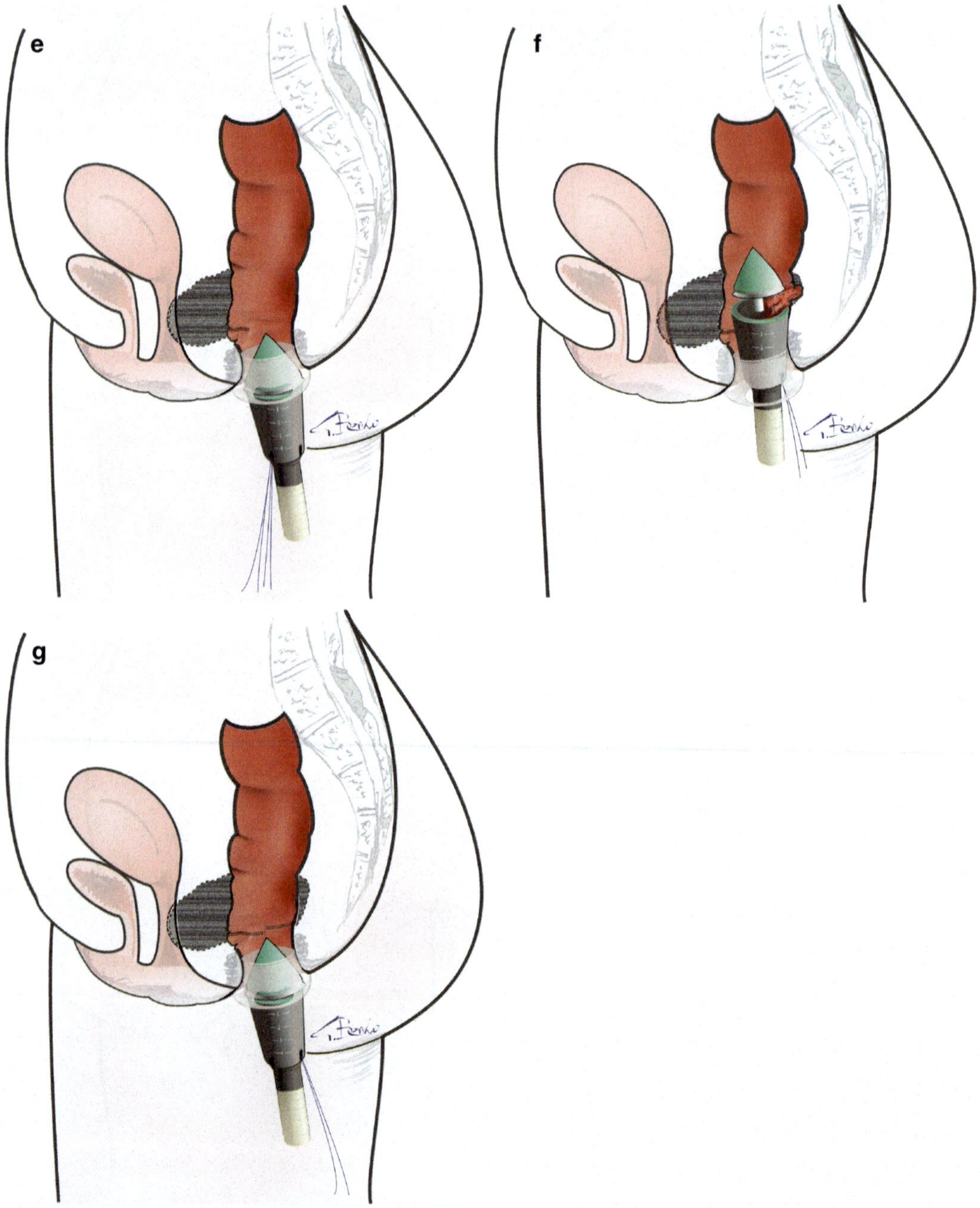

Fig. 18.8 (continued)

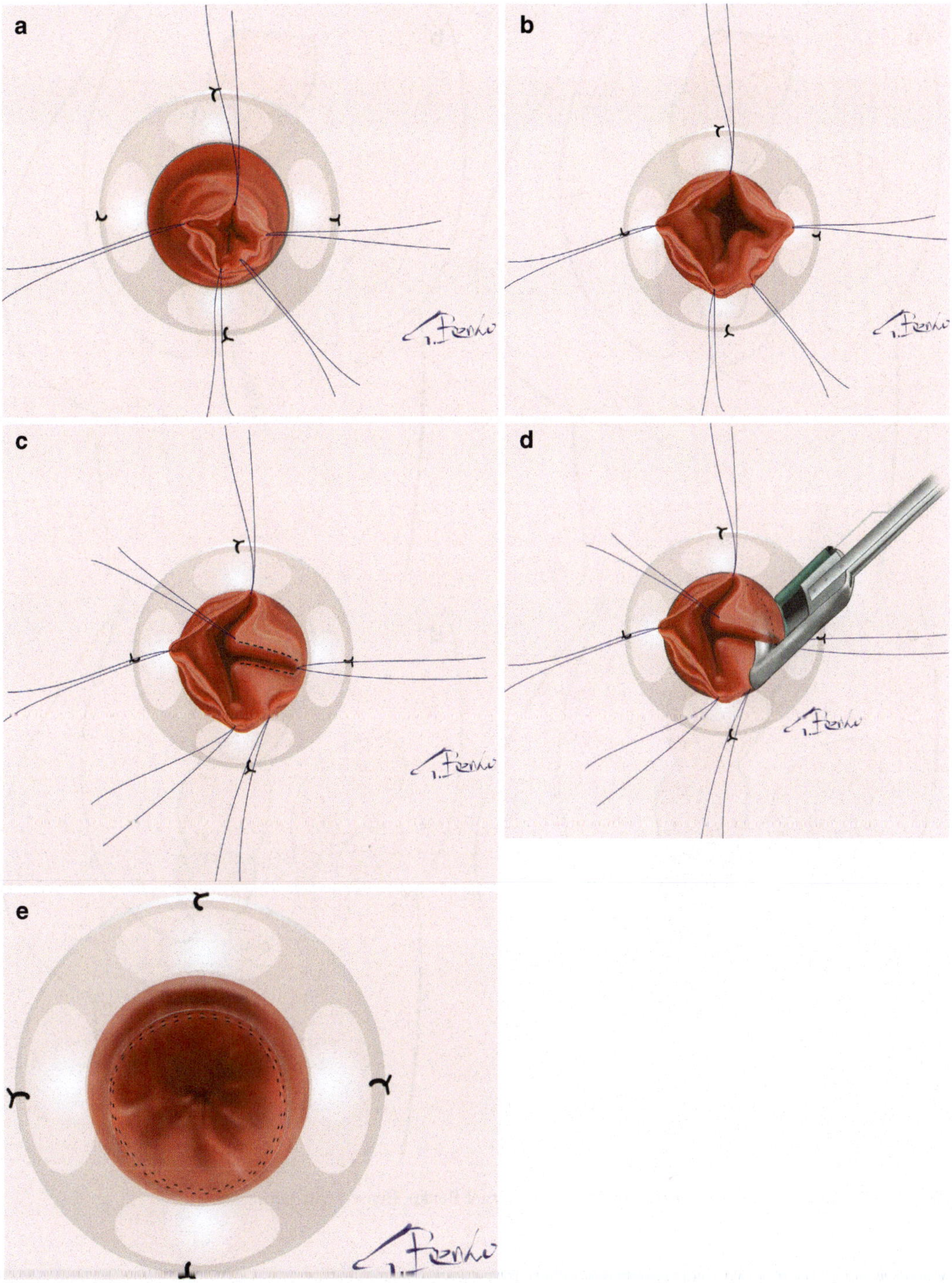

Fig. 18.9 (**a**–**e**) TRANSTAR procedure. (© Gernot Benko. Reused with permission)

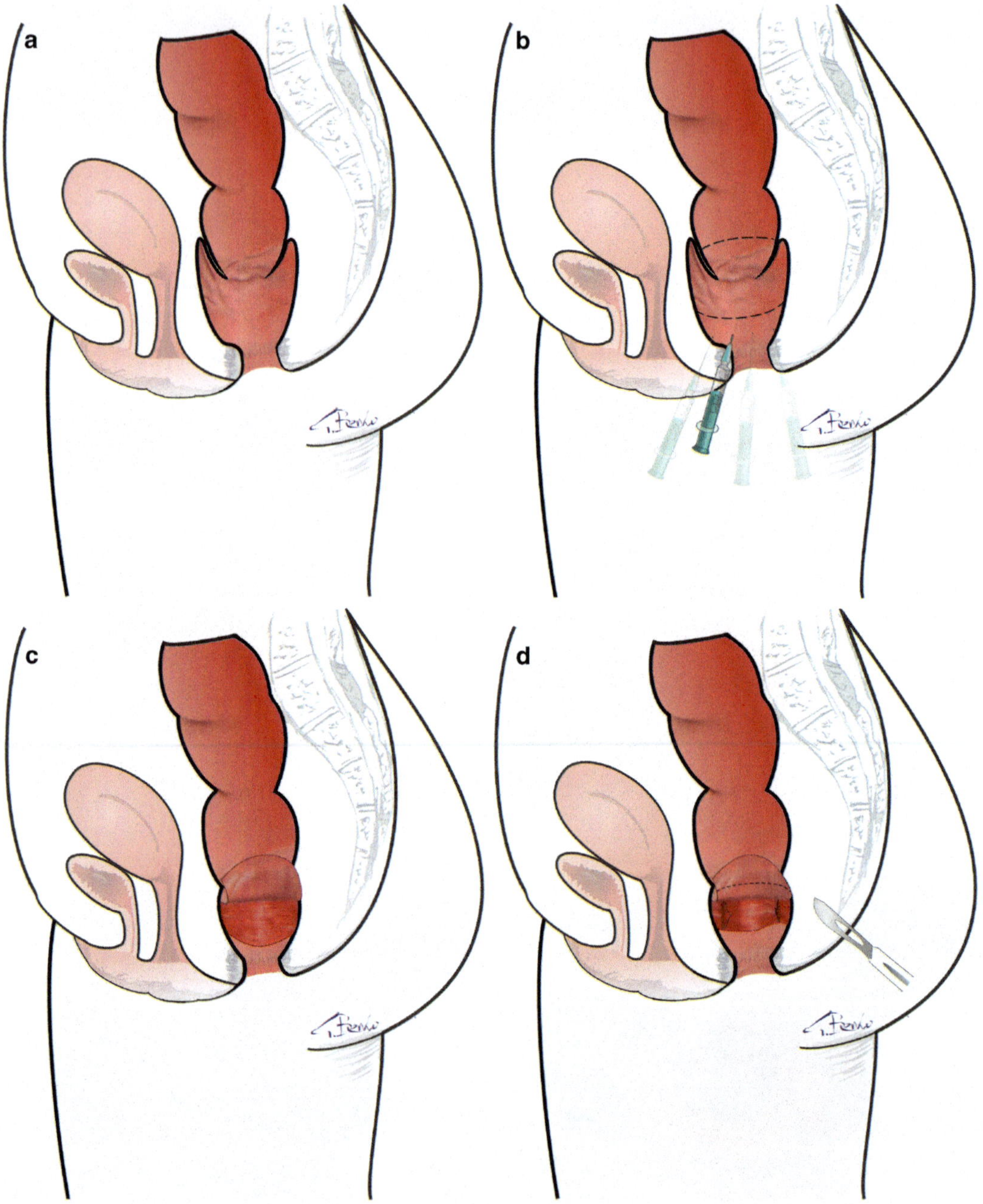

Fig. 18.10 (**a**–**e**) Intra-anal Delorme procedure. (© Gernot Benko. Reused with permission)

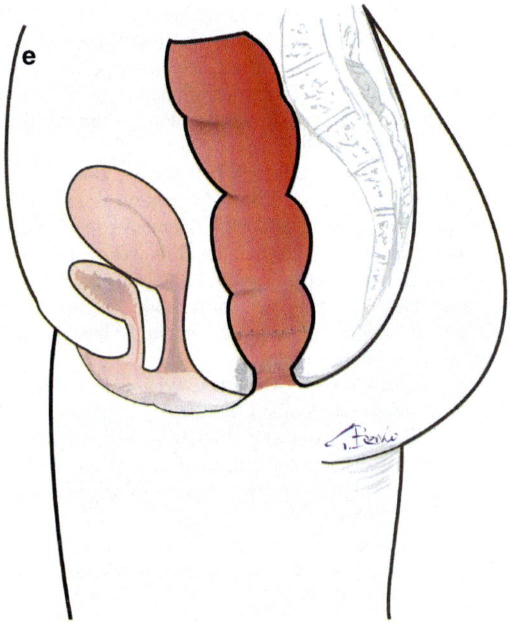

Fig. 18.10 (continued)

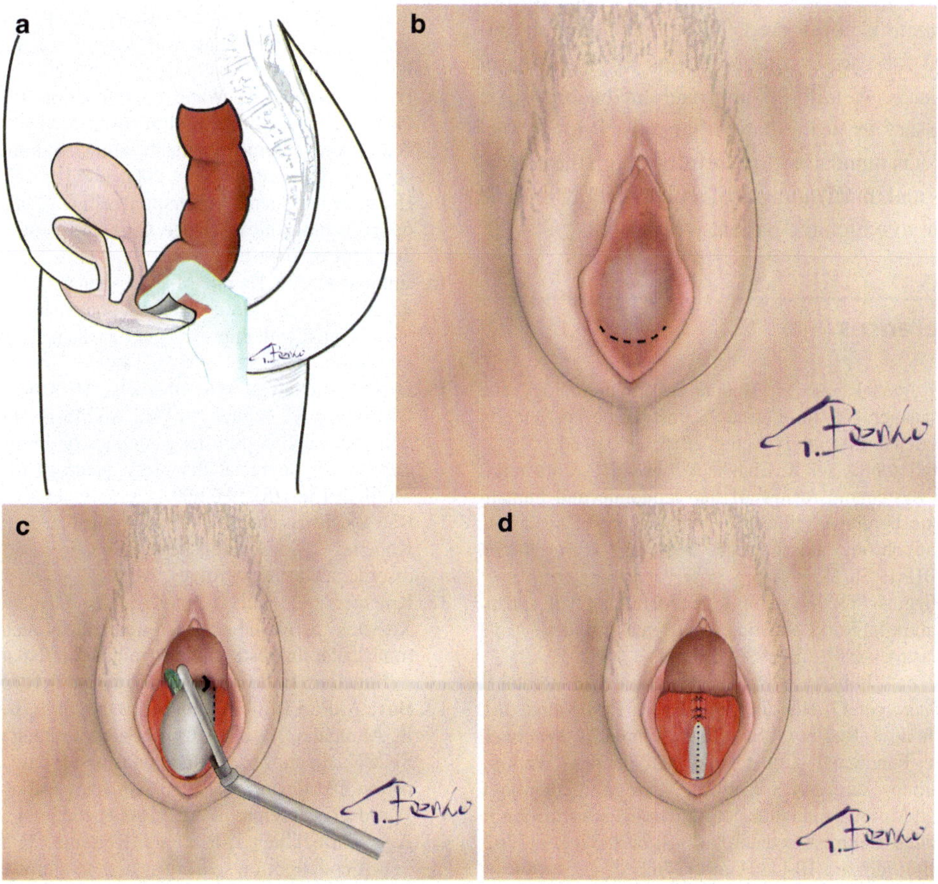

Fig. 18.11 (**a–e**) Transvaginal rectocoele resection. (© Gernot Benko. Reused with permission)

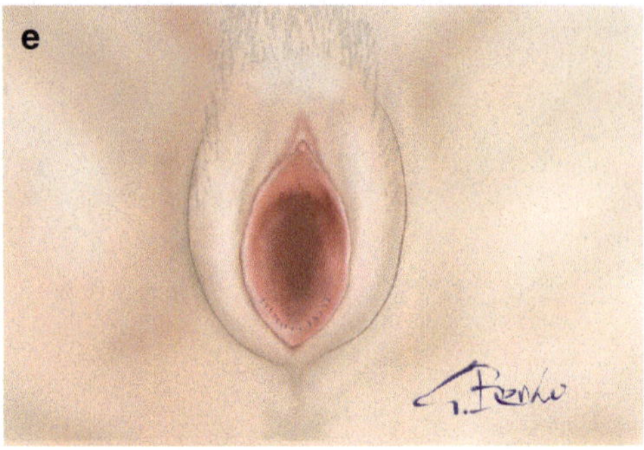

Fig. 18.11 (continued)

Summary

It cannot be stressed enough that surgery is the last possible option in the treatment of chronic constipation. Extensive diagnostic testing should be performed to accurately determine the pathophysiology of each patient before any further treatment is considered. There is not enough evidence for surgical interventions to give general guidance. A tailored approach in each case is necessary to fit the patient's needs. Prior to surgery, it is mandatory to carefully weigh the advantages and disadvantages of any procedure and the extent to which the patient will benefit.

References

1. Vazquez Roque M, Bouras EP. Epidemiology and management of chronic constipation in elderly patients. Clin Interv Aging. 2015;10:919–30.
2. Nellesen D, Yee K, Chawla A, Lewis BE, Carson RT. A systematic review of the economic and humanistic burden of illness in irritable bowel syndrome and chronic constipation. J Manag Care Pharm. 2013;19(9):755–64.
3. Higgins PD, Johanson JF. Epidemiology of constipation in North America: a systematic review. Am J Gastroenterol. 2004;99(4):750–9.
4. Serra J, Pohl D, Azpiroz F, Chiarioni G, Ducrotté P, Gourcerol G, Hungin APS, Layer P, Mendive JM, Pfeifer J, Rogler G, Scott SM, Simrén M, Whorwell P; Functional Constipation Guidelines Working Group. European Society of Neurogastroenterology and Motility guidelines on functional constipation in adults. Neurogastroenterol Motil. 2020;32(2):e13762. https://doi.org/10.1111/nmo.13762.
5. Harris LA, Horn J, Kissous-Hunt M, Magnus L, Quigley EMM. The better understanding and recognition of the disconnects, experiences, and needs of patients with chronic idiopathic constipation (BURDEN-CIC) study: results of an online questionnaire. Adv Ther. 2017;34(12):2661–73.
6. Xiong L, Gong X, Siah KT, Pratap N, Ghoshal UC, Abdullah M, et al. Rome foundation Asian working team report: real world treatment experience of Asian patients with functional bowel disorders. J Gastroenterol Hepatol. 2017;32(8):1450–6.
7. Ghoshal UC. Chronic constipation in Rome IV era: the Indian perspective. Indian J Gastroenterol. 2017;36(3):163–73.
8. Schmulson MJ, Drossman DA. What is new in Rome IV. J Neurogastroenterol Motil. 2017;23(2):151–63.
9. Simren M, Palsson OS, Whitehead WE. Update on Rome IV criteria for colorectal disorders: implications for clinical practice. Curr Gastroenterol Rep. 2017;19(4):15–017.
10. Yu SW, Rao SS. Anorectal physiology and pathophysiology in the elderly. Clin Geriatr Med. 2014;30(1):95–106.
11. Mertz H, Naliboff B, Mayer E. Physiology of refractory chronic constipation. Am J Gastroenterol. 1999;94(3):609–15.
12. Lyford GL, He CL, Soffer E, Hull TL, Strong SA, Senagore AJ, et al. Pan-colonic decrease in interstitial cells of Cajal in patients with slow transit constipation. Gut. 2002;51(4):496–501.
13. He CL, Burgart L, Wang L, Pemberton J, Young-Fadok T, Szurszewski J, et al. Decreased interstitial cell of cajal volume in patients with slow-transit constipation. Gastroenterology. 2000;118(1):14–21.
14. Huizinga JD, Chen JH. Interstitial cells of Cajal: update on basic and clinical science. Curr Gastroenterol Rep. 2014;16(1):363–013.
15. Emmanuel AV, Kamm MA. Response to a behavioural treatment, biofeedback, in constipated patients is associated with improved gut transit and autonomic innervation. Gut. 2001;49(2):214–9.
16. Rao SS, Benninga MA, Bharucha AE, Chiarioni G, Di Lorenzo C, Whitehead WE. ANMS-ESNM position paper and consensus guidelines on biofeedback therapy for anorectal disorders. Neurogastroenterol Motil. 2015;27(5):594–609.
17. Rao SS, Bharucha AE, Chiarioni G, Felt-Bersma R, Knowles C, Malcolm A, et al. Functional anorectal disorders. Gastroenterology. 2016.
18. Kraemer M, Paulus W, Kara D, Mankewitz S, Rozsnoki S. Rectal prolapse traumatizes rectal neuromuscular microstructure explaining persistent rectal dysfunction. Int J Color Dis. 2016;31(12):1855–61.
19. Bove A, Pucciani F, Bellini M, Battaglia E, Bocchini R, Altomare DF, et al. Consensus statement AIGO/SICCR: diagnosis and treatment of chronic constipation and obstructed defecation (part I: diagnosis). World J Gastroenterol. 2012;18(14):1555–64.
20. Agachan F, Chen T, Pfeifer J, Reissman P, Wexner SD. A constipation scoring system to simplify eval-

uation and management of constipated patients. Dis Colon Rectum. 1996;39(6):681–85.
21. Varma MG, Wang JY, Berian JR, Patterson TR, McCrea GL, Hart SL. The constipation severity instrument: a validated measure. Dis Colon Rectum. 2008;51(2):162–72.
22. Altomare DF, Spazzafumo L, Rinaldi M, Dodi G, Ghiselli R, Piloni V. Set-up and statistical validation of a new scoring system for obstructed defaecation syndrome. Color Dis. 2008;10(1):84–8.
23. Wang JY, Hart SL, Lee J, Berian JR, McCrea GL, Varma MG. A valid and reliable measure of constipation-related quality of life. Dis Colon Rectum. 2009;52(8):1434–42.
24. Lindberg G, Hamid SS, Malfertheiner P, Thomsen OO, Fernandez LB, Garisch J, et al. World Gastroenterology Organisation global guideline: constipation--a global perspective. J Clin Gastroenterol. 2011;45(6):483–7.
25. Dukas L, Willett WC, Giovannucci EL. Association between physical activity, fiber intake, and other lifestyle variables and constipation in a study of women. Am J Gastroenterol. 2003;98(8):1790–6.
26. Wisten A, Messner T. Fruit and fibre (Pajala porridge) in the prevention of constipation. Scand J Caring Sci. 2005;19(1):71–6.
27. Rao SS, Yu S, Fedewa A. Systematic review: dietary fibre and FODMAP-restricted diet in the management of constipation and irritable bowel syndrome. Aliment Pharmacol Ther. 2015;41(12):1256 70.
28. Dupont C, Campagne A, Constant F. Efficacy and safety of a magnesium sulfate-rich natural mineral water for patients with functional constipation. Clin Gastroenterol Hepatol. 2014;12(8):1280–7.
29. Bharucha AE, Pemberton JH, Locke GR. American Gastroenterological Association technical review on constipation. Gastroenterology. 2013;144(1):218–38.
30. Manabe N, Cremonini F, Camilleri M, Sandborn WJ, Burton DD. Effects of bisacodyl on ascending colon emptying and overall colonic transit in healthy volunteers. Aliment Pharmacol Ther. 2009;30(9):930–6.
31. Tack J, van Outryve M, Beyens G, Kerstens R, Vandeplassche L. Prucalopride (Resolor) in the treatment of severe chronic constipation in patients dissatisfied with laxatives. Gut. 2009;58(3):357–65.
32. Camilleri M, Kerstens R, Rykx A, Vandeplassche L. A placebo-controlled trial of prucalopride for severe chronic constipation. N Engl J Med. 2008;358(22):2344–54.
33. Quigley EM, Vandeplassche L, Kerstens R, Ausma J. Clinical trial: the efficacy, impact on quality of life, and safety and tolerability of prucalopride in severe chronic constipation a 12-week, randomized, double-blind, placebo-controlled study. Aliment Pharmacol Ther. 2009;29(3):315–28.
34. Thomas RH, Allmond K. Linaclotide (Linzess) for Irritable Bowel syndrome With Constipation and For Chronic Idiopathic Constipation. PT. 2013;38(3):154–60.
35. Lacy BE, Levy LC. Lubiprostone: a novel treatment for chronic constipation. Clin Interv Aging. 2008;3(2):357–64. https://doi.org/10.2147/cia.s2938.
36. Cotelle O, Cargill G, Marty MM, Bueno L, Cappelletti MC, Colangeli-Hagege H, et al. A concomitant treatment by CO2-releasing suppositories improves the results of anorectal biofeedback training in patients with dyschezia: results of a randomized, double-blind, placebo-controlled trial. Dis Colon Rectum. 2014;57(6):781–9.
37. Rao SS, Seaton K, Miller M, Brown K, Nygaard I, Stumbo P, et al. Randomized controlled trial of biofeedback, sham feedback, and standard therapy for dyssynergic defecation. Clin Gastroenterol Hepatol. 2007;5(3):331–8.
38. Rao SS, Valestin J, Brown CK, Zimmerman B, Schulze K. Long-term efficacy of biofeedback therapy for dyssynergic defecation: randomized controlled trial. Am J Gastroenterol. 2010;105(4):890–6.
39. Chiarioni G, Salandini L, Whitehead WE. Biofeedback benefits only patients with outlet dysfunction, not patients with isolated slow transit constipation. Gastroenterology. 2005;129(1):86–97.
40. Heymen S, Scarlett Y, Jones K, Ringel Y, Drossman D, Whitehead WE. Randomized, controlled trial shows biofeedback to be superior to alternative treatments for patients with pelvic floor dyssynergia-type constipation. Dis Colon Rectum. 2007;50(4):428–41.
41. Chiarioni G, Whitehead WE, Pezza V, Morelli A, Bassotti G. Biofeedback is superior to laxatives for normal transit constipation due to pelvic floor dyssynergia. Gastroenterology. 2006;130(3):657–64.
42. Pilkington SA, Emmett C, Knowles CH, Mason J, Yiannakou Y, NIHR CapaCiTY working group, et al. Surgery for constipation: systematic review and practice recommendations: results V: sacral nerve stimulation. Color Dis. 2017;19(Suppl 3):92–100.
43. Meurette G, Lehur PA, Coron E, Regenet N. Long-term results of Malone's procedure with antegrade irrigation for severe chronic constipation. Gastroenterol Clin Biol. 2010;34(3):209–12.
44. Knowles CH, Grossi U, Chapman M, Mason J, NIHR CapaCiTY working group, Pelvic floor Society. Surgery for constipation: systematic review and practice recommendations: results I: colonic resection. Color Dis. 2017;19(Suppl 3):17–36.
45. Mercer-Jones M, Grossi U, Pares D, Vollebregt PF, Mason J, Knowles CH, et al. Surgery for constipation: systematic review and practice recommendations: results III: rectal wall excisional procedures (Rectal Excision). Color Dis. 2017;19(Suppl 3):49–72.
46. Giarratano G, Toscana C, Toscana E, Shalaby M, Sileri P. Stapled transanal rectal resection for the treatment of rectocele associated with obstructed defecation syndrome: a large series of 262 consecutive patients. Tech Coloproctol. 2019;23(3):231–7. https://doi.org/10.1007/s10151-019-01944-9.

47. Grossi U, Horrocks EJ, Mason J, Knowles CH, Williams AB, NIHR CapaCiTY working group, et al. Surgery for constipation: systematic review and practice recommendations: results IV: recto-vaginal reinforcement procedures. Color Dis. 2017;19(Suppl 3):73–91.

48. Grossi U, Knowles CH, Mason J, Lacy-Colson J, Brown SR, NIHR CapaCiTY working group, et al. Surgery for constipation: systematic review and practice recommendations: results II: hitching procedures for the rectum (rectal suspension). Color Dis. 2017;19(Suppl 3):37–48.

19 Obstructed Defecation Syndrome

Adam Studniarek and Anders Mellgren

Introduction

Constipation varies in presentation and severity among patients and will in varying degree affect patients' quality of life. The estimated prevalence of constipation in the general population in North America ranges from 2% to 27%, depending on the definition used [1]. Two types of constipation can be distinguished: prolonged bowel transit time and outlet difficulties (obstructed defecation). Some patients have a combination of both types.

Defecatory disorders can be a result of several different functional and/or anatomical abnormalities. The etiology is sometimes not clear and can be multifactorial. Common findings in patients with obstructed defecation syndrome (ODS) include inappropriate contraction of puborectalis, rectal intussusception, rectal prolapse, and rectocele [2]. As ODS can be a multifactorial condition, a detailed assessment is essential for further treatment planning. All patients should initially be offered conservative management and only a small subset of patients will require surgical intervention.

A. Studniarek, MD · A. Mellgren, MD, PhD, FACS, FASCRS (✉)
Division of Colon & Rectal Surgery, University of Illinois at Chicago, Chicago, IL, USA

Normal Physiology

Normal defecation is a sequence of events involving a number of pelvic reflexes that are controlled and coordinated by the brain stem. Newborns have the basic control mechanism present at birth. As the higher cortical activity develops with time, bowel control improves through subsequent "training."

Propagated colonic contractions move the stool into the rectum, and if the stool is large enough it causes rectal distention and stimulates a desire to defecate. This urge is usually associated with rectal contraction and relaxation of IAS which contributes to delivering stool down to the proximal anal canal. This generally increases the defecatory urge, which can be controlled and surpassed by an active contraction of external anal sphincter (EAS) and puborectalis. This contraction closes the anal sphincter and pushes the stool back into the rectum [3].

During the process of defecation, the individual usually sits or squats. The diaphragm, the abdominal muscles, and the levators contract, while the external anal sphincter and puborectalis relax [4]. Once defecation has started, it can continue without additional effort as long as the neurological function is not impaired. The anorectal angle, the angle between the rectum and the anal canal, plays a role in this process [4]. Relaxation of the puborectalis muscle allows for opening up of the anorectal angle and easy passage of stool to the exterior.

L. C. C. Oliveira (ed.), *Anorectal Physiology*, https://doi.org/10.1007/978-3-030-43811-1_19

Diagnostic Assessment

Symptoms of ODS can include straining, incomplete evacuation, hard and lumpy stools, and some patients may need to use digital manipulation to expel stool from the rectum [5]. As the causes of ODS vary, it is crucial to obtain an adequate history and perform a careful diagnostic work-up. Usually, there is not a single test that can be applied to make differential diagnosis easier and tests need to be tailored to each individual patient. Poor patient selection can contribute to suboptimal treatment outcomes [6, 7].

A detailed history is important in order to identify possible underlying causes, such as endocrine, neurological disorders, etc. Most patients need to undergo a colonoscopy to exclude anatomical abnormalities such as tumors or strictures. If there is a suspicion of Crohn's disease, the patient may also need to undergo assessment of the small bowel and upper endoscopy.

Clinical Assessment

The initial evaluation of a patient with defecatory dysfunction should include a thorough history and physical examination. Typical symptoms are documented, including frequency and consistency of bowel movements, rectal bleeding, soiling, anorectal pain, the need to strain during defecation, or a need for digital assistance in the vagina or beside the anus to assist with defecation.

The physical examination should be detailed. The pelvic floor muscle strength and support are evaluated and sensation is tested. An external rectal assessment should be performed with buttocks separated, with special attention to potential causes of pain such as anal fissure, infection, or thrombosed/incarcerated hemorrhoids. Anal asymmetry or decreased sensation may suggest a neurogenic etiology. If there is a suspicion of external rectal prolapse, an examination with the patient sitting on a commode should be considered. A careful digital rectal examination is performed, noting possible anatomical abnormalities. The anal anal resting tone, squeeze tone, and tone when patient is asked to bear down on the examiner's finger are documented. An evaluation of the rectovaginal septum should be performed to assess possible defects or weaknesses in the rectovaginal septum and/or perineal body and signs of possible rectocele and/or enterocele. Abnormalities at possible vaginal examination may be documented with the Pelvic Organ Prolapse Quantification System (POP-Q) [8].

Defecography

Dynamic defecography is a radiological examination used to objectively assess pelvic floor anatomy and function [9]. It can be performed with fluoroscopic or MRI technique (Figs. 19.1 and 19.2). Images are obtained during rest, squeezing, straining, and defecation. Defecography can evaluate the emptying of the rectum and if there are any signs of nonrelaxation of the pelvic floor. Defecography is useful to determine the level of the pelvic floor and the anorectal angle during defecation. The assessment also helps to identify anatomic abnormalities, including rectocele, rectal intussusception, external rectal prolapse, enterocele, and vaginal vault prolapse.

Fluoroscopic defecography evaluates the emptying process when the patient empties con-

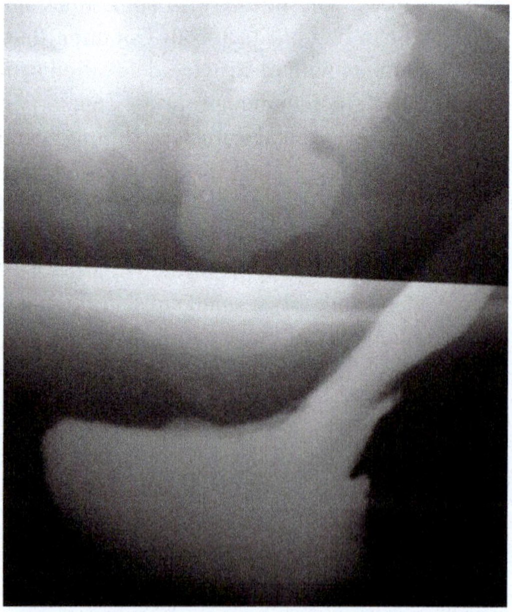

Fig. 19.1 Defecography showing a large anterior rectocele and a sigmoidocele

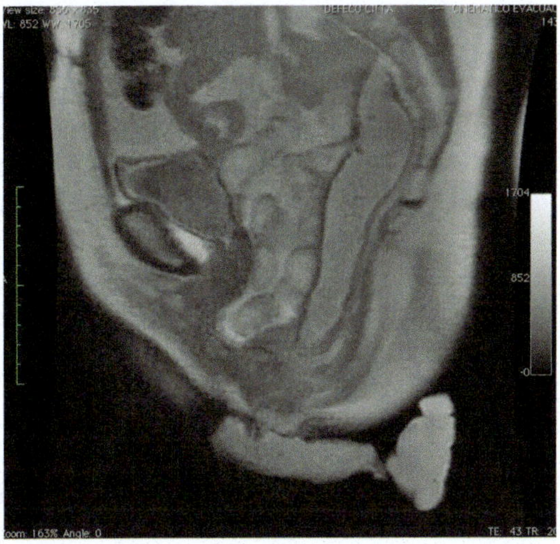

Fig. 19.2 MRI defecography showing a large enterocele

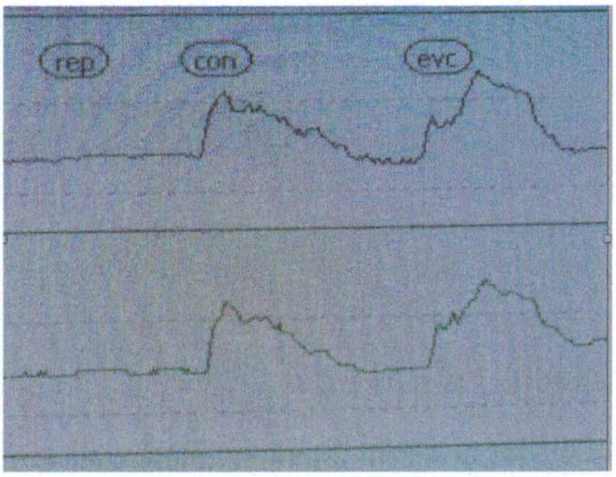

Fig. 19.3 Manometry showing nonrelaxing puborectalis during attempted evacuation

trast medium from the rectum in a sitting position. The examination provides more information if contrast medium is also placed in the vagina. Some centers also use oral contrast or contrast in the urinary bladder. Fluoroscopic defecography assesses the rectal shape and function during evacuation in the physiologic sitting position.

Dynamic MR defecography is usually performed with the patient emptying contrast medium lying in the supine position. It provides an assessment of the pelvic organs during evacuation. A drawback is that the supine position is less physiologic than the sitting position.

It has been difficult to correlate the feeling of complete evacuation with the amount of retained contrast on defecography [6]. The interpretation of radiological findings should be done by experienced clinicians and not all findings will need surgical correction.

Anorectal Manometry

Anorectal manometry may be valuable in the assessment of ODS. It can discriminate between central nervous system causes, anismus, hemorrhoids, and internal anal sphincter abnormalities. Anorectal manometry uses pressure-sensitive catheters to measure anorectal function, including anal pressures, rectal sensation, and compliance. It allows measuring pressures and reaction to squeeze and push motions when a small balloon attached to a catheter is inserted into the rectum and patient is asked to squeeze, relax, and push. These findings can help in diagnosis of nonrelaxation (Fig. 19.3). Anorectal manometry can also help exclude Hirschsprung's disease and identify decreased rectal sensation and nonrelaxing pelvic floor. In patients with decreased rectal sensation, a tolerance to increasing volumes of rectal content has been observed [10].

Electromyography

Anorectal electromyography (EMG) can evaluate muscle activity at different physiological phases. EMG can assess sphincter relaxation or contraction during voiding or defecation and can be assessed in conjunction with urodynamics. EMG is usually performed by using surface electrodes, and assessment of muscle activity is obtained during squeezing, pushing, and at rest. As the patient bears down, an unchanged activity suggests a nonrelaxing pattern, and an increase in electrical activity suggests a paradoxical contraction.

Ultrasound

Ultrasound has become an increasingly popular diagnostic modality for evaluation of pelvic floor disorders due to its relatively low cost, accessibility, and good patient tolerance. A variety of different probes can be used for imaging and will enhance the evaluation of anatomic and physiologic abnormalities, including anal sphincter injuries, rectal procidentia, nonrelaxation, rectocele, vaginal vault prolapse, and enterocele.

Colonic Transit Time

Transit time studies can be used in patients with infrequent bowel movements and who are refractory to laxatives or other conservative treatments. The goal of this study is to evaluate for slow transit constipation, which is defined as long transit time through the colon [11].

The Sitzmarks transit study, described by Hinton et al. in 1969 [12], remains the most commonly used method. The patient is asked to stop taking laxatives, enemas, or suppositories. The patient is then asked to ingest a capsule containing radiopaque markers. An x-ray of the abdomen is obtained at 1 and 5 days later. The examination is considered abnormal when more than 20% of markers are retained on day 5. Retained markers scattered throughout the colon support a diagnosis of colonic inertia, while accumulation of markers in the rectosigmoid area may suggest ODS (Fig. 19.4) [13].

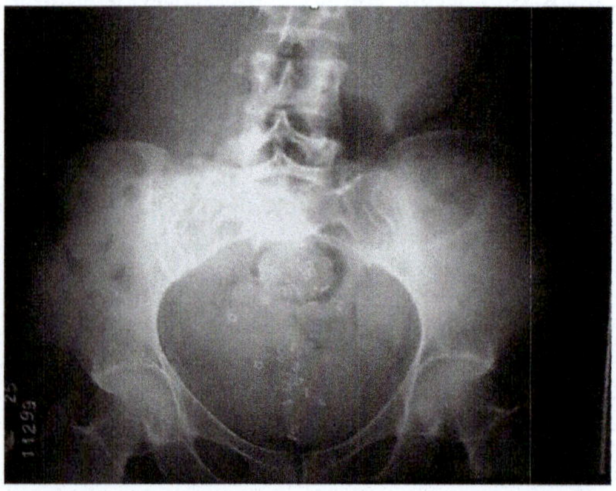

Fig. 19.4 Colonicic transit time showing the retained markers in the rectum in patients with ODS

Common Disorders in Patients with ODS

As indicated above, patients with ODS can have several different disorders, and some of them can only be identified with appropriate specialized work-up.

Nonrelaxing Pelvic Floor

Nonrelaxing pelvic floor is one of the most common findings in patients with ODS. The condition was first described by Preston and Lennard-Jones in 1985 [14] when they studied a group of women with severe constipation who could not properly expel stool. Their findings indicated that there was a paradoxical contraction of the puborectalis and the external anal sphincter muscles during defecation.

The inability to inhibit the activity of the puborectalis and the external anal sphincter can be related to impaired rectal sensation. In a study by Read et al. [15], they found that the perception of a desire to defecate was blunted in young women with severe constipation, whereas the perception of pain and distention during balloon inflation was unaffected. Nonrelaxation of the pelvic floor may sometimes be combined with colonic dysmotility. In a study by Miller et al. [16] with 277 patients with severe constipation symptoms, 27 had pelvic floor dysfunction, and 37 had both pelvic floor dysfunction and impaired colonic propulsion.

Failure of the internal anal sphincter to relax during rectal distention can also be a contributing factor to ODS, especially in patients with Hirschsprung's disease. In patients with short-segment Hirschsprung's disease, anorectal manometry will demonstrate a high anal tone that fails to relax in response to rectal distention. Some patients with megarectum may require abnormally large distension volumes in order to mediate an anal relaxation [17].

Levator ani syndrome is a type of nonrelaxing pelvic floor dysfunction that can contribute to the development of ODS. Patients typically have a constant, dull anal pain in levator muscles. This pain is usually caused by an increased tone in the levator muscles and is associated with a paradoxical contraction or nonrelaxation of the levator muscles during defecation [18]. This condition should be differentiated from proctalgia fugax, which is characterized by sudden and severe pain in the anal area that lasts for seconds to minutes and disappears between the attacks [19].

Slow Transit Constipation

Slow transit time constipation was described in 1986 by Preston et al. [14] in a group of women who displayed slow gut transit time with normal caliber of the colon. The exact mechanism of dysmotility of the colon is poorly understood. Patients often present with symptoms of abdominal bloating, discomfort, and infrequent urge to defecate. These symptoms are usually not relieved with fiber supplementation or other conservative measures.

Rectal Procidentia

Rectal procidentia is a "telescoping" of the rectum on itself. If the telescoped bowel does not reach beyond the anal verge, the prolapse is usually named internal prolapse or rectal intussusception. If the telescoped bowel reaches beyond the anal verge, the prolapse is usually named external rectal prolapse or sometimes just rectal prolapse.

Patients with external rectal prolapse can usually feel the protruding bowel and some patients may think it is "large hemorrhoids." At clinical examination, the prolapsing bowel contains the full thickness of the rectal wall and is usually characterized by circular folds.

Patients with internal and external prolapse may have a variety of symptoms, including rectal mucous secretion, rectal emptying difficulties, and incomplete rectal evacuation symptoms. It is unclear whether the ODS symptoms are caused by the prolapse itself or they are related to increased pushing and decreased relaxation [20, 21]. Some patients also develop a frequent urge to defecate and a solitary rectal ulcer, and some patients develop weakened anal sphincter tone and associated fecal incontinence [22].

Enterocele and Vaginal Vault Prolapse

Enterocele is defined as bowel, usually small bowel and sometimes the sigmoid colon, descending down into the lower pelvic cavity between the vagina and the rectum. It is found more commonly if the patient has previously undergone a hysterectomy, but this is not required. During clinical examination it is usually possible to palpate a bulge in the rectovaginal septum. Typical symptoms may include a feeling of fullness in the pelvis, lower back pain, and bulging in the vagina.

Vaginal vault prolapse occurs when the top part of the vagina prolapses into the vaginal canal. Vaginal vault prolapse occurs more often after previous vaginal deliveries and previous hysterectomy, since the uterus helps supporting the upper vagina. Other contributing factors include obesity, constipation, previous surgeries, and chronic coughing. On physical exam, a protruding bulge in vaginal canal may be present. Typical symptoms may include sensation of a vaginal bulge, pressure in the vaginal canal, urinary emptying difficulties, vaginal discharge or bleeding, as well as frequent urinary tract infections.

Rectocele

Rectocele is an outpocketing of the anterior rectal wall into the vagina. Rectoceles can be caused by pelvic floor relaxation and/or structural defects in the rectal wall following obstetric trauma. Women can sometimes relieve the rectocele by pushing on the posterior vaginal wall at rectal emptying. On defecography, rectocele is frequently associated with rectal intussusception and/or pelvic floor nonrelaxation. Typical symptoms include a bulge in the posterior vaginal wall, rectal emptying difficulties, rectal pain, or dyspareunia.

Nonsurgical Management of Obstructed Defecation

Treatment of almost all patients with ODS starts with conservative management, including a high-fiber diet, high water intake, and laxatives.

Colonic lavage, consisting of retrograde large bowel irrigation with warm water, can have a positive role. Some favor rectal irrigation, which appears to be especially successful in patients with intestinal dysfunction [23–26].

Injection of botulinum toxin can provide usually short-term relief, especially in patients with nonrelaxing pelvic floor. This treatment is usually safe; a few studies have reported transient mild anal incontinence [27–32].

Pelvic Floor Exercises with Biofeedback

Nonrelaxation of the pelvic floor is traditionally treated with biofeedback training. Under supervision, patients undergo pelvic floor retraining. Patients can usually follow the activity in the pelvic floor muscles on a screen, and exercises are usually overseen by a specially trained physical therapist or a nurse. The therapist also helps patients with lifestyle management, dietary recommendations, and can provide advice on other important matters.

A new psychological approach is called a psych-echo-biofeedback, and it involves breathing exercises, hypnotic words, and pelvic floor exercises and is usually performed with a psychologist. Patient is encouraged to watch the contraction-relaxation sequence on a screen using transanal or transvaginal ultrasound [30].

Psychological support and counseling have been shown to be helpful in patients with depression and/or anxiety. It is important to remember that up to one-third of females complaining of severe ODS and proctalgia have been exposed to sexual abuse/trauma during childhood and/or adolescence [31].

Transanal electrostimulation, using a small probe inserted into the anus, has been described to be of some benefit in patients with pudendal neuropathy and/or rectal hyposensation [32].

Surgical Management of Obstructed Defecation

A variety of different surgical techniques for treatment of ODS have been proposed. All techniques have their advantages and disadvantages. In our opinion, a good functional outcome can only be achieved by evaluating each patient individually and tailoring the surgical intervention to his/her specific condition.

The Malone Antegrade Continence Enema (MACE) procedure can relieve constipation symptoms, but does not address the underlying etiology. This procedure consists of creating a stoma with the appendix or a piece of the terminal ileum and enables antegrade enemas [33]. This procedure is especially popular in children.

Treatment of Rectal Procidentia

For treatment of rectal procidentia, most colorectal surgeons would agree that younger medically fit patients benefit from a surgical repair using an abdominal approach. On the other hand, in older patients with multiple comorbidities, a perineal approach is considered safer. Perineal approach is associated with higher recurrence rates.

For fit patients, surgeons traditionally agreed that posterior rectopexy is a method of choice for correcting the prolapse. Sometimes it is combined with bowel resection (resection rectopexy). In 2004, D'Hoore et al. [34] published the results of a novel technique using a mesh stabilizing the anterior part of the rectum (ventral rectopexy) and this technique has become increasingly popular.

Both rectopexy methods commonly provide relief of symptoms in a majority of patients. Ventral rectopexy (VR) corrects the intussusception in the anterior rectal wall [35] and seems to have a superior functional outcomes in some studies. Direct comparisons in a randomized fashion are however lacking. In Europe, VR has become the more popular method. In the USA, posterior rectopexy has remained popular, but VR is increasingly considered as a treatment alternative.

Ventral Rectopexy

Patients with OD symptoms and rectal procidentia can be operated with VR. A peritoneal incision is made to the right of the sacral promontory and extended along the rectum. Lateral and posterior dissection should be avoided to prevent damage to the autonomic nerves. The rectovaginal septum (recto-prostatic septum in men) is mobilized to the pelvic floor, approximately 2–3 cm above the dentate line. A mesh is sutured to the ventral aspect of the distal rectum and the upper part of the mesh is fixed to the sacral promontory.

This method was first described by Andre D'Hoore et al. [34] in 2004. The initial idea for the VR repair derived from the cinegraphic data of Broden and Snellman [35], who demonstrated that the intussusception of the rectum starts usually in the anterior aspect. VR aims to correct the descent of the posterior and middle compartment by supporting the rectovaginal septum down to the pelvic floor between the rectum and the vagina [36].

VR was initially recommended mainly for patients with external rectal prolapse, and the method has gained widespread acceptance for this indication. However, already in the first publication [34], it was noted that functional results were excellent using this method. In the original study, 19 out of the 42 patients operated for external prolapse had OD symptoms. Postoperatively, OD symptoms resolved in 16 of these 19 patients.

Indications for VR are expanding, and the method can be useful not only in patients with external rectal prolapse, but also in patients with internal rectal prolapse. Subsequent studies have demonstrated improvement of OD symptoms in patients with both internal or external rectal prolapse. In a recent systematic review [37] of patients with internal prolapse undergoing rectopexy, the authors found that 502 of 658 patients (77%) operated with VR had improved OD symptoms postoperatively.

Complications after VR have been a concern, since the technique includes implantation of mesh close to the rectum. Minor complications have been reported ranging from 0% to 36% and major complications from 0% to 5% based on the available studies [36, 37]. Mesh erosions into the bowel or vagina are reported in less than 5% of patients. It should be noted that no study can provide significant long-term follow-up, since most patients have been operated after 2010.

Posterior Rectopexy

Rectal procidentia can also be treated with posterior rectopexy, which includes a posterior mobilization and anchoring of the rectum to the sacrum. Suture rectopexy uses sutures for anchoring of the rectum to the sacrum. Cutait et al. [38] described this approach for the first time in 1959, and the method usually results in a low recurrence rate (5–10%). Some reports show an improvement in fecal incontinence as well, but some patients will still frequently have ODS symptoms. There are also a variety of methods using mesh for anchoring the rectum to the sacrum, but these methods have largely been abandoned since it has not been shown that mesh adds any additional significant benefit.

Suture rectopexy combined with concomitant sigmoidectomy, resection rectopexy, was described by Frykman and Goldberg in 1969 [39]. This technique can diminish postoperative constipation symptoms in some patients and this method has therefore been a popular method for treating rectal prolapse in patients who have concomitant constipation symptoms. Some surgeons however prefer VR, since this method does not include a bowel resection and may offer better functional results. Studies with direct comparisons are limited. Formijne Jonkers et al. [40] recently compared the outcomes after resection rectopexy in one institution (28 US patients) with the outcomes after VR in another institution (40 European patients). The authors concluded that both resection rectopexy and VR offer improvements in functional symptoms. They found that continence may improve better after resection rectopexy, but this method also had a slightly higher complication rate than VR.

Concomitant Sacrocolpopexy and Rectopexy

Sacrocolpopexy can be added to posterior rectopexy in patients with concomitant vaginal vault prolapse. Watadani et al. [41] reported on 110 women, with external rectal prolapse ($n = 96$) or internal rectal prolapse ($n = 14$) and enterocele ($n = 86$) and/or vaginal prolapse ($n = 48$). No patient developed recurrent rectal prolapse. Preoperatively, 93% of patients reported constipation; after surgery 82% reported resolution or improvement. In a prospective study by Dulucq et al. [42] in 77 patients with full-thickness rectal prolapse, they found that fecal incontinence improved in 89% of patients and constipation improved in 36%.

Sacrocolpopexy can also be added to VR. Some surgeons think that this may not be necessary, since the mesh used during VR can also be anchored to the vagina. Other surgeons, however, believe that results are better if patients with vaginal vault prolapse also receive a concomitant sacrocolpopexy.

Perineal Operations for Rectal Prolapse

Perineal operations for rectal prolapse have a limited role in the treatment of obstructed defecation. These operations are usually reserved for patients with compromised health and external rectal prolapse. OD is usually not the primary indication.

Altemeier's procedure was first described by Miles in 1933 [43] and was subsequently popularized by Altemeier in 1971 [44]. It involves excising the prolapsing rectum with a perineal approach, creating a low end-to-end coloanal anastomosis.

Delorme's operation was first introduced by Delorme in 1900, and it consists of stripping the mucosa from the prolapsed segment, plication of the muscle layers, and re-approximation of the mucosa [45]. This procedure is typically reserved for patients with a short segment of full-thickness rectal prolapse.

Stapled Transanal Rectal Resection

Stapled transanal rectal resection (STARR) uses a circular stapler to resect portions of the rectal wall and is usually used in patients who have internal rectal prolapse and/or rectocele [46]. Boccasanta et al. [47] reported excellent short-term results and a majority of patients had improved defecation. There were relatively low number of complications; some patients reported painful defecation and/or urgency. The postoperative proctalgia is thought to be caused by fibrosis around retained staples and can possibly improve after removal of staples [48]. STARR remains controversial in terms of the long-term outcomes and complications. In a more recent study by the Italian Society of Colo-Rectal Surgery [49], 55% of patients still had at least three symptoms of ODS at 18 months after STARR and 19% of patients required another intervention.

There have been a few studies comparing the STARR technique with VR for patients with OD. In a retrospective study, Altomare et al. [50] found that both STARR and VR improved defecation in patients with ODS and both techniques were associated with minimal complications. However, overall pelvic wellness evaluated by the TAPE score improved significantly only after VR. Madbouly and Mohii [51] recently reported on 112 patients who were prospectively randomized to either STARR ($n = 56$) or VR ($n = 56$) for treatment of ODS. Improvement was seen with both techniques, but VR had better long-term functional outcome, less complications, and less recurrences compared to STARR.

A more recent modification of the STARR technique uses a Contour® Transtar™ device (Ethicon Endo-Surgery, Cincinnati, Ohio, USA). The Transtar™ technique may achieve better results in terms of postoperative proctalgia [52]. A recent multicenter European trial has demonstrated short-term relief of obstructed defecation symptoms in patients after treatment with Transtar procedure, and no major postoperative complications or mortality was reported.

Rectocele Repair

There are a number of surgical repair options for the management of rectocele, each with a different approach. Transvaginal approach is common and allows repair of the rectocele as well as relaxed perineum and the vaginal opening. Transperineal approach can accommodate mesh placement, but mesh use is no longer common because of the complication risks. Transanal approach has an easy postoperative course, but allows only for limited repair of the rectocele.

In a prospective study by Mellgren et al. [53], they evaluated outcomes in patients who underwent posterior colporrhaphy and perineorrhaphy with a transvaginal approach. Constipation improved in 88% of patients and 52% of patients had no constipation at all on postoperative follow-up. Yamana et al. [54] assessed the clinical and physiological outcomes after transvaginal rectocele repair, and they demonstrated that difficult evacuation improved in 90% of patients and completely disappeared in 30% of patients. High rate of patient satisfaction was obtained in 83% of patients.

The transanal approach can be a successful method of treatment in selected patients. Williams et al. [55] evaluated 51 patients who underwent transanal rectocele repair and found that 91% had a resolution of rectocele-related symptoms and only 9% had recurrences on follow-up.

Transanal repair of rectocele with full rectal mucosectomy can be performed using a circular stapler device as described by Regadas et al. [56]. This technique was evaluated in a multicenter prospective trial reported by Cruz et al. [57]. This trial suggests that this method is safe and effective. The postoperative assessment demonstrated residual grade I anorectocele in 10.6% of patients and the Wexner constipation score decreased significantly from 16 to 4. Complications include bleeding from the staple line requiring additional hemostatic suture (17% of patients) and persistent postoperative pain (9% of patients). A small number of patients can develop a stricture at the staple line, which can be treated by stricturectomy and/or digital dilatation.

Pelvic Organ Prolapse Suspension

Pelvic organ prolapse suspension (POPS) is performed with laparoscopic approach with a mesh fixed to the vagina. The mesh is tunneled laterally and anchored to the abdominal wall laterally on each side creating a pelvic organ suspension.

Ceci et al. [58] evaluated the outcomes of this technique in 54 women. Postoperative assessment demonstrated improvement in symptoms, and only one patient with a residual recto-anal intussusception and a residual rectocele was found, whereas 16% of patients reported defecation urgency. This technique remains controversial, and further long-term follow-up studies are necessary to establish the long-term outcomes.

Sacral Nerve Stimulation

Sacral nerve stimulation (SNS) is used extensively in the treatment of fecal incontinence, urge incontinence, and urinary retention. However, SNS also has beneficial effects on constipation in some patients. Kamm et al. [59] reported in a prospective trial that SNS is effective in the treatment of some patients with constipation resistant to conservative management. They reported increased defecation frequency, increased days per week with evacuation, and a decrease in the Cleveland Clinic constipation score. Overall, 87% of the implanted patients achieved some degree of treatment success.

Subtotal Colectomy for Slow Transit Constipation

Surgical treatment of intractable slow transit constipation should be the last resort reserved for patients who have failed conservative management for severe, handicapping constipation. Different extents of colectomy have been used, but usually a subtotal colectomy with an ileorectal anastomosis is recommended. Long-term results are frequently disappointing, especially in patients with OD [60].

Summary

Surgery for ODS remains a challenging topic for both colorectal surgeons and patients. In order to provide the best care for specific symptoms related to ODS, careful patient assessment and selection should be carried out to achieve optimal results.

Conservative management should be attempted with every patient prior to surgical intervention. Thereafter, each individual patient should be treated according to the etiology of ODS. Surgical approaches vary and they should be tailored toward the specific condition causing ODS. There is no one standard surgical technique for the treatment of ODS, and each patient should be evaluated individually and treated based on the disease etiology, patient, and procedural risk factors.

References

1. Higgins PD, Johanson JF. Epidemiology of constipation in North America: a systematic review. Am J Gastroenterol. 2004;99:750–9.
2. Ming S, et al. Obstructed defecation. Review article. J Gastroenterol Hepatol. 1993;8:383–9.
3. Phillips SF, Edwards DA. Some aspects of anal continence and defecation. Gut. 1965;6(4):396–406.
4. Parks AG, Porter NH, Melzak J. Experimental study of the reflex mechanism controlling the muscle of the pelvic floor. Dis Colon Rectum. 1962;5:407–14.
5. Goh V, Halligan S, Kaplan G, Healy JC, Bartram CI. Dynamic MR imaging of the pelvic floor in asymptomatic subjects. AJR Am J Roentgenol. 2000;174:661–6.
6. Wijffels NA, Jones OM, Cunningham C, Bemelman WA, Lindsey I. What are the symptoms of internal rectal prolapse? Color Dis. 2013;15:368–73.
7. Janssen LW, van Dijke CF. Selection criteria for anterior rectal wall repair in symptomatic rectocele and anterior rectal wall prolapse. Dis Colon Rectum. 1994;37:1100–7.
8. Persu C, Chapple CR, Cauni V, Gutue S, Geavlete P. Pelvic Organ Prolapse Quantification System (POP-Q) - a new era in pelvic prolapse staging. J Med Life. 2011;4(1):75–81.
9. Murthy VK, Orkin BA, Smith LE, Glassman LM. Excellent outcome using selective criteria for rectocele repair. Dis Colon Rectum. 1996;39:374–8.
10. Ahn JY, et al. Effect of biofeedback therapy in constipation according to rectal 581 sensation. Gut Liver. 2013;7:157–62.
11. Bump RC, et al. The standardization of terminology of female pelvic organ prolapse and pelvic floor dysfunction. Am J Obstet Gynecol. 1996;175:10–7.
12. Hinton JM, Lennard-Jones JE, Young AC. A new method for studying gut transit times using radiopaque markers. Gut. 1969;10:842–7.
13. Ting KH, Mangel E, Eibl-Eibesfeldt B, Müller-Lissner SA. Is the volume retained after defecation a valuable parameter at defecography? Dis Colon Rectum. 1992;35:762–7.
14. Preston DM, Lennard-Jones JE. Severe chronic constipation of young women: idiopathic slow transit constipation. Gut. 1986;27:41–8.
15. Read NW, Timms J. Impairment of defecation in young women with severe constipation. Gastroenterology. 1986;90:53–60.
16. Miller R, Duthie GS, Bartolo DC, Roe AM, Locke-Edmunds J, Mortensen NJ. Anismus in slow transit constipation. Br J Surg. 1991;78:690–2.
17. Pemberton J. Evaluation and surgical treatment of severe chronic constipation. Ann Surg. 1991;214:403–1.
18. Read NW, Abouzekry L. Why do patients with fecal impaction have fecal incontinence? Gut. 1986;27:283–7.
19. Bordeianou LG, Camichael JC, Paquette IM. Consensus statement of definitions for anorectal physiology testing and pelvic floor terminology (revised). Dis Colon Rectum. 2018;61(4):421–7.
20. Bajwa A, Emmanuel A. The physiology of continence and evacuation. Best Pract Res Clin Gastroenterol. 2009;23:477–85.
21. Felt-Bersma RJ, Tiersma ES, Cuesta MA. Rectal prolapse, rectal intussusception, rectocele, solitary rectal ulcer syndrome, and enterocele. Gastroenterol Clin N Am. 2008;37:645–68.
22. Bharucha AE, Wald A, Enck P, Rao S. Functional anorectal disorders. Gastroenterology. 2006;130(5):1510–8.
23. Pizzetti D, Annibali R, Bufo A, Pescatori M. Colonic hydrotherapy for obstructed defecation. Color Dis. 2005;7(1):107–8.

24. Taffinder NJ, Tan E, Webb IG, McDonald PJ. Retrograde commercial colonic hydrotherapy. Color Dis. 2004;6(4):258–60.
25. Koch SM, Melenhorst J, van Gemert WG, Baeten CG. Prospective study of colonic irrigation for the treatment of defaecation disorders. Br J Surg. 2008;95(10):1273–9.
26. Christensen P, Krogh K, Buntzen S, Payandeh F, Laurberg SD. Long-term outcome and safety of transanal irrigation for constipation and fecal incontinence. Colon Rectum. 2009;52(2):286–92.
27. Pucciani F, Reggioli M, Ringressi MN. Colorectal obstructed defaecation: what is the role of rehabilitation? Colorectal Dis. 2012;14(4):474–9.
28. Bove A, Bellini M, Battaglia E, Bocchini R, Gambaccini D, Bove V, Pucciani F, Altomare DF, Dodi G, Sciaudone G, Falletto E, Piloni V. Consensus statement AIGO/SICCR diagnosis and treatment of chronic constipation and obstructed defecation (part II: treatment). World J Gastroenterol. 2012;18(36):4994–5013.
29. Dolk A, Holmström B, Johansson C, Frostell C, Nilsson BY. The effect of yoga on puborectalis paradox. Int J Color Dis. 1991;6(3):139–42.
30. Del Popolo F, Cioli VM, Plevi T, Pescatori M. Psycho-echo-biofeedback: a novel treatment for anismus--results of a prospective controlled study. Tech Coloproctol. 2014;18(10):895–900. Epub.
31. Devroede G. Early life abuses in the past history of patients with gastrointestinal tract and pelvic floor dysfunctions. Prog Brain Res. 2000;122:131–55.
32. Maria G, Sganga G, Civello IM, Brisinda G. Botulinum neurotoxin and other treatments for fissure-in-ano and pelvic floor disorders. Br J Surg. 2002;89(8):950–61.
33. Poirier M, Abcarian H, Nelson R. Malone antegrade continent enema: an alternative to resection in severe defecation disorders. Dis Colon Rectum. 2007;50(1):22–8.
34. D'Hoore A, Cadoni R, Penninckx F. Long-term outcome of laparoscopic ventral rectopexy for total rectal prolapse. Br J Surg. 2004;91:1500–5.
35. Broden B, Snellman B. Procidentia of the rectum studied with cineradiography. A contribution to the discussion of causative mechanism. Dis Colon Rectum. 1968;11(5):330–47.
36. Cullen J, Rosselli JM, Gurland BH. Ventral rectopexy for rectal prolapse and obstructed defecation. Clin Colon Rectal Surg. 2012;25(1):34–6.
37. Emile SH, Elfeki HA, Youssef M, Farid M, Wexner SD. Abdominal rectopexy for the treatment of internal rectal prolapse: a systematic review and meta-analysis. Color Dis. 2017;19(1):O13–24.
38. Cutait D. Sacro-promontory fixation of the rectum for complete rectal prolapse. Proc R Soc Med. 1959;52(suppl):105.
39. Frykman HM, Goldberg SM. The surgical treatment of rectal procidentia. Surg Gynecol Obstet. 1969;129:1225–30.
40. Formijne Jonkers HA, Maya A, Draaisma WA, Bemelman WA, Broeders IA, Consten EC, Wexner SD. Laparoscopic resection rectopexy versus laparoscopic ventral rectopexy for complete rectal prolapse. Tech Coloproctol. 2014;18(7):641–6.
41. Watadani Y, Vogler SA, Warshaw JS, Sueda T, Lowry AC, Madoff RD, Mellgren A. Sacrocolpopexy with rectopexy for pelvic floor prolapse improves bowel function and quality of life. Dis Colon Rectum. 2013;56(12):1415–22.
42. Dulucq JL, Wintringer P, Mahajna A. Clinical and functional outcome of laparoscopic posterior rectopexy (Wells) for full-thickness rectal prolapse. A prospective study. Surg Endosc. 2007;21(12):2226–30. Epub 2007 May 5.
43. Miles WE. Rectosigmoidectomy as a method of treatment for procidentia recti. Proc R Soc Med. 1933;26:1445–52.
44. Altemeier WA, Culbertson WR, Schwengerdt C, et al. Nineteen years' experience with the one-stage perineal repair of rectal prolapse. Ann Surg. 1971;173:993–1006.
45. Classic articles in colonic and rectal surgery. Edmond Delorme 1847–1929. On the treatment of total prolapse of the rectum by excision of the rectal mucous membranes or recto-colic. Dis Colon Rectum. 1985;28:544–553.
46. Pescatori M, Gagliardi G. Postoperative complications after procedure for prolapsed hemorrhoids (PPH) and stapled transanal rectal resection (STARR) procedures. Tech Coloproctol. 2008;12(1):7–19.
47. Boccasanta P, Venturi M, Stuto A, Bottini C, Caviglia A, Carriero A, Mascagni D, Mauri R, Sofo L, Landolfi V. Stapled transanal rectal resection for outlet obstruction: a prospective, multicenter trial. Dis Colon Rectum. 2004;47(8):1285–96; discussion 1296-7.
48. Petersen S, Jongen J, Schwenk W. Agraffectomy after low rectal stapling procedures for hemorrhoids and rectocele. Tech Coloproctol. 2011;15(3):259–64.
49. Gagliardi G, Pescatori M, Altomare DF, Binda GA, Bottini C, Dodi G, Filingeri V, Milito G, Rinaldi M, Romano G, Spazzafumo L, Trompetto M. Results, outcome predictors, and complications after stapled transanal rectal resection for obstructed defecation., Italian Society of Colo-Rectal Surgery (SICCR). Dis Colon Rectum. 2008;51(2):186–95; discussion 195.
50. Altomare DF, Picciariello A, Memeo R, Fanelli M, Digennaro R, Chetta N, De Fazio M. Pelvic floor function following ventral rectopexy versus STARR in the treatment of obstructed defecation. Tech Coloproctol. 2018;22(4):289–94.
51. Khaled M. Madbouly, Ahmed D. Mohii Transanal rectal resection for treatment of obstructed defecation in the elderly: long-term results of a prospective randomized study. Dis Colon Rectum. 2019;62:47–55.
52. Lenisa L, Schwandner O, Stuto A, Jayne D, Pigot F, Tuech JJ, Scherer R, Nugent K, Corbisier F, Espin-Basany E, Hetzer FH. STARR with Contour Transtar: prospective multicenter European study. Colorectal Dis. 2009;11(8):821–7.
53. Mellgren A, Anzén B, Nilsson BY, Johansson C, Dolk A, Gillgren P, Bremmer S, Holmström B. Results

of rectocele repair. A prospective study. Dis Colon Rectum. 1995;38(1):7–13.
54. Yamana T, Takahashi T, Iwadare J. Clinical and physiologic outcomes after transvaginal rectocele repair. Dis Colon Rectum. 2006;49(5):661–7.
55. Williams SB, Orkin BA, Holt BF, Drenon EA. Transanal rectocele repair: excellent intermediate outcome (review). Surgery. 2006;12(4):191–6.
56. Regadas FS, Regadas SM, Rodrigues LV, Misici R, Silva FR, Regadas Filho FS. Transanal repair of rectocele and full rectal mucosectomy with one circular stapler: a novel surgical technique. Tech Coloproctol. 2005;9(1):63–6.
57. Cruz JV, Regadas FS, Murad-Regadas SM, Rodrigues LV, Benicio F, Leal R, Carvalho CG, Fernandes M, Roche LM, Miranda AC, Câmara L, Pereira JC, Parra AM. Leal VM TRREMS procedure (transanal repair of rectocele and rectal mucosectomy with one circular stapler): a prospective multicenter trial. Arq Gastroenterol. 2011;48(1):3–7.
58. Ceci F, Spaziani E, Corelli S, et al. Technique and outcomes about a new laparoscopic procedure: the Pelvic Organ Prolapse Suspension (POPS). G Chir. 2013;34(5–6):141–4.
59. Kamm MA, Dudding TC, Melenhorst J, Jarrett M, Wang Z, Buntzen S, Johansson C, Laurberg S, Rosen H, Vaizey CJ, Matzel K, Baeten C. Sacral nerve stimulation for intractable constipation. Gut. 2010;59(3):333–40.
60. Knowles CH, Grossi U, Chapman M, Mason J. Surgery for constipation: systematic review and practice recommendations: results I: colonic resection. Color Dis. 2017;19(Suppl 3):17–36.

Defecation Disorders in Children and Adolescents

20

Carlos Zaslavsky and Lucia Camara Castro Oliveira

Introduction

Children and adolescents with defecatory disorders usually present with poor quality of life [1]. Moreover, the impact on the families of affected children is considerable [2]. Characterization of childhood and adolescent defecation disorders has evolved, and recently the disorders were included in Rome IV consensus. This consensus defines criteria and reviews the epidemiology pathophysiology and management of functional constipation and nonretentive fecal incontinence [3, 4]. Defecation disorders are common in children and adolescents. Functional constipation is distributed equally among different social classes with no relationship to family size, ordinal position child/adolescent in the family, or parental age. Boys with constipation had higher rates of fecal incontinence than girls [4, 5]. In general, consultations of defecatory disorders are motivated by constipation, and in 5% of cases, patients have organic etiology for dysfunction [1]. Prevalence studies in Brazil show that constipation occurs in up to 37% of children up to 12 years of age [6]. In North American children, 10–25% of consultations in pediatric gastroenterology are due to constipation [3], and 80% of constipated women have been diagnosed since childhood or adolescence [7]. Fecal incontinence is estimated to affect 0.8–4.1% of children in Western societies [4, 8]. In Brazil, in adolescents up to 18 years of age, the prevalence of fecal incontinence was 1.9% [9]. In children, fecal incontinence is a common problem that is usually associated with the presence of constipation and stool retention [10].

C. Zaslavsky
Santa Casa of Porto Alegre, Department of Digestive Motility, Porto Alegre, Brazil

L. C. C. Oliveira (✉)
Director Department of Anorectal Physiology of Rio de Janeiro and CEPEMED, Rio de Janeiro, Brazil

Normal Functions of the Colon and Rectum

The body's ability to control the passage of feces is undeveloped at birth and is achieved at an average age of 28 months. Continence is the body's ability to recognize when the rectum fills; to decide whether the contents are solid, liquid, or gas; and then to empty the rectum in a socially convenient time. To perform this difficult physiological feat, the body requires adequate coordination between anatomy and factors involved in continence control. Understanding normal function makes it easier to explain and treat the abnormalities that cause constipation and fecal incontinence [11]. Anal continence depends on many factors already discussed in previous Chapters. If fecal contents arrive in the rectum too quickly, they cannot be retained. Both anal sphincters, internal and external, have a resting tone. The internal anal sphincters produce 85%

L. C. C. Oliveira (ed.), *Anorectal Physiology*, https://doi.org/10.1007/978-3-030-43811-1_20

of the resting anal tone, and the external anal sphincter, a striated muscle, provides the remainder of the resting tone. The muscles of the pelvic floor can fatigue 60 seconds after a maximal contraction. Continence has developed when the child recognizes the sensation of rectal filling and realizes that is not acceptable to defecate immediately, but to find a socially convenient time and place (potty training). In a child, defecation can be summarized as rectal filling, rectal distension, relaxation of the internal anal sphincter and puborectalis, increased intra-abdominal pressure, expulsion of the feces, and return to the resting state, combined with parental encouragement for potty training [11, 12]. The mechanism of continence and defecation is complex and multifactorial. The main function of the rectum is to act as a reservoir of feces for a short period. The rectum has the capacity to adapt to the increase in rectal content (complacency), allowing the sensation of the first volume up to the maximum tolerated volume without a marked increase in intraluminal pressure. Rectal distention due to feces and gas or distension caused during an anorectal manometry examination causes a relaxation of the IAS, a phenomenon known as rectoanal inhibitory reflex (RAIR), mediated by ganglion cells (Fig. 20.1).

During this inhibition, fecal continence is maintained by the simultaneous contraction of the external anal sphincter (EAS). The external anal sphincter and the puborectalis muscle form a functional unit composed of striated muscle fibers, under voluntary control, which presents in tonic contraction even during rest. This contraction is responsible for 15–25% of the anal canal tone. EAS is innervated by the pudendal nerve (S2 to S4) and responds by contracting voluntary fecal retention and a sudden increase in abdominal pressure when the subject coughs or by relaxing on defecation. The puborectalis muscle is innervated directly by sacral branches (S3 and S4). When conditions are inadequate to defecation, EAS and the puborectalis muscle contract, avoiding it. In summary, defecation involves the sensation of increased rectal contents and their discrimination, intact fecal retention mechanisms, and voluntary control of the expulsion of fecal contents [11–13].

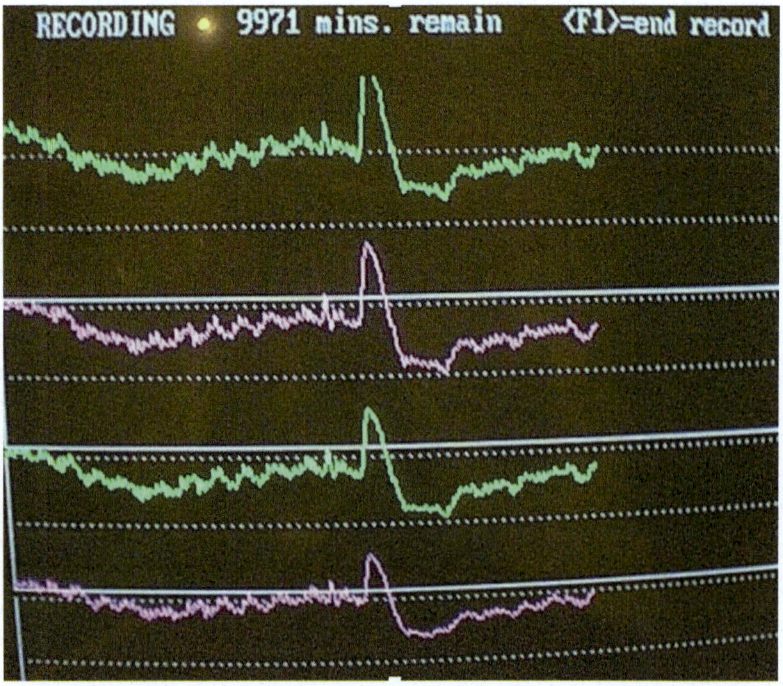

Fig. 20.1 Anorectal manometry showing RAIR with insufflation of 30 mL in to the balloon

Pathophysiology of Defecation Disorders in Children

The Rome IV defecatory disorders, in children and adolescents, are classified as functional constipation and nonretentive fecal incontinence. The diagnostic criteria for functional constipation must include two or more of the following occurring at least once per week for a minimum of 1 month with insufficient criteria for a diagnosis of irritable bowel syndrome: two or fewer defecations in the toilet per week in a child of a developmental age at least 4 years, at least one episode of fecal incontinence per week, history of fecal retention, history of painful or hard bowel movements, presence of large fecal mass in the rectum, and history of large-diameter stools that can obstruct the toilet [2, 4]. Functional constipation in children is often the result of repeated attempts of voluntary withholding of feces. Abnormal defecation or pelvic dyssynergia has been reported in 63% of children with chronic constipation. Progressive fecal accumulation in the rectum eventually leads to pelvic floor muscle fatigue and anal sphincter poor competence leading to fecal incontinence [14, 15]. Because functional constipation is equally common in both sexes and children with diverse socioeconomic backgrounds, dietary practices, and cultural influences, the triggering event is most likely the instinct to avoid defecation because of pain or social reason. As a consequence of withholding stool, the colonic mucosa absorbs water from the feces, and the retained stools become progressively more difficult to evacuate. This process leads to a vicious cycle of stool retention in which the rectum is increasingly distended, resulting in overflow fecal incontinence, loss of rectal sensation, and, ultimately, loss of normal urge to defecate. Increasing fecal accumulation in the rectum also causes decreased motility in the foregut, leading to anorexia, abdominal distention, and pain [4, 16–18]. These events lead to painful defecation such as toilet training, changes in routine or diet, stressful events, intercurrent illness, unavailability of toilets, or the child/adolescent postponing defecation because she or he is "too busy." The passage of large, hard stools that painfully stretch the anus may frighten the child/adolescent, resulting in fearful determination to avoid the defecation. They respond to the urge to defecate by contracting their anal sphincter and gluteal muscles, attempting to withhold stools. Eventually the rectum habituates to the stimulus of the enlarging fecal mass, and the urge to defecate subsides. With time, such retentive behavior becomes an automatic reaction [17, 19–21]. Childhood constipation is a chronic condition, which continues to affect 25–50% of youth into adulthood. Approximately one-half of children with constipation also experience fecal incontinence (Fig. 20.2). Although functional constipation continuing into adulthood has adverse effects on both physical and psychological health, fecal incontinence in a growing child/adolescent has additional, profound psychosocial consequences [14, 22, 23]. In children and adolescents with functional constipation, it is possible to identify the segment of the colon with altered motility. Using the technique of radiopaque markers to measure transit time in the colon, functional constipation can be divided into three subgroups: slow colonic transit, outlet obstruction or, and normal transit time. Slow transit in the colon consists of time delay in the right colon, the left

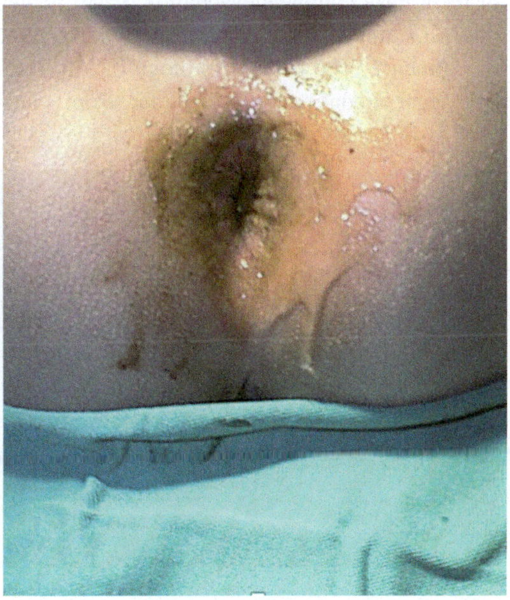

Fig. 20.2 Overflow incontinence in a child with fecal impaction and a huge fecaloma

colon, or both. When radiopaque markers remain in the rectal ampulla, constipation is considered to be outlet obstruction type (Fig. 20.3) [24, 25].

The Rome IV diagnosis of nonretentive fecal incontinence must include at least a 1-month history of the following symptoms in a child with a developmental age older than 4 years: defecation into places inappropriate to the sociocultural context; no evidence of fecal retention; and after appropriate medical evaluation, the fecal incontinence cannot be explained by another medical condition [2, 4]. In the literature, fecal incontinence in children is frequently described by the terms "encopresis" [26] and "fecal soiling" [27], but the Rome III [28] and the Paris Consensus on Childhood Constipation Terminology have favored the term fecal incontinence [29]. The pathophysiology of fecal incontinence is related to constipation in children. The presence of fecal incontinence is so well accepted as being a result of stool retention, that even the new Rome IV criteria for the diagnosis of functional constipation in children have one of the possible diagnostic characteristics as the presence of fecal incontinence greater than once per week [2, 4, 10]. Children and adolescents with nonretentive fecal incontinence pass stools into inappropriate places without evidence of stool retention. The majority of them have complete evacuation of the bowel, not just staining of the underwear as in retentive incontinence. The pathophysiology of nonretentive fecal incontinence is still far from clear. Patients with nonretentive fecal incontinence have normal defecation frequencies and colonic and anorectal motility parameters, differentiating this condition from functional constipation [4]. The diagnosis of nonretentive fecal incontinence should be based on clinical symptoms, such as normal defecation frequencies and absence of abdominal or rectal palpable mass, in combination of normal transit marker studies [30].

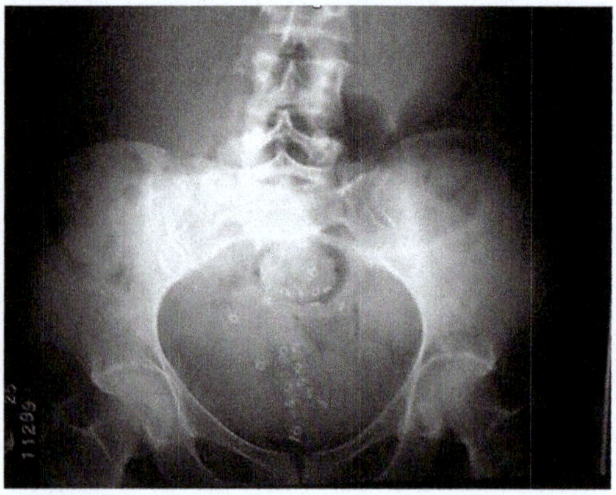

Fig. 20.3 Transit time suggestive of outlet obstruction syndrome

Etiology

Constipation may accompany numerous clinical and surgical organic diseases or ingestion of medications and, in the absence of clinical evidence suggesting them (90–95%), may be termed functional constipation (Table 20.1) [20, 28].

It is important to evaluate Hirschsprung's disease because of its importance in the differential diagnosis in children and adolescents with severe constipation. Hirschsprung's disease is a rare condition that occurs in 1 out of every 5000 births. It most commonly affects males and can be associated with Down syndrome, renal malformations, and intestinal atresia. The newborn presents with intestinal obstruction with large dilatation of the abdomen and absence of meconium elimination. The anal narrowing caused by aganglionosis prevents the natural elimination of gas, and the digital rectal examination, when performed, is accompanied by a characteristic explosive elimination of the congenital megacolon,

Table 20.1 Etiology of constipation

Functional constipation
Slow transit in the colon
Anorectal dysfunction
Normal colonic motility
Irritable bowel syndrome
Organic intestinal alteration
Stenosis and/or anal fissure
Anorectal malformation
Hirschsprung's disease
Neural dysplasia
Others
Hypothyroidism
Diabetes
CNS impairment or psychiatric illness

which may facilitate the diagnosis. In these cases, anorectal manometry can confirm the absence of the rectoanal inhibitory reflex (Fig. 20.4) [22, 23, 31, 32]. Rectal biopsy is the gold standard for diagnosing Hirschsprung's disease [4].

Fecal incontinence is a significant problem associated with functional constipation. The primary reason for fecal incontinence is fecal retention. Epidemiological and hospital studies have shown fecal incontinence in 75–90% of children with constipation [8, 33]. Of those with constipation, 95% of affected children have functional constipation, and the remaining 5% of children have organic etiologies such as anatomic abnormalities, Hirschsprung and imperforate anus, or spinal defects [34]. Nonretentive fecal incontinence might be a manifestation of an emotional disturbance in school-aged children and adolescents. Psychosocial factors and consequent deranged defecation dynamics may play a role in nonretentive fecal incontinence [4, 35]. Fecal incontinence may be triggered by an event in the life of a child or a family. Nonretentive fecal incontinence might be a manifestation of an emotional disturbance and has been described as a result of sexual abuse in childhood. Abuse history is common in gastroenterological practice, and patients with defecation disorders tend to have more severe abuse [36, 37]. Anatomic or neurologic causes account for up to 5% of cases: anterior location of the anus, anal stenosis, perineal fistula with anal atresia, anorectal trauma, post-surgical repair of anal atresia, meningomyelocele or tumor of the spinal cord, Hirschsprung's disease, cerebral palsy, neuromuscular disease, and laxative abuse [33].

Diagnosis

The diagnosis of functional constipation should be based on the pediatric Rome IV criteria [4]. Symptoms of functional constipation include infrequent, painful defecation, hard stools, fecal incontinence, and abdominal pain [38]. In addition, the Bristol Stool Scale is a valuable tool for diagnosis and post treatment evaluation [39–41]. If only one of the Rome criteria is present and the diagnosis is uncertain, a digital rectal examination is recommended [38]. Patients should undergo a complete physical examination to investigate clinical evidence of organic causes for constipation, the abdomen, perineum, and anus, including rectal examination and neurological evaluation of the extremities. Young patients with a history of constipation without other clinical problems and normal physical examination do not require complementary exams [28, 40]. Laboratory testing to screen organic causes for constipation, or for cow's

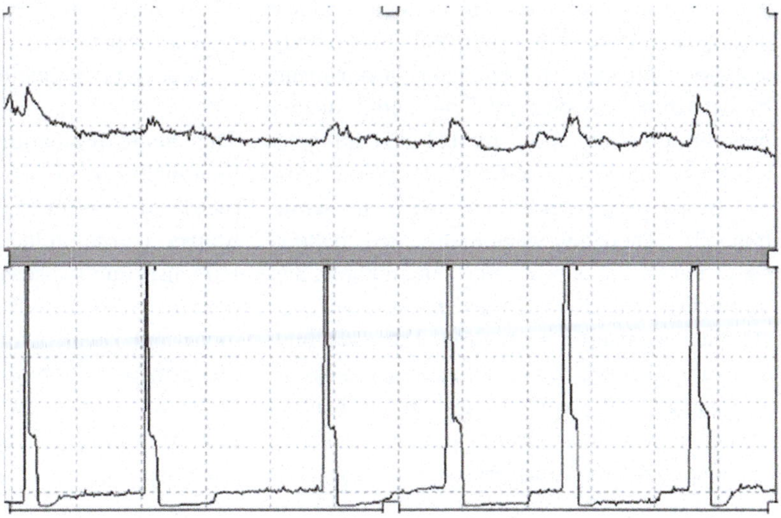

Fig. 20.4 Manometry revealing absence of inhibitory reflex in young patient with constipation and suspicion of congenital megacolon

milk allergy, is not recommended in the absence of suggestive symptoms [4]. There is no role for routine use of an abdominal x-ray to diagnose functional constipation. A plain abdominal radiograph may be used in a child/adolescent if fecal impaction is suspected but in whom physical examination is unreliable / or not possible [25]. A barium enema should not be used as an initial diagnostic tool for evaluation of functional constipation [4, 38]. The subjective aspect of the concept of constipation makes it necessary to establish an objective criterion to describe it. Some patients have a history of infrequent bowel movements but have no objective findings of constipation. In these patients an evaluation of colonic transit times with radiopaque markers may be helpful [42]. The total and segmental colonic transit time, traced by radiopaque markers, allows the identification of the colon segment that has a motility alteration that causes constipation, and helps in determining an effective treatment, depending on segmental transit times. Severe chronic constipation in children and adolescents may be due to slow colonic transit; different patterns of delayed transit have been described [43, 44]. Slow transit constipation was diagnosed when there was a transit delay through the right colon, the left colon, or both. Outlet obstruction was diagnosed when there was an exclusive delay in the rectosigmoid [24]. In rare and specific situations, colonic transit time can be utilized, especially when constipation is refractory to conventional clinical treatment [45, 46]. Anorectal manometry assesses continence and defecatory mechanisms by determining resting anal pressure, squeeze pressure, presence of an internal anal sphincter inhibitory reflex, relaxation of the pelvic muscles on attempted defecation, and rectal sensation [3]. Anorectal manometry and balloon expulsion testing appear to be safe, reliable, and useful tests in the evaluation and management of chronic constipation in children. There is a high correlation between a normal anorectal manometry and balloon expulsion test. If the balloon expulsion test is abnormal and the anorectal manometry does not identify a cause for the distal obstruction, additional studies may be needed [47]. Children with functional constipation and fecal incontinence who were able to defecate the rectal balloon were twice as likely to recover 12 months after the start of treatment [48]. Manometric studies are useful to differentiate between constipation-associated fecal incontinence and nonretentive fecal incontinence in children. It has been shown that children with constipation associated have higher threshold for rectal sensation with largest volume of balloon to provoke rectal sensation [8]. The finding of a prolonged duration of relaxation in children with fecal incontinence suggests that this abnormality in anorectal function may be, at least in part, responsible for this complication of constipation [49]. According to the North American Society for Pediatric Gastroenterology, Hepatology and Nutrition [50], the indications of the anorectal manometry include:

1. Evaluation of the internal anal sphincter relaxation in response to rectal balloon distention to help exclude Hirschsprung's disease (Fig. 20.3).
2. Evaluation of patients with anorectal malformations with persistent defecation problems after surgical repair.
3. To assess persistent defecation problems after surgery for Hirschsprung's disease.
4. To evaluate defecation dynamics in patients with chronic constipation.
5. To assess rectal sensation and sphincter tone in patients with fecal incontinence with neurogenic problems.

The main indication to perform anorectal manometry is to assess the presence of the rectoanal inhibitory reflex, but the rectal biopsy is the gold standard for diagnosing Hirschsprung's disease [4]. Anorectal manometry provides a noninvasive objective assessment and severity of constipation and fecal incontinence in children [51]. High-resolution anorectal manometry is an increasingly common procedure in pediatric patients to rule out Hirschsprung's disease as well as assess anorectal function and sensation. Despite the low risk of the conventional and high-resolution anorectal manometry, healthcare providers should be aware of the high levels of anxiety and distress that occur in patients and their parents associated with this procedure [52]. Rectal biopsy by aspiration or surgery is performed when there is a clinical suspicion of Hirschsprung's disease for the investigation of ganglion cells. Surgical biopsy is known as myectomy and consists of excision of an internal sphincter segment (Fig. 20.5). Rectal biopsies demonstrating the absence of ganglion cells in the submucosal plexus are diagnostic of Hirschsprung's disease [17].

Colon manometry is performed with the help of colonoscopy allowing the catheter to be inserted into the right colon and therefore obtaining the motility patterns of the colon. Although this information may be valuable in research, it has not yet proved important in the evaluation of patients with functional constipation or fecal incontinence [50]. The diagnosis of nonretentive fecal incontinence should be based on clinical symptoms, such as normal defecation frequency and absence of abdominal or rectal palpable mass in combination with normal transit marker studies [4]. Clinical severity should be assessed using the Cleveland Clinic Florida Fecal Incontinence Score (CCF-FIS), where 0 indicates perfect continence and 20 indicates total incontinence [53]. The physical examination should be complete, including the anorectal region and neurological evaluation. Rectal examination, during which the patient is asked to contract the sphincter anal, is considered an adequate evaluation procedure [4]. By the digital rectal examination, one can evaluate the resting and squeeze tone of the sphincters, but anorectal manometry allows a more precise evaluation, and should be utilized to guide treatment options [50]. Endoanal ultrasonography is a simple test for evaluation of structural damage of both the interna and external anal sphincters, which may be useful in adolescents and in the follow-up of children with anorectal malformation [54].

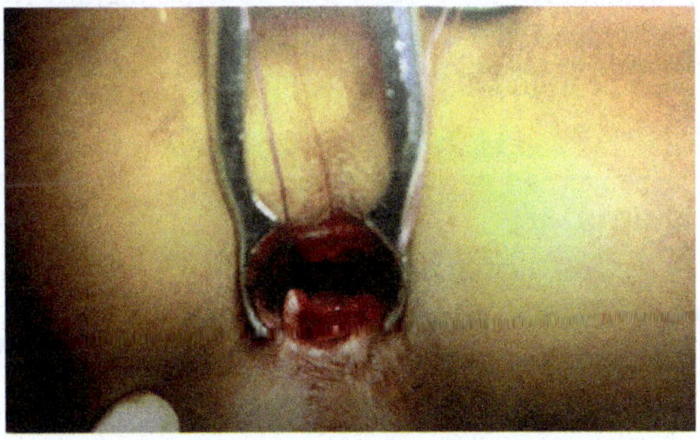

Fig. 20.5 Anal myectomy

Treatment

The initial management of patients with functional constipation is non-pharmacological treatment and consists of education, counseling the family and the patient, regular dietary advice (sufficient fiber and fluid intake), and in older children toilet training, and stool diary [2]. The pharmacology approach comprises two steps: rectal or oral disimpaction for children who present with fecal impaction and maintenance therapy to prevent further fecal retention. Evidence shows that polyethylene glycol and enemas are equally effective for fecal disimpaction [2, 31]. Table 20.2 describes the initial measures of the treatment of children and adolescents with functional constipation.

For maintenance therapy evidence shows that polyethylene glycol is more effective compared with lactulose, milk of magnesia, mineral oil, or placebo. More studies have been performed evaluating the effect of lactulose than studies evaluating the effect of milk of magnesia and mineral oil in children with constipation. More importantly, lactulose is considered to be safe for all ages. For these reasons, lactulose is recommended in case polyethylene glycol is not available [31, 55]. The goal of biofeedback training is to improve bowel function by restoring a normal pattern of defecation. Biofeedback therapy is an instrument-based learning process that is based on "operant conditioning" techniques. The governing principle is

Table 20.2 Treatment of children and adolescents with functional constipation

Initial treatment [2]	
	Education: counseling the family and/or the patient about constipation
	Rectal or rectal mass withdrawal by oral or rectal route; necessary step before maintenance treatment
Maintenance treatment to prevent recurrence	
	Dietary guidance with fiber enhancement
	Stimulate an intestinal habit
	Laxatives to regulate bowel habits
Referral to a specialist	
	Refractory cases
	Review of prescribed treatments
	Therapeutic measures according to functional evaluation
Other measures	
	Biofeedback
	Surgery
	Psychotherapy

[a]In isolation, changing the diet does not provide benefits to the constipated patient

that any behavior when reinforced repeatedly can be learned and improved. In patients with dyssynergic defecation, the goal of biofeedback training is threefold:

1. To correct the dyssynergia or incoordination of the abdominal, rectal, puborectalis, and anal sphincter muscles in order to achieve a normal and complete evacuation
2. To facilitate normal evacuation by simulated defecation training using balloons
3. To enhance rectal sensory perception in patients with impaired rectal sensation

Biofeedback therapy is recommended for the short-term and long-term treatment of constipation with dyssynergic defecation [3, 13, 40].

In cases of Hirschsprung's disease, anal myectomy is useful for confirmation of diagnosis, promotes relaxation of the anal canal, and can facilitate the normalization of the evacuation [56]. Chronic constipation is not initially a disease of a psychological nature, but there may be of benefit to refer children with constipation and behavioral abnormalities to a mental health provider [31]. Sacral neurostimulation has not yet proved useful in the treatment of constipation in children and adolescents [57]. If constipation is not treated early, it may lead to clinical and emotional repercussions for both the child and his or her family. There is a low adherence rate to drug treatment of constipation in children. It is necessary to seek new strategies to increase treatment adherence, while avoiding complications and reducing costs [58]. Treatment of anal incontinence should be directed to the specific etiology and the change of bowel habits. When associated with fecaloma, it should be removed by manual removal, with or without anesthesia. After removal, a program of laxatives and/or fibers and behavioral reinforcement should be started to prevent new accumulations [42]. As in adults, the first treatment of anal incontinence should be clinical, with dietary measures, pelvic exercises, and medications [59]. Because patients benefit from individual psychotherapy, one measure that can improve self-esteem and increase adherence to treatment is the use of motivational interviewing [60, 61]. Parents need to understand that psychological disturbances, learning difficulties, and behavioral problems are usually significant contributors to defecatory symptoms. Victims of sexual abuse must be identified and referred for appropriate counseling [4].

Biofeedback therapy with anorectal manometry can improve bowel movement dynamics in case of anal incontinence and is the treatment of choice for patients who do not respond to initial clinical treatment. Biofeedback is superior of pelvic exercises [3, 42]. The most successful approach to management of nonretentive fecal incontinence involves behavioral therapy [4]. Sacral neuromodulation is an effective therapy for individuals above 18 years [57]. Plug devices may also be useful in some patients with seepage [3].

Summary

Defecation disorders in children and adolescents are a challenge for physicions and the families. Fecal incontinence significantly decreases quality of life compared with functional constipation alone in children and adolescents. Strategies for early identification and treatment of constipation and/or fecal incontinence may mitigate the negative impact of these highly prevalent conditions [62]. Anal incontinence may not be reported by the family, which may result in children being abused in school and having antisocial behavior [63]. In an educational article, the International Society for Continence in Childhood suggests that the treatment is clinical and receptive to patient [42]. The systematic treatment of anal incontinence, as well as in adults, leads to the improvement of most patients [59]. The majority of cases of constipation and fecal incontinence are secondary to functional disorders, rather than organic causes, and result in behavioral problems, which affect the social life of the child, as well the family [21].

References

1. Di Lorenzo C. Pediatric anorectal disorders. Gastroenterol Clin N Am. 2001;30(1):269–87.
2. Zeevenhooven J, Koppen IJN, Benninga MA. The new Rome IV criteria for functional gastrointestinal disorders in children and toddlers. Pediatr Gastroenterol Hepatol Nutr. 2017;20(1):1–13.
3. Rao SCS, Bharucha AE, Chiaroni G, Felt-Bersma R, Knowles C, Malcolm A, Wald A. Anorectal disorders. Gastroenterology. 2016;150:1430–42.
4. Hyams JS, Di Lorenzo C, Saps M, Shulman RJ, Staiano A, van Tiburg M. Childhood functional gastrointestinal disorders: child/adolescent. Gastroenterology. 2016;150:1456–68.
5. Lewis LL, Palsson OS, Whitehead WE, Tilburg MAL. Prevalence of functional gastrointestinal disorders in children and adolescents. J Pediatr. 2016;177:39–43.
6. Zaslavsky C, Ávila ED, Araújo MA, Pontes MRN, Lima EN. Intestinal constipation in childhood- a prevalence study. Revista AMRIGS. 1988;32: 100–2.
7. Stewart WF, Liberman JN, Sandler RS, Woods MS, Stemhagena A, Chee E, et al. Epidemiology of constipation (EPOC) – study in the United States: relation of clinical subtypes to sociodemographic features. Am J Gastroenterol. 1999;94(12):3530–40.
8. Rajindrajith S, Devanarayana NM, Benninga MA. Review article: faecal incontinence in children: epidemiology, pathophysiology, clinical evaluation and management. Aliment Pharmacol Ther. 2013;37(1):37–4.
9. Zaslavsky C, Jurach MT, Barros CP. Epidemiology of anal incontinence, Porto Alegre-RS, Brasil. Revista AMRIGS. 2012;56(4):289–94.
10. Nurko S, Scott SM. Coexistence of constipation and incontinence in children and adults. Best Pract Res Clin Gastroenterol. 2011;25:28–41.
11. O'Brien MD, Phillips SF. Colonic motility in health and disease. Gastroenterol Clin N Am. 1996;25(1):147–63.
12. Palit S, Luniss PJ, Scott MS. Dig Dis Sci. 2012;57:1445–64.
13. Loening-Baucke V. Biofeedback therapy for fecal incontinence. Dig Dis Sci. 1990;8(2):112–24.
14. Loening-Baucke V. Constipation in early childhood: patient characteristics, treatment, and long term follow up. Gut. 1993;34(10):1400–4.
15. Griffiths DM. The physiology of continence: idiopathic fecal constipation and soiling. Semin Pediatr Surg. 2002;11(2):67–74.
16. Partin JC, Hamill SK, Fishel JE, Partin JS. Painful defecation and fecal soiling in children. Pediatrics. 1992;89(6 Pt 1):1007–9.
17. Baker SS, Liptak GS, Colletti RB, Croffie JM, Di Lorenzo C, Ector W, et al. Constipation in infants and children: evaluation and treatment. A medical position statement of the North American Society for Pediatric Gastroenterology and Nutrition. J Pediatr Gastroenterol Nutr. 1999;29(5):612–26.
18. Morais MB, Verena HV. Intestinal constipation. J Pediatr. 2000;76(2):S147–58.
19. Penberthy JK. Precipitants of constipation during early childhood. J Am Board Fam Pract. 2003;16: 213–8.
20. Hyman PE, Fleisher D. Functional fecal retention. Pract Gastroenterol. 1992;31:29–37.
21. Hyman PE, Milla PJ, Benninga MA, Davidson GP, Fleisher DF, Taminiau J. Childhood functional gastrointestinal disorders: neonate/toddler. Gastroenterology. 2006;130(5):1519–26.
22. Van Ginkel R, Reitsma JB, Büller MP, Van Wijk MP, Jan AJ, Taminiau M, Benninga MA. Childhood constipation: longitudinal follow-up beyond puberty. Gastroenterology. 2003;125(2):358–63.
23. Kovacic K, Sood MR, Mugie S, Di Lorenzo C, Nurko S, Heinz NBS, Ponnambakam A, Beesley C, Sanghavi R, Silverman AH. A multicenter study on childhood constipation and fecal incontinence: effects on quality of life. J Pediatr. 2015;166:1482–7.
24. Zaslavsky C, Silveira TR, Maguilnik I. Total and segmental colonic transit time with radiopaque markers in adolescents with functional constipation. J Pediatr Gastroenterol Nutr. 1998;27(2):138–42.
25. Benninga MA, Tabbers MM, Rijn V. How to use plain abdominal radiograph in children with functional defecation disorders. Arch Dis Child Educ Pract. 2016;101:187–93.
26. Loening-Baucke V. Encopresis. Curr Opin Pediatr. 2002;14:570–5.
27. Partin JC, Hamill SK, Fischel JE, Partin JS. Painful defecation and fecal soiling in children. Pediatrics. 1992;89(6):1007–9.
28. Rasquin A, Di Lorenzo C, Forbes D, Guiraldes E, Hyams JS, Staiano A, et al. Childhood

functional gastrointestinal disorders: child/adolescent. Gastroenterology. 2006;130(5):1527–37.
29. Benninga M, Candy DC, Catto-Smith AG, Clayden G, Loening-Baucke V, Di Lorenzo C, Nurko S, Staiano A. The Paris Consensus on Childhood Constipation Terminology (PACCT) Group. J Pediatr Gastroenterol Nutr. 2005;40(3):273–5.
30. Bongers ME, Voskuijl WP, van Rijn RR, Benninga MA. The value of the abdominal radiograph in children with functional gastrointestinal disorders. Eur J Radiol. 2006;59(1):8–13.
31. Tabbers MM, DiLorenzo C, Berger MY, Faure C, Langendam MW, Nurko S, et al. Evaluation and treatment of functional constipation in infants and children: evidence-based recommendations from ESPGHAN and NASPGHAN. J Pediatr Gastroenterol Nutr. 2014;58(2):258–74.
32. Rodriguez L, Sood M, Lorenzo D, Saps M. Na ANMS-NASPGHAN consensus document on anorectal and colonic manometry. Neurogastroenterol Motil. 2017;29:1–8.
33. Loening-Baucke V. Fecal incontinence in children. Am Fam Physician. 1997;55(6):2229–35.
34. Muddasani S, Moe A, Semmelrock C, Gilbert CL, Enemuo V, Chiou WH, Chumpitazi BP. Physical therapy for fecal incontinence with pelvic floor dyssynergia. J Pediatr. 2017;11:1–5.
35. Rajindrjith S, Devanarayana NM, Benninga MA. Fecal incontinence in adolescents is associated with child abuse, somatization, and poor health-related quality of life. J Pediatr Gastroenterol Nutr. 2016;62(5):698–703.
36. Mellon MW, Whiteside SP, Friedrich WN. The relevance of fecal soiling as an indicator of child sexual abuse: a preliminary analysis. J Dev Behav Pediatr. 2006;27(1):25–32.
37. Drossman DA. Abuse, trauma, and GI illness: is there a link? Am J Gastroenterol. 2011;106(1):14–25.
38. Tabbers MM, Boluyt N, Berger MY, Benninga MA. Clinical practice: diagnosis and treatment of functional constipation. Eur J Pediatr. 2011;170(8):955–63.
39. Vriesman MH, Velasco-Benitez CA, Ramirez CR, Benninga MA, Di Lorenzo C, Saps M. Assessing children's report of stool consistency: agreement between the pediatric Rome III questionnaire and the Bristol Stool Scale. J Pediatr. 2017 Nov;190:69–73.
40. Santos JD, Lopes RI, Koyle MA. Bladder and bowel dysfunction in children: an update on the diagnosis and treatment of a common, but underdiagnosed pediatric problem. J Can Urol Assoc. 2017;11(Suppl 1):S64–72.
41. Zaslavsky C, Guerra TC. The Bristol Form Scale in the clinical diagnosis of constipation in in children. Rev AMRIGS. 2016;60(2):127–31.
42. Koppen IJN, Vriesman MH, Tabbers MM, Di Lorenzo C, Benninga MA. Awareness and implementation of the 2014 ESPGHAN/NASPGHAN guideline for childhood functional constipation. J Pediatr Gastroenterol Nutr. 2018;66(5):732–7.
43. Benninga MA, Buller HA, Tytgat GNJ, Akkermans MA, Bossuyt PM, Taminiau JAJM. Colonic transit time in constipated children: does pediatric slow-transit constipation exist? J Pediatr Gastroenterol Nutr. 1996;23(3):241–51.
44. Bautista Casasnovas A, Varela Cives R, Villanueva Jeremias A, Castro-Gago M, Cadranel S, Tojo Sierra R. Measurement of colonic transit time in children. J Pediatr Gastroenterol Nutr. 1991;13(1):42–5.
45. Velde SV, Notebaert A, Meersschaut V, Herregods N, Van Winckel M, Van Bierliet S. Colon transit time in healthy children and adolescents. Int J Colorectal Dis. 2013;28(12):1721–4.
46. Zaslavsky C, De Barros SG, Gruber AC, MacIel AC, Da Silveira TR. Chronic functional constipation in adolescents: clinical findings and motility studies. J Adolesc Health. 2004;34(6):517–22.
47. Belkind-Gerson J, Goldstein AM, Kuo B. Balloon expulsion test as a screen for outlet obstruction in children with chronic constipation. J Pediatr Gastroenterol Nutr. 2013;56(1):23–6.
48. Loening-Baucke V. Balloon defecation as a predictor of outcome in children with functional constipation and encopresis. J Pediatr. 1996;128(3):336–40.
49. Raghunath N, Glassman MS, Halata MS, Berezin SH, Stewart JM, Medow MS. Anorectal motility abnormalities in children with encopresis and chronic constipation. J Pediatr. 2011;158(2):293–6.
50. Rodriguez L, Sood M, Di Lorenzo C, Saps M. Na ANMS-NASPGHAN consensus document on anorectal and colonic manometry in children. Neurogastroenterol Motil. 2016;10:1–8.
51. Keshtgar AS, Ward HC, Clayden GS. Pathophysiology of chronic childhood constipation: functional and morphological evaluation by anorectal manometry and endosonography and colonic transit study. J Pediatr Surg. 2013;48(4):806–12.
52. Lamparyk K, Mahajan L, Debeljak A, Steffen R. Anxiety associated with high-resolution anorectal manometry in pediatric patients and parents. J Pediatr Gastroenterol Nutr. 2017;65(5):e98–e100.
53. Jorge JMN, Wexner SD. Etiology and management of fecal incontinence. Dis Colon Rectum. 1993;36(1):77–97.
54. Caldaro T, Romeo E, De Angelis P, Gambita RA. Three-dimensional endoanal ultrasound and anorectal manometry in children with anorectal malformations: new discoveries. J Pediatr Surg. 2012;47(5):956–63.
55. Infante Pina D, Segarra Cantón O, Vilalta Casas R, Pardina JC, Linan MJL, Busoms CM. Eficacia, tolerancia y seguridad a largo plazo del polietilenoglicol 3350 con eletrolitos en el tratamiento del estreñimiento functional en niños. Annales de Pediatrie. 2014;80(5):278–84.
56. Heikkinen M, Lindahl H, Rintala RJ. Long-term outcome after internal sphincter myectomy for internal sphincter achalasia. Pediatr Surg Int. 2005;21(2):84–7.

57. Sulkowski JP, Nacion MK, Deans KJ, Minneci PC, Levitt MA, Mousa HM, et al. Sacral nerve stimulation for fecal and urinary incontinence and constipation in children. J Pediatr Surg. 2015;50(10): 1644–7.
58. Steiner SA, Torres MR, Penna FJ, Gazzinelli BF, Corradi CG, Costa AS, Ribeiro IG, de Andrade EG, do Carmo Barros de Melo M. Chronic functional constipation in children: adherence and factors associated with drug treatment. J Pediatr Gastroenterol Nutr. 2014;58(5):598–60.
59. Zaslavsky C, Jurach MT, Coronel AL. Anal incontinence: clinical treatment, Porto Alegre, Brazil. Rev AMRIGS. 2014;58(3):220–4.
60. Zaslavsky C, Loureiro RG, Jurach MT. Motivational interview in the treatment of anal incontinence. Rev AMRIGS. 2013;57(3):180–4.
61. MacNeily AE. Should psychological assessment be a part of incontinence management in children and adolescents? J Urol. 2016;195(5):1327–8.
62. Kovacic K, Sood MR, Di Lorenzo C, Nurko S, Heinz N, et al. A multicenter study on childhood constipation and fecal incontinence: effects on quality of life. J Pediatr. 2015;166(6):1482–7.
63. Joinson C, Heron J, Butler U, von Gotard A. Psychological differences between children with and without soiling problems. Pediatrics. 2006;117(5): 1575–84.

21 Anal Manometry and Chagasic Colopathy

Helio Moreira Jr.

Historical Aspects of Chagas Disease

Chagasic disease remains an endemic condition in South and Central American countries, particularly in the Brazilian midwest. Carlos Chagas identified the etiologic agent in 1909, describing the evolutionary cycle of the parasite in the human body as well as the epidemiological aspects of the disease [1]. Later, he correlated dysphagia in patients presenting an acute form of the disease [2].

Amorim and Correa Netto (1932) [3] and Etzel (1934) [4] described esophageal and colonic histological abnormalities in patients with Chagasic disease. They emphasized the myenteric neural plexus lesions as a common finding and corroborated with the achalasia theory described by Hertz in 1915 [5].

Since the first manuscripts and case reports from Carlos Chagas up to 1955, this neglected disease was placed in second plane, until Koeberle, a prominent pathologist and professor at the Medical School of Ribeirão Preto/SP, demonstrated a clear association of trypanosome infection (*Tripanossoma cruzi*, a protozoan parasite), followed to myenteric plexus lesions, and colonic and/or esophageal dilatation [6].

H. Moreira Jr. (✉)
Surgical Department, Colorectal Unit,
School of Medicine, Federal University of Goiás,
Goiânia, Goiás, Brazil

Epidemiological and Pathological Aspects of Chagasic Colopathy

Chagas disease is related to poor socioeconomical conditions. The higher prevalence of the disease in the Brazilian midwest area was related to rudimental infrastructure, mainly in the rural areas. More specifically, the houses where people lived were commonly infested with contaminated triatominae, a hematophagous insect responsible for the infection, as it likes to nest in cracks and holes in substandard housing.

The highest incidence of infected patients used to be from the states of Minas Gerais, Bahia, Goiás, and north of São Paulo, according to a serologic inquiry performed between 1970 and 1980 [7]. In 1989, Zicker and colleagues [8] evaluated rural laborers from Goiás and found an impressive prevalence of 13.1% of seroprevalence for *Trypanosoma cruzi*. However, due to efficient programs of vector control initiated in the seventies, Brasil was internationaly certified in 2006 that infections by *Triatoma infestans*, an exotic bug species responsible for the majority vectorial transmition in the past, no longer has been reported. Despite of these successful programs, it is estimated that there are still aproximately 12 millions of infected patients in the Americas, being 1 million in Brazil.

Although T cruzi infection is more commonly acquired through the triatomine bug, it may also be trasmitted transplacentally, through transfusion of contaminated blood products, from a

L. C. C. Oliveira (ed.), *Anorectal Physiology*, https://doi.org/10.1007/978-3-030-43811-1_21

transplanted organ of an infected donor, or rarely from contaminated food or laboratory accidents. This is very relevant in the US scenario as only after 2012 that donated blood at all US blood centers underwent screening for T cruzi infection. It has also been reported the presence of the triatomine bug in the southern United States. However, the vast majority of persons infected with T cruzi living in the United States acquired the infection while living in Latin American countries, specially imigrants. Fortunately the US is not considered a endemic country for Chagas disease. However, it has been estimated that 240.000 to 400.000 (considering undocumented imigrants) persons living in the United States are infected with the T cruzi parasite. Therefore, medical facilities serving latin american imigrantcommunities need to be aware of the disease prevalence in this population [9, 10]. As trypomastigote parasites reach the harbored blood flow, there is a tropism of this flagellate to the muscular cells of the heart and/or to the muscular visceral layers, more commonly to the esophagus and colon. Inside these cells, the parasite becomes into an amastigote form and rapidly multiplies until the cell ruptures. It then returns to a trypomastigote form as it again reaches the blood flow, perpetrating the cycle of infecting new cells, every 5–10 days. From this point, a still obscure immunobiological reaction between harbored and parasite is established, characterizing the chronic phase of the disease.

In the digestive tract, the infection promotes focal irregular myositis, with inflammatory infiltration surrounding Meissner and Auerbach neural plexuses, despite the absence of the parasite itself. This fact is supported by the theory of a similar antigenic component between the *T. cruzi* and the cells structure, determining an autoimmune inflammatory response.

This polymorphic acute phase makes an accurate diagnosis quite difficult to establish. Moreover, it is almost impossible to determine if, or when a symptomatic megacolon will develop. Other unknown factors influence on the harbored-parasite relation, as the majority will become asymptomatic Chagasic patients, named as the indeterminate form of the disease, and only 8–10% will develop the digestive form (megacolon or megaesophagus) [11]. It is however well known that the degree of neuronal destruction on the bowel wall is a crucial factor for megaviscera development. Koeberle [6] observed that the cutoff point is at least 55% of neural organ destruction.

Chagasic colopathy is particularly rare in individuals younger than 10 years of age. The highest incidence is between the fourth and sixth decade of life, resulting in a high social cost. Isolated radiological megacolon is rare (14.9%), but associated megaesophagus is quite common (53.8%). Both megaesophagus and Chagasic cardiopathy occur in approximately one quarter of cases (26.4%). Chagasic cardiopathy alone is the rarest event (4.9%) [8].

In 1939, Cannon [12] demonstrated that an enervated organ amplifies the response to physiological and pharmacological stimuli. During the chronic phase of Chagasic colopathy, a dyskinesia and motor hyperreactivity of the bowel associated with internal anal sphincter achalasia occurs, developing constipated symptoms that worsen over time [13–15]. In order to overcome bowel transit obstacles, the colon increases the amplitude and frequency of peristalsis, which will eventually lead to irregular bowel habits. This will ultimately result in bowel muscular hypertrophy and dilatation.

Santos Jr. and colleagues showed that rectal sensitivity is also impaired in patients with Chagasic megacolon. It is frequently diminished, necessitating a higher intrarectal stimulus to arouse evacuation [16]. It remains unclear, however, if this is caused by rectal wall denervation or if it is due to rectal dilatation.

Clinical Findings and Diagnosis of Chagasic Colopathy

Chagasic colopathy presents as chronic constipation that progressively worsens. Abdominal distention and bloating is a common complain, and advanced cases may present a bowel movement every 20–30 days (Fig. 21.1). There are reports of patients without a bowel movement for up to

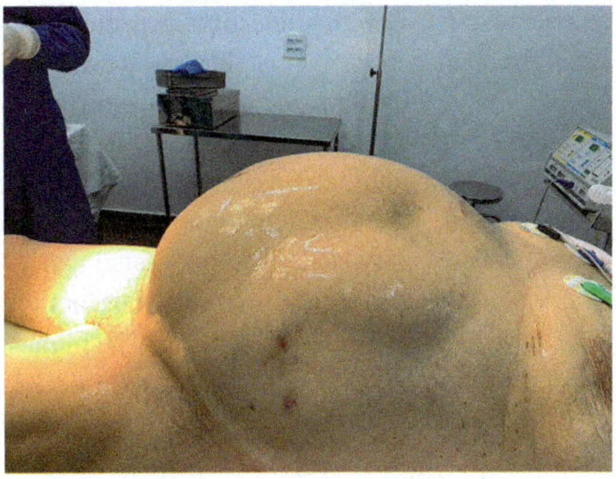

Fig. 21.1 Patient with distended abdomen due to huge Chagasic megacolon

90 days [7]. Constipation is insidious at the onset, but progressively worsens over time to become a persistent condition despite changing dietary habits or the use of oral or retrograde laxatives. Fecaloma is a common complication, reported in approximately half of the patients. Another complication is volvulus of the sigmoid colon. This acute scenario is present in approximately 25% of patients and may initially be treated with endoscopic decompression. However, surgical intervention may be advised due to an ischemic or even necrotic sigmoid, preferably before progressing to bowel perforation (Fig. 21.2).

Clinical suspicion for Chagasic colopathy is considered when the patient presents with a compatible medical history and positive epidemiological data for Chagas disease. Physical examination varies from normal to a voluminous palpable sigmoid colon. Fecaloma may present as an abdominal mass in the left lower quadrant associated with abdominal bloating.

Chagas disease is confirmed by serological testing [17, 18] with an accuracy that varies from 91.4% to 99.4%, depending on the used serological test (complement fixation, indirect immunofluorescence, direct hemagglutination, indirect hemagglutination, and ELISA) or the association of more than one method [19]. Barium enema is the gold standard exam to identify a dilated and/or elongated colon associated with the absence of haustras, mainly in the sigmoid and rectal segment (Fig. 21.3).

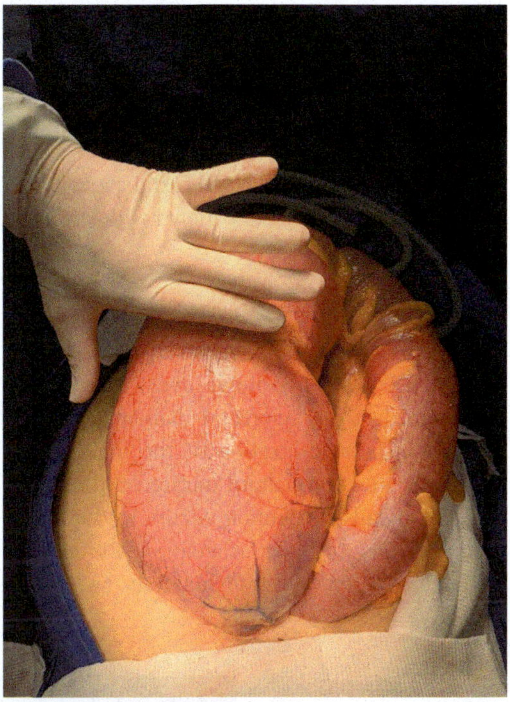

Fig. 21.2 The same patient as in Fig. 21.1, showing the large distended loops of bowel

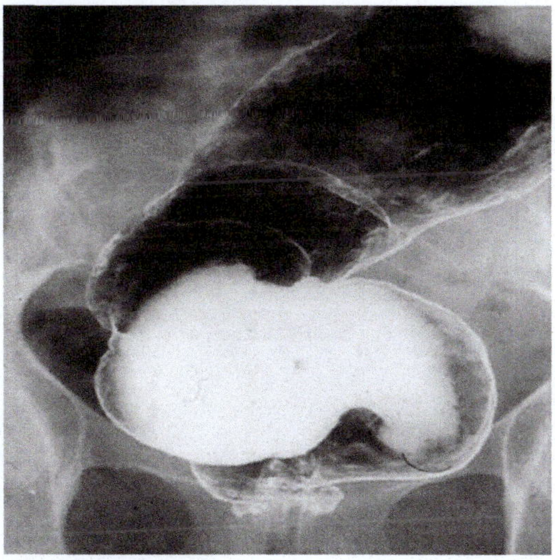

Fig. 21.3 Barium enema with a large Chagasic megacolon

Anorectal manometry may play an important role in the diagnosis of Chagasic colopathy, specifically in patients with positive serology and symptoms of constipation but without colonic dilatation.

Anorectal Manometry

Anorectal manometry allows objective confirmation of anal pressure during resting, squeeze, and pushing phases, elicitation of the rectoanal inhibitory reflex (RAIR), and rectal capacity and compliance for Chagasic patients with suspected colonic involvement.

Evaluation of the RAIR and Chagasic Colopathy

The RAIR is a physiologic event that may be reproduced during anorectal manometry. RAIR is elicited upon relaxation of the internal anal sphincter after lower third intrarectal distention. Physiologically, distention occurs by fecal content. In order to defer evacuation, the external sphincter is contracted to maintain anal continence. After a few seconds, rectal contents are accommodated in the upper third of the rectum, and continence is guaranteed due to restored contraction of the internal anal sphincter.

Anorectal manometry is mandatory in some specific clinical scenarios that are suspicious for Chagasic colopathy such as internal anal achalasia, which are well demonstrated using this test (Figs. 21.4, 21.5, and 21.6).

Rectal Capacity

Rectal capacity is a subjective measure wherein rectal sensation is the key factor. Patients with Chagasic colopathy commonly present with megarectum, and as such, rectal sensory threshold also tends to be altered. The authors recommend routinely determining rectal sensory threshold and capacity before eliciting the RAIR,

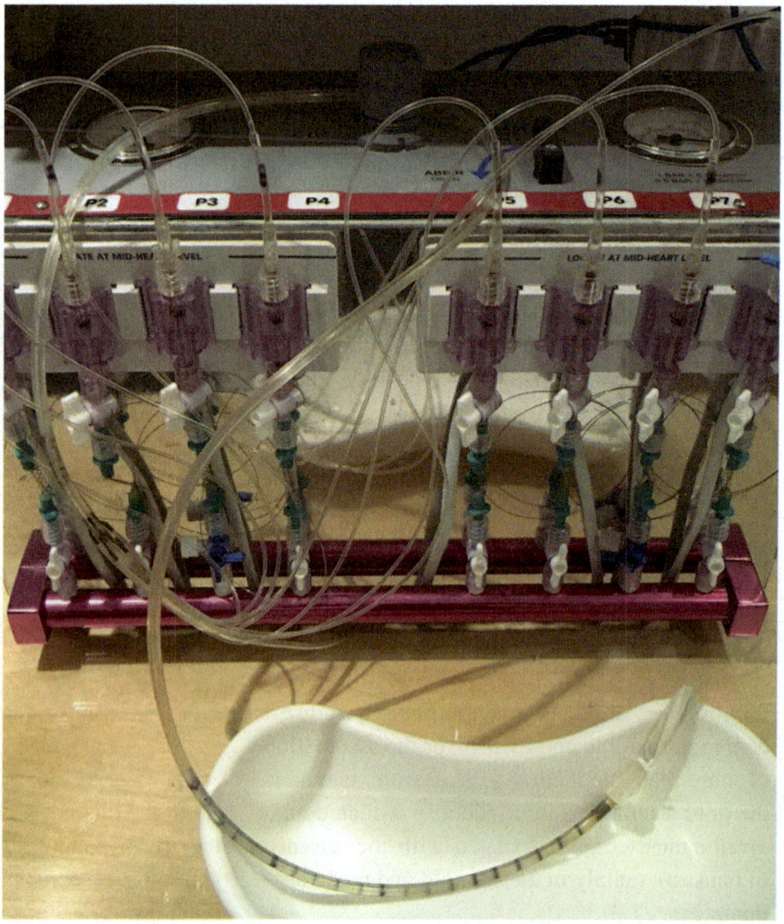

Fig. 21.4 Manometry equipment

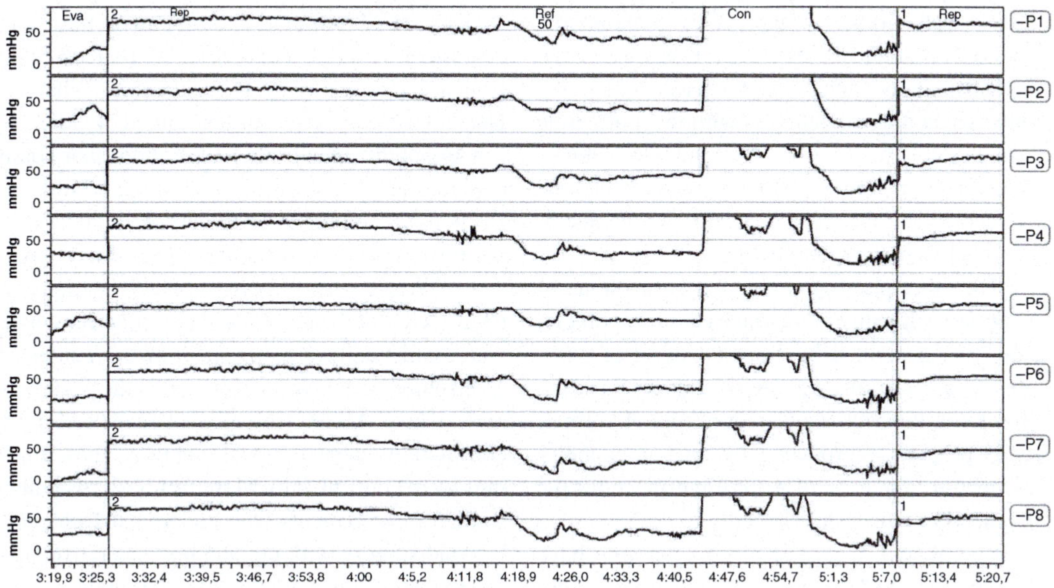

Fig. 21.5 Normal rectoanal inhibitory reflex (RAIR)

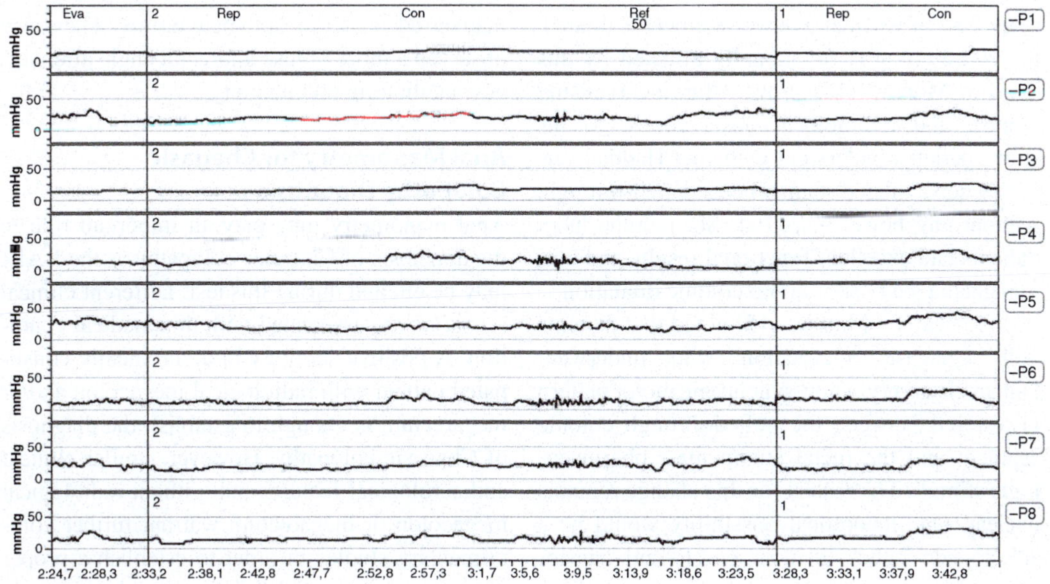

Fig. 21.6 Absent rectoanal inhibitory reflex (RAIR)

as it may need a larger intrarectal ballon distention to properly elicit RAIR.

Anorectal Manometry in Chagasic Patients

Until the mid-1960s, only histopathological parameters were established for Chagasic colopathy. Functional disorders were initially described in 1964, when Vieira and colleagues [20] observed colonic muscular layer hyperactivity after subcutaneous administration of methacholine (cholinergic drug) on manometry. Moreover, an uncoordinated activity of the sigmoid colon and rectum was noticed by simultaneous evaluation of the intraluminal pressure, which ultimately aggravates constipation. This uncoordinated motor pattern has also been described in patients with Chagasic colopathy under physiologic conditions without cholinergic stimulation. This pharmaco-manometric abnormality is not seen in patients with congenital megacolon and thus may be useful to help distinguish between these two entities.

The other physiopathological cornerstone of Chagasic colopathy is the internal anal sphincter achalasia, described by Habr-Gama et al. [14]. This entails the absence of sphincter relaxation after lower third rectal distention. This abnormality is properly reproduced by anal manometry and therefore may be of great importance for the diagnosis of Chagasic colopathy.

Patients with Chagasic megacolon who have undergone bowel resection of the dilated segment followed by an end-to-end colorectal anastomosis remains with postoperative bowel motor synchronism. This observation shows that, regardless of whether a bowel segment is dilated, the entire colon is functionally affected by this disease. Moreira [15] proved this theory using clinical and radiological observations. In his study, patients underwent Duhamel-Haddad surgery when a dilated segment was pulled-through, without any bowel resection. Six months later, patients had excellent functional results and barium enema study showed no colonic dilatation.

Furthermore, Moreira [15] noted that patients with Chagasic megacolon who underwent Duhamel surgery, a non-synchronic motor pattern is restored between the pulled-through colonic segment and the rectal stump after pharmacologic stimuli. The reason for this change remains unclear. One theoretical possibility would be a side-to-side rather than an end-to-end anastomosis. Similar results were reported in patients with Chagasic megacolon who underwent Hartmann's surgery. Anal manometry showed a non-synchronic bowel contraction between the rectal stump and the descending colon after pharmacological stimuli. However, under physiological circumstances, without cholinergic stimuli, data remains controversial about bowel contraction patterns between these two segments.

Teixeira-Moreira [21] compared pre- and postoperative anal manometry in patients with Chagasic colopathy who underwent Duhamel-Haddad surgery. The author aimed to identify functional changes in the pelvic floor that could explain the good clinical outcome after surgery. He noted a significant symmetrical reduction in the postoperative resting and squeeze anal pressures. He also reported better rectal sensation and decreased rectal capacity and compliance, which were statistically significant. However, internal anal sphincter achalasia persisted in the postoperative course. All physiological changes definitely contribute to normal bowel function. These findings were confirmed by Moreira Jr [22], who also studied these patients who underwent Duhamel-Haddad surgery with pre and post-operative anal ultrassound a cinedefecography. The authors concluded that the diminushed resting pressure were due to internal shincter tapering in the posterior hemi-circumference partial (ost commonly) or to inadverted posterior sphynctetotomy. They also observed a more obtuse anorectal angle after surgey, facilitating rectal empty.

Anal Manometry for Chagasic Colopathy Diagnosis

Anal manometry may play an important role in the diagnosis of Chagasic colopathy as the RAIR may be elicited during this test. Different clinical scenarios are associated with this medical condition. A positive serology in a chronically constipated patient with radiological megacolon and/or megarectum is enough to establish the diagnosis of Chagasic colopathy. However, similar clinical and serological features but without radiological megacolon or megarectum warrant further complimentary studies in order to establish a proper diagnosis, due to the following considerations:

1. Serological prevalence of Chagas disease in the midwest of Brazil is high (up to 8%).
2. Approximately 40% of Chagasic patients will ultimately develop some isolated or associated clinical manifestation of the disease, including Chagasic colopathy, esophageal disease, or myocardiopathy.

LEITE ACA [23] studied 64 chagasic constipated patients: 23 with megacolon/megarectum (G1), 21 without megacolon/megarectum (G2) and 20 non-Chagasic constipated patients without megacolon/megarectum (G3). Rectoanal Inhibitory Reflex was absent in 91.3% of patients in G1, 47.29% in G2 and present in all patients in G3. There was a significant difference in the absence of the Rectoanal Inhibitory Reflex when comparing the groups (G1 vs. G2: $p = 0.002$, G1 vs. G3: $p < 0.001$, G2 vs. G3: $p < 0.001$).The significant presence of the RAIR in G2 patients (Chagas disease without megacolon/megarectum) indicates that about half of this sub-group is constipated due to different causes other than chagasic colopathy. Thus, careful analysis of cause and effect is crucial as chronic constipation is a very common symptom. Anorectal manometry is an important complimentary test as the RAIR may be elicited. If achalasia is present, Chagasic colopathy is then confirmed as the primary diagnosis. Otherwise, other causes of chronic constipation must be ruled out.

Summary

Although Chagasic colopathy remains a prevalent disease in Latin American countries, especially in central areas of Brazil, improved socioeconomic factors, including migration of rural populations to the larger cities, campaigns to control the transmitting agent, and improved knowledge of the natural history of this disease, have resulted in a reduced incidence of new cases in the last few decades. Despite these improvements, patients with severe symptoms of the disease are still being routinely diagnosed. Patients who have undergone surgery vary in age, with the majority in their 60s. Occasionally, younger patients are seen (the youngest in the author's experience was 25 years of age), which may indicate that the disease is not yet completely under control.

A better understanding of the physiopathology of this disease represented a great advancement in its diagnosis and treatment. Careful clinical analysis associated with complimentary tests yields an accurate diagnosis. Anorectal manometry may play an important role for an accurate diagnosis in patients with chronic constipation, those with positive serologic tests for Chagas disease, and those in whom radiologic study shows a non-dilated colon. If anorectal manometry elicits an absent RAIR, Chagasic colopathy can be diagnosed. If anorectal manometry shows a positive RAIR, other causes of constipation should be considered.

A definitive diagnosis of this disease requires multidisciplinary findings including clinical, radiological, serological, histological, and functional outcomes. To date, Chagas disease is still challenging to diagnose and treat, which has likely contributed to the fact that this entity still arouses great curiosity and scientific interest.

References

1. Chagas C. Nova espécie mórbida do homem produzida por um trypanosoma *(Trypanosoma cruzi)*. Nota prévia Brasil – Méd. 1909;16:161.
2. Chagas C. Tripanossomíase americana – forma agudada moléstia. Mem Instituto Oswaldo Cruz. 1916;8:37.
3. Amorim M, Netto C. Histopatologia e pathogenese do megaesophago e megarecto – considerações em torno de um caso de "mal do engasgo". Ann Fac Méd Univ São Paulo. 1932;8:102.
4. Etzel E. Neuropatologia do megaesôfago e megaccolo – estudo de 5 casos. Ann Fac Méd Univ São Paulo. 1934;10:383.
5. Hertz AF. Achalasia of the cárdia. Q J Med. 1915;8:300–8.
6. Köberle F. Patogênese dos megas. Rev Goiana de Med. 1956;2:101–6.
7. Moreira H, Rezende JM. Megacólon chagásico – Clínica, diagnóstico e tratamento. In: Moreira H, editor. Coloproctologia – Conceitos: Ed. Escaleno; 1993. p. 15–60.
8. Zicker F, Oliveira RM, Luquetti AO, Oliveira OS, Smith PG. Seroprevalence of Chagas disease among unskilled urban workers in central Brazil. Trans R Soc Trop Med Hyg. 1989;83:511–3.
9. Bern C, Montgomery SP. An estimate of the burden of Chagas disease in the United States. Clin Infect Dis. 2009;49:e52–e54.
10. Manne-Goehler J, Chukwuemeka AU, Montgomery SP, Wirtz VJ. Estimating the burden of Chagas disease in the United States [published online

November 7, 2018]. PLOS Neglected Trop Dis. https://doi.org/10.1371/journal.pntd.0005033.
11. Moreira H, de Rezende JM. Chagasic Megacólon. In: Wexner SD, Bartolo DC, editors. Constipation: etiology, evaluation & management. Oxford: Butterworth-Heinemann; 1995. p. 189–98.
12. Cannon WB. A law of denervation. Am J Med Sci. 1939;198:737–50.
13. Meneguelli UG, Godoy RA, Macedo JFS, Oliveira RB, Dantas RO. Basal motility of dilated and non-dilated sigmoid colon and rectum in Chagas disease. Arq Gastroenterol. 1982;19:127.
14. Habr-Gama A. Motilidade do cólon e reto (contribuição à fisiopatologia do megacólon chagásico) Tese Fac Med da Univ de São Paulo 1966.
15. Moreira H. Estudo eletromanométrico da atividade motora do coto retal e colo descendente em pacientes submetidos às operações de hartman e Duhamel. Rev Goiana Med. 1974;20:125.
16. Santos Jr JCM, Guimarães AS, Aprili F, Gomes DR, Rocha JJR. Colite isquêmica como complicação do megacólon. XXXVI Congresso Brasileiro de Coloproctologia – Foz do Iguaçu – 1977 - Resumos pp. 84–85.
17. Guerreiro C, Machado A. Da reação de Bordet e Gengou na moléstia de Carlos Chagas, como elemento diagnóstico. Brasil-Med. 1913;27(23): 225–6.
18. Freitas JLP de. Contribuição para o estudo do diagnóstico da moléstia de Chagas por processos de laboratório. Tese Fac de Med Univ de São Paulo – Ribeirão Preto 1947.
19. de Rezende JM, Luquetti AO. Chagasic megavisceras. Pan American Health organization. In: Chagas's disease and the nervous system. Washington, D.C.: PAHO; 1994. p. 354.
20. Vieira CB, de Godoy RA, Carril CF. Hipersensibilidade do intestino grosso de pacientes com doença de Chagas e Megacólon aos agentes colinérgicos. Rev Bras Gastroenterol. 1964;16:41.
21. Teixeira-Moreira JP. Avaliação eletromanométrica no pré e pós-operatório de pacientes portadores de megacólon chagásico submetidos à operação de Duhamel. Tese Univ Federal Goiás 2001.
22. Moreira Júnior H, Moreira JPT, Isaac RR, de Almeida AC, Moreira H, Klug WA. Anatomical and physiological changes in pelvic diaphragm in patients with chagasic megacolon submitted to Duhamel surgery. J Coloproctol. 2013;33(2):62–9.
23. Leite, Arminda Caetano de Almeida e, Moreira, Marise Amaral Rebouças, Barbosa, Maria Alves, Moreira Júnior, Hélio, Leite, Paula Chrystina Caetano Almeida, Moreira, José Paulo Teixeira. Clinical and manometric investigation in constipated Chagasic patients with and without megacolon. J Coloproctol (Rio de Janeiro). 2019;39(2);145–52.

New Functional Concepts of Chagas Disease

22

Enio Chaves de Oliveira, Mauro Bafutto, Salustiano Gabriel Neto, and Alexandre Barcelos Morais da Silveira

Introduction

Different theories were proposed to explain the morphological and functional findings in the patient with Chagas disease [1]. Early after the description of Chagas disease, the digestive involvement was reported, and either Chagasic megacolon or Chagasic megaesophagus was compared to congenital megacolon and idiopathic megaesophagus, respectively [2–4].

After a short acute phase, patients present with a chronic phase. Initially asymptomatic, after 10–20 years the patient may develop visceral lesions. Approximately 60% will not have any type of involvement (indeterminate form), 20–30% will have heart disease (cardiac form), and 10–15% will present with digestive disease along the digestive tract, mainly megaesophagus and/or megacolon (digestive form). Different forms may also occur [1].

Professor Fritz Köberle [5–7] was the first to report the association between neuronal destruction in the myoenteric and submucous plexuses of the digestive tract and dilation of some segments, therefore establishing the pathophysiology of acquired megacolon and megaesophagus in chronic Chagas disease. Early studies had demonstrated neuronal lesions in the digestive tract of Chagasic patients [8, 9]. Late studies reinforced the link between neuronal loss in the enteric nervous system and mega formation in chronic Chagasic patients [10–12]. The "digestive form" was a term first used by Professor Joffre Marcondes Rezende in 1959 [13] who worked in Central Brazil in an endemic area of Chagas disease.

Chronic Chagasic patients presented with megaesophagus with dysphagia or megacolon with long-lasting constipation. Therefore, clinicians thought that dilation was responsible for symptoms. This theory gained wide acceptance due to comparisons between Chagasic megacolon and congenital megacolon (Hirschsprung) [2, 4, 14]. Some reports associated achalasia of the internal anal sphincter as a potential contributor to congenital megacolon [15, 16].

Motility studies enhanced understanding of the pathophysiology of acquired megacolon. Chagasic patients presented with a higher sensitivity to methacholine with an increase in tone and activity as a result of denervation with and without colonic enlargement [17–19].

In 1966, Habr-Gama [20] studied the colon and rectum of 55 patients by manometry: 15 were normal controls, 25 had megacolon, 15 had megaesophagus, and 15 were operated for

E. C. de Oliveira (✉) · S. G. Neto
School of Medicine, Universidade Federal de Goiás, Department of Surgery, Goiania, Brazil

M. Bafutto
School of Medicine, Universidade Federal de Goiás, Internal Medicine Department, Goiania, Brazil

A. B. M. da Silveira
Universidade Federal de Uberlândia, Anatomy Department, Uberlandia, Brazil

L. C. C. Oliveira (ed.), *Anorectal Physiology*, https://doi.org/10.1007/978-3-030-43811-1_22

symptomatic megacolon. The authors highlighted four important aspects:

1. Patients with Chagasic megacolon have a higher motor activity in the colon and rectum with more contraction waves with higher amplitude and shorter duration.
2. Patients with megaesophagus have similar rectosigmoid activity to patients with megacolon.
3. Motor activity in the sigmoid colon and rectum is independent in normal subjects, while in Chagasic patients rectosigmoid motility is synchronous showing altered motor wave coordination.
4. Patients who undergo a pull-through operation to treat megacolon showed the same pattern of motility in the distal colon as patients with megacolon. Alteration in motility occurred despite a dilated colon. Enlargement of the colon was not responsible for altered motility.

In 1970 and again in 1974, Habr-Gama et al. [21, 22] identified two findings that explained the pathophysiology of Chagasic megacolon: increased colonic motor activity and non-relaxation of the internal anal sphincter with a full rectum (achalasia). Patients with altered motility and achalasia of internal anal sphincter present constipation and develop colon enlargement.

In 1968, Meneghelli [19] reported an increased motor activity in the colon of Chagasic patients to methacholine. In 1977, Meneghelli [23] studied the colonic motility of 31 individuals: 11 non-Chagasic, 9 Chagasic without megacolon, and 11 Chagasic with megacolon and megarectum. The author found a decreased motor colonic activity in Chagasic patients independent of colon and rectal enlargement. These results diverged from the studies of Habr-Gama et al. [20–22]. Meneghelli [23] detailed radiographic characterization of the colon. Patients were grouped according to dilation of the colon and/or rectum.

Santos Jr [24] in a painstaking study with 30 patients (7 normal controls, 10 with Chagasic megacolon without megarectum, and 13 with Chagasic megacolon and megarectum) evaluated the rectal sensibility and anorectal inhibitory reflex. He concluded that 40% of Chagasic patients with megacolon without megarectum have rectal sensibility similar to normal subjects.

In 2008, Professor Joao Gomes Netinho and his group [25] demonstrated that 43.6% of patients with Chagasic megacolon had anorectal inhibitory reflex, contrary to the theory about achalasia as a key factor in the pathophysiology of Chagasic megacolon (Fig. 22.1).

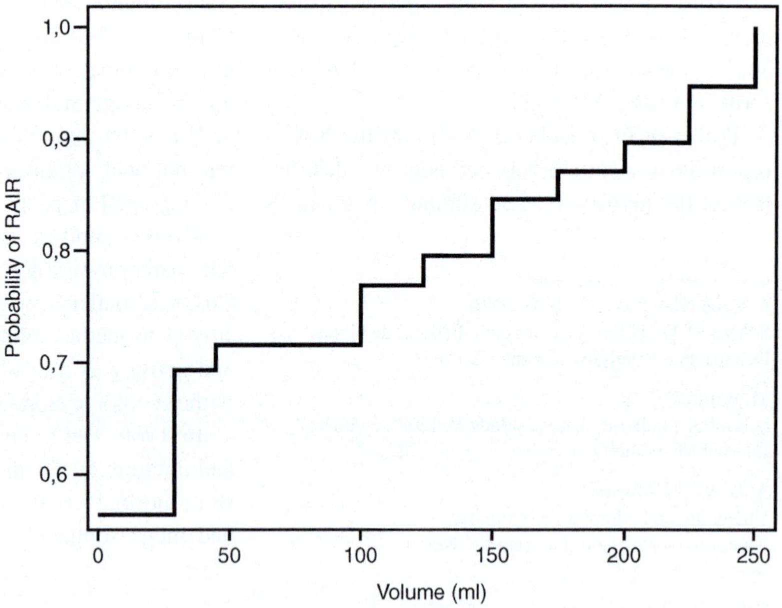

Fig. 22.1 Kaplan-Meier curve showing the probability of inhibitory anorectal reflex occurrence in Chagasic patients with megacolon and increasing inflated volumes in the rectum [25]

For many years, the predominant theory about acquired megacolon was based on the destruction of myoenteric and submucosal plexuses leading to dysmotility and achalasia of the internal anal sphincter. Fecal stasis (constipation) occurred in the posterior colon and, after a variable period, resulted in enlargement of the sigmoid colon.

Constipation and Dilation in the Chagasic Megacolon

Constipation is a very common symptom in the general population, and positive serology for Chagas disease does not make anyone different from the normal population relative to gastrointestinal symptoms, including constipation.

Our group has been working for many decades in an area of Brazil endemic of Chagas disease where urban laborers have an estimated 14% positive serology for Chagas disease [26]. Different clinical-radiological presentations may be found in patients who are grouped in the digestive form of Chagas disease:

- Patients without megacolon/megarectum and without constipation
- Patients without megacolon/megarectum and with constipation
- Patients with megacolon/megarectum and without constipation
- Patients with megacolon/megarectum and with constipation

The pathophysiology theory, based on colonic dysmotility and achalasia of the internal anal sphincter, clearly does not explain all four clinical-radiological groups.

Ximenes et al. [27] studied patients with positive serology for Chagas disease using barium enema and reported that up to 35% (7/20) of patients with megacolon had normal bowel movements. Rassi et al. [28] reported that 57.1% of patients referred to cardiology consultation for heart Chagas disease as having megacolon but with normal bowel movements. Hernandez [29], in a field survey in an endemic area of Central Brazil, reported that 75% of patients with megacolon had normal bowel movements.

There is clear evidence in hospital population-based studies associating Chagasic megacolon and constipation. In addition, the pathophysiology theory to explain all the clinical-radiological groups is inadequate. Another very important factor to the old and incomplete pathophysiology theory is the regional variability of the disease (Fig. 22.2) [30].

Chagas disease is found from Mexico to Chile and Argentina with varying clinical presentations [30]. Central Brazil has a high prevalence of the digestive form, while the southwest is predominantly cardiac and the northwest is indeterminate [1].

Chagasic Megacolon and Internal Anal Sphincter

Anorectal manometry was developed to evaluate the physiology of anorectal muscles and was first used in Hirschsprung [16, 22] and then in Chagasic megacolon. A parallel was made with the two diseases and this has influenced the incomplete pathophysiology theory [21]. There are some basic differences between the two diseases. However, the key point to understand why anorectal manometry is inappropriately used to evaluate Chagasic megacolon is that many patients have associated megarectum. These studies do not identify patients with and without megarectum [23, 24, 28]. Anatomical and physiological characteristics of the rectum may play an important role in the results of anorectal manometry even in non-Chagasic patients [31, 32].

Our group (Núcleo de Estudo de Doença de Chagas at the Universidade Federal de Goiás) has collected clinical and serological data of more than 28,000 Chagasic patients in the last four decades. Of the 1500 patients with megacolon, approximately 48% have megarectum. Megarectum is a rectal diameter greater than 6.5 cm on pelvic x-ray with barium enema [33, 34]. Megarectum may influence testing for the rectoanal inhibitory reflex (RAIR) and rectal compliance (Figs. 22.2, 22.3, and 22.4). If

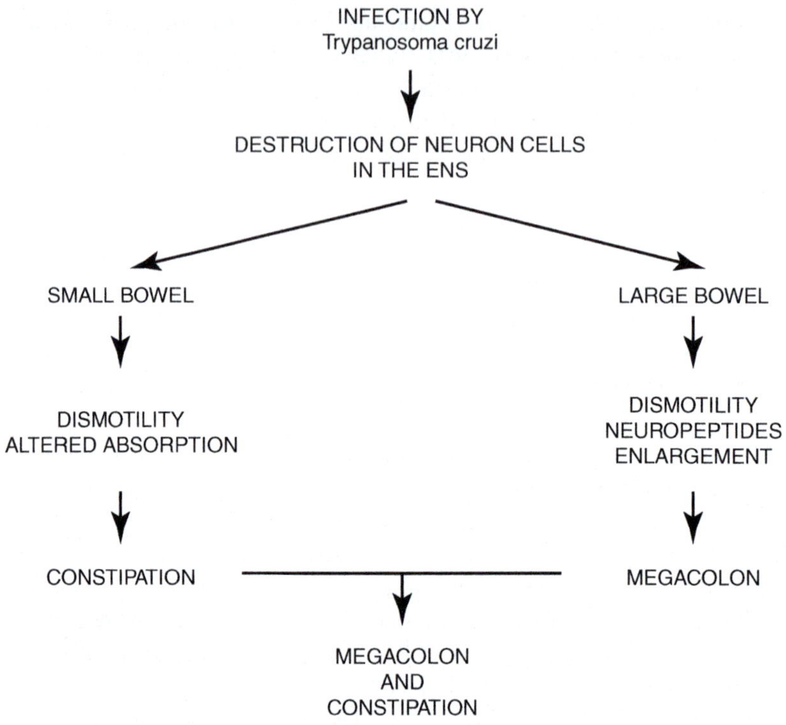

Fig. 22.2 Pathophysiology of megacolon in chronic Chagas disease

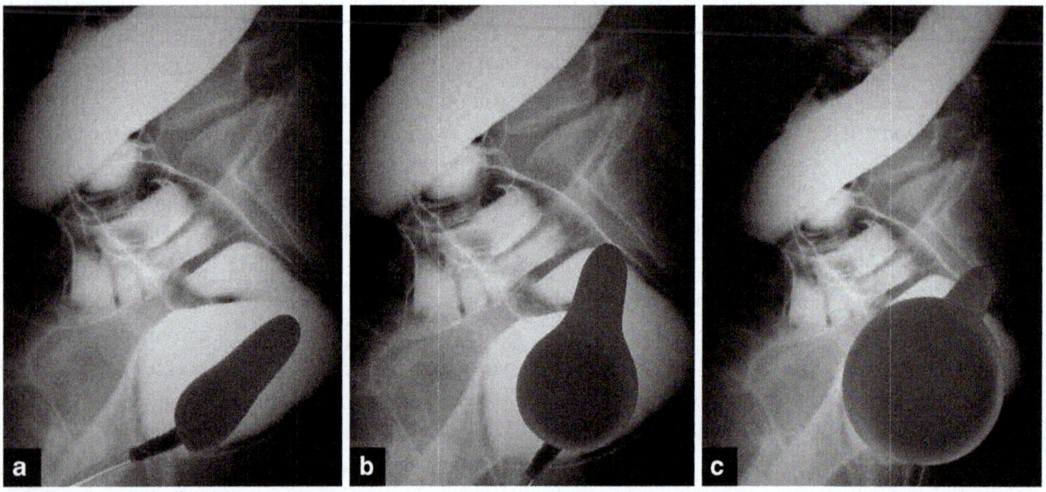

Fig. 22.3 (**a–c**) Barium enema of Chagasic patient without megarectum. Schematic demonstration of an inflated balloon with volumes 10 mL, 50 mL, and 150 mL could make pressure on rectal wall and elicit the inhibitory anorectal reflex

patients with megarectum are tested with a small volume of air in the rectal balloon, the RAIR may not be elicited due to the small volume (usually up to 50 mL) (Figs. 22.2 and 22.3). Patients with megacolon without megarectum may be evaluated by anorectal manometry (Figs. 22.4, 22.5, and 22.6). Anorectal manometry may overestimate the prevalence of megarectum in constipated patients, while imaging exams such as barium enema may underestimate the prevalence

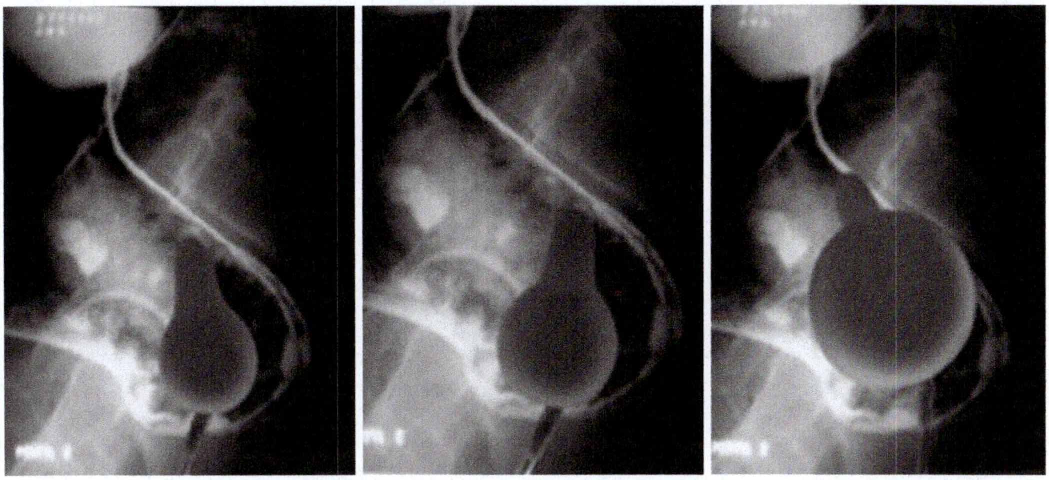

Fig. 22.4 Barium enema of Chagasic patient with megarectum. Schematic demonstration of an inflated balloon with volumes 10 mL, 50 mL, and 150 mL could not make pressure on rectal wall and not elicit the inhibitory anorectal reflex

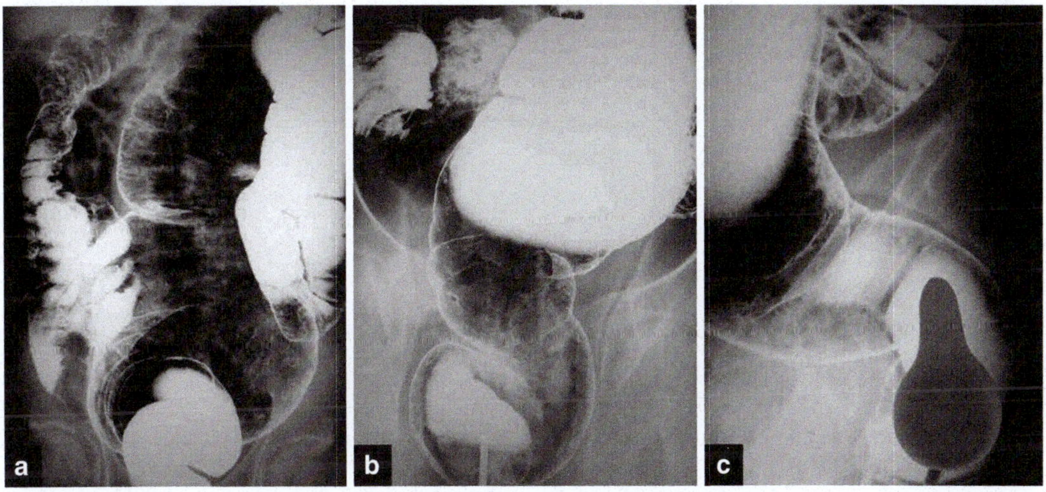

Fig. 22.5 Barium enema of Chagasic patient showing megasigmoid without megarectum. Schematic demonstration of an inflated balloon with volume of 50 mL could elicit the inhibitory anorectal reflex. (**a**) megasigmoid; (**b**) Normal rectum (front view); (**c**) Normal rectum (side view) with schematic demonstration of an inflated balloon with 50 mL of air that could stimulate the rectal wall

[34]. Ideally, measuring the diameter of the rectum should take into account the lumen rectal pressure on the distended rectal wall.

Constipation and Chagasic Megacolon

Constipation in Chagasic patients with megacolon may be associated with a dilated colon, but this is not the cause of the constipation. Historically, authors have used hospital samples to report these data. When more than 50% of patients with megacolon have normal bowel movements, colonic enlargement per se is not the cause of the constipation. In addition, Chagasic patients without bowel complaints had no colonic evaluation. The pioneering work of Professor Anis Rassi and Carlos Antônio Ximenes from the School of Medicine at the Universidade Federal de Goiás [27, 28] evaluated the Chagasic patient with a holistic

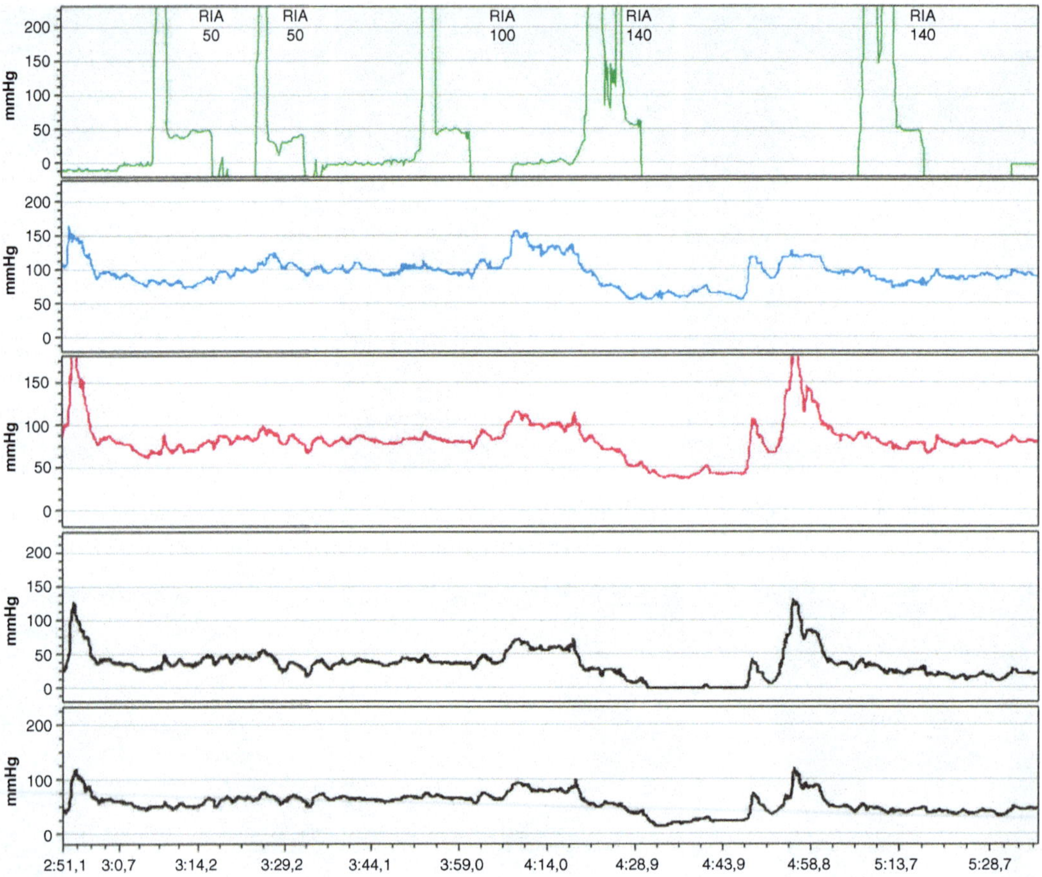

Fig. 22.6 Anorectal manometry showing the rectoanal inhibitory reflex in Chagasic patient with megacolon without megarectum with a balloon positioned in the rectum and inflated with volumes greater than 100 mL

approach (heart, esophagus, and colon) regardless of the patients' complaints.

Analyzing 2446 patients referred to the Núcleo de Estudos em Doença de Chagas da Universidade Federal de Goiás and who underwent barium enema, 1174 had megacolon and 679 (63.2%) had constipation. However, we observed that most patients with megacolon referred from cardiology and gastroenterology clinics did not have constipation (Table 22.1).

Bafutto et al. [35, 36] studied two groups of Chagasic patients with megacolon: patients with normal bowel movements and constipated patients using H2 breath test to address orocecal transit time. The authors reported an increased orocecal transit time in Chagasic constipated patients regardless of colonic enlargement. It was demonstrated that Chagasic patients have small bowel involvement and slow transit leading to constipation. Clearly the dilated colon may also

Table 22.1 Patients referred to the Núcleo de Estudos em Doença de Chagas, Universidade Federal de Goiás, showing the initial clinic or complaint, 2014

Clinic	With megacolon	Without megacolon	Total
Cardiology			
Constipated	36 (44.3%)	104 (45.4%)	140
Non-constipated	45 (55.6%)	125 (54.6%)	170
Gastroenterology			
Constipated	198 (43%)	128 (14.0%)	326
Non-constipated	263 (57%)	785 (86.0%)	1048
Digestive surgery			
Constipated	445 (83.7%)	54 (41.5%)	499
Non-constipated	87 (26.3%)	76 (59.5%)	163
Total	1174	1272	2446

represent a physical obstacle to feces and worsening bowel transit. These findings are corroborated by basic knowledge on Chagas disease that shows denervation along the entire digestive tube [1]. Further studies regarding the role of the small bowel in constipation are needed to understand the pathophysiology of constipation.

Destruction of the enteric nervous system appears to be selective, and symptoms will depend on quantity and type of neuron loss in the plexus submucosal and myoenteric plexus. More extensive destruction of the submucosal plexus is associated with constipation, while destruction of the myoenteric plexus leads to colonic enlargement [37].

Dilation Versus Constipation in Chagasic Patients: A Molecular Approach

The gastrointestinal tract has two components responsible for its innervation: an extrinsic component of neurons originating from the central nervous system (CNS) and an intrinsic component represented by the enteric nervous system (ENS). The extrinsic innervation of the gastrointestinal tract consists of sympathetic and parasympathetic neurons. In the sympathetic nervous system, noradrenaline is the most common neurotransmitter in postganglionic neurons that innervate the intestine [38]. The ENS presents neurons and support cells (glial cells or enteroglial cells) grouped into small groups called enteric ganglia interconnected by nerve fibers.

Enteric ganglia, although small, are so numerous that the system as a whole has millions of neurons. Another important aspect to understand the ENS is the study of support cells or enteroglial cells, whose main function is the protection and maintenance of neuronal tropism [39]. Currently, a range of pathologies interfere in the functioning of the ENS, either functionally or structurally. Some of these diseases are classified as functional of the gastrointestinal tract, and among the acquired ones, the digestive form of Chagas disease can be identified as the most important [40]. It has recently been demonstrated that in Chagasic patients, there is a decrease in the number of enteroglial cells, generally identified by the expression of the S-100 protein, in the colon's nervous plexus region. Moreover, it is known that while neurons are rarely parasitized by *T. cruzi*, glial cells are commonly attacked by this parasite, and neuronal changes are mainly a consequence of the destruction of enteric glia and the inflammatory process triggered by cellular parasitism [41, 42]. Another important marker for enteroglial cells is the expression of the GFAP protein. This is a non-constitutive structural protein of the intermediate filament class. It is interesting to note that while in Chagasic patients with megacolon, the number of enteroglial cells expressing GFAP does not change in relation to non-Chagasic individuals, a significant increase in the number of these cells is observed in the group of non-mega-Chagasic patients. To speculate about the possible repercussions of this finding on the pathophysiology of Chagasic infection, we must first consider the importance of GFAP in the biology of enteroglial cells. Being a constituent of the intermediate filaments, one of the functions of GFAP is possibly to increase the cohesion between the glial cells, thereby creating a protective barrier for neuronal bodies. Thus, the increase of GFAP expression in non-megacolon patients could represent an attempt to protect components of the ENS against harmful factors inherent to the inflammatory process or even to the parasite itself [42, 43].

Based on this assumption, patients who fail to increase the expression of this protein will have a greater loss of neuronal bodies of the ENS and consequently greater possibility of developing the megacolon [44]. This hypothesis is in agreement with the ultrastructural alterations described by Tafuri (1975): "In the same ganglion and in the bundles of extraganglionic nerve fibers are nerve fibers deeply altered next to normal ones." It is believed that this discrepancy is due to the differentiated expression of GFAP by enteric glial cells after *T. cruzi* infection.

Chagasic patients with megacolon and no clinical involvement of the organ (without constipation, complaints of stasis, or difficulties of evacuation) are occasionally observed, with megacolon being discovered for other reasons (exams of routine or complaints in the genitourinary system). Histopathological analysis of these samples reveals a dilated organ with a large decrease in enteroglial cells (S-100 and immunoreactive GFAP), but with integral neuronal part of ENS. Likewise, it is not difficult to find Chagasic patients without evidence of the presence of megacolon (opaque enema or colonoscopy) but with several complaints regarding intestinal constipation and evacuation difficulties. Histopathological analysis of samples from these patients demonstrates a preservation of the enteroglial component with intense destruction of the neurons of the myenteric plexus of the rectum and the sigmoid [45, 46]. These data lead us to conclude that constipation and dilation are situations of distinct pathophysiological etiologies that may or may not be associated in the same patient, but because they are independent events, they need different management and approaches. While the non-dilated colon with motility compromised by denervation finds definitive solution only in the excision surgery of the affected portion, the dilated colon that has preserved motility can be clinically conducted with the use of low-potency steroid anti-inflammatory drugs and periodically reevaluated (Figs. 22.6, 22.7, and 22.8). Nevertheless, there is another key component that seems to influence these clinical forms of Chagasic megacolon: the expression of serotonin in the gut.

Recently many studies have been undertaken to identify the role of serotonin in the small and large bowel and gastrointestinal disease as Chagasic megacolon and irritable bowel syndrome. Our group demonstrated that serotonin mRNA transcription is reduced during inflammation in the intestine due to chronic disease and consequently a decrease in the production of serotonin mainly in the mucosa. The reduction in serotonin signaling is associated with chronic constipation, irritable bowel syndrome, and other symptoms as observed in Chagasic megacolon. These diseases have a common factor – the inflammatory process reinforcing the hypothesis that the reduction of serotonin expression is directly related to constipation independent

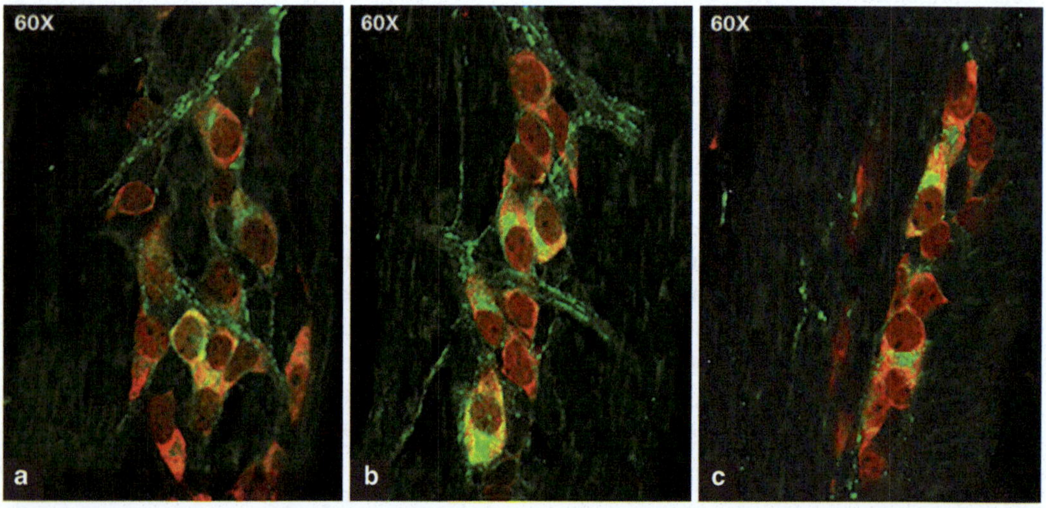

Fig. 22.7 Identification of neuronal bodies through immunohistochemistry by expression of Hu C/D protein (red) and neuronal fibers by the expression of peripherin (green). (**a**) Myenteric ganglion of non-Chagasic individual; (**b**) myenteric ganglion of Chagasic patient with dilated/non-constipated megacolon. Note that there is no difference in the destruction of neuronal bodies in between Chagasic patient with megacolon and noninfected individual. Despite the dilation observed in the colon from Chagasic patient, its function still preserved. (**c**) Myenteric ganglion of Chagasic patient with non-dilated/constipated megacolon. This patient presents normal gauge of the colon in relation to the non-Chagasic; however the neuronal destruction led to the functional loss of the organ, characterized by intestinal constipation

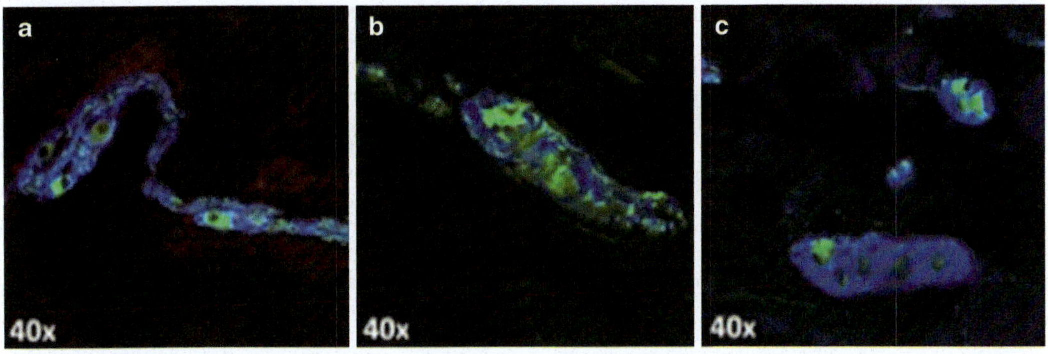

Fig. 22.8 Identification through immunohistochemistry of neuronal bodies by expression of PGP 9.5 (green) and enteroglial cells by expression of the immunoreactive S-100 protein (blue). (**a**) Myenteric ganglion of the non-Chagasic individual; (**b**) myenteric ganglion of a Chagasic patient with dilated/non-constipated megacolon. Note that despite presenting a number of neurons equivalent to the non-Chagasic individual, this patient presents a very low expression of enteroglial cells (blue). (**c**) Myenteric ganglion of the colon of a Chagasic patient with non-dilated/constipated megacolon. This patient has a low number of neurons although the expression of enteroglial cells is high (blue), thus ensuring that the colon does not dilate despite presenting its motor part compromised

of viscera enlargement [47]. These findings may present new opportunities to treatment.

Summary

The pathophysiology of Chagasic megacolon is firmly established, and colonic dilation alone is not the only explanation for constipation. Internal anal sphincter achalasia seems to be a false-negative result in patients with megarectum.

Submucosal and myoenteric plexuses play different roles in the pathophysiology of Chagasic megacolon. The small bowel should be addressed in Chagasic patients with constipation. Local factors such as neuropeptides are involved in the process of colonic enlargement. Constipation and colonic enlargement are independent factors in Chagasic patients but nevertheless may be associated.

References

1. Rezende JM, Luquetti AO. Chagasic megaviscera. In: Pan American Health Organization, editor. Chagas disease and the nervous system, Scientific Publication 547. Washington: Pan American Health Organization; 1994. p. 160–83.
2. Amorim M, Correa NA. Histopathologia e pathogenes do megaesôfago e megareto: considerações em torno de um caso de "mal do engasgo". An Fac Med Univ São Paulo. 1932;8:101–27.
3. Etzel E. Distribuição geográfica do megaesôfago-megacolon: estado atual da teoria etiológica da avitaminose B1-estudo de 626 casos. Revista da Associação Paulista de Medicina. 1939;15:103–58.
4. Fonseca L. Alguns aspectos clínico-radiológicos do intestino delgado e do cólon na acalásia. Arq Hosp S Casa S Paulo. 1955;1:63–91.
5. Koberle F. Megacolon. J Trop Med Hyg. 1958;61(1):21–4.
6. Koberle F. Enteromegalie. Zentralbl Alleg Path. 1957;96:244–6.
7. Koberle F. Patologia y anatomia patológica de la enfermedad de Chagas. Boletin de la Oficina Sanitaria Panamericana (OSP). 1961;51(5):404–28.
8. Almeida MO. Contribuição ao estudo experimental do mal de engasgo. Mem Inst Oswaldo Cruz. 1929;9:146–51.
9. Vampre E. Terceira contribuição ao estudo do mal de engasgo. B Soc Med Cir S Paulo. 1923;6(5):75–88.
10. Okumura M, Correa Netto A. Produção experimental de megas em animais inoculados com Trypanosoma cruzi. Rev Hosp Clin Fac Med S Paulo. 1961;16(5):338–41.
11. Koberle F. Enteromegaly and cardiomegaly in Chagas disease. Gut. 1963;4(4):399–405.
12. Alcantara FG. Molestia de Chagas experimental: manifestações viscerais. O Hospital. 1964;66(3):625–33.
13. Rezende JM. Forma digestiva da moléstia de Chagas. Rev Goiana Med. 1959;5:193–227.
14. Etzel E. Neuropatia do megaesôfago e megacolo. An Fac Med Univ S Paulo. 1934;10(3):383–95.
15. Hurst AF. The sphincters of the alimentary canal and their clinical significance. Brist M J. 1925;1:145–51.
16. Hurst AF. Anal achalasia and megacolon. Guy's Hosp Rep. 1934;84:317–50.
17. Vieira CB, De Godoy RA, Carril CF. Hypersensitivity of the large intestine to cholinergic agents in patients

with Chagas' disease and megacolon. Rev Bras Gastroenterol. 1964;16(2):41–8.
18. Vieira CB, Godoy RA, Menghelli UG, Carril CF. Resposta do cólon sigmóide não ectásico a metacolina na forma crônica da molestia de Chagas. Arq Gastroenterol. 1966;3(1):21–6.
19. Meneghelli UG. Estudos farmacológicos no megacólon chagásico. Rev Goiana Med. 1968;14(1):61–7.
20. Habr-Gama A. Motilidade do cólon sigmóide e do reto. Contribuição a fisiopatologia do megacólon chagásico. Tese de Doutorado Universidade de São Paulo. 1966.
21. Habr-Gama A, Costa Curta L, Raia A. Anatomia e fisiologia do esfincter interno do ânus. Rev Soc Bras Proctol. 1970;3:21–30.
22. Habr-Gama A, Haberkorn S, Rodrigues JJG, Raia A, Betarello A. Manometria ano-reto-colica. Arq Gastroenterol. 1974;11:201–16.
23. Meneghelli UG. Motilidade do sigmóide e do reto de portadores da moléstia de 'chagas em condições basia e sob ação da pentagastrina Tese de Livre Docência Faculdade de Medicina de Ribeirão Preto, Universidade de São Paulo. 1977:110 p.
24. Santos Jr JCM. Estudo comparativo da sensibilidade retal e do reflexo reto-esfincteriano entre pacientes chagásicos com megacólon e reto não dilatado, com megacólon e megarreto e em pacientes normasi. Tese de Doutorado Faculdade de Medicina de Ribeirão Preto, Universidade de São Paulo. 1977:68 p.
25. Cavenaghi S, Felicio OCS, Ronchi LS, Cunrath GS, Melo MMC, Netinho JG. Prevalence of recto-anal inhibitory reflex in chagasic megacolon. Arq Gastroenterol. 2008;45(2):128–31.
26. Zicker F, Smith PG, Luquetti AO, Oliveira OS. Mass screening for Trypanosoma cruzi infections using the immunofluorescence, ELISA and haemagglutination tests on serum samples and on blood eluates from filter-paper. Bull World Health Organ. 1990;68(4):465–71.
27. Ximenes CA, Rezende JM, Moreira H, Vaz MGM. Técnica simplificada para o diagnóstico radiológico do megacólon chagásico. Rev Soc Bras Med Trop. 1984;17(suppl)
28. Rassi A, Rezende JM, Moreira H, Ximenes CA, Luquetti AO, Lousa L, et al. Associação da cardiopatia, megaesôfago e megacolon na fase crônica da doença de Chagas. Rev Soc Bras Med Trop. 1986;19(supl II):29.
29. Hernandez EBR. Estudo clínico e radiológico do reto-sigmóide em indivíduos de uma área endêmica de doença de Chagas, Mambaí, Goiás. Tese de Doutorado Faculdade de Medicina, Universidade de Brasília. 2002:112 p.
30. Souza DH, Vaz Mda G, Fonseca CR, Luquetti A, Rezende Filho J, Oliveira EC. Current epidemiological profile of Chagasic megaesophagus in Central Brazil. Rev Soc Bras Med Trop. 2013;46(3):316–21.
31. Barleben A, Mills S. Anorectal anatomy and physiology. Surg Clin North Am. 2010;90(1):1–15, Table of Contents.
32. Lee JM, Kim NK. Essential anatomy of the anorectum for colorectal surgeons focused on the gross anatomy and histologic findings. Ann Coloproctol. 2018;34(2):59–71.
33. Lopes GP, Ferreira-Silva MM, Ramos AA, Moraes-Souza H, Prata A, Correia D. Length and caliber of the rectosigmoid colon among patients with Chagas disease and controls from areas at different altitudes. Rev Soc Bras Med Trop. 2013;46(6):746–51.
34. Gladman MA, Dvorkin LS, Scott SM, Lunniss PJ, Williams NS. A novel technique to identify patients with megarectum. Dis Colon Rectum. 2007;50(5):621–9.
35. Bafutto M. Avaliação do tempo de trânsito orocecal e do teste de tolerância a glicose em pacientes chagásicos e não chagásicos com ou sem constipação intestinal. Tese de Doutorado Faculdade de Medicina, Universidade de Federal de Goiás. 2014:62 p.
36. Bafutto M, Luquetti AO, Gabriel Neto S, Penhavel FAS, Oliveira EC. Constipation is related to small bowel disturbance rather than colonic enlargement in acquired Chagasic megacolon. Gastroenterology Res. 2017;10(4):213–7.
37. Oliveira EC, Menezes JG, Cardoso VK, Luquetti AO, Neto SG, Garcia SB. The relationship between megacolon and constipation in Chagas' disease. Neurogastroenterol Motil. 2009;21(Suppl 1):5.
38. Furness JB. The enteric nervous system and neurogastroenterology. Nat Rev Gastroenterol Hepatol. 2012;9(5):286–94.
39. von Boyen G, Steinkamp M. [The enteric glia and neurotrophic factors]. Z Gastroenterol. 2006;44(9):985–90.
40. da Silveira AB, Lemos EM, Adad SJ, Correa-Oliveira R, Furness JB, D'Avila RD. Megacolon in Chagas disease: a study of inflammatory cells, enteric nerves, and glial cells. Hum Pathol. 2007;38(8):1256–64.
41. Jabari S, de Oliveira EC, Brehmer A, da Silveira AB. Chagasic megacolon: enteric neurons and related structures. Histochem Cell Biol. 2014;142(3):235–44.
42. Moreira MD, Brehmer A, de Oliveira EC, Neto SG, Luquetti AO, Bueno LL, et al. Regenerative process evaluation of neuronal subclasses in chagasic patients with megacolon. Hum Immunol. 2013;74(2):181–8.
43. da Silveira AB, Freitas MA, de Oliveira EC, Neto SG, Luquetti AO, Furness JB, et al. Neuronal plasticity of the enteric nervous system is correlated with chagasic megacolon development. Parasitology. 2008;135(11):1337–42.
44. von Boyen G, Steinkamp M. The role of enteric glia in gut inflammation. Neuron Glia Biol. 2010;6(4):231–6.

45. Jabari S, da Silveira AB, de Oliveira EC, Quint K, Wirries A, Neuhuber W, et al. Interstitial cells of Cajal: crucial for the development of megacolon in human Chagas' disease? Colorectal Dis. 2013;15(10):e592–8.
46. Jabari S, da Silveira AB, de Oliveira EC, Quint K, Neuhuber W, Brehmer A. Preponderance of inhibitory versus excitatory intramuscular nerve fibres in human chagasic megacolon. Int J Color Dis. 2012;27(9):1181–9.
47. de Freitas MA, de Oliveira EC, de Oliveira FC, Jabari S, Brehmer A, da Silveira AB. Is the increased presence of CD8 T-lymphocytes related to serotonin levels in Chagas disease? Colorectal Dis. 2015;17(3):268–9.

23 Functional Aspects of Diabetes and Collagenosis

Lucia Camara Castro Oliveira
and Henrique Sarubbi Fillmann

Introduction

The digestive system is commonly affected in patients with diabetes mellitus (DM) and in those with collagen diseases. These diseases lead to important motor and sensory changes in the gastrointestinal tract. DM is a chronic disease, considered one of the major health problems in the world, and can determine changes in multiple organs and systems [1].

The impairment in the digestive tract is relevant, estimating that more than 76% of the patients with type I DM present with some type of symptoms [2, 3]. The entire digestive system can be affected, including the anal sphincter muscles, developing *diabetic anal incontinence* [3–5].

The high incidence of diabetic neuropathy in patients with a long history is associated with important alterations in norectal manometry. These changes may occur alone or may be associated with others. We now know that biochemical changes such as elevated nitric oxide levels in the patient with DM may also be responsible for changes of anal pressure levels in these individuals [6]. Collagen diseases are also referred to as connective tissue diseases and collagenous or vascular collagen disease. Their clinical presentation includes scleroderma, periarteritis nodosa, dermatomyositis, malignant nephrosclerosis, thrombangiitis obliterans, some nephritis and subacute endocarditis, systemic lupus erythematosus, rheumatic fever, rheumatoid arthritis, Sjögren's syndrome, mixed connective tissue disease, and antiphospholipid antibody syndrome. In this chapter we will present the manometric changes in DM and scleroderma.

Manometry in Diabetes Mellitus

Anorectal dysfunction in the patient with DM is common and manifests primarily as constipation and anal incontinence. These changes are more frequent in patients with evidence of peripheral or autonomic neuropathy, representing about 24–30% of patients with type I DM [7]. Diabetic microangiopathy damages nerve conduction and the transmission of stimuli at synapses, and thus the striated musculature of the pelvic floor may also be affected. The diabetic microangiopathy also damages the sympathetic colonic innervation, propulsive waves occur in greater intensity, and the rectum receives the colon content still liquid, causing a rapid distension of the ampola and a relaxation of the internal involuntary anal sphincter with episodes of incontinence [8, 9]. In the elderly patient with pudendal neuropathy, this mechanism may be sufficient to cause anal incontinence.

L. C. C. Oliveira
Director Department of Anorectal Physiology of Rio de Janeiro and CEPEMED, Rio de Janeiro, Brazil

H. S. Fillmann (✉)
Hospital Sao Lucas-PUC-RS, Department of Surgery, Porto Alegre, Brazil
e-mail: henrique@fillmann.com.br

L. C. C. Oliveira (ed.), *Anorectal Physiology*, https://doi.org/10.1007/978-3-030-43811-1_23

There is considerable evidence suggesting that oxidative stress plays a very important role in the pathogenesis of DM complications including anal sphincter hypotonia and, consequently, fecal incontinence. Several studies have demonstrated the increase of the reactive species of oxygen and nitrogen in individuals with DM. Among these is nitric oxide. This molecule is one of the most important agents responsible for the relaxation of the internal anal sphincter muscle. We know that this anal hypotonia is independent of diabetic neuropathy and is even detected earlier [6].

The incidence of anal incontinence is still underestimated, including in patients with DM. Incontinence is often not reported, and often the general practitioner does not include this item in the patient's medical history, making accurate prevalence difficult to confirm. In one study including 136 patients with DM, the authors found a 20% incidence of incontinence [9]. Similarly, diabetic autonomic neuropathy (DAN) is associated with constipation symptoms and can occur in up to 60% of patients with DM.

In general, long-standing diabetic patients have decreased anal resting pressure, voluntary striated muscle hypotonia with muscle fatigue, and changes in sensorial rectal perception [10, 11].

In a manometric study including 11 patients with DM and 20 controls, Sun et al. [12] studied the pathophysiology of fecal incontinence. They concluded that the mechanism is multifactorial but found important changes in the internal anal sphincter such as short waves during resting, with a range of 10–40 cmH_2O, rectal contractions in response to rectal distension, and reduced rectal compliance. Alterations in the electrical activity of the internal anal sphincter would be an important trigger for episodes of incontinence.

Epanomeritakis et al. [13] assessed 38 patients with DM and 25 controls who underwent anal manometry in order to establish the relationship between the degree of anorectal dysfunction and the duration of disease. When compared to the control group, all DM patients had decreased resting and contraction pressures, altered reactive rectoanal inhibitory reflex (RAIR), and reduced rectal perception or rectal sensitivity to distension.

A simple test that may also help in the assessment of incontinent patients with DM is the introduction of volume into the rectal ampulla to assess the duration of maintaining volume without leakage. This method was employed by Schiller et al. [8], who observed that diabetic patients presented with incontinence episodes with smaller volumes and with shorter duration than nondiabetic patients.

Russo et al. [14] performed an evaluation of anorectal function in a group of diabetic patients with and without hyperglycemia. They observed that hyperglycemia causes a decrease in anal sphincter contraction pressure and rectal compliance.

Alterations in the RAIR profile are common in diabetic patients (Fig. 23.1) [15], with a slower recovery time. In a study that included 30 diabetic patients and 22 control subjects, Deen et al. [16] compared the RAIR findings and reported a weakened reflex in most diabetic patients and normal reflex in the control group. This may suggest an alteration of the intrinsic neural function. RAIR was absent in all diabetic patients who complained of anal incontinence.

In a study of 35 diabetic patients who underwent anal manometry, Jorge et al. [17] found that resting pressure was significantly higher in individuals complaining of incomplete evacuation.

Constipation is also a frequent complaint in the diabetic patient, although the genesis of the disorder is unclear. Neuropathic dysfunction has been proposed as the main factor involved in the constipated diabetic patient [18]. There are numerous experimental studies demonstrating the effect of DM on the anal sphincter muscles. Wistar rats with streptozotocin-induced diabetes showed a significant decrease in anal sphincter pressure when compared to a control group 8 weeks after induction. It is interesting to note that autonomic neuropathy secondary to diabetes only occurred in these animals after 16 weeks of DM induction. This highlights the role of biochemical changes that are much more common than neuropathies, especially the increase of reactive nitrogen species [19].

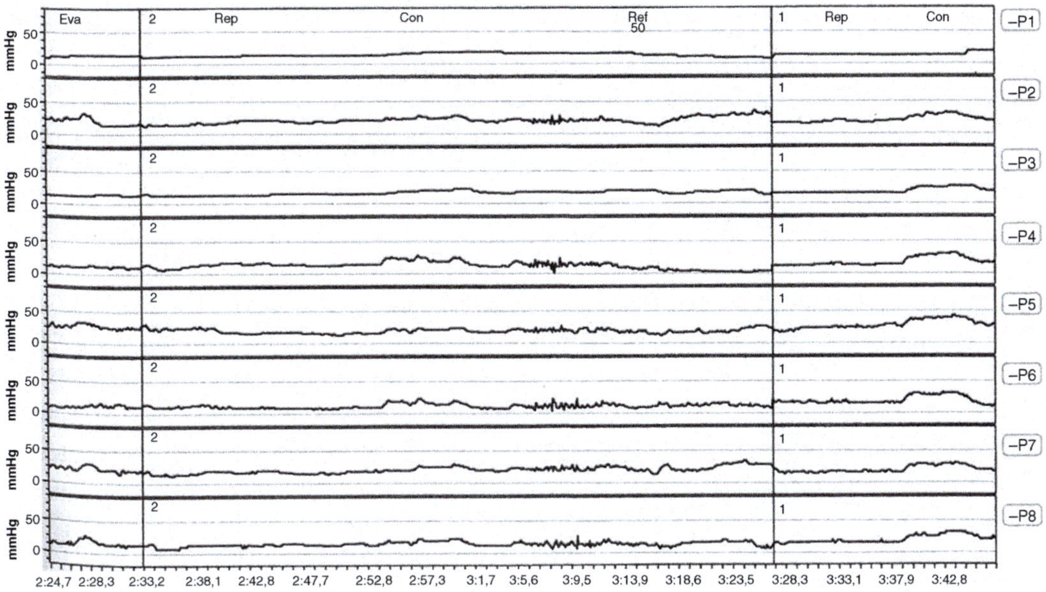

Fig. 23.1 Manometry showing inhibited RAIR in diabetic patients

Manometry in Systemic Sclerosis

Systemic sclerosis generates disorders of high and low gastrointestinal motility, such as decreased or absent pressure of the lower esophageal sphincter, low amplitude of contractions in the distal esophagus, and reduction or absence of migrant motor complexes in the smooth muscle of the gastrointestinal tract. These anomalies can lead to symptoms and signs of heartburn, reflux, dysphagia, early satiety, belching, abdominal distension, bacterial overgrowth, abdominal pain, weight loss, diarrhea, constipation, and anal incontinence. The internal anal sphincter may also be impaired by degeneration and fibrosis, leading to concomitant anal incontinence in scleroderma patients. The use of anorectal manometry in these patients has the objective of providing important data in the diagnosis and also influencing the treatment.

Decreased pressure of the internal anal sphincter can be seen in patients with scleroderma, especially in those with constipation. Most of these patients exhibit a decrease or even absence of the RAIR. Rectal prolapse is related to decreased resting pressures of the anal sphincter and decreased rectal capacity and compliance and may also be associated with fecal incontinence [18–20].

Fecal incontinence is an underestimated condition in patients with scleroderma. Ten incontinent patients were evaluated by anorectal manometry and compared to 20 incontinent patients without scleroderma. Those in the scleroderma group had higher voluntary contraction pressures and similar resting pressures. The threshold for the first rectal sensation in the scleroderma group was lower. Episodes of incontinence, anal canal length, and maximum tolerated volume did not vary between the two groups. The RAIR was abnormal in 80% of patients with systemic and normal sclerosis in 70% of the control group (Fig. 23.2). In addition, patients with scleroderma had more watery stools. Treatment with a combination of diet and constipating medications may prove helpful in these patients, since the presence of watery stools is common in addition to anal resting hypotonia [21, 22]. In a prospective study including 13 female patients with systemic sclerosis, it was demonstrated that more than 70% of the patients had hypotonia

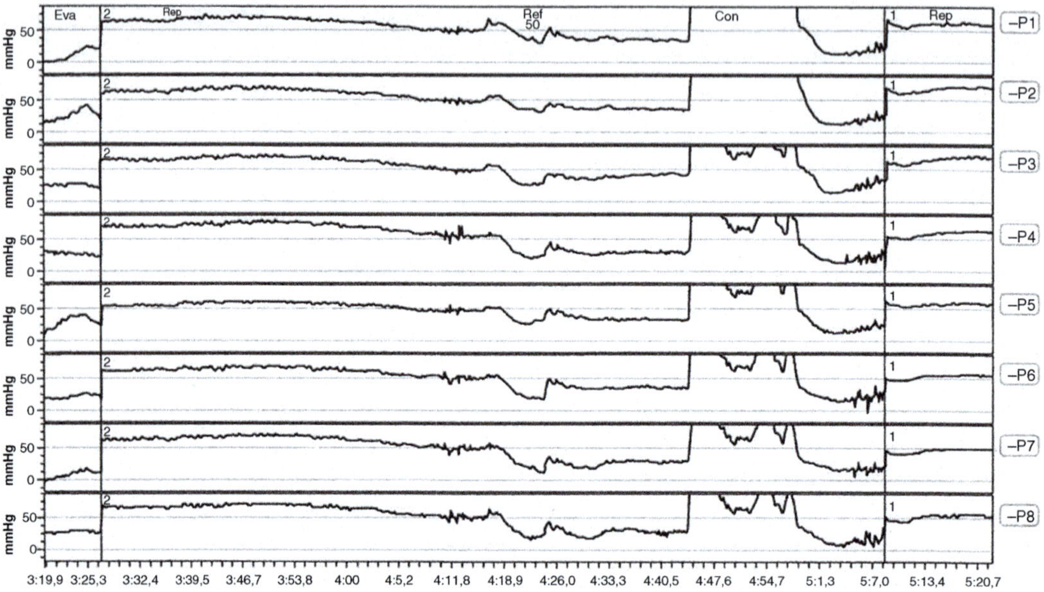

Fig. 23.2 Manometry in a patient with scleroderma. During investigation of the rectoanal inhibitory reflex (RAIR), we noticed relaxation of the internal anal sphincter after a contraction of the external anal sphincter. There was also resting anal hypertonia and short waves

of the internal anal sphincter [23]. In order to understand the underlying mechanisms associated with fecal incontinence in patients with system sclerosis, 25 patients underwent neurophysiological evaluation with EMG and fiber density as well as anal manometry and ultrasound. The authors concluded that lower voluntary squeeze pressures in incontinent patients and sonographic abnormalities in the external anal sphincter were present only in patients with incontinence, suggesting that the external anal sphincter is more important in maintaining fecal continence in patients with systemic sclerosis than has previously been reported. The finding of increased fiber density in most patients further supports involvement of the external anals sphincter function in systemic sclerosis and could indicate previous nerve injury with consequent incomplete reinnervation [24]. In another study comparing a control group with 44 patients with systemic sclerosis, it was demonstrated that patients with systemic sclerosis (both symptomatic or asymptomatic) have a thin and atrophied internal anal sphincter, suggesting that internal anal sphincter atrophy develops even in asymptomatic patients [25]. These authors suggested that sacral neuromodulation could be an option in these patients. However, before embarking on a more invasive treatment, biofeedback therapy should first be performed. In a study including 13 patients with systemic sclerosis underwent biofeedback therapy for management of fecal incontinence [26]. After biofeedback treatment, a significant improvement in symptoms and quality of life was observed.

Summary

Diabetes and collagenosis affect innervation of the pelvic floor and, through neuropathy, may cause symptoms of incontinence or obstructed defecation. Anorectal manometry is a simple method that can be used for diagnosis followed by treatment with biofeedback therapy or other treatment modalities.

References

1. Llanos G, Libman I. Diabetes in the Americas. Bull Pan Am Health Organ. 1994;28:285–301.
2. Feldman M, Schiller LR. Disorders of gastrointestinal motility associated with diabetes mellitus. Ann Intern Med. 1983;98:378–84.
3. Rothstein RD. Gastrointestinal motility disorders in diabetes mellitus. Am J Gastroenterol. 1990;85:782–5.
4. Goyal RK, Spiro HM. Gastrointestinal manifestations of diabetes mellitus. Med Clin North Am. 1971;55:1031–44.
5. Janatuinen E, Pikkarainen P, Laakso M, Pyörälä K. Gastrointestinal symptoms in middle-aged diabetic patients. Scand J Gastroenterol. 1993;28:427–32.
6. Fillmann HS, Llessuy S, Marroni CA, Fillmann LS, Marroni NP. Diabetes mellitus and anal sphincter pressures: an experimental model in rats. Dis Colon Rectum. 2007;50(4):517–22.
7. Rosztóczy A, Róka R, Várkonyi TT, Lengyel C, Izbéki F, Lonovics J, Wittmann T. Regional differences in the manifestation of gastrointestinal motor disorders in type 1 diabetic patients with autonomic neuropathy. Z Gastroenterol. 2004;42(11):1295–300.
8. Vinik AI, Maser RE, Mitchell BD, Freeman R. Diabetic autonomic neuropathy. Diabetes Care. 2003;26(5):1553–79.
9. Schiller LR, Ana CAA, Schmulen AC, Hendler RS, Harford WV, Fordtran JS. Pathogenesis of fecal incontinence in diabetes mellitus. N Engl J Med. 1982;307:1666–71.
10. Azvar E, Ersoz O, Karisik E, et al. Hyperglycaemia-induced attenuation of rectal perception depends on pattern of rectal balloon inflation. Dig Dis Sci. 1997;42:2206–12.
11. Chey WD, Kim M, Hasler WI, Owyang C. Hyperglycaemia alters perception of rectal distension and blunts the rectoanal inhibitory reflex in healthy volunteers. Gastroenterology. 1998;108:1700–8.
12. Sun WM, Katsinelos P, Horowitz M, Read NW. Disturbances in anorectal function in patients with diabetes mellitus and faecal incontinence. Eur J Gastroenterol Hepatol. 1996;8(10):1007–12.
13. Epanomeritakis E, Koutsoumbi P, Tsiaoussis I, Ganotakis E, Vlata M, Vassilakis JS, Xynos E. Impairment of anorectal function in diabetes mellitus parallels duration of disease. Dis Colon Rectum. 1999;42(11):1394–400.
14. Russo A, Botten R, Kong M-F, Chapman IM, Fraser RJL, Horowitz M, Sun W-M. Effects of acute hyperglycaemia on anorectal motor and sensory function in diabetes mellitus. Diabet Med. 2004;21:176–82.
15. Thiruppathy K, Bajwa A, Kuan KG, Murray C, Cohen R, Emmanuel A. Gut symptoms in diabetics correlate with components of the rectoanal inhibitory reflex, but not with pudendal nerve motor latencies or systemic autonomic neuropathy. J Dig Dis. 2015;16(6):342–9.
16. Deen KI, Premaratna R, Fonseka MM, De Silva HJ. The recto-anal inhibitory reflex: abnormal response in diabetics suggests an intrinsic neuroenteropathy. J Gastroenterol Hepatol. 1998;13(11):1107–10.
17. Jorge JX, Borges CI, Panão EA, Delgado FJ, Simões MA, Coelho ÁC, Silva AL, Almeida CC. Recto-anal manometric characteristics of type 2 diabetic patients who have sensation of incomplete defecation. J Diabetes Complications. 2013;27(2):167–70.
18. Falchuk KR, Conlin D. The intestinal and liver complications of diabetes mellitus. Adv Intern Med. 1993;38:269–87.
19. Fillmann HS, Tieppo J, Kretzmann Filho NA, Rodrigues G, Picada JN, Panès J, Marroni NAP. Superoxide dismutase ameliorates the bowel alterations induced by diabetes mellitus - an experimental study. Free Radicals Antioxid. 2012;26:378–81.
20. Sjogren RW. Gastrointestinal motility disorders in scleroderma. Arthritis Rheum. 1994;37:1265–82.
21. Battle WM, Snape WJ Jr, Wright S, et al. Abnormal colonic motility in progressive systemic sclerosis. Ann Intern Med. 1981;94:749–52.
22. Jaffin BW, Chang P, Spiera H. Fecal incontinence in scleroderma. Clinical features, anorectal manometric findings, and their therapeutic implications. J Clin Gastroenterol. 1997;25(3):513–7.
23. Pinto RA, Corrêa Neto IJF, Nahas SC, Bustamante Lopes LA, Sobrado Júnior CW, Cecconello I. Functional and anatomical analysis of the anorectum of female scleroderma patients at a center for pelvic floor disorders. Arq Gastroenterol. 2018;55Suppl 1(Suppl 1):47–51.
24. Franck-Larsson K, Graf W, Eeg-Olofsson KE, Axelson HW, Rönnblom A. Physiological and structural anorectal abnormalities in patients with systemic sclerosis and fecal incontinence. Scand J Gastroenterol. 2014;49(9):1076–83.
25. Thoua NM, Abdel-Halim M, Forbes A, Denton CP, Emmanuel AV. Fecal incontinence in systemic sclerosis is secondary to neuropathy. Am J Gastroenterol. 2012;107(4):597–603.
26. Collins J, Mazor Y, Jones M, Kellow J, Malcolm A. Efficacy of anorectal biofeedback in scleroderma patients with fecal incontinence: a case-control study. Scand J Gastroenterol. 2016;51(12):1433–38.

Solitary Rectal Ulcer Syndrome

24

Andre da Luz Moreira and Mitchell A. Bernstein

Introduction

Solitary rectal ulcer syndrome (SRUS) is an uncommon benign disorder of the anterior wall of the rectum. It was first described by Cruveilhier [1, 2] in 1829, and subsequently it was presented as a clinical entity by Madigan and Morson [3] in 1969. This condition is often associated with defecation disorders [4–6]. The term "solitary rectal ulcer" is inadequate, since the ulceration may be multiple, not confined to the rectum, and the lesions may present as hyperemic mucosa or polypoid rather than ulcerated [3, 6]. Internal prolapse or intussusception most often precedes the endoscopic presentation. Excessive straining at defecation and spastic pelvic floor are associated with SRUS [5, 7].

SRUS is rare, with annual incidence of 1–3.6 per 100,000 [8, 9]. The syndrome affects men and women in approximately equal proportions; however, there is a slight predominance among females [6, 8–10]. Although typically described as affecting young adults, the condition is also seen in children, adolescents, and the elderly. The mean age of presentation was 49 years in one series with up to 25% of patients presenting over the age of 60 years [11].

A. da Luz Moreira
University Hospital Pedro Ernesto, Department of Colorectal Surgery, Rio de Janeiro, Brazil

M. A. Bernstein (✉)
NYU Langone Medical Center, Division of Colon and Rectal Surgery, Department of Surgery, New York, NY, USA
e-mail: Mitchell.Bernstein@nyumc.org

Pathophysiology

The cause of SRUS is unknown. Physiological and histopathological studies suggest a spectrum of disease, raising the possibility of a variety of causes in different patients. In clinical practice there is a subgroup of patients who strain excessively at defecation, while in others such a behavioral disorder is not clinically apparent. In some patients, ulceration occurs as part of an external prolapse. It is generally agreed that occult or external rectal prolapse and paradoxical contraction of the pelvic floor muscles are among the factors involved in the development of SRUS. Opposing forces on the rectal mucosa have been described, with the downward force of defecation being opposed by the paradoxical contraction of the pelvic floor. The hypertonic anal canal with the presence of the prolapsed mucosa can result in ulceration [5, 7, 12]. In some patients, however, especially those with low rectal ulceration, direct trauma from a finger or other object can be identified as a cause. Unfortunately, there are patients who do not respond to biofeedback therapy for the treatment of paradoxical puborectalis contraction or surgical correction of mucosal prolapse [13].

L. C. C. Oliveira (ed.), *Anorectal Physiology*, https://doi.org/10.1007/978-3-030-43811-1_24

Presentation and Diagnosis

SRUS is a benign and rare condition and is often underdiagnosed. Patients usually have a long history of symptoms before the diagnosis can be established. SRUS commonly presents with passage of blood and mucus from the rectum associated with straining, chronic constipation, and symptoms of obstructive defecation (Table 24.1).

Digital manipulation to assist defecation and rectal emptying is reported in up to 60% of the cases [9, 11, 14, 15]. Massive rectal bleeding requiring blood transfusion is very rare [16–18]. Evidence of rectal prolapse, either internal or external, with or without abnormal perineal descent on defecography can be seen.

Diagnosis can usually be made by sigmoidoscopy (Fig. 24.1). Ulceration is not universally present, with polypoid, non-ulcerated lesions and erythematous areas also seen. The lesions are usually located 5–10 cm from the anal verge, almost always on the anterior rectal wall [6, 11]. These are commonly solitary but may be multiple. Although most commonly on the anterior rectal wall, they may be more extensive and even circumferential. Ulcers are seen in approximately 57% of the cases, but in 30% can present with multiple or synchronous lesions. The polypoid appearance may be present in 25% of the cases and it may suggest or mimic malignancy (Fig. 24.2) [1, 9, 19–22].

The differential diagnosis of SRUS includes inflammatory bowel disease, chronic ischemic colitis, malignancy, and rectal endometriosis [6]. Less frequently, amebiasis, lymphogranuloma venereum, secondary syphilis, and colitis cystica profunda may be considered [1, 15, 19]. There are reports of inadequate diagnoses and treatments in up to 26% of patients with SRUS [14–22].

A biopsy should always be taken. The histological features of SRUS are well established. There is a thickening of the mucosa with elongation and distortion of the glands, particularly

Table 24.1 Symptoms of SRUS

Constipation
Straining
Pain and pressure in the perineal region
Incomplete evacuation
Digital manipulation
Rectal bleeding
Mucous discharge
Association with rectal prolapse

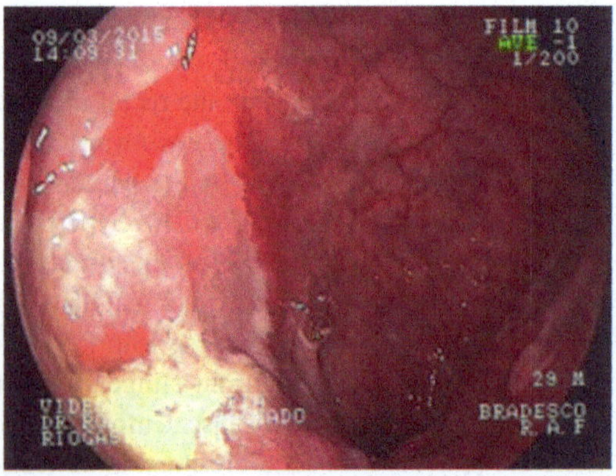

Fig. 24.1 Solitary rectal ulcer syndrome

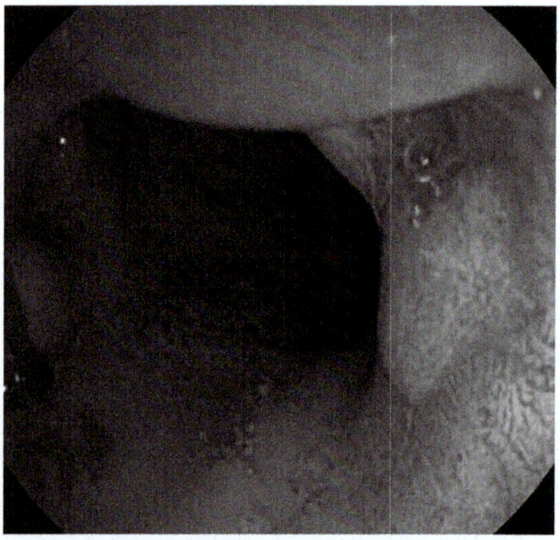

Fig. 24.2 SRUS presenting with a polypoid lesion mimicking a rectal malignancy

at their bases. When the glands are displaced into the submucosa, superficial bleeding can occur. The lamina propria is edematous and contains a variable proliferation of fibroblasts. There is thickening of the muscularis mucosa and extension of the smooth muscle fibers from the muscularis mucosa, vertically upward between the crypts [3, 23]. Features which SRUS or mucosal prolapse share with ischemic colitis are degeneration of the epithelium of the upper parts of the crypts, surface erosion, engorgement of superficial capillaries, and a lack of inflammatory features. Decussation and thickening of the two rectal muscle layers, nodular induration of the inner circular layer, and grouping of the outer longitudinal layer are unique to SRUS and are not seen in patients with isolated rectal prolapse. Diffuse excessive collagen deposition is characteristic of SRUS and is a valuable method of differentiating it from inflammatory bowel disease. Mucin staining patterns may also be used to distinguish this syndrome from other nonmalignant conditions, with sialomucin present in all cases and sulfomucin present in most [3, 6, 17, 21]. In all cases, the histopathologic evaluation of the inflammatory process is mandatory to distinguish SRUS from dysplasia and malignancy.

Although defecography can provide insights regarding pathophysiology, it has a limited role in establishing diagnosis of SRUS. Internal or external rectal prolapse has been reported in 45–80% of cases. Defecation was often delayed or incomplete. A study using defecography before and after rectopexy showed almost invariable correction of the prolapse and alteration of rectal configuration at rest but failed to identify any preoperative features which could be used to predict clinical outcome. The presence of prolonged evacuation time after surgery correlated with a poor symptomatic outcome [24].

Barium enema is not a reliable test in the diagnosis of SRUS. Nodularity of the rectal mucosa, thickening of the rectal folds, stricture formation, polypoid lesions, and ulceration may be seen but cannot always be differentiated from other conditions. Barium enema has been reported to be normal in 40–50% of cases [24, 25].

Endoanal and endorectal ultrasonography reveals thickening of the muscularis propria and a lack of distinction between the mucosa and the muscularis propria. The lack of relaxation of the puborectalis during straining may also be seen. Although not a universal finding, marked thickening of the internal sphincter is the most striking feature of endoanal ultrasonography. This is most likely to be due to repeated trauma, but an associated primary abnormality cannot be excluded. One study found a thick internal anal sphincter to be predictive of high-grade rectal intussusception in patients with SRUS. This could be used as an alternative to defecography, thus avoiding irradiation. There is often an associated thickening of the external sphincter and the submucosa, the latter sometimes also losing its homogeneous appearance [26, 27].

Physiological studies have been performed in patients with SRUS, but the results are variable, and the tests do not contribute to making the diagnosis or to predicting therapeutic response. The maximum resting pressure is reduced when the ulceration is associated with external prolapse. Patients may present with a lower tolerance to the volumes of the balloon during manometry. These raised anal canal and rectal electrosensory thresholds have been reported and may relate to the effect of trauma on nerve function, to tissue disruption, or may be part of the primary pathology. The rectal distension threshold is often reduced, which may reflect a reduction in rectal compliance. The reported high rectal voiding pressure is likely to relate to excessive straining. There is general agreement on a high incidence of rectoanal incoordination, excessive perineal descent, and pudendal neuropathy in patients with solitary rectal ulcer.

Treatment

There is no specific cure for SRUS. Symptoms may be improved by treatment, but it is uncommon to achieve endoscopic and histological normality. It is important to reassure patients about the benign nature of the rectal lesions. The initial treatment is conservative (e.g., diet adjustments, physiotherapy, and local treatment) and can prevent disease progression. Application of local agents, such as sucralfate enemas, to the area of ulceration fails to address the underlying defecatory disorder and is usually unsuccessful. Topical steroids and sulfasalazine enemas are not effective as seen in inflammatory bowel disease. A high-fiber diet can help but by itself is insufficient to obtain healing. The response rate ranges from 19% to 70%, and patients with an associated prolapse seem to benefit least from this intervention alone. In clinical practice some patients with SRUS appear to have a dominant behavioral disorder with excessive straining, while in others there is no history of straining. Biofeedback has been used both as the sole therapy for solitary rectal ulcer and as an adjunct to surgical therapy. Improvement in symptoms of SRUS after biofeedback retraining can deteriorate in some patients with time. Biofeedback for these patients included correction of pelvic floor defecatory behavior; regulation of toileting habits; encouragement to stop laxatives, suppositories, and enemas; and an attempt to address any psychological factors that may have been relevant [5, 11, 20, 28].

The surgery is reserved for patients when the conservative treatment fails. The least invasive operative procedures are suturing, or local excision of the area affected by SRUS. These local approaches fail to address the possible underlying etiology of the condition, and thus the long-term benefit is uncertain.

If a rectal prolapse is identified on defecography, its surgical repair should be considered [17]. The best results have been reported after posterior or anterior rectopexy. The long-term results of surgery have recently been addressed in 66 patients who had 72 procedures over a 10-year period, including rectopexy, Delorme's procedure, and resection. From patients who had had a rectopexy, 18% had complete resolution of symptoms, 36% had marked improvement, 9% had only an initial improvement, and 36% had no improvement or felt worse after the operation. No patient had a complete resolution of symptoms after a Delorme's procedure, but 56% were greatly improved, 11% had only initial improvement, and 33% were unchanged or worse. Resection with coloanal anastomosis or anterior resection improved only one of five patients. Overall, 52% of patients who did not have a stoma were improved after surgery. Patients with rectal prolapse seemed to fare better after abdominal rectopexy than those without rectal prolapse. One study noted successful treatment of SRUS by abdominal rectopexy in 83% of patients with vs 25% of patients without full-thickness prolapse. Resection may be used in conjunction with rectopexy in an attempt to decrease the incidence of postoperative constipation or can be performed as the primary operation to remove the affected segment. In a recent study of resection and coloanal anastomosis for benign rectal lesions, two patients with SRUS had poor results, both requiring a colostomy. For some patients a stoma, either temporary or permanent, is required. Many, however, continue to experience functional symptoms [29–33].

Summary

Solitary rectal ulcer syndrome is a benign disorder with complex mechanisms. Its is a rare condition that, unfortunately, frequently goes unrecognized. In addition, it is often challenging as there is no specific treatment modalities for a definitive cure.

References

1. Cruveilhier J. Ulcere chronique du rectum. Anatomie pathologique du corps humain. Paris: JB Bailliere; 1829.
2. Saul SH, Sollenberger LE. Solitary rectal ulcer syndrome. Am J Surg Pathol. 1985;9(6):411–21.

3. Madigan MR, Morson BE. Solitary ulcer of the rectum. Gut. 1969;10(11):871–81.
4. Kuijpers HC, Schreve RH, ten. Cate Hoedemakers H. Diagnosis of functional disorders of defecation causing the solitary rectal ulcer syndrome. Dis Colon Rectum. 1986;29(2):126–9.
5. Womack NR, Williams NS, Mist JH, Morrison JF. Anorectal function in the solitary ulcer syndrome. Dis Colon Rectum. 1987;30(5):319–23.
6. Felt-Bersma RJF, Cuesta MA. Rectal prolapse, rectal intussusception, rectocele, and solitary rectal ulcer syndrome. Gastroenterol Clin N Am. 2001;30(1):199–222.
7. Snooks SJ, Nicholls RJ, Henry MM, Swash M. Electrophysiological and manometric assessment of the pelvic floor in the solitary rectal ulcer syndrome. Br J Surg. 1985;72(2):131–3.
8. Haray PN, Morris-Stiff GJ, Foster ME. Solitary rectal ulcer syndrome – an underdiagnosed condition. Int J Color Dis. 1997;12(5):313–5.
9. Martin CJ, Parks TG, Biggard JD. Solitary ulcer in Northern Ireland 1971-1980. Br J Surg. 1981;68(10):744–7.
10. Rotholtz NA, Wexner SD. Surgical treatment of constipation and fecal incontinence. Gastroenterol Clin N Am. 2001;30(1):131–66.
11. Tjandra N, Fazio VW, Church JM, Lavery IC, Oakley JR, Milsom JW. Clinical conundrum of solitary rectal ulcer. Dis Colon Rectum. 1992;35(3):227–34.
12. Kim DJ, Bonis PAL. Solitary rectal ulcer syndrome. UpToDate. 2001. Disponível em: http://www.uptodate.com/contents/solitary-rectal-ulcer-syndrome.
13. Halligan S, Nicholls RJ, Bartram CI. Evacuation proctography in patients with solitary rectal ulcer syndrome: anatomic abnormalities and frequency of impaired emptying and prolapse. AJR Am J Roentgenol. 1995;164(1):91–5.
14. Tjandra JJ, Fazio VW, Petras RE. Clinical and pathologic factors associated with delayed diagnosis in solitary rectal ulcer syndrome. Dis Colon Rectum. 1993;36(2):146–53.
15. Vaizey CJ, van den Bogaerde JB, Emmanuel AV, Talbot IC, Nicholls RJ, Kamm MA. Solitary rectal ulcer syndrome. Br J Surg. 1998;85(12):1617–23.
16. Binderow SR, Mayer R, Freed JS. Massive hemorrhage from solitary rectal ulcer: toward a definitive treatment. Mt Sinai J Med. 1995;62(4):308–11.
17. Levine DS. "Solitary" rectal ulcer syndrome. Are "solitary" rectal ulcer syndrome and "localized" colitis cystica profunda analogous syndrome caused by rectal prolapse? Gastroenterology. 1987;92(1):243–53.
18. Hull TL, Milsom JW. Pelvic floor disorders. Surg Clin N Am. 1994;74(6):1399–413.
19. Sondheimer JM, Slagle TA, Bryke CR. Solitary rectal ulcer syndrome in a teenaged boy. J Pediatr Gastroenterol Nutr. 1985;4(5):835–8.
20. Evans A, Bishop PR. Clinical quiz. Solitary rectal ulcer. J Pediatr Gastroenterol Nutr. 2001;32(2):214–31.
21. Abid S, Khawaja A, Bhimani SA, Ahmad Z, Hamid S, Jafri W. The clinical, endoscopic and histological spectrum of the solitary rectal ulcer syndrome: a single-center experience of 116 cases. BMC Gastroenterol. 2012;14:12–72.
22. Blanco F, Frasson M, Flor-Lorente B, Minguez M, Esclapez P, García-Granero E. Solitary rectal ulcer: ultrasonographic and magnetic resonance imaging patterns mimicking rectal cancer. Eur J Gastroenterol Hepatol. 2011;23(12):1262–6.
23. Kang YS, Kamm MA, Engel AF, Talbot IC. Pathology of the rectal wall in solitary rectal ulcer syndrome. Gut. 1996;38(4):587–90.
24. Goei R, Baeten C, Janevski B, van Engelshoven J. The solitary rectal ulcer syndrome: diagnosis with defecography. Am J Roentgenol. 1987;149(5): 933–6.
25. Millward SF, Bayjoo P, Dixon MF, Williams NS, Simpkins KC. The barium enema appearances in solitary rectal ulcer syndrome. Clin Radiol. 1985;36(2):185–9.
26. Halligan S, Sultan A, Rottenberg G, Bartram CI. Endosonography of the anal sphincters in solitary rectal ulcer syndrome. Int J Color Dis. 1995;10(2):79–82.
27. Van Outryve MJ, Pelckmans PA, Fierens H. Transrectal ultrasound study of the pathogenesis of solitary rectal ulcer syndrome. Gut. 1993;34(10):1422–6.
28. Binnie NR, Papchrysostomou M, Clare N, Smith AN. Solitary rectal ulcer: the place of biofeedback and surgery in the treatment of the syndrome. World J Surg. 1992;16(5):836–40.
29. Keighley MRB, Shouler P. Clinical and manometric features of the solitary rectal ulcer syndrome. Dis Colon Rectum. 1984;27(8):507–12.
30. Uhlig BE, Sullivan ES. The modified Delorme operation: its place in surgical treatment for massive rectal prolapse. Dis Colon Rectum. 1979;22(8):513–21.
31. Stavorovsky M, Weintroub S, Ratan J, Rozen P. Successful treatment of a benign solitary rectal ulcer by temporary diverting sigmoidostomy: report of a case. Dis Colon Rectum. 1977;20(4): 347–50.
32. Sitzler PJ, Kamm MA, Nicholls RJ, McKee RF. Long-term clinical outcome of surgery for solitary rectal ulcer syndrome. Br J Surg. 1998;85(9):1246–50.
33. Niv Y, Bat L. Solitary rectal ulcer syndrome – clinical, endoscopic, and histological spectrum. Am J Gastroenterol. 1986;81(6):486–91.

Multiple Approaches for the Treatment of Rectoceles

25

Sergio Larach, Lawna Hunter,
and Beatriz Martin-Perez

Introduction

Rectoceles involve herniation of the anterior rectal wall through the rectovaginal septum and into the posterior vaginal wall. Outpouching of the rectal wall into the vagina is typically the result of a weakened rectovaginal septum, which may have occurred due to age, parturition, post-menopausal status, collagen disorders, or other currently unknown processes [1]. They are often found in females, rarely in men, and may or may not be related to evacuatory disorders. Since the immediate cause of rectoceles is not completely known, the baseline prevalence has not been well established, and due to the concurrence of both vaginal and rectal symptomatology, cases are under the care of either gynecologists, colorectal surgeons, or both.

Rectoceles are often asymptomatic, but when symptoms do arise, it can be debilitating, affecting women's daily living. A working knowledge of anatomy, symptomatology, medical therapies, and surgical options are necessary in the proper diagnosis and treatment of a rectocele. The ultimate goal in any surgery is restoration of function; techniques for accomplishing this will be discussed.

S. Larach (✉)
Endosurgical Center of Florida, Orlando, FL, USA

L. Hunter
San Diego, CA, USA

B. Martin-Perez
Hospital Clinic, Gastrointestinal Surgery,
Barcelona, Spain

Anatomy

The existence and integrity of the rectovaginal septum continues to be an area of debate, even though current studies are supporting the existence of this plane, as has been demonstrated by the work of Fritsch (2002), Berglas (1953), Bock (1982), and Ludwikowski (2002) [2–5]. Anatomically, the rectovaginal septum is the female equivalent of Denonvilliers' fascia within a male's rectovesical septum [6]. It is a fusion of the endopelvic fascia, which is comprised of a fibrous layer of dense collagen, smooth muscle, elastin, and neurovascular bundles. Thickness of the rectovaginal septum varies. In some patients, it is a thick fascial layer. In other patients, the septum is a thin translucent structure.

A healthy vagina has three different layers of support and stabilization: superior, lateral, and inferior. The first layer of support occurs where the superior vaginal apical endopelvic fascia is attached to the cardinal-uterosacral ligament complex [7]. The second layer of stabilization occurs laterally. Here, the endopelvic fascia is connected to the arcus tendineus fascia pelvis, and the lateral posterior vagina attaches to the fascia which overlies the levator ani muscles. The final site of support occurs inferiorly where

L. C. C. Oliveira (ed.), *Anorectal Physiology*, https://doi.org/10.1007/978-3-030-43811-1_25

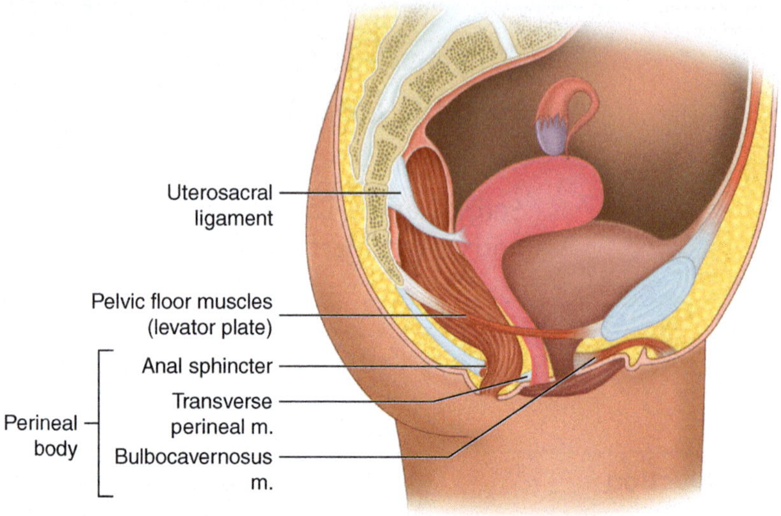

Fig. 25.1 Most common locations of rectovaginal tears originate in rectoceles. Note the inferior tear's connection with the perineal body

the low, posterior vagina attaches to the perineal body (Fig. 25.1).

During childbirth or as a result of an improperly healed episiotomy, many women develop transverse defects in the rectovaginal septum or have a separation of its attachments [8]. Even though rectoceles most often present in childbearing women, they can also occur in nulliparous women. The most common defect arises inferiorly, just above the connection with the perineal body. Less common defects can occur with lateral, midline, or high transverse fascial defects (Fig. 25.2). The presence of defects within the septum are anatomically evident upon observance of compromised integrity of the rectovaginal fascia through the transvaginal approach to rectoceles, reaffirming the importance of what is described as defect-specific rectocele repairs [1]. This type of defect-specific repair has been the standard for transanal repairs and is becoming more popular among gynecological approaches.

Symptomatology

The most common complaint for women with rectocele is perineal and vaginal pressure. Oftentimes, there are added symptoms of constipation, obstructive defecation, fecal incontinence, pelvic pain, dyspareunia, as well as rectal and/or vaginal prolapse [1, 9, 10]. Some patients will also present with concurrent urological symptoms, such as cystocele or urinary incontinence. It must be known that many patients display a multitude of arrangements of the previously listed symptoms; all are not necessary for diagnosis. Diagnosis is further complicated by the lack of correlation between degree of prolapse and severity of symptoms [1, 11, 12].

Porter (1999) found that 20–80% of women referred to pelvic floor clinics were diagnosed with rectoceles [6, 13]. Defecographic studies used in the initial workup showed that 80% of the women and 13% of men with rectoceles of at least 1 cm depth were asymptomatic. Larger rectoceles are associated with difficulty in evacuation, rectal pain and/or bleeding, and the sensation of having a vaginal mass. Of patients presenting with rectocele, 12–70% of patients reported rectal pain. Chronic cases of rectocele commonly present with fecal incontinence, pudendal neuropathy, and tenesmus [6]. While the etiology for pudendal neuropathy is not completely known, it is believed to result from constant or severe compression. Risk factors for developing pudendal neuropathy include chronic obstructive pulmonary disease (COPD), obesity, and chronic constipation.

Of the patients with enlarging rectoceles, many have further separation of the levator muscles with an increase in vaginal caliber [14]. Such anatomical changes are correlated with increas-

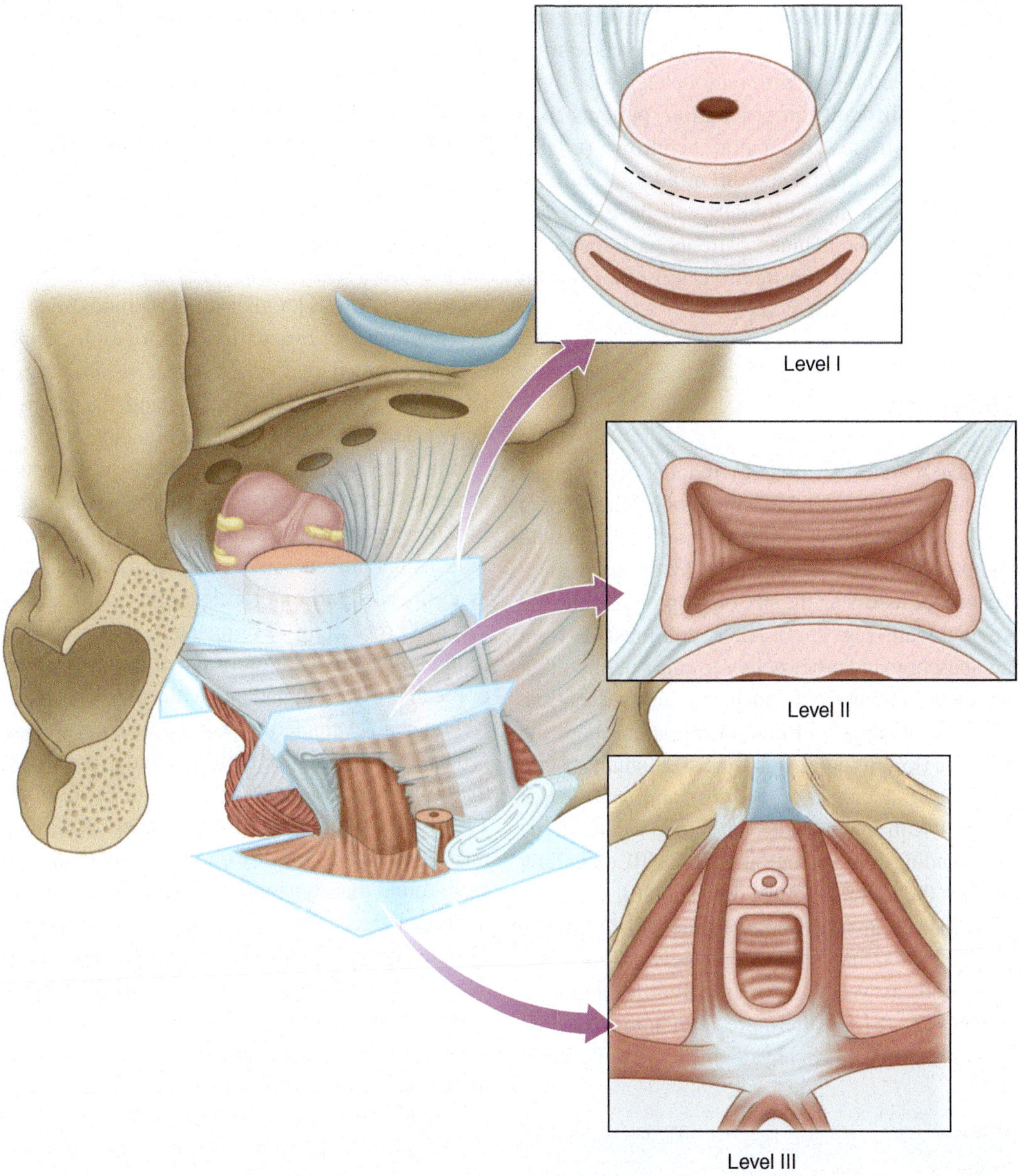

Fig. 25.2 Levels of vaginal support. Note that the endopelvic fascia is the main stabilizing component within the three layers

ing degrees of prolapse and developed sexual dysfunction due to enlargement of the genital hiatus.

During the evacuatory process, many women describe symptoms of vaginal bulging and the sensation of having a vaginal mass. Patients who develop a vaginal mass during defecation typically have a difficult time evacuating completely. Therein, patients may develop obstructive defecation syndrome, and there is often a need for digital splinting to aid in the defecatory process [6]. The need for manual assistance in initiated defecation is common, but not always present, occurring in 20–75% of cases [9, 10, 15]. Patients presenting with a large vaginal mass often develop hemorrhoids due to intense straining during the defecatory process. In fact, hemorrhoids are the most common coexisting colorectal pathology with rectoceles [16].

Due to difficulty in evacuation, patients may develop obstructive defecatory syndrome (ODS), although you must keep in mind that not all patients presenting with ODS will have a rectocele. In the same manner, patients with rectocele may present with constipation, but other etiologies must be investigated [6].

Diagnosis

Clinical diagnosis is done by manual or bimanual examination via digital examination of the vaginal wall; this should occur prior to and concurrent with the patient's Valsalva maneuver. With this maneuver, we are assessing the size of the prolapse into the vaginal wall (Fig. 25.3). After, you must perform an anoscopy and proctoscopy to rule out any other anorectal pathologies, such as intussusception. This examination will give information about the position (low, middle, high), the size (small <2 cm, medium 2–4 cm, large >4 cm), and the degree (Type I with bulging into the upper vagina, Type II extending into the introitus, Type III extending beyond the introitus) of the rectocele [6].

Dynamic magnetic resonance (MR) allows for the best visualization of the pelvic floor compartments, as well as the pelvic muscles and ligaments [17, 18]. Furthermore, dynamic MR (MR defecography) allows physicians to identify rectoceles and assess its size, location, failure to empty during simulated defecation, and the presence of a nonrelaxing puborectalis. This imaging displays a picture of the behavior of the three compartments during defecatory efforts, thus allowing the physician to evaluate combined pathologies [17, 18]. Enhancement of this technique is obtained by injecting contrast into the rectum and vaginal areas with T1 and T2 scanning modalities (Fig. 25.4) [19–21]. Other imaging modalities are defecography, or evacuation proctography, which is an X-Ray where contrast is placed into the rectal cavity and the vagina, showing the changes during defecation and Valsalva maneuvers.

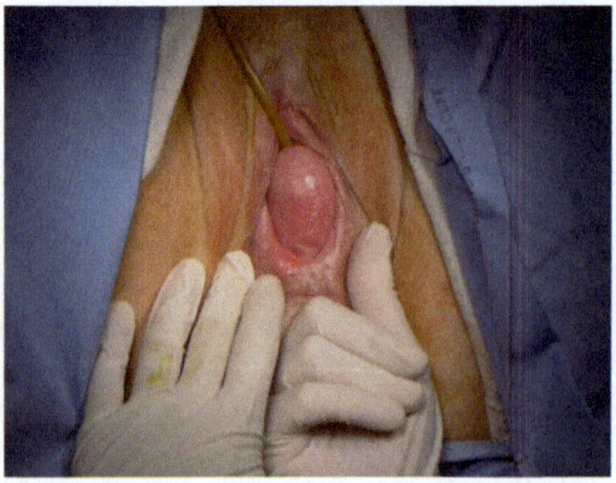

Fig. 25.3 Clinical examination of a level 3 rectocele

Imaging is an imperative part of the diagnosis process. Not only does it allow the physician to confirm the presence of a rectocele, but it will also provide insight into any other pelvic pathologies, such as cystoceles and enteroceles, and could change surgical proceedings. For instance, Mellgren et al. found a 93% association of rectal intussusception or rectal prolapse in patients with enteroceles [22]. Kelvin et al. evaluated 74 women with pelvic floor prolapse, using a four-contrast study, and found that 19% had associated enteroceles [23]. Of the 19% of newly diagnosed enteroceles, 50% were missed during physical examination. Attenberger et al. found that dynamic MR was essential to the treatment decisions in 22 out of 50 cases [17]. Thirteen of the 22 cases had a change in treatment due to MRI findings, and 12 out of 50 patients were diagnosed with an enterocele that was not diagnosed through physical examination. In 4 out of 50 cases, enteroceles were suspected due to history and symptomatology; MRI findings excluded the diagnosis and changed treatment from surgical to conservative [17].

Lefavre et al. stated that while 80% of colorectal surgeons use MR defecography, only 6% of gynecologists use this imaging in preoperative evaluations [1, 24, 25]. It is imperative that defecography be used in the screening process because it could shed insight into the need for a different surgical approach.

Additional studies to be used for proper diagnosis of a rectocele are dynamic transperineal ultrasonography, manometry, electromyography (EMG), colonoscopy, and transit studies.

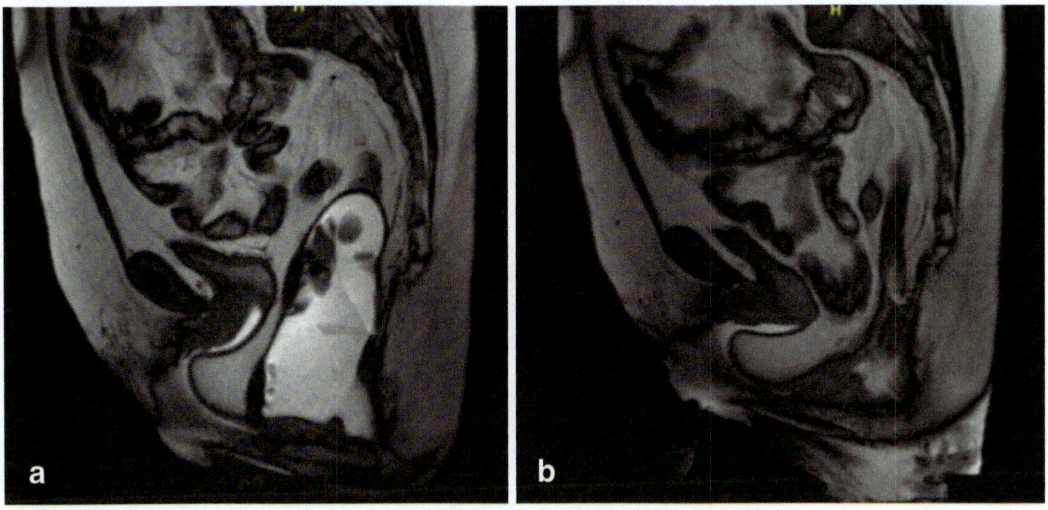

Fig. 25.4 MRI is a valuable evaluation tool for the three pelvic compartments (anterior, middle, and posterior) and any associated pathologies. (**a**) Expulsion of rectal contrast revealing an anterior rectocele. (**b**) Valsalva maneuver demonstrating an enterocele with vaginal prolapse

Dynamic ultrasound is advantageous due to the lack of pelvic radiation, and it allows for visualization of thinning of the rectal wall via protrusion of the rectum into the vagina [17–21]. Manometry and electromyography (EMG) are suggested for select patients with suspicious symptoms of fecal incontinence. Colonoscopies are recommended as a tool for essential evaluation of the lower GI tract. Transit studies will be indicated in patients evaluated with concomitant constipation.

Upon completion of physical examination and imaging studies, physicians will be able to clearly determine the position of rectoceles; such knowledge effects the surgical decision-making process, and it is important to classify the rectocele for proper management. Rectoceles are classified via the Pelvic Organ Prolapse Quantification system (POP-Q) along with imaging studies [1]. POP-Q provides physicians with a reproducible and objective testing for determining site-specific defects in pelvic support. High rectoceles correspond to degradation of the uterosacral ligament and cardinal ligaments associated with deep cul-de-sac and may result in an enterocele [6]. Medium rectoceles are the most classical herniation of the rectovaginal septum. Low rectoceles are more frequently a consequence of a prior episiotomy and the destruction of the tendinous portion of the perineum [26]. High rectoceles are usually treated with a transabdominal technique while the mid and low retoceles may be treated for a perineal technique or a transabdominal approach. Additionally, transabdominal techniques are recommended if there is a multicompartimental prolapse.

Medical Management

It is recommended that physicians try nonsurgical bowel management program for patients with rectocele prior to determining if surgery is completely necessary. Treatments such as high-fiber diets, psyllium supplements, increasing water consumption, prescribing stool softeners, placement of vaginal pessaries for pelvic organ support, topical vaginal estrogens, and recommending the avoidance of active laxatives and prolonged straining are all viable nonsurgical therapy options (Fig. 25.5) [1, 6, 27].

Surgical Treatment

Each surgical indication and approach should be individualized due to the fact that there is no gold standard technique for rectocele repair.

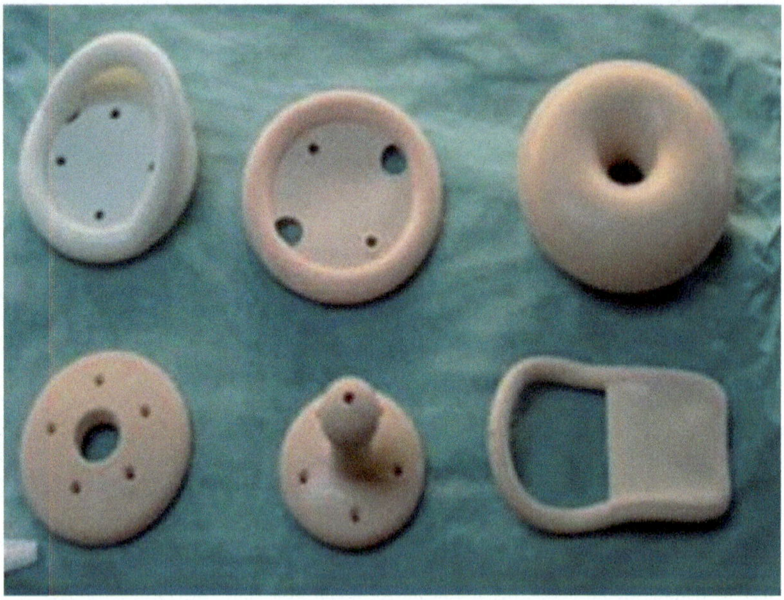

Fig. 25.5 An assortment of pessaries. Pessaries are most commonly used in the treatment of uterine prolapse but are also helpful in the treatment of other pelvic floor pathologies

Approaches for rectocele repair include transvaginal, transperineal, endorectal, and transabdominal. In view of all of the different types of rectoceles, gynecologists and colorectal surgeons will attest to the difficulty of deciding which surgical procedure is warranted for each individual patient. The surgical procedure which will best provide for a more complete treatment of rectocele is chosen according to the rectocele's anatomical position and presence, or lack thereof, of evacuatory outlet syndrome [6].

Transvaginal Repair

Posterior colporrhaphy is commonly performed simultaneously with a perineoplasty to address a relaxed perineum and widened genital hiatus [8]. Posterior colporrhaphy was an operation devised in the early nineteenth century largely to deal with perineal tears incurred during vaginal delivery [6]. Surgically, the dissection is driven laterally to the lateral vaginal sulcus and medial margins of the puborectalis muscles [1, 6, 8]. Plication of the rectovaginal fascia, with or without the levator ani muscles, occurs using interrupted sutures while depressing the anterior rectal wall. The repair also includes approximation of the levator muscles. An isolated transvaginal approach, or combined with an abdominal technique, can be used for reconstruction of the attenuated rectovaginal septum in cases with associated enterocele or cystocele [6].

Resolution of constipation occurs in approximately 80% of patients [6, 28]. Excellent results for resolution of constipation were studied by Mellgren et al. study [22]. He reported on 25 patients prospectively for an average of 1 year and found that constipation was present in 88% preoperatively and relieved in 84% postoperatively. While constipation is positively improved, Zbar et al. report a rate of 36% for postoperative fecal incontinence and a 25% rate of dyspareunia [6, 8].

Site-Specific Fascial Defect Repair Technique

The goal of site-specific fascial defect repair of rectoceles is to identify the fascial tear and reapproximate the edges [1]. The surgical dissection is similar to the traditional transvaginal exposure, identifying the defect and repairing with interrupted sutures. It is advised through the procedure to insert the fingers through the rectum to identify the weak point in the fascia. This technique is not standardized and is operator dependent.

Transperineal Repair

Transperineal approach is more common among colorectal surgeons. A perineal U-shaped incision is created and a dissection is completed lifting the vaginal wall. This allows the surgeon to complete a colporrhaphy and a sphincter repair if needed, with the option of combined procedures like hemorrhoidectomy and/or fissurectomy. During the dissection, the plane between the external anal sphincter and vaginal mucosa is entered and dissected superiorly up to the posterior cul-de-sac, with special care taken to not enter the pouch of Douglas [1]. Plication of the rectal wall with interrupted sutures should start on the apex of the rectocele (Fig. 25.6). For reinforcement, this technique has also been used with the addition of biological or synthetic mesh [8]. A cautious word should be said, even though high success rates for mesh implants have been reported, erosion of the synthetic mesh has been associated with these repairs and has resulted in litigation [1].

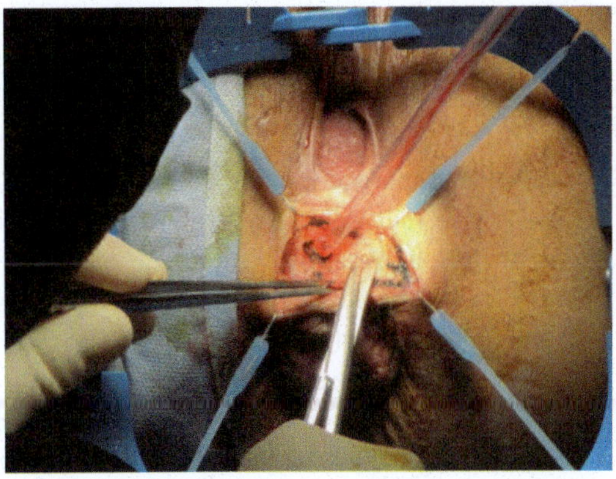

Fig. 25.6 The transperineal approach allows for repair of the rectocele, as well as reinforcement of the rectovaginal septum with a biological or synthetic mesh if necessary

Transanal Repair

In 1967, Marks described a transanal repair of rectoceles. The plication of the rectal wall and removal of the excess anterior mucosa were the fundamentals of this procedure, forming a scar to add support to the anterior wall of the rectum [8, 28].

The transanal approach has been a preferred method by colorectal surgeons, for the addition of concurrent anorectal surgeries to the repair (hemorrhoidectomy, fissurectomy, etc.). Patient is placed in the jacknife position. In this procedure, the defect in the rectovaginal septum is denuded from rectal mucosa [6], resecting the mucosal flap until you reach the apex of the rectocele. The rectal wall is approximated with interrupted sutures and finally a mucosal layer is approximated over (Fig. 25.7). A modification of this procedure has been described by Block (1986) using a suture without resection of the redundant mucosa [29]. This technique has also been modified with the usage of a linear stapling GIA-disposable endoanal stapler with the same success [6, 8]. A common postoperative symptom is tenesmus; necrosis and fistulas have also been described as a postoperative complication.

Given the deleterious effects on internal sphincter function and resting anal pressure, this approach should not be used when an option of a sphincter repair is indicated in patients with concomitant fecal incontinence [30]. This approach does not allow for reapproximation of

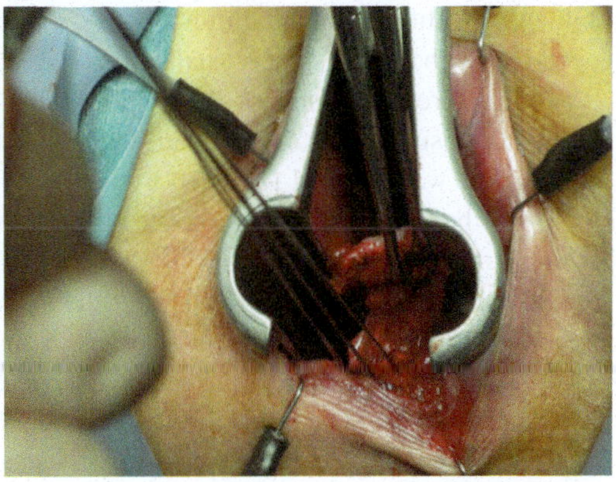

Fig. 25.7 Plication of the rectal wall after creating of a flap of the rectal mucosa during a transanal repair

the levator muscles and exposure may be limited for high rectoceles, so other approaches should be considered. The only randomized trial comparing transvaginal with endorectal repair is by Arnold (1990), which displayed a significantly high incidence of postoperative dyspareunia in the sexually active cohorts after transvaginal repair [6, 30].

Stapled Transanal Rectal Resection (STARR)

Patient's presenting with rectocele and symptoms of ODS are best treated with a stapled transanal rectal resection (STARR) procedure. It is pertinent that patients presenting with chronic constipation or any change in their bowel habits are properly diagnosed with either pelvic floor motility disorder or obstructive defecation disorders. Due to continued obstructive symptoms, patients with clinically diagnosed internal prolapse and rectoceles associated with pelvic dyssynergy should be treated conservatively with biofeedback therapy and/or dietary modifications. Surgery is often not required and, if performed, will result in continued abnormal defecation and may be responsible for high rectocele recurrence rate [1, 31].

In 2003, Longo described a new technique for treatment of ODS caused by rectocele and rectal intussusceptions: stapled transanal rectal resection (STARR) [32]. In this procedure, the rectocele was identified, and circumferential resection occurred without constant visualization due to the circumferential shape of the stapler, which obliterates visualization of the canal. The circular anal dilator was inserted followed by the placement of the anterior half purse-string sutures with full thickness of the rectal wall. Endovaginal digital control allows you to verify that the vaginal wall has been spared. Upon stapling of the anterior hemicircumference, the same procedure is repeated for the posterior half. This technique is primarily used by colorectal surgeons because gynecologists do not typically have a strong working knowledge of the symptoms of defecatory dysfunction [1].

Since its origin, complications have been the biggest concern of the STARR procedure. Ribaric's (2014) multi-institutional study found 1-year real-world outcomes of postoperative complications in 11% of patients, including bleeding (5%), staple line complications (3%), urinary retention (2%), and persistent pain (1%); there were no major complications or mortalities [33]. Other studies found postoperative complications of perianastomotic abscess, hemorrhoidal thrombosis, anal fissure, rectal stenosis and sepsis with retropneumoperitoneum and intra-abdominal bowel injury [32, 33].

In response to the need to cut down on postoperative complications, the TRANSTAR™ stapler was created with a hemicircumferential stapler. It allows for constant visualization throughout the entire procedure because the integrity of visualization is maintained during stapling. The CONTOUR® TRANSTAR™ stapling kit (Ethicon Endo-Surgery Inc., Cincinnati, OH, USA) is opened, and a circular anal dilator is introduced through the anus; it is then fixed to the perianal skin with four cardinal sutures [33]. Upon visualization of the defect, parachute suture placement occurs, which allows for opening of the prolapse. Under constant visualization, the redundant tissue is circumferentially resected through the use of the TRANSTAR™ stapler, thus completing the procedure.

Transabdominal Laparoscopic Rectocele Repair

The use of laparoscopy-assisted rectal mobilization for rectocele and enterocele repair is in evolution. Adequate mobilization of the rectum up to the level of the levator muscles must be achieved. Placement of mesh has been described for the reinforcement of the rectovaginal septum [6]. Laparoscopy is indicated when multiple compartment dysfunctions have to be addressed, as it allows for the combination of uterosacral suspension, levatorplasty, sacral colpopexy for vault prolapse, or any other indicated procedures.

Summary

The relationship of the pathogenesis of constipation and rectocele is unclear. Since rectoceles are usually found in the setting of other pelvic problems, it is argued that they may share a common etiology. The physician must observe all of the different implications of the etiology of the different pelvic floor disorders for evaluation and therapy [27].

Preoperative evaluation should include the clinical assessment from patient history and physical examination, in association with complementary studies of MR defecography and manometry. Colorectal surgeons must be aware of associated defecatory disturbances that may alter surgical procedures. Due to the fact that patients with ODS or constipation present primarily to colorectal surgeons, gynecologists predominantly address vaginal anatomical changes with limitations for the treatment of associated defecatory dysfunction. Patients who are presenting with concomitant obstructive defecation symptomatology should have a transanal repair, preferably the STARR procedure.

Due to the global transformation of the pelvic floor and the lack of knowledge of true etiology of rectoceles, patients may receive surgical treatments that are successful for a limited period of time [34]. Further research is needed in order to provide every patient with a more definitive intervention and better overall outcomes.

References

1. Lefevre R, Davila GW. Functional disorders: rectocele. Clin Colon Rectal Surg. 2008;21(2):129–37.
2. Fritsch H, Brenner E, Lienemann A, Ludwikowski B. Anal sphincter complex-reinterpreted morphology and its clinical relevance. Dis Colon Rectum. 2002;45:188–94.
3. Berglas B, Rubin IC. Histologic study of the pelvic connective tissue. Surg Gynecol Obstet. 1953;97:277–89.
4. De Blok S. The connective tissue of the female fetal pelvic region: a dissectional analysis. Acta Morphol Neerl Scand. 1982;20:191–212.
5. Ludwikowski B, Hayward IO, Fritsch H. Rectovaginal fascia: an important structure in pelvic visceral surgery? About its development, structure and function. J Pediatr Surg. 2002;37:634–8.
6. Zbar AP, Lienemann A, Fritsch H, Beer-Gabel M, Pescatori M. Rectocele: pathogenesis and surgical management. Int J Color Dis. 2003;18:369–84.
7. DeLancey JOL. Anatomic aspects of vaginal eversion after hysterectomy. Am J Obstet Gynecol. 1992;166(6 pt 1):1717–28.
8. Beck DE, Nechol LA. Rectocele. Clin Colon Rectal Surg. 2010;23:90–8.
9. Capps WF. Rectoplasty and perineoplasty for the symptomatic rectocele: a report of fifty cases. Dis Colon Rectum. 1975;18:237–44.
10. Sarles JC, Arnaud A, Selezneff I, Olivier S. Endorectal repair of rectocele. Int J Color Dis. 1989;4:167–71.
11. Fenner DE. Diagnosis and assessment of sigmoidoceles. Am J Obstet Gynecol. 1996;175:1438–42.
12. Altman D, Lopez A, Kierkegaard J, et al. Assessment of posterior vaginal wall prolapsed: comparison of physical findings to cystodefecoperitoneography. Int Urogenecol J Pelvic Floor Dysfunct. 2005;16: 96–103.
13. Porter WE, Steele A, Walsh P, Kohli N, Karram MM. The anatomic and functional outcomes of defect-specific rectocele repairs. Am J Obstet Gynecol. 1999;181:1353–9.
14. Kahn MA, Stanton SL. Techniques of rectocele repair and their effects of bowel function. Int Urogynecol J. 1998;9:37–47.
15. Blatchford GJ, Cali RL, Christensen MA. Surgical treatment of rectocele. In: Wexner SD, Bartolo DCC, editors. Constipation: etiology evaluation and management. Oxford/Boston: Butterworth-Heinemann; 1995. p. 199–209.
16. Pitchford CA. Rectocele: a cause of anorectal pathological changes in women. Dis Colon Rectum. 1967;10:464–6.
17. Attenberger IU, Morelli JN, Budjan J, et al. The value of dynamic magnetic resonance imaging in interdisciplinary treatment of pelvic floor dysfunction. Abdom Imaging. 2015. Advance online publication. https://doi.org/10.1007/s00261-015-0476-y.
18. Garcia del Salto L, Criado JD, Aguilera del Hoyo LF, et al. MR imaging-based assessment of the female pelvic floor. Radiographics. 2014;34(5):1417–39.
19. Flusberg M, Sahni VA, Erturk SM, Mortele KJ. Dynamic MR defecography: assessment of the usefulness of the defecation phase. AJR Am J Roentgenol. 2011;196(4):394–9.
20. Foti PV, Farina R, Riva G, et al. Pelvic floor imaging: comparison between magnetic resonance imaging and conventional defecography in studying outlet obstruction syndrome. Radiol Med. 2013;118(1):23–9.
21. Pannu JK, Kaufman HS, Cundiff GW, Genadry R, Bluemke DA, Fishman EK. Dynamic MR imaging of pelvic organ prolapse: spectrum of abnormalities. Radiographics. 2000;20(6):1567–82.
22. Mellgren A, Anzen B, Nilsson BY, et al. Results of rectocele repair. A prospective study. Dis Colon Rectum. 1995;38:7–13.

23. Kelvin FM, Maglinte DD, Hornback JA, Benson JT. Pelvic prolapse: assessment with evacuation proctography (defecography). Radiology. 1992;184:547–51.
24. Mizrahi N, Kapoor D, Nogueras JJ, et al. A gynecologic perspective of posterior compartment defects. Colorectal Dis. 2002;4(Suppl):H68.
25. Davila GW, Ghoniem GM, Kapoor DS, Contereras-Ortiz O. Pelvic floor dysfunction management practice patterns: a survey of members of the international urogynecological association. Int Urogynecol J. 2002;13:319–25.
26. Cortesse A, Haab F, Costa P, Delmas V. Traitement des rectoceles par voie basse. Prog Urol. 2009;19:1080–5.
27. Pemberton JK, Swash M, Henry L. The pelvic floor: its function and disorders. China: Harcourt Publishers Limited; 2002.
28. Marks MM. The rectal side of the rectocele. Dis Colon Rectum. 1967;10(5):387–8.
29. Block IR. Transrectal repair of rectocele using an obliterative suture. Dis Colon Rectum. 1986;29:707–11.
30. Arnold MW, Stewart WRC, Aguilar PS. Rectocele repair; four year's experience. Dis Colon Rectum. 1990;33:684–7.
31. Brubaker L. Rectocele. Curr Opin Obstet Gynecol. 1996;8:376–9.
32. Longo A. Obstructed defecation because of rectal pathologies. Novel surgical treatment: stapled transanal rectal resection (STARR). XIV Colorectal Symposium. Ft Lauderdale; 2003.
33. Ribaric G, D'Hoore A, Schiffhorst G, Hempel E. STARR with CONTOUR® TRANSTAR™ device for obstructed defecation syndrome: one-year real-world outcomes of the European TRANSTAR registry. Int J Color Dis. 2014;29:611–22.
34. Lee PYH, Meurette G. Pelvic floor disorders. In: Beck DE, Roberts PL, Saclarides TJ, Senagore AJ, Stamos MJ, Nasseri Yl. The ASCRS textbook of colon and rectal surgery: Second Edition. New York: Springer Science +Business Media LLC. 2011:323–335.

Rectal Prolapse

26

Janet T. Lee, Sarah A. Vogler, and Robert D. Madoff

Introduction

Rectal prolapse, also known as procidentia, is defined by full-thickness protrusion of the rectum through the anal canal, beyond the anal verge. This is a benign condition but can cause symptoms such as discomfort, fecal incontinence, pelvic pain, and rectal bleeding that can greatly affect a patient's quality of life. Studies have estimated that 0.5% of the population is affected by rectal prolapse, although the exact incidence is unknown and the rates are much higher in specific patient populations [1]. Women are up to 6 times more likely to be affected by rectal prolapse compared with men [1]. Rectal prolapse also is more likely to occur in the elderly, with the peak incidence in women aged 70.

The exact cause of rectal prolapse is unknown. Most authorities think that rectal prolapse initially begins as internal rectal intussusception which then progresses to full-thickness prolapse with straining [2]. Others consider rectal prolapse to represent a sliding hernia that develops through a defect in the pelvic floor. Anatomically, rectal prolapse is associated with laxity of rectal attachments, a deep pouch of Douglas, and a redundant or mobile rectum or sigmoid colon. Multiparity and connective tissue disorders such as Ehlers-Danlos are risk factors, but a smaller subset of patients may be nulliparous females or male [3]. Concomitant uterine or vaginal prolapse, enterocele, rectocele, or cystocele may also be found in female patients with rectal prolapse, indicating a generalized weakness of pelvic floor support.

Although typically associated with the elderly, younger patients can also be affected by this problem. Sun et al. [4] performed a retrospective review of young rectal prolapse patients who underwent surgery between 1994 and 2012. The study included 44 patients with mean age of 23 and 72% female. One of the major features of this patient population was high incidence of chronic psychiatric disease requiring medications (41%). Laxative use was also common (56%) among the patients with psychiatric disease. Other studies have also shown that younger patients are more likely to have disordered defecation, bowel dysmotility, autism, psychiatric comorbidities, and developmental delays [5]. Substance abuse also may be a risk factor for development of prolapse [5].

Other conditions associated with rectal prolapse include internal rectal intussusception and mucosal prolapse. Internal rectal intussusception involves telescoping of the rectal wall, which descends to the pelvic floor but does not pass

J. T. Lee (✉)
University of Minnesota, Department of Surgery, Saint Paul, MN, USA

S. A. Vogler
University of Minnesota, Department of Surgery, Edina, MN, USA

R. D. Madoff
University of Minnesota, Department of Surgery, Minneapolis, MN, USA

L. C. C. Oliveira (ed.), *Anorectal Physiology*, https://doi.org/10.1007/978-3-030-43811-1_26

through the anal canal. Rectal mucosal prolapse occurs when the mucosa of the rectum comes out past the anal verge while the muscular wall of the rectum remains in place. Female patients with rectal prolapse may also have prolapse of organs in the middle and anterior compartments of the pelvis, such as uterovaginal prolapse or cystocele.

Symptoms

Patients with rectal prolapse most often present when they note a bulge or mass protrude from the anus with defecation or straining. The mass may or may not reduce spontaneously or the patient may describe manually reducing the prolapse. A particularly strenuous Valsalva may precede an episode of prolapse. As the condition progresses, the prolapse may emerge after coughing, sneezing, or minor exertion, and in some patients the rectum is chronically prolapsed and promptly re-emerges after reduction. Patients sometimes present to the emergency room or urgent care with an incarcerated prolapse that cannot be reduced.

Oftentimes, patients may describe a long history of constipation or difficulty evacuating bowel movements. Studies have shown that up to 25–50% of patients have constipation, colonic dysmotility, or pelvic floor dyssynergia [6, 7]. Other common symptoms include mucous discharge or seepage, rectal bleeding, pelvic pain, and tenesmus. In addition to chronic symptoms, rectal prolapse may also present acutely with ulceration, bleeding, incarceration, and ischemia. Repeated trauma from rectal prolapse can lead to overstretching of the anal sphincter, which, combined with traction injury to the pudendal nerve from pelvic floor descent, can cause incontinence and a patulous anus. Up to 88% of patients with rectal prolapse have associated fecal incontinence, which may or may not improve with repair of the prolapse [8]. The recurrent mechanical trauma from prolapse can also lead to ulceration and inflammation of the rectal mucosa and solitary rectal ulcer syndrome.

Mucosal prolapse and internal rectal intussusception can present with some of the same symptoms of rectal prolapse. Constipation is also frequently seen, as well as feeling of incomplete evacuation, seepage, or fecal incontinence. With mucosal prolapse, patients will also feel a mass or extra tissue protruding from the anus with bowel movements, although the size is usually much smaller than full-thickness rectal prolapse. Internal rectal intussusception is also associated with mechanical trauma to the rectum and may lead to ulceration but may also be asymptomatic. However up to 20–50% of asymptomatic volunteers have been shown to have rectal intussusception on defecography [9], strongly suggesting that this condition is not always pathologic.

Evaluation and Diagnosis

Patients with suspected rectal prolapse or internal rectal intussusception should be evaluated with a thorough history and physical examination. Eliciting a careful history is essential to determining the correct diagnosis and optimal treatment plan for each patient. Patients should be asked about bowel habits, consistency of stools, straining, and fecal incontinence. Clustering of bowel movements may be suggestive of a rectocele or internal rectal intussusception with stool trapping. Presence of pelvic pressure or difficulty with urination may suggest presence of multi-compartment pelvic organ prolapse.

On physical examination, rectal prolapse may be obvious with visual inspection in left lateral decubitus or prone position. Other times, however, the patient may need to be examined while straining on a commode in order to recreate the prolapse. Patients may also resort to taking pictures of the prolapse at home and showing the images to the examiner in the clinic.

The appearance of full-thickness rectal prolapse is characterized by rectal mucosa with smooth concentric folds, as opposed to rectal mucosal prolapse, which is characterized by radially oriented grooves separating columns or clusters of hemorrhoids (Fig. 26.1). The amount of prolapse can vary from a few centimeters to several inches. The mucosa may appear edematous if frequent prolapse occurs. The perianal skin may

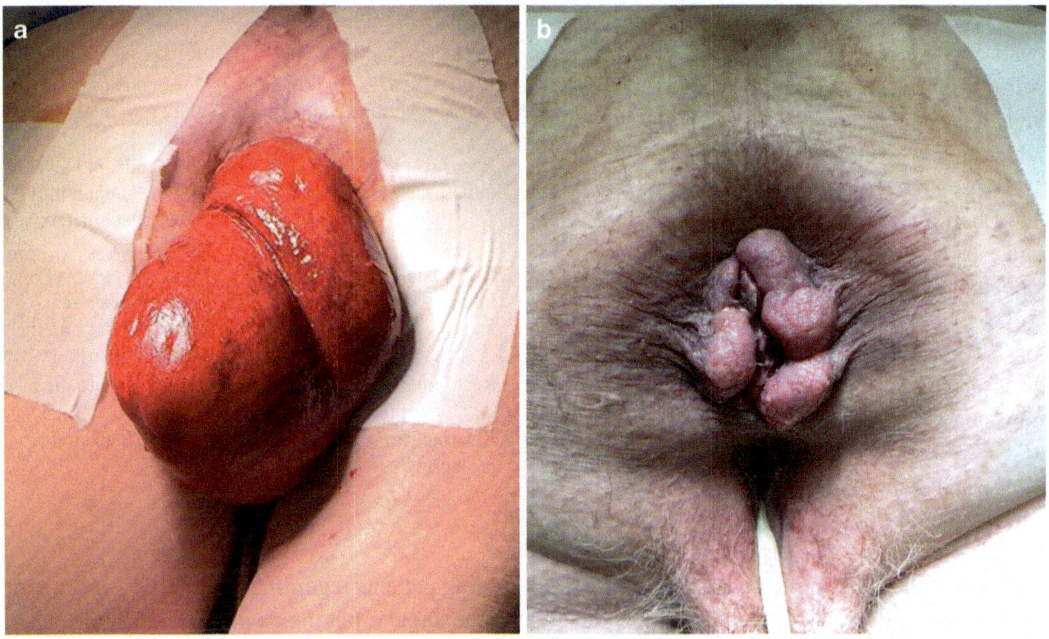

Fig. 26.1 (**a**) Full-thickness rectal prolapse and (**b**) Mucosal prolapse

appear excoriated or macerated from frequent seepage and discharge. In the most extreme presentation, the prolapse may be incarcerated and become ischemic (Fig. 26.2).

Digital rectal examination should be performed with the prolapse reduced and sphincter function should be assessed. Sphincter muscles are usually weak on examination. The ability to squeeze and relax should also be assessed at the time of examination. Internal rectal intussusception may also be diagnosed with digital rectal examination as circular intussusception can sometimes be palpable. Rectocele, pelvic floor descent, enterocele, cystocele, and uterine prolapse may also be identified with proper physical examination.

Once rectal prolapse has been identified, other diagnostic tests are recommended. Colonoscopy should be performed to rule out colonic masses that can potentially serve as a lead point for prolapse. This can also clear the colon of any polyps, malignancy, or inflammatory disorders prior to surgical repair of the prolapse. Solitary rectal ulcer syndrome (SRUS) may be seen in up to 10–15% of patients with rectal prolapse [10]. Patients with SRUS may present with rectal bleeding, difficult defecation, tenesmus, mucous discharge, and rectal or anal pain [3]. The ulcer

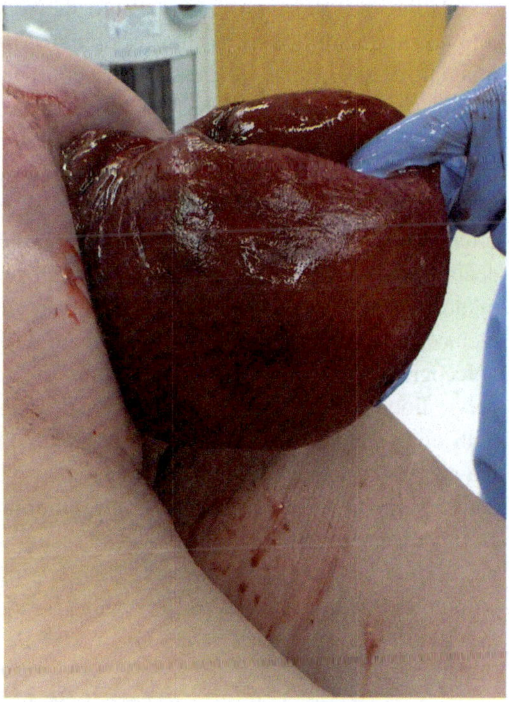

Fig. 26.2 Ischemic, incarcerated prolapse

is classically seen on the anterior or anterolateral wall of the rectum about 5–10 cm from the anal verge with hyperemic margins and a pale base. It may appear as a small, shallow lesion with either ulceration or a polypoid appearance on endoscopy. Edema and thickening of the anterior wall may be noticeable as well. Solitary rectal ulcer may also appear as multiple small lesions. The appearance of SRUS can often be mistaken for a polyp or malignancy. The histologic appearance of SRUS is characterized by fibrous obliteration of the lamina propria and thickening of the muscularis mucosa [3]. Colitis cystica profunda (CCP) is a related condition that has a similar etiology and symptoms as SRUS. CCP may have a similar endoscopic appearance but histologically is characterized by mucous cysts lined by columnar epithelium deep in the muscularis mucosa.

Fluoroscopic or MRI defecography is recommended to evaluate the extent of multicompartment pelvic organ prolapse and coordination of the pelvic floor muscles. This study involves instillation of radiopaque contrast into the rectum and vagina followed by fluoroscopic or MR imaging of the patient evacuating the contrast. Fluoroscopic defecography has the advantage of always being performed with the patient seated upright on a commode, as they are when they defecate, versus MRI defecography, which most often is performed with the patient in the lateral decubitus position. Abnormal pelvic floor movement and coordination, pelvic organ prolapse, rectocele, and RI can also be identified with defecography. Patients with nonrelaxation of the pelvic floor may be at risk for continued issues with constipation after anatomic correction of the prolapse and may benefit from preoperative and postoperative biofeedback therapy. Internal rectal intussusception identified on defecography should be graded as low grade (telescoping within the rectum, Grades I–II) or high grade (telescoping down to anal canal, Grades III–IV) (Table 26.1) [11].

Anal manometry, balloon expulsion, and electromyography can be performed also as part of a full pelvic floor evaluation prior to surgical repair of prolapse. These tests can help in counseling patients regarding continence expectations after surgery, but the results do not have a strong impact on surgical decision-making. Colonic transit study is another test that can be considered in patients with severe constipation refractory to medical management, as it may help guide surgical decision-making.

Table 26.1 Oxford grading system for rectal prolapse [11]

Type of prolapse	Grade	Radiologic features
Recto-rectal intussusception	I (high rectal)	Descends no lower than proximal limit of rectocele
Recto-rectal intussusception	II (low rectal)	Descends into level of rectocele but not to level of sphincters/anal canal
Recto-anal intussusception	III (high anal)	Descends to level of the sphincter/anal canal but not into the canal
Recto-anal intussusception	IV (low anal)	Descends into the anal canal
External rectal prolapse	V (overt rectal prolapse)	Protrudes from the anus

Treatment Approaches: Nonoperative

Conservative measures can help alleviate some of the symptoms of rectal prolapse. Most of these efforts involve avoidance of straining, dietary modification, and treatment of underlying constipation. Patients are taught to manually reduce prolapse on their own. Patients with infrequent prolapse that occurs with severe straining and reduces spontaneously may be able to prevent progression to chronic, frequent prolapse through dietary changes and behavioral modification. A high-fiber diet with adequate water intake and avoidance of caffeine can help with stool consistency, making stools easier to pass. Biofeedback therapy can reduce straining and improve relaxation of the pelvic floor. Without any changes, however, prolapse will likely progress over time leading to worsening nerve damage, sphincter damage, and fecal incontinence. In general, most patients with rectal prolapse do not improve without surgical intervention.

Treatment Approaches: Perineal Approach

Surgery is the only curative treatment for rectal prolapse. Several different procedures have been described, but the optimal "best procedure" has yet to emerge from the literature. Surgeons often tailor the surgery to the patient based on patient condition, previous surgical history, and their own personal experience. In general, rectal prolapse procedures can be divided into two major categories: perineal approaches and abdominal approaches. Patients that are not an acceptable risk for an abdominal operation may best be served by a perineal approach. Patients undergoing perineal procedures typically have less pain, and the procedures can theoretically be done without general anesthesia.

Anal Encirclement (Thiersch Procedure)

Anal encirclement, also known as the Thiersch procedure, involves narrowing the anal canal by placement of a subcutaneous suture or material, theoretically making it more difficult for the rectum to prolapse. The goal of the procedure is to physically prevent the rectum from prolapsing outside the body. However, it does not correct the internal rectal intussusception and laxity that ultimately causes prolapse. The Thiersch procedure was first described in 1891 using a silver wire to encircle the anus and can be performed with just local anesthesia [12]. More modern materials used for anal encirclement include silicone, various types of mesh, sutures, fascia, or tendons. The procedure can cause fecal impaction, sepsis, erosion of the material into the skin or anal canal, or further problems with obstructive defecation. Recurrence rates are high (33–44%), and the procedure is rarely used today and is reserved for patients that are too ill to undergo any other kind of perineal procedure [13] .

Mucosal Sleeve Resection (Delorme Procedure)

Patients with a short segment of rectal mucosal prolapse or prolapse of only a portion of the rectal wall may be candidates for a mucosal sleeve resection known as the Delorme procedure. The Delorme procedure is performed by lifting and resecting a circumferential mucosal sleeve, imbricating of the underlying muscularis layer, and performing a mucosal anastomosis at the dentate line.

The procedure is typically performed in the prone position, although it can also be performed in lithotomy position, under general, spinal, or local anesthesia. The rectal mucosa is injected with local anesthetic containing epinephrine and incised circumferentially 1–2 cm proximal to the dentate line and dissected down through the mucosa and submucosa only, leaving the underlying muscularis propria intact. A tube of rectal mucosa is then dissected proximally for the full length of the prolapsing segment. The exposed muscle layer is then plicated longitudinally with approximately eight circumferentially arrayed absorbable sutures, and a mucosal anastomosis is performed to cover the reefed muscle.

Perineal Rectosigmoidectomy (Altemeier Procedure)

Perineal rectosigmoidectomy, also known as the Altemeier procedure, comprises a full-thickness resection of the redundant rectum and sigmoid colon with creation of a coloanal anastomosis [14]. It is the procedure of choice for patients with incarcerated prolapse and evidence of ischemia or necrosis as the compromised bowel can be completely resected. The procedure is typically performed in the prone position, although it can be performed in lithotomy or left lateral decubitus position. General or spinal anesthesia can be used. A full bowel preparation is recommended prior to the procedure. The rectum is prolapsed and then injected with local anesthetic containing epinephrine circumferentially, approximately 1–2 cm proximal to the dentate line. A full-thickness incision of the outer rectal wall is made circumferentially at this location. Dissection proceeds proximally by systematically dividing the mesorectum to release tension and allow for prolapse of additional rectum and sigmoid colon. As dissection continues, the peritoneal cavity is entered anteriorly and this can help guide dissection. Redundant colon and rectum is gently pulled out of the pelvis and the mesorectum and mesentery ligated until there is no longer excess length in the bowel and mesentery. Use of an energy device can help with this portion of the procedure. A levatorplasty, plication of the levatores ani, can next be performed anteriorly, posteriorly, or in both locations. The proximal portion of the colon or rectum is then transected, taking care that the level of resection does not lead to undue tension on the anastomosis. Control of the proximal bowel can be maintained by serially dividing its circumference and placing full-thickness anastomotic sutures at the cardinal positions as the transection proceeds. Alternatively, the anvil of a circular stapler can be secured in the proximal bowel segment if a stapled anastomosis is preferred. The sutured anastomosis is completed with interrupted stitches placed at intervals between the quadrant sutures.

Treatment Approaches: Abdominal

Several different variations of transabdominal procedures have been described to treat rectal prolapse. These procedures may involve mobilization of the rectum, fixation of the rectum to the sacrum, resection of redundant colon or rectum, placement of mesh, or a combination of these elements. In general, incarcerated prolapse is not amenable to a transabdominal approach.

Suture Rectopexy

Suture rectopexy involves fixation of the rectum to the presacral fascia after circumferential rectal mobilization. Posterior rectal mobilization alone without the rectopexy is not recommended for treatment of rectal prolapse [15]. The critical steps in this procedure include rectal mobilization, either circumferentially or posteriorly, down to the level of the pelvic floor. The rectum is then pulled cephalad and secured to the presacral fascia between the level of the third sacral foramen and the sacral promontory using permanent suture. In addition to the suture fixation, mobilization of the rectum theoretically creates adhesions that also help fix the rectum in place. Some studies have shown that division of the lateral ligaments with rectal mobilization may increase the risk of postoperative constipation [16]. Because of this observation, at least unilateral preservation of the lateral ligaments is recommended. Suture rectopexy can be performed open, laparoscopically, or robotically.

Resection Rectopexy

Resection rectopexy involves resection of redundant sigmoid colon in combination with the suture rectopexy procedure and was first described by Frykman in 1955 [17]. The procedure can also be performed open, laparoscopically, or robotically. The procedure begins with rectal mobilization down to the level of the levator muscles, with attempts to leave the lateral stalks intact. The rectum is then pulled cephalad and secured to the presacral fascia prior to performing resection of the redundant sigmoid colon. A stapled or handsewn anastomosis is created after sigmoid colon resection. In general, this procedure is used more frequently in patients with significant constipation or diverticular disease and is supported by moderate-quality evidence [15].

Anterior Sling Rectopexy (Ripstein Procedure)

The anterior sling rectopexy, also known as the Ripstein procedure, is largely of historical interest. The procedure involves circumferential mobilization of the rectum and placement of mesh around the anterior wall of the rectum at the level of the peritoneal reflection. Once the mesh is secured to the mobilized rectum, the mesh and rectum is pulled cephalad and secured to the presacral fascia. Because kinking of the rectum at the sling is believed to have caused significant constipation in some patients, the anterior sling rectopexy is seldom used today.

Posterior Mesh Rectopexy (Wells Procedure)

Mobilization of the rectum with partial posterior mesh placement is known as the Wells procedure. Initially, a prosthetic sponge was used instead of mesh, but due to a high complication rate, the procedure now typically utilizes a synthetic mesh. After rectal mobilization with preservation of the lateral ligaments, mesh is wrapped only partially around the rectum, leaving the anterior wall of the rectum mesh-free. The rectum is elevated out of the pelvis and the mesh is then fixed to the presacral fascia, typically on the right side of the pelvis. The peritoneum is closed over the mesh to exclude it from the peritoneal cavity.

Ventral Mesh Rectopexy

Initially described using fascia lata by Orr and Loygue [18], ventral mesh rectopexy (VMR) has gained increasing popularity since its modification and reintroduction by D'Hoore in 2004 [19]. The procedure is characterized by a limited dissection of the anterior portion of the distal rectum in the rectovaginal septum. Dissection is carried down to the level of the pelvic floor. Permanent or biologic mesh can be used for the repair. The distal end of the mesh is sutured along the ventral distal rectum (Fig. 26.3). The proximal tail of the mesh is secured to the presacral fascia at the sacral promontory (Fig. 26.4). This proce-

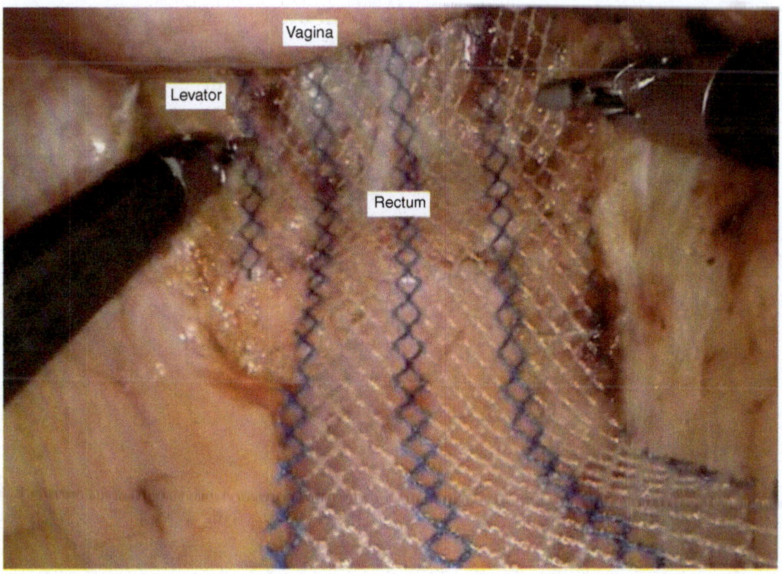

Fig. 26.3 Ventral mesh rectopexy distal mesh placement

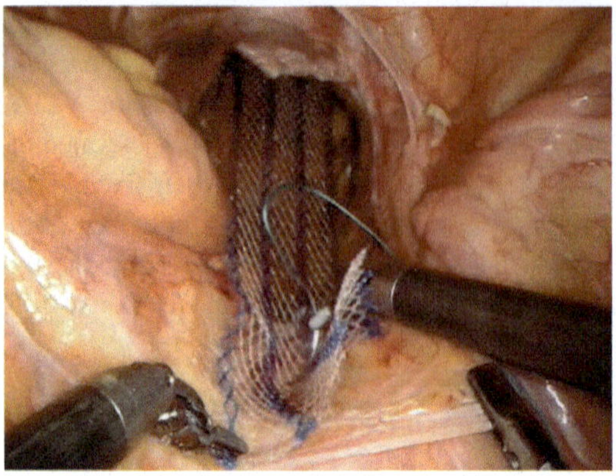

Fig. 26.4 Ventral mesh rectopexy anchoring to sacral promontory

dure differs from other rectopexy procedures as it only mobilizes the anterior portion of the rectum, leaving the posterior and lateral portions of the rectum intact. Variants of the procedure also include anchoring the mesh to the levator muscles at the level of the pelvic floor and anchoring the midportion of the mesh to the posterior vagina. The peritoneum is closed over the mesh to isolate the mesh from the peritoneal cavity. VMR may be used for full-thickness rectal prolapse, rectocele, and rectal intussusception and can also be performed in combination with urogynecologic procedures to address vaginal vault or bladder prolapse. Unlike posterior rectopexy, VMR avoids potential injury to autonomic nerves and division of lateral ligaments. VMR can be performed robotically, laparoscopically, or open.

Comparison of Procedures

Data from the literature are inconclusive when determining the optimal procedure for rectal prolapse. Tou et al. [20] performed a Cochrane database review of all randomized controlled trials (RCTs) addressing full-thickness rectal prolapse in adults. The study included 15 RCTs with a total of 1007 participants. Unfortunately, the authors ran into difficulty comparing data and drawing significant conclusions as many of the trials compared different procedures, had different primary outcomes, and had poor methodology. A few main conclusions could be drawn from their review of the literature [20]. Division of the lateral ligaments was associated with fewer recurrences of prolapse but more constipation. Laparoscopic rectopexy was also associated with fewer complications and a shorter length of stay compared with open rectopexy. Resection rectopexy was associated with a lower rate of postoperative constipation. Procedures that did not include rectopexy had a higher rate of recurrent prolapse. The authors highlight the need for larger more rigorous trials in order to identify clinically important differences among the surgical treatments for rectal prolapse. VMR was not included in any of the studies in this Cochrane review.

Altemeier Versus Delorme

Both the Altemeier and Delorme procedures are typically performed in patients that are at higher risk for undergoing abdominal surgery. In general, both procedures are considered very safe with relatively low rate of complications. Anastomotic leak rates following perineal rectosigmoidectomy are generally lower than coloanal anastomosis for malignant conditions. According to the American Society of Colon and Rectal Surgery (ASCRS) clinical practice guidelines, there is level 1C evidence that the Delorme procedure and Altemeier procedure can be used to treat patients with short-segment full-thickness rectal prolapse or prolapse of any length, respectively [15] .

Emile et al. [21] performed a systematic review of all perineal approach procedures for rectal prolapse. The study reviewed 39 articles including 1748 patients who underwent Altemeier and 712 who underwent Delorme procedures. The vast majority (90.3%) were female with a mean age of 69.1 (range of 32–81.5). Overall, recurrence rate was 16.6% with individual rates of 11.4% for Altemeier and 14.4% for Delorme. The majority of patients in both groups had improvement in symptoms of fecal incontinence (Altemeier 61.4%, Delorme 69%). When combined with the Delorme procedure, levatorplasty had a higher observed improvement in fecal incontinence (72.9% versus 66.6%) although statistical significance was not reported. Based on differences in reporting, these rates could not be determined when levatorplasty was performed in Altemeier procedures. Overall, there were 350 complications reported (13.2%) [21]. Median complication rate for Altemeier procedures was 11.1% and for Delorme was 8.7%. The most common complication after Altemeier was anastomotic leak (1.88%). Bleeding from the suture or staple line was the most common complication in the Delorme and stapled procedures. Mortality rate was 0.64% for all procedures and was generally similar for both the Altemeier and Delorme procedures (0–3.8% versus 0–5.2%).

Perineal Versus Abdominal Approaches

A randomized comparison of surgical treatments for rectal prolapse was published in 2013 referred to as the PROlapse Surgery: Perineal or Rectopexy (PROSPER) trial [22]. Patients were randomized between abdominal and perineal surgery or were randomized to a specific procedure if the approach (perineal vs abdominal) was already elected. Patients either had an Altemeier or Delorme procedure for perineal procedures or suture rectopexy or resection rectopexy for abdominal procedures. Initially, the study aimed to accrue 950 patients, but the target was revised to 300 due to slow recruitment of subjects. Ultimately only 293 patients were included in the study with 213 patients undergoing perineal procedures and 78 undergoing abdominal procedures. Patients were followed for a median time of 36 months. The primary outcomes were recurrent prolapse and impact on fecal incontinence, bowel function, and quality of life scores. There were no significant differences in baseline characteristics between groups. Recurrence rate was not significantly different between the Altemeier and the Delorme groups (24% vs 31%, $p = 0.4$). Likewise, there was no significant difference in recurrence rates between the resection rectopexy and suture rectopexy groups (13% vs 26%, $p = 0.2$). Recurrence rates were also similar when patients randomized to perineal vs abdominal approaches were compared (20% versus 26%, $p = 0.8$). The recurrence rates for all perineal procedures and all abdominal procedures were also compared (nonrandomized comparison), and these differences were not statistically significant either (28% vs 19%, $p = 0.2$). There were no significant differences in change in fecal incontinence scores, bowel function, or quality of life scores between any intergroup comparisons. There were four anastomotic leaks in the study and they all came from the Altemeier group. This study may not have been large enough to detect clinically significant differences between groups, as they had to change their target subject numbers drastically due to slow recruitment.

Fang et al. [23] used the National Surgical Quality Improvement Program (NSQIP) database to compare the safety of abdominal versus perineal procedures for rectal prolapse. Overall, 1569 patients were included in this study with an overall mortality rate of 0.5%. Emergent cases were excluded. The mortality rate for perineal procedures was lower than abdominal procedures (0.9% vs 0.13%, $p = 0.033$). When stratified by American Society of Anesthesiologists (ASA) score, however, there was no significant difference in mortality between perineal and abdominal groups for patients with ASA 3 and 4 (1.3% versus 0.35%, $p = 0.19$). The authors conclude that abdominal procedures can also be considered when treating patients with rectal prolapse, previously thought to be "high risk" for surgery.

The ASCRS guidelines simply recommend that patients with an "acceptable risk for surgery" should undergo an abdominal procedure with rectal fixation based on only moderate-quality evidence (2B) [15].

Suture Versus Resection Versus Ventral Rectopexy

Review of the literature shows high-quality evidence (level 1A) that rectopexy is a key component in the abdominal approach to rectal prolapse [15], although the optimal type of rectopexy procedure is debatable. As mentioned previously, there was no significant difference in recurrence rates between resection rectopexy and suture rectopexy in the PROSPER trial [22]. Various other studies have made comparisons between different methods of rectopexy.

Formijne Jonkers et al. [24] performed a retrospective review of patients who underwent laparoscopic resection rectopexy (LRR) or laparoscopic VMR at two different institutions. The study included a total of 68 patients with 28 in the LRR group and 40 in the VMR group. The primary outcome was incidence of constipation/obstructed defecation syndrome. Fecal incontinence, postoperative complications, length of stay, and recurrences were also measured. Both groups had significant improvement in constipation and incontinence scores postoperatively. Change in score was not significantly different between groups. Complication rate was significantly higher in the LRR group (32.1% versus 7.5%, $p < 0.05$). The most common complication in the LRR group was prolonged ileus ($n = 4$) followed by wound infection ($n = 2$). The most common complication in the VMR group was myocardial ischemia not requiring intervention ($n = 2$). No recurrences were found in either group.

Lundby et al. [25] performed a double-blind randomized single-center study comparing bowel function after laparoscopic suture rectopexy with laparoscopic VMR for patients with full-thickness rectal prolapse. They excluded patients younger than 18, pregnant or breastfeeding patients, dementia or psychiatric patients, and patients with recurrent prolapse. Polypropylene mesh was used for the VMR with multifilament sutures to anchor the mesh to the rectum, while tacks were used for the sacral promontory. The primary outcome of the study was change in obstructive defecation score (ODS) before and after surgery at 12-month follow-up. Of note, suture rectopexy was the standard procedure for the institution, while VMR was only performed 10 times between the three surgeons involved in the study. Overall 75 patients were enrolled in the study with 37 in the suture rectopexy group and 38 in the VMR group. Operative time was significantly shorter in the PSR group (90 min vs 125 min, $p < 0.0001$), while other intraoperative characteristics were similar. Reduction in ODS, fecal incontinence, and constipation scores were similar between groups. Total gastrointestinal transit time increased from baseline in both groups, but the increase was significantly shorter in the VMR group compared with suture rectopexy group (0.14 days versus 1.25 days, $p > 0.006$). Postoperative complications, hospital length of stay, and mucosal prolapse postoperatively were similar between groups as well. In the suture rectopexy group, there was a 5% rectal prolapse recurrence rate and 11% mucosal prolapse rate within 12 months. In the VMR group, there were no rectal prolapse recurrences but 5% had mucosal prolapse at 12 months. Patients were assessed with defecography at the 12-month mark. In the suture rectopexy group, 53% had invagination of the rectum versus 28% in VMR group ($p = 0.037$). Ability to evacuate the rectum, as determined by defecography, was similar between groups. There were no mesh complications reported during the study period.

Mäkelä-Kaikkonen et al. [26] looked at long-term results of laparoscopic VMR from 508 consecutive patients treated with VMR for either rectal prolapse or internal rectal intussusception from 2005 to 2013. Questionnaires were sent to patients with a median follow-up time of 44 months. Subjective symptom relief was high at 76% and was more frequently reported in patients who underwent surgery for rectal prolapse compared with RI (86 vs 68%, $p < 0.001$). The overall complication rate was 11.4% and recurrence rate was 7.1%. Overall, the study showed durable long-term outcomes of VMR.

Variations on Surgical Approach for Ventral Mesh Rectopexy

The ASCRS consensus guidelines recommend VMR as an alternative approach to repair of rectal prolapse with acceptable complication rates based on level 1C data [15]. Several authors have reported their experience with VMR and variations of VMR using biologic grafts, synthetic mesh, and laparoscopic and robotic approaches [19, 26–37]. Proponents of the robotic approach may cite increased visualization from greater field magnification, three-dimensional imaging, and improved dexterity from the multiarticulated instruments as advantages of robotic VMR [28, 30] . Improved ergonomics may also be a potential benefit.

Van Iersel et al. [29] performed a retrospective review of their 5-year experience with robotic VMR. The study included 258 patients with a mean follow-up of 23.5 months (range 0.2–65.1 months). They had no conversions and five intraoperative complications. Overall mortality was 0.4% and major complication rate was low (1.95%). There was one mesh-related complication: an asymptomatic vaginal mesh erosion. Patients had a significant improvement in obstructed defecation and fecal incontinence scores from baseline. Actuarial analysis of 5-year rectal prolapse and internal rectal intussusception recurrence rates were 12.9% and 10.4%.

Rondelli et al. [38] performed a systematic review and meta-analysis of robotic versus laparoscopic approaches for VMR. Six stud ies were included in meta-analysis representing 340 patients. Operative time was longer with the robotic approach ($p = 0.04$), but intraoperative blood loss, postoperative complications, and postoperative length of stay significantly favored the robotic group. There was no significant difference in recurrence rate, conversion rate, or reoperation rate between groups. Bowel function, sexual function, and procedural costs could not be determined in this study. In general, the authors report a trend favoring robotic surgery in their meta-analysis but this was not statistically significant.

Variations on Mesh

VMR is most commonly performed using a synthetic mesh made of polypropylene. Relative contraindications to mesh placement include morbid obesity, previous radiation, high-grade endometriosis, and previous sigmoid diverticulitis. Absolute contraindications include pregnancy, no demonstrable pelvic anatomical problem, a hostile abdomen, severe proctitis, psychological instability, and anismus [39]. Biologic mesh has been investigated as an alternative to synthetic mesh. McLean et al. [31] reviewed the short- and long-term outcomes for laparoscopic VMR using biologic mesh in a cohort of 224 patients with either rectal prolapse, internal rectal intussusception, SRUS, or vaginal vault prolapse. The overall complication rate was 10.7% of patients but only 0.45% mesh morbidity. Recurrence rate was 11.4% with 5% recurring within 12 months and 10.7% recurring at 5 years. The vast majority of patients reported a significant improvement in constipation and incontinence outcomes (>90%). Vaginal vault suture erosion occurred in three patients (1.3%) requiring removal. Sacral osteomyelitis occurred in one patient (0.4%). One patient had erosion of the mesh via a previous transperineal wound 12 weeks after the procedure.

Evans et al. [40] performed a retrospective review from data from five pelvic floor databases to determine potential safety-related issues related to mesh when used for VMR. The study included 2203 patients with the vast majority (80.1%) using synthetic mesh, while the remainder used a biologic mesh. Mesh erosion occurred in 45 patients (2%), with the following distribution: vagina [20], rectum [17], rectovaginal fistula [7], and perineum [1]. Approximately half of the patients (51%) were treated with local excision of the mesh. Some of the patients with mesh erosion (40%) had a major morbidity from the mesh erosion and required an invasive procedure to remove the mesh, resect a portion of the rectum, and/or create a colostomy. Erosion rate was higher in the synthetic mesh group (2.4%) compared with the biologic mesh group (0.7%). Use of polyester mesh was associated with an increased risk of erosion compared with polypropylene mesh (HR 4.09, 95% CI 2.16–7.73). Median time to identification of mesh erosion was 27.0 ± 18.1 months for synthetic mesh and 2.5 ± 6.1 months for biologic mesh. Overall, the authors conclude that use of polypropylene mesh is safe when performing VMR, although the use of an international mesh registry may help to monitor potential mesh complications and best practices.

Current Practice for Primary Rectal Prolapse

Gunner et al. [41] performed a survey to find out current trends in treating rectal prolapse. Questionnaires regarding current surgical practice were administered in both 1997 and 2014 to surgeons from the Association of Coloproctology of Great Britain and Ireland. There were 153 respondents in 1997 and 122 respondents in 2014. More surgeons in 2014 compared with 1997 preferred the abdominal approach for medically fit patients with rectal prolapse (81.7% vs 63.5%, $p < 0.010$). The Delorme procedure was the most common perineal procedure, but the proportion of Altemeier procedures increased from 14.9% to 39.3% between the two time periods. The most common abdominal procedure in 2014 for rectal prolapse was VMR (48.6%) followed by suture rectopexy (45.9%) and resection rectopexy (9.9%). The overwhelming majority of VMR procedures were performed laparoscopically (96.3%). Suture rectopexy declined substantially between time periods (92.6% vs 45.95%, $p < 0.01$).

Recurrent Rectal Prolapse

In general, abdominal prolapse procedures have a lower recurrence rate than perineal procedures. Recurrent prolapse may occur because of technical or patient-specific factors. Treatment of recurrent rectal prolapse can be more difficult and have a high rate of recurrent prolapse.

Risk Factors for Recurrent Prolapse

Fu and Stevensen [42] performed a retrospective cohort analysis of patients undergoing laparoscopic VMR to determine risk factors for recurrence. The study included 231 patients treated by a single surgeon. Patients with previous prolapse surgery were not excluded. Recurrence was defined as either full-thickness prolapse, mucosal prolapse, or internal rectal intussusception. The overall recurrence rate was 11.7%. Prolonged pudendal nerve terminal motor latency (HR 5.57, 95% CI 1.13–27.42; $p = 0.04$) and use of synthetic mesh (HR 4.24, 95% CI 1.27–14.2; $p = 0.02$) were significant risk factors identified in multivariate analysis. Age >70 years and poor preoperative continence were associated with risk of recurrent prolapse in univariate analysis. Technical failures included mesh detachment from the sacral promontory and inadequate midrectal mesh fixation.

Bishawi et al. [43] also examined recurrent rectal prolapse in a pooled group of 532 patients treated with rectopexy (not including VMR). Recurrence was defined as full-thickness rectal prolapse only and mucosal prolapse was not included. The overall recurrence rate was 8.6% at a median follow-up of 60 months. Risk factors identified on univariate analysis included incontinence, constipation, and extent of rectal mobilization. With multivariate analysis, degree of mobilization was independently associated with recurrence. Patients that had circumferential mobilization in combination with rectopexy had a lower risk of recurrent prolapse compared with those that had posterior or anterior only.

Ding et al. [44] looked at outcomes for both primary and recurrent prolapse in a retrospective cohort study of patients treated with Altemeier procedures. The study included 113 patients with primary rectal prolapse and 23 with recurrent prolapse. There were no significant differences between groups for operative time, blood loss, length of stay, follow-up, baseline demographics, or complication rates. No significant risk factors for recurrence could be identified. Mean interval to recurrence was 16 months in the recurrent group versus 21.5 months in the primary group. The authors concluded that patients with recurrent rectal prolapse are more likely to have recurrent prolapse after surgery than patients undergoing primary repair (39% versus 18%, $p = 0.007$) [44].

Treatment of Recurrent Prolapse

Surgical options for recurrent rectal prolapse are somewhat limited by patient-specific risk factors and the initial procedure performed. Patients that have had a previous resection with anastomosis, either transabdominal or perineal approach, are usually not candidates for a repeat resection procedure due to the potential for devascularization unless the entire anastomosis is resected. For example, a patient with a resection rectopexy should not undergo an Altemeier procedure unless the entire anastomosis from the previous procedure can be entirely resected and only healthy proximal colon is brought down to the anus. Perineal procedures have been reported to have a significantly higher rate of re-recurrence after reoperation compared with abdominal procedures ($p = 0.03$) [45]. Patients that are high risk for an abdominal procedure, however, may be limited to perineal procedures only.

Implantation of a mesh with a primary procedure can potentially make a reoperative surgery more difficult and limit surgical options for recurrent prolapse. Attempted removal of mesh for a procedure involving resection can potentially perforate the rectum and is not recommended. Options for recurrent prolapse after VMR include suture rectopexy, reattachment of the mesh to the sacrum (if that has pulled away), reinforcement of existing mesh, or repeat VMR.

Hotouras et al. [46] performed a systematic review of the literature specifically for management of recurrent rectal prolapse. A total of 14 studies were included in analysis with the majority of studies being retrospective. The authors found that most studies had small patient numbers and lacked systematic preoperative and postoperative assessment of function. The authors separated outcomes based on perineal procedures and abdominal procedures for recurrent rectal prolapse. Recurrence rates after perineal procedures for recurrent rectal prolapse varied from 0% to 50% with follow-up ranging from 8.8 to 81 months (median 35 months). Complications were also widely variable with outcomes that were difficult to compare. The authors note that at least four of the studies reviewed had anastomotic leak with Altemeier procedures with an overall leak rate of 4–6%. Abdominal approaches for recurrent prolapse included suture rectopexy and resection rectopexy. Recurrence rate ranged from 0% to 15% in the studies included, although follow-up duration ranged from <1 year to 23 years. Morbidity rates also were widely variable, ranging from 0% to 32%. The authors concluded that insufficient data is available to create a treatment algorithm for recurrent rectal prolapse and that larger high-quality studies are necessary to guide practice.

Summary

Rectal prolapse is characterized by full-thickness protrusion of the rectum beyond the anal verge and can lead to significant morbidity and impact on quality of life. Prolapse is more common in women and older patients, although there is a subset of younger patients and male patients. Patients should be evaluated with careful history and physical examination. Presence of nonrelaxing pelvic floor, colonic dysmotility, and concomitant pelvic organ prolapse should be considered in development of treatment plans. Surgery is the only curative treatment for rectal prolapse and a wide variety of procedures exist to treat prolapse. Both perineal and abdominal approaches are options to treat prolapse, although there are differences in recurrence rates and complication rates that are not clearly delineated in the literature. Most patients may be candidates for abdominal procedures involving rectopexy. Ventral mesh rectopexy has recently gained popularity, particularly in Europe, as a potentially safe and durable treatment of rectal prolapse. There is a paucity of data regarding surgery and best practice recommendations for recurrent rectal prolapse. Future large-scale randomized controlled trials and use of registries are needed to help guide best practices for rectal prolapse and recurrent rectal prolapse management.

References

1. Kairaluoma M, Kellokumpu I. Epidemiologic aspects of complete rectal prolapse. Scand J Surg. 2005;94:207–10.
2. Wijeffels N, Collinson R, Cunningham C, Lindsey I. What is the natural history of internal rectal prolapse. Colorectal Dis. 2010;12:822–30.

3. Gurland B, Zutshi M. Rectal prolapse. In: The ASCRS textbook of colon and rectal surgery; 2016. p. 1064–89.
4. Sun C, Hull T, Ozuner G. Risk factors and clinical characteristics of rectal prolapse in young patients. J Visc Surg [Internet]. 2014;151(6):425–9. Available from: https://doi.org/10.1016/j.jviscsurg.2014.07.013.
5. Marceau C, Parc Y, Debroux E, Tiret E, Parc R. Complete rectal prolapse in young patients: psychiatric disease a risk factor of poor outcome. Color Dis. 2005;7:360–5.
6. Kim D, Tsang C, Wong W, Lowry A, Goldberg S, Madoff R. Complete rectal prolapse: evolution of management and results. Dis Colon Rectum. 1999;42:441–50.
7. Madoff R, Mellgren A. One hundred years of rectal prolapse surgery. Dis Colon Rectum. 1999;42:466–9.
8. Wallenhorst T, Bouguen G, Brochard C, Cunin D, Desfourneaux V, Ropert A, et al. Long-term impact of full-thickness rectal prolapse treatment on fecal incontinence. Surgery. 2015;158:104–11.
9. Shorvon P, McHugh S, Diamant N, Somers S, Stevenson G. Defecography in normal volunteers: results and implications. Gut. 1989;30:1737–49.
10. Cannon J. Evaluation, diagnosis, and medical management of rectal prolapse. Clin Colon Rectal Surg [Internet]. 2016;30(1):016–21. Available from: http://www.thieme-connect.de/DOI/DOI?10.1055/s-0036-1593431.
11. Collinson R, Cunningham C, D'Costa H, Lindsey I. Rectal intussusception and unexplained faecal incontinence: findings of a proctographic study. Color Dis. 2009;11:77–83.
12. Goldman J. Concerning prolapse of the rectum with special emphasis on the operation by Thiersch. Dis Colon Rectum. 1988;31:154–5.
13. Kuijpers HC. Treatment of complete rectal prolapse: to narrow, to wrap, to suspend, to fix, to encircle, to plicate or to resect? World J Surg. 1992;16:826–30.
14. Altemeier W, Culbertson W, Schowengerdt C, Hunt J. Nineteen years' experience with the one-stage perineal repair of rectal prolapse. Ann Surg. 1971;173:993–1006.
15. Bordeianou L, Paquette I, Johnson E, Holubar SD, Gaertner W, Feingold DL, et al. Clinical practice guidelines for the treatment of rectal prolapse. Dis Colon Rectum. 2017;60(11):1121–31.
16. Speakman C, Madden M, Nicholls R, Kamm M. Lateral ligament division during rectopexy causes constipation but prevents recurrence: results of a prospective randomized study. Br J Surg. 1991;78:1431–3.
17. Frykman H. Abdominal rectopexy and primary sigmoid resection for rectal procidentia. Am J Surg. 1955;90:780–9.
18. Loygue J, Nordlinger B, Cunci O, Malafosse M, Huguet C, Parc R. Rectopexy to the promontory for the treatment of rectal prolapse. Report of 257 cases. Dis Colon Rectum. 1984;27:356–9.
19. D'Hoore A, Cadoni R, Penninckx F. Long-term outcome of laparoscopic ventral rectopexy for total rectal prolapse. Br J Surg. 2004;91(11):1500–5.
20. Tou S, Brown SR, Nelson RL. Surgery for complete (full-thickness) rectal prolapse in adults. Cochrane Database Syst Rev. 2015;11(11):CD001758.
21. Emile SH, Elfeki H, Shalaby M, Sakr A, Sileri P, Wexner SD. Perineal resectional procedures for the treatment of complete rectal prolapse: a systematic review of the literature. Int J Surg [Internet]. 2017;46(2017):146–54. Available from: https://doi.org/10.1016/j.ijsu.2017.09.005.
22. Senapati A, Gray RG, Middleton LJ, Harding J, Hills RK, Armitage NCM, et al. PROSPER: a randomised comparison of surgical treatments for rectal prolapse. Color Dis. 2013;15(7):858–68.
23. Fang SH, Cromwell JW, Wilkins KB, Eisenstat TE, Notaro JR, Alva S, et al. Is the abdominal repair of rectal prolapse safer than perineal repair in the highest risk patients? An NSQIP analysis. Dis Colon Rectum. 2012;55(11):1167–72.
24. Formijne Jonkers HA, Maya A, Draaisma WA, Bemelman WA, Broeders IA, Consten ECJ, et al. Laparoscopic resection rectopexy versus laparoscopic ventral rectopexy for complete rectal prolapse. Tech Coloproctol. 2014;18(7):641–6.
25. Lundby L, Iversen LH, Buntzen S, Wara P, Høyer K, Laurberg S. Bowel function after laparoscopic posterior sutured rectopexy versus ventral mesh rectopexy for rectal prolapse: a double-blind, randomised single-centre study. Lancet Gastroenterol Hepatol. 2016;1(4):291–7.
26. Mäkelä-Kaikkonen J, Rautio T, Kairaluoma M, Carpelan-Holmström M, Kössi J, Rautio A, et al. Does ventral rectopexy improve pelvic floor function in the long term? Dis Colon Rectum [Internet]. 2018;61(2):230–8. Available from: http://insights.ovid.com/crossref?an=00003453-201802000-00016.
27. Mäkelä-Kaikkonen J, Rautio T, Pääkkö E, Biancari F, Ohtonen P, Mäkelä J. Robot-assisted vs laparoscopic ventral rectopexy for external or internal rectal prolapse and enterocele: a randomized controlled trial. Color Dis. 2016;18(10):1010–5.
28. Inaba CS, Sujatha-Bhaskar S, Koh CY, Jafari MD, Mills SD, Carmichael JC, et al. Robotic ventral mesh rectopexy for rectal prolapse: a single-institution experience. Tech Coloproctol [Internet]. 2017;21(8):667–71. Available from: http://link.springer.com/10.1007/s10151-017-1675-z.
29. van Iersel JJ, Formijne Jonkers HA, Paulides TJC, Verheijen PM, Draaisma WA, Consten ECJ, et al. Robot-assisted ventral mesh rectopexy for rectal prolapse: a 5-year experience at a tertiary referral center. Dis Colon Rectum. 2017;60(11):1215–23.
30. Mehmood RK, Parker J, Bhuvimanian L, Qasem E, Mohammed AA, Zeeshan M, et al. Short-term outcome of laparoscopic versus robotic ventral mesh rectopexy for full-thickness rectal prolapse. Is robotic superior? Int J Color Dis. 2014;29(9):1113–8.

31. McLean R, Kipling M, Musgrave E, Mercer-Jones M. Short- and long-term clinical and patient-reported outcomes following laparoscopic ventral mesh rectopexy using biological mesh for pelvic organ prolapse: a prospective cohort study of 224 consecutive patients. Colorectal Dis. 2018;20(5):424–36.
32. Trilling B, Sage P-Y, Reche F, Barbois S, Waroquet P-A, Faucheron J-L. Early experience with ambulatory robotic ventral rectopexy. J Visc Surg [Internet]. 2018. Available from: http://linkinghub.elsevier.com/retrieve/pii/S1878788617300486.
33. Emile SH, Elbanna H, Youssef M, Thabet W, Omar W, Elshobaky A, et al. Laparoscopic ventral mesh rectopexy vs Delorme's operation in management of complete rectal prolapse: a prospective randomized study. Color Dis. 2017;19(1):50–7.
34. Emile SH, Elfeki H. Symptom improvement, recurrence and infection after laparoscopic ventral rectopexy using biologic mesh. Color Dis. 2017;19(10):942.
35. D'Hoore A, Penninckx F. Laparoscopic ventral recto(colpo)pexy for rectal prolapse: surgical technique and outcome for 109 patients. Surg Endosc Other Interv Tech. 2006;20(12):1919–23.
36. D'Hoore A, Vanbeckevoort D, Penninckx F. Clinical, physiological and radiological assessment of rectovaginal septum reinforcement with mesh for complex rectocele. Br J Surg. 2008;95(10):1264–72.
37. Ramage L, Georgiou P, Tekkis P, Tan E. Is robotic ventral mesh rectopexy better than laparoscopy in the treatment of rectal prolapse and obstructed defecation? A meta-analysis. Tech Coloproctol. 2015;19(7):381–9.
38. Rondelli F, Bugiantella W, Villa F, Sanguinetti A, Boni M, Mariani E, et al. Robot-assisted or conventional laparoscopic rectopexy for rectal prolapse? Systematic review and meta-analysis. Int J Surg [Internet]. 2014;12(S2):S153–S159. Available from: https://doi.org/10.1016/j.ijsu.2014.08.359.
39. Mercer-Jones M, Brown S, Knowles C, Williams A. Position statement by the Pelvic Floor Society on behalf of the Association of Coloproctology of Great Britain and Ireland on the use of mesh in ventral mesh rectopexy (VMR). Color Dis. 2017. https://doi.org/10.1111/codi.13893.
40. Evans C, Stevenson ARL, Sileri P, Mercer-Jones MA, Dixon AR, Cunningham C, et al. A multicenter collaboration to assess the safety of laparoscopic ventral rectopexy. Dis Colon Rectum. 2015;58(8):799–807.
41. Gunner CK, Senapati A, Northover JMA, Brown SR. Life after PROSPER. What do people do for external rectal prolapse? Color Dis. 2016;18(8):811–4.
42. Fu CWP, Stevenson ARL. Risk factors for recurrence after laparoscopic ventral rectopexy. Dis Colon Rectum [Internet]. 2017;60(2):178–86. Available from: http://insights.ovid.com/crossref?an=00003453-201702000-00008.
43. Bishawi M, Foppa C, Tou S, Bergamaschi R. Recurrence of rectal prolapse following rectopexy: a pooled analysis of 532 patients. Color Dis. 2016;18(8):779–84.
44. Ding JH, Canedo J, Lee SH, Kalaskar SN, Rosen L, Wexner SD. Perineal rectosigmoidectomy for primary and recurrent rectal prolapse: are the results comparable the second time? Dis Colon Rectum. 2012;55(6):666–70.
45. Steele SR, Goetz L, Minami S, Madoff R, Mellgren A, Parker S. Management of recurrent rectal prolapse: surgical approach influences outcome. Dis Colon Rectum. 2006;49:440–5.
46. Hotouras A, Ribas Y, Zakeri S, Bhan C, Wexner SD, Chan CL, et al. A systematic review of the literature on the surgical management of recurrent rectal prolapse. Colorectal Dis [Internet]. 2015;17(8):657–64. Available from: http://doi.wiley.com/10.1111/codi.12946.

Enterocele: Diagnosis and Treatment

27

Gonzalo Hagerman Ruíz-Galindo,
José Arturo Aragón López,
and Jorge Daniel Silva-Velazco

Introduction

Enterocele is a herniation of the peritoneal cavity forming a sac that protrudes between the utero-sacral ligaments at the vaginal apex, which may then descend distally through the rectovaginal septum (Fig. 27.1) [1–5]. It may contain small bowel (enterocele) or sigmoid colon (sigmoido-cele) [1, 2, 6, 7]. Its incidence in asymptomatic patients is unknown, and according to some authors, it can range from 11% to 45% in patients with pelvic floor dysfunction and can also be seen in up to 10% of healthy females on defecography [1–3, 6, 8]. It occurs more frequently in older or multiparous women and in those with a history of hysterectomy [1, 2, 4, 7, 9].

The onset of enterocele may be multifactorial [9]. It is associated with pelvic floor injury, obstetric trauma, straining from constipation, increased intra-abdominal pressure, prior hysterectomy, and the effect of hormonal changes in the connective tissue of the pelvis [2, 4, 7, 9]. The most frequent type of enterocele occurs when a pelvic organ prolapses (usually the uterus or rectum) pulling down the cul-de-sac, thus creating a traction enterocele [1, 3]. On the other hand, pulsion enterocele may result from chronic abdominal pressure increase leading to protrusion or eversion of the rectal or vaginal wall [3]. Acquired cases are seen after pelvic surgery especially after hysterectomy (18–25% of patients develop enterocele at some point after hysterectomy) in which the cul-de-sac is widened by anterior displacement of the vagina [1, 3, 7, 10]. Finally, congenital enterocele represents less than 1% of cases and results from abnormal development of the rectovaginal septum [1, 3].

Symptoms at presentation are usually nonspecific and include false rectal tenesmus, evacuation difficulty, pelvic pressure, lower abdominal pain, and fecal incontinence [1, 3, 4, 7, 9, 11, 12]. It may be suspected with the presence of pelvic pressure or vaginal bulging, which has a clinical sensitivity for the diagnosis of only 50% [11, 12]. These symptoms may be exacerbated by increasing intra-abdominal pressure especially in the upright position [9]. Some authors have pointed out that the subjective "bearing down" symptoms are often related to mesenteric traction due to the presence of the bowel or omentum in the hernia sac [7]. According to some authors, it is hard to define the typical clinical features of enterocele because most patients have concomitant pelvic floor abnormalities, such as rectocele or intussusception [9].

G. H. Ruíz-Galindo (✉) · J. D. Silva-Velazco
Centro Medico ABC, Department of Colorectal Surgery/General Surgery, Mexico City, Mexico

J. A. A. López
Hospital Espanol de Veracruz, Department of Surgery, Veracruz, Mexico

L. C. C. Oliveira (ed.), *Anorectal Physiology*, https://doi.org/10.1007/978-3-030-43811-1_27

Fig. 27.1 Enterocele

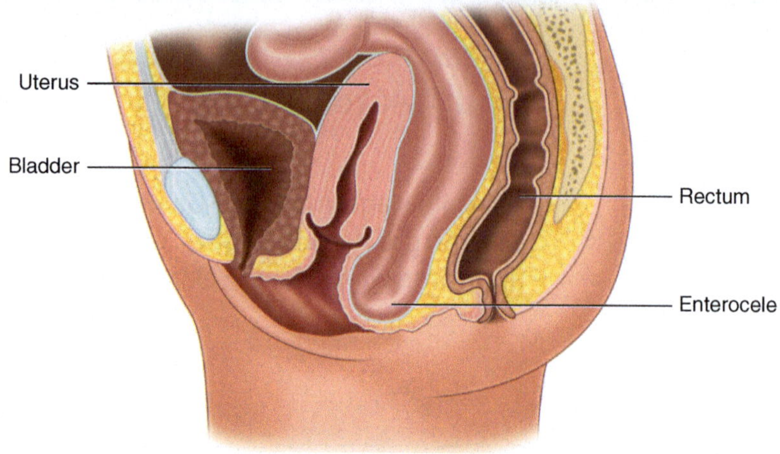

Diagnosis

The diagnosis of enterocele is based on identifying a hernia sac between the vagina and rectum [7]. This can be achieved by a combined vaginal and rectal digital examination (preferably in the standing position) utilizing a bivalve vaginal speculum ideally with transillumination of the rectovaginal septum with a light source through the rectum, where the small bowel between the rectum and vagina will block the transmission of light [7]. It has been recently described that a pelvic exam showing vaginal bulging has a sensitivity of 50%; however it is often difficult to distinguish rectocele from a true enterocele protruding at the level of the vaginal vault [1, 5, 9, 12, 13].

Solitary presentation is rare, and most cases are associated with rectocele or prolapse of other pelvic organs and/or abnormal descent of the pelvic floor [1, 6]. Most cases are diagnosed during pelvic floor evaluation of other pathologies with conventional defecating proctogram (defecography), dynamic magnetic resonance imaging defecography, or cystoscopy/ureteroscopy [1, 4, 9, 12].

Dynamic Evacuation Proctography (Defecography)

It remains as the gold-standard diagnostic procedure, even though it may miss up to 20% of enteroceles [12, 14]. This study offers a dynamic view of the rectum, pelvic floor muscles, and anal sphincters by filling the rectum with contrast and simulating defecation [15]. It may show the direct compression of the rectum by enterocele which makes rectal emptying difficult [16]. This study has the advantages of being an easy to perform cheap, and readily available diagnostic tool. Nevertheless, it has some limitations including that it is observer dependent, requires usage of ionizing radiation, provides images only in the lateral plane, and does not accurately assess soft tissues [11, 14, 17].

Enterocele is classified according to the depth of small bowel descent determined by defecography: Grade I above pubococcygeal line, Grade II located below the pubococcygeal line and over the ischiococcygeal line, and Grade III when it is below the ischiococcygeal line [3].

Dynamic Magnetic Resonance Imaging Defecography

It is an alternative study with high sensitivity (100%) for diagnosing enterocele [12, 18, 19]. In contrast to conventional defecography, this study does not use ionizing radiation, is not observer dependent, and allows assessment of soft tissue and the three pelvic compartments simultaneously with multiplanar imaging [13, 14, 18]. Nonetheless, it is expensive and not as frequently available as conventional defecography [12, 14].

Dynamic Transperineal Ultrasound

Recently, the use of dynamic transperineal ultrasound has been described. It offers the advantage of being a noninvasive and nonirradiating office procedure with sensitivity comparable to that of dynamic defecography for the diagnosis of a range of pelvic floor disorders [12, 20]. Several studies have shown that ultrasound is better tolerated than defecation proctography and it is considerably less expensive [21–24]. It is thus likely that ultrasound will replace defecography as the initial investigation of choice in women with posterior pelvic floor symptoms [20].

Cystoscopy

Cystoscopy is useful only to differentiate an enterocele from cystocele [13].

Treatment

Surgical treatment is reserved for symptomatic cases or vaginal or rectal ulceration [1]. The approach may be vaginal or abdominal [3–5, 7, 25]. The goal of enterocele surgical repair is excision or obliteration of the peritoneal sac with approximation of the uterosacral ligaments in the midline [1, 4, 5, 9, 13, 26]. This obliteration may be carried out synchronously with an abdominal procedure being performed for any coexistent pathology or, more commonly, accompanying a vaginal approach for hysterectomy or cystocele/rectocele repair [3, 4, 25, 27]. When it is performed at the time of vaginal hysterectomy (McCall culdoplasty), the uterosacral ligaments are incorporated into the closure of the peritoneum and upper vagina after the uterus is removed; this aims at preventing subsequent enterocele recurrence [7, 26–29].

Because enterocele presents due to weakened vault support, the vaginal vault must also be resuspended [13]. Vaginal vault suspension can be performed through transvaginal reattachment of the uterosacral ligaments (McCall culdoplasty), the sacrospinous ligament, or the iliococcygeus fascia and/or muscle to the vaginal apex [13].

In sacrospinous fixation, a posterior vaginal incision is made and extended to the top of the vagina [13, 27]. The sacrospinous ligament running from the ischial spine to the sacral bone is identified, and two sutures are placed through the ligament and secured to the top of the vagina resulting in increased support to the upper vagina without vaginal shortening; this can be performed unilaterally or bilaterally (Fig. 27.2)

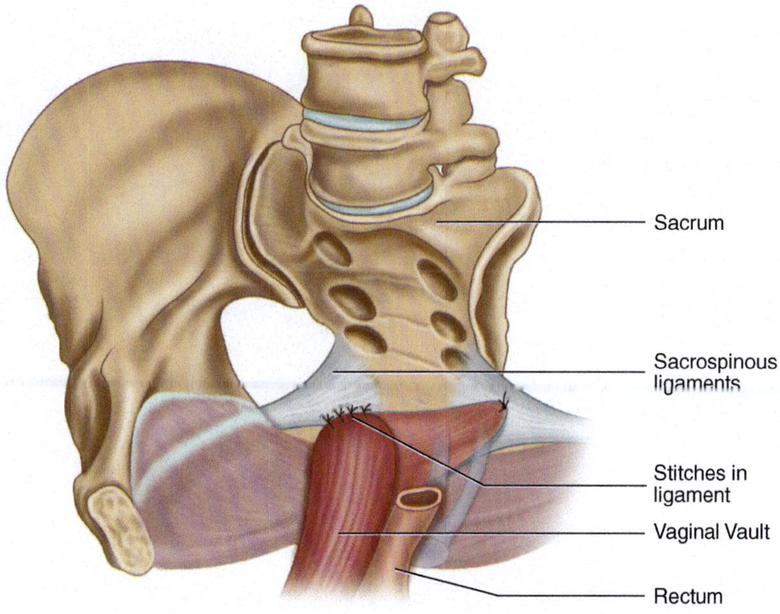

Fig. 27.2 Sacrospinous fixation

[13, 27–29]. This technique should be considered at the time of vaginal hysterectomy, when the vault descends to the introitus during closure [29]. Several systematic reviews have shown that sacrospinous fixation is a highly effective procedure with low recurrence and complication rates and good patient satisfaction [30–34]. This procedure carries the risk of a high incidence (8–30%) of postoperative anterior compartment prolapse and stress urinary incontinence, presumably due to posterior fixation of the upper vagina which predisposes the anterior compartment to excessive intra-abdominal pressure [29]. There is no evidence that bilateral sacrospinous fixation or fixation using permanent suture material is associated with lower recurrence rates [29]. Long-term follow-up studies have reported that prolapse symptoms are present in up to 16% of women at 2–15-year follow-up [29]. This procedure may not be the appropriate therapeutic choice for women with a short vagina and should be carefully considered in women with preexisting dyspareunia [29, 35].

For fixation of the vaginal apex to the iliococcygeus fascia and/or muscle, one or two sutures are placed into the iliococcygeus fascia and/or muscle just anterior to the ischial spine [13]. In patients not sexually active, it is performed without the need of creating a vaginal incision, by placing a monofilament permanent suture at full thickness through the vaginal wall into the muscle [13]. The use of a polypropylene mesh to support vaginal vault suspensions has been reported with high success rates [13].

Abdominal sacrocolpopexy is considered the gold-standard procedure for vaginal vault prolapse treatment [7, 13]. It is performed through an incision in the lower abdomen or laparoscopically [27]. Suspension of the vaginal apex to the sacral promontory with or without a mesh can be performed [13]. The mesh is secured to the sacrum and the peritoneum is sutured over the mesh (Fig. 27.3) [27, 29]. A systematic review of observational studies reported long-term success rates of 78–100%, with intestinal mesh erosion in 2–11% of the cases [29]. Although this procedure requires an abdominal incision and there is a risk for bleeding from the sacral promontory and postoperative ileus may occur, the resultant anatomy carries the lowest recurrence and least risk of sexual dysfunction and dyspareunia [13].

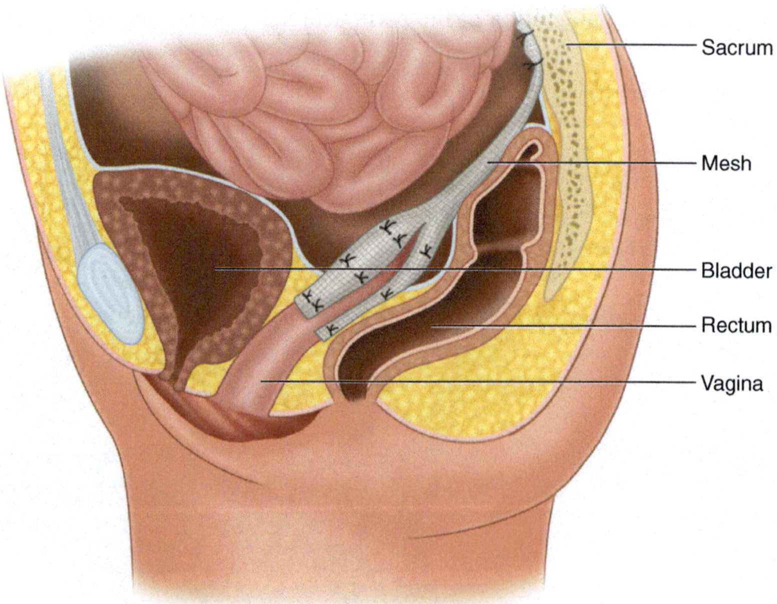

Fig. 27.3 Abdominal sacrocolpopexy

Vaginal Versus Abdominal Approach for the Surgical Repair of Enterocele

According to recent literature, although the vaginal approach offers less morbidity, it is documented that up to 20% of these patients present recurrence and/or dyspareunia and is associated with greater risk of ureteral injury as well [1, 9, 13, 16, 29]. Nevertheless, most studies report that both approaches offer successful results in about 80% of the cases [1, 29]. Some studies comparing abdominal sacral colpopexy versus vaginal sacrospinous colpopexy report that the former was associated with significantly lower rates of recurrent vault prolapse, less postoperative stress urinary incontinence and less postoperative dyspareunia at the price of longer operative and recovery times, and higher cost than for vaginal surgery [29]. On the other hand, they report no statistically significant differences in patient satisfaction, number of women reporting prolapse symptoms, reoperation rates for stress urinary incontinence, and reoperation rates for prolapse [29]. Vaginal sacrospinous colpopexy resulted in a reduction in operative time, it was less expensive, and patients had an earlier return to their daily activities [27, 29].

Laparoscopic Versus Open Abdominal Sacrocolpopexy

Laparoscopic sacrocolpopexy can be as equally effective as open abdominal sacrocolpopexy in selected women and may require mesh application. It can be performed in combination with other vaginal procedures to correct prolapse of other organs [29]. One multicenter randomized controlled trial (RCT) compared open abdominal sacrocolpopexy versus laparoscopic sacrocolpopexy and showed that the potential advantages of laparoscopic sacrocolpopexy were significantly less intraoperative blood loss and shorter hospital stay [29]. In contrast, a single multicenter RCT compared open and laparoscopic sacrocolpopexy in the treatment of pelvic organ prolapse without significant differences found for operative time, adverse events, or quality of life among the study patients [27]. Other studies concur with these findings as well [36–38].

Mesh/Graft Repair

Some techniques require the routine use of grafts or a mesh to bridge the gap between the vaginal cuff and the concavity of the sacrum [27]. This can be a synthetic mesh (e.g., a permanent polypropylene or absorbable polyglactin mesh) or a biological graft [27]. Biological grafts can be autologous (such as fascial sheath), alloplastic (e.g., porcine dermis), or homologous (e.g., cadaveric fascia lata) [27].

In one trial, abdominal sacral colpopexy with either absorbable cadaveric fascia lata graft (Tutoplast) or nonabsorbable (permanent) monofilament polypropylene mesh (Trelex) were compared. There were no recurrences in either group; however, the failure rate (recurrence at any other vaginal site) was significantly higher (32%) in the fascial graft group than in the mesh group (9%) There were no vaginal erosions in the fascial graft group, but 3.7% had mesh-related erosion in the nonabsorbable mesh group [27]. De Ridder et al. compared two types of absorbable mesh, polyglactin (Vicryl) inlay versus porcine dermis graft (Pelvicol). The failure rate at 25 months of follow-up was significantly greater in the Vicryl group (31%) when compared with the Pelvicol group (9.5%) [27]. In another RCT, Natale et al. compared polypropylene mesh (Gynemesh) repair with porcine dermis graft (Pelvicol) repair. At 2 years, significantly fewer women had anterior vaginal wall recurrence (28%) in the mesh group, whereas 44% of the porcine graft group recurred [39].

In a RCT that compared laparoscopic sacrocolpopexy vs. transvaginal mesh repair in women with vaginal vault prolapse, the women in the laparoscopic sacrocolpopexy group had longer operative time, shorter hospital stay, and quicker return to daily activities with significantly greater patient satisfaction at 2-year follow-up [40]. Another RCT compared transvaginal sacrospinous fixation repair with and without a mesh with 12-month mean follow-up and showed higher recurrence rates in the non-mesh vaginal sacrospinous fixation group without significant difference in quality of life between groups [41]. The limited evidence available on transvaginal mesh kits does not support their use as first-line treatment, and if considered, women should be fully informed of the permanent nature of the mesh and about potential mesh use complications, some of which are serious and have long-term effects that can be difficult to treat [27, 29].

Summary

Enterocele is a herniation of the peritoneal cavity associated with pelvic floor injuries and surgical procedures. Surgical treatment is reserved for symptomatic cases. The goal of the repair is to provide obliteration of the peritoneal sac with approximation of the uterosacral ligaments in the midline using multiple available abdominal techniques.

References

1. Ellis CN, Essani R. Treatment of obstructed defecation. Clin Colon Rectal Surg. 2012;25(1):24–33.
2. Oom DM, Gosselink MP, Schouten WR. Enterocele: diagnosis and treatment. Gastroenterol Clin Biol. 2009;33(2):135–7.
3. Enríquez-Navascués JM, Elósegui JL, et al. Surgical treatment of rectocele and enterocele: an integrated view of disorders of the posterior vaginal compartment. Cir Esp. 2005;78(Suppl 3):66–71.
4. Mehta D, Deval, Shaam MB, Yadav JK. Rare cause of intestinal obstruction – vaginal enterocele. Indian J Surg. 2010;72(2):104–6.
5. Reynolds RK. Gynecology. In: Doherty GM, editor. Current diagnosis and treatment: surgery. 14th ed. New York: McGraw-Hill Education; 2015.
6. Mellgren A, Johansson C, Dolk A, et al. Enterocele demonstrated by defecography is associated with other pelvic floor disorders. Int J Colorec Dis. 1994;3:1221–124.
7. Cronje HS, De Beer JA, Bam R. The pathophysiology of an enterocele and its management. J Obstet Gynaecol. 2004;24(4):408–13.
8. Mellgren A, Dolk A, Johansson C, Bremmer S, Anzén B, Holmström B. Enterocele is correctable using the Ripstein rectopexy. Dis Colon Rectum. 1994;37(8):800–4.
9. Delest A, Cosson M, Doutrelant C, Querleu D, Crepin G. Enterocele. Retrospective study of 134 cases: risk factors and comparison between abdominal and perineal routes. J Gynecol Obstet Biol Reprod. 1996;25(5):464–70.
10. Chaliha C, Stanton S. Complications of surgery for genuine stress incontinence. Br J Obstet Gynaecol. 1999;106(12):1238–45.
11. Kaiser AM. Chapter 2. Evaluation tools. McGraw-Hill Manual: colorectal surgery. New York: The McGraw-Hill Companies; 2009.
12. Beer-Gabel M, Assoulin Y, Amitai M, Bardan E. A comparison of dynamic transperineal ultrasound (DTP-US) with dynamic evacuation proctography (DEP) in the diagnosis of cul de sac hernia (enterocele) in patients with evacuatory dysfunction. Int J Color Dis. 2008;23(5):513–9.
13. Deng DY, Shindel AW. Chapter 40. Female urology and female sexual dysfunction. In: JW MA, Lue TF, editors. Smith and Tanagho's General Urology. 18th ed. New York: The McGraw-Hill Companies; 2013.
14. Martin G, Garcia J, Roig JV, et al. Analysis of experience in the use of dynamic pelvic magnetic resonance imaging in the assessment of obstructive defaecation syndrome. Cir Esp. 2012;90(5):292–7.
15. Aarons CB, Sentovich SM. Anorectum. In: Doherty GM, editor. CURRENT diagnosis and treatment: surgery. 14th ed. New York: McGraw-Hill Education; 2015.
16. Gosselink MJ, van Dam JH, Huisman WM, Ginai AZ, Schouten WR. Treatment of enterocele by obliteration of the pelvic inlet. Dis Colon Rectum. 1999;42(7):940–4.
17. Lin AY, Fleshman JW. Chapter 39. Benign disorders of the anorectum (pelvic floor, fissures, hemorrhoids, and fistulas). In: Zinner MJ, Ashley SW, editors. Maingot's abdominal operations. 12th ed. New York: The McGraw-Hill Companies; 2013. p. 803–32.
18. Law YM, Fielding JR. MRI of pelvic floor dysfunction: review. AJR. 2008;191(6 Suppl):S45–53.
19. Lienemann A, Anthuber C, Baron A, Reiser M. Diagnosing enteroceles using dynamic magnetic resonance imaging. Dis Colon Rectum. 2000;43(2):205–12.
20. Dietz HP, Beer-Gabel M. Ultrasound in the investigation of posterior compartment vaginal prolapse and obstructed defecation. Ultrasound Obstet Gynecol. 2012;40(1):14–27.
21. Konstantinovic ML, Steensma AB, Domali E, et al. Correlation between 3D/4D translabial ultrasound and colpocystodefecography in diagnosis of posterior compartment prolapse. Ultrasound Obstet Gynecol. 2007;30(4):448.
22. Steensma AB, Oom DMJ, Burger CW, Schouten WR. Assessment of posterior compartment prolapse: a comparison of evacuation proctography and 3D transperineal ultrasound. Color Dis. 2010;12(6):533–9.

23. Perniola G, Shek C, Chong CCW, Chew S, Cartmill J, Dietz HP. Defecation proctography and translabial ultrasound in the investigation of defecatory disorders. Ultrasound Obstet Gynecol. 2008;31(5):567–71.
24. Grasso RF, Piciucchi S, Quattrocchi CC, Sammarra M, Ripetti V, Zobel BB. Posterior pelvic floor disorders: a prospective comparison using introital ultrasound and colpocystodefecography. Ultrasound Obstet Gynecol. 2007;30(1):86–94.
25. Eilber KS, Rosenblum N, Gore J, Raz S, Rodriguez LV. Perineocele: symptom complex, description of anatomic defect, and surgical technique for repair. Urology. 2006;67(2):265–8.
26. Miklos JR, Kohli N, Lucente V, Saye WB. Site-specific fascial defects in the diagnosis and surgical management of enterocele. Am J Obstet Gynecol. 1998;179(6 Pt 1):1418–22. discussion 822-3
27. De Ridder D, Claehout F, Verleyen P, Boulanger S, Deprest J. Procine dermis xenograft as reinforcement for cystocele stage III repair: a prospective randomized controlled trial (**Abstract**). Neurourology and Urodynamics 2004;23:435–6.
28. Rogers RM, Colborn GL, Weidman TA, Skandalakis JE, Mirilas PS, Galloway NTM. Chapter 26. Female genital system. In: Skandalakis JE, Colburn GL, Weidman TA, Foster RS, Kingsworth AN, Skandalakis LJ, et al., editors. Skandalakis' surgical anatomy. New York: The McGraw-Hill Companies; 2004.
29. El Naqa AM, Guerrero KL, Abdel MS. Post-hysterectomy vaginal vault prolapse (Green top Guideline No. 46). RCOG/BSUG. 2015;46:1–21.
30. Sze E, Karram MM. Transvaginal repair of vault prolapse: a review. Obstet Gynecol. 1997;89:466–75.
31. Beer M, Kuhn A. Surgical techniques for vault prolapse: a review of the literature. EJOG. 2005;119(2):144–55.
32. Morgan DM, Rogers MA, Huebner M, Wei JT, Delancey JO. Heterogeneity in anatomic outcome of sacrospinous ligament fixation for prolapse: a systematic review. Obstet Gynecol. 2007;109(6):1424–33.
33. Cvach K, Dwyer P. Surgical management of pelvic organ prolapse: abdominal and vaginal approaches. World J Urol. 2012;30(4):471–7.
34. Holley RL, Varner RE, Gleason BP, Apffel LA, Scott S. Recurrent pelvic support defects after sacrospinous ligament fixation for vaginal vault prolapse. J Am Coll Surg. 1995;180(4):444–8.
35. Tulikangas PK, Piedmonte MR, Weber AM. Functional and anatomic follow-up of enterocele repairs. Obstet Gynecol. 2001;98(2):265–8.
36. D'Hoore A, Vanbeckevoort D, Penninckx F. Clinical, physiological and radiological assessment of rectovaginal septum reinforcement with mesh for complex rectocele. Br J Surg. 2008;95(10):1264–72.
37. Solomon ER, Muffly TM, Hull T, Paraiso MF. Laparoscopic repair of recurrent lateral enterocele and rectocele. Int Urogynecol J. 2015;26(1):145–6.
38. Cook J, Seman E, O'Shea R. Laparoscopic treatment of enterocele: a 3-year evaluation. Aust N Z J Obstet Gynaecol. 2004;44:107–10.
39. Cervigni M, Natale F, Weir J, et al. Prospective randomized trial of two new materials for the correction of anterior compartment prolapse: Pelvicol and Prolene Soft. Neurourol Urodyn. 2005;24: 585–6.
40. Maher C, Feiner B, DeCuyper E, Nichlos C, Hickey K, O'Rourke P. Laparoscopic sacral colpopexy versus total vaginal mesh for vaginal vault prolapse: a randomized trial. Am J Obstet Gynecol. 2011;204(360):e1–7.
41. Halaska M, Maxova K, Sottner O, et al. A multicenter, randomized, prospective, controlled study comparing sacrospinous fixation and transvaginal mesh in the treatment of posthysterectomy vaginal vault prolapse. Am J Obstet Gynecol. 2012;207(301):e1–7.

Ventral Rectopexy: Indications, Surgical Considerations, and Outcomes

28

Paul Cavallaro and Liliana Bordeiano u

Introduction

Laparoscopic ventral mesh rectopexy (LVR) was first described by D'Hoore [1] in 2004 as an alternative approach to the standard abdominal rectopexy for external rectal prolapse. This operation takes advantage of a critical innovation in surgical technique for rectal prolapse – sparing the sacral nerves by limited posterior rectal dissection. This novel approach avoids posterolateral rectal dissection, thereby avoiding rectal denervation and minimizing postoperative de novo constipation. Furthermore, the dissection allows for correction of middle compartment prolapse, elevation of the pouch of Douglas, and reinforcement of the rectovaginal septum. Since its description, LVR has been widely adopted by colorectal surgeons for treatment of external prolapse and other pelvic floor disorders such as internal rectal intussusception. In fact, the proportion of laparoscopic operations for rectal prolapse has increased from 10% to 40% since 2005 with very favorable outcomes [2]. In this chapter, we will describe the spectrum of rectal prolapse and indications for laparoscopic ventral mesh rectopexy, detail operative technique, and review the literature with a focus on outcomes and complications of this operation.

P. Cavallaro
Department of General Surgery, Massachusetts General Hospital, Boston, MA, USA

L. Bordeianou (✉)
Center for Pelvic Floor Disorders, Section of Colon and Rectal Surgery, Massachusetts General Hospital, Boston, MA, USA
e-mail: lbordeianou@mgh.harvard.edu

The Spectrum of Rectal Prolapse

External rectal prolapse refers to a full-thickness intussusception of the rectal wall with evisceration through the anus. This disease is felt to represent the final stage in a series of progressive stages of prolapse, preceded by intrarectal and intra-anal intussusception as described in the Oxford rectal prolapse staging system [3]. While many agree that internal intussusception and external prolapse represent a disease on a spectrum, the true incidence of progression from intussusception to external prolapse is unknown.

Symptoms of external prolapse generally include incomplete rectal evacuation, incontinence of mucous and/or stool, and sensation of a mass that has prolapsed through the anus. On physical exam, the surgeon will find a full-thickness prolapse of the rectal mucosa, classically with a concentric ring of folds. If not evident on exam, the prolapse can be induced by the patient squatting and bearing down. A thorough examination of the perineum and digital rectal exam should also be completed to evaluate the integrity of the anal sphincter.

The clinical significance of internal intussusception has long been debated by the pelvic floor community. Many colorectal surgeons,

L. C. C. Oliveira (ed.), *Anorectal Physiology*, https://doi.org/10.1007/978-3-030-43811-1_28

particularly in Europe, feel that internal intussusception is causative of symptoms consistent with obstructive defecation syndrome (ODS) including constipation and incomplete evacuation due to the telescoping of the intussuscepted rectum causing a mechanical obstruction [4]. In contrast, internal intussusception has been identified in 20–50% of asymptomatic volunteers on defecography [5, 6]. Furthermore, radiographic findings of intussusception have not been shown to correlate with rectal emptying, constipation severity, or balloon expulsion. Interestingly, increasing grades of intussusception have been shown to be associated with increasing severity of fecal incontinence [7].

Given the unclear relationship between internal intussusception and rectal prolapse, we recommend limiting your choice of patients for ventral rectopexy to those who have overt rectal prolapse or patients with nearly visible internal intra-anal intussusception. In those with internal intussusception, additional evaluation with radiographic studies (defecography, dynamic pelvic MRI), physiologic studies (manometry, anal sphincter EMG), colonoscopy, and sitz marker colonic transit studies is important to ensure that intussusception is not an incidental finding in the context of other coexisting pelvic floor disorders. These studies may reveal other anatomic disorders, such as rectocele or enterocele, and functional disorders such as anismus (paradoxical non-relaxation of the anal sphincters). A thorough history is a requisite, including screening for confounding disorders, such as irritable bowel syndrome, prior to any surgery.

Indications

At this time, the only indication for LVR that is agreed upon by the global colorectal surgery community is external rectal prolapse. As the operation's popularity has flourished since 2004, its use has been extended to other pelvic floor disorders on an institution and surgeon-dependent basis. In fact, a consensus statement from a group of European pelvic floor specialists listed high-grade internal intussusception and solitary rectal ulcer syndrome (SRUS) as relative indications for LVR [8]. These additional indications are intriguing but have not been generally accepted by the international community as of yet, and there are an increasing number of studies examining the outcomes. As previously stated, we continue to feel that the clinical significance of internal rectal intussusception is debatable. Furthermore, there is an unfortunate lack of high-quality evidence supporting use of LVR for internal intussusception, and no studies have been able to document a clear correlation between surgical correction of anatomic abnormalities and improvement in obstructed defecation [9]. This same European consensus statement listed specific contraindications for LVR: pregnancy, no pelvic anatomical problems, severe adhesions, active proctitis, psychologic instability, and anismus resistant to conventional treatment.

Technique

We follow the principles in surgical technique initially described by D'Hoore in 2004. Briefly, peritoneal access is obtained at the umbilicus. One working 12 mm port is placed in the right lower quadrant. Two additional 5 mm posts are in the LLQ and RUQ. The uterus is retracted anteriorly with suture. The rectosigmoid is retracted to the left side and out of the pelvis to expose the sacral promontory. The peritoneum over the sacral promontory is incised with an energy device, adjacent to the mesorectum and rectum. This peritoneal incision is carried down distally along the right side of the rectum and then extended transverse across the deepest portion of the pouch of Douglas (Fig. 28.1). At this point, we pay special attention to avoid damage to the hypogastric nerves. Next, Denonvilliers' fascia is incised, and the rectovaginal septum is opened using the energy device (Fig. 28.2). Contrary to the standard abdominal rectopexy, there is no lateral or posterior mobilization of the rectum. Once the ventral rectum is mobilized, we place either a Prolene (permanent) or a Biodesign®

(biologic) mesh into the abdomen. The mesh is sutured to the ventral aspect of the distal rectum using nonabsorbable 2-0 Vicryl or 2-0 PDS sutures (Fig. 28.3). Additional sutures are placed to fix the mesh to the lateral borders of the rectum more proximally in rows of two. We usually create about three to four rows of sutures to each side of the rectum. The mesh is then fixed to the sacral promontory with a nonabsorbable 2-0 Gore-Tex suture (Fig. 28.4). It is key to have minimal traction on the rectum after placement of the mesh. Next, the posterior vaginal fornix may be sutured to the anterior aspect of the mesh to correct a coexisting middle-compartment prolapse, if present. Lastly, the edges of incised peritoneum are closed over the mesh which elevates the new pouch of Douglas, restoring anatomy and covering the foreign material of the mesh with peritoneum (Fig. 28.5).

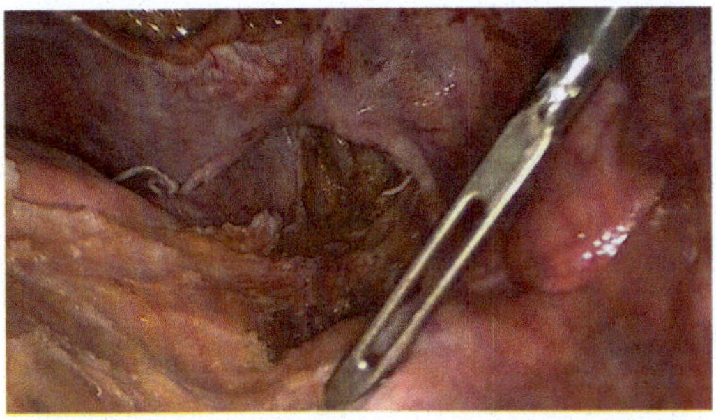

Fig. 28.1 Incision of the peritoneum along the right border of the rectum/mesorectum. The laparoscopic grasper indicates the sacral promontory

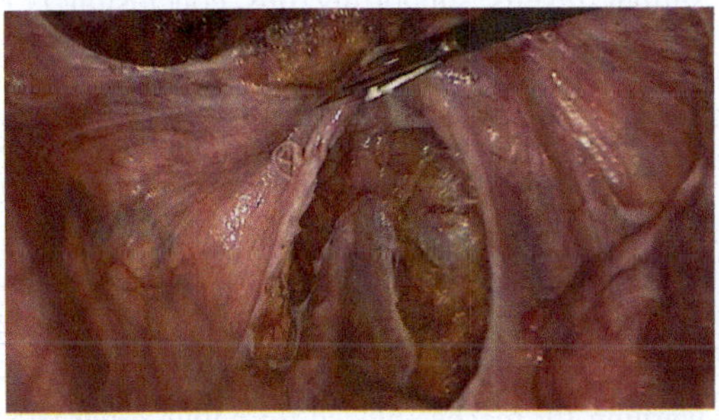

Fig. 28.2 Incising of the peritoneum carried anteriorly along the pouch of Douglas with the dissection of Denonvilliers' fascia. There is no posterior rectal dissection

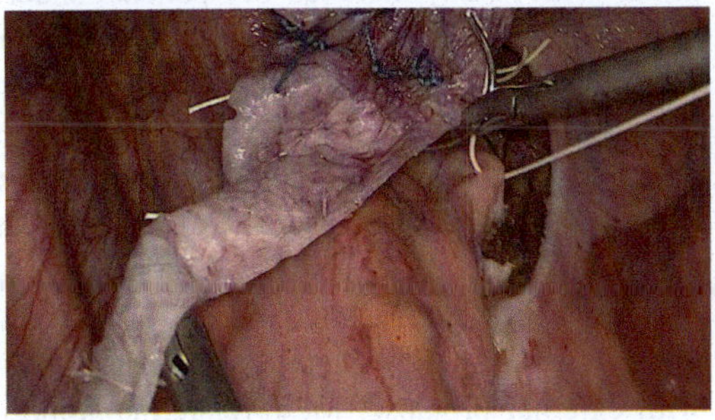

Fig. 28.3 Laparoscopic suturing of the mesh to the ventral rectum

Learning Curve

There have been two studies describing the learning curve for proficiency with the LVR. Mackenzie et al. [10] evaluated operative technique, as well as outcomes and improvement in quality of life after 636 LVR operations performed by a single senior colorectal surgeon. They developed proficiency gain curves and determined that the learn-

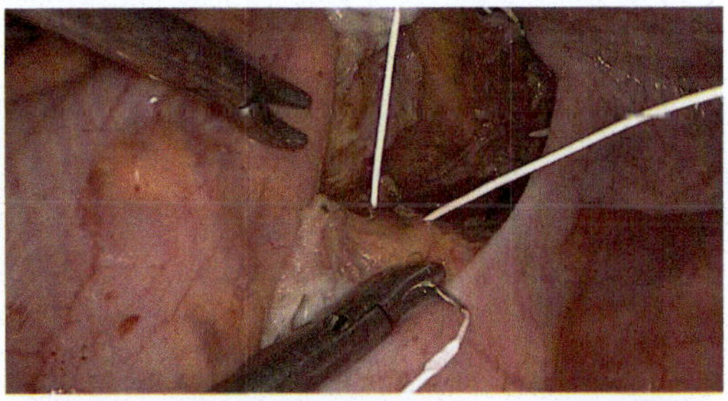

Fig. 28.4 Fixation of the proximal portion of the mesh to the sacral promontory with permanent suture

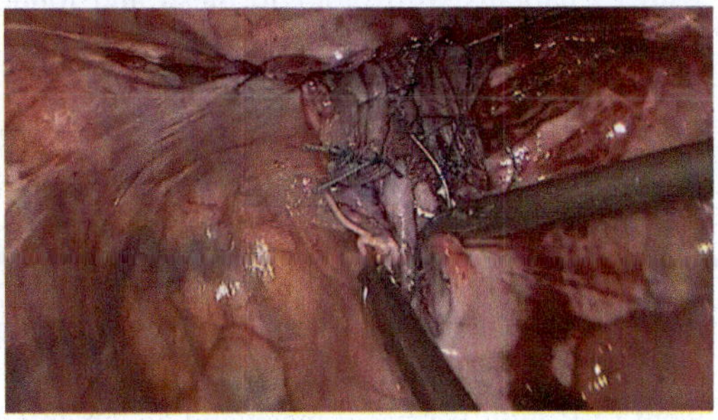

Fig. 28.5 Closure of the peritoneum over the mesh

ing curve for operative time was 54 cases but for other clinical and quality-of-life outcomes was between 82 and 105 cases. A more recent study in 2017 [11] looked at 311 LVRs performed at two district hospitals by two surgeons in the United Kingdom. Cumulative sum curve analysis suggested a learning curve of between 25 and 30 cases based on operative times and length of stay, and this was similar between both surgeons. They did not find a significant change point for morbidity or mortality. In our experience, technical proficiency with this operation was felt to be achieved at approximately 10 cases, but surgeons have been routinely performing laparoscopic suture rectopexies using posterior approach for decades prior to learning the LVR technique.

Types of Meshes

Synthetic mesh is typically used to fix the rectum to the sacral promontory. However, given the concern for mesh complications that have been reported with similar procedures in the pelvic floor, some authors have studied the feasibility of using biologic mesh. A systematic review [12] of 13 observational studies comprising 866 patients in 2013 (11 studies with 767 synthetic mesh, 2 studies with 99 biologic mesh) found no difference in recurrence (3.7% vs 4.0%, $p = 0.78$) or mesh complications (0.7% vs 0%, $p = 1.0\%$) between synthetic and biological mesh repair.

Three more recent studies have further detailed the use of biologic mesh for LVR. Ogilvie et al. [13] matched 29 patients with permanent mesh with 29 patients with biologic mesh and found no difference in symptom resolution, recurrence, or mesh-related complication. Albayati et al. [14] studied 51 patients that underwent LVR with biologic mesh and reported a complication rate of 13.7% with an overall 25% reduction in obstructed defecation symptoms and 20% reduction in incontinence symptoms. Lastly, McLean et al. [15] reported on 224 patients that underwent LVR with Permacol mesh and documented a complication rate of 10.7%; however, mesh-related morbidity was only 0.5%. Recurrence was 10.7% at 5 years, and there was significant improvement in patient-related constipation and incontinence symptoms.

Robotic-Assisted Surgery

After LVR proved to be a generally safe operation, some colorectal surgeons sought to utilize the benefits of robotic surgery to further enhance surgical technique. Perrenot [16] performed robot-assisted LVR on 77 patients between 2002 and 2010. After a learning curve of 18 cases, morbidity was found to be acceptably low (10%) with 13% recurrence at an average of 52 months. Similar to laparoscopic surgery, there was a 50% reduction in constipation. There were five conversions to open surgery. The authors concluded that robot-assisted LVR was safe with acceptable outcomes, warranting further studies. Two subsequent retrospective single-center studies found similar results supporting this conclusion [17, 18].

Outcomes

Patients undergoing LVR have had favorable outcomes, and while a concern for long-term mesh-related complications exists based on the FDA's warning for meshes placed for pelvic organ prolapse, there has been an acceptable morbidity profile with current follow-up data. The largest study of long-term outcomes after LVR was performed by Consten et al. [19] in 2015. This observational cohort study included 919 patients undergoing LVR for either external rectal prolapse or Oxford grade III/IV internal rectal prolapse with symptoms of fecal incontinence or obstructed defecation. Patients were followed for a median of 34 months, and there were 68 recurrences at a median 24 months. Using Kaplan-Meier methods, they estimated a 14.3% risk of 10-year recurrence for all patients and an 8.2% risk for patients with external prolapse. Mesh-related complications occurred in 4.6% of patients, including 7 mesh erosions into the vagina (5 of which had an associated perineotomy). Patients reported improvements in both fecal incontinence (11.1% vs 37.5%) and obstructed defecation (15.6% vs 54.0%). A 2018 study [20] analyzing long-term outcomes in pelvic floor function studied 508 patients with either external rectal prolapse or symptomatic internal rectal prolapse with a median follow-up time of 44 months. Subjective symptom severity was quantified with Wexner score, obstructive defecation score, and quality-of-life scores. Approximately 76% of patients experienced subjective symptom relief, with higher rates of relief in patients with external prolapse compared to internal prolapse/intussusception (86% vs 68%). Complications occurred in 11% of patients, and mesh-related complications occurred in 7 patients – 5 of which were mesh erosions into the vagina and 2 of which were rectovaginal fistulas. Of note, three of the five mesh erosion complications occurred in patients with intraoperative vaginal perforation. The overall recurrence rate was 7% for external prolapse during the study period. Interestingly, de novo symptoms were reported in 124 patients – two-thirds of these patients reported an urge sensation, and this was more common in patients with internal intussusception, while 13 patients reported loss of a sensation to defecate. The authors of both studies similarly concluded that LVR was a safe and effective treatment for both external and internal rectal prolapse with an acceptable rate of mesh-related complications. However, based on these studies, there does seem to be some heterogeneity in outcome based on indication. The literature regarding outcomes for specific indications will be reviewed in subsequent sections.

While recurrence rates for LVR have been acceptable, they are not negligible. To date there has been one study attempting to decipher risk factors for recurrence. Fu et al. [21] studied 231 consecutive patients undergoing LVR by a single surgeon for either external rectal prolapse, internal intussusception, SRUS, or rectocele. Despite the heterogenous population, they reported a complication rate of 5.2% over a median follow-up time of 47 months and a recurrence rate of 11.7%. All but two of the recurrences occurred in patients with full-thickness external rectal prolapse. On univariate analysis, predictors of recurrence included age >70 years, worse preoperative Cleveland Clinic Incontinence Score, prolonged pudendal nerve terminal motor latency (PNTML), and the use of synthetic mesh. On multivariate analysis, only prolonged PNTML and the use of synthetic mesh were independently found to be associated with recurrence. On reoperation for recurrence, the most common findings during laparoscopy were that either the mesh/graft used for LVR had detached from the sacral promontory or the mesh/graft had come off the mid-rectum. This highlights the importance of technical proficiency for positive results after LVR.

Most patients spend one night in the hospital after their operation; however, we often discharge patients to home on the day of surgery if they are hemodynamically stable with their pain well controlled and after passing a trial of void. This practice is supported in the literature: Powar et al. [22] reported that 23% of LVRs are discharged home on the day of surgery with no increase in complications or readmissions.

In line with improvements in constipation and incontinence symptoms, patients similarly report higher scores on quality-of-life instruments. In a 2016 study [23], patients documented significantly higher scores on the Short Form 36 Health Survey scales (including physical functioning, bodily pain, health perception, social functioning, emotional and mental health) at 3, 6, and 12 months postoperatively. Similarly, all of the Fecal Incontinence Quality of Life and Patient Assessment of Constipation-Quality of Life scales significantly improved after LVR. A separate study [24] specifically evaluated sexual function after LVR for either external prolapse, internal intussusception, rectocele, or enterocele and found that the number of patients being satisfied with their sexual function was similar before and after surgery (91% vs 85%). Approximately 13% of respondents felt that sexual function decreased after surgery.

Complications

Driving much of the surge in volume of LVR since its description is the fact that patients have enjoyed relatively few complications. Laparoscopy-associated complications such as port site hernia, hematoma, and iatrogenic bowel injury seldom do occur. It is important to keep in mind, however, that although rare, serious procedure-specific complications have been reported after this operation, specifically for mesh-related complications. These procedure-specific complications have significant consequences for long-term functional outcomes, and patients must be counseled appropriately.

Considering the high-profile reports of complications from transvaginal meshes placed into the pelvis, many authors have focused on mesh complications after LVR. Evans et al. [25] followed 2203 patients at multiple centers undergoing LVR. Approximately 80% of meshes placed were synthetic. Mesh erosion into the rectum or vagina occurred in 2% of cases over a 14-year study period, and 40% of these patients underwent reoperation for major mesh morbidity (12 laparoscopic mesh removal, 3 mesh removals with colostomy, and 3 anterior resections). The remainder required minor revisions with local excision of a stitch or exposed mesh. A 2017 meta-analysis [26] compiled eight studies with close to 4000 patients and similarly found mesh-related erosion rates of 1.87% in synthetic meshes and 0.22% in biologic meshes. There was a range in time of diagnosis from 2 to 124 months. When a mesh complication is encountered, it is crucial that colorectal surgeons are comfortable with management. Amoudi et al. [27] proposed specific principles of management for variations in mesh-related complications. In all patients, dissection should be carried down to the pelvic floor with removal of the original mesh. A new lightweight Teflon-coated polypropylene mesh may be used to replace the original mesh. If the complication is due to mesh detachment or poor fixation, the detached site can be fortified using a new mesh tacked to the promontory and then sutured to the old mesh. Rectal injury/erosion should be managed with anterior resection and a limited LVR with a new mesh to prevent recurrent prolapse. Rectovaginal fistulae following mesh erosion have been described and reported to be difficult to manage. Some have argued for control of the local sepsis with a defunctioning ostomy and removal of the mesh [28], in addition to transabdominal or transvaginal repair of the rectum depending on the site of the fistula. Others argued that biologic mesh can be considered in reoperation in a contaminated field to manage the initial prolapse symptoms [27]. We have not yet had to deal with any of these complications at our institution but are monitoring the literature on the topic with significant attention.

A variety of other rare complications have been reported after LVR and should be mentioned. Lumbar discitis has been described in several instances [29, 30]. This seems to occur at the site of fixation of the mesh into the promontory, leading some to believe that bacteria are translocating at the site of rectal fixation onto the mesh. High-grade hemorrhoids requiring surgical intervention may be common with an actuarial 5-year estimated incidence of 24% in one study [31]. Other rare complications include SBO [32] and RP fibrosis [33].

External Prolapse

In their original description of LVR, D'Hoore et al. [1] operated on 42 patients with full-thickness external rectal prolapse with favorable results. There were no postoperative mortalities and two postoperative complications, both urinary tract infections. Two patients developed recurrent rectal prolapse at 54 and 91 months – both of these patients had prior failed Delorme procedure. Before surgery, 31 patients were incontinent, and 28 of these patients reported improved continence after LVR. Sixteen patients achieved normal continence. Likewise, of the 19 patients with preoperative obstructed defecation, 16 reported that their symptoms resolved. As expected, the four patients with slow transit constipation observed no improvement. Importantly, only two patients reported de novo mild ODS symptoms. Surprisingly, even in this initial description, conversion to laparotomy was rare, occurring in two cases.

Since then, LVR has been adopted by the colorectal surgeon community, and a number of single-institution studies sharing outcomes for external prolapse have been published, all with similar results. Boons et al. [34] performed LVR on 65 consecutive patients with external rectal prolapse. At a median follow-up of 19 months, there was one recurrence. Outcomes were assessed at 3 months, and constipation was improved in 72% with a decrease in median Wexner score from 9 to 4. De novo constipation was noted in one patient. Similarly, continence was improved in 83% of patients. Improvements in these functional parameters continued at 24 months of follow-up. Importantly, there were no mesh infections, erosions, or other mesh-related complications in this cohort. Faucheron et al. [35] operated on 175 patients with external prolapse and followed them for a median of 74 months. Recurrence occurred in two female patients – one at month 6 and one at month 24. The overall complication rate in this population was 5%, including urinary tract infection, transient brachial plexus palsy, and small bowel perforation from adhesiolysis. One patient presented with erosion of the mesh into the rectum at 9 months – she had a reportedly uneventful transanal removal of the mesh. Randall et al. [36] published their results from 190 LVRs for external rectal prolapse. Of their patients, 120 had follow-up for >5 years and 16 had follow-up for >10 years. Incontinence scores improved by a median of 8 points ($p < 0.001$) with a 93% improvement overall. QOL scores assessed at 1 and 4 years improved by 46%. Sexual function was improved in 37% of patients. Five patients developed a partial recurrence limited to the left side, and the overall recurrence rate was 3%. Four additional patients developed posterior lateral intussusception. Seven patients developed mesh complications – four meshes eroded into the vagina, two meshes eroded into the rectum, and one rectovaginal fistula was noted. Three of these were treated by transvaginal mesh removal, and three were managed by laparoscopic mesh removal. The authors of the three aforementioned studies independently concluded that LVR is a safe approach to manage full-thickness external rectal prolapse with favorable long-term improvements in constipation, incontinence, and quality of life. Recurrence rates are comparable to other transabdominal approaches. The complication rate is low – importantly however, there are some documented mesh complications, and additional studies on long-term outcomes are needed to quantify the true risk of mesh complications, which can be devastating. Interestingly, a 2016 study [37] showed that one-third of LVR patients have postoperative internal intussusception on defecography, which was associated with less improvement in functional measures.

There have been two randomized control trials comparing LVR with other common operations for external rectal prolapse. One trial [38] randomized 50 patients to either LVR or Delorme procedure, the most common perineal approach in the country the study was performed (Egypt). The majority of patients (66%) had fecal incontinence. Given the small sample size, there were no noted differences in recurrence rates or outcomes; however, recurrent prolapse was observed in 16% of the LVR patients and 8% of the Delorme patients. Postoperative incontinence and constipation scores were similarly improved from preoperative scores in both groups. Lundby et al. [39] randomized 75 patients to either LVR or laparoscopic posterior sutured rectopexy. Interestingly, in this study, LVR and posterior sutured rectopexy had similar postoperative ODS scores and de novo constipation despite the sacral nerve-sparing approach. Lastly, a 2014 study (24500726) retrospectively compared laparoscopic resection rectopexy (the most common abdominal approach in the United States) with LVR (the most common approach in Europe). Each operation was performed at a single center, either in the United States or the Netherlands. In all, there were 28 resection rectopexy patients and 40 LVR patients. The resection rectopexy group was younger; however, the groups were otherwise similar. A significant reduction in constipation and incontinence occurred in both groups. A comparison of the two operations showed a trend to significance favoring resection rectopexy for improvement of incontinence ($p = 0.09$). The complication rate was significantly higher after resection rectopexy compared to LVR (9 vs 3, $p < 0.05$). The authors concluded that LVR and laparoscopic resection rectopexy are safe options with acceptable outcomes for external rectal prolapse; however, further prospective, randomized controlled trials are needed to compare the two operations.

Traditionally, transabdominal approaches are often reserved for younger patients with limited comorbidities, while older patients were managed with perineal approaches. However, perineal procedures suffer from significantly higher recurrence rates and worse functional outcomes. With the technical advances provided by laparoscopy in terms of reduction of comorbidity, surgeons began to reappraise the appropriateness of transabdominal approaches for elderly patients. Wijffels et al. [40] examined a prospectively collected database from two tertiary pelvic floor centers and evaluated outcomes in patients over the age of 80 with external rectal prolapse. In this age group, the median LOS was 3 days. There were no mortalities, and there was a complication rate of 13% (3 pneumonias, 3 UTIs, 3 port site hernias, 1 SBO, 1 MI, 1 wound infection, and 1 fluid overload). Recurrence occurred in 3 patients at a mean of 23 months. Similarly, Bjerke [41] studied 46 patients with a median age of 83 – of these patients, 14 had previously undergone a prolapse operation and 12 had perineal procedures. Median LOS was 2 days and the 30-day complication rate was 15%. There were four major complications that were intraoperative complications – one trocar bladder perforation, one thermal rectal injury, one hematoma, and one small bowel thermal injury. Two patients died within 30 days – one 93-year-old woman died from cardiac arrest on POD3 and one 85-year-old who underwent reoperation for small bowel thermal injury and died on POD10 from cardiac arrest. Functional outcomes were favorable with a significant reduction in incontinence scores at 2 months and 1 year. There were 2 recurrences at a median follow-up of 1.5 years. Gultekin et al. [42] retrospectively compared 1263 patients over the age of 70 to younger patients undergoing LVR and found no significant difference in mortality or complications between groups on multivariate analysis. Taken together, the data suggest that LVR is safe, well-tolerated, and efficacious in older, frail patients and may be an alternative to perineal approaches with more durable results.

While LVR seems to be an excellent therapeutic option for external rectal prolapse, some have identified patients in which results may not be optimal. Gurland [43] studied 108 LVRs, 36 of which were on patients with recurrent prolapse. When comparing patients with primary repair vs repair of recurrent prolapse, prolapse recurrence rates for primary repairs were significantly lower: 1.4%, 6.9%, and 9.7% compared to 13.9%, 25%, and 25% at 1, 3, and 5 years. Time to recurrence was significantly shorter in patients undergoing LVR for recurrent prolapse, 8.8 vs 30.7 months. The authors noted that the majority of recurrent prolapse occurred secondary to technical errors, primarily with failure to adequately fix the mesh to the sacral promontory. While they concluded that LVR still had reasonable outcomes to repair recurrent prolapse, they emphasized that patients should be counseled that they are at increased risk for prolapse recurrence. Additionally, the importance of technical proficiency was stressed, as many of the recurrences were due to inadequate fixation. Two studies have specifically studied outcomes in men (25175930, 27641548). Both studies reported that LVR is a safe and effective operation for external prolapse in men; however, Rautio et al. [44] found that men were at higher risk for reoperation in the postoperative period (33%). The majority of reoperations were for recurrent prolapse and persistent postoperative mucosal anal prolapse symptoms. Importantly, LVR did not impact sexual function and did not cause any voiding or urinary symptoms.

Internal Intussusception

As discussed above, the role of surgical management for internal rectoanal intussusception is not clear, at least in management algorithms in the United States. This practice is much more common in Europe, with 31 of 32 European colorectal surgeons reporting acceptance of ODS from internal intussusception as an indication for LVR [45]. As such, many studies evaluating the outcomes of LVR for internal intussusception have come from European centers. While the majority of these studies highlight very favorable outcomes for both ODS and incontinence due to internal intussusception, we recommend caution in regard to adoption of LVR for internal intussusception, given the clouded understanding of how this radiologic anatomic finding impacts pelvic floor function and whether or not it represents a pathologic abnormality that needs to be corrected or a variant of normal anatomy. Our algorithm for management of ODS stresses the importance of maximum medical therapy, biofeedback, and recognition/management of confounding risk factors such as IBS. Only when these measures fail in high-grade internal intussusception do we consider offering surgery. Despite the controversy regarding surgical management of internal intussusception and lack of adoption of this practice in the United States, this section will summarize the existing literature detailing outcomes of LVR for intussusception.

One of the first studies to examine the role of LVR in ODS came in 2008 [46]. Seventeen patients with ODS were included, most of which were secondary to rectocele or internal intussusception. Of these patients, 15 had improvement in constipation in the short term. However, at a mean follow-up of 38 months, there was no significant difference in ODS scores. In fact, 12 patients had higher scores postoperatively than preoperatively, and one-third of patients complained of continued straining, incomplete evacuation, and digitation. A number of subsequent studies documented favorable outcomes, at least in the short term. Collinson [47] prospectively studied 75 patients undergoing LVR for high-grade rectoanal intussusception that failed medical therapy. Preoperative constipation and fecal incontinence both significantly improved at 3 and 12 months, and no patients reported worse function. Sileri et al. [48] similarly studied 34 patients undergoing LVR for high-grade internal intussusception with incontinence or constipation refractory to conservative management of an aggressive bowel regimen, laxative, and biofeedback from a pelvic floor therapist. Preoperative constipation and incontinence were significantly improved at 3 months. Two patients experienced persistent or recurrent prolapse, and the complication rate was comparable to that in the literature. Borie et al. [49] retrospectively compared LVR with stapled transanal rectal resection for ODS secondary to intussusception or rectocele. STARR was performed in 27 patients and LVR was performed in 25 patients. After surgery, ODS symptoms were significantly reduced in 56% undergoing LVR and 59% undergoing STARR in the short term. Approximately 80% of patients were very or moderately satisfied after LVR. Complication rates were similar between groups. A 2017 meta-analysis [50] of rectopexy for internal intussusception reviewed a total of 14 studies comprising 1300 patients, 1147 of which had an LVR and the remainder had resection rectopexy. Approximately 77% of patients undergoing LVR reported an improvement in ODS symptoms, and 63% reported an improvement in fecal incontinence. Recurrence occurred in 6.5% of patients and the overall complication rate was 13.6%. The mesh-related complication rate was 1.1% and included mesh detachment, erosion, fistula formation, and intestinal obstruction. A single study by Tsunoda et al. [51] performed postoperative defecography on 26 patients that underwent LVR for intra-anal internal intussusception and found that the high-grade intussusception was eliminated in all patients, although 8 developed intrarectal intussusception. These patients had an associated 50% reduction in ODS symptoms, possibly linking correction of the anatomy with improvement in symptoms. Adopters of LVR for internal intussusception cite these data for their satisfactory results in the majority of patients, low recurrence rate, and low morbidity rate.

Fecal incontinence has been shown to correlate with worsening internal intussusception grade, and several studies have focused on incontinence as a primary endpoint. In a 2013 study [52] of 72 patients undergoing LVR for high-grade intussusception causing fecal incontinence refractory to medical management, the median fecal incontinence score 1 year after surgery was significantly lower than preoperative scores. A follow-up study [53] of 50 patients with incontinence and high-grade internal intussusception compared to 41 patients undergoing LVR for external prolapse showed that incontinence and quality-of-life scores were similarly reduced in both groups.

Data are relatively lacking in regard to long-term outcomes. A recent study in 2018 [20] attempted to answer this question. This study included 508 consecutive patients treated with laparoscopic ventral rectopexy for either external prolapse or internal intussusception. Symptomatic IRP was present in 214 patients, 79% of whom had obstructed defecation, 17% had incontinence, and 20% had combined symptoms. The median follow-up length was 44 months. Fewer patients with internal intussusception had relief of obstructed defecation symptoms compared to external rectal prolapse, 68% vs 86%; however, the raw rate of long-term relief in this population was still viewed as favorable. Of patients being operated on for internal intussusception, 6.1% required reoperation which was consistent with previous reports. De novo defecatory urge occurred more commonly after LVR for internal intussusception than external prolapse. From these data, the authors concluded that the results of LVR for internal intussusception were long-lasting, although they benefitted less than patients with external prolapse.

Rectocele

Parallel to the debate on the clinical significance of internal intussusception is a similar debate on rectocele. The majority of rectoceles is asymptomatic and found incidentally. Those that are symptomatic are reported to result in symptoms ranging from obstructed defecation to the sensation of a lump in the vagina. To the contrary, a prospective study of patients evaluated in a pelvic floor center found that rectoceles were not associated with worsening ODS severity, anorectal abnormalities, or pelvic floor dyssynergia [54]. In this context, after some surgeons observed favorable outcomes of LVR for internal intussusception, the operation was then applied to patients with rectocele to avoid the complications of incontinence and dyspareunia common to conventional rectocele repair. Wong et al. [55] were the first to report on LVR performed exclusively for rectocele. At a median follow-up of 29 months, they found a significant decrease in vaginal discomfort (86–20%) and ODS symptoms (83–46%), with no change in fecal incontinence or de novo symptoms. A subsequent study [56] included patients with rectocele in addition to external prolapse and internal intussusception as candidates for LVR. Patients with rectocele or internal prolapse had a significant reduction in incontinence and constipation postoperatively. Their complication rate was 4.6% and included two mesh infections complicated by discitis at the site of mesh fixation. The group from the former study went on to study the impact on anorectal and sexual function in two separate studies [57, 58]. They reported a significant relief in the predominant symptoms of vaginal bulge and sexual dysfunction with no de novo dyspareunia. These data are limited to a single center and should be generalized to other centers with caution until further studies are available.

Solitary Rectal Ulcer Syndrome

Solitary rectal ulcer syndrome (SRUS) is often associated with ODS and internal rectal intussusception with symptoms consisting of bleeding, mucous discharge, pain, and difficulty evacuating stool. There is little, if any, consensus on management of this entity. One study [59] reported an 86% improvement in SRUS symptoms after placement of a ventral mesh during a standard posterior open mesh rectopexy. This finding led to a subsequent study evaluating the efficacy of LVR for SRUS [60] – in 48 patients with SRUS refractory to biofeedback, LVR led to epithelial healing of the lesion at 3 months with improvement in ODS and QOL scores at 2 years. Recurrent lesions occurred in two patients on long-term follow-up. A similar study [61] showed healing of the ulcer in 90% of patients with significant improvements in functional scores.

Summary

Laparoscopic ventral mesh rectopexy is an innovative operative technique to avoid excessive rectal dissection and de novo constipation symptoms. Outcomes for management of external rectal prolapse are favorable and comparable to traditional abdominal approaches. The limited dissection and minimally invasive approach have made this the operation of choice for many surgeons for this indication. Recurrence is low in this population (about 5%) and functional outcomes are excellent. Furthermore, there is less physiologic insult compared to open procedures, and it has been shown to be safe for older, frail patients who are a significant constituent of this population. Some surgeons have translated LVR to other indications including internal intussusception, rectal prolapse, and SRUS. Prospective data evaluating outcomes in these patients is limited and therefore this practice should be adopted with caution. Mesh complications are rare but have been shown to occur in the follow-up period documented in the current literature. Many of these mesh-associated complications require re-intervention, and surgeons must have an armamentarium of approaches for management. Fortunately, the overall complication rate is low, and LVR has reproducibly been shown to have a more than adequate safety profile.

References

1. D'Hoore A, Cadoni R, Penninckx F. Long-term outcome of laparoscopic ventral rectopexy for total rectal prolapse. Br J Surg. 2004;91(11):1500–5.
2. Rogers AC, McCawley N, Hanly AM, Deasy J, McNamara DA, Burke JP. Trends in the treatment of rectal prolapse: a population analysis. Int J Color Dis. 2018;33(4):459–65.
3. Collinson R, Cunningham C, D'Costa H, Lindsey I. Rectal intussusception and unexplained faecal incontinence: findings of a proctographic study. Color Dis. 2009;11(1):77–83.
4. Dvorkin LS, Gladman MA, Scott SM, Williams NS, Lunniss PJ. Rectal intussusception: a study of rectal biomechanics and visceroperception. Am J Gastroenterol. 2005;100(7):1578–85.
5. Shorvon PJ, McHugh S, Diamant NE, Somers S, Stevenson GW. Defecography in normal volunteers: results and implications. Gut. 1989;30(12):1737–49.
6. Palit S, Bhan C, Lunniss PJ, et al. Evacuation proctography: a reappraisal of normal variability. Color Dis. 2014;16(7):538–46.
7. Hawkins AT, Olariu AG, Savitt LR, et al. Impact of rising grades of internal rectal intussusception on fecal continence and symptoms of constipation. Dis Colon Rectum. 2016;59(1):54–61.
8. Mercer-Jones MA, D'Hoore A, Dixon AR, et al. Consensus on ventral rectopexy: report of a panel of experts. Color Dis. 2014;16(2):82–8.
9. Lundby L, Laurberg S. Laparoscopic ventral mesh rectopexy for obstructed defaecation syndrome: time for a critical appraisal. Color Dis. 2015;17(2):102–3.
10. Mackenzie H, Dixon AR. Proficiency gain curve and predictors of outcome for laparoscopic ventral mesh rectopexy. Surgery. 2014;156(1):158–67.
11. Pucher PH, Mayo D, Dixon AR, Clarke A, Lamparelli MJ. Learning curves and surgical outcomes for proctored adoption of laparoscopic ventral mesh rectopexy: cumulative sum curve analysis. Surg Endosc. 2017;31(3):1421–6.
12. Smart NJ, Pathak S, Boorman P, Daniels IR. Synthetic or biological mesh use in laparoscopic ventral mesh rectopexy–a systematic review. Color Dis. 2013;15(6):650–4.
13. Ogilvie JW, Stevenson AR, Powar M. Case-matched series of a non-cross-linked biologic versus non-absorbable mesh in laparoscopic ventral rectopexy. Int J Color Dis. 2014;29(12):1477–83.
14. Albayati S, Morgan MJ, Turner CE. Laparoscopic ventral rectopexy for rectal prolapse and rectal intussusception using a biological mesh. Color Dis. 2017;19(9):857–62.
15. McLean R, Kipling M, Musgrave E, Mercer-Jones M. Short- and long-term clinical and patient-reported outcomes following laparoscopic ventral mesh rectopexy using biological mesh for pelvic organ prolapse: a prospective cohort study of 224 consecutive patients. Color Dis. 2018;20(5):424–36.
16. Perrenot C, Germain A, Scherrer ML, Ayav A, Brunaud L, Bresler L. Long-term outcomes of robot-assisted laparoscopic rectopexy for rectal prolapse. Dis Colon Rectum. 2013;56(7):909–14.
17. Mäkelä-Kaikkonen J, Rautio T, Klintrup K, et al. Robotic-assisted and laparoscopic ventral rectopexy in the treatment of rectal prolapse: a matched-pairs study of operative details and complications. Tech Coloproctol. 2014;18(2):151–5.
18. van Iersel JJ, Formijne Jonkers HA, Paulides TJC, et al. Robot-assisted ventral mesh rectopexy for rectal prolapse: a 5-year experience at a tertiary referral center. Dis Colon Rectum. 2017;60(11):1215–23.
19. Consten EC, van Iersel JJ, Verheijen PM, Broeders IA, Wolthuis AM, D'Hoore A. Long-term outcome after laparoscopic ventral mesh rectopexy: an observational study of 919 consecutive patients. Ann Surg. 2015;262(5):742–7; discussion 747–748.
20. Mäkelä-Kaikkonen J, Rautio T, Kairaluoma M, et al. Does ventral rectopexy improve pelvic floor function in the long term? Dis Colon Rectum. 2018;61(2):230–8.
21. Fu CW, Stevenson AR. Risk factors for recurrence after laparoscopic ventral rectopexy. Dis Colon Rectum. 2017;60(2):178–86.
22. Powar MP, Ogilvie JW, Stevenson AR. Day-case laparoscopic ventral rectopexy: an achievable reality. Color Dis. 2013;15(6):700–6.

23. Tsunoda A, Takahashi T, Ohta T, Kusanagi H. Quality of life after laparoscopic ventral rectopexy. Color Dis. 2016;18(8):O301–10.
24. Formijne Jonkers HA, Poierrié N, Draaisma WA, Broeders IA, Consten EC. Impact of rectopexy on sexual function: a cohort analysis. Int J Color Dis. 2013;28(11):1579–82.
25. Evans C, Stevenson AR, Sileri P, et al. A multicenter collaboration to assess the safety of laparoscopic ventral rectopexy. Dis Colon Rectum. 2015;58(8):799–807.
26. Balla A, Quaresima S, Smolarek S, Shalaby M, Missori G, Sileri P. Synthetic versus biological mesh-related erosion after laparoscopic ventral mesh rectopexy: a systematic review. Ann Coloproctol. 2017;33(2):46–51.
27. Badrek-Al Amoudi AH, Greenslade GL, Dixon AR. How to deal with complications after laparoscopic ventral mesh rectopexy: lessons learnt from a tertiary referral centre. Color Dis. 2013;15(6):707–12.
28. Ouaïssi M, Cresti S, Giger U, et al. Management of recto-vaginal fistulas after prosthetic reinforcement treatment for pelvic organ prolapse. World J Gastroenterol. 2010;16(24):3011–5.
29. Draaisma WA, van Eijck MM, Vos J, Consten EC. Lumbar discitis after laparoscopic ventral rectopexy for rectal prolapse. Int J Color Dis. 2011;26(2):255–6.
30. Vujovic Z, Cuarana E, Campbell KL, Valentine N, Koch S, Ziyaie D. Lumbosacral discitis following laparoscopic ventral mesh rectopexy: a rare but potentially serious complication. Tech Coloproctol. 2015;19(4):263–5.
31. van Iersel JJ, Formijne Jonkers HA, Verheijen PM, Draaisma WA, Consten EC, Broeders IA. High-grade hemorrhoids requiring surgical treatment are common after laparoscopic ventral mesh rectopexy. Tech Coloproctol. 2016;20(4):235–42.
32. Salminen HJ, Tan WS, Jayne DG. Three cases of small bowel obstruction after laparoscopic ventral rectopexy using the V-Loc(®) suture. Tech Coloproctol. 2014;18(6):601–2.
33. Leow S, Voyvodic F, Murphy E, de Fontgalland D. Unprecedented complication of biologic mesh use in ventral rectopexy. ANZ J Surg. 2017;88(11):E811–2.
34. Boons P, Collinson R, Cunningham C, Lindsey I. Laparoscopic ventral rectopexy for external rectal prolapse improves constipation and avoids de novo constipation. Color Dis. 2010;12(6):526–32.
35. Faucheron JL, Voirin D, Riboud R, Waroquet PA, Noel J. Laparoscopic anterior rectopexy to the promontory for full-thickness rectal prolapse in 175 consecutive patients: short- and long-term follow-up. Dis Colon Rectum. 2012;55(6):660–5.
36. Randall J, Smyth E, McCarthy K, Dixon AR. Outcome of laparoscopic ventral mesh rectopexy for external rectal prolapse. Color Dis. 2014;16(11):914–9.
37. Tsunoda A, Takahashi T, Ohta T, Fujii W, Kusanagi H. New-onset rectoanal intussusception may not result in symptomatic improvement after laparoscopic ventral rectopexy for external rectal prolapse. Tech Coloproctol. 2016;20(2):101–7.
38. Emile SH, Elbanna H, Youssef M, et al. Laparoscopic ventral mesh rectopexy vs Delorme's operation in management of complete rectal prolapse: a prospective randomized study. Color Dis. 2017;19(1):50–7.
39. Lundby L, Iversen LH, Buntzen S, Wara P, Høyer K, Laurberg S. Bowel function after laparoscopic posterior sutured rectopexy versus ventral mesh rectopexy for rectal prolapse: a double-blind, randomised single-centre study. Lancet Gastroenterol Hepatol. 2016;1(4):291–7.
40. Wijffels N, Cunningham C, Dixon A, Greenslade G, Lindsey I. Laparoscopic ventral rectopexy for external rectal prolapse is safe and effective in the elderly. Does this make perineal procedures obsolete? Color Dis. 2011;13(5):561–6.
41. Bjerke T, Mynster T. Laparoscopic ventral rectopexy in an elderly population with external rectal prolapse: clinical and anal manometric results. Int J Color Dis. 2014;29(10):1257–62.
42. Gultekin FA, Wong MT, Podevin J, et al. Safety of laparoscopic ventral rectopexy in the elderly: results from a nationwide database. Dis Colon Rectum. 2015;58(3):339–43.
43. Gurland B, Carvalho ME, Ridgeway B, Paraiso MF, Hull T, Zutshi M. Should we offer ventral rectopexy to patients with recurrent external rectal prolapse? Int J Color Dis. 2017;32(11):1561–7.
44. Rautio T, Mäkelä-Kaikkonen J, Vaarala M, et al. Laparoscopic ventral rectopexy in male patients with external rectal prolapse is associated with a high reoperation rate. Tech Coloproctol. 2016;20(10):715–20.
45. Kim M, Meurette G, Ragu R, Lehur PA. Current surgical treatment of obstructed defecation among selected European opinion leaders in pelvic floor surgery. Tech Coloproctol. 2016;20(6):395–9.
46. van den Esschert JW, van Geloven AA, Vermulst N, et al. Laparoscopic ventral rectopexy for obstructed defecation syndrome. Surg Endosc. 2008;22(12):2728–32.
47. Collinson R, Wijffels N, Cunningham C, Lindsey I. Laparoscopic ventral rectopexy for internal rectal prolapse: short-term functional results. Color Dis. 2010;12(2):97–104.
48. Sileri P, Franceschilli L, de Luca E, et al. Laparoscopic ventral rectopexy for internal rectal prolapse using biological mesh: postoperative and short-term functional results. J Gastrointest Surg. 2012;16(3):622–8.
49. Borie F, Bigourdan JM, Pissas MH, Guillon F. Laparoscopic ventral rectopexy for the treatment of outlet obstruction associated with recto-anal intussusception and rectocele: a valid alternative to STARR procedure in patients with anal sphincter weakness. Clin Res Hepatol Gastroenterol. 2014;38(4):528–34.
50. Emile SH, Elfeki HA, Youssef M, Farid M, Wexner SD. Abdominal rectopexy for the treatment of internal rectal prolapse: a systematic review and meta-analysis. Color Dis. 2017;19(1):O13–24.

51. Tsunoda A, Ohta T, Kiyasu Y, Kusanagi H. Laparoscopic ventral rectopexy for rectoanal intussusception: postoperative evaluation with proctography. Dis Colon Rectum. 2015;58(4):449–56.
52. Gosselink MP, Adusumilli S, Gorissen KJ, et al. Laparoscopic ventral rectopexy for fecal incontinence associated with high-grade internal rectal prolapse. Dis Colon Rectum. 2013;56(12):1409–14.
53. Gosselink MP, Joshi H, Adusumilli S, et al. Laparoscopic ventral rectopexy for faecal incontinence: equivalent benefit is seen in internal and external rectal prolapse. J Gastrointest Surg. 2015;19(3):558–63.
54. Hicks CW, Weinstein M, Wakamatsu M, Pulliam S, Savitt L, Bordeianou L. Are rectoceles the cause or the result of obstructed defaecation syndrome? A prospective anorectal physiology study. Color Dis. 2013;15(8):993–9.
55. Wong M, Meurette G, Abet E, Podevin J, Lehur PA. Safety and efficacy of laparoscopic ventral mesh rectopexy for complex rectocele. Color Dis. 2011;13(9):1019–23.
56. Formijne Jonkers HA, Poierrié N, Draaisma WA, Broeders IA, Consten EC. Laparoscopic ventral rectopexy for rectal prolapse and symptomatic rectocele: an analysis of 245 consecutive patients. Color Dis. 2013;15(6):695–9.
57. Wong MT, Abet E, Rigaud J, Frampas E, Lehur PA, Meurette G. Minimally invasive ventral mesh rectopexy for complex rectocoele: impact on anorectal and sexual function. Color Dis. 2011;13(10):e320–6.
58. Abet E, Lehur PA, Wong M, Rigaud J, Darnis E, Meurette G. Sexual function and laparoscopic ventral rectopexy for complex rectocoele. Color Dis. 2012;14(10):e721–6.
59. Nicholls RJ, Simson JN. Anteroposterior rectopexy in the treatment of solitary rectal ulcer syndrome without overt rectal prolapse. Br J Surg. 1986;73(3):222–4.
60. Badrek-Amoudi AH, Roe T, Mabey K, Carter H, Mills A, Dixon AR. Laparoscopic ventral mesh rectopexy in the management of solitary rectal ulcer syndrome: a cause for optimism? Color Dis. 2013;15(5):575–81.
61. Evans C, Ong E, Jones OM, Cunningham C, Lindsey I. Laparoscopic ventral rectopexy is effective for solitary rectal ulcer syndrome when associated with rectal prolapse. Color Dis. 2014;16(3):O112–6.

Pudendal Neuropathy and Pudendal Canal Decompression

29

Ali A. Shafik

Introduction

Pudendal neuropathy or pudendal canal syndrome are synonymous for proctalgia. Proctalgia is a sudden onset of severe anal, rectal, or perineal pain without specific cause. It is chronic, intermittent, and cramp-like. It usually lasts for a few minutes and disappears (rarely > 30 min) [1–3]. It may follow straining at defecation (90% of cases), sudden explosive bowel action, or ejaculation. It is aggravated by sitting, relieved by standing, and absent when recumbent or when sitting on a toilet seat; it is more common among young men [3–6].

One of the features for clinical diagnosis is pain that may be produced by palpating the lateral aspect of the pelvic floor and relieved by levator massage [7]. This term is widely used in North America, and the syndrome appears to be more common in women than in men. Furthermore, symptoms are often worse at night and frequently waken the patient from sleep [8]. Numerous causes have been suggested including neuralgia, neuroses, infections, allergy, vasospasm, venous stasis, mechanical factors, and psychiatric disorders, but none can be supported by conclusive evidence [2, 4, 9–11].

A. A. Shafik (✉)
Chairman and Professor of Colorectal Surgery Department, Faculty of Medicine, Cairo University, Cairo, Egypt
e-mail: ali@alishafik.com

Nevertheless, proctalgia treatment has many modalities, but no treatment of proven efficiency is available [1, 3]. Sitz baths [1], digital anal dilatation [7], biofeedback [12], lidocaine injection [13], botulinum A toxin injection [14], superior hypogastric block [15], behavioral treatment, and psychotherapy [16] have been involved in the treatment of proctalgia.

In 1991, Shafik [17] described a new syndrome called "pudendal canal syndrome" (PCS) in seven females. Diagnosis of this syndrome was based on the following:

1. Pain, numbness, and tingling localized to the anal region and not related to defecation
2. Absence of proctological lesions
3. Tenderness over both pudendal canals (PCs)
4. Motor and sensory changes localized to distribution of the pudendal nerve (PN)
5. Low rectal neck pressure at rest and on voluntary squeeze
6. Increased pudendal nerve terminal motor latency (PNTML)

Shafik reported that the increased intra-abdominal pressure beyond the normal physiologic limits, as occurs in chronic straining at defecation, urination, or prolonged second stage of labor, would eventually result in subluxation and sagging of the levator ani (LA) muscle. This will lead to traction on the inferior rectal nerve which pulls on and stretches the pudendal nerve resulting in entrapment of its

L. C. C. Oliveira (ed.), *Anorectal Physiology*, https://doi.org/10.1007/978-3-030-43811-1_29

distal portion in the pudendal canal and pudendal neuropathy.

Surgical Anatomy of the Pudendal Nerve

The Roots of the Pudendal Nerve

The pudendal nerve is the principal nerve of the perineum derived usually from the second, third, and fourth ventral division of the sacral plexus but sometimes takes fiber from S1, ventral division. It arises from the ventral division with nerves to the quadratus femoris and gemellus inferior (L4, 5 S1), nerve to the obturator internus and gemellus superior (L5 S1, 2), and sciatic nerve which gives two branches, medial and lateral popliteal. The medial popliteal takes fibers from the ventral division (L4, 5 S1, 2, 3), while the lateral popliteal takes fibers from the dorsal division (L4, 5 S1). Jünemann et al. [18] described that the pudendal nerve is formed just proximal to the ischial spine. Sometimes the inferior rectal nerve arises directly from the roots of the sacral plexus. Jünemann found that the pudendal nerve at the ischial spine gives a significant branch splits away and penetrates the dorsum of the levator ani to enter the pelvic cavity and subdivided into numerous branches that supply the inner surface of the levator ani.

As the pudendal nerve roots leave the pelvis through the greater sciatic foramen between the piriformis and the coccygeus muscles, the pudendal nerve is formed just proximal to the sacrospinous ligament. The pudendal nerve lies on the sacrospinous ligament medial to the internal pudendal vessels. It enters the pudendal canal through the lesser sciatic foramen. Some somatic fibers coming from S2 and S3 run close to the pelvic plexus to innervate the levator ani and the urethral sphincter. The nerve then travels caudally into a small space "clamp" between the sacrospinous ligament and the sacrotuberous ligament very near the ischial spine. Just inferior to the ischial spine, the nerve gives its first branch, the dorsal nerve of the penis or the clitoridal nerve [18]. These nerves are separated from the main trunk by the pudendal vein and artery. Then, it enters Alcock's canal formed by a division of the obturator muscle aponeurosis. In the canal the nerve crosses the sharp edge of the sacrotuberous ligament (falciform process) [19, 20]. Caudally, at the level of the anus, the nerve gives medially the inferior rectal nerves (usually two branches) which innervate the anal sphincter (and probably the puborectalis) and the skin of the posterior perineum and antero-laterally the transversus perinei branch (for this muscle, for the ischiocavernosus muscle, and maybe for the urethral sphincter) [18]. The remaining part of the nerve is usually called the perineal nerve. This nerve gives a bulbocavernosus branch and finally divides into a sphincteric branch (innervation of the urethra) and a branch which innervates the skin of the anterior perineum (Fig. 29.1) [21].

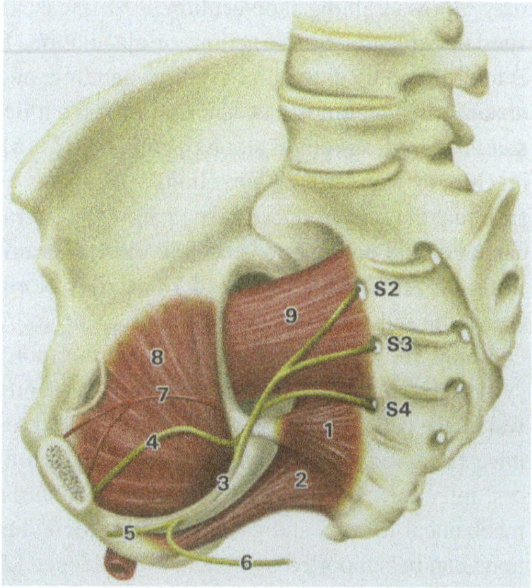

Fig. 29.1 Pelvic nerve anatomy. 1) Sacrospinal ligament. 2) Sacrotuberous ligament 3) Alcock's canal with the pudendal nerve. 4) Nerve of the clitoris (not in Alcock's canal). 5) Perineal branch of the pudendal nerve. 6) Inferior rectal nerve (separated from the pudendal nerve between the ligaments in 50% of the cases, going through the sacrospinal ligament in 15% of the cases). 7) Arcus tendineus fascia pelvis. 8) Obturator muscle. 9) Piriformis muscle – S2, S3, and S4: sacral roots forming the pudendal nerve

Pudendal Canal (Alcock's Canal)

Sinnatamby described that the sacrotuberous ligament is attached to the medial half of the lower part of the tuber ischii [22]. Its upper edge is prolonged forward on the medial surface of the ischium as the falciform ligament. There is dense fascia on the obturator internus called obturator fascia. The canal contains internal pudendal artery, vein, and pudendal nerve. Pudendal canal is formed from lateral prolongation of delicate perineal fascia which thickens and splits to enclose pudendal neurovascular bundles; the pudendal canal connects lesser sciatic foramen to the posterior edge of the perineal membrane. As the pudendal nerve enters the pudendal canal, it gives off inferior rectal nerve in its posterior part. Shortly afterward it divides into the perineal nerve and the dorsal nerve of the penis or clitoris.

Inferior Rectal Nerve

This nerve usually arises as the pudendal nerve enters the pudendal canal. It crosses the ischiorectal fossa inferomedially with the inferior rectal vessels. It supplies levator ani muscle, and reaching the anus innervates the skin and fascia around it and the sphincter ani externus muscle. Cutaneous branches of the inferior rectal nerve communicate with the perineal branch of the posterior femoral cutaneous nerve.

Perineal Nerve

It is the inferior larger terminal branch of the pudendal nerve. It runs forward below the internal pudendal artery. It divided into posterior scrotal (labial) and muscular branches. The posterior scrotal or labial nerve gives medial and lateral branches that pierce or pass over the inferior fascia of the urogenital diaphragm and runs forward in the lateral part of the urethral triangle with scrotal (labial) branches of the perineal artery [22]. They supply the scrotal skin or that of the labium majus, connecting with the perineal branch of the posterior femoral cutaneous and inferior rectal nerve. Muscular branches supply the transversus perinei superficialis, bulbospongiosus, ischiocavernosus, transversus perinei profundus, external urethral sphincter, external anal sphincter, and levator ani muscle. In males, the nerve to the bulb of the urethra leaves the nerve to the bulbospongiosus muscle by piercing the bulbospongiosus muscle to supply the corpus spongiosum of the penis end in the mucosa of the urethral bulb.

Dorsal Nerve of the Penis or Clitoris

This nerve passes forward through the deep perineal space with the internal pudendal artery. It lies close to medial surface of the inferior pubic ramus superior to the perineal membrane and crus of the penis or clitoris and inferior to the sphincter urethrae muscle. It sends a twig through the perineal membrane to the crus and corpus cavernosum of the penis or clitoris. It then pierces the perineal membrane near its anterior border and passes on the dorsal surface of the penis or clitoris lateral to the dorsal artery. It distributed to distal two-thirds of the penis sending branches around the side to reach the inferior surface of that organ. The nerve is much smaller in female than in males (Figs. 29.2, 29.3, 29.4, and 29.5).

Perineum

The perineum is an approximately diamond-shaped region which lies below the inner aspect of the thighs and anterior to the sacrum and coccyx. It is usually described as if from the position of an individual lying supine with the hip joints in abduction and partial flexion. The surface projection of the perineum and the form of the skin covering it varies considerably depending on the position of the thighs, but the deep tissues themselves occupy relatively fixed positions. The perineum is bounded anteriorly by the pubic symphysis and its arcuate ligament, posteriorly by the coccyx, anteriorly by the ischiopubic rami and the ischial tuberosities, and posterolaterally by the sacrotuberous ligament. The deep limit of the perineum is the inferior surface of the pelvic

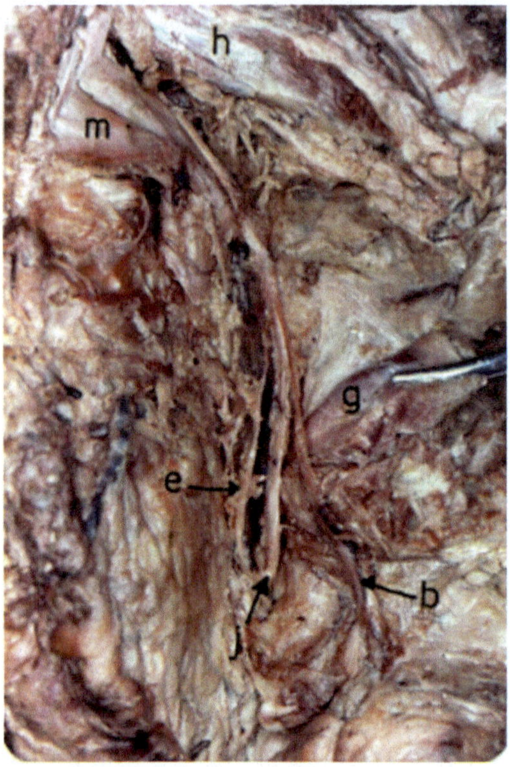

Fig. 29.2 A photograph of female cadaver shows the course of the right pudendal nerve from the sacrospinous ligament to lesser sciatic foramen and the pudendal canal. The pudendal nerve gives the inferior rectal nerve and then divides into two terminal branches (perineal nerve and dorsal nerve of the clitoris). It shows piriformis muscle, sacrospinous ligament, and obturator internus muscle. h) Piriformis muscle. m) Sacrospinous ligament. g) Obturator internus muscle. e) Inferior rectal nerve. b) Dorsal nerve of the clitoris. j) Perineal nerve

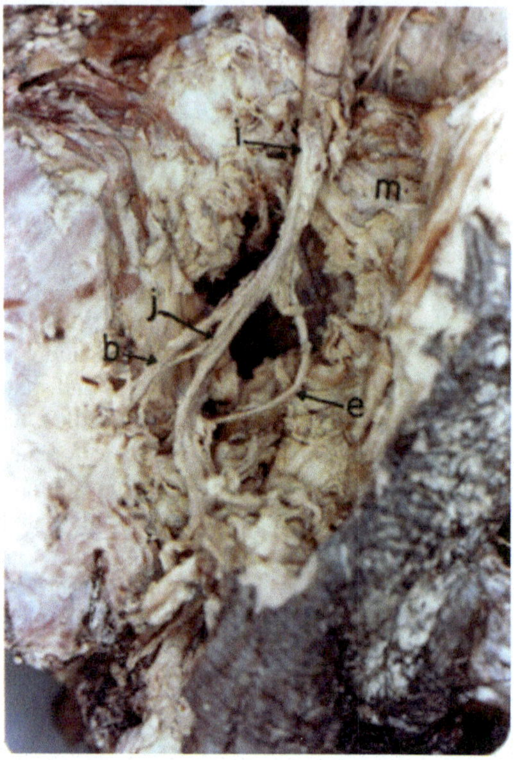

Fig. 29.3 A photograph of female cadaver shows the branches of the left pudendal nerve. The first branch is the inferior rectal, and the two terminal branches are dorsal nerve of the clitoris and perineal nerve. i) Pudendal nerve. m) Sacrospinous ligament. j) Perineal nerve. b) Dorsal nerve of the clitoris. e) Inferior rectal nerve

diaphragm, and its superficial limit is the skin which is continuous with that over the medial aspect of the thighs and the lower abdominal wall. An arbitrary line joining the ischial tuberosities (the interischial line) divides the perineum into an anterior urogenital triangle and a posterior anal triangle. The urogenital triangle faces downward and forward, whereas the anal triangle faces downward and backward.

The male urogenital triangle contains the bulb and attachments of the penis, and the female urogenital triangle contains the mons pubis, the labia majora, the labia minora, the clitoris, and the vaginal and urethral orifices.

Ischiorectal (Ischioanal) Fossa

It is an approximately horseshoe-shaped region filling the majority of anal triangle. The "arms" of the horseshoe are triangular in cross section because the levator ani slops downward toward the anorectal junction [23]. Although it is often referred to as a space, it is filled with loose adipose tissue and occasional blood vessels. The anal canal and its sphincter lie in the center of the horseshoe [23]. The anal canal and the sloping levator ani muscles form the medial wall of the fossa, while the lateral wall is formed by the ischial tuberosity below with obturator internus above. At the base the anterior boundary is the posterior border of the perineal body and urogenital diaphragm, and the posterior boundary

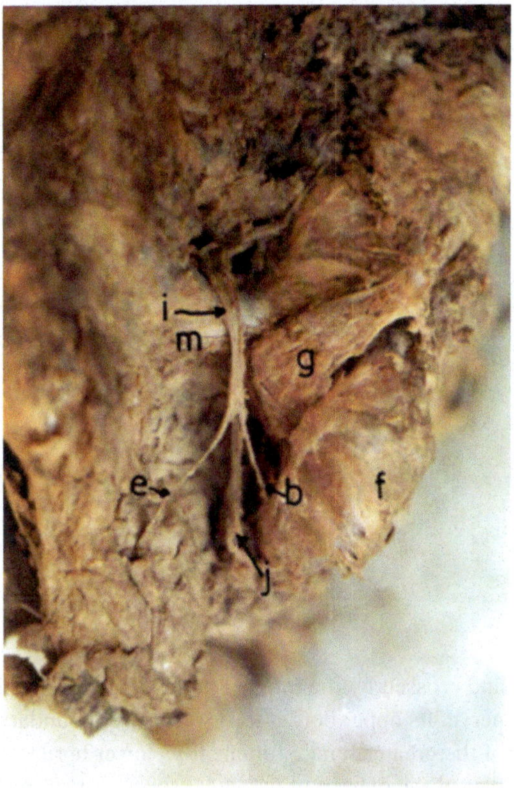

Fig. 29.4 A photograph shows the pudendal nerve crosses the sacrospinous ligament and the inferior rectal nerve and its two terminal branches (perineal and dorsal nerve of clitoris). i) Pudendal nerve. g) Obturator internus muscle. b) Dorsal nerve of the clitoris. e) Inferior rectal nerve. j) Perineal nerve. f) Ischial tuberosity

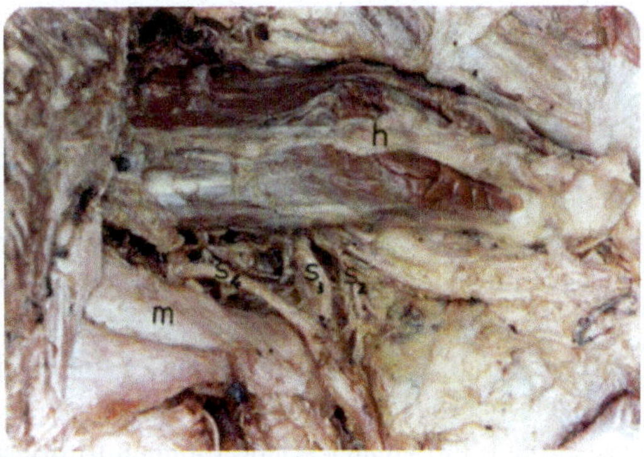

Fig. 29.5 A photograph of female cadaver shows the piriformis muscle, and at its lower border, the roots of the pudendal nerve is presented. The roots of the pudendal nerve extended till the upper border of the sacrospinous ligament. h) Piriformis muscle. m) Sacrospinous ligament. S2, 3, 4 = numbers of sacral foramen that the roots arise

is the sacrotuberous ligament overlapped by the lower border of gluteus maximus [22].

Posteriorly, the fossa contains the attachment of the external anal sphincter to the tip of the coccyx: above and below this, the adipose tissue of the fossa is uninterrupted across the midline. These continuations mean that infections, tumors, and fluid collections within may not only enlarged relatively freely to the side of the anal canal but may also spread with little resistance to the opposite side and deep to the perineal membrane. The internal pudendal vessels and accompanying nerves lay in the lateral wall of the ischiorectal fossa, enclosed in fascia to form the pudendal canal [23]. The inferior rectal vessels and nerves cross the fossa from the pudendal canal and often branch within it.

Each fossa contains the ischiorectal fat pad, the pudendal canal, and a number of vessels and nerves. The ischiorectal fat pad allows for dilatation of the anal canal during defecation and of the vagina during parturition when the passage of fetal head virtually obliterates the space [22].

The fossa is an important surgical plane during resections of the anal canal and anorectal junction for malignancy. It provides an easy plane of dissection with relatively few vessels encountered, which encompasses all of the muscular structures of the anal canal. It leads to the inferior surface of the levator ani through which the dissection is carried (Fig. 29.6) [23].

Sacrotuberous Ligament

The sacrotuberous ligament (great or posterior sacrosciatic ligament) is situated at the lower and back part of the pelvis. It is flat and triangular in form, narrower in the middle than at the ends. The attachments of the sacrotuberous ligament include the lower transverse sacral tubercles, the inferior margins sacrum, and the upper coccyx [24]. The upper edge of the ischial attachment is prolonged forward and attached to a curved ridge of bone. This prolongation is the falciform process; it lies just below the pudendal canal [19, 20]. The membranous falciform process of the

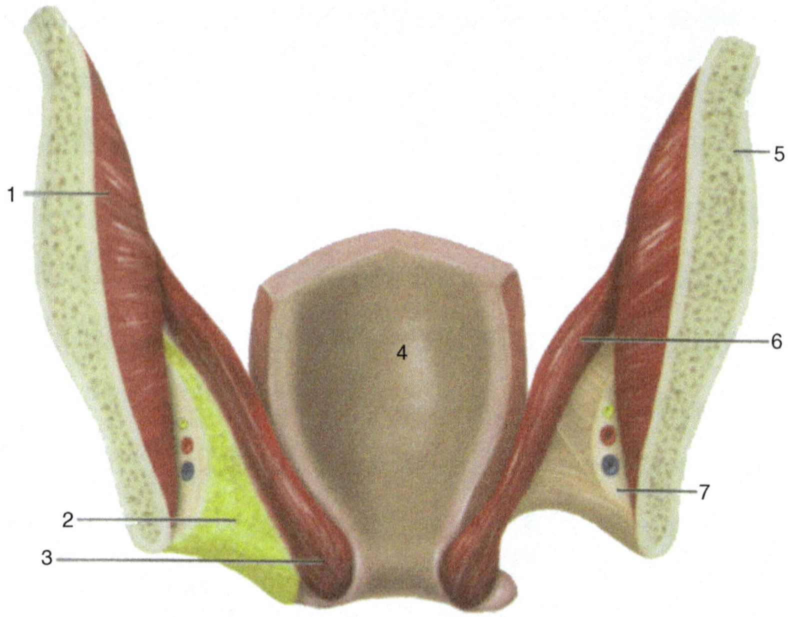

Fig. 29.6 Frontal section shows the walls of the ischiorectal fossa and are as follows: medial, external anal sphincter; lateral, obturator internus muscle; superior, levator ani; inferior, superficial fascia and skin
Legend: 1, fat; external anal sphincter; 3, rectum; 4, basin bone; 5, anal muscle lift; 6, pudendal canal

sacrotuberous ligament was found to be absent in 13% of cadavers. When present it extends toward the ischioanal fossa traveling along the ischial ramus and fusing with the obturator fascia. The sacrotuberous ligament contains the coccygeal branch of the inferior gluteal artery [23].

The lower border of the ligament was found to be directly continuous with the tendon of origin of the long head of the biceps femoris in approximately 50% of subjects [25]. Biceps femoris could therefore act to stabilize the sacroiliac joint via the sacrotuberous ligament.

Sacrospinous Ligament

The sacrospinous ligament (small or anterior sacrosciatic ligament) is thin and triangular in form; it is attached by its apex to the spine of the ischium and, medially, by its broad base, to the lateral margins of the sacrum and coccyx, in front of the sacrotuberous ligament with which its fibers are intermingled [23].

Relations

It is in relation, anteriorly, with the coccygeus muscle, to which it is closely connected; posteriorly, it is covered by the sacrotuberous ligament and crossed by the internal pudendal vessels and nerve. Its upper border forms the lower boundary of the greater sciatic foramen; its lower border is part of the margin of the lesser sciatic foramen.

Its main function is to prevent posterior rotation of the ilia with respect to the sacrum. Laxity of this ligament along with the sacrotuberous ligament allows for this posterior rotation to occur. Stresses to these ligaments occur most often when leaning forward or getting out of a chair.

Piriformis Muscle

The piriformis (from Latin piriformis = "pear shaped") is a muscle in the gluteal region of the lower limb. It runs from the base of the spine to the top of the femur, or thigh bone.

More precisely, it originates from the anterior part of the sacrum, the part of the spine in the gluteal region, and from the gluteal surface of the ilium (as well as the sacroiliac joint capsule and the sacrotuberous ligament) [23]. It exits the pelvis through the greater sciatic foramen to insert on the greater trochanter of the femur; it passes over the sciatic nerve in the majority of cases. However, variations in this arrangement have been reported with the nerve crossing above or through the muscle belly itself [26, 27]. Its tendon often joins with the tendons of the superior gemellus, inferior gemellus, and obturator internus muscles prior to insertion.

The piriformis is a flat muscle, pyramidal in shape, lying almost parallel with the posterior margin of the gluteus medius. It is situated partly within the pelvis against its posterior wall and partly at the back of the hip joint. It arises from the front of the sacrum by three fleshy digitations, attached to the portions of bone between the first, second, third, and fourth anterior sacral foramina and to the grooves leading from the foramina: a few fibers also arise from the margin of the greater sciatic foramen and from the anterior surface of the sacrotuberous ligament. The muscle passes out of the pelvis through the greater sciatic foramen, the upper part of which it fills, and is inserted by a rounded tendon into the upper border of the greater trochanter behind, but often partly blended with, the common tendon of the obturator internus and gemelli [23].

Variations

It is frequently pierced by the common perineal nerve and thus divided more or less into two parts. It may be united with the gluteus medius or send fibers to the gluteus minimus or receive fibers from the superior gemellus. It may have only one or two sacral attachments or be inserted in to the capsule of the hip joint. It may be absent. The piriformis is an external rotator of the hip and functions in conjunction with the quadratus femoris, obturator externus, obturator internus, and the gemellus superior and inferior. The rotary component of this muscle group has been reported to decrease with flexion of the hip. At 90° hip flexion, this group of muscles has a significant abductor component. Some report that the piriformis functions as an internal rotator in hip flexion [28, 29].

Neuroanatomy of Pudendal Nerve

El Badawi A and Schink EA [30] assumed that a triple innervation with sympathetic, parasympathetic, and somatic elements innervates the external urethral sphincter. However it is believed mainly that the pudendal nerve is the most important nerve for innervation and control of the external urethral sphincter mechanism [31]. Jünemann et al. [18] established the exact neuroanatomy of pudendal nerve as well as its neurophysiology relationship to the urethral closure mechanism.

It is well known that the pudendal nerve is a mixed nerve carrying motor and sensory fibers. It is a part of the pelvic plexus, yet its fibers are derived from the sacral roots as they leave the spinal canal through the sacral foramen. Once the roots traverse the sacral foramen, they divide first into somatic and autonomic component (S2 to S4). Autonomic branches originating from S2 to S4 traverse more ventrally and form the pelvic plexus, which is primarily the parasympathetic autonomic supply to the pelvic organs, especially the detrusor muscle and urethral smooth musculature. From the somatic component of the S2 to S4 nerve roots, branches combine to form one major trunk of the pudendal nerve superior to the sacrospinous ligament on the coccygeal muscle and lateral to the coccygeal bone. Further caudal, the pudendal nerve enters laterally to the ischiorectal fossa at the medial side in a fascial sheath (Alcock's canal) close to the obturator internus muscle, where it is protected by an overhang of the gluteal musculature dorsally. The internal pudendal artery and vein emanating from the internal iliac artery and vein, respectively, join the pudendal nerve on its way into the ischiorectal fossa. Directly inferior to the ischial spine, these blood vessels run between the dorsal nerve of the penis and the pudendal nerve, thus separating the underlying dorsal nerve of the penis from the mother trunk, the pudendal nerve.

The dorsal nerve of the penis travels dorsolaterally in relation to the main trunk, above the obturator internus muscle and underneath the levator ani muscle. It perforates the transversus perinei muscle laterally to enter the dorsum of the penis ventromedially. The nerve runs lateral to the dorsal artery and deep dorsal vein of the penis to its final destination the glans penis. Further down in the ischiorectal fossa, the pudendal nerve splits into several motor and sensory fibers.

At the level of the anus, the rectal nerve fibers take off medially to target the final destination, the external anal sphincter. The remaining part of the pudendal nerve trunk, often called the perineal nerve, continues further caudally in an intermediate position between the penile dorsal nerve (which is deep at this level) and the more superficial anal sphincteric branches of the pudendal nerve; the most caudally located nerve branch supplies the bulbospongiosus muscle and provides the ventral portion of the penis and scrotum with sensory fibers.

Dorsolaterally, topographically at the level of the anal sphincteric branches, another single branch emerges from the main pudendal nerve trunk. This branch to the pelvic floor provides the innervation of the transversus perinei and ischiocavernosus muscles. The nerve divides into the transversus perinei muscle from the dorsomedial side wall of the pelvis, and some fibers perforate it to supply the underlying ischiocavernosus muscle also.

At the level of the ischial spine where the trunk of the pudendal nerve is located, splitting of the outer sheath will reveal separate dorsal bundles to different destinations as outlined previously. These bundles will constitute the motor supply to the anal sphincter, levator ani, transversus perinei, and urethral sphincter, as the initiation point of the pudendal nerve proximal to the ischial spine. A significant branch splits away and penetrates the dorsum of the levator ani; it enters the pelvic cavity and subdivides into numerous branches that supply the inner surface of the levator ani muscle. This neuroanatomical interrelationship becomes important for providing continence and an adequate urethral sphincter mechanism.

Neurophysiology of the Pudendal Nerve

Previous studies on the neuronal pathways to the lower urinary tract in animals by Araujo [32] and in humans by Brindley et al. [33] have shown that neurological control of this area is provided mainly by the sacral segments S2 to S4 and that in human S3 is the principal innervator of the pelvis and controls the micturition reflex. The neural control of the bladder, which is derived from the autonomic nervous system (primarily parasympathetic), and the somatic neural control of the external urethral sphincter are located in the S2 to S4 sacral segments mainly S2 and S3 [32, 34]. From anatomical dissections it becomes clear that the external urethral sphincter, the intrinsic rhabdosphincter of the membranous urethra, and the extrinsic levator ani and transversus perinei muscles (pelvic floor) receive its neural supply through the pudendal nerve via the somatic nerve fibers that emanate from the S2 and S3 sacral roots [35, 36]. Jünemann et al. [18] proved these morphological findings. They stimulated the S2 and S3 roots and the pudendal nerve in five patients before and after selective neurotomy or neural blockade with lidocaine to determine the contribution of each nerve and sphincteric muscular component to the external urethral closure pressure.

Pressure recording the bladder, bladder neck urethral sphincter, and rectum was performed with a 3-balloon membrane catheter connected to a Statham transducer.

They represent the neural stimulation responses in several patients who underwent sacral wire and nerve root electrode implantations for neurogenic lower urinary tract dysfunction. Pudendal nerve stimulation shows a typical response of the pelvic floor stimulation of the main trunk of the pudendal nerve after selective neurotomy.

A 29-year-old woman with transverse myelitis and severe urinary dysfunction underwent neurotomy of the transversus perinei branch on the left side and of the anal sphincteric branches on the right side. An electrode was placed around each pudendal nerve, and each side was stimulated separately as clearly shown; stimulation of the left pudendal nerve resulted in a normal anal sphincteric response with a pressure increase to 50 cm water, whereas the intraurethral pressure did not change. This result shows a nonresponsive extrinsic urethral rhabdosphincter transversus perinei muscle to neurostimulation of the main trunk of the pudendal nerve. Due to the selective neuroanatomy of the transversus perinei branch, stimulation of the right pudendal nerve showed reverse response and there was no pressure change within the anal sphincter branch. The urethral pressure was increased to a maximum of 60 cm water and showed response to stimulation of the S3 sacral root in a male paraplegic who underwent bilateral selective S3 somatic neurotomy and left pudendal neurotomy. Stimulation of the left S3 ventral root induced strong bladder contraction (80 cm water) and no urethral sphincteric response. However on the right bladder, response was relatively weak, but the urethral sphincteric pressure increases to 80 cm water. Both studies show clearly that the pelvic floor particularly the transversus perinei muscle has a major role in the external urethral sphincteric mechanism. Furthermore it is obvious that the nerve supply to the transversus and ischiocavernosus muscle is derived from the pudendal nerve and its transversus perinei branch, respectively. Additionally it becomes obvious that the S2 and S3 ventral roots are the most important nerves for external urethral sphincteric mechanism (vide infra). Neurostimulation of the S3 ventral root after selective sacral and pudendal neurotomy neutralizes the response of intrinsic and extrinsic striated external sphincter levator ani and transversus perinei muscle to stimulation of the S3 sacral root, whereas selective sacral root neurotomy alone is not sufficient to abolish the response of the external urethral sphincter to S3 ventral root stimulation (vide infra).

Sacral Nerve Stimulation

A 27-year-old male quadriplegic with a C5 injury suffered from detrusor hyperreflexia with sphincteric dysreflexia and pelvic pain. Six months after right pudendal neurotomy, the patient underwent unilateral selective S3 somatic neurotomy on the left side.

Preoperative S3 ventral root stimulation on the left side produced a strong sphincteric contraction. A weaker response was noticed after left S2 ventral root stimulation. The maximal intraurethral pressure increase was about 50% of that with S3 stimulation. This phenomenon is related to clear neuroanatomical separation of the S2 and S3 somatic nerve fibers in this patient.

The S2 somatic branches mainly supply the transversus perinei and ischiocavernosus muscle via the pudendal nerve; the S3 somatic branches provide the major fiber innervating the intrinsic rhabdosphincter of the external urethral sphincter and the levator ani muscle. However, in addition, the S3 nerve branches to the pudendal nerve innervate the extrinsic urethral sphincter (pelvic floor) which accounts for the increased urethral closure pressure upon stimulation of S3. In this patient stimulation of S3 on the right side increased the intraurethral pressure to 50% of response to stimulation on the left side because of the right pudendal neurotomy (no response to stimulation of the right S2 ventral root).

The stimulation results after unilateral selective S3 somatic neurotomy verified the previous statement. Left S3 ventral root stimulation after neurotomy showed an external urethral sphincteric response that was reduced by almost 70% owing to the failure of the rhabdosphincter and levator ani to contract. Stimulation of the left S2 ventral root alone showed an intraurethral pressure reduction of 30% and stimulation of S2 and S3 ventral roots on the right side showed no change at all.

From these findings, it is obvious that in this patient the S2 ventral root is the major branch that forms the pudendal nerve, whereas the S3 ventral root primarily controls the intrinsic striated external urethral sphincter and the levator ani muscle via sacral somatic nerve fibers. However, both nerves clearly crossover to innervate the other sphincteric component partially.

To determine if the remaining sphincteric response to stimulation of the left S3 ventral root is owing to contraction of the pelvic floor and transversus perinei muscle, they blocked the left pudendal nerve with lidocaine. Repeated stimulation of S3 and S2 ventral roots showed clearly that the external urethral was abolished completely owing to selective somatic neurotomy of the S3 sacral nerve and additional pudendal nerve block. It becomes obvious that stimulation responses before lidocaine blockade were mediated via the pudendal nerve.

An hour after pudendal nerve blockade, complete restoration of the remaining of external urethral sphincter response could be observed. The pelvic floor and transversus perinei muscle contract in response to sacral root stimulation.

Pudendal nerve damage can be assessed by measurement of pudendal nerve terminal motor latency (PNTML) [37]. Abnormal descent of the pelvic floor occurs in patient with pudendal neuropathy which is itself found in idiopathic fecal incontinence and urinary stress incontinence. The relationship between perineal descent and prolonged pudendal nerve terminal motor latency (PNTML) has been studied statistically in relation to patients with constipation [38], patients with fecal incontinence [39]. and patients following vaginal delivery.

Pathophysiology

Study of the course of the pudendal nerve, as we have just seen, stresses possible conclusions [19]:

- In the ligamentous pinch at the level of the ischial spine, the nerve is fixed between the sacrotuberal and sacrospinal ligaments, sometimes even ensheathed by ligamentous expansions which constitute a perineural compartment. Traveling in the midst of ligamentous layers, it then assumes a flattened appearance. We have even seen it traverse the sacrospinal ligament in its thickness or override a sharp sacrospinal ligament which may even be calcified. It is also at this level that the piriformis muscle, situated cranial to the nerve, may come into contact with it and sometimes possess a fibrous sheet endangering the nerve trunk.
- The falciform process of the sacrotuberal ligament may arise high up and come into contact with the nerve, which literally straddles it.
- The vessels are often of considerable size. The pudendal artery may describe perineural curves or constrict the nerve trunk with its collateral branches. The pudendal vein is often tortuous and dilated, leaving little room for the nerve component within the vascular sheath that is often encountered.

The course of the nerve is thus very special. It successively traverses three very different regions and describes a curve which drags it around the region of the ischial spine, which it straddles like a violin string on its bridge. The seated position was simulated in the cadaver. It is associated with an ascent of the ischioanal fat, which becomes applied laterally to the falciform process of the sacrotuberal ligament, so that this is elevated and approximated to the nerve trunk. The latter, contained in the aponeurosis of the obturator internus muscle, has no means of escape, any more than it has in the gluteal region in the ligamentous grip described above, since the piriformis muscle bars any possibility of upward escape.

There is an obvious parallel here with other peripheral nerve entrapments: the median nerve in the carpal tunnel, the common perineal nerve over the head of the fibula, and the ulnar nerve at the elbow, to mention only the most well-known of the entrapment syndromes. These anatomic findings determine the diagnostic, clinical, and neurophysiologic approaches and lead to decisions on treatment.

The exact mechanism of nerve dysfunction and damage is dependent on its etiology. For patients with nerve entrapment and compression, an inflammatory response is engendered. This results in venous stasis, increased vascular permeability, and eventually demyelination. This can result in scar formation and, in cases of severe injury, permanent nerve damage.

For patients with nerve tension injury, the inflammatory effect is not as severe, and demyelination is not a factor. However, neuronal function is impaired. For patients with fixation along the nerve's course, an injury will be more common because the nerve lacks mobility and is more readily stretched. Also, pelvic floor dysfunction itself may cause pain along the pudendal nerve distribution.

Abnormal descent of the pelvic floor with excessive stretching of the pudendal nerve occurred in many patients with pudendal neuropathy and with idiopathic fecal incontinence and urinary stress incontinence. The length of the pudendal nerve from its roots to its termination was measured and compared with the direct distance from the roots of the nerve to the rectum as measured inside the pelvis. The whole length of the pudendal nerve ranged from 1.8 to 2 cm in stillbirths and from 3.3 to 3.8 cm in adult cadavers [40]. The direct distance from the nerve roots to the rectum in stillbirths ranged from 0.8 to 1 cm and in adult cadavers from 1.7 to 1.9 cm. This distance equals nearly half the length of the pudendal nerve. On these bases Shafik's pudendal nerve decompression operation was done by dividing the sacrospinous ligament and slitting the opening of the pudendal canal. The pudendal nerve was freed to pass directly from the pelvis to the perineum without crossing the gluteal region, thus saving nearly half of the distance crossed before by the nerve.

The cause of control disorders in CRP is disputed [41, 42]. The present study could shed some light on the genesis of these disorders. The patients of group A, who had fecal incontinence in association with CRP, showed diminished EAS and levator EMG activity as well as significantly prolonged nerve conduction time to the EAS, compared with controls and patients with no fecal incontinence. They also showed perianal hypo- or anesthesia. These investigative results point to pudendal neuropathy as the cause of these manifestations, the motor and sensory manifestations being localized to pudendal nerve distribution.

Patients with complete rectal prolapse (CRP) without fecal incontinence (group 11) showed normal nerve conduction to the EAS in 77.3%; in the remaining 22.7%, the conduction was prolonged. The EMG activity of the EAS was normal in patients with a normal PNTML, and that of the levator ani muscle was reduced in all the patients. These results indicate that pudendal neuropathy had not occurred in the majority of this group. However, the ten patients with prolonged nerve conduction and with reduced EMG activity of the EAS seemed to have an early stage of pudendal neuropathy with subclinical partial denervation of the EAS. These patients had no fecal incontinence at the time of presentation, but we hypothesize that they may develop it later or when pudendal neuropathy becomes more marked. Also, the CRP patients with normal PNTML may develop pudendal neuropathy and fecal incontinence after a variable period of time. The cause of pudendal neuropathy in CRP needs to be clarified.

Pudendal Canal Syndrome and Fecal Incontinence

Fecal incontinence is a distressing proctologic problem. The cause may, for example, be the result of a fistula operation or obstetric injury to the sphincters. However, many patients have no obvious cause and incontinence is considered idiopathic.

Pudendal neuropathy [39], secondary to levator dysfunction syndrome [6] with perineal descent, is one of the causes of fecal incontinence.

The prolonged conduction time in the innervation of the EAS occurring in patients with CRP and fecal incontinence, and not in patients with only CRP, postulates a relationship between incontinence and the prolonged conduction time. The prolonged PNTML seems to result from pudendal neuropathy caused by nerve entrapment in the pudendal canal.

The cause of pudendal neuropathy in CRP needs to be discussed. Under normal physiologic conditions, straining at defecation or during delivery effects levator ani contraction; the muscle becomes elevated and laterally retracted with a resulting widening of the levator hiatus and evacuation of the pelvic organ's contents [43]. An increase in intra-abdominal pressure beyond the physiologic limits, as in chronic constipation or in repeated deliveries with a prolonged second stage, would throw its brunt on the levator ani muscle and the anococcygeal raphe and eventually results in levator subluxation and sagging [6, 43]. This seems to explain the reduced EMG activity of the levator ani muscle in all the patients with CRP, whether continent or not. Likewise, this would support the concept that levator dysfunction could be the primary event in the genesis of CRP [44]. The subluxated and sagged levator ani muscle, lying at a level lower than normal, seems to pull on the pudendal nerve. The stretch involves the distal portion of the nerve that extends from the ischial spine to the muscle as evidenced by the prolonged PNTML. The major part of this portion of the nerve lies in the pudendal canal. The winding of the nerve around the sacrospinous ligament seems to fix the nerve at this point and to expose only the distal portion to stretching. Continuous levator activity would lead to pudendal nerve stretch and traumatization with a resulting neurapraxia or axonotmesis. Being entrapped in the pudendal canal and suffering continuous stretching, the pudendal nerve may develop edema with subsequent nerve compression inside the canal, leading to nerve damage. Eventually entrapment neuropathy and pudendal canal syndrome will occur with a resulting fecal incontinence.

Diagnosis

Main Symptoms of Pudendal Canal Syndrome

Perineodynia

For perineodynia, four situations were encountered: no pain, proctalgia, unilateral pain, and bilateral pain. The effect of surgery was estimated by the patient using one of the following proposals: cured, improved, unchanged, or worsened [20, 45]

Anal Incontinence

For anal incontinence, a four-level ordinal scale was used: no incontinence, gas incontinence, liquid incontinence, and solid incontinence. "Cured" was defined as "no incontinence." The patient was considered "improved" if there was a change of at least one level in the scale going from "solid" to "gas" incontinence. The patient was defined as "worsened" if there was a change of at least one level in the opposite direction [46–48].

Urinary Incontinence

For urinary incontinence, a four-level ordinal scale was used: no incontinence, mild incontinence, moderate incontinence, and severe incontinence. The two types of urinary incontinence, stress and urge incontinence, were evaluated separately even if both were present in the same patient (mixed incontinence). "Cured" was defined as "no incontinence." The patient was considered "improved" if there was a change of at least one level in the scale going from "severe" to "mild" incontinence. If the change observed was in the opposite direction, the patient was considered "worsened" [47, 49].

Erectile Dysfunction Syndrome

Shafik A [47], in 1995, described the results of the treatment of seven patients with neurogenic erectile dysfunction (ED) by pudendal canal decompression that are presented. Patients had penile, perineal, and scrotal hypoesthesia or anesthesia. EMG of the external urethral sphincter and levator ani muscle revealed diminished activity. There were increased bulbocavernosus and pudendal nerve terminal motor (PNTML) latencies. Nocturnal penile tumescence activity was absent. These findings pointed to neurogenic ED due to pudendal canal syndrome (PCS). It is suggested that chronic straining at stool in these patients led to levator subluxation and sagging and to pulling on the pudendal nerve with a resulting entrapment in the pudendal canal, pudendal neuropathy, and PCS. ED results from involvement of the penile and perineal branches of the pudendal nerve. PCS causes ED, which improves with pudendal canal decompression.

Three Clinical Signs of the PCS (Examinations were performed in lithotomy position)

Abnormal Anal or Vulvar Sensibility

Sensibility was tested with a needle comparing the left and the right sides of the vulva and of the skin 2 cm lateral to the anus. The interpretations of the results were done using a four-level ordinal scale: 0 = total anesthesia; 1 = reduced sensibility; 2 = normal sensibility; and 3 = hypersensibility. The levels 0, 1, and 3 were considered as "abnormal sensibility."

Painful Alcock's Canal on Rectal Examination

The pain induced by the palpation of the pudendal canal by rectal examination was evaluated using a seven-level ordinal scale: 0 = no pain; 1 = mild pain; 2 = mild pain with Tinel sign (irradiation of the pain); 3 = moderate pain; 4 = moderate pain with Tinel sign; 5 = severe pain; and 6 = severe pain with Tinel sign. Alcock's canal was considered "painful" if the pain was 4 or more.

Painful "Skin Rolling Test"

Beginning from 5 cm behind the level of the anus, the skin was pinched and then rolled to the front until the skin fold was at the level of the clitoris. The skin rolling test was considered "painful" if it induced a severe pain at least at one level (Fig. 29.7) [50].

The site of the pain is in the perineum and may be anterior (urogenital), posterior (anal), or mixed. Situated in the territory of the pudendal nerve, it is uni- or bilateral and to be distinguished from other regional pains with which it must not be confused (coccydynia, located more posteriorly, neuralgia of the ilioinguinal, iliohypogastric, or genitofemoral nerves). In two-thirds of the cases, women are affected. The character of the pain consists of sensations of burning, of torsion or heaviness, and also of foreign bodies in the rectum or vagina. The pain is piercing and very comparable to acute toothache. The mode of onset is often gradual, but a fall is sometimes provocative; sometimes it is postoperative, especially after orthopedic procedures where a traction table has been used. Pain after repetitive energetic bicycling has led to the term "cyclist's syndrome" [51]. It may be much more indolent and develop gradually over time without a definite provoking factor. Lastly, it may be exacerbated by a regional surgical procedure: proctologic, urologic, or gynecologic. The exacerbation of the pain is then only the patient's awareness of a therapeutic failure.

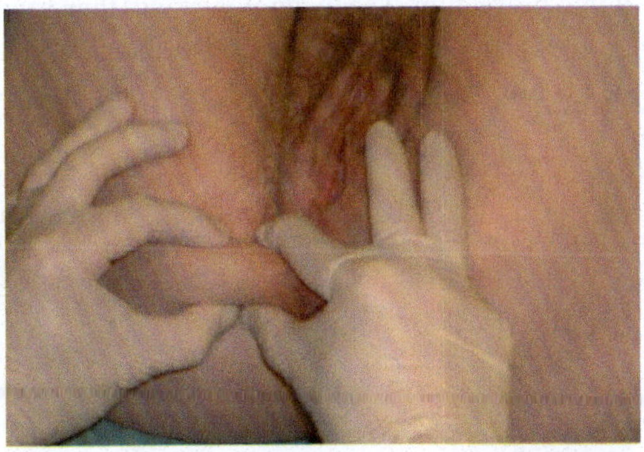

Fig. 29.7 Skin rolling test: the skin of the perineum is pinched just beneath the level of the anus and then rolled to the front searching for a sharp pain at one level. This sign is well known in the diagnosis of neuralgia

The positional nature of the pain is very suggestive. At a certain point in the case history, the seated position provokes or exacerbates the pain. These patients have no pain at night and are comfortable when standing or lying on the non-painful side especially. It is an important point that they have no pain when on the lavatory seat, i.e., when the painful zone is relieved from pressure. The main daily activities requiring the seated position (work, meals, driving, theaters, etc.) are no longer available to these patients, whose mental attitude is one of chronic pain sufferers so obsessed with their miserable state as to be rapidly regarded by their doctors as psychiatric cases.

Clinical Assessment

Perineal sensation is preserved for long, as is muscular trophicity. Urinary disturbances are usually absent, and sexual problems are related to loss of libido resulting from the pain. Rectal examination is painful opposite the ischial spine. Pressure at this level quite often elicits the same type of pain as that felt spontaneously. The striking subjective manifestation is contrast with the sparse examination findings. Supplementary radiological examination certainly excludes any other pathology, especially tumoral. Thus, the neurophysiologist has to attempt to derive organic conclusions from an essentially subjective pathology. History and physical as well as other diagnostic tests will help differentiate between pudendal nerve entrapments versus nerve dysfunction.

For patients with pudendal neuralgia, the patient will describe pain or other nerve dysfunction in accordance with the distribution of the pudendal nerve. The patient may or may not give a history of common triggering factors (i.e., pelvic surgery, trauma, delivery, etc.). Patients will state that sitting increases symptoms and standing decreases symptoms somewhat. On exam, altered skin sensitivity will be noted. Pressure on the pudendal trunk will produce pain (equivalent to Tinel sign). This can be performed both transvaginally and transrectally.

A pudendal nerve block may produce significant or complete pain relief for several hours to several weeks. The block may be used as a diagnostic tool; resultant pain relief demonstrates at least some of the symptoms are stemming from an inflamed nerve. The block may be performed via the transperineal, transvaginal, or transgluteal route with or without radiographic assistance. Finally, electrophysiological evaluation can help confirm the site of entrapment and the type of nerve damage. The studies consist of EMG testing of the external sphincter, sacral reflex, pudendal nerve terminal motor latency (PNTML), and somatosensory-evoked potential studies.

In addition to entrapment, pudendal neuralgia can also be caused by compression or tension dysfunctions. On exam, a patient will still present with a positive Tinel sign and often pelvic floor dysfunction. Specialized physical therapy in conjunction with pudendal nerve blocks can result in significant reductions in pain and can improve function. When the pudendal neuralgia is caused by an actual nerve entrapment, physical therapy and injections alone are often not successful in completely eradicating the problem. In the event conservative management is failing, sacral reflex testing is indicated to confirm or rule out an entrapment. This will determine if the patient requires further conservative management or a surgical decompression.

Pudendal Nerve Terminal Motor Latency (PNTML)

The most widely used method of electrophysiological testing of pudendal nerve function is that described by Kiff and Swash 1984 at St. Mark's Hospital in London. They used a rubber fingerstall that has two stimulating electrodes at the tip and two surface electrodes for recording mounted 3 cm proximally at its base. The index finger, mounted with the device, is inserted into the rectum and placed on the ischial spine. Electrical stimulation is then initiated, and the latency of the response to the anal sphincter is recorded on surface or needle electrodes. The normal mean terminal latency is 2.0 ± 0.3 msec. It must be pointed out that the pudendal nerve terminal motor latency (PNTML) test is solely a motor study and is of importance only if the study is abnormal. In other words, the sensory nerve fiber component of the nerve more peripherally located can be compromised without involving the motor fibers. This anatomical situation can result in a patient with sensory fiber compression and pain having a negative PNTML test. In addition, the test does not indicate the extent of injury or entrapment but only if the nerve is responding abnormally. A comprehensive examination should include sensory nerve tests, as well as testing of the components of motor function and EMG of the pelvic floor.

Patient will be asked to lie on a stretcher, turn to his left side, and bend his knees. An electrode pad (similar to an EKG pad) will be placed on his buttock or thigh. The technician will then put on a rubber glove with an electrode on the index finger (the St. Mark electrode) (Fig. 29.8). After lubricating his index finger, he will gently insert it into his rectum. The technologist will then send a mild, electrical stimulus through the electrode on his finger to the pudendal nerve. This stimula-

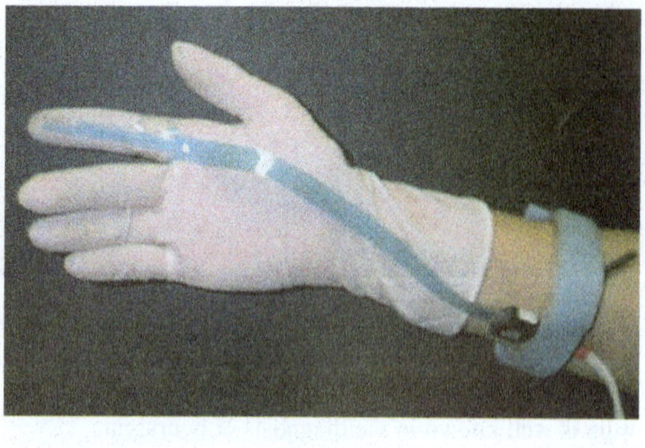

Fig. 29.8 The St. Mark electrode

tion may cause the muscles of the thigh to twitch involuntarily. The technologist will then gently rotate his finger to repeat the test on the opposite branch of the nerve. A computer will record the response of the pudendal nerve to the stimulation. A physician will interpret the results and determine if any nerve conduction delays exist. The actual procedure will take 15–20 minutes.

Rectal Sensation Test

The rectal sensation was done by Lubowski and Nicholls [52]. The rectal balloon was inflated with air with increments of 50 ml every 15 second until its presence was perceived by the patient; this volume was measured by continuing the inflation to a point which produced pain or marked pelvic discomfort.

Rectoanal Inhibitory Reflex (RAIR)

The rectoanal inhibitory reflex (RAIR) was tested by placing a 5 cm latex balloon in the rectum with its distal end 5–6 cm above the anal verge. The balloon was inflated with air in 10 ml increments, while simultaneously recording anal pressure. The minimum volume required to induce anal sphincter relaxation of 20 cm H_2O or more was recorded and called the threshold volume initiating the rectoanal reflex [52].

Anorectal Manometry

Anorectal manometry measures intraluminal pressures generated by the internal and external anal sphincters. The examination is typically performed in the left lateral decubitus position. The catheter is inserted approximately 8 cm into the anal canal, and the rectal pressure is recorded. The catheter is then slowly withdrawn until a rise is noted. This is the beginning of the anal canal high-pressure zone. The distance from the anal verge is recorded, allowing determination of the length of the anal canal. The maximum resting pressure is the maximum pressure generated in the anal canal with the patient at rest. The internal anal sphincter is responsible for 80–85% of the resting tone, with the remainder due to tonic action of the external anal sphincter. Maximum voluntary contraction is determined by asking the patient to squeeze and represents action of the external anal sphincter. In patients who are unable to squeeze in a laboratory setting, the pressure generated during a cough usually accurately reflects squeeze pressure. The presence of the rectoanal inhibitory reflex is determined by inflating a balloon in the rectum, while simultaneously measuring intraluminal pressure in the anal canal. The expected response is a decrease in the resting tone of at least 50%.

Absolute manometric values overlap considerably between "normal" patients and those with disorders of defecation. The values are useful, however, when taken in the clinical context. They are also useful in predicting and measuring responses to therapy.

Defecography

The dynamics of defecation can be assessed radiographically using cinedefecography [53]. The patient is placed on a radiolucent commode and the rectum filled with a thick barium paste. The patient is then asked to squeeze, cough, strain, and then empty the rectum while lateral fluoroscopic views of the pelvis are obtained. Static images are useful for determining length of the anal canal; the anorectal angle at rest, during squeezing, and during evacuation; and degree of perineal descent.

Defecography qualitatively assesses puborectalis function and adequacy of rectal emptying. It also permits real-time visualization of rectocele, enterocele, and internal rectal nerve. A paradoxical contraction of the puborectalis muscle, leading to obstructive defecation, is identified as anismus.

Electromyography

Concentric needle EMG was done at rest and during voluntary contraction on both sides of the external anal sphincter and on each bulbocavernosus muscle. The richness of the EMG was grossly evaluated using a six-level scale: 1 = simple; 2 = poor; 3 = intermediate poor; 4 = intermediate; 5 = intermediate rich; and 6 = rich.

Electromyography (EMG) of the external anal sphincter and pelvic floor can be performed using either needle or surface electrodes. Needle EMG can be either concentric needle EMG or single-fiber EMG. Concentric needle EMG records the summed electrical activity of approximately 30 motor units. It has been used primarily to map sphincter defects, such as after obstetrical injury. Single-fiber EMG uses a fine, 25-(Mu) m electrode. It is able to record the action potential from a single muscle fiber and is able to quantify fiber density. An increase in fiber density occurs with denervation due to sprouting from neighboring nerve fibers. Single-fiber EMG can also be used to measure neuromuscular "jitter," which is an electromyographic indication of denervation and reinnervation [54]. EMG has the disadvantage of being uncomfortable for the patients, and it requires considerable technical expertise on the part of the examiner.

Treatment

Pudendal Nerve Perineal Injections (PNPI)

Several authors describe the use of pudendal blocks to relieve chronic, nonmalignant pelvic pain. Use the technique of Bensignor, giving a series of three monthly transgluteal, pudendal nerve perineural injections (PNPI) of bupivacaine 0.25%, 6 ml, and triamcinolone 40 mg, 1 ml. Perineural anesthesia may be diagnostic or therapeutic. Two PNPI are given into the interligamentary space at ischial spine, and one is given into the pudendal canal. Two hours after injection, examination of six sites with pinprick detects analgesia or hypalgesia (clitoris (glans), labia (posterior scrotum), or perianal area, bilaterally).

Injections may be guided by palpation, fluoroscopy, EMG stimulation ultrasound, or CT guidance. Complete pudendal anesthesia correlates with a good therapeutic response. Our patients monitor symptom indices weekly for 14 weeks.

Results

- Symptom relief after PNPI may last hours, days, or weeks. They may completely resolve after one, two, or three PNPI. Bensignor indicated control of neuralgia in 70% of patients at 6 months. Amarenco reported a 15% response at 12 months. One of early patients continues to have durable relief for over 4 years including relief of serious pain, urinary retention, and erectile dysfunction.
- Symptom changes include improved erections, decrease of pain following ejaculation, increased vaginal lubrication, pain-free intercourse, improved orgasms, improved defecation, and other suitable improvements. One group with persistent pain is a subset of patients, with previous pelvic surgery complicated by urine leakage.
- Injections into Alcock's may not anesthetize the inferior anal nerve because that branch exits the main trunk proximal to Alcock's canal in 50% of cadavers.
- PNPI can be repeated and have successful responses 2 years after the second series.

Complications of PNPI are infrequent and include pain "flares" that may last several days; bleeding through the needle requires repositioning of the needle; sciatic nerve anesthesia may cause transient gait disturbance. Penetration of the pudendal nerve by the needle apparently occurred in three of our patients causing significant aggravation of pudendal pain requiring up to 3 months for pain resolution. Incontinence of urine or flatus may occur for 1 or 2 hours after PNPI (Fig. 29.9).

Physical Therapy

Musculoskeletal dysfunctions can cause pudendal neuralgia as well as other painful pelvic syndromes. Physical therapy is an effective method of minimizing or eliminating the concurrent pain

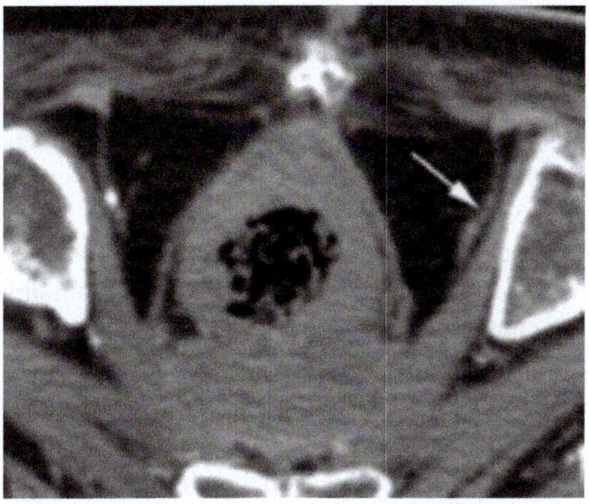

Fig. 29.9 Arrow identifies the site of PNPI injection

generators that occur when the pudendal nerve is irritated (i.e., pelvic floor hypertonicity and myofascial trigger points, extrapelvic hypertonicity and trigger points, adverse neural tension, sacroiliac joint dysfunctions, connective tissue restrictions, and faulty neuromuscular recruitment patterns). It is important to acknowledge this interaction between musculoskeletal and neural dysfunction as it is unusual that one exists without the other.

Physical therapists require special training to treat pudendal neuralgia. The therapist should have a strong manual therapy bias and an extensive working knowledge of pudendal neuralgia. The program should emphasize restoring normal length to the pelvic floor (through internal myofascial release) and pelvic floor relaxation techniques. Typically, the shortened pelvic floor/pudendal neuralgia will become symptomatically exacerbated with Kegel exercises, and these should be avoided until otherwise instructed by a professional. The program should also include connective tissue mobilization, neural mobilization, and a home exercise program.

Surgical Treatment

It is currently available (worldwide) for nerve decompression. The theory is similar to other nerve decompression procedures performed for nerve entrapments in other regions of the body (i.e., carpal tunnel release). The procedures differ in their approach to the area of entrapment and have never been compared head to head.

Minimal criteria for surgery:

At least one of the three following symptoms resistant to conservative treatments (physiotherapy, drugs, infiltrations, modification of diet or behavior):

1. Anal incontinence
2. Perineodynia
3. Urinary incontinence

Associated with at least two of the five following criteria:

1. Increased anal or perineal PNTML
2. Pathological EMG of the anal sphincter or bulbocavernosus muscles (neurogenic trace; reduced activity, richness "poor" or "simple").
3. Painful Alcock's canal on rectal examination (at least on one side)
4. Abnormal perineal sensibility (at least at one level)
5. Painful "skin rolling test" (at least on one side)

Surgical Techniques

Perineal Approach

In the perineal approach described by Prof Ahmad Shafik, this involves a vertical para-anal incision. The anesthesia was either spinal or general and the patient was put in the lithotomy position. A vertical para-anal incision, 2 cm from the anal orifice, was done, and the ischiorectal fossa was entered and the pad of fat pushed medially. Keeping close to the lateral wall of the fossa, dissection proceeded along the obturator fascia identifying the pudendal canal backward to the ischial spine. The neurovascular bundle was palpated with the forefinger along its course in the pudendal canal. The inferior rectal nerve was identified as it passes transversely across the ischiorectal fossa, stretched by the index finger and followed to the pudendal canal. The fascia of the pudendal canal was incised along its whole length, and the pudendal nerve was freed away from the canal till the sacrospinous ligament which was divided in some cases. Hemostasis was secured and the wound was closed after being drained. The same procedure was repeated on the pudendal nerve in the other side (Figs. 29.10, 29.11, 29.12, 29.13, 29.14, and 29.15).

Transgluteal (TG) Approach

First described by Robert et al., this approach involves a gluteal incision overlying the ischial tuberosity. The sacrotuberous ligament is identified and windowed over a 2–3 cm segment. The neurovascular bundle is then identified and traced to the entrance to Alcock's canal. This overlies the sacrospinous ligament, which is also incised at the level of the ischial spine. The overlying fascia of Alcock's canal is also incised as is the remnants of the falciform process if necessary. The nerve is then transposed in front of the ischial spine.

The procedure has the advantage of good access and visualization. All areas of possible entrapment are dealt with. The procedure is limited by the necessary sacrificing of both the sacrospinous and sacrotuberous ligament in some patients. There have been reports indicating that the transgluteal approach has been modified in some patients so that the sacrotuberous ligament is not compromised. Particularly for those patients with sacroiliac joint dysfunction, it is imperative that the sacrotuberous ligament is intact. If a patient is considering the transgluteal technique, sacroiliac joint dysfunction should be

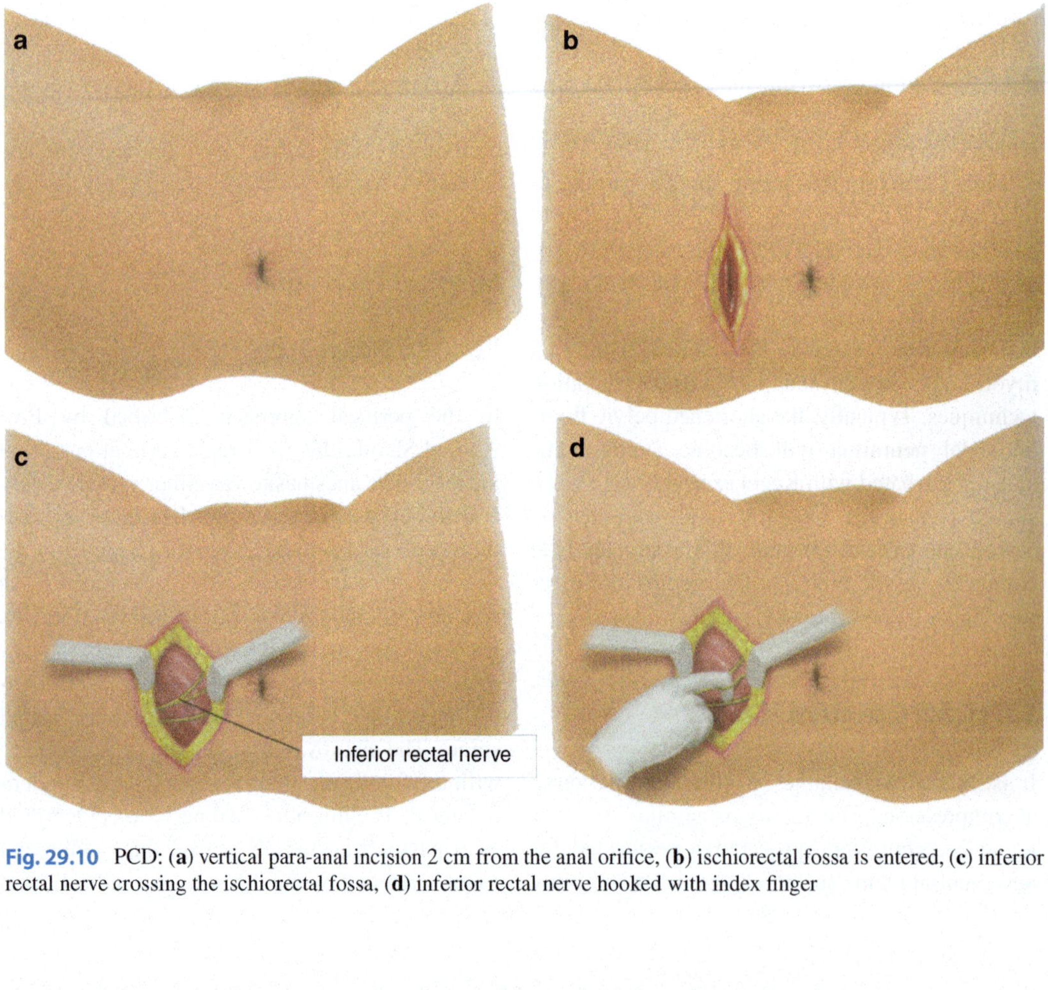

Fig. 29.10 PCD: (**a**) vertical para-anal incision 2 cm from the anal orifice, (**b**) ischiorectal fossa is entered, (**c**) inferior rectal nerve crossing the ischiorectal fossa, (**d**) inferior rectal nerve hooked with index finger

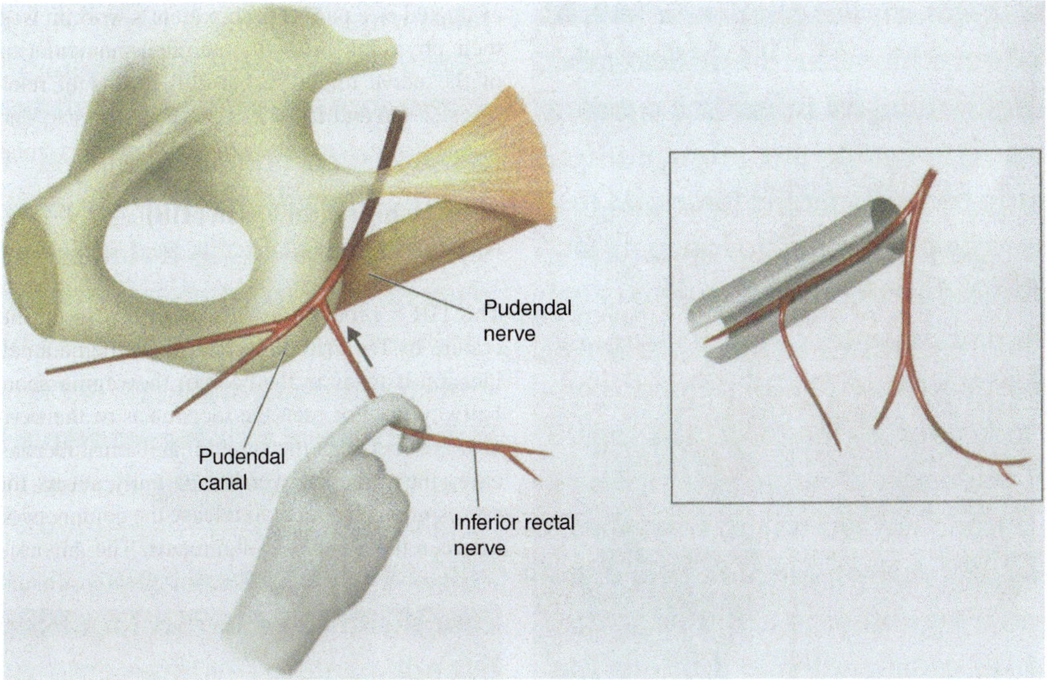

Fig. 29.11 Inferior rectal nerve stretched by index finger and followed to PN. Inset: PN in- and outside the PC and the fascia of PC incised

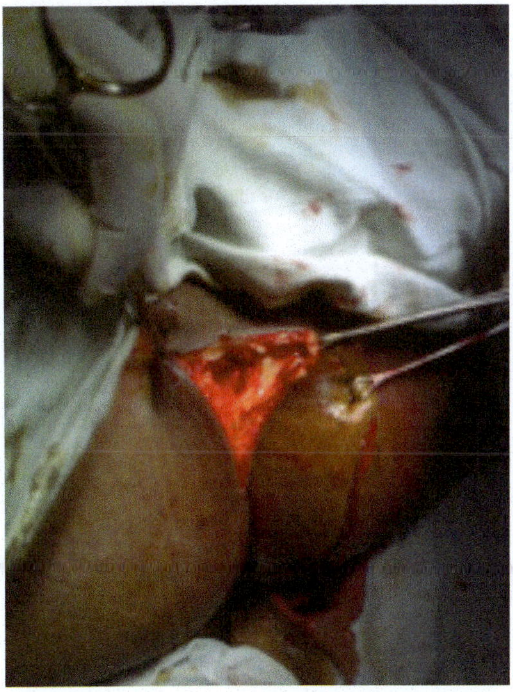

Fig. 29.12 Vertical para-anal incision 2 cm midway between ischial tuberosity and anal verge

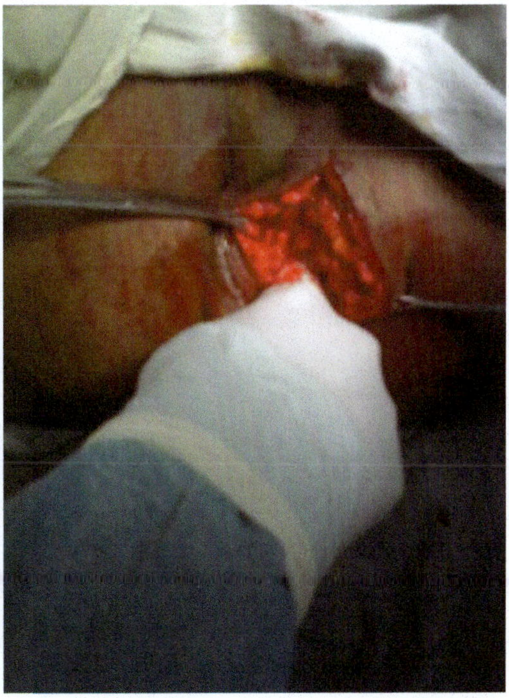

Fig. 29.13 Exposure of ischiorectal fossa and digital dissection to identify the inferior rectal nerve

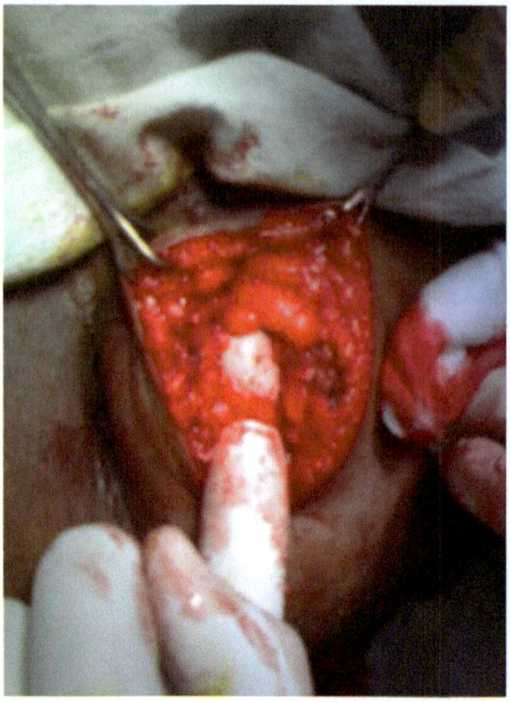

Fig. 29.14 Inferior rectal nerve hooked with the index finger

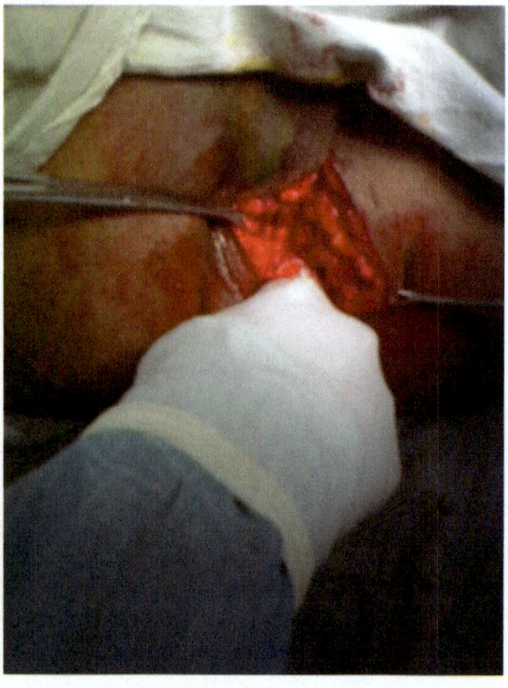

Fig. 29.15 Digital following of the inferior rectal nerve to pudendal nerve and blunt dissection of Alcock's canal

evaluated as a part of each patient's workup with their physician. Also, the surgical manipulation of the nerve might be harmful in and of itself (Figs. 29.16 and 29.17).

Transischiorectal Fossa (TIR) Approach

The TIR approach was first described in the literature by Dr. Eric Bautrant; for women a small incision is made in the back of the vagina about halfway up. For men the incision is in the perineal area between the scrotum and anus. In most cases the surgeon severs or partially severs the sacrospinous ligament to release the compression between the ST and SS ligaments. The Alcock's

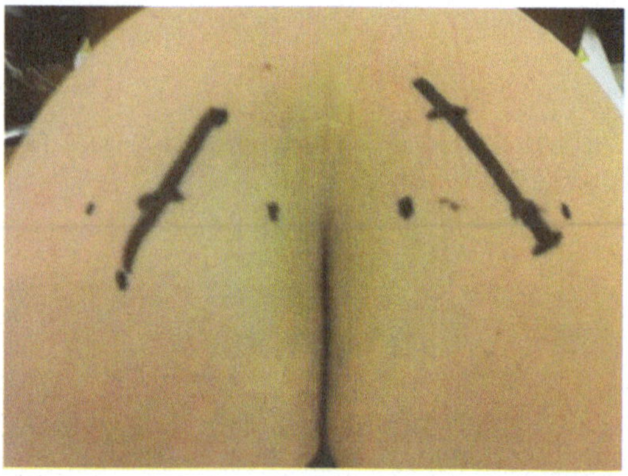

Fig. 29.16 Skin marks for the incisions

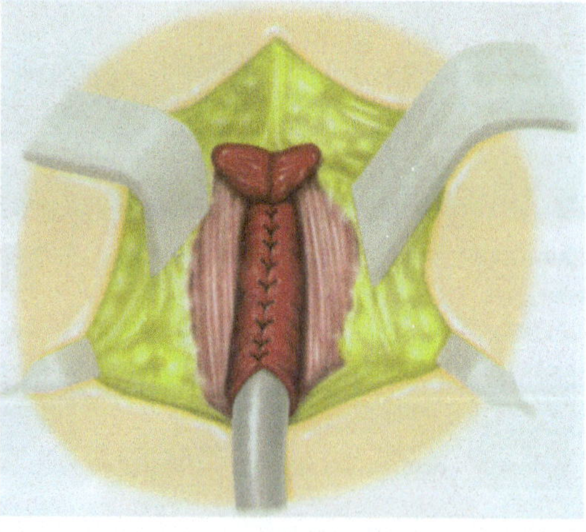

Fig. 29.17 Operative field for the transgluteal approach

canal is explored by the surgeon's finger and the nerve released from any fascia that might be tethering it.

The rectum is retracted medially and the sacrospinous ligament is identified. The pudendal nerve is tested using a stimulating electrode on a fingerstall above the ischial spine. A needle electrode is placed in the external anal sphincter. PNTML and electrical potential surface area are measured. The sacrospinous ligament is divided progressively. Bipolar cautery is employed for hemostasis. If necessary, the falciform extension of the sacrotuberous ligament is partially divided. The operator then can digitally explore the canal caudally to ensure that the nerve is completely free. Nerve testing is repeated several times during the procedure. Once a normal result is returned, the dissection is complete. A drain is placed and the incision is closed.

The procedure has the advantage of good access to all of the surgically amenable areas of entrapment. The complication rate is low (1–2.5% for hemorrhage and surgical site infection). The intraoperative nerve testing limits dissection to what is necessary to relieve entrapment. This increases the safety and efficacy of the procedure. The procedure avoids direct manipulation of the nerve, thus reducing the risk of surgical nerve damage. Finally, complete sectioning of the sacrotuberous ligament is avoided. Disadvantages include a smaller field of vision during the procedure. Operators must be familiar with the approach via the vagina or perineum (in men). Expertise in intraoperative nerve testing is not widespread.

Laparoscopic Pudendal Nerve Decompression and Transposition (LaPNDT) Approach

Erdogru et al. [55] described laparoscopic pudendal nerve decompression and transposition (LaPNDT). Ports are inserted as shown in Figs. 29.18 and 29.19. The external iliac vein is identified, and the peritoneum is incised between the ureter and external iliac vein. Blunt dissection is used to create a peritoneal window

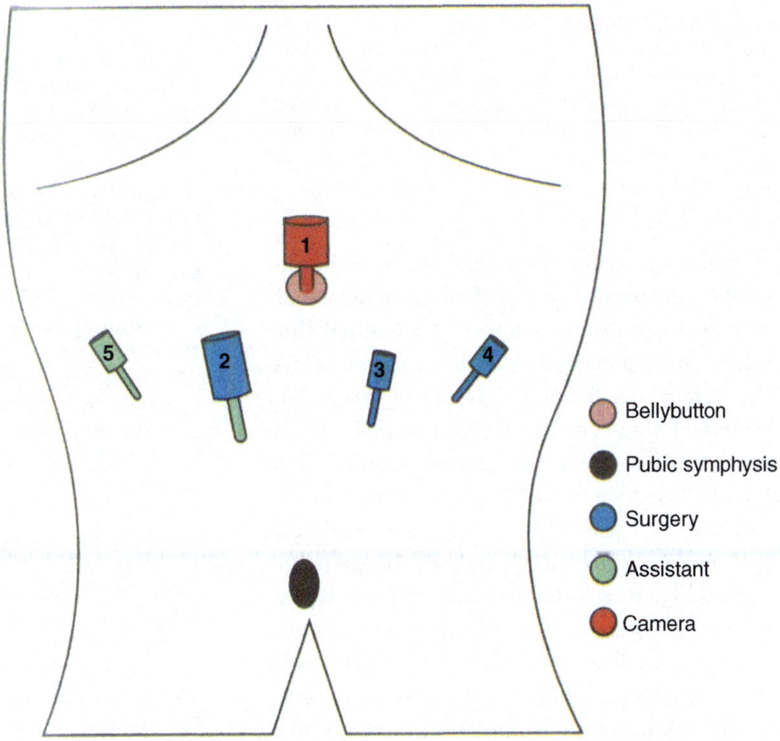

Fig. 29.18 Locations of the trocars as assistance, surgeon, and camera of the laparoscopic pudendal nerve decompression and transposition

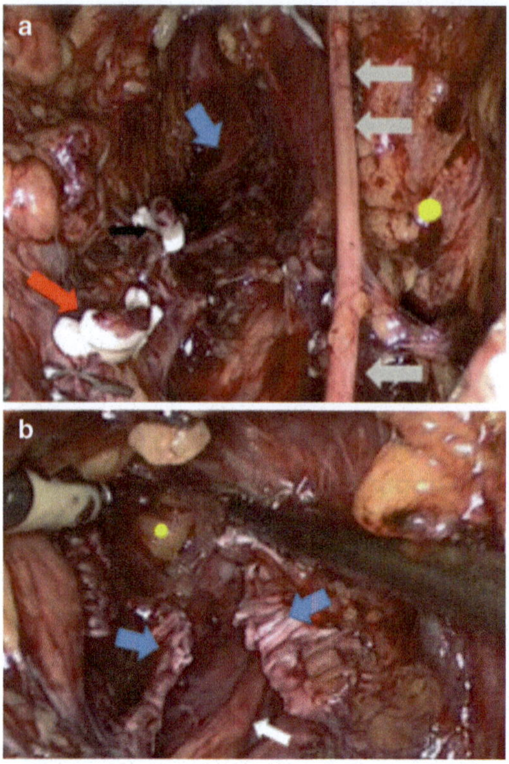

Fig. 29.19 (**a**) Exposure of the sacrospinous ligament (SSL) from an internal perspective and the lateral deep pelvic topographic anatomy. Blue arrow, the right SSL. Black arrow, divided right obturator vein. Red arrow, divided medial umbilical ligament. Gray arrows, right obturator nerve. (**b**) With division of SSL, pudendal nerve clearly identified beneath the ligament and the fatty tissue in front of Alcock's canal entrance. Blue arrows, divided right SSL. White arrow, the right pudendal nerve. Yellow dot, the fatty tissue in front of Alcock's canal entrance

medial to the obturator nerve. The inner border of the peritoneal layer is then retracted medially to allow visualization of the internal iliac artery, vein, and arcus tendineus fascia pelvis. For better visualization, division of the medial umbilical ligament as the first branch of the internal iliac artery and partial internal iliac lymph dissection might be required; all fatty tissue has to be removed over the arcus tendineus fasciae. The arcus tendineus fascia of the pelvis is incised and retracted medially to fully expose the lateral border of both the internal iliac artery and the vein. These vessels might then be mobilized and traced distally, to the posterior border of the ischiococcygeus muscle, if there is a narrow space such as in an obese patient. Following these structures distally allows exposure of the sacrospinous ligament and the lateral deep pelvic topographic anatomy from an internal perspective. Using 5 mm scissors, the SSL is then completely divided, with retraction and protection of the pudendal nerve with an endo dissector. Dissection of the pudendal nerve continues to the proximal entrance to Alcock's canal, with splitting of the inner side of the levator ani muscles to reach the fatty tissue in front of the canal entrance and inner part of the aponeurosis of the internal obturator muscle. An omental flap to protect the nerve against scar tissue development around it,

Summary

Pudendal neuropathy is a rare and challenging condition. Treatment should be based on a detailed evaluation, selecting the best option for each specific case.

References

1. Whitehead WE. Functional anorectal disorders. Semin Gastrointest Dis. 1996;7:230–6.
2. Staude G. Proctalgia fugax. Differential diagnosis and therapy of fleeting anal cramp. Fortschr Med. 1992;110:278–80.
3. Peery WH. Proctalgia fugax: a clinical enigma. South Med J. 1988;81:621–3.
4. Ger GC, Wexner SD, Jorge JM, Lee E, Amaranath LA, Heymen S, Nogueras JJ. Evaluation and treatment of chronic intractable rectal pain-a frustrating endeavor. Dis Colon Rectum. 1993;36:139–45.
5. Whitehead WE, Wald A, Diamant NE, Enck P, Pemberton JH, Rao SS. Functional disorders of the anus and rectum. Gut. 1999;45 Suppl 2:1155–9.
6. Shafik A. A new concept of the anatomy of the anal sphincter mechanism and the physiology of defecation. XVIII. The levator dysfunction syndrome. A new syndrome with report of 7 cases. Coloproctology. 1983;5:158–65.
7. Rockefeller R. Digital dilatation for relief of proctalgia fugax. Am Fam Physician. 1996;54:72–6.
8. Eckardt VF, Dodt O, Kanzler G, Bernhard G. Anorectal function and morphology in patients with sporadic proctalgia fugax. Dis Colon Rectum. 1996;39:755–62.

9. Barros P, Pamo O. Digestive disorders in the adult population of Lima. Rev Gastroenterol Peru. 1989;9:131–8.
10. White WB, Hanna M, Stewart JA. Systemic herpes simplex virus type 2 infection. Proctitis, urinary retention, arthralgia, and meningitis in the absence of primary mucocutaneous lesions. Arch Intern Med. 1984;144:826–7.
11. Pilling LF, Swenson WM, Hill JR. The psychologic aspects of proctalgia fugax. Dis Colon Rectum. 1965;8:372–6.
12. Gilliland R, Hevman JS, Altomare DF, et al. Biofeedback for intractable rectal pain: outcome and predictors of success. Dis Colon Rectum. 1997;40:190–6.
13. Peleg R, Shvartzman P. Low-dose intravenous lidocaine as treatment for proctalgia fugax. Reg Anesth Pain Med. 2002;27:97–9.
14. Katsinelos P, Kalomenopoulou M, Christodoulou K, Katsiba D, Tsolkas P, Pilpilidis I, Papagiannis A, Kapitsinis I, Vasiliadis I, Souparis T. Treatment of proctalgia fugax with botulinum A toxin. Eur J Gastroenterol Hepatol. 2001;13:1371–3.
15. Waldman SD, Wilson WL, Kreps RD. Superior hypogastric plexus block using a single needle and computed tomography guidance: description of a modified technique. Reg Anesth. 1991;16:286–7.
16. Whitehead WE. Behavioral medicine approaches to gastrointestinal disorders. J Consult Clin Psychol. 1992;60:605–12.
17. Shafik A. Pudendal Canal Syndrome. Description of a new syndrome and its treatment. Report of 7cases. Coloproctology. 1991;13:102–9.
18. Jünemann KP, Jecht E, Müller-Mattheis V, Weiske WH, Wetterauer U, Baccouche M, Lühmann R, Zentgraf M, Bähren W, Altwein JE. [Open multicenter study of the differential diagnosis of erectile dysfunction with a papaverine-phentolamine combination (BY023)]. Urologe. 1988;27:2–7.
19. Robert R, Brunet C, Faure A, Lehur PA, Labat JJ, Bensignor M, Leborgne J, Barbin JY. Surgery of the pudendal nerve in various types of perineal pain: course and results. Chirurgie. 1993;119:535–9.
20. Robert R, Prat-Pradal D, Labat JJ, Bensignor M, Raoul S, Rebai R, Leborgne J. Anatomic basis of chronic perineal pain: role of the pudendal nerve. Surg Radiol Anat. 1998;20:93–8.
21. Shafik A, Doss SH. Pudendal canal: surgical anatomy and clinical implications. Am Surg. 1999;65:176–80.
22. Sinnatamby CS. Last's anatomy, regional and applied, eleventh edition. New York: Churchill Livingstone; 2006.
23. Williams PL, Warmic R. True pelvis, pelvic floor and premium editors. Gray's anatomy. 39th, ed. New York: Churchill Livingstone; 2005.p 1357–1371.
24. Loukas M, Louis RG Jr, Hallner B, Gupta AA, White D. Anatomical and surgical considerations of the sacrotuberous ligament and its relevance in pudendal nerve entrapment syndrome. Surg Radiol Anat. 2006;28:163–9.
25. Vleeming A, Stoeckart R, et al. The sacrotuberous ligament: a conceptual approach to its dynamic role in stabilizing the sacroiliac joint. Clin Biomech. 1989;4:200–3.
26. Goodgold J, Eberstein A. Electro diagnosis of neuromuscular diseases. Baltimore: Williams & Wilkins Company; 1987. p. 129.
27. Hollinshead WH, Rosse C. Textbook of anatomy. 4th ed. Philadelphia: Harper and Row Publishers; 1985. p. 365–86.
28. Kapandji IA. The physiology of joints. 2nd ed. London: Churchill Livingstone; 1970. p. 68.
29. Steindler A. Kinesiology of the human body. 4th ed. Springfield, II: Charles C. Thomas Publisher; 1976. p. 296.
30. Elbadawi A, Schenk EA. A new theory of the innervation of bladder musculature. 4. Innervation of the vesicourethral junction and external urethral sphincter. J Urol. 1974;111:613–5.
31. Bradley WE. Innervation of the male urinary bladder. Urol Clin North Am. 1978;5:279–93.
32. de Araujo CG, Schmidt RA, Tanagho EA. Neural pathways to lower urinary tract identified by retrograde axonal transport of horseradish peroxidase. Urology. 1982;19:290–5.
33. Brindley GS, Polkey CE, Rushton DN. Sacral anterior root stimulators for bladder control in paraplegia. Paraplegia. 1982;20:365–81.
34. Thüroff JW, Bazeed MA, Schmidt RA, Wiggin DM, Tanagho EA. Functional pattern of sacral root stimulation in dogs. II. Urethral closure. J Urol. 1982;127:1034–8.
35. Smith AR, Gilpin SA, Gosling JA, Warrell DW. The pathogenesis of genitourinary prolapse and stress urinary incontinence of urine: a histological and histochemical study. Br J Obstet Gynaecol. 1989;96:15–23; Schmidt RA, Chan HP, Kodera Y, Doi K, Chen CT. Evaluation of cassette performance: physical factors affecting patient exposure and image contrast. Radiology. 1983;146:801–6.
36. Snooks SJ, Swash M. Perineal nerve and transcutaneous spinal stimulation: new methods for investigation of the urethral striated sphincter musculature. Br J Urol. 1984;56:406–9.
37. Kiff ES, Swash M. Slowed conduction in the pudendal nerves in idiopathic (neurogenic) fecal incontinence. Br J Surg. 1984;71:614–6.
38. Kiff ES, Barnes PR, Swash M. Evidence of pudendal neuropathy in patients with perineal descent and chronic straining at stool. Gut. 1984;25:1279–82.
39. Kiff ES, Swash M. Normal proximal and delayed distal conduction in the pudendal nerves of patients with idiopathic (neurogenic) faecal incontinence. J Neurol Neurosurg Psychiatry. 1984;47:820–3.
40. Shafik A. The posterior approach in the treatment of pudendal canal syndrome. Coloproctology. 1992;14:310–5.
41. Shafik A. A new concept of the anatomy of the anal sphincter mechanism and the physiology of defeca-

tion. VIII. Levator hiatus and tunnel: anatomy and function. Dis Colon Rectum. 1979;22:539–49.
42. Porter NH. A physiological study of the pelvic floor in rectal prolapse. Ann R Coll Surg Engl. 1962;31:379–404.
43. Shafik A. A new concept of the anatomy of the anal sphincter mechanism and the physiology of defecation. XII. Anorectal mobilization: a new surgical access to rectal lesions. Preliminary report. Am J Surg. 1981;142:629–35.
44. Shafik A. Urethral discharge, constipation, and hemorrhoids. New syndrome with report of 7 cases. Urology. 1981;18:155–60.
45. Miloiu A. In: Georges Abraham CS, editor. Perineodynia concept. Bucarest (Roumania); 2002.
46. Keighley MR, Fielding JW. Management of faecal incontinence and results of surgical treatment. Br J Surg. 1983;70:463–8.
47. Shafik A. Stress urinary incontinence: an alternative concept of pathogenesis. Int Urogynecol J. Pelvic Floor Dysfunction. 1994;5:3–11.
48. Ganio E, Ratto C, Masin A, Luc AR, Doglietto GB, Dodi G, Ripetti V, Arullani A, Frascio M, Berti Riboli E, Landolfi V, Del Genio A, Altomare DF, Memeo V, Bertapelle P, Carone R, Spinelli M, Zanollo A, Spreafico L, Giardiello G, de Seta F. Neuromodulation for fecal incontinence: outcome in 16 patients with definitive implant. The initial Italian Sacral Neurostimulation Group (GINS) experience. Dis Colon Rectum. 2001;44:965–70.
49. Snooks SJ. Faecal incontinence. Geriatr Nurs (Lond). 1985;5(4):4–5.
50. Maigne R. Low back pain of thoracolumbar origin. Arch Phys Med Rehabil. 1980;61:389–95.
51. Amarenco G, Le Cocquen-Amarenco A, Kerdraon J, Lacroix P, Adba MA, Lanoe Y. Perineal neuralgia. Presse Med. 1991;20:71–4.
52. Lubowski DZ, Nicholls RJ. Faecal incontinence associated with reduced pelvic sensation. Br J Surg. 1988;75:1086–8.
53. Broden B, Snellman B. Procidentia of the rectum studied with cineradiography: a contribution to the discussion of causative mechanism. Dis Colon Rectum. 1968;11:330–47.
54. Gee AS, Durdley P. Neurophysiology of the pelvic floor. Semin Colon Rectal Surg. 1996;7:129–36.
55. Erdogru T, Avci E, Akand M. Laparoscopic pudendal nerve decompression and transposition combined with omental flap protection of the nerve (Istanbul technique): technical description and feasibility analysis. Surg Endosc. 2014;28:925–32.

Anorectal and Pelvic Floor Pain Syndromes

30

Mauro Pereira de Azevedo

Introduction

Pain is a very important symptom that works toward the preservation of vital integrity. It advises against threats to the body and enables a response to avoid them, decreasing the potential harm. Pain also promotes psychological and physiological reactions, variable according to the intensity and duration of injury [1]. Prompt pain control is very important to avoid adverse outcomes and the perpetuation of pain as a disease. Intense or continuous stimulation of the nervous system can elicit an altered response in nervous cells, by inducing modifications in the function or the structure of nervous cells, also at peripheral and central levels, a phenomenon called neuroplasticity. This elicits pain memory, pain independent of the existence of an actual lesion [2]. This is chronic pain.

The International Association for the Study of Pain (IASP) defines pain as "an unpleasant sensory and emotional experience associated with actual or potential tissue damage or described in terms of such damage." According to IASP, chronic pain is defined as pain which remains for 3 months or more after a clinical or biological event, regardless of the event, basically, pain that remains beyond the end of the healing process [3, 4]. Chronic pain is a serious public health problem, with a prevalence of over 37% of the world's population suffering from some kind of chronic pain, according to a study by the WHO (World Health Organization) [5].

Pain physiology has been largely studied, and we know that the initiation of pain occurs in the peripheral nociceptors. Nociceptors are silent receptors or receptors with a high threshold for activation, located in peripheral and profound tissues, and they are the basic unit for the initiation of pain stimulus. This nociceptor activation occurs after a painful stimulation, which can be mechanical, thermal, electrical, or chemical in nature. After stimulation, chemical mediators (protons, chemokines, prostaglandins, bradykinin, and others) are activated, which sensitize the nociceptors, reducing its threshold for activation. The chemical stimulus is transformed into an electrical potential, which is conducted through a nerve fiber to the spinal cord, where it will undergo the process of modulation. Modulation aims to control the influx of stimulus and its response. This process initially takes place in the spinal cord and is influenced by higher structures of the brain – serotoninergic and noradrenergic systems of pain modulation. After modulation in the spinal cord, the nociceptive stimulus ascends via the spinothalamic and spinoreticular pain tracts in the spinal cord to areas where it will be modulated, integrated, and localized, after which the feeling of pain will be expressed with

M. P. de Azevedo (✉)
Marcilio Dias Navy Hospital - Brazil, Rio de Janeiro, Brazil

Rio de Janeiro Society of Anesthesiology, Rio de Janeiro, Brazil

Brazilian Society of Anesthesiology, Rio de Janeiro, Brazil

L. C. C. Oliveira (ed.), *Anorectal Physiology*, https://doi.org/10.1007/978-3-030-43811-1_30

an organic response to the trauma (affective, discriminative, and endocrine-metabolic response). These responses serve to restore damaged tissues. If the stimulus is strong or long enough, pain can be modified into a pathological pain, which exists independent of the existence of a visible lesion.

A thorough evaluation of pain characteristics and the knowledge of its physiopathology and, if possible, the origin of pain is of paramount importance. After an appropriate diagnosis, the most suitable treatment can be selected, including:

1. *Behavioral and lifestyle changes* – sedentary lifestyle, alimentation, overweight or obesity, and ergonomic aspects (in work and daily life)
2. *Physiotherapy methods* – massages, perineal muscle strengthening and stretching exercises, shock waves, biofeedback retraining, and electrogalvanic stimulation
3. *Medications*, including common analgesics, anti-inflammatory and muscle relaxants for low-intensity pain, and weak opioids for moderate pain and stronger opioids for strong pain, according to the OMS analgesic ladder
4. *Use of adjuvant drugs*, including antidepressant and anticonvulsant drugs, particularly in neuropathic pain
5. *Invasive procedures* when the above methods do not yield relief
6. *Spinal cord modulation or implant of an infusion pump with analgesic drugs* for refractory cases

It is quite rare to use all these options. We must be very careful with the use of opioid medication when dealing with benign pain, mostly because of the fear of addiction, a current public health epidemic in North America [6].

Pain is a personal experience, and its treatment must be individualized, in accordance with certain basic principles: [1]

- It must always be *multimodal therapy* [7], that is, the use of a combination of drugs and analgesic techniques (pharmacological or not) that acts through diverse pathways producing a synergetic or addictive effect.
- It must always be *preventive therapy* [8, 9], that is, no matter when the drug or therapy (before or after the lesion) is used, we must endeavor to prevent the onset of pain.
- It must always be *multidisciplinary*, because pain is always multifactorial.
- It must avoid the use of opioid analgesics, because there is a risk of addiction and worsening pain (collateral effects of constipation, nauseas, and vomits) [10, 11].

In dealing with perianal syndrome disorders, the purpose is to find a treatable cause. If this is not possible, symptom relief with medications and physiotherapy methods is the goal.

Anorectal and Pelvic Floor Pain Syndromes

In the human pelvis, several anatomical structures and organs can generate pain of organic or functional nature. When dealing with these conditions, the approach must be multidisciplinary. The IASP definition of chronic pelvic pain states that it can have multiple origins: muscular, neurological, urological, gynecological, or anorectal, and, depending on the patient, can have different diagnoses or different management plans. According to the IASP, chronic anal pain is included in the group *chronic pelvic pain syndromes (CPPS)*, in the axis of gastrointestinal pain origins, which includes irritable bowel syndrome, beyond the chronic pain of anorectal origin [12, 13].

CPPS is defined as a "chronic or persistent pain perceived* in structures related to the pelvis of either men or women. It is often associated with negative cognitive, behavioral, sexual, and emotional consequences as well as with symptoms suggestive of lower urinary tract, sexual, bowel, pelvic floor, or gynecological dysfunction."

*"Perceived indicates that the patient and clinician, to the best of their ability from the history, examination, and investigations (where appropriate), have localized the pain as being perceived in the specified anatomical pelvic area."

Andromanakos et al. [14] noted that, "chronic perineal pain is not a disease but an idiopathic multifactorial vague disorder, a syndrome of a complex interaction between neurological, musculoskeletal, and endocrine systems that is further affected by behavioral and psychological factors."

Chronic or recurrent pain in the anal canal, rectum, or other pelvic organs occurs in 7–24% of the general population. In 85% of cases, there is no identifiable organic cause. These chronic painful syndromes course with a hypervigilant state, due to central sensitization. There is an increased response of nociceptors to painful stimulation, producing the urgency to defecate, postprandial fullness, gastric acid reflux, and reduction in pain threshold, in such a way that non-painful stimulation becomes painful, like the act of defecation (allodynia) [15]. Rectum or sigmoid distention provokes abnormal visceromuscular reflexes. Hyperactivity of the pelvic floor can lead to a visceral dysfunction, with urinary urgency, intestinal constipation, or dyspareunia. One common characteristic of CPPS is the *excessive tension (spasm) of striated musculature in the pelvic floor* [16].

When dealing with patients complaining of perianal pain, the first step is to define an organic cause, such as hemorrhoids, anal fissure, anal fistula, rectal ulceration, neoplastic disease, sacrococcygeal cyst, and pain from other origins like the urogenital tract, lumbar and sacral spine, pudendal neuralgia, piriformis syndrome, or gynecological disease. If an organic cause cannot be identified, there should be a high suspicion for

Table 30.1 Rome criteria for functional gastrointestinal disorders [17]

A. Esophageal disorders	
A1. Functional chest pain	A4. Globus
A2. Functional heartburn	A5. Functional dysphagia
A3. Reflux hypersensitivity	
B. Gastroduodenal disorders	
B1. Functional dyspepsia	B3. Nausea and vomiting
B1a. Postprandial distress syndrome (PDS)	B3a. Chronic nausea vomiting syndrome (CNVS)
B1b. Epigastric pain syndrome (EPS)	B3b. Cyclic vomiting syndrome (CVS)
B2. Belching disorders	B3c. Cannabinoid hyperemesis syndrome (CHS)
B2a. Excessive supragastric belching	B4. Rumination syndrome
B2b. Excessive gastric belching	
C. Bowel disorders	
C1. Irritable bowel syndrome (IBS)	C2. Functional constipation
IBS with predominant constipation (IBS-C)	C3. Functional diarrhea
IBS with predominant diarrhea (IBS-D)	C4. Functional abdominal bloating/distension
IBS with mixed bowel habits (IBS-M)	C5. Unspecified functional bowel disorder
IBS unclassified (IBS-U)	C6. Opioid-induced constipation
D. Centrally mediated disorders of gastrointestinal pain	
D1. Centrally mediated abdominal pain syndrome (CAPS)	
D2. Narcotic bowel syndrome (NBS)/opioid-induced GI hyperalgesia	
E. Gallbladder and sphincter of Oddi (SO) disorders	
E1. Biliary pain	
E1a. Functional gallbladder disorder	
E1b. Functional biliary SO disorder	
E2. Functional pancreatic SO disorder	
F2: Functional Anorectal Pain	
F2a. Levator Ani Syndrome	
F2b. Unspecified Functional Anorectal Pain	
F2c. Proctalgia Fugax	

a functional anorectal disorder. Thus, this is an exclusion diagnosis.

In 2016, the Rome III criteria for functional gastrointestinal disorders diagnosis was reviewed and updated as the Rome IV criteria, as can be seen in Table 30.1. The criteria for diagnosis are the same as version III.

F2. Functional Anorectal Pain

F2a. Levator Ani Syndrome

Diagnostic Criteria

Symptoms of chronic proctalgia associated with painful discomfort during posterior traction of the puborectalis muscle

F2b. Unspecified Functional Anorectal Pain

Diagnostic Criteria

Symptoms of chronic proctalgia but without painful discomfort during posterior traction of the puborectalis muscle

F2c. Proctalgia Fugax

Diagnostic Criteria

Must include *all items* below:

1. Recurrent episodes of localized pain in the anus or lower rectum.
2. These episodes last seconds to few minutes.
3. Absence of anorectal pain between the painful episodes.

Chronic anorectal pain can also be an associated symptom in patients with advanced prolapse of the posterior compartment, which often presents as a functional defecation disorder such as obstructed defecation or fecal incontinence, according to Hompes et al. [18]. Some authors include other pain diseases when studying perianal pain, given the similarity in presentation and frequency, including coccygodynia, myofascial syndromes, and piriformis syndrome [14].

Pelvic floor muscle dysfunction is the leading cause of chronic perineal painful syndromes [19]. The pelvic floor is formed by several muscles, ligaments, and fascia that provide the support to the pelvic viscera and maintain fecal and urinary continence, defecation, and sexual function and assist in childbirth. The pelvic diaphragm is shaped by the levator ani and coccygeus muscles. The levator ani muscle is a complex formed by three muscles: pubococcygeus, puborectalis, and iliococcygeus that are innervated by branches from the sacral plexus (S2–S4). The piriformis and obturator internus muscles lay in the inner portion of the pelvis, whose function is related to the hip joint. Dysfunction of these muscles, such as in piriformis syndrome, can cause perianal pain that must be considered in the differential diagnosis of chronic perianal syndromes [19].

The levator ani muscle and the muscles that compose the anus and urethral sphincter are in a unique condition of continuous tonic activity, relaxing only during defecation or urination; this requires complex sensorial and motor coordination. Excessive tension (spasm) in the striated musculature of the pelvic floor seems to be a common feature in pelvic painful syndromes [16].

The pelvic organs and musculature work together; their innervation arises from the same spinal segment and conveys to the same superior centers. This innervation is accompanied by fibers from the autonomic nervous system, which regulates organ function. Therefore, when the pelvic viscera are stimulated, autonomic reactions occur.

Chronic pain, regardless of the cause, is accompanied by central sensitization, a phenomenon that increases and tends to perpetuate the disease. It does not necessarily begin with a peripherical lesion; it may be derived from a deviation in the function of the central nervous system, which can lead to persistent hyperexcitability (neuroplasticity in the dorsal horn of the spinal cord). Pain becomes the disease, regardless of whether there is a causative factor. If this abnormal behavior is not interrupted, emotional aspects become more marked, and a vicious cycle is established (Fig. 30.1).

Chronic anorectal pain is typically localized on the medial aspect of the anal canal, generally deep and intense, and worsens when the individ-

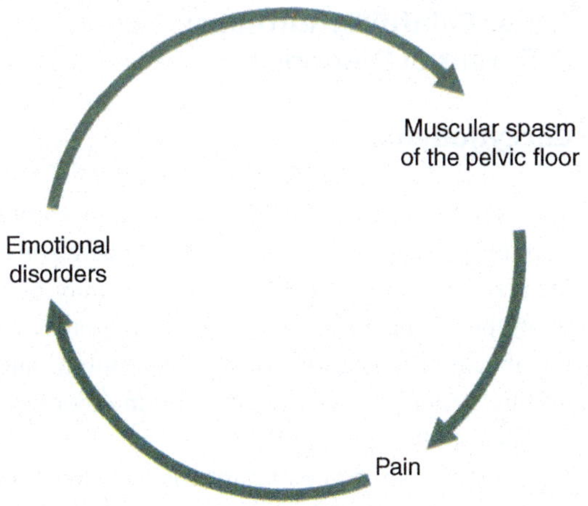

Fig. 30.1 Anorectal pain cycle

ual sits; pain can be irradiated to adjacent areas. Usually there is no disturbance in the sleep pattern, because pain is typically relieved at bedtime. Again, it is very important to exclude other causes of anorectal pain prior to defining it as a functional pain.

Levator Ani Syndrome (LAS)

LAS is characterized by chronic or recurrent burning anorectal pain that radiates to the gluteal region. It appears episodically and lasts from at least 20 minutes to several hours or even days. Pain worsens in the seated position or during defecation and can be relieved in the lying position. Patients report a sensation of pressure or foreign body in the anal region and may also complain of fecal or urinary incontinence [20].

LAS is usually idiopathic but can be associated with other causes including pelvic lesions, childbirth, physical activities, prolonged sitting position, obesity, and surgical or diagnostic procedures. The incidence of LAS is higher in women, as are other dysfunctional diseases such as fibromyalgia. There are usually associated physiological alterations such as depression and anxiety. Some patients have reported a history of sexual abuse.

The mechanism of LAS is linked to an alteration in the coordination between anorectal muscles and pelvic floor muscles, which can be demonstrated by electromanometry and defecography. Digital rectal examination causes a painful stimulation of the levator ani muscle, which can be relieved by anesthetizing the pudendal nerve. This confirms the engagement of this nerve and striated musculature in the physiopathology of pain. It also contributes to hormonal changes and the aging process that leads to the failure of the pelvic floor and accentuates perineal prolapse. Some patients may complain of pain in sites other than the anorectal region such as in the perineum, coccyx, or sacrum.

LAS is a diagnosis of exclusion: alterations noted during digital rectal examination and exclusion of other causes of pain by appropriate investigations, especially imaging tests.

Treatment aims at muscle relaxation by digital massage and sitz baths with warm water. Analgesics, antidepressant, anticonvulsant, anxiolytic, corticosteroids, and laxative medications also help to control pain. Biofeedback has shown promising results.

Sacral neuromodulation is an option when other methods of analgesia fail [21]. Botulinum toxin injected into the internal anal sphincter can relieve symptoms in some patients with functional anal pain [22, 23]. Surgical treatment is not indicated due to the high risk of complications, especially fecal incontinence [14].

Unspecified Functional Anorectal Pain

The Rome classification defines this kind of pain as that with similar characteristics to LAS but without pain to the posterior traction of the puborectalis muscle. According to Chiarioni et al. [16], this group can be heterogenous as some patients may develop constipation while others may develop pelvic pain.

Proctalgia Fugax (PF)

PF is a benign pain that originates from spastic contractions of the puborectalis muscle (paroxysmal), with an acute characteristic (oppressive) and with a sudden onset and end. The pain is located in the anus or lower rectum, lasts a few minutes, and is often related to a spasm of the anal sphincter. Sometimes the pain is so intense that the patient feels faint.

PF is more commonly seen in females, with a prevalence of 8–18% in the general population. It can be precipitated by sexual activity, masturbation, stress, intestinal constipation, defecation, or menstruation, but more often there is no identifiable precipitating factor. It is associated with other functional pathologies such as anxiety disorder or irritable bowel syndrome [24]. More recently, electromyographic evaluation of the pelvic floor has demonstrated a potential association with a lesion on the distal portion of the pudendal nerve [25].

Two aspects must be observed in patients with PF: one is its association with stress, emotional or physical, which leads to consequent augmentation of the intrinsic activity of the smooth muscle of internal anal sphincter; the other aspect is its high prevalence (about 52%) in patients with irritable bowel syndrome, another functional pathology. According to Jeyarajah et al. [24], PF is, by nature, a pathology with a high psychoneurotic component precipitated by anxiety. These patients present with a higher incidence of depression, neuroticism, psychological disorders, and pain, when compared with patients who have an organic pathology such as anal fissure.

Diagnosis is made by exclusion criteria, and treatment aims to control the precipitating factor and uses methods of pelvic floor muscle relaxation. Pharmacological and non-pharmacological methods can be used, including antispasmodics, topical nitroglycerin, oral diazepam, calcium channel blockers, salbutamol inhalation, local injection of type A botulinum toxin, neural blockade, antidepressants, sitz baths with warm water, biofeedback, internal anal sphincter massage, and psychotherapy. It is important the therapy must be individualized to each patient. There is little evidence to indicate surgical treatment, which is restricted to intervention of the internal anal sphincter but only when it is more than 3.5 mm, confirmed by endoanal ultrasonography. As most painful episodes last only a few minutes, most therapies are unable to act in a reasonable timeframe, except in more severe cases. Therapy must be driven to avoid recurrence (reduction of the frequency) by offering physiotherapy such as biofeedback (high success rate of 85%) and psychotherapy. In some cases, the eventual use of antidepressant medication may be helpful [26–28].

Other Painful Syndromes: Differential Diagnosis

Coccygodynia

This syndrome is related to spastic pain and/or burning sensation accompanied by edema and localized primarily in the coccyx but may also be in the perineum, gluteus, lumbar region, anorectal region, posterior aspect of the thigh, and urinary system. It is more common in older age, females, and obese people.

Etiology may be post-traumatic – dislocation and/or distention of the sacrococcygeal vertebra or ligaments, fracture by fall, labor, prolonged sitting position, pelvic floor musculature spasm, congenital disease, neoplasia, infection, and degenerative arthritis, among others. One third of these cases are idiopathic.

Pain worsens with standing or sitting for long periods of time and is relieved when laying down. Sexual activity and defecation may also worsen the pain.

Digital rectal examination with pressure over the coccyx (reproduce the pain) and X-ray can be used for diagnosis [29, 30].

Initially, treatment must be conservative with analgesics; anti-inflammatory medications and, eventually, opioid medications in more severe cases; and adjuvant medications (muscle-relaxing agents, antidepressants, or anticonvulsants). It is indicated to use padding when sitting, massage therapy, physiotherapy, ultrasound, sitz baths with warm water, and pelvic floor muscle-relaxing techniques. In some cases, infiltration with local anesthetics or corticosteroids can help to relieve the pain. When all types of treatment have failed and pain intensifies, surgical removal of the coccyx may be indicated [31].

Pudendal Neuralgia

Also known as pudendal channel syndrome, pudendal neuralgia is derived from compression (long time in the sitting position, constant use of bikes or motorcycles), stretch (vaginal delivery, severe intestinal constipation), or direct trauma over the pudendal nerve in the trajectory between the sacrotuberal and sacrospinal ligaments, which causes a burning sensation over the gluteal, perineal, or genital region. It is a neuropathic pain, similar to carpal tunnel syndrome.

According to the Nantes Group in France, the essential features to diagnose this syndrome include the following:

1. Pain must be limited to the region innervated by the pudendal nerve.
2. Pain occurs especially in the sitting position.
3. Pain rarely wakes up the patient during rest.
4. During clinical examination, there is no sensitive alteration identifiable even in the presence of paresthesia.
5. Pain improves after pudendal nerve blockade.

The treatment is pudendal nerve blockade (sometimes more than one treatment is needed) with concomitant use of analgesics, anti-inflammatory drugs, anticonvulsants, or antidepressants. Eventually the use of a spinal neurostimulator may be needed (Fig. 30.2) [15, 32].

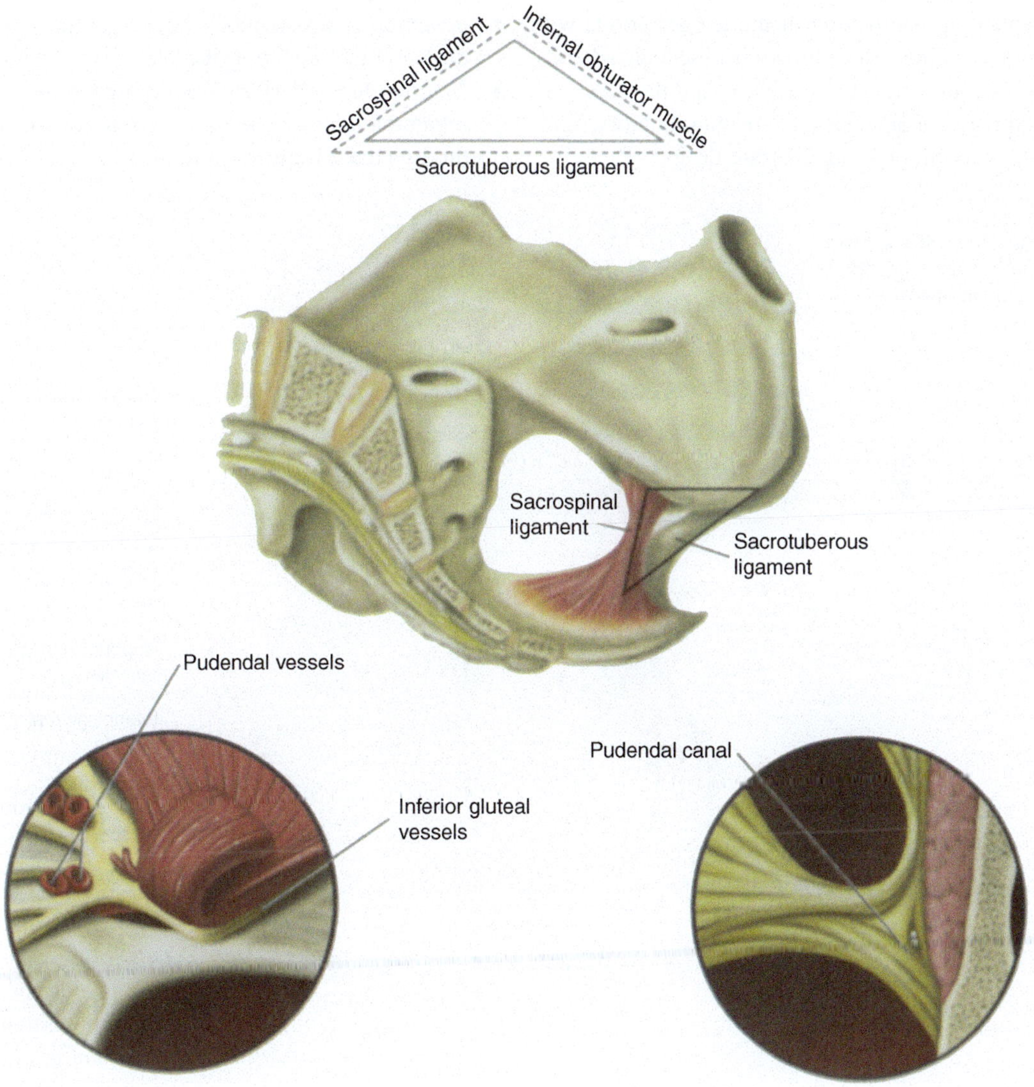

Fig. 30.2 Pudendal nerve path [33]

Myofascial Syndrome

Myofascial syndrome is characterized by a stabbing or deep muscle pain, with sensitive trigger points in the muscles, which deflagrate the pain. This pain is exacerbated by movement and a sitting position. It is derived from a state of tension or chronic or repeated contraction of the pelvic floor muscles, which leads to development of such trigger points (microtraumas in the muscle). Diagnosis is made by palpation and identification of painful areas in the anal lifter musculature and the presence of taut bands, reproducing the pain feature [15].

Treatment must be multidisciplinary and is aimed at inactivation of the trigger points with dry needling or needling with local anesthetics, digital massage, biofeedback, and the concomitant use of analgesic, anti-inflammatory, anticonvulsant, or antidepressant drugs.

Piriformis Syndrome

This syndrome is characterized by pain in the gluteal or perineal region or rectal pain during defecation, with radiation to the ipsilateral leg or hip [34]. It is due to sciatic nerve trapping as it traverses the piriformis muscle (Fig. 30.3), which is linked to hip movement. It can resemble carpal tunnel syndrome.

Pain usually occurs after a muscle trauma in the gluteal region and develops a piriformis muscle lesion [35].

Piriformis syndrome has two components:

- *Somatic* – corresponds to a myofascial syndrome of the piriformis muscle
- *Neuropathic* – derived from compression or inflammation of the sciatic nerve in its course through the infrapiriform foramen

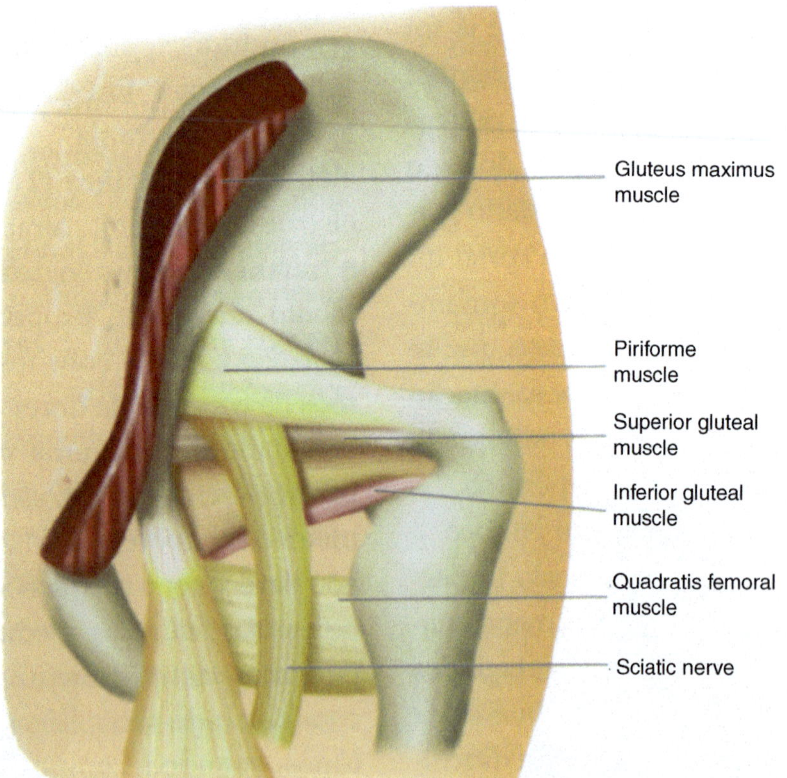

Fig. 30.3 Sciatic nerve, piriformis muscle, and adjacent structural anatomy

Clinical Approach to the Pain in the Chronic Anorectal Syndromes

Frequently, the origin of pain is known or suspected as the patient may likely have previously sought care. It is important to evaluate the intensity of pain and, more importantly, the impact that pain has had in the patient's daily life. When dealing with chronic pain, especially with a high degree of emotional/psychological involvement, one of the main focuses of treatment is to restore the patient's social and professional life. There is typically a vicious cycle of depression, pain, and social isolation that must be disrupted. Of course, pain treatment is of utmost importance, either by pain killers, adjuvant medications, or physiotherapy. Psychological therapy is very helpful to alleviate depression and stress symptoms.

Opioid medication in chronic benign pain is very debatable. The fear of addiction, tolerance, hyperalgesia, and other side effects like severe constipation (narcotic bowel syndrome) [36, 37] can worsen pain syndrome. In rare cases, when pain is too severe or incapacitating and when all therapies have failed, sacral neurostimulation may control the pain. Ablative procedures in the sacral plexus or the ganglion impar may also be an option.

Summary

Chronic pain syndrome is complex and multifactorial and demands a multidisciplinary team to deal with it. It presents as an anatomical, psychological, physiological, and traumatic entity. It is difficult to diagnose and demands a complex and prolonged treatment. It is important that patients be treated on an individual basis, using a physiological and physiotherapeutic approach.

References

1. Bajwa ZH. Evaluating the patient with chronic pain. In: Bajwa ZH, Warfield CA, editors. Principles & practice of pain medicine. 2nd ed. New York: McGraw-Hill; 2004. p. 61–8.
2. Scott MF, Jane CB, James PR, eds. Philadelphia: Wolters Kluwer/Lippincott, Williams and Wilkins; 2010.
3. IASP Taxonomy Pain. Available from: https://www.iasp-pain.org/Taxonomy#Pain.
4. Apkarian AV, Baliki MN, Geha PY. Towards a theory of chronic pain. Prog Neurobiol. 2009;87(2):81–97.
5. Tsang A, Von Korff M, Lee S, Alonso J, Karam E, Angermeyer MC, et al. Common chronic pain conditions in developed and developing countries: gender and age differences and comorbidity with depression-anxiety disorders. J Pain. 2008;9(10):883–91.
6. Soelberg CD, Brown RE Jr, Du Vivier D, Meyer JE, Ramachandran BK. The US opioid crisis: current federal and state legal issues. Anesth Analg. 2017;125(5):1675–81.
7. Kohring JM, Orgain NG. Multimodal analgesia in foot and ankle surgery. Orthop Clin North Am. 2017;48(4):495–505.
8. Katz J, Clarke H, Seltzer Z. Review article: preventive analgesia: quo vadimus? Anesth Analg. 2011;113(5):1242–53.
9. Pogatzki-Zahn EM, Zahn PK. From preemptive to preventive analgesia. Curr Opin Anaesthesiol. 2006;19(5):551–5.
10. Nicol AL, Hurley RW, Benzon HT. Alternatives to opioids in the pharmacologic management of chronic pain syndromes: a narrative review of randomized, controlled, and blinded clinical trials. Anesth Analg. 2017;125(5):1682–703.
11. Vallejo R, Barkin RL, Wang VC. Pharmacology of opioids in the treatment of chronic pain syndromes. Pain Physician. 2011;14(4):E343–60.
12. IASP. Classification of chronic pain, second ed. (Revised). Available from: https://www.iasp-pain.org/PublicationsNews/Content.aspx?ItemNumber=1673. Available from: https://s3.amazonaws.com/rdcms-iasp/files/production/public/Content/ContentFolders/Publications2/ClassificationofChronicPain/Part_II-F.pdf.
13. Treede RD, Rief W, Barke A, Aziz Q, Bennett MI, Benoliel R, et al. A classification of chronic pain for ICD-11. Pain. 2015;156:1003–7.
14. Andromanakos NP, Kouraklis G, Alkiviadis K. Chronic perineal pain: current pathophysiological

aspects, diagnostic approaches and treatment. Eur J Gastroenterol Hepatol. 2011;23(1):2–7.
15. Messelink B, Baranowski A, Hughes J. Abdominal and Pelvic Pain, from definition to best practice. In: IASP, editor. Abdominal and Pelvic Pain, from definition to best practice. Washington, DC: IASP Press; 2015. p. 213.
16. Chiarioni G, Asteria C, Whitehead WE. Chronic proctalgia and chronic pelvic pain syndromes: new etiologic insights and treatment options. World J Gastroenterol. 2011;17(40):4447–55.
17. Drossman DA. Functional gastrointestinal disorders: history, pathophysiology, clinical features and Rome IV. Gastroenterology. 2016;150(6):1262–79.
18. Hompes R, Jones OM, Cunningham C, Lindsey I. What causes chronic idiopathic perineal pain? Color Dis. 2011;13(9):1035–9.
19. Finamore PS, Goldstein HB, Whitmore KE. Pelvic floor muscle dysfunction: a review. Female Pelvic Med Reconstr Surg. 2008;14(6):417–22.
20. Mazza L, Formento E, Fonda G. Anorectal and perineal pain: new pathophysiological hypothesis. Tech Coloproctol. 2004;8(2):77–83.
21. Govaert B, Melenhorst J, van Kleef M, van Gemert WG, Baeten CG. Sacral neuromodulation for the treatment of chronic functional anorectal pain: a single center experience. Pain Pract. 2010;10(1):49–53.
22. Zhang Y, Smith CP. Botulinum toxin to treat pelvic pain. Toxicon. 2017;147:129–33.
23. Bibi S, Zutshi M, Gurland B, Hull T. Is Botox for anal pain an effective treatment option? Postgrad Med. 2016;128(1):41–5.
24. Jeyarajah S, Chow A, Ziprin P, Tilney H, Purkayastha S. Proctalgia fugax, an evidence-based management pathway. Int J Color Dis. 2010;25(9):1037–46.
25. Damphousse M, Jousse M, Verollet D, Guinet A, Le Breton F, Lacroix P, et al. Evidence of pudendal neuropathy in Proctalgia Fugax: perineal neurophysiological assessment in 55 patients. Prog Urol. 2012;22(4):220–4.
26. Atkin GK, Suliman A, Vaizey CJ. Patient characteristics and treatment outcome in functional anorectal pain. Dis Colon Rectum. 2011;54(7):870–5.
27. Rao SS, Bharucha AE, Chiarioni G, Felt-Bersma R, Knowles C, Malcolm A, et al. Functional anorectal disorders. Gastroenterology. 2016;130(5):1510–8.
28. Armananzas L, Arroyo A, Ruiz-Tovar J, Lopez A, Santos J, Moya P, et al. Chronic idiopathic anal pain. Results of a diagnostic-therapeutic protocol in a colorectal referral unit. Cir Esp. 2015;93(1):34–8.
29. Fogel GR, Cunningham PY 3rd, Esses SI. Coccygodynia: evaluation and management. J Am Acad Orthop Surg. 2004;12(1):49–54.
30. Kleimeyer JP, Wood KB, Lonne G, Herzog T, Ju K, Beyer L, et al. Surgery for refractory coccygodynia: operative versus nonoperative treatment. Spine (Phila Pa 1976). 2017;42(16):1214–9.
31. Hanley EN, Ode G, Jackson Iii BJ, Seymour R. Coccygectomy for patients with chronic coccydynia: a prospective, observational study of 98 patients. Bone Joint J. 2016;98-b(4):526–33.
32. Carmel M, Lebel M, et al. Pudendal nerve neuromodulation with neurophysiology guidance: a potential treatment option for refractory chronic pelvi-perineal pain. Int Urogynecol J. 2010;21(5):613–6.
33. Zundert V, Ostheimer GW. Pain relief and anesthesia in obstetrics. New York: Churchill Livingstone; 1996. p. 381.
34. Jankovic D, Peng P, van Zundert A. Brief review: piriformis syndrome: etiology, diagnosis, and management. Can J Anaesth. 2013;60(10):1003–12.
35. Miller TA, White KP, Ross DC. The diagnosis and management of piriformis syndrome: myths and facts. Can J Neurol Sci. 2012;39(5):577–83.
36. Farmer AD, Gallagher J, Bruckner-Holt C, Aziz Q. Narcotic bowel syndrome. Lancet Gastroenterol Hepatol. 2017;2(5):361–8.
37. Drossman D, Szigethy E. The narcotic bowel syndrome: a recent update. Am J Gastroenterol Suppl. 2014;2(1):22–30.

Neurogenic Bowel and Bladder Dysfunction

31

Lucia Camara Castro Oliveira and
Eliane Garcez da Fonseca

Introduction

Neurogenic bowel and bladder dysfunction (NBBD) is the broad term used to describe alterations in the fecal and urinary continence process secondary to a damage in the central nervous system (CNS), autonomic nervous system (ANS), and/or peripheral nervous system (PNS) [1]. The neurological conditions that most frequently lead to NBBD in adults can be seen in Table 31.1. In children NBBD is predominantly associated with congenital neural tube defects but can also be related to acquired causes (Table 31.2).

In the last decades, the importance of early investigation and proactive management of neurogenic bladder (NBl) has been well established and emphasized. Neurogenic bladder can cause severe and irreversible renal damage. The early institution of clean intermittent catheterization combined with anticholinergics has led to renal damage prevention and made urinary continence possible. The bowel and bladder have similar neurologic control, and neurogenic dysfunctions of these tracts are often associated. Neurogenic bowel (NBo) patients present with constipation and fecal incontinence (FI). Constipation is associated with recurrent urinary tract infection and worsening in bladder function. In addition, fecal incontinence is a debilitating condition, associated with decrease in quality of life, self-esteem, and social interaction [2, 3].

Historically, constipation and fecal incontinence have been left in a second plan during the management of patients with NBBD [4]. However, because the associated cost to manage these conditions is considerable and increasing, it has gained more importance in the last years. New approaches

Table 31.1 Causes of neurogenic bowel

Diabetes mellitus
Parkinson disease
Trauma with spinal cord injury
Amyotrophic lateral sclerosis
Spina bifida
Myelomeningocele
Multiple sclerosis
Stroke

Table 31.2 Causes of neurogenic bowel and bladder in children

Myelomeningocele
Lipomeningocele
Sacral agenesis
Tethered cord
Anorectal malformation
Spinal cord tumors
Trauma with spinal cord injury
Transverse myelitis

L. C. C. Oliveira (✉)
Director Department of Anorectal Physiology of Rio de Janeiro and CEPEMED, Rio de Janeiro, Brazil

E. G. da Fonseca
Polyclinic Piquet Carneiro, Department of Pediatrics, Rio de Janeiro, Brazil

L. C. C. Oliveira (ed.), *Anorectal Physiology*, https://doi.org/10.1007/978-3-030-43811-1_31

and techniques to control constipation and fecal incontinence have been developed, creating new therapeutic options for these patients. In this chapter, we will discuss the clinical aspects of neurogenic bowel and bladder dysfunction with focus on neurogenic bowel.

Pathophysiology

The structures of the lower urinary tract are in close relation to the anus and rectum, being in close proximity to each other. The embryological origin of the bladder and rectum is the same, and both are innervated by autonomic and somatic nerves with similar central neurological control [5–7]. The difference between the lower bowel tract and the lower urinary tract is that the bowel tract has an enteric nervous system [8]. Interactions between the two organ systems have been increasingly recognized, and there is a coordinated activity between them. Voiding can occur without defecation, and the initiation of micturition often precedes that of defecation, even if both organs are considered equally full. The filling status of the bladder influences sensation in the rectum and vice versa, and the potential for mutual influence in pathology is emerging. Many severe pathophysiologic mechanisms are involved in neurogenic bowel, affecting patients with motor incomplete spinal cord injury similarly to those of patients with motor complete lesions with spinal sacral reflexes [9]. Upper motor neuron damage results in loss of voluntary control, maintained reflex activity in the anorectum, increased colonic transit time, and constipation. In these cases, the anal sphincter muscles can be affected (hypertonia) or not. Lower motor neuron damage results in loss of voluntary control, loss of reflex activity in the anorectum, prolonged transit time, constipation and rectal impaction, and a hypotonic anal sphincter. Emotional and behavioral changes can occur with cerebral damage, impacting on the individual's ability to manage their bowel function; systemic changes can affect bowel function through spasticity, fatigue, muscle weakness, and reduced mobility [10].

Clinical Presentation

Patients usually present loss of bladder and bowel control: urinary and fecal incontinence or urinary retention, constipation and fecal impaction, fecal accidents, and urinary tract infections. In addition, neurogenic bowel dysfunction can lead to skin and decubitus ulcers, rectal bleeding, or prolapse with a great impact in quality of life. Patients can present with hemorrhoids or anal fissures. These conditions can be very painful and lead to trigger autonomic dysreflexia in susceptible patients. Autonomic dysreflexia can be commonly observed and manifested by rapidly developing severe headache and other signs such as flushing, sweating, and blotchiness above the lesion. It is a medical emergency, as it can lead to potentially life-threatening hypertension in individuals with an injury above the sixth thoracic vertebrae.

Classification

Supraconal Disorder: Upper Motor Neuron Bowel Syndrome or Hyperreflexic (Spastic) Bowel (Table 31.3)

Patients who have disease/injury above the conus medullaris involving the loss of supraspinal inhibitory input present with spastic pelvic floor and non-relaxing muscles with associated constipation as a result of a non-coordinated defecation process. The increase in colonic wall, pelvic

Table 31.3 Bedside signs to differentiate supraconal and infraconal disorders

Supraconal disorder (hyperreflexic bowel)	Infraconal disorder (areflexic bowel)
High resting anal tone	No/low resting anal tone
Anal reflex present – reflex contraction of anus in response to stroking of anal skin	Anal reflex absent
Bulbocavernosus reflex present – reflex anal contraction on squeezing the glans penis/clitoris	Bulbocavernosus reflex absent

floor, and anal tone results in reduced colonic compliance, overactive segmental peristalsis, and underactive propulsive peristalsis. Due to the disruption of the afferent and efferent signals between the colon and the brain, patients loose the perception of the presence of feces in the rectum, and bowel accidents can occur. The spastic constricted state of the external anal sphincter (EAS) worsens the situation further by causing retention of stool [11–13].

Table 31.4 Clinical assessment of patients with neurogenic bowel

Previous medical history
Pre-injury bowel habit
Episodes of urinary and fecal incontinence
Time spent in the toilet
Diet and meal frequency
Medications
Levels of activity
Ability to communicate – cognitive status
Home circumstances and psychological factors

Infraconal Disorder: Lower Motor Neuron Type or Areflexic (Flaccid) Bowel

There is a reduction in the motility of the colon, and peristaltic movements are very slow or absent. Patients also present with constipation and absence of desire to defecate. Due to the lack of sensibility, patients can also have loose stools [14].

Patients should have a detailed evaluation (Table 31.4). Abdominal X-rays can help to confirm the retention of stool and the presence of a dilated colon (Fig. 31.1).

Transabdominal ultrasound can help to confirm the dilated rectum in children [15].

Paradoxical contraction and anorectal sensation can be demonstrated by anal manometry. Magnetic resonance defecography can demonstrate anismus.

Nevertheless, in most cases, a digital exam and the assessment of perineal sensitivity by the pinprick reflex are sufficient.

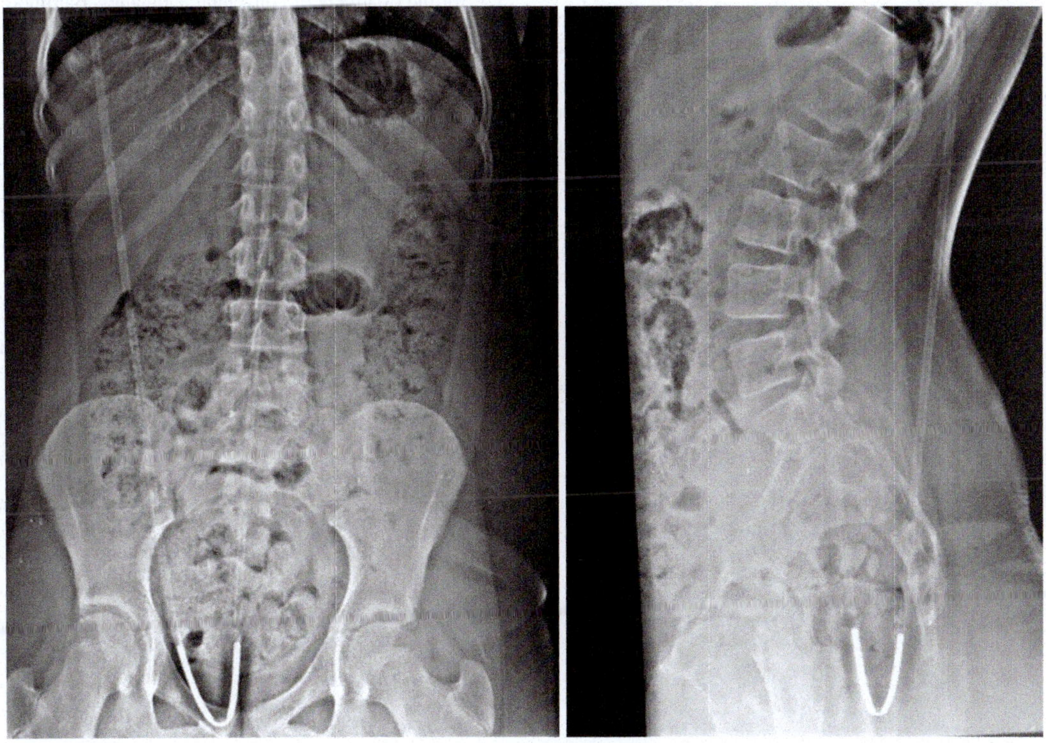

Fig. 31.1 Abdominal plain film in the AP and sagittal plane showing a dilated and redundant colon

Treatment

The goal of the treatment is to improve quality of life and reduce morbidity from decubitus ulcers and urinary tract infections. A multidisciplinary team approach is very important. Usually, a non-surgical management is successful in up to 90% of the cases [14–17].

Treatment options include conservative management with diet and lifestyle modifications, oral laxatives, rectal interventions for evacuation using suppositories, enemas, transanal irrigation, and anal plug. Surgical interventions can be utilized in a stepwise algorithm, including antegrade colonic irrigation with the Malone surgery, sacral neuromodulation, and as a last option, fecal diversion through a colostomy or ileostomy (Table 31.5).

In fact, fecal diversion should be utilized as a last option in these patients as there is an increased risk of complications when compared to other patients [18, 19]. In addition, although stomas can be quite effective in reducing the time spent on bowel care and controlling FI, they do not correct the colonic transit time, and as such, there might be a continual requirement for stoma irrigation or laxative use.

Mosiello et al. proposed a stepped approach to the treatment of bowel dysfunction, based in the invasiveness of the different approaches [20].

Table 31.5 Treatment options for neurogenic bowel patients

Diet and fluid management/lifestyle alterations
Laxatives, lubricants, or constipating and bulking agents
Rectal enemas and suppositories
Biofeedback therapy with electrostimulation
Transanal irrigation
Stimulation of tibial nerve
Sacral neuromodulation
Antegrade colonic conduit – Malone procedure
Sacral anterior root stimulation
Fecal diversion

Summary

Neurogenic bowel and bladder dysfunction is a challenge for both the pediatrician and colorectal surgeon. As the neurogenic damage can have multiple consequences, a multidisciplinary approach is usually required.

References

1. Coggrave M, Norton C, Wilson-Barnett J. Management of neurogenic bowel dysfunction in the community after spinal cord injury: a postal survey in the United Kingdom. Spinal Cord. 2009;47(4):323–30.
2. Lie HR, Lagergren J, Rasmussen F, Lagerkvist B, Hagelsteen J, Borjeson MC. Bowel and bladder control of children with myelomeningocele: a Nordic study. Dev Med Child Neurol. 1991;33:1053.
3. Krogh K, Lie HR, Bilenberg N, Laurberg S. Bowel function in Danish children with myelomeningocele. APMIS. 2003;109:S81.
4. Woodhouse CR. Myelomeningocele: neglected aspects. Pediatr Nephrol. 2008;23:1223–31.
5. Skandalakis JE, Gray SW, Ricketts R. The colon and rectum. In: Skandalakis JE, Gray SW, editors. Embryology for surgeons. The embriological basis for the treatment of congenital anomalies. Baltimore: Williams e Wilkins; 1994. p. 242–81.
6. Godlewski G, Prudhomme M. Embryology and anatomy of the anorectum. Surg Clin N Am. 2000;80:319–43.
7. Davies MRQ. Anatomy of the nerve supply of the rectum, bladder and internal genitalia in anorectal dysgenesis in the male. J Ped Surg. 1997;4:536–41.
8. Goldstein A, Hofstra R, Burns A. Building a brain in the gut: development of the enteric nervous system. Clin Genet. 2013;83:307–16.
9. Vallès M, Mearin F. Pathophysiology of bowel dysfunction in patients with motor incomplete spinal cord injury: comparison with patients with motor complete spinal cord injury. Dis Colon Rectum. 2009 Sep;52(9):1589–97.
10. Hawker K, Frohman E. Bladder, bowel and sexual dys- function in multiple sclerosis. Curr Treat Options Neurol. 2001;3:207–14.
11. Krogh K, Nielsen J, Djurhuus JC, et al. Colorectal function in patients with spinal cord lesions. Dis Colon Rectum. 1997;40:1233–9.
12. Leduc BE, Spacek E, Lepage Y. Colonic transit time after spinal cord injury. J Spinal Cord Med. 1997;20:416–21.

13. Nino-Murcia M, Stone JM, Chang PJ, Perkash I. Colonic transit in spinal cord-injured patients. Investig Radiol. 1990;25:109–12.
14. Kumar L, Athanasakos E, Emmanuel AV. Evaluation and treatment of neurogenic bowel dysfunction – a review european neurological review. Eur Neurol Rev. 2016;11(2):109–15.
15. Joensson IM, Siggard C, Rittig S, Hagstroem S, Djurhuus JC. Transabdominal ultrasound of rectum as a diagnostic tool in children constipation. J Urol. 2008;179:1997–2002.
16. Stiens SA, Bergman SB, Goetz LL. Neurogenic bowel dysfunction after spinal cord injury: clinical evaluation and rehabilitative management. Arch Phys Med Rehabil. 1997;78:86–102.
17. Correa GI, Rotter KP. Clinical evaluation and management of neurogenic bowel after spinal cord injury. Spinal Cord. 2000;38:301–8.
18. Gor RA, Katorski JR, Elliott SP. Medical and surgical management of neurogenic bowel. Curr Opin Urol. 2016 Jul;26(4):369–75.
19. Preziosi G, Emmanuel A. Neurogenic bowel dysfunction: pathophysiology, clinical manifestations and treatment. Expert Rev Gastroenterol Hepatol. 2009;3:417–23.
20. Mosiello G, Marshall D, Rolle U, Crétolle C, Santacruz BG, Frischer J, Benninga MA. Consensus review of best practice of transanal irrigation in children. J Pediatr Gastroenterol Nutr. 2017 Mar;64(3):343–52.

Anorectal Physiology in Low Rectal Resection Syndrome

32

Nouf Y. Akeel and Tracy Hull

Introduction

Over the past decades, many advances in the management of rectal cancer have led to significant improvement of oncological outcomes. However, the functional outcomes and the quality of life may still be impaired. Among the patients who underwent sphincter-saving operations, 85% reported quality of life impairment related to low anterior resection syndrome (LARS) with 40% reported major impairment [1, 2]. LARS score is a simple system based on the symptoms and their impact on quality of life (QoL) to evaluate bowel dysfunction after LAR [3]. More details are discussed in Chaps. 4 and 33. Symptoms of LARS consist of diarrhea, fecal incontinence, stool frequency, urgency, and/or difficult evacuation. Patients may report constipation, bowel fragmentation (having multiple bowel movements in a short period of time), and feelings of incomplete emptying. Some patients reach a stable function 1.5 years post-operation [2].

Therefore, in addition to the oncological outcomes, bowel function must be considered when dealing with rectal cancer. Insight into the pathophysiology of LARS is crucial to improving the patient's quality of life.

N. Y. Akeel
Department of Surgery, Faculty of Medicine, King Abdulaziz University, Jeddah, Saudi Arabia

T. Hull (✉)
Cleveland Clinic, Department of Colorectal Surgery, Cleveland, OH, USA
e-mail: hullt@ccf.org

Physiology of Normal Bowel Function

The process of defecation and anal continence is complex; it is an interplay between the central nervous system, intestinal motility, an adequate rectal reservoir, and functioning anal sphincter complex. Propagating and nonpropagating contractions start in the proximal colon and result in transporting the stool to the rectum. As stool and gas fill the rectum, a transient relaxation of the internal anal sphincter (IAS) occurs which permits the upper anal canal to sample the rectal content. This is termed the rectoanal inhibitory reflex (RAIR); it allows the discrimination of flatus from solid and liquid stool. RAIR is mediated via the neurotransmitter nitric oxide. When the rectum becomes distended to its threshold capacity, the intraluminal pressure increases as well as the phasic rectal contractions, and the urge to defecate starts. The parasympathetic pelvic splanchnic nerves lying in the rectal wall or surrounding the levator ani musculature are responsible for transmitting the change in the intraluminal pressure. If it is not convenient to defecate, a bowel movement may be postponed by the voluntary contraction of the external anal sphincter (EAS) and puborectalis muscle. The urge will disappear

L. C. C. Oliveira (ed.), *Anorectal Physiology*, https://doi.org/10.1007/978-3-030-43811-1_32

as the rectal wall relaxes to accommodate the volume and recovers the baseline pressure.

An intact anal sphincter is not the only factor maintaining continence. To prevent stool leakage, the anal pressure is higher than the rectal pressure until defecation is initiated. The intrarectal pressure is affected by colonic motility and rectal wall compliance such that if the rectal wall is stiff (noncompliant) and cannot stretch to reduce the pressure, leakage may occur. Conversely, for the evacuation process, a higher intrarectal pressure must be generated simultaneously with relaxation of the EAS and puborectalis muscle [4, 5].

Pathophysiology of Low Anterior Resection Syndrome

There are multiple mechanisms that may contribute to the development of LARS. Internal anal sphincter (IAS) dysfunction, decrease in anal canal sensation, absent RAIR, disruption in local reflexes between the anus and the neorectum, and reduction in rectal reservoir capacity and compliance are frequently described in the literature [6]. These are the result of the multimodal therapy utilized to effectively treat rectal cancer which includes total mesorectal excision (TME) and pelvic radiotherapy. Also, technical aspects of creating the colorectal or coloanal anastomosis may affect the bowel and anal sphincter function. Long-term data indicate that these changes are permanent and persist beyond the adaptation period [7].

Anal Sphincter Dysfunction

Multiple studies have demonstrated a structural and functional defect in the IAS after a low anterior resection (LAR). Nakahara et al. evaluated the anorectal function in eight patients who have undergone a rectal resection and low stapled coloanal anastomosis. They found a significant decrease in the postoperative resting anal pressure, the maximum squeeze pressure, and the high-pressure zone. Additionally, they noted an absent RAIR with no tendency to recover at 1 year [8].

Another author evaluated the sphincter complex with endorectal ultrasound and found IAS disruption in 18% of patients who underwent LAR with stapled anastomosis; however, the external anal sphincter (EAS) was not damaged after surgery [9]. A marked drop in the maximum resting anal pressure without significant recovery at 2 years after LAR was demonstrated in other studies [7, 10]. It has been suggested that anal manipulation and stretch during the insertion of the circular stapler may be the etiology of the IAS injury. The IAS on ultrasound in some studies appeared thin and patchy which resembled a similar pattern of injury noted after anal dilatation [6, 9, 11]. Ho et al. did not find a significant drop in the maximum squeeze pressure or EAS defects in patients who had sigmoid resection using stapled transanal anastomosis vs. biofragmentable anastomotic ring group [11]. However, it is known that fecal incontinence is multifactorial such that an IAS injury does not always lead to incontinence as the presence of an adequate rectal capacity/reservoir may compensate for the weakened sphincter [11]. Therefore a combination of IAS injury with an inadequate reservoir may be required for dysfunction.

Tomita and Igarashi evaluated the anorectal functions in patients with soiling 5 or more years after LAR. Patients with soiling had significantly lower resting anal pressure, maximum squeeze pressure, maximum rectal-tolerated threshold volume, and rectal compliance. Also those with soiling had reduced RAIR compared to the non-soiling group [12].

Injury to the Pelvic Plexus

Denervation during pelvic dissection is theorized to be another mechanism of sphincter dysfunction [13–15]. Kinugasa and his group showed that the major autonomic nerve input to the IAS appeared to originate from the inferior rectal branches of the pelvic plexus located on the anterolateral side of the distal rectum rather than the nerve fibers from Auerbach's plexus. These nerves encompassed both sympathetic and parasympathetic fibers; however, they have a sympathetic-dominant nerve supply. Injury to these nerves during surgery seemed to result in denervation to the major part of the IAS which results in ineffective contraction [16].

Impaired Anal Sensation and Disappearance of RAIR

After rectal transection, the communication from the proximal intermural innervation is severed to the internal sphincter. This results in an inability to discriminate the character of the rectal contents. The loss of anal sensation and RAIR has been associated with incontinence [12, 17]. Forty-six patients who underwent LAR with stapled anastomosis were evaluated preoperatively and up to 2 years. During preoperative assessment, RAIR was present in 93%; however only 18% had an intact RAIR on postoperative day 10. There was a significant recovery in 85% of the patients at 2 years follow-up [10]. Regeneration of nerves and activation of an alternate reflex pathway through stretch receptors in the pelvic floor muscles could explain the recovery of RAIR [4]. Conversely, other studies found poor RAIR recovery in these patients 1 year after surgery [8, 18].

Rectal Capacity and Compliance

The neorectum (segment of the colon) has different reservoir capacity, sensation, and motility compared with the native rectum [19]. Moreover, colonic mobilization may lead to significant extrinsic denervation with destruction of inhibitory sympathetic innervations and an increase in motility in the distal colon [20]. The combination of these factors can produce urgency, frequency, and incontinence [15, 19, 21–23].

Several reconstructive techniques other than a straight coloanal (or ultralow colorectal) anastomosis (SCAA) (Fig. 32.1) have been introduced to compensate for the loss of capacity. These include the colonic J-pouch (CJP) (Fig. 32.2), the transverse coloplasty (TC) (Fig. 32.3), and the side-to end anastomosis (STEA) (Fig. 32.4). Fazio et al. compared the complications, long-term functional outcomes, and QoL of patients with coloanal anastomosis with or without a reservoir for low rectal cancer. After the intraoperative assessment, the patients were randomized to TC or SCAA if CJP was not feasible. Alternately, those who were CJP eligible were randomized to CJP or TC. CJP was found to be superior to TC and SCAA regarding the number of bowel movements, the Fecal Incontinence Severity Index (FISI), and clustering. There was no significant difference in the complications or the quality of life across all the groups up to 2 years after surgery. This study did not have a sufficient power to demonstrate a significant difference between TC and SCAA [24].

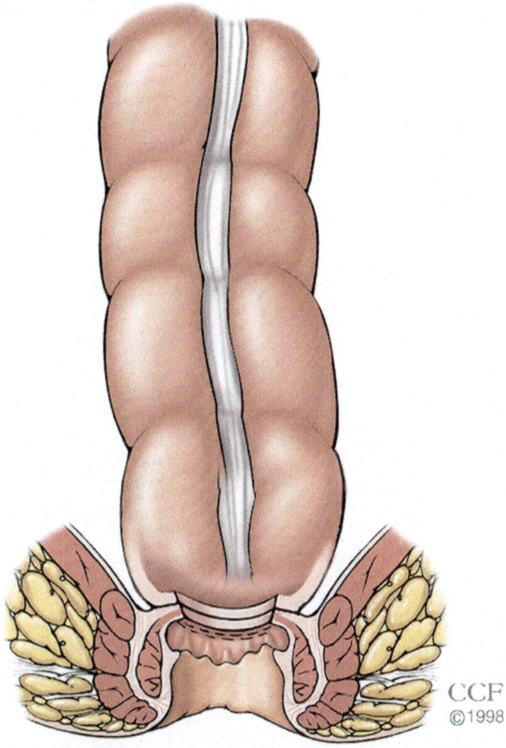

Fig. 32.1 Straight coloanal anastomosis. (Reprinted with permission, Cleveland Clinic Center for Medical Art & Photography ©1998–2018)

Long-term data up to 5 years showed superior functional outcomes (frequency, clustering, use of constipating drugs, diet modification) of CJP over SCAA [25]. Hida et al. evaluated the func-

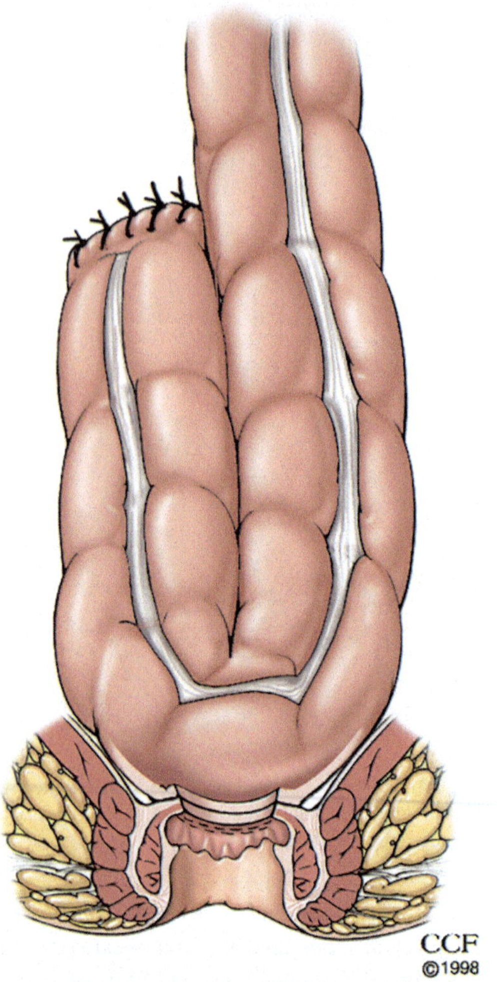

Fig. 32.2 Colonic J-pouch. (Reprinted with permission, Cleveland Clinic Center for Medical Art & Photography ©1998–2018)

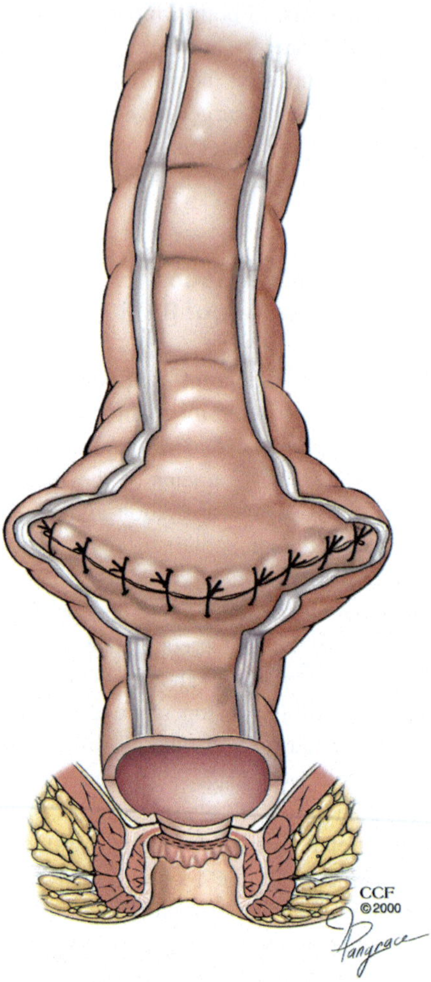

Fig. 32.3 Transverse coloplasty. (Reprinted with permission, Cleveland Clinic Center for Medical Art & Photography ©1998–2018)

tional and physiological outcomes 3 years post-LAR. Patients with CJP showed better continence and satisfaction in bowel function, less urgency, and frequency compared to those with SCAA. The manovolumetric assessment showed greater threshold volume, maximum tolerated volume, and compliance in the CJP group [26].

A comparison of CJP and TC by Ho et al. in a randomized controlled trial showed a surprisingly high anastomotic leak rate in the coloplasty group (16% vs. 0%, $p = 0.01$). All the leaks were anterior at the level of the coloanal anastomosis. The high leak rate was attributed to the compromised blood supply in an end-to-end anastomosis. The patients in the TC group had less urgency and nocturnal leaks but more stool fragmentation at 4 months. However, both groups were comparable at 1 year in terms of bowel function, continence, and quality of life [27]. Machado et al. found comparable functional and surgical outcomes between CJP and SECAA at 1- and 2-year follow-up [28, 29]. A meta-analysis (1636 patients) evaluated the reconstruction techniques after low anterior resection for rectal cancer. The

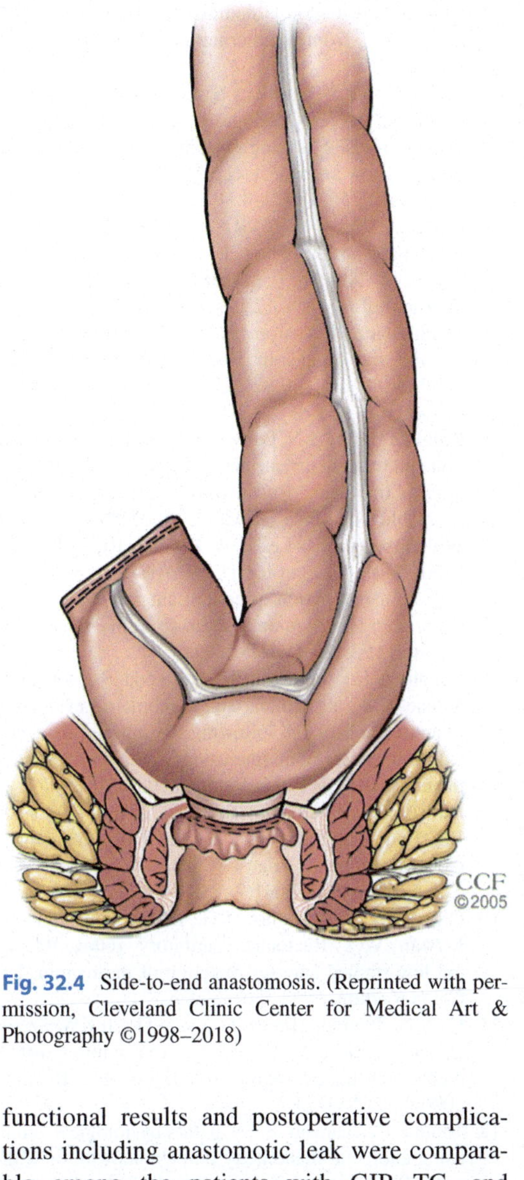

Fig. 32.4 Side-to-end anastomosis. (Reprinted with permission, Cleveland Clinic Center for Medical Art & Photography ©1998–2018)

functional results and postoperative complications including anastomotic leak were comparable among the patients with CJP, TC, and SECAA. The results were favoring CJP over SCAA when looking at stool frequency and the use of antidiarrheal medication up to 1 year after surgery [30]. Improvement in functional outcomes was not reflected when assessing quality of life in some trials [24, 27, 30–32].

Many authors have concluded that the neorectal capacity has a minor significance in improving bowel function, as the increase in rectal capacity did not translate into better functional outcomes [29, 31, 33, 34]. Therefore it has been suggested that the decrease in the peristaltic waves and motility may be more significant than the neorectal capacity [29, 31, 34].

Predisposing Factors for Low Anterior Resection Syndrome

Neoadjuvant Chemoradiotherapy

Several studies revealed a deleterious effect of radiation on anorectal function [1, 2, 35–41]. Bregendahl et al. evaluated the biomechanical wall properties, sensory function and perception, and postprandial response of the neorectum as well as anal sphincter function in 16 patients with major LARS who received neoadjuvant radiation. They compared this group to 23 patients with major LARS treated with TME alone. The patients in the radiation group had impaired neorectal sensitivity illustrated by a reduced sensation to thermal and mechanical stimulations, reduced resting anal pressure, and an altered perception of defecatory urge. They speculated that these changes are possibly due to the damaged afferent nerves and loss of IAS integrity and denervation [42]. Similarly, results of a meta-analysis showed that fecal incontinence was more often noted in irradiated patients (risk ratio (RR) = 1.67; 95% confidence interval (CI), 1.36, 2.05; $p = 0.0001$), and manometric results were significantly worse [32]. Moreover, preoperative radiotherapy was associated with an increased risk of anastomotic leak and stenosis [43].

Level of the Anastomosis

As discussed previously, the loss of the rectal reservoir capacity is one of the factors associated with LARS, and a low anastomosis has been associated with an increase in fecal incontinence [29]. Patients with an anastomosis 5 cm or less from the anal verge have a higher risk of major LARS [40]. However, the benefit of a longer rectal remnant was lost in the patients treated with neoadjuvant therapy [38]. The combination of neoadjuvant radiation and a low lying tumor was associated with the highest risk of developing LARS [2].

Anastomotic Leak and Pelvic Sepsis

An anastomotic leak/pelvic sepsis also may contribute to LARS. Neorectal volume, compliance, urge to defecate, and maximum tolerated volume were significantly reduced in patients with leakage. Clinically, those patients had more frequent bowel movements and tendency to have urgency, incontinence, and impaired evacuation [44]. Ashburn et al. published that patients with anastomotic leak reported a worse quality of life, more frequent daytime and nighttime bowel movements, and more incontinence for solid stool in comparison with those without a leak [45].

Summary

The surgeon must consider LARS and its impact on the QoL when dealing with a rectal cancer patient. The development of LARS has multiple contributing mechanisms which are (IAS) dysfunction, decrease in anal canal sensation, absent RAIR, injury to the pelvic plexus, and reduction in rectal reservoir capacity and compliance. The level and reconstructive techniques of the anastomosis, pelvic sepsis, and anastomotic leak and the radiation therapy have an impact on the postoperative bowel function.

References

1. Battersby NJ, Juul T, Christensen P, Janjua AZ, Branagan G, Emmertsen KJ, et al. Predicting the risk of bowel-related quality-of-life impairment after restorative resection for rectal cancer: a multicenter cross-sectional study. Dis Colon Rectum. 2016;59(4):270–80. Available from: internal-pdf://161.87.246.188/Predicting.
2. Qin Q, Huang B, Cao W, Zhou J, Ma T, Zhou Z, et al. Bowel dysfunction after low anterior resection with neoadjuvant chemoradiotherapy or chemotherapy alone for rectal cancer: a cross-sectional study from China. Dis Colon Rectum. 2017;60(7):697–705. Available from: internal-pdf://145.144.205.231/Bowel.
3. Emmertsen KJ, Laurberg S. Low anterior resection syndrome score: development and validation of a symptom-based scoring system for bowel dysfunction after low anterior resection for rectal cancer. Ann Surg. 2012;255(5):922–8.
4. Hallbook O, Sjodahl R. Surgical approaches to obtaining optimal bowel function. Semin Surg Oncol. 2000;18(3):249–58. Available from: internal-pdf://94.198.175.212/Surgical.
5. Szmulowicz U. Anatomy and physiology of the pelvic digestive organs. In: Hull T, editor. Posterior pelvic floor abnormalities. Philadelphia: Elsevier; 2011. p. 1–40.
6. Ridolfi TJ, Berger N, Ludwig KA. Low anterior resection syndrome: current management and future directions. Clin Colon Rectal Surg. 2016;29(3):239–45. Available from: internal-pdf://209.155.80.13/Current.
7. Williamson MER, Lewis WG, Finan PJ, Miller AS, Holdsworth PJ, Johnston D. Recovery of physiologic and clinical function after low anterior resection of the rectum for carcinoma: myth or reality? Dis Colon Rectum. 1995;38(4):411–8.
8. Nakahara S, Itoh H, Mibu R, Ikeda S, Oohata Y, Kitano K, et al. Clinical and manometric evaluation of anorectal function following low anterior resection with low anastomotic line using an EEA™ stapler for rectal cancer. Dis Colon Rectum. 1988;31(10):762–6.
9. Farouk R, Duthie GS, Lee PWR, Monson JRT. Endosonographic evidence of injury to the internal anal sphincter after low anterior resection. Dis Colon Rectum. 1998;41(7):888–91.
10. O'riordain MG, Molloy RG, Gillen P, Horgan A, Kirwan WO. Rectoanal inhibitory reflex following low stapled anterior resection of the rectum. Dis Colon Rectum. 1992;35(9):874–8.
11. Ho Y-H, Tsang C, Tang CL, Nyam D, Eu KW, Seow-Choen F. Anal sphincter injuries from stapling instruments introduced transanally. Dis Colon Rectum. 2000;43(2):169–73.
12. Tomita R, Igarashi S. A pathophysiological study using anorectal manometry on patients with or without soiling 5 years or more after low anterior resection for lower rectal cancer. Hepato-Gastroenterology. 2008;55(86–87):1584–8.
13. Carlstedt A, Nordgren S, Fasth S, Appelgren L, Hulten L. Sympathetic nervous influence on the internal anal sphincter and rectum in man. Int J Color Dis. 1988;3(2):90–5.
14. Hirano A, Koda K, Kosugi C, Yamazaki M, Yasuda H. Damage to anal sphincter/levator ani muscles caused by operative procedure in anal sphincter-preserving operation for rectal cancer. Am J Surg. 2011;201(4):508–13.
15. Bryant CL, Lunniss PJ, Knowles CH, Thaha MA, Chan CL. Anterior resection syndrome. Lancet Oncol. 2012;13(9):e403–8. Available from: internal-pdf://121.226.169.227/Anterior.

16. Kinugasa Y, Arakawa T, Murakami G, Fujimiya M, Sugihara K. Nerve supply to the internal anal sphincter differs from that to the distal rectum: an immunohistochemical study of cadavers. Int J Color Dis. 2014;29(4):429–36.
17. Saigusa N, Belin BM, Choi H-J, Gervaz P, Efron JE, Weiss EG, et al. Recovery of the rectoanal inhibitory reflex after restorative proctocolectomy. Dis Colon Rectum. 2003;46(2):168–72.
18. Pedersen IK, Christiansen J, Hint K, Jensen P, Olsen J, Mortensen PE. Anorectal function after low anterior resection for carcinoma. Ann Surg. 1986;204(2):133–5.
19. Van Duijvendijk P, Slors JFM, Taat CW, Van Tets WF, Van Tienhoven G, Obertop H, et al. Prospective evaluation of anorectal function after total mesorectal excision for rectal carcinoma with or without preoperative radiotherapy. Am J Gastroenterol. 2002;97(9):2282–9.
20. Lee WY, Takahashi T, Pappas T, Mantyh CR, Ludwig KA. Surgical autonomic denervation results in altered colonic motility: an explanation for low anterior resection syndrome? Surgery. 2008;143(6):778–83. Available from: internal-pdf://83.41.153.56/Surgical.
21. Ryu Y, Akagi Y, Yagi M, Sasatomi T, Kinugasa T, Yamaguchi K, et al. Fecoflowmetric analysis of anorectal motor function in postoperative anal-preserving surgery patients with low rectal cancer comparison with the Wexner score and anorectal manometry. Int Surg. 2015;100(1):29–37.
22. Kye BH, Kim HJ, Kim G, Yoo RN, Cho HM. The effect of biofeedback therapy on anorectal function after the reversal of temporary stoma when administered during the temporary stoma period in rectal cancer patients with sphincter-saving surgery: the interim report of a prospective rand. Medicine. 2016;95(18):e3611. Available from: internal-pdf://228.60.152.105/The.
23. Gosselink MP, Zimmerman DD, West RL, Hop WC, Kuipers EJ, Schouten WR. The effect of neo-rectal wall properties on functional outcome after colonic J-pouch-anal anastomosis. Int J Colorectal Dis. 2007;22(11):1353–60. Available from: http://link.springer.com/10.1007/s00384-007-0326-7.
24. Fazio VW, Zutshi M, Remzi FH, Parc Y, Ruppert R, Furst A, et al. A randomized multicenter trial to compare long-term functional outcome, quality of life, and complications of surgical procedures for low rectal cancers. Ann Surg. 2007;246(3):481–90.
25. Dehni N, Tiret E, Singland JD, Cunningham C, Schlegel RD, Guiguet M, Parc R. Long-term functional outcome after low anterior resection: comparison of low colorectal anastomosis and colonic J-pouch-anal anastomosis. Dis Colon Rectum. 1998;7(41):817–22.
26. Hida J-I, Yoshifuji T, Okuno K, Matsuzaki T, Uchida T, Ishimaru E, et al. Long-term functional outcome of colonic J-pouch reconstruction after low anterior resection for rectal cancer. Surg Today. 2006;36(5):441–9.
27. Ho Y-H, Brown S, Heah S-M, Tsang C, Seow-Choen F, Eu K-W, et al. Comparison of j-pouch and Coloplasty pouch for low rectal cancers. Ann Surg. 2002;236(1):49–55.
28. Machado M, Nygren J, Goldman S, Ljungqvist O. PMC1422690; Similar outcome after colonic pouch and side-to-end anastomosis in low anterior resection for rectal cancer: a prospective randomized trial. Ann Surg. 2003;238(2):214–20. Available from: internal-pdf://215.148.120.254/Similar.
29. Machado M, Nygren J, Goldman S, Ljungqvist O. Functional and physiologic assessment of the colonic reservoir or side-to-end anastomosis after low anterior resection for rectal cancer: a two-year follow-up. Dis Colon Rectum. 2005;48(1):29–36. Available from: internal-pdf://241.243.36.84/Functional.
30. Hüttner FJ, Tenckhoff S, Jensen K, Uhlmann L, Kulu Y, Büchler MW, et al. Meta-analysis of reconstruction techniques after low anterior resection for rectal cancer. Br J Surg. 2015;102(7):735–45.
31. Fürst A, Burghofer K, Hutzel L, Jauch KW. Neorectal reservoir is not the functional principle of the colonic J-pouch: the volume of a short colonic J-pouch does not differ from a straight coloanal anastomosis. Dis Colon Rectum. 2002;45(5):660–7. Available from: http://www.springerlink.com/index/AU1WE1BYNUQ97UCT.pdf.
32. Doeksen A, Bakx R, Vincent A, van Tets WF, Sprangers MAG, Gerhards MF, et al. J-pouch vs side-to-end coloanal anastomosis after preoperative radiotherapy and total mesorectal excision for rectal cancer: a multicentre randomized trial. Color Dis. 2012;14(6):705–13. Available from: http://doi.wiley.com/10.1111/j.1463-1318.2011.02725.x.
33. Joo JS, Latulippe JF, Alabaz O, Weiss EG, Nogueras JJ, Wexner SD. Long-term functional evaluation of straight coloanal anastomosis and colonic J-pouch: is the functional superiority of colonic J-pouch sustained? Dis Colon Rectum. 1998;41(6):740–6. Available from: http://content.wkhealth.com/linkback/openurl?sid=WKPTLP:landingpage&an=00003453-199841060-00011.
34. Tsunoda A, Kamiyama G, Narita K, Watanabe M, Nakao K, Kusano M. Prospective randomized trial for determination of optimum size of side limb in low anterior resection with side-to-end anastomosis for rectal carcinoma. Dis Colon Rectum. 2009;52(9):1572–7. Available from: http://journals.lww.com/dcrjournal/Abstract/2009/09000/Prospective_Randomized_Trial_for_Determination_of.6.aspx.
35. Bregendahl S, Emmertsen KJ, Lous J, Laurberg S. Bowel dysfunction after low anterior resection with and without neoadjuvant therapy for rectal cancer: a population-based cross-sectional study. Color Dis. 2013;15(9):1130–9. Available from: internal-pdf://144.156.249.46/Bowel.
36. Jimenez-Rodriguez RM, Segura-Sampedro JJ, Rivero-Belenchon I, Diaz Pavon JM, Garcia Cabrera AM, Vazquez Monchul JM, et al. Is the interval from surgery to ileostomy closure a risk factor for low anterior resection syndrome? Color Dis. 2017;19(5):485–90. Available from: internal-pdf://243.123.141.220/Is.

37. Hughes DL, Cornish J, Morris C. Functional outcome following rectal surgery-predisposing factors for low anterior resection syndrome. Int J Color Dis. 2017;32(5):691–7. Available from: internal-pdf://122.109.141.60/predisposing.
38. Bondeven P, Emmertsen KJ, Laurberg S, Pedersen BG. Neoadjuvant therapy abolishes the functional benefits of a larger rectal remnant, as measured by magnetic resonance imaging after restorative rectal cancer surgery. Eur J Surg Oncol. 2015;41(11):1493–9. Available from: internal-pdf://81.234.211.107/Neoadjuvant.
39. Loos M, Quentmeier P, Schuster T, Nitsche U, Gertler R, Keerl A, et al. Effect of preoperative radio(chemo) therapy on long-term functional outcome in rectal cancer patients: a systematic review and meta-analysis. Ann Surg Oncol. 2013;20(6):1816–28. Available from: internal-pdf://103.99.132.94/Effect.
40. Ekkarat P, Boonpipattanapong T, Tantiphlachiva K, Sangkhathat S. Factors determining low anterior resection syndrome after rectal cancer resection: a study in Thai patients. Asian J Surg. 2016;39(4):225–31. Available from: internal-pdf://67.83.208.205/Factors.
41. Parc Y, Zutshi M, Zalinski S, Ruppert R, Furst A, Fazio VW. Preoperative radiotherapy is associated with worse functional results after coloanal anastomosis for rectal cancer. Dis Colon Rectum. 2009;52(12):2004–14. Available from: internal-pdf://0943947739/Preoperative.
42. Bregendahl S, Emmertsen KJ, Fassov J, Krogh K, Zhao J, Gregersen H, et al. Neorectal hyposensitivity after neoadjuvant therapy for rectal cancer. Radiother Oncol. 2013;108(2):331–6. Available from: http://www.sciencedirect.com/science/article/pii/S0167814013003319.
43. Qin Q, Ma T, Deng Y, Zheng J, Zhou Z, Wang H, et al. Impact of preoperative radiotherapy on anastomotic leakage and stenosis after rectal cancer resection: post hoc analysis of a randomized controlled trial. Dis Colon Rectum. 2016;59(10):934–42. Available from: http://www.ingentaconnect.com/content/wk/dcr/2016/00000059/00000010/art00016.
44. Hallböök O, Sjödahl R. Anastomotic leakage and functional outcome after anterior resection of the rectum. Br J Surg. 1996;83(1):60–2. Available from: http://doi.wiley.com/10.1002/bjs.1800830119.
45. Ashburn JH, Stocchi L, Kiran RP, Dietz DW, Remzi FH. Consequences of anastomotic leak after restorative proctectomy for cancer: effect on long-term function and quality of life. Dis Colon Rectum. 2013;56(3):275–80. Available from: http://journals.lww.com/dcrjournal/Abstract/2013/03000/Consequences_of_Anastomotic_Leak_After_Restorative.2.aspx.

33 Treatment of the Low Anterior Resection Syndrome

Soren Laurberg, Katrine J. Emmertsen, T. Juul, Hossam Elfeki, and Peter Christensen

Introduction

Within the last decades, there have been substantial improvements in the outcome for rectal cancer with a reduction in 30-day mortality and risk of local recurrence, improved long-term survival, and reduction in the use of permanent stoma. Today, the majority of patients are cured for their cancer without a permanent stoma. Therefore, there is an urgent need for focus on the long-term consequences of the treatment. The most prominent sequela after reconstructive surgery for rectal cancer is bowel dysfunction – often termed low anterior resection syndrome (LARS). It will affect about 50% of all patients [1], but generally, there has been little focus on prevention and treatment of LARS.

Symptoms

In brief, most patients with major LARS will have a combination of urgency, clustering, change in bowel frequency, and incontinence of flatus and/or feces. The symptom burden decreases within the first 1–2 years after surgery, but major LARS at 3 months posttreatment is a predictor of long-term problems [2].

S. Laurberg (✉) · K. J. Emmertsen · T. Juul
H. Elfeki · P. Christensen
Aarhus University Hospital, Aarhus, Denmark
e-mail: Soeren.Laurberg@aarhus.rm.dk

Risk Factors and Pathophysiology

The two main risk factors for LARS are a low anastomosis and use of neoadjuvant radiotherapy. The pathophysiology is multifactorial. Several studies suggest that denervation of the left colon during mobilization plays a key role due to augmentation of the gastrocolic reflex. Also, following surgery alone, there is a substantial reduction in risk of LARS if just a short remnant of the rectum can be left in situ possibly due to spared afferent nerve fibers. Radiotherapy leads to major sensory disturbances in the pelvic cavity and hence to control of the bowel. In addition, loss of a rectal reservoir and lesion of the internal anal sphincter may play a role.

Information and Surveillance

Generally, there has been little focus on information of risk of major LARS to individual patients prior to treatment. However, recently a simple predictive model for risk of LARS was developed [3].

There is a substantial variation in the surveillance programs after curative treatment, but generally, the guidelines have focused on detection of recurrence. In the majority of guidelines, there is no advice regarding detection and treatment of LARS [4].

L. C. C. Oliveira (ed.), *Anorectal Physiology*, https://doi.org/10.1007/978-3-030-43811-1_33

Diagnosis

Previously, numerous scores have been developed to describe bowel dysfunction including the St Mark's incontinence score and the Cleveland Clnic Florida-Fecal Incontinence Score (CCF-FIS). Yet, these scores are not sufficient to cover the more complex symptomatology of LARS and have never been validated in patients with rectal cancer. The LARS score [5] is a 5-question score that has been translated into more than 35 languages and has been validated internationally. Patients' symptoms can be categorized into major, minor, or no LARS. It is simple to use and reflects the impact on quality of life of symptoms from a patient perspective. Overall, close to 50% of all rectal cancer patients will have major LARS after a resection, and it will be present in more than 80% of patients following neoadjuvant therapy and resection.

However, it must be taken into account that bowel symptoms are common in the general population, and in 50- to 75-year-olds; about 15% will be categorized as having major LARS. Therefore, some patients may have major LARS not related to the treatment, and only a proportion of patients with major LARS may be interested in referral for treatment.

Prerequisite Before Referral for Treatment of Major LARS

The LARS symptoms are unspecific and may be due to other conditions including complications and recurrence. It is therefore mandatory that relevant investigations including endoscopy and rectal examination have been performed before referral.

Treatment

In the literature, there are no studies systematically evaluating the treatment of the LARS [6–9]. At present, only a few small observational studies suggest that diet intervention, pelvic floor training, transanal irrigation (TAI), Gatekeeper™, serotonin receptor antagonists, and sacral nerve modulation may play a role [10–18]. The treatment of LARS should be performed in a dedicated pelvic floor unit with experience in the treatment of functional bowel problems. The majority of patients with fecal incontinence referred to a pelvic floor unit can be managed successfully by dedicated trained nurses with conservative treatment. The treatment of major LARS should initially be conservative and run by dedicated nurses following a well-defined algorithm with prospective monitoring of the intervention and outcomes.

Treatment Algorithm

There is no generally accepted treatment algorithm for major LARS. The following describes our current treatment strategy in our national research center for sequelae to cancer in the pelvis.

1. At the *scheduled follow-up*, all radically operated rectal cancer patients are screened using the LARS score. If the patients have major LARS, they are offered treatment at our pelvic floor clinic. Before referral, they will have a clinical examination, endoscopy, and relevant imaging to exclude sign of recurrence, strictures, inflammation, or other pathology that could explain their symptoms.
2. *Before the first visit* in the clinic, they fill out additional questionnaires regarding dietary

habits, bowel function including the Bristol stool scale, and the impact of symptoms on quality of life.

3. *Personalized conservative management of LARS*

   First visit in the clinic. Initially, the patients will be asked to explain their bowel symptoms and impact on the quality of life. Then, the specialist nurses will go through the questionnaires with the patient and for each symptom ask about the impact. Subsequently, the patient will rank the most severe symptoms from the patient perspective based upon the symptom burden. Finally, after a clinical investigation of sphincter function and an inspection of the anal area, a personalized conservative intervention program is set up.

The basic intervention program may include dietary advice for diarrhea or constipation and selective addition of fibers and use of loperamide (Bristol 6–7), laxatives (Bristol 1–2), and mini enema. In case of constipation and evacuation difficulties, the patient receives written and oral instructions on toilet positioning. If there is a week sphincter, they receive a link to instructions on how to perform pelvic floor self-training. If the patient has perianal skin problems, they are instructed to use a skin barrier lotion.

Follow-up visits After 2 months, the effect of treatment is monitored with PROMs, and the intervention may be adjusted, and 3 new options can be introduced:

(a) Selectively, the patients may be introduced to *biofeedback training* in the pelvic floor clinic for incontinence or constipation.
(b) Transanal irrigation can be tried in patients with incontinence, urgency, or evacuation problems if there is a failure of mini enema.
(c) Referral to medical gastroenterologist for evaluation of persistent diarrhea – the threshold for referral must be very low in patients treated with radiotherapy, because if the small bowel is included in the radiation field, this may lead to bile acid malabsorption or bacterial overgrowth.

At the third visit, there might be additional adjustments, and the effect is evaluated and the outcome monitored with PROMs.

Satisfied patients will be dismissed from the clinic, but long-term outcome will be continuously monitored, and the patient can contact the clinic if symptoms recur.

If there is insufficient effect of treatment, patients will be referred to one of the surgeons in the clinic. The surgical option would primarily be a PNE test, and if the test is negative and SNS therefore is not an option, patients may be offered a permanent stoma.

Of the first 78 patients that followed this treatment strategy, 70 (89.74%) were managed by personalized conservative treatment which included biofeedback in 4 patients and TAI in 16 patients. While only four patients were referred to the medical gastroenterologist, another four patients had surgery (SNS or stoma). In case of interest in our setup for the diagnosis and treatment algorithm for sequelae to cancer surgery (Fig. 33.1), you are welcome to contact the authors for a more detailed description.

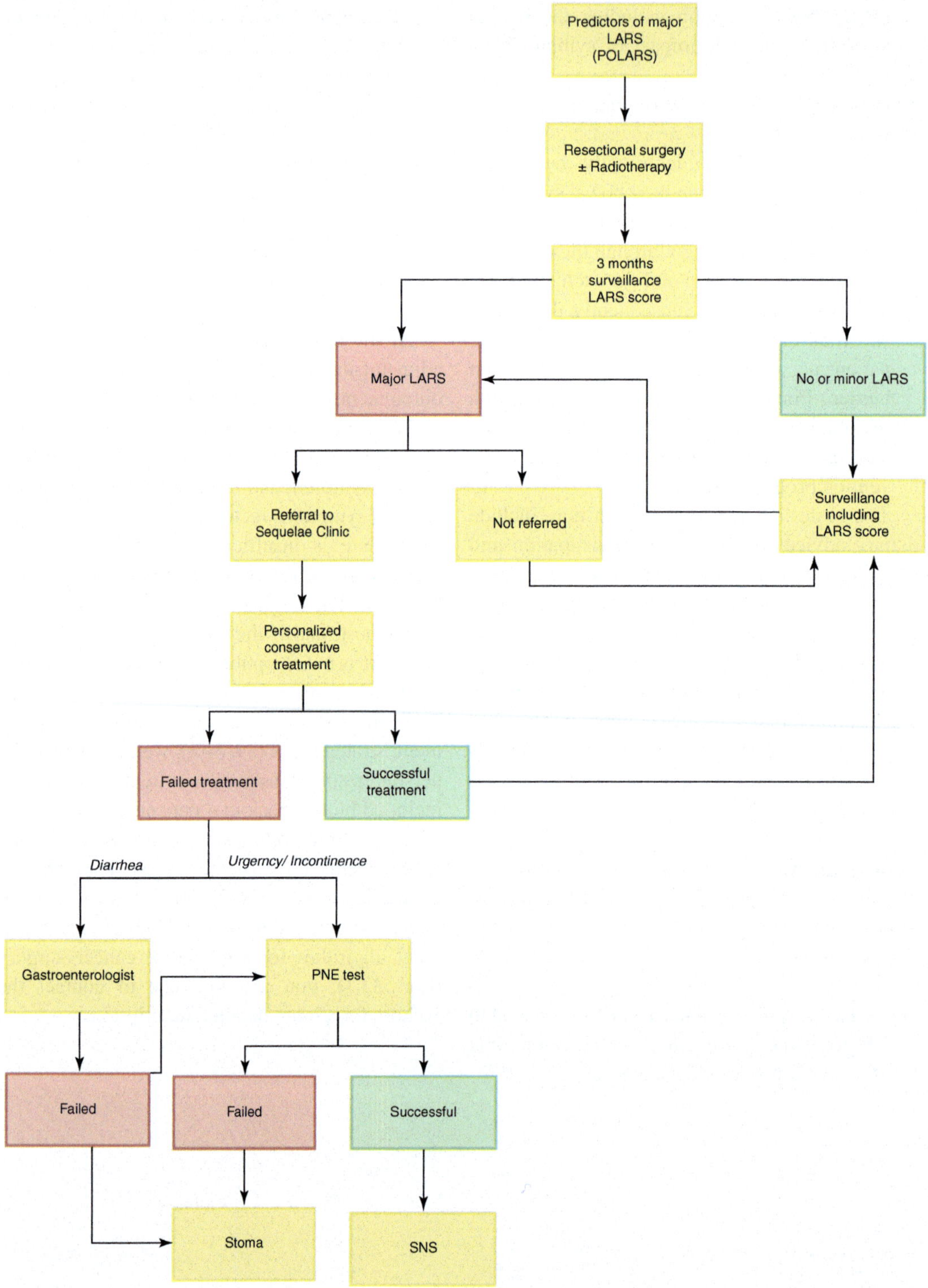

Fig. 33.1 Treatment algorithm for low anterior resection syndrome

Summary

LAR syndrome is frequently a challenge for the colorectal surgeon. Identification and treatment of late sequelae to cancer are unrecognized issues that need further focus. In the future, all patients treated for rectal cancer should have prospective monitoring of LARS and be offered treatment. There is an urgent need for prospective and randomized trials to optimize knowledge of treatment algorithms for LARS. However, late sequelae after rectal cancer resection also include bladder and sexual function and chronic pain, and PROMs should be applied, and evidence-based treatment strategies should be developed and validated. Such studies are presently ongoing in our national research center.

References

1. Juul T, Ahlberg M, Biondo S, et al. International validation of the low anterior resection syndrome score. Ann Surg. 2014;259(4):728–34. https://doi.org/10.1097/SLA.0b013e31828fac0b.
2. Emmertsen KJ, Laurberg S. Impact of bowel dysfunction on quality of life after sphincter-preserving resection for rectal cancer. Br J Surg. 2013;100(10):1377–87. https://doi.org/10.1002/bjs.9223.
3. Battersby NJ, Bouliotis G, Emmertsen KJ, et al. Development and external validation of a nomogram and online tool to predict bowel dysfunction following restorative rectal cancer resection: the POLARS score. Gut. 2018;67(4):688–96. https://doi.org/10.1136/gutjnl-2016-312695.
4. Bastiaenen VP, Hovdenak Jakobsen I, Labianca R, et al. Consensus and controversies regarding follow-up after curative intent treatment of non-metastatic colorectal cancer: a synopsis of guidelines used in countries represented in ESCP. Color Dis. 2018:1–25. https://doi.org/10.1111/codi.14503.
5. Emmertsen KJ, Laurberg S. Low anterior resection syndrome score. Ann Surg. 2012;255(5):922–8. https://doi.org/10.1097/SLA.0b013e31824f1c21.
6. Dulskas A, Smolskas E, Kildusiene I, Samalavicius NE. Treatment possibilities for low anterior resection syndrome: a review of the literature. Int J Color Dis. 2018;33(3):251–60. https://doi.org/10.1007/s00384-017-2954-x.
7. Bazzell A, Madsen LT, Dains J. Clinical management of Bowel dysfunction after low anterior resection for rectal cancer. J Adv Pract Oncol. 2016;7:618–29. https://doi.org/10.6004/jadpro.2016.7.6.4.
8. Ridolfi TJ, Berger N, Ludwig KA. Low anterior resection syndrome: current management and future directions. Clin Colon Rectal Surg. 2016;29(3):239–45. https://doi.org/10.1055/s-0036-1584500.
9. Martellucci J. Low anterior resection syndrome: a treatment algorithm. Dis Colon Rectum. 2016;59(1):79–82. https://doi.org/10.1097/DCR.0000000000000495.
10. Sun V, Grant M, Wendel CS, et al. Dietary and behavioral adjustments to control bowel function by long-term colorectal cancer survivors. J Clin Oncol. 2014;32(3_suppl):573. https://doi.org/10.1200/jco.2014.32.3_suppl.573.
11. Martellucci J, Sturiale A, Bergamini C, et al. Role of transanal irrigation in the treatment of anterior resection syndrome. Tech Coloproctol. 2018;22(7):519–27. https://doi.org/10.1007/s10151-018-1829-7.
12. Croese AD, Whiting S, Vangaveti VN, Ho YH. Using sacral nerve modulation to improve continence and quality of life in patients suffering from low anterior resection syndrome. ANZ J Surg. 2018;88(11):E787–91. https://doi.org/10.1111/ans.14871.
13. Sun V, Crane TE, Slack SD, et al. Rationale, development, and design of the Altering Intake, Managing Symptoms (AIMS) dietary intervention for bowel dysfunction in rectal cancer survivors. Contemp Clin Trials. 2018;68:61–6. https://doi.org/10.1016/j.cct.2018.03.010.
14. McCutchan GM, Hughes D, Davies Z, et al. Acceptability and benefit of rectal irrigation in patients with low anterior resection syndrome: a qualitative study. Color Dis. 2018;20(3):O76–84. https://doi.org/10.1111/codi.13985.
15. Trenti L, Biondo S, Noguerales F, et al. Outcomes of Gatekeeper™ prosthesis implantation for the treatment of fecal incontinence: a multicenter observational study. Tech Coloproctol. 2017;21(12):963–70. https://doi.org/10.1007/s10151-017-1723-8.
16. Eftaiha SM, Balachandran B, Marecik SJ, et al. Sacral nerve stimulation can be an effective treatment for low anterior resection syndrome. Color Dis. 2017;19(10):927–33. https://doi.org/10.1111/codi.13701.
17. Mege D, Meurette G, Vitton V, et al. Sacral nerve stimulation can alleviate symptoms of bowel dysfunction after colorectal resections. Color Dis. 2017;19(8):756–63. https://doi.org/10.1111/codi.13624.
18. Itagaki R, Koda K, Yamazaki M, et al. Serotonin (5-HT3) receptor antagonists for the reduction of symptoms of low anterior resection syndrome. Clin Exp Gastroenterol. 2014;7(1):47–52. https://doi.org/10.2147/CEG.S55410.

Pelvic Floor Rehabilitation

34

Silvana Uchôa, Bary Berghmans,
and Maura R. Seleme

Introduction

Disorders of the gynecological, urinary, and gastrointestinal systems are often treated by medical and surgical options. However, physiotherapeutic procedures have become increasingly helpful in the rehabilitation of patients with such disorders, due to the optimal outcomes achieved [1]. The rehabilitation of the pelvic floor comprises a group of techniques used to aid the reacquisition of the sphincter control. Among these, there are the so-called behavioral therapies and pelvic physiotherapeutic procedures. Behavioral therapy is defined as a group of specific, low-cost interventions performed to alter the relationship between the patient's signs and symptoms and their environment. This goal can be achieved by modifying the behavior and/or the environment in which it is found. Behavioral techniques help the patient learn ways to control the urethral and anal sphincters (especially of the bladder and bowel control muscles). These techniques are safe and have no side effects.

S. Uchôa (✉)
Universidade Catolica de Pernambuco, Recife, Brazil

B. Berghmans
Pelvic Care Center Maastricht, Maastricht University Medical Centre, Maastricht, The Netherlands

M. R. Seleme
Faculdade Inspirar, Department of Pelvic Physiotherapy, Curitiba, Brazil

Table 34.1 Types of therapeutic modes

Biofeedback
Electrostimulation
Electroanalgesia
Kinesiotherapy or pelvic floor muscle training (PFMT)
Vaginal cones
Rectal balloon training (RBT)

The physiotherapist must first obtain basic knowledge about the anatomy and physiology of the pelvic floor muscles (PFM) to better understand the objectives and effects of the therapeutic modalities used in the rehabilitation of the pelvic floor, as well as the biophysical and biological properties of electrical stimulation, with its indications and contraindications, and also the choice of the correct exercise modality, emphasizing that all therapeutic planning should only be elaborated through careful evaluation and physiotherapeutic diagnosis.

With regard to the rehabilitation of the PFM, biofeedback plays an important role and can be assisted by electrostimulation, either excitatory or analgesic. Similarly, a pelvic floor rehabilitation program includes different types of therapeutic modalities (Table 34.1), emphasizing the kinesiotherapy or pelvic floor muscle training (PFMT), vaginal cones, and also the use of a technique very important for sensorimotor training, which is rectal balloon training (RBT) [2, 3].

L. C. C. Oliveira (ed.), *Anorectal Physiology*, https://doi.org/10.1007/978-3-030-43811-1_34

Pelvic floor dysfunction can cause poor quality of life, especially among women, and the aforementioned physiotherapeutic procedures can act in a curative manner or promote a marked improvement in the function, symptomatology, and quality of life of patients. The different disorders that may benefit from the rehabilitation of the pelvic floor are listed below:

- Urinary incontinence: physiotherapy treatment is most effective in stress, urgency, and mixed incontinence.
- Anal incontinence: postsurgical (as in hemorrhoidectomies, fistulectomies, etc.), underactivity of the external sphincter of the anus, or traumatic origin (obstetrics, accidents, impalation).
- Constipation: by outlet obstruction, anismus or paradoxical contraction of the puborectal muscle, or spastic pelvic floor syndrome.
- Anorectal pain: as in proctalgia fugax and levator ani syndrome, among others.
- Posterior vaginal wall prolapse (rectocele).
- Anterior vaginal prolapse (cystocele).
- Chronic pelvic pain.
- Descending perineum syndrome.
- Sexual dysfunctions.

Pelvic Floor Muscles (PFM)

The pelvic floor covers the base of the abdominopelvic cavity and closes the cavity below the pelvic bone, provides support for the pelvic viscera, and actively participates in maintaining the normal function of these organs. It consists of three layers: endopelvic fascia, pelvic diaphragm, and urogenital diaphragm (Fig. 34.1).

The endopelvic fascia, also known as visceral pelvic fascia, has an important role to suspend the viscera mechanically and to provide support for pelvic structures [4].

The pelvic diaphragm is formed by layers of muscle and fascia within the pelvis, which are supported as a mesh extending from the pubis to the coccyx. The pelvic diaphragm muscles surround the openings of the urethra, vagina, and rectum. They are constituted by the elevators of

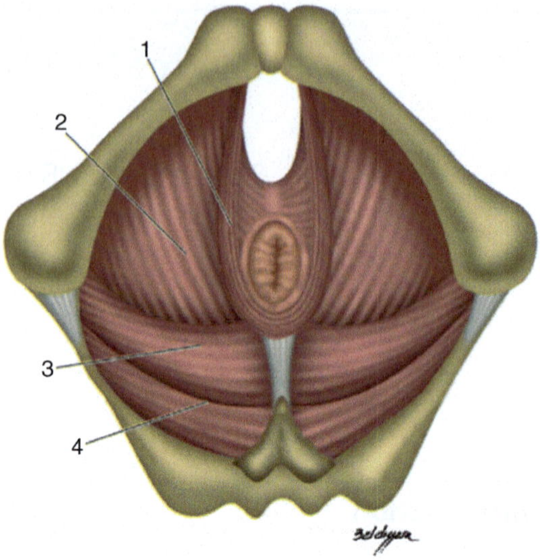

Fig. 34.1 Pelvic floor muscles: (1) puborectal muscle; (2) pubococcygeus or pubovisceral; (3) iliococcygeus; (4) Ischiococcygeus

the anus (formed by the pubococcygeal, puborectal, and iliococcygeus) and ischiococcygeus muscles. The urogenital diaphragm is formed by the deep transverse muscle of the perineum, the urethral sphincter, and the superficial perineal muscles, which include the bulbospongiosus, the ischiocavernosus, the superficial transverse muscle of the perineum, and the external anal sphincter. The perineum is the set of soft tissues that close the pelvis and hold the viscera in upright position. It is delimited by osteofibrous structures, containing in the anterior portion the pubic symphysis and the ischiopubial branches and, in the posterior portion, the sacrum, the coccyx, and the great sacroischial ligament [2].

In women, the perineum is crossed by the urethra, vagina, and anus; it is considered a passage zone, as well as a zone that supports part of the lower trunk. As these two functions are very different, they also require two distinct properties: great sensory and motor coordination, acting intermittently to allow the outputs with great flexibility, and great strength, to ensure firm submission during the remaining time.

These muscles are innervated by the pudendal nerve and by direct branches of the motor roots of S3 and S4 and are composed of type I (slow) and type II (fast) fibers and play an important role in maintaining continence. The muscle levator ani or levators are the most important and essential musculature for the therapist's approach, since it is through it that the muscular qualities of the pelvic floor are developed [2, 5]. The levators, by means of their ramifications, wrapped the proximal urethra and may therefore represent a second sphincter.

Levator Ani Functions

During labor, the levator ani muscles play an essential role in position changes, particularly in the rotation and orientation of the fetal head, which comes into contact with the homolateral levator which, by its tonicity and direction of the passage, places it in contact with the opposite levator. The tonicity of the two bundles, in turn, causes the head to advance forward and downward. On the other hand, the levator represents the last obstacle to presentation, since overcoming this muscular ring requires maximum relaxation of these muscles, which are heavily distended [2]. The physiotherapist aims at the entire postpartum approach to increase awareness of this musculature, contributing to the prevention of perineal complications of childbirth. For sexual function the levator muscles are responsible for vaginal dynamics involving the vagina and favoring the container/content adaptation. Strengthening of these muscles can improve the quality of sexual intercourse.

Use of Physiotherapeutical Modalities

A positive attitude is an essential component of any medical treatment program, especially in the case of pelvic floor rehabilitation. It is essential that the patient takes part in the rehabilitation process and that the patient is motivated to participate in the sessions.

A record of emptying the bladder or the use of bowel and urinary diaries should be performed for at least 2 weeks before starting a behavioral program. The International Continence Society (ICS) recommends completing a diary for three consecutive days. It should include the number of incontinence accidents, activities associated with accidents, sensation during involuntary loss (urgency, level of perceived sensation), periods of regular micturition, and also defecation as well as fluid intake.

The initial assessment of the patient has been addressed already in previous chapters and should include detailed medical history as well as a rectal or vaginal examination for evaluation of bladder, urethra, and rectum prolapses, as well as muscle strength and the patient's ability to control pelvic muscles [6]. Next, the physiotherapeutic modalities used in the rehabilitation of the pelvic floor will be described.

Biofeedback Therapy

Biofeedback is a process of guiding patients so that they can learn to control some of the physiological events that are not under their control. One of the great difficulties in muscle training, especially in the PFM, is to make the individual perceive if he is contracting and/or relaxing the correct muscles and with sufficient strength, either in magnitude or duration, and with coordination and adequate relaxation. Basmajian defines biofeedback as:

"A technique that uses equipment, usually electronic, to reveal to humans beings continuously and instantaneously some of their normal and abnormal internal physiological events in the form of visual and/or auditory signals in order to teach them to manipulate these events (involuntary or unconscious) through the manipulation of the signs represented." [7]

Biofeedback devices monitor and demonstrate muscle activity on the monitor (screen). Special electrodes are adhered to the perineum, around the anus, or to intracavitary regions (vaginal or anal sensors) to measure PFM activity through the capture of electrical signals from muscular activity that, when processed by a computer, show the motor unit recruitment in a graph, which can be compared to the magnitude and the contraction strength duration performed, but *biofeedback* does not measure strength directly, only motor unit recruitment (Fig. 34.2). An audible signal can also be listened when, for example, the amplitude of the contraction reaches the levels or goals desired during the training.

Methods of using anal biofeedback are described in Chap. 35, biofeedback therapy. In general, it can be performed by electromyography (EMG) or manometry. In both cases, the introduction of an intra-anal probe or intrarectal balloon, respectively, is required. In the case of EMG – biofeedback – three sensors are also used on the abdomen (right lateral oblique muscle). This helps to confirm that the activity of the abdominal muscle is not interfering with the PFM, i.e., biofeedback allows the person to learn to perform a selective contraction of these muscles. This type of training allows greater control of muscle contraction and its duration, as well as the training of relaxation of these muscles. Muscle activity, that is, its relaxation and contraction, can be seen by the EMG tracings (Fig. 34.3).

Therefore, the purpose of biofeedback is to modify an inadequate physiological response or to provide the acquisition of a new physiological response. With the aid of biofeedback, learning will be done in three stages. The first step represents the detection and amplification of a function by the apparatus and translation into a visual and/or sound signal immediately available to the patient. The second step is to show the patient that he is able, by a voluntary act, to strengthen, attenuate, or maintain the function that is shown by the visual or auditory signals. This function, of which he has no knowledge (pelvic floor), is shown to him by signals that he can manipulate. It is the stage of awareness, an essential and even exclusive biofeedback role. The third step is the automation of this function, obtained by the manual work of perineal blockage under stress. In this phase, biofeedback, in particular the wireless device, will promote the integration of the PFM function into the patient's body image. This operant conditioning is a closed loop learning, and the feedback is permanent. At any time during the course of action, the patient can act in one direction or the other. In human learning the information represents the ideal reinforcement [2].

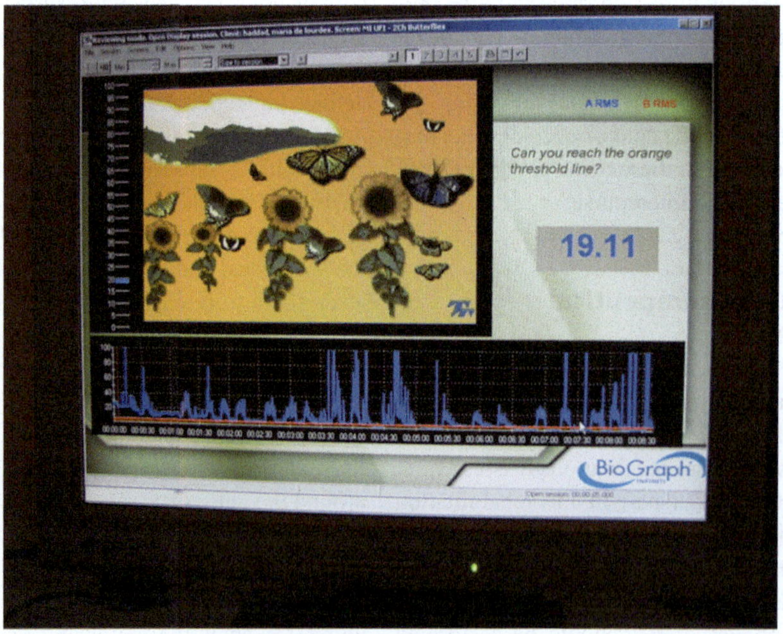

Fig. 34.2 Electromyographic biofeedback screen

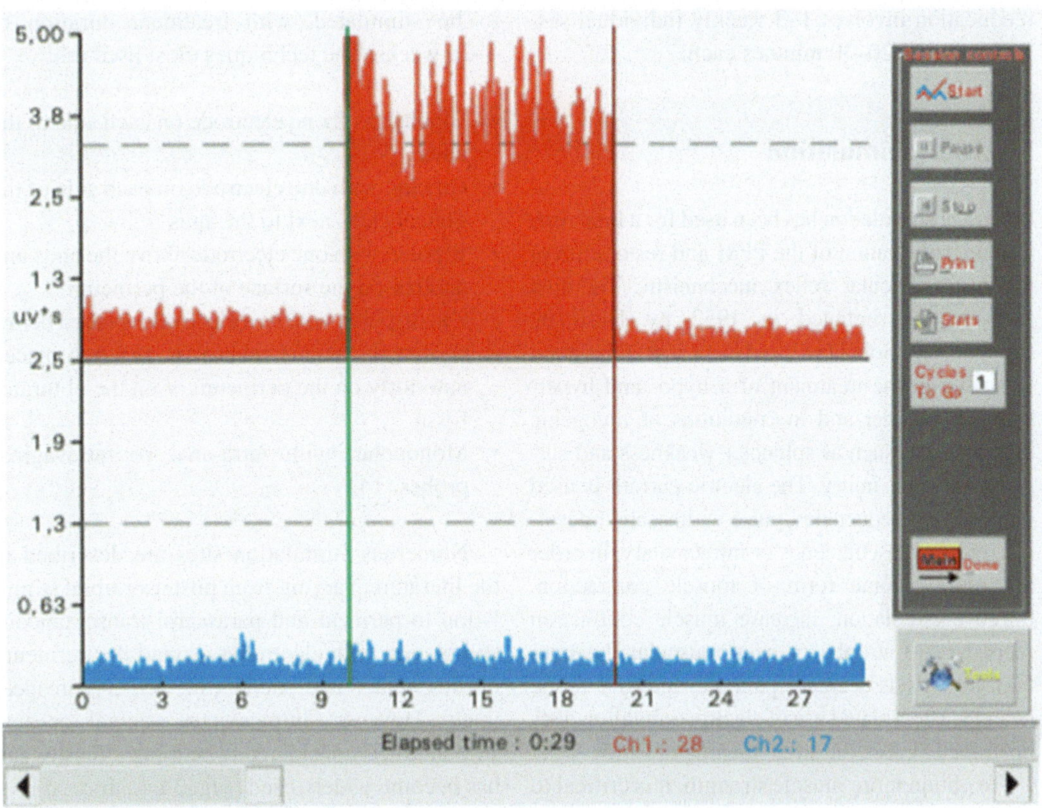

Fig. 34.3 Sustained contraction exercise for 10 seconds

Finally, the goal of using biofeedback as a therapeutic modality is to train the individual first to recognize the sensation of transient relaxation of the internal anal sphincter (IAS) and then to respond to that sensation with the closure of the external anal sphincter (EAS) strong enough to prevent fecal leakage. Some patients recognize the sensation of the IAS, but they are not able to contract the EAS with enough strength to prevent any loss. In such cases, the use of biofeedback should be added with PFMT to help strengthen the EAS response and increase its capacity.

In biofeedback training the skeletal muscles are being trained, mainly the EAS and the puborectalis muscle and the pelvic floor muscles in general. But functional training must be added when the patient has awareness of the right contraction, especially with the wireless device. Functional training of the PFM needs to be incorporated in the treatment by the pelvic physiotherapist to mimic daily life activities and situations in which the patients used to have incontinence, in order to realize fully automatic avoidance of leakage. If successful, the patient will be extremely motivated to adhere and continue the PFMT [8].

Reeducation may be useful in cases of sphincter incontinence, as well as when there is a change in the perception of defecating desire. There are no clinical, manometric, defecographic, or electromyographic predictive factors [8]. However, the presence of total denervation of the pudendal nerve with absence of objective sphincter contraction contraindicates reeducation by this technique [9–12].

Reeducation may be useful in cases of small sphincter injury (maximum 1 quadrant or 25% of the sphincter) [11] and particularly as a complement to repair surgery (sphincteroplasty). The reeducation involves 1–3 weekly individual sessions, lasting 20–30 minutes each.

Electrical Stimulation

Electrical stimulation has been used for a long time in the rehabilitation of the PFM and restoration of the neuromuscular reflex mechanism. The first studies were initiated in 1952 by Huffman, Osborne, and Sokol [13], and electrical stimulation was used in the treatment of a hypo- and hyper-reflexive bladder and in conditions of myogenic incontinence, such as sphincter weakness and surgical sphincter injury. The electric current is used with various frequencies, pulse widths, and intensities, either transcutaneous or intracavitary, in order to promote some form of muscle contraction, improve circulation, increase muscle contraction secondary to atrophy or neuromuscular dysfunction, as well as decrease pain and improve tissue healing. The isolated use of electrostimulation does not directly cause the strengthening of PFM.

To obtain more muscle strength, it is critical to combine electrical stimulation with PFMT during the treatment session. Electrotherapy is useful in cases where the patient has a weak PFM or has no or insufficient awareness of this contraction, which in clinical practice is rather frequent and a common finding. In such cases, in order to improve or restore awareness of PFM activity, electrotherapy can play an important role in neuromuscular information.

It is through direct stimulation of the pudendal nerves that the electrical stimulation of the pelvic floor produces contraction of the levator muscle and the urethral and anal sphincters, causing concomitantly an inhibitory reflex of the detrusor contraction. There are virtually no side effects, other than local discomfort. Contraindications to the procedure include patients with demand pacemakers, pregnancy, menstruation, neoplastic lesions, infections, and exposed metal implants.

Electrical stimulation is usually performed on a daily basis, three times a week, or at least two sessions per week. The session time varies from 15 to 30 minutes, depending on the type of fiber to be stimulated, with treatment duration of 4–20 weeks. The techniques most used are:

- Bipolar, with one electrode on each side of the anus
- Bipolar, with one electrode on each side of the gluteal cleft, next to the anus
- Bipolar, with one electrode above the anus and another on the surface of the perineum
- Quadripolar, with two electrodes placed below or into the sciatic tuberosity and two placed anteriorly on the perineum or on the obturator fossa
- Monopolar, with intra-anal or intravaginal probes [14]

Numerous stimulation sites are described in the literature, varying from posterior tibial stimulation to perineal and parasacral transcutaneous stimulation with electrodes around the perineum fibrous center and sacral (S2, S3, S4), respectively. However, intracavitary vaginal or anal stimulation is considered to be more effective and has become widely used (Fig. 34.4a, b).

In general, surface stimulation is reserved for cases where it is impossible to use intracavitary probes, such as fistulas, fissures, or hemorrhoids [12, 15–17]. The relative efficacy of vaginal and anal stimulation has led to numerous controversies [12, 18, 19]. However, some studies have shown that, from a theoretical point of view, anal stimulation is more effective than vaginal stimulation. The explanation is justified by the difference in impedance because, in the anal canal, there is a greater density of the nervous afferents, as well as the smaller distance between the electrodes and the pudendal nerve [19].

It is worth mentioning that, in order to use electrical stimulation as therapy, parameters such as frequency, pulse width, intensity, and working time-rest time are fundamental for effective application. The intensity of the current may vary between 2 and 80 mA, but it should be used with sufficient intensity to produce at least 65% of the maximum voluntary contraction, in order to result in increased muscle strength, i.e., motor level stimulation.

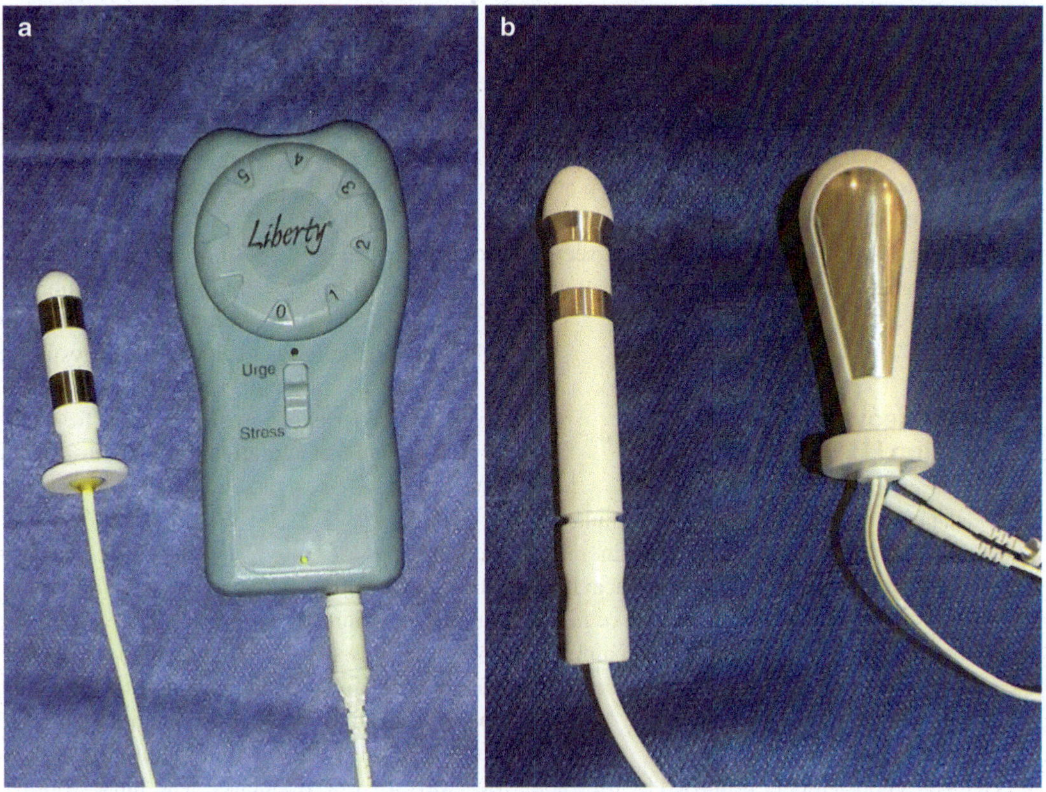

Fig. 34.4 (**a**) Electrostimulation device with a probe. (**b**) Anal and vaginal probes

The choice of electric current frequency is based on the type of muscle. The levator ani and external sphincter muscles consist of motor units of slow and fast fibers, which often respond best to frequencies from 10 to 20 Hz and 30 to 60 Hz, respectively, and pulse duration ranging from 250 μs to 1 ms.

Initially, in the case of a weak muscle, the resting period must be twice the stimulation period, progressively evolving until equal time is reached between the passage of the current and the rest time. Frequency and pulse width are two inseparable parameters in terms of amount of energy, pulse width being the determining factor for the desired type of recruitment. It can be stated that the smaller the pulse width, the greater the intensity of the current for its effectiveness. There is yet no consensus about whether maintenance treatment is necessary, so further studies should be carried out for this purpose, but what is currently used are biphasic, low-frequency currents and also, more recently, interferential current (medium frequency) with a patient-tolerable intensity and a period of treatment for 3–6 months [20, 21].

Recent data show that, in case of underactivity of the EAS leading to incontinence, the frequency of the electric current should be between 50 and 100 Hz, at a current intensity of 65–100 mA. For urgency incontinence, the ideal parameters may be 4 and 10–20 Hz, and current intensity ranges from 35 to 100 mA.

Although there is no agreement among these parameters in the specific literature, most scientific papers and publications consider these as ideal. As the stimulation of the pelvic floor is predominantly indirect, reflex, by the contributions of the internal pudendal nerve, to be effective, the probe should be positioned as close as possible to the afferents of the pudendal nerve [2]. Relevant literature regarding effects of electrical stimulation reports cure or improvement rates from 54%

to 77% of patients, with minimal reported side effects [20, 21].

In a recent systematic review, the authors concluded that there is sufficient evidence showing the efficacy of biofeedback associated with electrical stimulation in the treatment of anal incontinence. They also stated that the interference current associated with biofeedback seems to be the safest and most effective treatment modality [21].

Transcutaneous Electrical Nerve Stimulation (TENS)

Transcutaneous electrical nerve stimulation (TENS), which is used to achieve pain relief through cutaneous electrodes or intracavitary probe, will stimulate small-diameter nerve fibers and release endorphins or act upon the pain gate theory. The two most recognized theories to explain the effects of TENS on pain are:

1. Pain gate control theory proposed by Melzack and Wall [22]
2. The release of endogenous opioids during the application of acupuncture TENS for patients with chronic pain [23, 24]

TENS is a treatment modality with proven efficacy in controlling acute and chronic pain syndromes [16, 23].

Modes of Stimulation

The modes of stimulation are conventional, acupuncture-like, burst and brief-intense TENS, which are combinations of parameters chosen to influence pain relief. They are determined by adjusting the frequency parameters, the pulse width, and the current intensity of the TENS device. There are devices that also allow changes in frequency modulation. The most commonly used modes of stimulation will be described below.

Conventional Mode

It is the most commonly used TENS modality and aims to activate large alpha and beta proprioceptive myelinated nerve fibers. The parameter set for conventional mode consists of frequencies of 50–150 pulses per second (pps) and pulse duration of 20–100 μs, with sufficient intensity for the patient to feel a sensation of "tingling," but without muscle contraction.

Acupuncture-Like TENS

This mode produces muscle contraction in myotomes as a result of descending efferents inhibiting ascending nociceptive pathways in the dorsal horn of the spine at multiple segmental levels and consists of low frequencies, from 1 to 4 pps, with pulse duration of 150 to 250 μs. Intensity is based on patient tolerance.

Burst Mode

TENS burst is similar to *acupuncture-like* TENS, but the accommodation effect is less. This mode combines low-frequency (1–4 Hz) pulse frequency with high internal frequency (70–100 Hz), so that high pulse frequencies are emitted in fixed trains or bursts of 5–7 per pulse between the carrier wave of low frequency. The pulse width is 250 μs, and the intensity of current is high.

Brief-Intense TENS

This mode is the least tolerated by the patient. The parameters set for the brief-intense mode consist of about 100 pps, with a pulse width of 150–250 ms, and at intensity regulated based on patient tolerance, yet strong enough to promote tetanic muscle contraction or muscle fasciculation depending on the location of the electrode. According to Melzack, this mode of brief-intense stimulation can break the memory of pain, being the most effective mode of stimulation to induce short-term electroanalgesia [22].

Kinesiotherapy (PFMT)

Kinesiotherapy, or PFMT, is one of the treatment modalities most used by the physiotherapist in order to restore and improve the musculoskeletal system. It is important for the therapist to know the exact medical diagnosis so that the pelvic floor reeducation program is fine-tuned on that, indi-

Table 34.2 Oxford grading scale score for assessing muscular strength

0	No muscle activity
1	Minor muscle "flicker"
2	Weak muscle activity without a circular contraction
3	Moderate muscle contraction
4	Good muscle contraction
5	Strong muscle contraction

vidualized, and effective [23]. It is essential that the patient is aware of the PFM, and the physiotherapist should then make a complete functional assessment taking into account the so-called 4F program: this is "find," "feel," "force," and "follow through" so that at the onset of the PFMT all parameters for the training are correct and appropriate. Thus, extensive history taking, documentation of symptoms, identification of etiologic and prognostic factors, internal vaginal and rectal exams, and pressure assessment by manometric or electromyography examination should be performed. In clinical practice, assessment of pelvic floor muscle strength test is often performed using the Oxford scale, which evaluates the presence and intensity of voluntary contraction of the pelvic floor, graduating from 0 to 5 (Table 34.2).

The baseline score obtained by the patient will be recorded and will be used for later evaluations. This score provides information about the amount of lifting (support function) and closure (sphincter function) by the PFM. A good functional assessment of the PFM helps to determine the possible duration of rehabilitation, as well as the potential for rehabilitation. Another important parameter is endurance, 50% of the person's maximal voluntary contraction, which is the ability to maintain a moderate muscle contraction for 30 seconds or more. Therapists also determine how many fast muscle contractions with maximal intensity and complete relaxation can be performed and observe their quality as well as quantity of muscle relaxation, classifying it as delayed, absent, partial, or (in)complete. The resting activity (rest tone) between contractions is also evaluated, specifically looking for deficiencies that occur with an altered activity.

Muscle coordination and contraction of other muscle groups, especially the buttocks, adductors, and abdominals, are also observed and evaluated. There are also other problems, such as the presence of trigger points in the pelvic floor, reduced sensitivity, and myofascial scarring or adhesions, as they may promote an obstacle to muscle strengthening.

PFMT aims to improve urethral resistance through active exercise of the pubococcygeus muscle. These exercises strengthen the voluntary pelvic and periurethral muscles. The contraction exerts a closing force on the urethra and increases the muscular support for the visceral pelvic structures, so that the same occurs when these exercises are done for the anal region.

Some women lose cortical control of the pelvic floor muscles after severe pain in the perineum following delivery, and neuropraxia and even denervation have been observed [24, 25].

Pelvic floor muscle exercises are valuable for their strength properties and pain relief. They also speed up healing, reducing edema and stimulating good circulation. After delivery or surgery, these exercises can begin as early as the doctor allows [26]. Furthermore, they should be performed constantly, since any muscle in the body, without being used, weakens very quickly. For this, it is necessary to use a schedule with home exercises to maintain the strength achieved (Fig. 34.5).

The frequency and amplitude of activity of the motor unit supplying PFMs are increased by changes in posture and increases in intra-abdominal pressure that occur during activities such as coughing or sneezing and decrease during effort in preparing for defecation [27].

Conditions that promote chronic increase of intra-abdominal pressure, such as chronic constipation, chronic cough, and obesity, predispose to all kinds of pelvic floor dysfunctions, especially in already damaged tissues. For anal continence, the anorectal sphincters and pudendal nerves must be intact; the rectum with normal reservoir function and capacity, normal anorectal sensitivity, and anorectal angle should be maintained by tonic contraction of the puborectal muscles; the PFM should function normally. However, the largest contribution is due to the IAS muscle, which is responsible for 50–80% of the anal resting tone. In order to

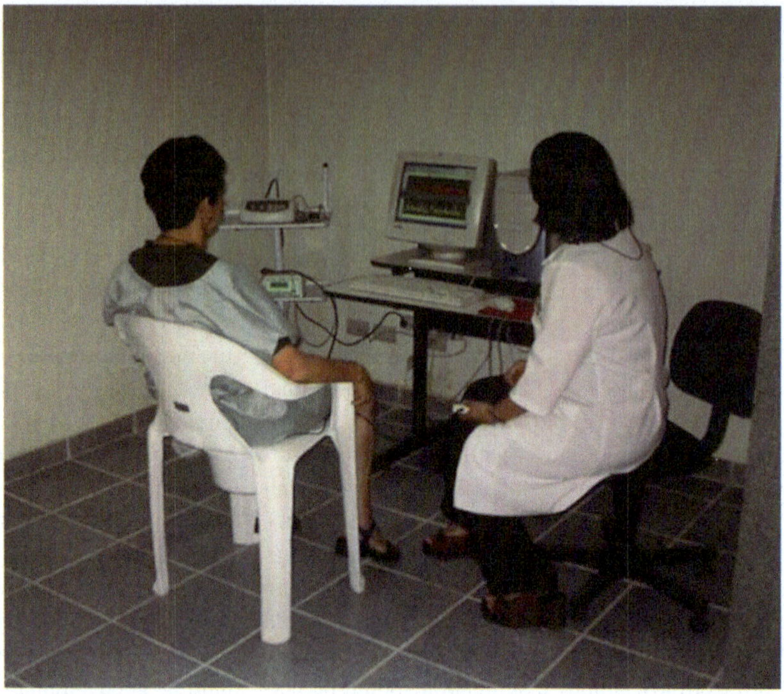

Fig. 34.5 Kinesiotherapy (PFMT)

obtain increase of PFM activity, PFMT should address both types of muscle fibers (consisting of 70% of type I, slow fibers, and 30% of type II, fast fibers). During PFMT, fast contractions should be alternated with sustained contractions and also ask for fast contractions at the end of sustained contractions. The objectives of PFMT are:

- Improve muscle coordination, strength, and endurance.
- Increase muscle cross-sectional area.
- Increase the closing pressure of the urethra and the anus.
- Increase the patient's ability to contract PFM with sudden increase in intra-abdominal pressure.
- Facilitate inhibition of the detrusor muscle by the pudendal-pelvic reflex.

PFMT should be executed in short sessions of intensive (maximal) contractions, from three times a week to every day. According to Bo [27], PFMT involves series of 8–12 maximal contractions holding for 3–8 seconds, alternated with 15 fast contractions between these series and one sustained contraction as long as the person can perform. They can be performed in any posture but preferably in the posture where loss of urine and/or stool occurs. Mild pain or discomfort in the area can be felt in the first few days, mainly because the individual is not used to train the pelvic floor muscles. These symptoms tend to disappear in the first week with continued training. However, if the pain persists or becomes more severe, the training should be discontinued, and the patient should consult a medical doctor for further advice.

It is important to emphasize that relevant literature shows that preferably PFMT should be guided by a skilled physiotherapist, who is experienced to offer the appropriate intensity and duration of PFMT.

Exercises with Vaginal Cones

Vaginal cones are a set of small capsules of anatomical shape, made up of inert, resistant, and relatively heavy material that when inserted in the vaginal canal provides the necessary input for the woman to contract the PFM during the exer-

Fig. 34.6 Complete kit of vaginal cones

cises with them. It was Plevnik [28] in 1985, the first to present the cone concept for strengthening the PFM, especially in urinary incontinence. The use of vaginal cones serves as an adjunct to PFMT. The patient uses a kit of cones, which are identical in shape and volume but have increasing weights. A kit is usually composed of five or six cones, with weights varying between 20 and 100 g (Fig. 34.6).

As part of the exercise program, the woman inserts the weighted cone into the vaginal canal with the conical portion resting on the upper surface of the perineal muscle and should seek to retain it by contracting the PFM for a period of 15–20 minutes. This procedure should be done twice a day. The sustained contraction required to retain the cone increases the strength of the pelvic muscles, and the cone weight serves to provide proprioceptive feedback to the contraction of the desired PFM [28–30].

The Swiss ball is used in pelvic floor rehabilitation because it provides proprioception and can be used to help pelvic movement in association with PFM contractions, promoting the perception and increased PFM strength.

Swiss ball exercises improve the sensory perception of PFM with each movement. Thus, spontaneous acceptance of new exercises is promoted, which are also self-motivating [29–34].

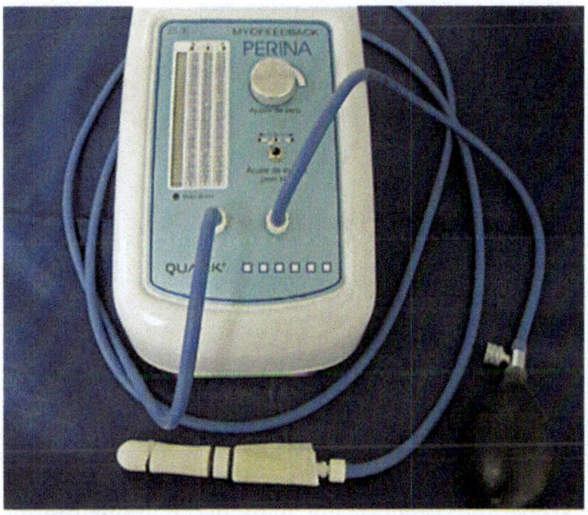

Fig. 34.7 Rectal balloon

Rectal Training with Balloon (RTB)

It is reported that rectal sensation may be more important than the sphincter strength in attempts to relieve the symptoms of constipation or anal incontinence. Rectal balloon training (RBT) (Fig. 34.7) is used to improve rectal sensitivity by gradual reductions in distension of the rectal balloon so that the individual can perceive and distinguish smaller rectal volumes or be able to inhibit urgency using progressive distention of the balloon, through a voluntary anal contraction to neutralize the action of the rectoanal inhibitory reflex (RAIR) in response to the rectal filling.

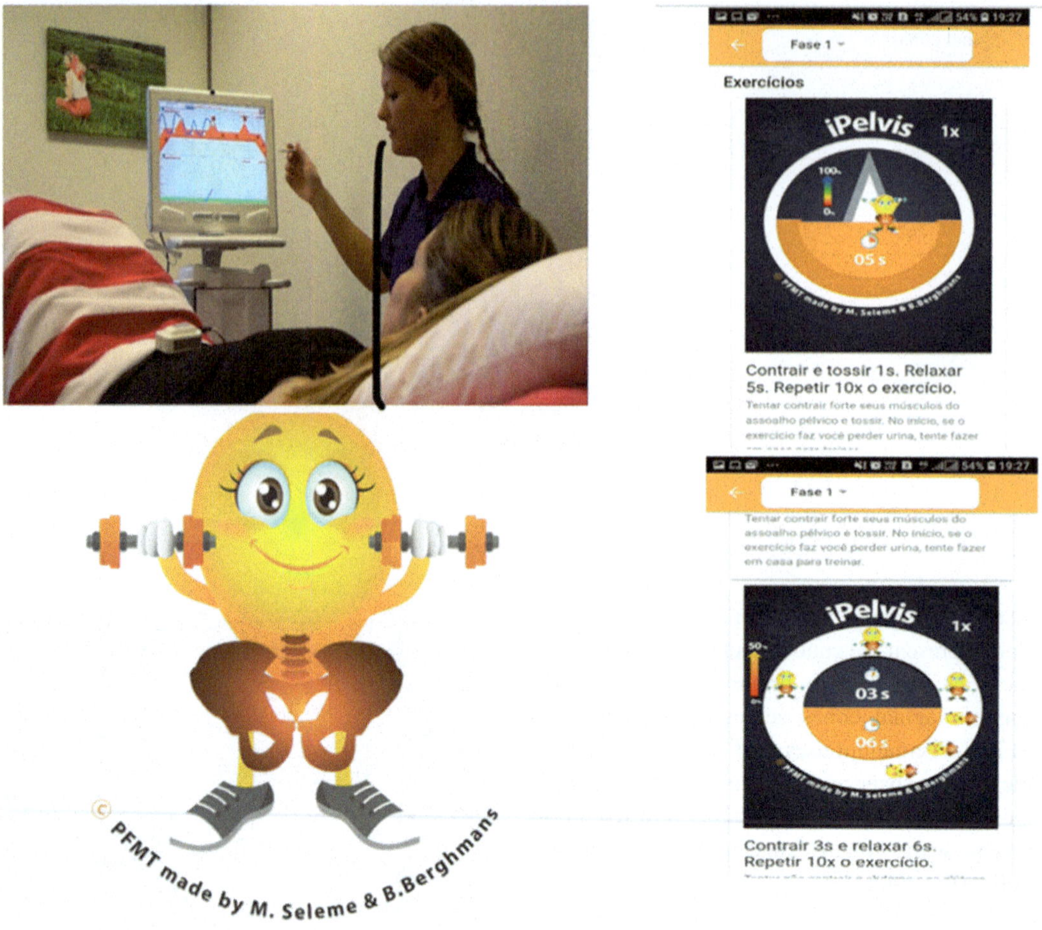

Fig. 34.8 iPelvis app

Since the precise mechanisms responsible for improvement after physiotherapeutic interventions like PFMT or biofeedback remain obscure, some researchers have argued that rectal sensitivity training is the most important element to provide feedback to the individual [35].

Finally, the necessity of prolonged therapy for patients that could be trained and maintained in home training gave impulse for the development of new information technology and apps such as the iPelvis (Fig. 34.8) [36–38]. Therefore, the concept of the "5 F" pelvic floor muscle training focusing on the restoration of continence during sport, work, and daily activities was developed. Motivation, adherence, compliance, and behavioral change are keywords for our clinic- and home-based pelvic floor muscle training. This innovative mApp iPelvis contains very attractive and useful components to realize optimal individual functional training. iPelvis connects the patient and healthcare providers; is easily accessible, funny, and attractive; and supports its users in many ways every day. iPelvis shows the patient's progress in an enthusiastic way, facilitates improvement of patient's quality of life, puts pressure on the patient to adhere and comply in an empathic way, and helps the patient to face health problems such as urinary incontinence, fecal incontinence, pre- and postoperative surgery, prolapses, and sexual dysfunctions, appealing to the patient's own responsibility to keep on training. The iPelvis system, built on behavior change theories, can serve any health professional to focus his patients on behavior change, using easy and ludic strategies to animate the patient, and does so in a controlled and structured way.

Summary

The beneficial effects of physiotherapy modalities to treat pelvic floor disorders have been demonstrated in different studies, with an improvement in up to 70–80% of patients. In addition, satisfaction rates can be achieved in up to 40–50%, resulting in a better quality of life [2, 29–31].

The use of biofeedback in combination with other physiotherapeutic modalities may promote an improvement in 55–87% of cases of anal incontinence. Moreover, electrical stimulation has been reported to cure or improve 54–77% of cases, with minimal side effects [3, 26, 29, 32, 39]. The results reported in the literature are described as successful in 70% of cases in the short term, whereas long-term follow-up shows satisfactory results in 50% of patients [33]. Pelvic floor rehabilitation may be used before or after surgical treatment, in cases of surgical failure, or as prevention for pelvic floor dysfunction Multidisciplinary treatment of pelvic floor disorders is associated with improvement in the expected outcomes. The role of the physiotherapist in rehabilitation of the pelvic floor is to help patients to improve symptoms as well as to improve the pre- and postoperative results. The ultimate goal is improvement in quality of life.

References

1. Shelly B. Assoalho pélvico. In: Hall MC, Brody LT, editors. Exercício terapêutico na busca da função. Rio de Janeiro: Guanabara Koogan; 2001. p. 356–90.
2. Grosse D, Sengler J. Reeducação perineal. 1st ed; 2002. p. 13–33/61–99.
3. Kegel AH. Physiologyc therapy for urinary stress incontinence. JAMA. 1951;146:915–7.
4. Bent AE, Ostergard DR, Cundiff GW, Swift SE. Ostergard's urogynecology and pelvis floor dysfunction. 5th ed: Lippincott Willians & Wilkins; 2003.
5. Calais Germain B. El periné femenino y el parto. Ed: Los Libros de la Liebre de Margo, SL; 1998.
6. Corços J, Davis MB, Drew S, West L, Incontinência urinária e fecal: o uso do biofeedback eletromiográfico para treinar a musculatura do assoalho pélvico, 1997.
7. Basmajian JV. Biofeedback – principles and practice for clinicians, vol. 1. 3rd ed; 1989. p. 281–5.
8. Berghmans B. Pelvic floor muscle training: what is important? a mini-review. Obstet Gynecil Int J. 2017;6(4):00214.
9. Enck P, Van Der Voort IR, Klosterhalfen S. Biofeedback therapy in fecal incontinence and constipation. Neurogastroenterol Motil. 2009;21:1133–41.
10. Binder-Mcleod AS. Biofeedback eletromiográfico para melhorar o controle motor voluntário. In: Robinson AJ, Snyder-Mackler L, editors. Eletrofisiologia clínica. 2nd ed. Porto Alegre: Artmed; 2001. p. 386–92.
11. Kim YS, Weinstein M, Raizada V, Jiang Y, Bhargava V, Rajasekaran R, et al. Anatomical disruption & length-tension dysfunction of anal sphincter complex muscles in women with fecal incontinence. Dis Colon Rectum. 2013;56(11):1282–9.
12. Scott KM. Pelvic floor rehabilitation in the treatment of fecal incontinence. Clin Colon Rectal Surg. 2014;27:99–105.
13. Huffman JW, Osborne SL, Sokol JK. Electrical stimulation in the treatment of intractable stress incontinence: a preliminary report. Arch Phys Med Rehabil. 1952;33(11):674–6.
14. Pickles B, Compton A, Cott C, Simpson J, Vandervoort A. Fisioterapia na terceira idade. São Paulo: Sartos Livraria editora; 1998. p. 230–54.
15. Uchôa SM. Aplicações da Eletroestimulação nas Disfunções do assoalho Pélvico. In: Agne JE, editor. Eletrotermofototerapia. 1ª edição: Santa Maria; 2013. p. 507–11.
16. Uchôa SM, Matos M. Técnicas de reeducação anorretal. In: Amaro JL, Haddad JM, Trindade JC, Ribeiro R, editors. Reabilitação do assoalho pélvico nas disfunções urinárias e anorretais. ed. São Paulo: Segmento Farma; 2012. p. 311–27.
17. Plevnik S, Vodusek DB, et al. Optimization of pulse duration for electrical stimulation in treatment of urinary incontinence. World J Urol. 1988;4:22–3.
18. Davis HL. Is electrostimulation beneficial to denervated muscle? A review of results from basic research. Physiother Can. 1983:55–66.
19. Martellucci J. Electrical stimulation for pelvic floor disorders. Ed. London: Springer; 2015. p. 89–90.
20. Martins FG. Tratamento conservador da incontinência urinária feminina. Urologia Contemporânea. 2000;6(1):21–6.
21. Vonthein R, Heimerl T, Schwandner T, Ziegler A. Electrical stimulation and biofeedback for the treatment of fecal incontinence: a systematic review. Int J Color Dis. 2013;28:1567–77.
22. Melzack R, Wall PD. Pain mechanisms: a new theory. Science. 1965;150:1971.
23. Schüssler B. Aims of pelvic floor evaluation in: pelvic floor reeducation principles and practice: Springer; 1994. p. 40.
24. Snooks S, Barnes P, et al. Damage to the innervation pf the voluntary anal and periurethral musculature in incontinence: an electrophysiological study. J Neurol Neurosurg Psychiatry. 1984;47:1269–73.
25. Snooks S, Stechell M, et al. Injury to innervation of pelvic floor sphincter musculature in childbirth. Lancet. 1984;324:546–50.

26. Polden M, Mantle J. Fisioterapia em ginecologia e obstetrícia. Rio de Janeiro: Livraria Santos; 2002. p. 235.
27. Jundt K, Peschers U, Kentenich H. The investigation and treatment of female pelvic floor dysfunction. Dtsch Arztebl Int. 2015;112:564–74.
28. Plevnik S. New methods for testing and strengthening the pelvic floor muscles. In: Proceeding of 18o Annual Meeting of ICS. London; 1985. p. 267–8.
29. Carrière B. The Swiss ball. In: Theory, basic exercises and clinical application: Springer; 1998. p. 327–58.
30. Sartori DVB, Gameiro MO, Yamamoto HA, Kawano PR, Guerra R, Padovani CR, et al. Reliability of pelvic floor muscle strength assessment in healthy continent women. BMC Urol. 2015;15:29.
31. Grosse D, Sengler J, et al. Les techniques de stimulation dans la reeducation vesico-sphinctérienne. J Urol. 1993;99(5):229–42.
32. Janez J, Plevnik S, Suhel P. Urethral and bladder responses to anal electrical stimulation. J Urol. 1979;122:192–3.
33. Susset JG, Galea G, Read L. Biofeedback therapy for female incontinence due to low urethral resistance. J Urol. 1990;143:1205–8.
34. Guillemont F. Reeducation perineale et incontinence fecale. Explorations fonctionnelles digestives. 2 Décembre 2000.
35. Roongsirisangrat S, Rangkla S, Manchana T, Tantisiriwat N. Rectal balloon training as an adjunctive method for pelvic floor muscle training in conservative management of stress urinary incontinence: a pilot study. J Med Assoc Thail. 2012 Sep;95(9):1149–55.
36. GFS L, et al. An ideal e-health system for pelvic floor muscle training adherence: systematic review. Neurourol Urodyn. 2019;38(1):63–80.
37. Dumoulin C, Hay-Smith EC, Mac H-SG. Pelvic floor muscle training versus no treatment, or inactive control treatments, for urinary incontinence in women. Cochrane Database Syst Rev. 2014;14:CD005654.
38. Berghmans B. Pelvic floor muscle training: what is important? a mini-review. Obstet Gynecol Int J. 2017;6(4):00214. https://doi.org/10.15406/ogij.2017.06.00214.
39. Bo K. Pelvic floor muscle strength and response to the pelvic floor training for stress urinary incontinence. Neurourol Urolodyn. 2003;22:654–8.

Biofeedback

35

Lucia Camara Castro Oliveira
and Rebeca Kisel Laska

Introduction

The first report on biofeedback therapy dates back to 1969, with a description of the learning of visceral and glandular responses [1]. Since then, biofeedback therapy has gained more attention and has become a treatment option for a wide array of disorders [2]. In coloproctology, it has become one of the most utilized nonsurgical treatments for both fecal incontinence and constipation. Fundamentally, biofeedback offers a visual or audio feedback for the patient during any muscular training or contraction, which is usually performed unconsciously. In fact, we consider biofeedback as a process of helping the patient to understand the adequate movement or function of a certain muscular group – in coloproctology, that is the action of the anorectal sphincters. Moreover, in order to obtain a successful outcome, a number of measures should be established and followed by both the patient and the doctor (Table 35.1). We consider biofeedback therapy as a planned strategy that involves not only the physiotherapy itself but the items described in Table 35.1.

L. C. C. Oliveira (✉)
Director Department of Anorectal Physiology of Rio de Janeiro and CEPEMED, Rio de Janeiro, Brazil

R. Kisel Laska
Clinica del Piso Pelvico, Department of Urogynecology, Zapopan, Mexico

Table 35.1 Biofeedback program elements

Establish an optimal doctor-patient relation
Motivated doctor or therapist
Explain anatomy and physiological concepts to the patient
Diet and nutritional evaluation and counseling
Establish a bowel diary
Weekly supervision of medications
Instrumental biofeedback therapy with animation screens or visual stimulus
Home exercises with portable instruments when available

The first publication describing the use of biofeedback for a dysfunction of the pelvic floor was by Kohlenberg [3] in 1973, when he treated a young male patient with encopresis. The exact mechanism of biofeedback therapy is still unknown, but a number of theories have been proposed. Rectal awareness clearly improves, and rectal sensation and the striated muscles can improve contraction [4, 5]. There are controversies in the literature that are likely related to different systems and methodologies utilized and the selection of patients, which make comparisons between institutions unreliable. Nevertheless, according to the guidelines of the American Society of Colon and Rectal Surgeons (ASCRS) and the American College of Gastroenterology (ACG), biofeedback therapy is a first-line treatment for patients with fecal incontinence and dyssynergic defecation (Table 35.2) [6–9].

L. C. C. Oliveira (ed.), *Anorectal Physiology*, https://doi.org/10.1007/978-3-030-43811-1_35

Table 35.2 Indications for biofeedback

Anal incontinence	Weak sphincter – any etiologies After vaginal delivery After sphincter repair Before closure of a colostomy After resection of rectal prolapse
Urinary incontinence	
Dyssynergic defecation	With anismus With rectocele Pelvic pain Solitary rectal ulcer

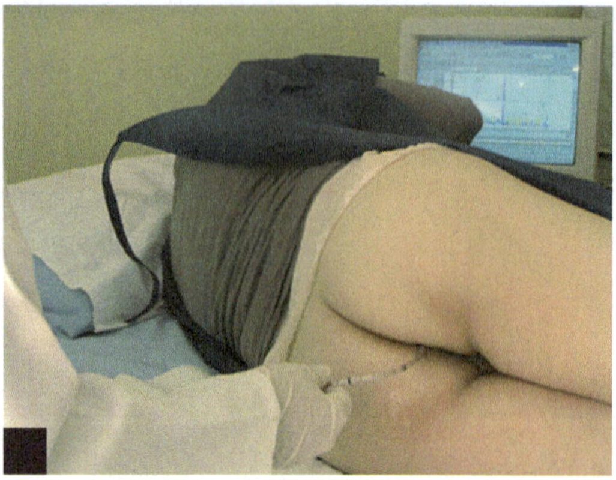

Fig. 35.1 Patient position for manometric and balloon biofeedback therapy

Methods and Technique

Different systems can be utilized for biofeedback therapy: manometry systems, balloon, electromyography, and even ultrasound systems [10–12]. Regardless of the method employed, biofeedback is a safe, minimally invasive, and inexpensive technique that requires a motivated therapist and patient. The doctor or therapist must understand the principles of anatomy and physiology of the pelvic floor. The room must be confortable and quiet and the patient must feel confident to perform the exercises. One of the few contraindications is hearing or visual impairment. A rectal preparation or special diet is typically not required. No matter what system is utilized, for fecal incontinence or dyssynergic pelvic floor, the patient is usually placed in the left lateral position with flexed legs (Fig. 35.1). This allows visualization of the screen by the patient, which is part of the process for effective biofeedback session.

In principle, biofeedback therapy can help the patient exercise the anorectal and pelvic floor muscles and also improve awareness of fecal contents [4, 5]. Kegel was the first to describe working and exercising the pelvic floor muscles [12] for the treatment of urinary incontinence. He utilized a periometer to develop a number of perineal exercises that were efficient for recovery of postpartum urinary incontinence. Kegel's exercises were initially questioned by the medical community, but, over time, his theories were recognized and incorporated by physiotherapists. In fact, one of his greatest contributions was the concept of utilizing an instrument to monitor the results, thereby providing feedback to patients.

There are four main methods for biofeedback therapy: manometric, balloon, with electromyography (EMG), and utilizing ultrasound (Figs. 35.2, 35.3 and 34.7). Ultrasonographic biofeedback has been utilized after prostatectomy and for urinary incontinence in the postpartum period [12, 13].

Patients are instructed to contract and relax the anorectal sphincter, while the probe or electrode is recording the muscle activity on a screen or scale. Physiological activity is then monitored, and unconscious physiological stimulus is provided by audio or by visual instruments to allow the patient to gain control over these sphincteric functions by the completion of treatment. Sessions typically last 40 minutes over a 10-week period, after which patients are instructed to perform home exercises. The treatment outcomes are followed by bowel diaries, and good and excellent results are expected when there is more than 50% and 70% reduction of incontinence episodes, respectively. Poor prognostic factors are low anterior resection syndrome (LARS), anal deformity, obesity, irritable bowel syndrome, and severe muscle damage.

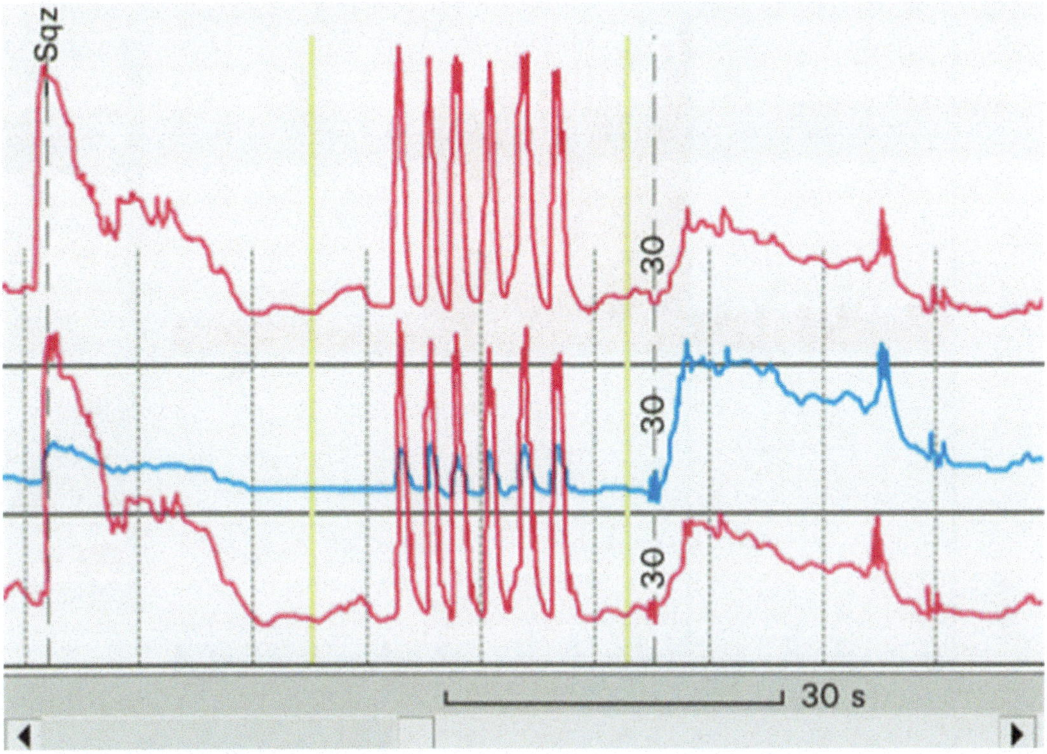

Fig. 35.2 Manometry screen for biofeedback

Biofeedback Results for Anal Incontinence

Among all the nonsurgical treatment modalities available for patients with incontinence, biofeedback has been the most widely used. Although the evidence is based on very heterogenic and non-randomized studies, satisfactory results can be seen in the majority of patients [14].

Incontinence experts from the Cochrane group conducted a recent review of 23 articles on biofeedback for treatment of fecal incontinence, demonstrating that there are in fact only 5 eligible randomized trials, including a total of 109 patients [15]. All studies included a small number of patients and used limited measures to evaluate the results. In addition, follow-up information was not consistently reported and only two studies provided data in suitable form for statistical analysis. Consequently, the limited number of clinical studies associated with a poor methodology did not allow an assessment of the possible role of anal biofeedback for the treatment of fecal incontinence. However, there are suggestions that biofeedback may have a therapeutic effect. In fact, biofeedback is one of the most widely used nonsurgical therapeutic methods for the treatment of fecal incontinence, with success rates reported between 40% and 98% (Table 35.3) [16–29]. In patients with pelvic floor dysfunction, the presence of both anal and urinary incontinence is not unusual. Biofeedback theraphy can therefore help both conditions that were discussed in Chap. 34.

One advantage of biofeedback therapy is that it is a low-cost procedure with minimal morbidity. Biofeedback is an interesting therapeutic option for incontinent patients who do not qualify for surgical treatment, especially in patients who did not have satisfactory sphincter function after successful surgical repair [30].

In a randomized study including 27 incontinent women due to obstetric injury, improvement

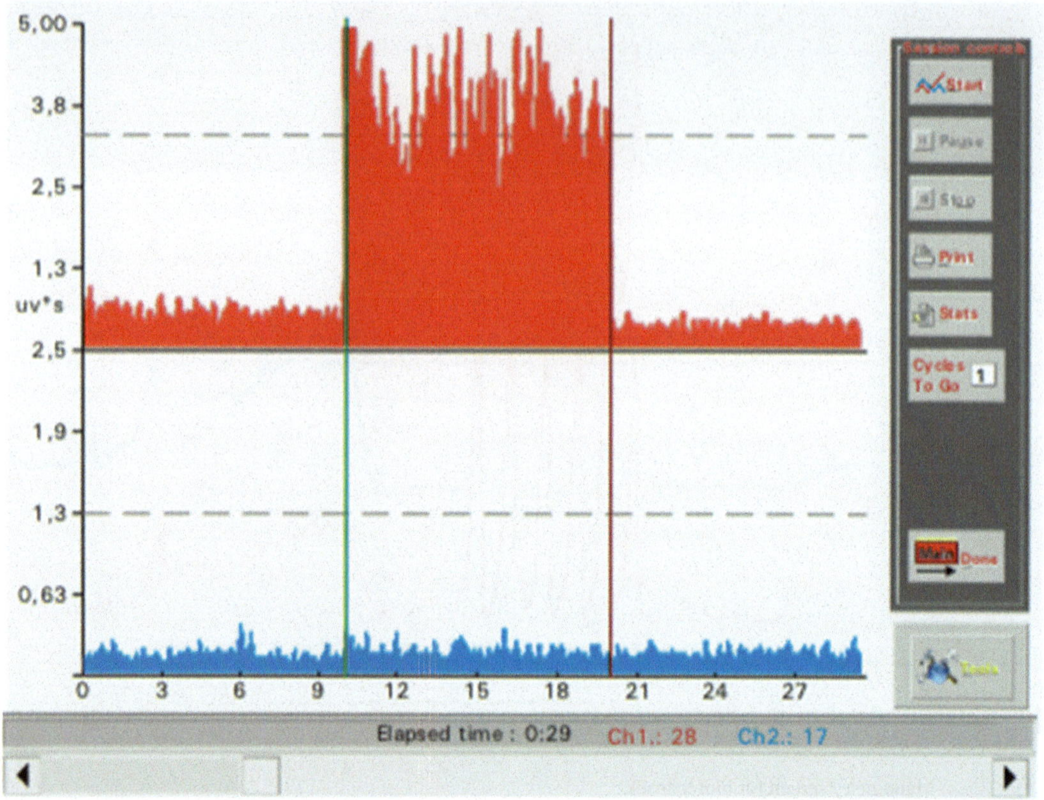

Fig. 35.3 EMG biofeedback screen

in the Cleveland Clinic Florida Fecal Incontinence Score (CCF-FIS) after treatment was significant [29]. Those patients were divided into three groups: I, biofeedback performed 3 months before and 6 months after surgery; II, biofeedback performed only 6 months after surgery; and III, no biofeedback therapy. Patients who underwent biofeedback (groups I and II) presented statistically significant better results. The exact mechanism of action remains debatable and not completely understood. However, some of the proposed mechanisms are:

1. Improvement in muscular strength
2. Improvement in rectal perception and awareness to control rectal contents
3. Improvement in rectal capacity, allowing accommodation of rectal feces

Many investigators have been trying to study the mechanism of biofeedback therapy, but results are contradictory.

For example, some authors have reported an increase in sphincter pressure in patients with satisfactory results, while others such as McHugh [31] Wald [32] and Loening-Baucke [33] found no change in pressure. Others have reported an improvement in rectal sensation that correlates with functional response. Success rates reported in the literature range from 40% to 98%. Despite overall enthusiasm, it has been shown that good results in the first 6 months can deteriorate within 2 years; therefore this data could be useful to reinforce this treatment after the first 6 months. A recent study including 17 children with myelomeningocele reported an improvement in patients' perception and anal sensation after biofeedback therapy [34]. Patients should be motivated. In fact, motivation and interaction with the doctor or therapist are factors associated with the best results. In addition, the patient should be able to understand the muscular movement that he/she is being asked to perform. The most optimal results of this therapy are usually obtained in patients with muscle weakness after

Table 35.3 Biofeedback series for fecal incontinence. Results and outcomes

Author, year	Patients (*n*)	Average length of treatment (weeks)	Control group	BF system	Clinical improvement
Heyman et al. (2000) [16]	40	36	No	EMG	Great improvement
Solomon et al. (2000) [10]	44	8	No	US	40% improvement
Mahony et al. (2004) [18]	54	12	–	EMG	Significant improvement
Ilnycki et al. (2005) [19]	23	–	–	NR	61% improvement
Norton et al. (2003) [20]	171	52	–	Manometry	53% improvement
Naimy et al. (2007) [21]	49	8	No	EMG	NR
Sun et al. (2008) [22]	126	24	No	EMG	98% improvement
Heyman et al. (2009) [23]	108	12	No	Manometry	76% improvement
Bartlett et al. (2015) [24]	75	–	–	Balloon	No improvement in the FISI scale
Leite et al. (2013) [25]	52	–	No	Balloon	No improvement in quality of life
Damon et al. (2014) [26]	157	–	Yes	–	57% improvement
Sjodahl et al. (2015) [27]	64	–	Yes	EMG	Reduced number of leaks (from 6 to 3)
Kuo et al. (2015) [28]	21	–	No	EMG	Improvement in the CCF-FIS (17 to 12)

FISI fecal incontinence severity index, *EMG* electromyography, *CCF-FIS* Cleveland Clinic Florida Fecal Incontinence Score, *NR* not reported

anorectal surgery. Conversely, in patients with neurological impairment (diabetic neuropathy, multiple sclerosis, myelomeningocele), suboptimal results can be expected, since rectal sensation is greatly impaired or completely absent in 70% of patients. This observation was also made by van Tets and colleagues [35] who studied 12 patients with neurogenic incontinence who underwent biofeedback treatment. After 12 weeks of training, no patient improved anal incontinence. Although it may be slow, biofeedback has no adverse effects, causes minimal discomfort to the patient, and is a good choice for incontinent patients with altered sphincter function.

In a long-term study in patients who underwent biofeedback several years ago, Enck et al. [36] reported that improvement in continence was observed in 19 patients not only during treatment but also for several years following therapy. Another long-term study demonstrated that biofeedback therapy improved continence at 6 and 30 months posttreatment in 83 incontinent patients. The authors have also shown that, for many patients, improvement was maintained after treatment [37].

In our personal initial experience with 120 incontinent patients selected for biofeedback therapy, only 66 completed the entire program [38]. This series included 56 females and 10 males with a median age of 66 years. These patients had a median of three biofeedback (1-8) sessions. The overall success was 84%. In an average follow-up period of 12.5 (1-43) (months, the CCF-FIS improved from 11.8 to 5 ($p < 0.0001$). An increase in pressures was also noted after biofeedback therapy from an average of 63 to 81 mmHg ($p = 0.0016$). One parameter related to a successful outcome was the absence of muscle fatigue (good results in 61%), whereas the presence of a severe sphincter defect corresponded with poor results. There were no complications among our patients. These initial satisfactory results made biofeedback therapy our preferred option for the treatment of mild-to-moderate fecal incontinence. We have already treated more than 400 patients with good results in 80% of the selected cases [39]. Our patients are evaluated clinically by the CCF-FIS and FIQL questionnaires before and after ten biofeedback sessions.

In a prospective study, Heymen et al. [23] randomized incontinent patients to biofeedback ($n = 45$) or only pelvic floor exercises ($n = 63$). After a period of 3 months, they observed that 76% of the group treated with biofeedback had improvement as compared with only 41% improvement in the pelvic floor exercise group.

In Brazil, Melão et al. [40] and Leite et al. [25] also reported their experience with biofeedback in 85 and 52 incontinent patients, respectively, utilizing an EMG system with good results.

In a systematic review by Vonthein et al. [41], 13 randomized trials including biofeedback and electrostimulation were analyzed. In 12 trials, at least 1 therapy group received biofeedback alone and/or in combination with electrostimulation, while electrostimulation alone was evaluated in 7 trials. Superiority of biofeedback + electrostimulation over any monotherapy was demonstrated in several trials. Amplitude-modulated medium-frequency (AM-MF) stimulation, also termed pre-modulated interferential stimulation, combined with BF was superior to both low-frequency electrostimulation and biofeedback alone, and 50% of the patients were continent after 6 months of treatment. In fact, the better results when combining biofeedback with electrostimulation were already demonstrated by Schwandner et al. [42] in 2010 in a study including 158 patients.

In a recent a randomized, controlled, superiority trial of 98 incontinent patients in Denmark, patients were randomly assigned to groups that received supervised pelvic floor muscle training (PFMT) and biofeedback plus conservative treatment or attention-control treatment plus conservative treatment [43]. In the intention-to-treat analysis, patients who received supervised PFMT had fivefold higher odds of reporting improvements in fecal incontinence symptoms and a significant improvement as observed by the Vaizey scoring system. In a recent long-term follow-up study including 108 consecutive female patients with fecal incontinence who completed an instrumented biofeedback course, the results of 61 of 89 contactable patients was presented [44]. The authors reported that long-term symptom improvement was observed in more than half of fecal incontinence patients at a 7-year post-biofeedback follow-up. However, they observed that patients improving during the initial biofeedback program had a higher chance of long-term improvement, while patients who did not respond to biofeedback should be considered early for other therapies.

Another recent study compared the effectiveness of four different biofeedback treatment regimes [45]. Three hundred and fifty patients were randomized and divided in four groups: group 1 received four monthly face-to-face biofeedback treatments, groups 2 and 3 received one face-to-face biofeedback followed by telephone biofeedback, and group 4 received a one-off face-to-face biofeedback treatment. All groups had significant improvements in fecal incontinence, quality of life, incontinence score, and mental status ($p < 0.001$ for all). Currently, this is a recommended treatment option largely consistent with society guidelines and consensus statements [6–9]. According to the American Society of Colon and Rectal Surgeons (ASCRS) it should be considered as an initial treatment for patients with incontinence and some preserved voluntary sphincter contraction with a strong grade of recommendation (1B) based on moderate-quality evidence [7].

Results of Biofeedback for Constipation

Patients who present with symptoms of obstructed defecation or outlet obstruction syndrome represent the subgroup of constipated patients that benefit most from biofeedback therapy. In these situations, a non-relaxed, spastic, paradoxically contracted puborectalis muscle or dyssynergic muscle, also known as anismus, can be managed by sphincter retraining through biofeedback [46, 47]. The patients are predominantly female and young. Many of them have a history of sexual abuse and different associated psychiatric and eating disorders. Clinical diagnosis is established through history of major defecation, several attempts to evacuate throughout the day, need for digital assistance, and chronic use of laxatives. The most commonly used diagnostic methods are cinedefecography

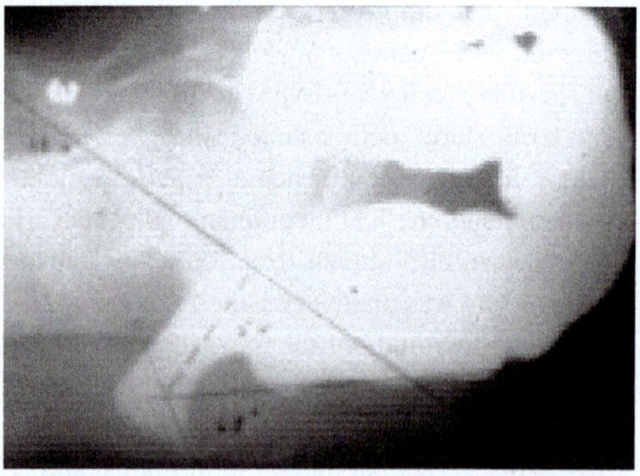

Fig. 35.4 Paradoxical contraction of the puborectalis or anismus on cinedefecography

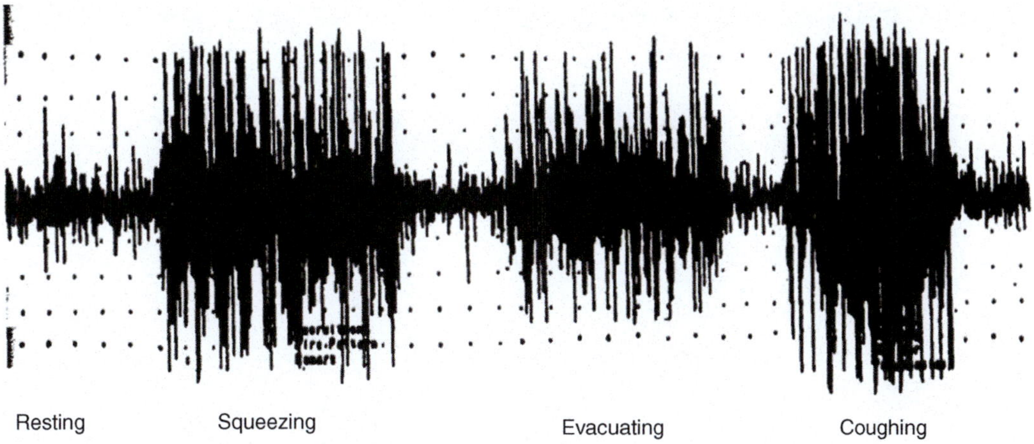

Fig. 35.5 Anismus by electromyography

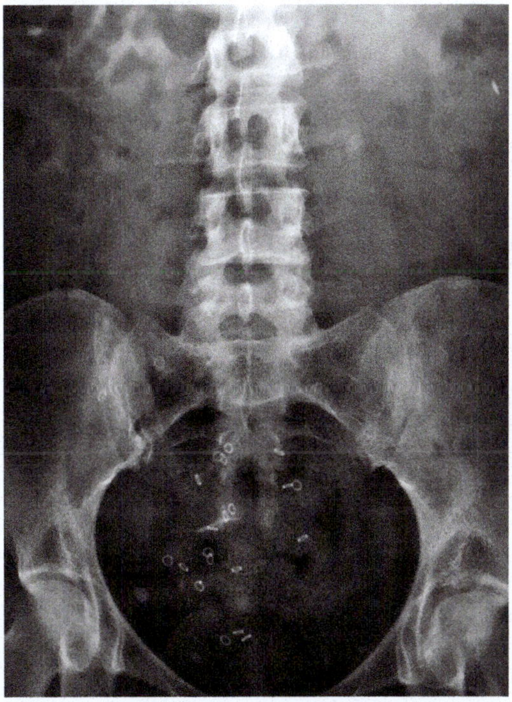

Fig. 35.6 Colonic transit time showing the markers in the rectum in a patient with obstructed defecation syndrome

(Fig. 35.4), EMG, (Fig. 35.5), and colonic transit time (Fig. 35.6). Cinedefecography shows maintenance of the closed anorectal angle, the absence of the lowering of the pelvic floor, and the opening of the anal canal, in addition to the retention of the barium paste, thus confirming the partial emptying of the rectal contents. Electromyography is considered the gold standard for the diagnosis of anismus (Fig. 35.5) but is currently seldom used.

The colonic transit time may exclude the association of ODS with colonic inertia (Fig. 35.6). Other tests that may also be utilized include anorectal manometry, balloon expulsion test, and, more recently, MRI defecography and echodefecography (see Chaps. 8 and 10). Anorectal manometry may demonstrate hypertonic sphincter and no relaxation of striated muscles during a voluntary simulation of evacuation (Fig. 35.7).

The balloon expulsion test is very useful, in that it is a simple and inexpensive method: a rectal balloon filled with 50–100 mL of liquid is introduced, and the patient is instructed to expel it.

The treatment, in these cases, should be performed by an integrated team, including proctologist, nutrologist, psychiatrist, psychologist, gastroenterologist, and physiotherapist. Patients are evaluated weekly and complete an evacuation diary for weekly monitoring purposes. Efforts should be made to discontinue the use of irritant laxatives, while stimulating the introduction of bulking agents such as fiber. Biofeedback therapy can be performed by means of EMG, anal manometry or balloon systems.

As for cases of incontinence, a good doctor-patient relationship and professional and patient motivation are necessary for successful outcomes. Regardless of the system utilized, clinical improvement can be observed depending on the number of sessions performed. In fact, the number of sessions is the most important predictive factor, which was well documented by Gilliland et al. [48] in a study that

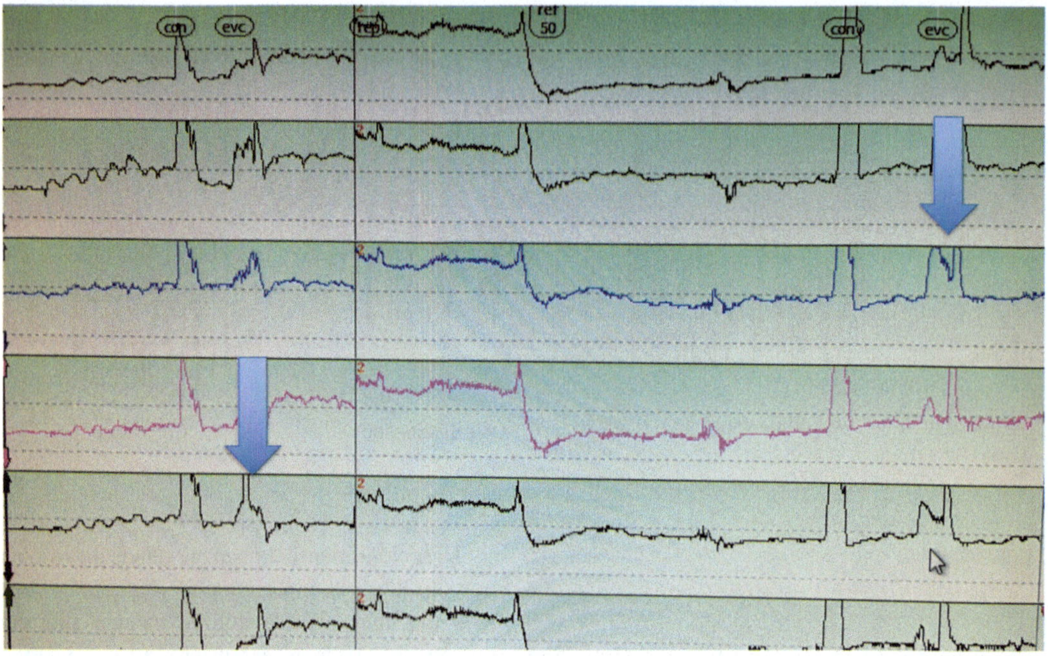

Fig. 35.7 Manometry with non-relaxing pelvic floor

Table 35.4 Biofeedback series for obstructed defecation. Results and outcomes

Author, year	Patients (*n*)	Manometry	EMG	Success rate (%)
Bleijenberg and Kuijpers (1987) [49]	10	No	Yes	70
Weber et al. (1987) [50]	42	Yes	No	48
Lester et al. (1991) [51]	16	Yes	No	44
Kawimbe et al. (1991) [52]	15	Yes	Yes	40
Dahl et al. (1991) [53]	14	No	Yes	93
Wexner et al. (1991) [54]	18	No	Yes	89
Fleshman et al. (1992) [55]	9	Yes	Yes	78
Turnbull and Ritvo (1992) [56]	7	Yes	No	NR
Keck et al. (1994) [57]	12	Yes	No	8
Papachrysostomou and Smith (1994) [58]	22	Yes	Yes	57
Bleijenberg and Kuijpers (1994) [59]	20	Yes	Yes	65
Koutsoumanis et al. (1994) [60]	NR	Yes	No	NR
Siproudhis et al. (1995) [61]	27	Yes	No	52
Leroi et al. (1996) [62]	15	Yes	No	53
Ho et al. (1996) [63]	62	Yes	Yes	90
Park et al. (1996) [64]	68	No	Yes	86
Ko et al. (1997) [65]	NR	Yes	Yes	80
Rao et al. (1997) [66]	25	Yes	No	76
Glia et al. (1997) [67]	20	Yes	Yes	90
Karlbohm et al. (1997) [68]	28	Yes	Yes	43
Rieger et al. (1997) [69]	19	Yes	Yes	23
Patankar et al. (1997) [70]	30	No	No	96
Chiotakakou et al. (1998) [71]	100	No	Yes	55
Kairaluoma et al. (2004) [72]	52	No	Yes	64
Farid et al. (2009) [73]	48	Yes	Yes	25
Heymen et al. (2007) [74]	117	No	Yes	70

NR not reported, *EMG* electromyography

included 194 constipated patients submitted to biofeedback. The authors did not observe differences in the results regarding the use of EMG or manometry and suggested at least five sessions of BF for a successful outcome in the treatment of anismus.

The different series in the literature report success rates between 40 and 80% of the cases (Table 35.4) [49–74].

A prospective, randomized study comparing biofeedback to other conservative alternatives for anismus treatment showed very promising results in 117 treated patients: the use of biofeedback was superior to that of diazepam ($p < 0.001$) and placebo ($p = 0.017$) with an increase in the number of spontaneous evacuations and an improvement in electromyography during evacuation attempts [74].

According to the results of published series, it can be said that the role of biofeedback in the treatment of anismus is well established, with a recommendation according to the American Society of Gastroenterology, Grade IA [6]. In a study analyzing the effects of long-term biofeedback (44 months) in a group of 347 patients with anismus treated with a mean of 5 sessions, the authors reported the follow-up of 103 patients. The initial results were maintained during the 44 months in 85 patients (82.5%) [75].

In 2016, Murad-Regadas et al. [76] evaluated 116 patients submitted to laxative and biofeedback due to anismus, with good results reported in 59%.

Currently, the author's experience includes more than 200 constipated patients treated with biofeedback with manometry and EMG. It is important to emphasize that, in cases of constipation, all measures that are included in what we call the biofeedback program (see Table 35.1) must be strictly followed. It is essential that patients are monitored based on dietary, drugs, and psychological guidelines.

Biofeedback Therapy for Pelvic Pain

Biofeedback therapy can also be utilized for the treatment of painful pelvic floor syndromes, chronic rectal pain, and levator syndrome, already described in Chaps. 29, 30, and 34.

In addition, there are some reports for the treatment of solitary rectal ulcer [77].

In these cases, the mechanism of action of biofeedback is also obscure, and the difficulty in demonstrating possible physiological parameters associated to the improvement consists in the fact that most of the published series, in general, only includes a small number of cases, which cannot be used for comparison.

One of the works with the highest number of cases included 86 patients (55 females) with a mean age of 68 years. Eleven patients performed only one biofeedback session and were excluded. Of the 75 patients who completed the sessions with EMG system, 26 (34.7%) reported improvement in rectal pain [78].

In another study published in the same year, 16 patients underwent biofeedback with manometry and followed for 1 year. Patients reported reduced pain and the need for analgesic and anti-inflammatory drugs [79]. Cornel et al. [80] used biofeedback with EMG for the treatment of chronic pelvic pain syndrome in 33 men who had chronic bacterial prostatitis, 2 of whom were excluded from the study and the remaining 31 presented significant improvement of pain and prostatitis symptoms.

Another recent study, applying biofeedback with EMG in 66 patients with chronic pelvic pain syndrome, demonstrated the improvement of symptoms in 60 patients after 3 courses of biofeedback [81]. All studies have demonstrated improvement of symptoms, although a specific mechanism of action has not yet been demonstrated. However, the use of biofeedback for the treatment of pelvic floor dysfunctions is a very attractive option, mainly because it is free from side effects and contributes in some way to the improvement of the presented symptom, whether it is related to rectal or pelvic pain, bowel problems, or anal incontinence. A recent meta-analysis including very selected studies demonstrated that, within the 11 trials included, they all reported benefits from biofeedback therapy through a variety of clinical and physiological outcomes [82]. The weight of evidence from this systematic review and meta-analysis favors the efectiveness of biofeedback therapy against other forms of therapy for a select group of patients with dyssynergic defecation.

Summary

Biofeedback therapy is an important therapeutic modality for the treatment of anal and also urinary incontinence. For patients with obstructed defecation, it is a valuable option with scientific evidence validated in the literature, being safe and effective.

References

1. Miller NE. Learning of visceral and glandular responses. Science. 1969;163(866):434–45.
2. Wong AM, Lee MY, Kuo JK, et al. The development and clinical evaluation of a standing biofeedback trainer. J Rehabil Res Dev. 1997;34(3):322–7.
3. Kohlenberg RJ. Operant conditioning of human anal sphincter pressure. J Appl Behav Anal. 1973;6:201–8.
4. Latimer PR, Campbell D, Kasperski J. A component analysis of biofeedback in the management of fecal incontinence. Biofeedback Self Regulat. 1984;9:311–24.
5. Miner PB, Donelly TC, Read NW. Investigation of the mode of action of biofeedback in the treatment of fecal incontinence. Dig Dis Sci. 1990;35:1291–8.
6. Wald A, Bharucha AE, Cosman BC, Whitehead WE. ACG clinical guideline: management of benign anorectal disorders. Am J Gastroenterol. 2014;109:1141.
7. Paquette IM, Varma MG, Kaiser AM, et al. The American Society of Colon and Rectal Surgeons Clinical Practice Guideline for the treatment of fecal incontinence. Dis Colon Rectum. 2015;58:623.
8. Bharucha AE, SSC R, Shin AS. Surgical interventions and the use of device-aided therapy for the treatment of fecal incontinence and defecatory disorders. Clin Gastroenterol Hepatol. 2017;15:1844.
9. Tjandra JJ, Dykes SL, Kumar RR, Ellis CN, Gregorcyk SG, Hyman NH, et al. Practice parameters for the treatment of fecal incontinence. Dis Colon Rectum. 2001;50(10):1497–507.
10. Solomon MJ, Rex J, Eyers AA, Stewart P, Roberts R. Biofeedback for fecal incontinence using ultrasonography: novel approach. Dis Colon Rectum. 2000;43(6):788–92.
11. Sangwan YP, Coller JA, Barrett RC, Roberts PL, Murray JJ, Schoetz DJ Jr. Can manometric parameters predict response to biofeedback therapy in fecal incontinence? Dis Colon Rectum. 1995;38(10):1021–5.
12. Kegel A. The nonsurgical treatment of genital relaxation. Ann West Med Surg. 1948;2:213.
13. Kegel A. Progressive resistance exercise in the functional restoration of the perineal muscles. Am J Obstet Gynecol. 1948;56:242–5.
14. Lacima G, Pera M, Amador A, Escaramís G, Piqué JM. Long-term results of biofeedback treatment for faecal incontinence: a comparative study with untreated controls. Color Dis. 2010;12:742–9.
15. Norton C, Hosker G, Brazelli M. Biofeedback and/or sphincter exercises for the treatment of faecal incontinence in adults. Cochrane Database Syst Rev. 2000;2:CD002111.
16. Heymen S, Pikarsky AJ, Weiss EG, Vickers D, Nogueras JJ, Wexner SD. Prospective randomized trial comparing four biofeedback techniques for patients with faecal incontinence. Color Dis. 2000;2(2):88–92.
17. Solomon MJ, Rex J, Eyers AA, Stewart P, Roberts R. Biofeedback for fecal incontinence using transanal ultrasonography: novel approach. Dis Colon Rectum. 2000;43(6):788–92.
18. Mahony RT, Malone PA, Nalty J, Behan M, O'connell PR, O'herlihy C. Randomized clinical trial of intra-anal electromyographic biofeedback physiotherapy with intra-anal electromyographic biofeedback augmented with electrical stimulation of the anal sphincter in the early treatment of postpartum fecal incontinence. Am J Obstet Gynecol. 2004;191(3):885–90.
19. Ilnyckyj A, Fachnie E, Tougas G. A randomized controlled trial comparing an educational intervention alone vs education and biofeedback in the management of faecal incontinence in women. Neurogastroenterol Mot. 2005;17:58–63.
20. Norton C, Chelvanayagam S, Wilson-Barnett J, Redfern S, Kamm MA. Randomized controlled trial of biofeedback for fecal incontinence. Gastroenterology. 2003;125(5):1320–9.
21. Naimy N, Lindam AT, Bakka A, et al. Biofeedback vs. electrostimulation in the treatment of postdelivery anal incontinence: a randomized, clinical trial. Dis Colon Rectum. 2007;50:2040–6.
22. Sun D, Zhao P, Jia H, Wang D, Zhang W. Results of biofeedback therapy together with electrical stimulation in faecal incontinence with myogenic lesions. Acta Chir Belg. 2008;108(3):313–7.
23. Heymen S, Scarlett Y, Jones K, Ringel Y, Drossman D, Whitehead WE. Randomized controlled trial shows biofeedback to be superior to pelvic floor exercises for fecal incontinence. Dis Colon Rectum. 2009;52(10):1730–7.
24. Bartlett L, Sloots K, Nowak M, Ho YH. Supplementary home biofeedback improves quality of life in younger patients with fecal incontinence. J Clin Gastroenterol. 2015;49(5):419–28.
25. Leite FR, Lima MJ, Lacerda-Filho A. Early functional results of biofeedback and its impact on quality of life of patients with anal incontinence. Arq Gastroenterol. 2013;50(3):163–9.
26. Damon H, Siproudhis L, Faucheron JL, Piche T, Abramowitz L, Eléouet M, Etienney I, Godeberge P, Valancogne G, Denis A, Mion F, Schott AM, Oralia Trial Group. Perineal retraining improves conservative treatment for faecal incontinence: a multicentre randomized study. Dig Liver Dis. 2014;46(3):237–42.
27. Sjödahl J, Walter SA, Johansson E, Ingemansson A, Ryn AK, Hallböök O. Combination therapy with biofeedback, loperamide, and stool-bulking agents

is effective for the treatment of fecal incontinence in women – a randomized controlled trial. Scand J Gastroenterol. 2015;50(8):965–74.
28. Kuo LJ, Lin YC, Lai CH, Lin YK, Huang YS, Hu CC, Chen SC. Improvement of fecal incontinence and quality of life by electrical stimulation and biofeedback for patients with low rectal cancer after intersphincteric resection. Arch Phys Med Rehabil. 2015;96(8):1442–7.
29. Ghahramani L, Mohammadipour M, Roshanravan R, Hajihosseini F, Bananzadeh A, Izadpanah A, Hosseini SV. Efficacy of biofeedback therapy before and after sphincteroplasty for fecal incontinence because of obstetric injury: a randomized controlled trial. Iran J Med Sci. 2016;41(2):126–31.
30. Jensen LL, Lowry AC. Biofeedback improves functional outcome after sphincteroplasty. Dis Colon Rectum. 1997;40(2):197–200.
31. MacHugh S, Walma K, Diamant NE. Fecal incontinence. A controlled trial of biofeedback. Gastroenterology. 1990;91:1060.
32. Wald A. Biofeedback therapy for fecal incontinence. Ann Intern Med. 1981;99:146–9.
33. Loening-Baucke V. Efficacy of biofeedback training in improving fecal incontinence and anorectal physiological function. Gut. 1990;31:1395–402.
34. Zubiri E, Bigliardi R, Miculan S, Nanfito G. OP-21 quality of life before and after biofeedback in myelomeningocele patient. J Pediatr Gastroenterol Nutr. 2015;61(4):518–0.
35. Van Tets WF, Kuijpers JHC, Bleijenberg G. Biofeedback treatment is ineffective in neurogenic fecal incontinence. Dis Colon Rectum. 1996;39:992–4.
36. Enck F, Daublin C, Lubke HJ, Strohmeyer G. Long-term efficacy of biofeedback training for fecal incontinence. Dis Colon Rectum. 1994;37:997–1001.
37. Fernández-Fraga X, Azpiroz F, Aparici A, Casaus M, Malagelada JR. Predictors of response to biofeedback treatment in anal incontinence. Dis Colon Rectum. 2003;46(9):1218–25.
38. Oliveira L, Mello AV. Biofeedback therapy for fecal incontinence. Is it a valuable option? (abstr). Dis Colon Rectum. 2003;46(5):A37.
39. CCL O, editor. Biofeedback anal. In: Fisiologia Anorretal- segunda edição Rio de Janeiro: Editora Rubio; 2017. p. 469–86.
40. Melão S, Maria dos Reis LD, Ferreira RR, Kurachi G, Rizzardi K, Schuster M, et al. Effectiveness of treatment using fecal incontinence biofeedback isolated or associated with electrical stimulation. J Coloproctol. 2014;34(4):240–4.
41. Vonthein R, Heimerl T, Schwandner T, Ziegler A. Electrical stimulation and biofeedback for the treatment of fecal incontinence: a systematic review. Int J Color Dis. 2013;28(11):1567–77.
42. Schwandner T, König IR, Heimerl T, Kierer W, Roblick M, Bouchard R, Unglaube T, Holch P, Ziegler A, Kolbert G. Triple target treatment (3T) is more effective than biofeedback alone for anal incontinence: the 3T-AI study. Dis Colon Rectum. 2010;53(7):1007–16.
43. Ussing A, Dahn I, Due U, Sørensen M, Petersen J, Bandholm T. Efficacy of supervised pelvic floor muscle training and biofeedback vs attention-control treatment in adults with fecal incontinence. Clin Gastroenterol Hepatol. 2018. pii: S1542-3565(18)31391-0.
44. Mazor Y, Ejova A, Andrews A, Jones M, Kellow J, Malcolm A. Long-term outcome of anorectal biofeedback for treatment of fecal incontinence. Neurogastroenterol Motil. 2018;1:e13389.
45. Young CJ, Zahid A, Koh CE, Young JM, Byrne CM, Solomon MJ, Rex J, Candido J. A randomized controlled trial of four different regimes of biofeedback programme in the treatment of faecal incontinence. Color Dis. 2018;20(4):312–20.
46. Rao SS, Patcharatrakul T. Diagnosis and treatment of dyssynergic defecation. J Neurogastroenterol Motil. 2016;22:423–35.
47. Rao SS, Tuteja AK, Vellema T, et al. Dyssynergic defecation: demographics, symptoms, stool patterns, and quality of life. J Clin Gastroenterol. 2004;38:680–5.
48. Gilliand R, Heymen S, Vickers D, Wexner SD. Biofeedback in the treatment of fecal incontinence (abstr). Br J Surg. 1997;12:183.
49. Bleijenberg G, Kuijpers HC. Treatment of the spastic pelvic floor syndrome with biofeedback. Dis Colon Rectum. 1987;30:108–11.
50. Weber J, Ducrotte P, Touchais JY, Roussignol C, Denis P. Biofeedback training for constipation in adults and children. Dis Colon Retum. 1987;30:844–6.
51. Lestar B, Penninckx F, Kerremans R. Biofeedback defaecation training for anismus. Int J Colorect Dis. 1991;6:202–7.
52. Kawimbe BM, Papachrysostomou M, Binnie NR, Clare N, Smith AN. Outlet obstruction constipation (anismus) managed by biofeedback. Gut. 1991;32:1175–9.
53. Dahl J, Lindquist BL, Tysk E, Leissner P, Philipson L, Jãrnerot G. Behavioral medicine treatment in chronic constipation with paradoxical anal sphincter contraction. Dis Colon Rectum. 1991;34:769–76.
54. Wexner SD, Cheape JD, Jorge JMN, Heyman SR, Yesgelman DG. Prospective assessment of biofeedback for the treatment of paradoxical puborectalis contraction. Dis Colon Rectum. 1992;35:145–50.
55. Fleshman JW, Dreznik Z, Meyer K, Fry RD, Carney R, Kodner IJ. Outpatient protocol for biofeedback therapy of pelvic floor outlet obstruction. Dis Colon Rectum. 1992;35:1–7.
56. Turnbull GK, Ritvo PG. Anal sphincter biofeedback relaxation treatment for women with intractable constipation symptoms. Dis Colon Rectum. 1992;35:530–6.
57. Keck JO, Staniunas RJ, Coller YES, et al. Biofeedback training is useful in fecal incontinence but disappointing in constipation. Dis Colon Rectum. 1995;37:1271–6.

58. Papachrysostomou M, Smith AN. Effects of biofeedback on obstructed defecation — reconditioning of the defecation reflex. Gut. 1994;35:252–6.
59. Bleijenberg G, Kuijpers FIE. Biofeedback treatment of constipation: comparison of two methods. Am J Gastroenterol. 1994;89(7):1021–6.
60. Koutsomanis D, Leonnard-Jones JE, Kamm MA. Prospective study of biofeedback treatment for patients with slow and normal transit constipation. Eur J Gastroenterol Hepatol. 1994;6:131–7.
61. Siproudhis L, Dautreme S, Ropert A, et al. Anismus and biofeedback: who benefits? Eur J Gastroenterol Hepatol. 1995;7:547–52.
62. Leroi AM, Duval V, Roussignol C, Berkelmans I, Reninque P, Denis P. Biofeedback for anismus in 15 sexually abused women. Int J Colorect Dis. 1996;11:187–90.
63. Ho YH, Chiang JM, Tan M, Low JY. Biofeedback therapy for excessive stool frequency and incontinence following anterior resection or total colectomy. Dis Colon Rectum. 1996;39:1289–92.
64. Park UC, Choi SK, Piccirillo MF, Verzaro R, Wexner SD. Patterns of anismus and the relation to biofeedback therapy. Dis Colon Rectum. 1996;39:768–73.
65. Ko CY, Tong J, Lehman RE, Shelton AA, Schrock TR, Welton ML. Biofeedback is effective therapy for fecal incontinence and constipation. Arch Surg. 1997;132:829–34.
66. Rao SSC, Welcher KD, Pelsang RE. Effects of biofeedback therapy on anorectal function in obstructed defecation. Dis Dis Sci. 1997;42:2197–205.
67. Glia A, Gylin M, Gullberg K, Lindberg G. Biofeedback retraining in patients with functional constipation and paradoxical puborectalis contraction. Dis Colon Rectum. 1997;40:889–95.
68. Karlbohm U, Hallden M, Eeg-Olofsson, Pahlman L, Graf W. Result of biofeedback in constipated patients. A prospective study. Dis Colon Rectum. 1997;40:1149–55.
69. Rieger NA, Wattchow DA, Sarre RG, et al. Prospective study of biofeedback for treatment of constipation. Dis Colon Rectum. 1997;40:1143–8.
70. Patankar SK, Ferera A, Larach SW, et al. Electromyographic assessment of biofeedback training for fecal incontinence and chronic constipation. Dis Colon Rectum. 1997;40:907–11.
71. Chiotakakou-Faliakou E, Kamm MA, Roy AJ, Storrie JB, Turner IC. Biofeedback provides long term benefit for patients with intractable, slow and normal transit constipation. Gut. 1998;42:517–21.
72. Kairaluoma M, Raivio P, Kupila J, Aarnio M, Kellokumpu I. The role of biofeedback therapy in functional proctologic disorders. Scand J Surg. 2004;93(3):184–90.
73. Farid M, El Monem HA, Omar W, El Nakeeb A, et al. Comparative study biofeedback retraining and botulinum neurotoxin in the treatment of anismus patients. Int J Color Dis. 2009;24(1):115–20.
74. Heymen S, Scarlett Y, Jones K, Ringel Y, Drossman D, Whitehead WE. Randomized, controlled trial shows biofeedback to be superior to alternative treatments for patients with pelvic floor dyssynergia-type constipation. Dis Colon Rectum. 2007;50(4):428–41.
75. Lee HJ, Boo SJ, Jung KW, Han S, Seo SY, Koo HS, Yoon IJ, Park SH, Yang DH, Kim KJ, Ye BD, Byeon JS, Yang SK, Kim JH, Myung SJ. Long-term efficacy of biofeedback therapy in patients with dyssynergic defecation: results of a median 44 months follow-up. Neurogastroenterol Motil. 2015;27(6):787–95.
76. Murad-Regadas SM, Regadas FS, Bezerra CC, de Oliveira MT, Regadas Filho FS, Rodrigues LV, Almeida SS, da Silva Fernandes GO. Use of biofeedback combined with diet for treatment of obstructed defecation associated with paradoxical puborectalis contraction (anismus): predictive factors and short-term outcome. Dis Colon Rectum. 2016;59(2):115–21.
77. Forootan M, Shekarchizadeh M, Farmanara H, Esfahani ARS, Esfahani MS. Biofeedback efficacy to improve clinical symptoms and endoscopic signs of solitary rectal ulcer syndrome. Eur J Transl Myol. 2018;28(1):7327.
78. Gilliland R, Heymen JS, Altomare DF, Vickers D, Wexner SD. Biofeedback for intractable rectal pain: outcome and predictors of success. Dis Colon Rectum. 1997;40(2):190–6.
79. Heah SM, Ho YH, Tan M, Leong AF. Biofeedback is effective treatment for levator ani syndrome. Dis Colon Rectum. 1997;40(2):187–9.
80. Cornel EB, van Haarst EP, Schaarsberg RW, Geels J. The effect of biofeedback physical therapy in men with chronic pelvic pain syndrome Type III. Eur Urol. 2005;47(5):607–11.
81. Ye ZQ, Cai D, Lan RZ, Du GH, Yuan XY, Chen Z, Ma YZ, Hu YM, Zeng GY. Biofeedback therapy fo1.r chronic pelvic pain syndrome. Asian J Androl. 2003;5(2):1155–8.
82. Moore D, Young CJ. A systematic review and meta-analysis of biofeedback therapy for dyssynergic defaecation in adults [published online ahead of print, 2020 May 5]. Tech Coloproctol. 2020. https://doi.org/10.1007/s10151-020-02230-9.

Sacral Neuromodulation

36

Ezio Ganio

Introduction

Sacral neuromodulation has effectively revolutionized the management of some intractable cases of fecal incontinence (FI) and those that are unresponsive to other more complex procedures.

Urologists first began to study the possibilities of an electrical stimulation to control bladder dysfunction. Initial attempts to provoke artificial micturition involved direct stimulation of the spinal cord [1, 2]. It was suggested as a technique for electrical stimulation of the bladder by Boyd in 1954, with the use of electrical stimulation of the detrusor muscle in the 1970s [3] and the use of intra-anal stimulation electrodes in 1972 by Hopkinson.

None of these methods produced satisfactory bladder voiding. Research was then focused on electrical stimulation of the sacral nerve roots in order to treat serious bladder voiding dysfunctions [4–6, 7]. Tanagho and Schmidt [8] from the University of California, San Francisco, first applied the principles of sacral nerve stimulation (SNM) to patients affected by voiding dysfunction or incontinence due to bladder instability. In 1981, they performed the first sacral nerve stimulation implant.

Currently, stimulation of the sacral nerve roots is used successfully to control voiding difficulties such as urge incontinence, urinary retention, frequency-urgency syndromes, and bowel dysfunction.

Matzel first reported its use in the treatment of FI in Lancet in 1995 [9]. It became an accepted first-line treatment for patients who have not benefited from medical and behavioral therapies [10, 11]. SNM has been approved for FI by the Food and Drug Administration (FDA) in 2011 and was supported by the UK Clinical Practice Guidelines in 2004, where it showed equal benefit in mild, moderate, and severe FI.

Encouraging results have also emerged from its use in other conditions. It was observed that, in some patients with FI, there was also a subjective effect on defecation. Evidence for a possible role in constipation initially came from urological patients. In a series of 48 patients with coexisting constipation, intestinal frequency increased in 78%. Two studies then reported the effects of temporary stimulation. One showed improvement in two of eight patients [12], and the second showed subjective improvement [13]. This led to the world's first implant of a sacral nerve stimulator for intractable idiopathic constipation.

Although the exact mechanism by which SNM works still remains unknown, it is likely multifactorial. The stimulated target is a mixed nerve-carrying efferent/somatomotor and afferent/sensory nerves as well as autonomic nerves [14]. It seems to be an effect on several nerves within the sacral plexus: the somatic pudendal nerves and the efferent nerves directed to the pelvic floor muscles appear to be involved with

E. Ganio (✉)
Studio Medico Aurora, Milan, Italy

L. C. C. Oliveira (ed.), *Anorectal Physiology*, https://doi.org/10.1007/978-3-030-43811-1_36

increased function of the external anal sphincter. However, studies on delay between stimulation and effect show a latency ten times greater than expected, suggesting a more complex, multisynaptic pathway [15].

There seems to be an effect on sensory afferent nerves with an acute sensation, but there is little effect on intrinsic enteric neurons, and the anorectal reflex is not affected [13]. This, however, is a crude indicator of the function of the enteric nervous system, in a nervous system that has the proven ability to adapt and regenerate. The balance between the autonomic nervous system, parasympathetic and sympathetic nerves, is the determining factor in the motility of the colon and the function of the internal sphincter. The modulation of these nerves can be an important part of the physiological mechanism. The new evidences highlight a facilitation effect on afferent pathways.

Technique of Sacral Neuromodulation

Classically, SNM consists of two stages: percutaneous nerve evaluation (PNE) in the diagnostic stage and permanent implant in the therapeutic stage.

Percutaneous nerve evaluation (PNE) of the sacral roots (S2, S3, and S4) is divided into two phases: an acute phase to test the functional relevance and integrity of each sacral spinal nerve to striated anal sphincter function [9] and a chronic phase to assess the therapeutic potential of sacral spinal nerve stimulation in individual patients.

Percutaneous nerve evaluation: With the patient in prone position, the three sacral foramina S2, S3, and S4 are located using bony landmarks (Fig. 36.1). The sacral foramen S2 is typically found just under the projection of the posterior superior iliac spines and about one finger lateral to the median line. When the sciatic notches, which correspond to level S3, are identified, S4 is about 2 cm under foramen S3. Foramina S3 and S4 are also positioned about one finger across from the median line.

The acute phase test is performed under local anesthesia using a 20-gauge spinal insulated needle (Medtronic™ #041828-004) and an external neurostimulator (Medtronic™ Model 3625 Screener). The needle is inserted perpendicular to the sacrum, with an inclination to the skin of 60–80 degrees (Fig. 36.2). After the needle is positioned in the chosen foramen, it is connected to the external neurostimulator. The stimulation parameters used in the acute phase are pulse width (PW) of 210 μsec, frequency of 5–25 Hz, and an amplitude which resulted in an increased contraction of the pelvic floor and a deepening and flattening of the buttock muscle. This usually occurred between 1 and 3 volts. Stimulation of specific sacral nerves typically results in specific movements of the perineum, anal sphincter, and ipsilateral lower extremity. This ensures correct lead placement. Stimulation of S2 causes some movement of the perineum and the external sphincter along with a lateral rotation of the leg and contraction of the toes and foot. Stimulation of S3 causes a contraction of the pelvic floor and the external sphincter, the "bellows" contraction, and a plantar flexion of the big toe. Stimulation of S4 causes a contraction of the anus with a clamp-like perineal movement with no leg or foot movement. Vesicle, vaginal (or scrotal), and rectal paresthesia may be perceived by the patient during sacral nerve stimulation. A radiological check of the electrode position is mandatory.

Temporary SNM: Once an adequate muscular response is obtained, a temporary stimulator lead (Model 3065 U Medtronic™, Minneapolis, MN, USA, or Model 3057-1, Minneapolis, MN, USA) is inserted through the needle, following which the needle is removed. The lead is connected to an external stimulator (Screener Model 3625, Minneapolis MN, USA) to allow evaluation of the functional responses to the test, both subjectively with regard to continence and objectively using rectoanal physiology. Ten to fourteen days of stimulation is the minimum period needed for the test.

To evaluate the functional results of PNE, patients completed a clinical diary of fecal incontinence and bowel movements episodes in the 2 weeks preceding, during PNE, and in the 2 weeks following the PNE.

Surgical technique for permanent implant: Only selected patients who achieve an improve-

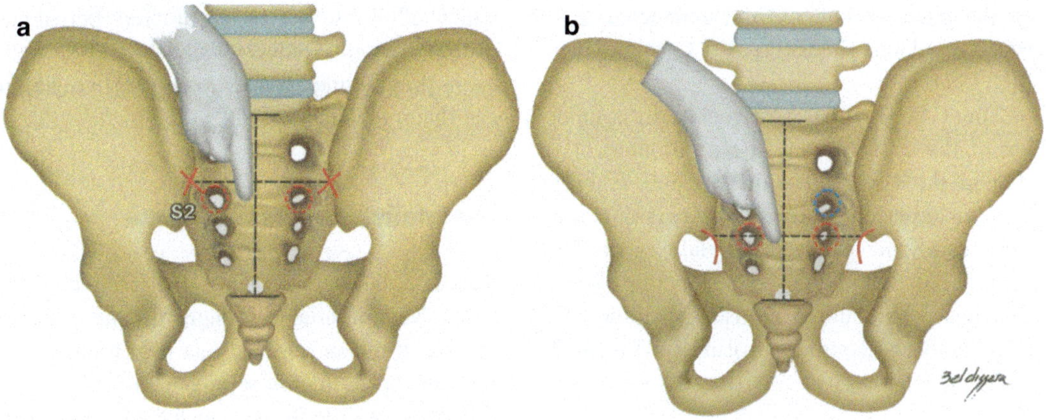

Fig. 36.1 The dorsal sacral foramina are positioned approximately 2 cm laterally to the sacral crest. S2 is about 1 cm medially and 1 cm below the posterior superior iliac spine, and S3 is positioned on a level with the upper border of the sciatic notch

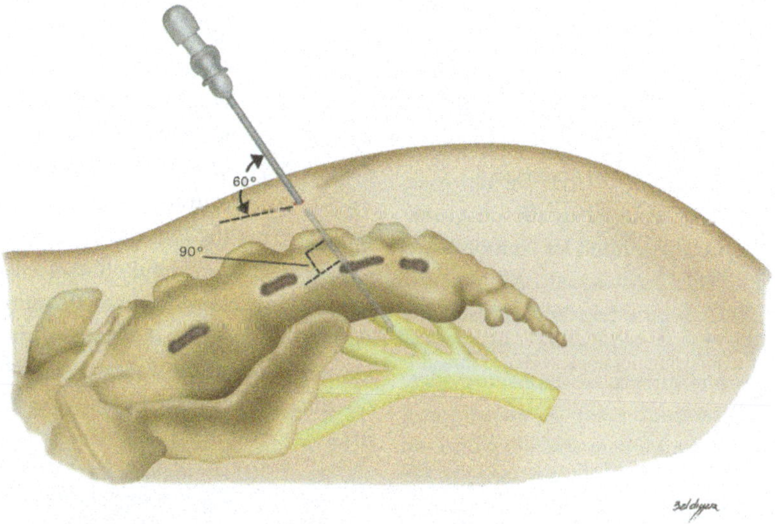

Fig. 36.2 The needle is inserted parallel to the foramina axes with an inclination to the skin of 60–80 degrees

ment of at least 50% compared to the previous clinical situation (reduction of 50% of days with incontinence and/or reduction of 50% of weekly incontinence episodes) reach this stage.

Classical Open Surgical Technique

Before incision, the sacral foramen is checked with an isolated needle. Once the sacral foramen is confirmed, an incision is made along the median line above the sacral spinous process, up to the level of the underlying lumbodorsal fascia. The lumbodorsal fascia is cut longitudinally, about one finger width from the median line. The paraspinous muscles underneath are divided sharply along the length of the fibers. The sacral foramen is checked, and the definitive electrode is inserted (model 3080, Minneapolis, MN, USA) and anchored to the periosteum (Fig. 36.3).

Each electrode is composed of four electrodes that can be selected individually through the programming of the neurostimulator. Once the tip is anchored, the rest of the electrode is channeled, with the aid of a tunneling tool, through the subcutaneous tissue layer, into a small incision made on the patient's buttock and connected to the neurostimulator (Interstim 3023, Minneapolis, MN, USA).

This procedure has now replaced the tunneling of the electrode and the positioning of the IPG in the lower abdominal region. The IPG abdominal placement requires a longer operative time, and some patients complain of displacement or pain at the IPG site postoperatively [16]. The neurostimulator (impulse generator: IPG) can be activated using a control unit (N'VisionTM) which allows to set all parameters percutaneously via a radio frequency signal. Each stimulator is programmed in the most effective way to suit that individual patient.

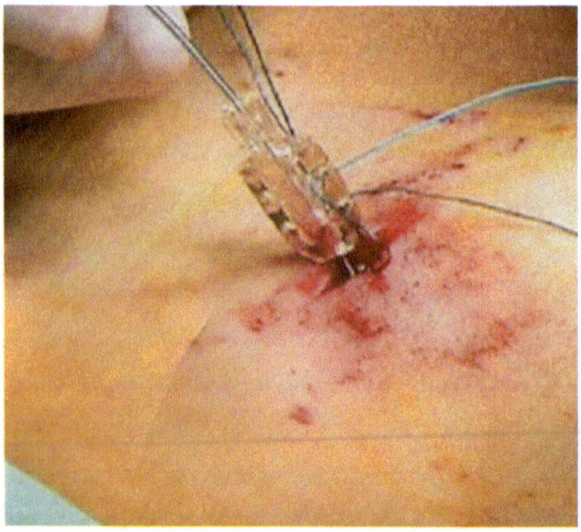

Fig. 36.3 The definitive electrode is anchored to the periosteum using not nonabsorbable thread

Minimally Invasive Technique

Recently, the introduction of the "tined lead" has made an important change in the surgical approach; the sacral electrode is now implanted with an approach only percutaneous. After insertion of the needle in the selected sacral foramen and test for nerve responses, a metal stylet (directional guide) is inserted through the needle. The needle is removed, two small incisions on either side of the guide are made, and a dilator is inserted on the guide of the directional guide and advanced into the sacral foramen. Leaving the introducer sheath in place, the chronic tined lead is inserted and advanced under fluoroscopic control (Fig. 36.4). Once the responses of the various electrodes are confirmed, the introducer sheath is removed, thereby deploying the tines and anchoring the lead [17].

Finally, the classical PNE and one-stage permanent implant could now be replaced by a two-stage procedure. Once the permanent lead is implanted, a percutaneous extension is used to

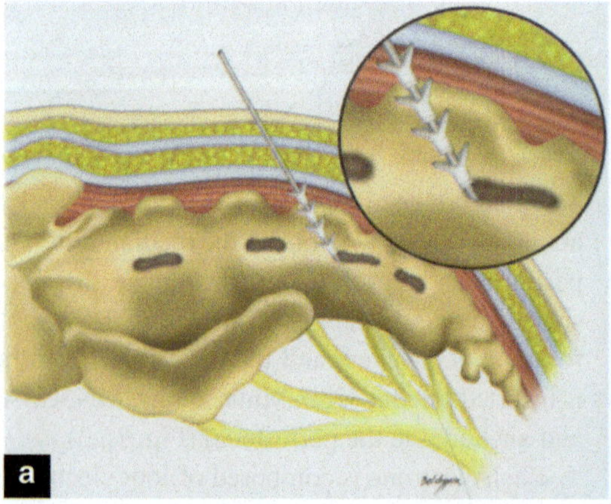

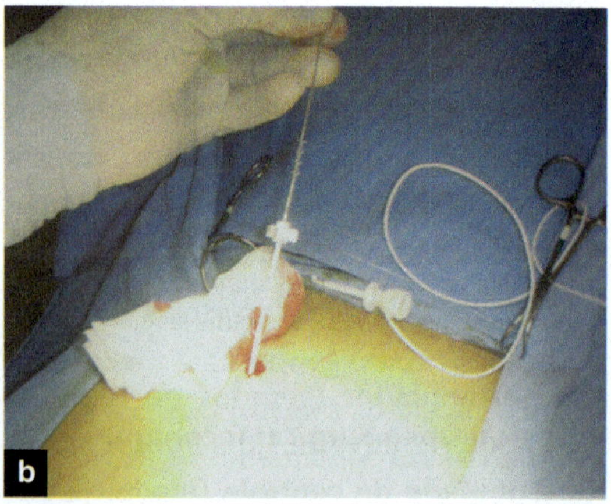

Fig. 36.4 (**a**) Quadripolar tined lead; the electrodes are shown. (**b**) After sacral foramen needle is inserted and location is verified by electrical stimulation to the needle and fluoroscopy, the plastic dilator is positioned. (**c**) The quadripolar lead is introduced through the dilator plastic sheath into position. Once the responses of the various electrodes are confirmed, the introducer sheath is removed, thereby deploying the tines and anchoring the lead

connect it to an external stimulator (Model 3065 U Medtronic™, Minneapolis, MN, USA, or Medtronic™ Model 3531 Verify) allowing a long period of evaluation (1–2 months) of the effectiveness of sacral neuromodulation. If the response is confirmed, the percutaneous extension will be removed and the lead directly connected to the IPG (Interstim 3023, Minneapolis, MN, USA). With the introduction of the minimally invasive technique, this two-stage modality has been proposed as alternative to the PNE itself.

Type of quadripolar lead stimulation is not standardized but should take the form of external settings of the stimulator around a pulse width of 210 ms, speed of 14–20 Hz, and an amplitude of <3 V for all four positions of the conductor answering by answer.

Prospective randomized data show that the two-stage implantation technique of SNM has a higher success rate when compared to the single-stage method despite a previous positive PNE response and effectively reduces the reoperation rate and overall procedural costs [18–19]. Spinelli reports that the success rate of this technique in patients selected for the permanent implant was significantly improved over a two-step technique initially using a temporary lead placement [20]. This allowed a longer test period with the permanent electrode before proceeding with the implantation of a definitive neurostimulator (IPG). In our experience with the use of the "tined lead" and the two-step technique, the percentage of successes to complete the procedure increased from 26.8% to 84.5%. In summary, the SNM technique has become somewhat standardized in recent years and has shifted from the placement of a temporary lead to a permanent one but remains a two-stage process of initial temporary stimulation pending a decision regarding the second-stage implantation of permanent pacemaker.

Indications

The lack in the knowledge of the exact mechanism makes difficult to give precise indications on the eligible patients. Two orders of considerations go advised: clinical and anatomical. It is a shared belief that the indication for neuromodulation is a severe incontinence not amenable to standard drugs or biofeedback therapy or that has failed conventional surgical management. Incontinence to solid or liquid feces at least once each week during the last 2 months, as reflected in a defecation diary kept by the patient, is a good practical criterion. Patients with only gas incontinence or minor staining are not a good candidate. If we consider the type of incontinence, patients with an urge incontinence (fecal loss at the first signs of the urge to defecate) show a better improvement if compared with passive incontinent patients (inadvertent and unpredictable fecal loss) [21].

Among anatomical considerations from the initial assertion of the need for integrity of the anal sphincters, an increasing worldwide application has shown that it is considered suitable for many cases of passive and urge AI as well as in those cases both with and without a disrupted anal sphincter ring [22]. More data are available to extend the use of neuromodulation in those patients normally destined for sphincteroplasty due to the presence of an EAS defect [23, 24], as well as in those cases with isolated IAS deficiencies in which a plant would normally be considered [25], in the incontinence and in the urgency associated to the syndrome of the low anterior resection (with or without the construction of a neorectal pouch) [26], and in those with partial lesion of the spinal cord [27, 28]. Patients with complete spinal cord lesion or complete peripheral nervous lesions such as spina bifida or iatrogenic nerve lesion are not candidate for sacral neuromodulation [21–22].

Patient's Evaluation and Impact on Gastrointestinal Physiology

Initial assessment included a complete clinical history and physical examination. Before applying the lead, patients usually perform anorectal physiological evaluation and an anal ultrasound to evaluate anal sphincter. Symptoms evaluated include the number of incontinence episodes, fecal urgency, use of pads, and impact on lifestyle. To establish baseline function, all patients completed a 14-day fecal incontinence diary of episodes of fecal incontinence and bowel movements prior to PNE or first-stage implant. The same diary is used during the test period and eventually in the 2 weeks following PNE. The evaluation of the clinical diary is the only parameter that currently allows selecting the patients for definitive implant. Usually, PNE test or first-stage implant is considered positive if there is >50% improvement in FI symptoms compared with baseline and a rapid return to pre-PNE conditions when stimulation is turned off.

Other definitions and outcomes used included intention-to-treat (ITT) analysis which is based on measuring outcome based on the number of patients initially enrolled in the treatment as opposed to per-protocol analysis (PPA) which only measures the final outcome based on the number of patients who had a successful PNE or first stage and then went on to receive a permanent implant.

Primary failure is defined as those who never had a clinical response to PNE, while secondary failure refers to those patients who had a successful response to PNE but failed to subsequently achieve therapeutic benefit from the permanent implant. Physiological investigation includes anorectal manometry, specifically anal pressures (maximum resting and squeeze pressure (mmHg)), rectal sensory thresholds (balloon volumes in milliliters of air or water), small/large bowel motility, and neurophysiological study of the pelvic floor (including pudendal SEP, sacral reflexes, and PNTML).

Clinical Outcome

Incontinence

At a follow-up of 6 months, Matzel [9] in 1995 reported a complete recovery of continence in two cases and soiling in one patient with definitive implant of a sacral electrode with an improvement of resting and squeezing anal pressure. In July 2000, Malouf [29] reported an improvement on the Wexner incontinence scale from 16–20 to 3–6 in 5 patients followed for at least 16 months after definitive implant of a sacral electrode. The resting anal tone showed consistent improvement, but no variation was observed with the squeezing anal pressure. In 2001, Rosen [30] published the results of 16 permanent implant selected out of 20 (80%) incontinent patients tested over a period of 10–14 days with PNE. Three patients had their electrodes removed because of infection. All the 13 (81.2%) functioning implant had a significant improvement in fecal continence. Resting and squeezing anal pressure improved significantly only in patients with neurologic incontinence but not in those with idiopathic incontinence. For the first time, a dedicated quality of life questionnaire was used, showing clear improvement in all the four items investigated (lifestyle, coping behavior, depression, embarrassment).

In our first experience [21] with five patients implanted out of 23 PNE tests (22%), definitive electrical stimulation of sacral roots was associated with an improvement in fecal continence from a mean of 4.8 episodes/week to a complete cessation, reproducing at a median follow-up of 19.2 months (range 5–37) the clinical effect of the PNE test. A significant increase of the resting but not squeezing anal pressure was observed, and an earlier rectal sensation to balloon distension was observed in this first series of patients. Using isobaric rectal distention, the pressure applied for the first sensation threshold decreased significantly ($p = 0.012$) as did the pressure for the urge threshold ($p = 0.008$). The distension pressure decreases for the first sensation, and urge threshold is very important, because it states a better sensibility or a facilitation of the rectal receptors.

The first medium-term results of SNM for fecal incontinence have been published by Kenefick [31] in 2002, reporting good results in 15 incontinent patients followed up for a mean of 24 months with 11 patients fully continent. Episodes of fecal incontinence decreased from 11 (2–30) per week before stimulation to 0 (0–4) per week after permanent stimulation ($p < 0.001$). Urgency and ability to defer improved in all patients. Resting and squeezing anal pressure significantly increased, and the volume requested for rectal sensitivity to initial distension was significantly lower ($p < 0.05$) than before SNM. There were no major complications. In these group of patients, the quality of life questionnaire [36-item Short Form Healthy Survey (SF36)] was administered before and after stimulation: "social function" and "role-physical" subscales of the SF36 improved significantly.

Medium-term results of a substantial series were presented by the GINS [32] group in the spring of 2002. Thirty-one patients had permanent implant out of 116 (27%) PNE tests, and PNE results were reproduced in all patients at a mean follow-up of 25.6 (range 1–56) months. The mean number of incontinence episodes for solid or liquid stools (per 14 days) decreased from 15 (range 2–22) at baseline to 3.2 (range 0–10) at 3 months follow-up ($p = 0.02$), to 2.9 (range 0–13) at 6 months, and to 0.3 (range 0–4) at 12 months follow-up. Again, anorectal manometry shows a positive trend in increasing sphincter pressure and rectal sensitivity. No local sepsis occurred. One patient complained of pain at the implant site when IPG case was used as anode (unipolar impulse) and another necessitated electrode repositioning for displacement after 3 months. In one patient, interruption of the electrode caused decreased effectiveness at 11 months post implant; the lead was changed and the patient recovered continence. SF36 was used in 18 of these patients before and after SNM. Improvement of continence had a positive impact on the health state, particularly in the reduction of physical limitations or disabilities. An overall analysis showed a significant improvement in patient's physical ($p < 0.05$) and mental health ($p < 0.05$) after implant.

Long-term results were reported first by Matzel in 2003 [33]. Functional improvement was achieved in 94% of 16 patients. At a median follow-up of 32.5 months (3–99), treatment was successful in 81%. Two of the electrodes were removed after 5 and 45 months for intractable pain. Mean squeeze pressure increased, but maximum squeezing pressure improved only in three of them. Resting pressure, perception, urge threshold, and maximum tolerable volume were not significantly changed. Using the disease-specific quality of life instrument (FIQL-ASCRS) before and during stimulation, the quality of life index was improved in all categories.

In a systematic review of the impact of sacral neuromodulation on clinical symptoms, Mirbagheri [34] with data obtained from 63 studies, the results demonstrated overall improvement in subjective and objective measures of FI in all studies, regardless of the design of the study. The PNE success rate, defined as >50% reduction in clinical symptoms over the evaluation period, ranged from 51.5% to 100%, with a median value of 81% on a per-protocol basis. The reported rates of "perfect continence" after permanent implant ranged from 13% to 88% (Table 36.1). Notwithstanding the inevitable heterogeneity of patient characteristics, pooling of these results ($n = 608$) gave a perfect continence rate of 36.5% on an ITT basis and 42.9% on a PPA.

The possibility of a placebo effect was investigated. Six trials assessed the effects of SNS for FI. In the crossover trial by Leroi [49], 24 participants while still blinded chose the period of stimulation they had preferred. Outcomes were reported separately for 19 participants who preferred the "on" and five who preferred the "off" period. For the group of 19, the median episodes of fecal incontinence per week fell from 1.7 during the "off" period to 0.7 during the "on" period; for the group of five, however, the median rose from 1.7 during the "off" period to 3.7 during the "on" period. In the crossover trial by Vaizey [50], participants reported an average of six and one episodes of fecal incontinence per week during the "off" and "on" periods, respectively, in two participants with FI. In another case crossover study by Kahlke [51], 14 participants with FI experienced significantly lower episodes of FI per week during the stimulator "on" (1 (SD, 1.7)) compared with the "off" period (8.4 (SD, 8.7)).

From 1996 to 2003, 94 patients affected by FI underwent the peripheral nerve evaluation (PNE) test for SNS in six Italian colorectal units [37]. Sixty of them (64%) had a good response to temporary SNS and therefore underwent a definitive

Table 36.1 Details of patients achieving full continence in 18 studies

Study	Sample size	Sacral neuromodulation (n)	% full continence	Full continence (per protocol) (n)
Leroi et al. [35][a]	34	34	5	15
Leroi et al. [36]	9	8	1	13
Altomare et al. [37][a]	52	38	9	24
Oom et al. [38]	46	37	8	22
Boyle et al. [39]	50	37	13	35
Hull et al. [40]	72	64	26	41
Oz-Duyos et al. [41]	47	28	14	50
Matzel et al. [42]	37	37	12	32
Jarret et al. [43]	59	46	19	41
Tjandra et al. [44]	59	54	25	46
Ganio et al. [13]	25	22	11	50
George et al. [45]	25	23	12	52
Matzel et al. [9]	3	3	2	67
Santoro et al. [46]	28	28	19	68
Kenefick et al. [31]	15	15	11	73
Kenefick [47]	19	19	14	74
Ganio et al. [21]	19	17	14	82
Vaizey et al. [48]	9	8	7	88
Total	**608**	**518**	**222**	**Pooled: 36.5[b]** **Range: 13–88[b]**

[a]Data after permanent implant only. [b]Intention-to-treat analysis (patient with perfect continence/total sample size). Per protocol analysis = 42.9%

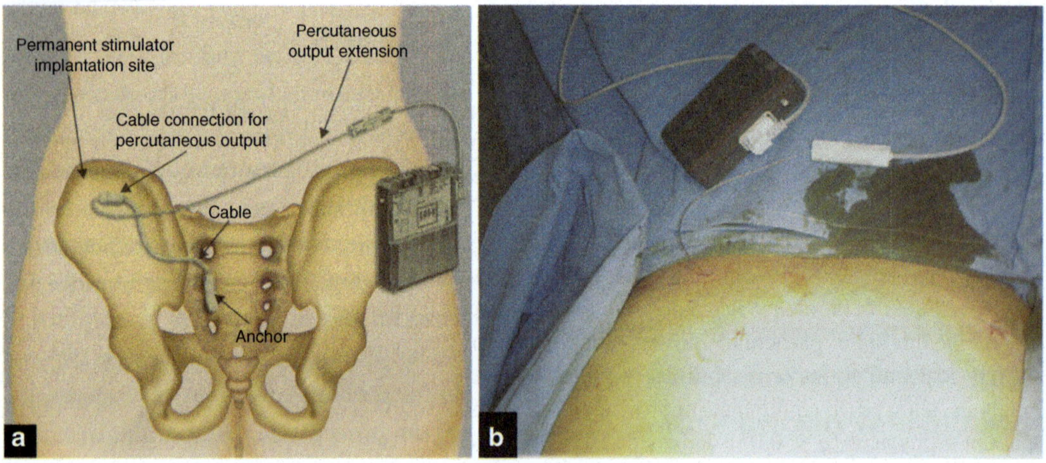

Fig. 36.5 Patient data of incontinent patients tested between 1999 and 2003

electrode(s) and electrostimulator implant. During a mean follow-up period of 74 ± 14 months (range 60–122 months), no patient was lost, but two died (3 and 4 years after the SNS implant) of diseases not related to FI or SNS, and six (10%) had the device removed because of complications or progressive failure of the therapeutic efficacy within the first 2 years of follow-up. Thus, 52 patients (86.7%) were available for the long-term evaluation (Fig. 36.5). Complications were reported in 15 patients (28.8%): pain at the site of implant in six cases (11.5%), electrode displacement (all implanted with the old technique) in eight patients (15.4%; all reimplanted, three using a tined lead), and early battery rundown in one case (1.9%). Pain was managed by reducing the stimulation voltage or by repositioning the implantable pulse generator in another place, while electrode displacements and battery rundown required substitution of the devices.

In the 52 patients available for long-term evaluation, the Cleveland Clinic Florida Fecal Incontinence Score (CCF-FIS) decreased significantly compared with baseline (from 15 ± 4 to 5 ± 5, $p < 0.001$). At least 50% improvement in continence was achieved in 74% of the patients, and at least 70% improvement (median value) was achieved in 50%. Full continence was achieved in 17% of the patients. The mean number of solid/liquid incontinence episodes decreased significantly from 0.5 (±0.5) to 0.1 (±0.3) per day ($p = 0.004$). Quality of life improved in all domains. The overall mean improvement in SF-36 scores was 39.8%. Both mean resting and squeeze anal pressures increased significantly, and maximum volume tolerated decreased significantly (Fig. 36.6).

A survey to review prospectively recorded data on all consecutive patients undergoing temporary testing for SNS from ten European centers with long-standing experience of SNS for FI was presented in 2015 [52]. From January 1998 to December 2006, a total of 407 patients underwent temporary stimulation, of whom

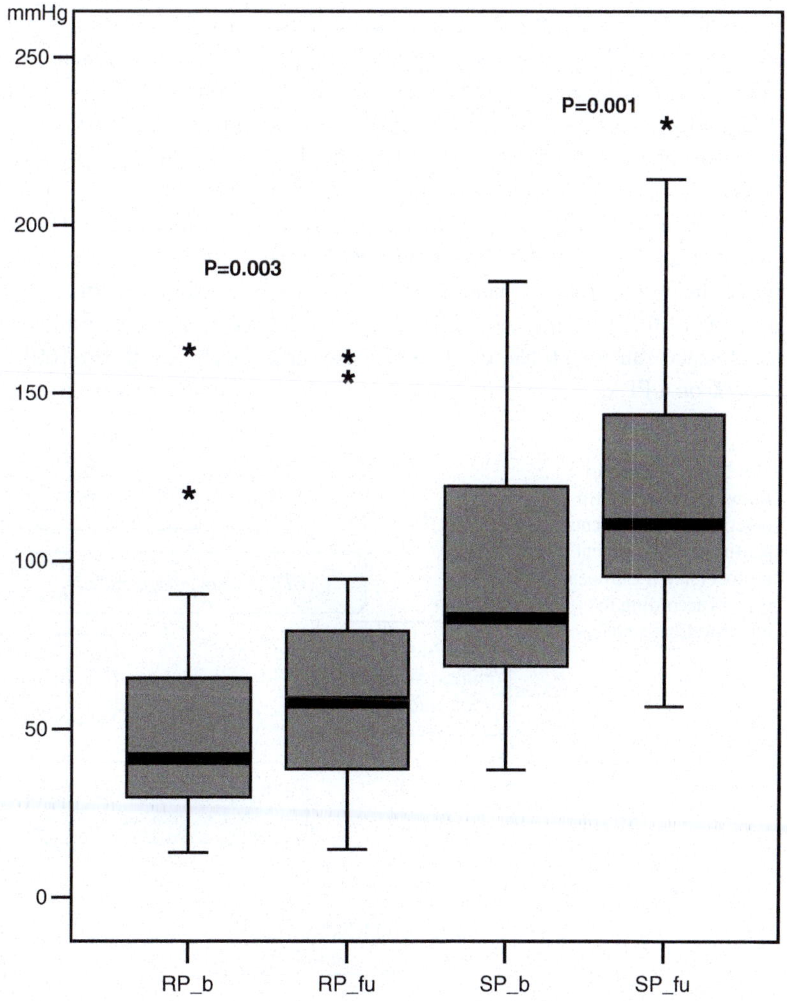

Fig. 36.6 Comparison between baseline (b) and last follow-up (fu) manometric data. RP resting pressure, SP squeezing pressure. ∗Outlier value

272 (66.8%) had an impulse generator implanted; 228 (56.0%) were available for long-term follow-up at a median of 84 (i.q.r. 70–113) months (Fig. 36.7).

Significant reductions in the number of FI episodes per week (from median 7 to 0.25) and summative symptom scores (median CCF-FIS from 16 to 7, St Mark's score from 19 to 6) were recorded after implantation (all $p < 0.001$) and maintained in long-term follow-up. In per-protocol analysis, long-term success was maintained in 71.3% of patients, and full continence was achieved in 50.0%; respective values based on intention-to-treat analysis were 47.7% and 33.4%.

A recent systematic review of data from published studies, although with differences in endpoints and reporting [34], reports overall an improvement in subjective and objective measures of FI across all studies, irrespective of study design. The PNE success rate, defined as >50% reduction in clinical symptoms over the evaluation period, ranged from 51.5% to 100%, with a median value of 81% on a per-protocol basis. The reported rates of "perfect continence" ranged from 13 to 88%, with a mean rate of complete continence of 36.5%. Notwithstanding the inevitable heterogeneity of patient characteristics, pooling of these results ($n = 608$) gave a perfect continence rate of 36.5% on an ITT basis and 42.9% on a PPA.

Incontinence and Sphincter Lesion

Initially, SNM was used only in FI of neurogenic origin, but subsequently, the indications have been extended to other conditions including incontinence in the presence of a sphincter defect, that represent the major cause of fecal incontinence, particularly in women.

Sphincteroplasty with overlap is the traditional treatment, but a significant reduction in benefits within 5 years of surgery has been reported. In a literature review [53] of SNM for FI in the presence of a sphincter defect, ten reports (119 patients) satisfied the inclusion criteria. All reported a lesion of the external anal and/or internal anal sphincter on endoanal ultrasound. A definitive implant was performed on 106 (89%) of the 119 patients who underwent a peripheral nerve evaluation test. The weighted average number of incontinent episodes per week decreased from 12.1 to 2.3, the weighted average CCF-FIS decreased from 16.5 to 3.8, and the ability to defer defecation, when evaluated, increased significantly. The features at anorectal manometry did not change. The quality of life improved significantly in almost all studies [54–59].

Similar resultas were observed in nine publications, where studies with less than 25 pactients were excluded [60]. All studies demonstrated highly improved function across all outcome

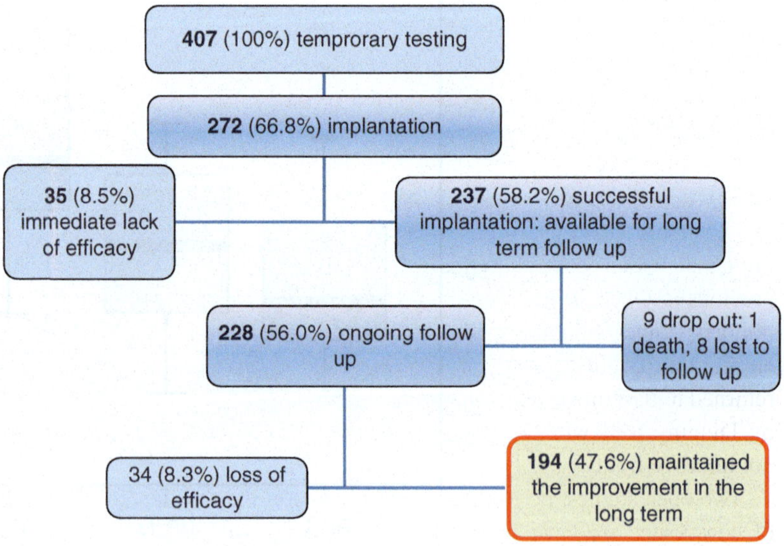

Fig. 36.7 Flow chart showing the long-term outcomes of all patients starting SNS treatment protocol in ten European centers according to intent-to-treat analysis

measures, and improvement was statistically significant in all, with sphincter gaps ranging from 17° to 180°. Outcomes remain stable at long-term follow-up. The size of the gap appears to have no impact on outcome.

Constipation

More recently, applications for sacral neuromodulation have been found in the treatment of chronic, intractable severe constipation. The Rome III criteria [61, 62] distinguish between functional constipation and constipation-predominant irritable bowel syndrome (IBS). The former is defined by the presence of two or more of the following symptoms, originating at least 6 months before diagnosis and currently active for 3 months: infrequent bowel movements (i.e., less than three stools/week), hard stools, excessive straining, a sensation of anorectal blockage, the use of manual maneuvers during evacuation, and a sensation of incomplete evacuation after defecation.

Two subtypes of functional constipation can be categorized: slow transit constipation (STC) and obstructed defecation (OD). However, in some cases, these two conditions overlap.

Kenefick [47] in 2006 reported of four women (aged 27–36 years) with severe, resistant idiopathic constipation for 8–32 years. Symptoms improved in all with temporary and in three with permanent stimulation at 8 months (range 1–11 months). Bowel frequency increased: 1–5 versus 6–28 evacuations/3 weeks. Symptom scores and quality of life improved.

A double-blind, crossover study was also performed to examine placebo effect and efficacy in two patients aged 36 years who had been implanted with a permanent stimulator 12 months previously. Once stimulation was removed in a blinded manner, symptoms, physiological parameters, and quality of life measures rapidly returned to baseline levels. In contrast, in the trial by Dinning [63] with 59 participants, SNS did not improve frequency of bowel movements.

Two open studies, performed on larger cohorts of constipated patients, have reported relatively satisfactory results of the SNM if we consider that the affected patients have a chronic, severe, and treatment-resistant pathology to usual medical conditions. The first, a multicenter prospective study including 62 constipated patients (81% with transit constipation) was published in 2010 [64]. Forty-five patients (73%) had a positive test after 3 weeks of stimulation and were implanted. Patients were followed for a median of 28 (1–55) months. Thirty-nine (87%) of the 45 implanted were significantly improved with respect to stool frequency, thrust efforts, incomplete defecation sensations, abdominal pain, and bloating. The CCF-FIS decreased from 18/30 to 10/30 at the last follow-up visit with a visual analogue scale of digestive symptoms that increased from 8 to 66/100 (0 was worst and 100 was best). Four out of 8 domains of the SF-36® quality of life score were also significantly improved after implantation.

The second study was retrospective and reported the results of the SNM in a population of 117 constipated patients [65]. Sixty-eight patients (58.1%) had a positive test and were implanted. At the last follow-up visit (median 37 months), 61 of the 68 (88%) implanted patients were still treated with SNM. The improvement appeared independent of the type of transit or distal constipation of the patients. After implantation, the results of the SNM seem to be maintained in the long term. Ratto et al. [66] evaluated 61 constipated patients (17 had transit constipation and 25 had distal constipation). Forty-two patients (68.9%) had a positive temporary test and were implanted. The average duration of follow-up was 51 ± 15 months. At the last follow-up visit, 47% of implanted patients had a significant improvement in their constipation scores. This improvement involved 64% of patients with distal constipation compared to 17% of patients with transit constipation suggesting that patients with evacuation difficulties would be better candidates than others for SNM treatment.

A prospective, open-label, multicenter study up to 5 years has been published so far [67]. Sixty-two patients (7 male, median age 40 years) underwent test stimulation, and 45 proceeded to permanent implantation. Twenty-seven patients exited the study and only 18 patients (29%)

attended 60-month follow-up. In 14 patients (23%) with Cleveland Clinic Florida Constipation Score, improvement was sustained at 60 months [17.9 ± 4.4 (baseline) to 10.4 ± 4.1, $p < 0.001$]. Benefit from sacral neuromodulation in the long term was observed in a small minority of patients with intractable constipation.

Finally, the results of a randomized, double-blind, French-controlled study were recently presented at the European Congress of Gastroenterology (UEGW 2015, Barcelona). This study mainly involved patients with distal constipation. Thirty-six patients underwent a temporary stimulation test and, in the case of a test response, were implanted. After implantation, two fairly long periods of 8 weeks of active or simulated stimulation were organized. Twenty patients (56%) were considered responders (improvement of symptoms by more than 50%) during the test period and were implanted. There was no significant difference between the percentages of responding patients during active stimulation versus simulated stimulation (60% vs. 55%, p = ns). After 1 year of follow-up, 11 of the 20 patients implanted (55%) were still responders.

In conclusion, although it seems that some patients are answering the SNM, without further details on the profile of its "good candidates," it seems difficult to validate this type of treatment in the management of constipated patients considering the limited success rate and the cost of this treatment.

Low Anterior Resection Syndrome

The proportion of rectal cancer patients undergoing sphincter-sparing operations ranged between 71% and 90%. Low anterior resection with end-to-end anastomosis is the most frequent procedure after mesorectal excision. Severe low anterior resection syndrome (LARS) developed in less than 40% of patients. The most important factor related to defecatory function impairment is the distance from the anal margin to anastomosis. Other factors thought to be involved were anastomotic leakage, preoperative radiation therapy, age, and postoperative radiotherapy. Lifestyle changes and dietary measures associated with or without drug treatment were the modalities of choice.

In a retrospective review, 12 patients (50% men) of a mean age of 67.8 (±10.8) years underwent sacral nerve test stimulation [68], and 10 patients (83%) proceeded to permanent implantation. Median time from anterior resection to stimulator implant was 16 (range 5–108) months. At a median follow-up of 19.5 (range 4–42) months, there were significant improvements in CCF-FIS and low anterior resection syndrome (LARS) scores ($p < 0.001$). In a systematic review, seven papers were identified including one case report and six prospective case series [69]. These included 43 patients with a median follow-up of 15 months. After peripheral nerve evaluation, definitive implantation was carried out in 34 (79.1%) patients. Overall, 32 (94.1%) of the 34 patients experienced improvement of symptoms which, based on intention-to-treat, was 32/43 (74.4%). The review suggests that SNS for fecal incontinence in LARS has success rates comparable to its use for other forms of FI.

Complications and Troubleshooting

A systematic analysis of published data on side effects of SNM reported adverse events and reoperation rates for 1954 patients, followed for 27 (1–117) months [70]. The majority of adverse events were reported within the first 2 years after stimulator implantation. Complications may broadly be divided into test-stimulation-related and implantation-related problems. Most relate to lead migration (about 12%), pain (3%), and infection (10%), with a 15% of reoperations for a combination of events including attenuated response, infection, IPG site pain, and lead migration [71].

Even though response to temporary stimulation is a prerequisite for permanent stimulator implantation, most of the concerns focused on lack or loss of benefit, which accounted for half of the primary problems described. Conceptually, one may question whether lack or loss of benefit is truly an adverse event. Adjustment of stimulation parameters effectively resolved many of the reported problems, which could thus be seen as analogous to dosing changes in pharmacotherapy (Table 36.2). Similarly, any treatment based on electrical stimulation will require energy and will, therefore, deplete the battery over time. Unless the need for battery replacement surfaces very early after stimulator implantation, it may also be considered routine maintenance of electrotherapy.

Pain or paresthesia accounted for 14.9% of the complaints, with 35.1% of these reports specifically referring to the generator site as affected area. Lead-related problems accounted for 10.7% of the reports. Lead migration is usually resolved by reprogramming and usually does not require a new lead to be inserted.

In some cases, there is an accommodation to stimulation, which does not respond to an increase in stimulation amplitude, and this may ultimately require a repeat insertion or a contralateral lead insertion. Problems relating to response may occur as a result of impedance resistance, with attenuation of electron flow through the circuit; impedance describes the resistance to the flow of electrons through a circuit. Impedance measurement can act as a troubleshooting technique, checking the system's integrity in patients who lose SNM efficacy. In this setting, high-resistance levels (>4 K) indicate an open circuit, which is usually due to a fractured lead, loose connections, or both.

The pooled rate of infection was 5.1% (4.1–6.4). Device explants were largely due to infection but were also caused by generator erosion through the skin or other local complications at the pocket site and lack of benefit, thus leading to a higher rate of reoperation. A total of 39 studies, covering 1810 patients, provided information about explant rates at the end of their follow-up period, with an average of 10.0% (7.8–12.7) (I2 = 54.0%) and a significant increase with the duration of follow-up. Lead complications, battery depletion, or pain all contribute to additional intervention, with an overall reoperation rate of 18.6% (14.2–23.9) (I2 = 80.5%) based on cohorts with a total of 1784 patients. Reoperation rates rose with longer follow-up times. Overall, data would suggest that, when SNM is used for functional bowel disease, about half of the patients will experience at least one device- or treatment-related adverse event [72].

Impact on Anorectal Physiology

In evaluating the impact on anorectal physiological parameters, a consistent trend was noted, with an increase in both maximum resting pressure and squeeze pressure after SNM with a median difference of the mean of 5.9 (−11.8–21) and 14.8 mmHg (−12.5–96), respectively [34]. No correlation could be made between manometric findings and clinical symptoms after stimulation. Rectal sensitivity, as measured by the volume required to elicit sensory thresholds, tended to improve (as evidenced by a reduction in sensory threshold volumes) after SNM. Uludağ [73] used an isobaric phasic distension protocol to evaluate

Table 36.2 The most commonly described corrective actions for key concerns from Manufacturer and User Facility Device Experience (MAUDE) on adverse events related to the Interstim device

Primary concern	Sample	Conservative therapy	Operative therapy
Lack of benefit	325	Stimulation adjusted: 160	Explant: 22
	Medication: 3	Replacement: 17	
	System check: 7	Pocket revision: 3	
Pain or discomfort	97	Stimulation adjusted: 34	Explant: 8
	Medication: 4	Replacement: 1	
	System check: 2	Pocket revision: 7	
Lead problem	70	Stimulation adjusted: 3	Explant: 1
	System check: 2	Replacement: 36	
Programming problems	30	Stimulation adjusted: 11	
		Replacement: 1	

rectal filling sensations of first sensation (FS), earliest urge to defecate (EUD), and irresistible, painful urge to defecate (maximum tolerable volume (MTV)). Rectal wall tension and compliance could be calculated from these recordings. During stimulation, median volume thresholds decreased significantly ($p < 0.01$) for FS (98.1 vs. 44.2 ml), EUD (132.3 vs. 82.8 ml), and MTV (205.8 vs. 162.8 ml). The median reductions of the mean values for sensory volumes were 11.9, 16.4, and 6.6 ml for first sensation, sensation of urge, and maximum tolerated volume, respectively. Pressure thresholds tended to be lower for all filling sensations, and median rectal wall tensions decreased significantly ($p < 0.01$) for all filling sensations.

The effect of SNM on rectal compliance was measured in seven studies [21, 36, 48, 73–76], but none of these showed any statistically significant changes, although the sample size in each study was small ranging from 11 to 23 patients. Other rectal physiological parameters such as rectal stool retention test, rectoanal angle, and rectal motility were not affected by SNM [73, 77]. However, Michelsen [78] demonstrated a significant decrease in postprandial rectal tone during stimulation.

Mechanism of Action

Debate as to the mechanism of action for sacral stimulation in patients with FI is still ongoing.

Action on the striated sphincters and a facilitation of voluntary contraction have been suggested and attributed to direct alpha motor fiber stimulation [14]. Several studies have tried to show an improvement of the external anal sphincter during neuromodulation, but results are controversial.

In a systematic review of the impact of sacral neuromodulation on clinical symptoms, only a small number of factors were associated with outcome. Notably, age was a significant variable in more than one study [79, 80], and the younger the patient (<70 years old), the more likely a successful response to SNM. Anal sphincter defects and multiple PNE procedures were correlated with failures of SNM in two studies [81, 82]. The variables that were not predictive of outcome included baseline anorectal physiological parameters and colonic transit study, body mass index, gender, stimulation parameters, etiology of FI (idiopathic vs. organic), baseline quality of life, duration and severity of FI, and presence of anxiety or depression.

However, according to observations by Fowler [15], studies on the latencies of the pelvic floor contraction during peripheral nerve evaluation show that the muscle response is reflexly mediated with a minimum latency ranging from 50 to 57 milliseconds instead of the 4–5 milliseconds observed with sacral root magnetic depolarization [48]. Are these reflexes originated from a segmental level within the sacral spinal cord or from supraspinal neuronal centers involving spino-bulbospinal pathways? Schurch [83] recorded a reflex response of 41.2 ms (range 33.3–62 ms) which corresponds to a segmental reflex, similar to the pudendo-anal reflex, in three patients with complete spinal cord injury (SCI). The findings confirm that the anal contractions observed during peripheral nerve evaluation are reflex responses mediated by afferent pathways of spinal origin, since they were obtained in complete SCI patients in whom all spino-bulbospinal loops are supposed to be interrupted. The finding that neuromodulation is working in non-neurogenic patients but is not successful in complete SCI patients could give evidence that preserved spino-bulbospinal loops contribute to the positive effects of neuromodulation.

The reported trends toward an improvement in rectal sensitivity with a reduction in the threshold of perception of rectal distension are of particular interest, and an effect at the level of the central nervous system by afferent stimulation may be hypothesized.

Some experimental animal studies seem to confirm the hypothesis that neuromodulation has an effect on the central nervous system via afferent sensory fibers. A double-blind randomized study with spinally transected rats has evaluated the role of neuromodulation on C-afferent fibers that form the afferent arc of the pathological reflex responsible of bladder hyperreflexia after spinal cord trauma. T10 spinal transection developed bladder hyperreflexia after 3 weeks associated to an increase in the neuropeptide content (substance P, neurokinin A, and calcitonin gene-related peptide (CGRP)) in L6 dorsal root ganglions. The electrical stimulation of S1 reduces the increase of neuropeptide in L6 and abolished bladder hyperreflexia suggesting that the blockade of C-afferent fibers is one of the mechanisms of action of sacral neuromodulation [84].

Recently, Chan [84] showed an increase of nerve fibers immunoreactive to vanilloid receptor subtype 1 (VR1) in the mucosal, submucosal, and muscle layers of patients with rectal hypersensitivity and fecal urgency. The VR1, present in A-delta and C-fibers and postsynaptic sites within the spinal cord dorsal horn, is known as an integrator of noxious stimuli. VR1 is activated by heat, protons, and capsaicin (an alkaloid, extractable from red pepper) and induces a flow of cations (especially Ca+ and Na+). Intravenous injection of capsaicin has produced dose-dependent sensations in the rectum of healthy people, indicating a high density of functional VR1 in this organ [27]. An increased density of VR1 fibers could lead to hyperexcitability of the dorsal spinal cord that results in a dysregulation of the sacral reflexes. These efferent reflexes include neurogenic inflammation and increase in the sympathetic tone which produces vasospasm, tissue hypoxia, and reflexive striated muscular spasticity [28].

An interesting contribution to the comprehension of the mechanism of action of the SNM comes from Hamdy [85]. He showed that the anal sphincter contraction induced by magnetic cortical stimulation was facilitated when this stimulation was preceded by repetitive stimulation of the pudendal or sacral nerve, suggesting that repetitive stimulation of a sacral nerve could cause sensory-motor interactions with better control of the sphincter function. Specific action of SNM on the primary sensory cortical area was evaluated by Malaguti [86] using somatosensory-evoked potentials (SEPs) of the pudendal and posterior tibial nerves in patients implanted with a monolateral permanent quadripolar electrode. In all patients, SNM produced a significant decrease in pudendal SEP latency at different pulse rates at the ipsilateral and contralateral implant sites. This finding was evidence of the effect of SNM on the cortical sensory area.

In a prospective trial, the latency (ms) of somatosensory-evoked cerebral potentials (SEP) induced by stimulation of the pudendal nerve was compared before (T0) and at 1 month during peripheral nerve evaluation (PNE) of SNM at frequencies of 21 Hz (T1) and 40 Hz (T2) [87] in patients with constipation or fecal incontinence. The results were correlated with the clinical outcome at 6 months. Twenty-eight (66.7%) of 42 patients had a good clinical result ("success") at 6 months.

In 16 (69.6%) of 23 incontinent patients with clinical "success" from SNM at 6 months (CCF-FIS ≤ 7), there was a significant difference between P40 latency at T0 and T2 (38.81 ms T0, 37.49 ms T2, $p = 0.049$). In the seven incontinent patients with "failure" at 6 months, there was no change between T0 and T2.

In 12 (63.2%) of 19 constipated patients with "success" at 6 months (Wexner constipation score ≤ 15), there was no difference between baseline (T0) and T2 P40 latency (39.28 ms T0, 38.25 ms T2, $p = 0.374$). In the seven constipated patients with "failure," there was a significant fall in P40 latency (41.20 ms T0, 39.30 ms T2, $p = 0.047$) but not into normal range.

The T0 P40 latency in incontinent patients having "success" was significantly higher than in the normal population ($p = 0.044$), and success was also associated with a fall in the SEP P40 latency after SNM at 40 Hz at 1 month to within the normal range (Fig. 36.8). In constipated patients, success of SNM appears to be associated with an SEP P40 latency at T0 marginally at the upper limit of normal but not falling significantly after SNM at 40 Hz at 1 month. Failure of SNM was associated with an SEP P40 latency at T0 very significantly higher than the normal value (approximately 2 SD) and falling significantly on SNM at 1 month at 40 Hz but not to within the normal range.

These results can be interpreted to indicate that clinical success in incontinent patients is associated with a reduction of P40 latency from elevated

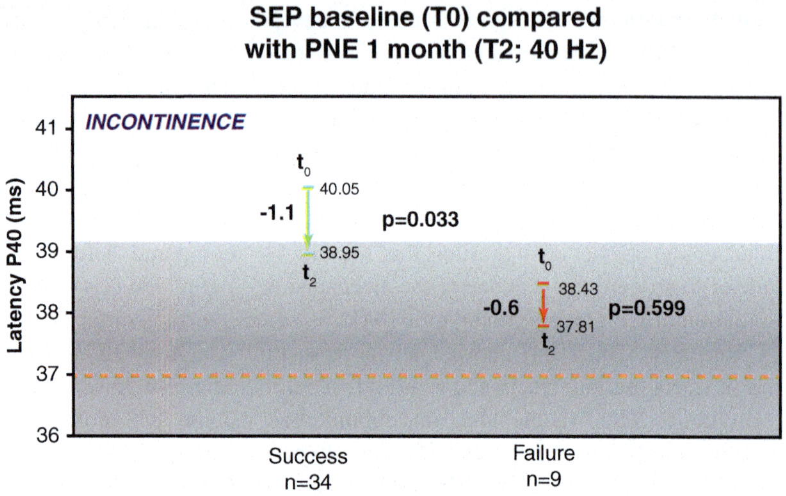

Fig. 36.8 SEP baseline (T0) compared with PNE 1 month (T2; 40Hz) for fecal incontinent patients

values to within the normal range. In the case of constipation, there was clear statistically significant evidence that those who failed had high P40 latencies at T0, and despite a reduction on SNM, they continued to have P40 values which still remained above the upper limit of normal.

These results support that SNM acts on the cortical level via the afferent pathway. Furthermore, the modifications to SEP induced by SNM seemed to be a prognostic factor for the clinical outcome.

Laurberg [88] used positron emission tomography to evaluate regional cerebral blood flow before and after 30 minutes of continuous stimulation and repeated this procedure after 2 weeks of continued stimulation before and 30 minutes after arrest of the stimulation in nine women and one man. The initial stimulation activated a region of the contralateral frontal cortex that normally is active during focused attention. After 2 weeks of stimulation, this activation had been replaced by activity in parts of the ipsilateral caudate nucleus, a region of the brain thought to be specifically involved in learning and reward processing. These changes may contribute to the improved continence, which is an acquired result of the stimulation.

A recent review of relevant studies on the central mechanism of SNM in FI confirms that the initial assumption of peripheral motor neurostimulation is not supported by increasing evidence, which reports effects of SNM outside the pelvic floor [89]. The new hypothesis states that afferent signals to the brain are essential for a successful therapy. In a total of eight studies on the central mechanism of SNM for FI, a variety of (sub)cortical and spinal changes after induction of SNM are described, and the corticoanal pathways, brainstem, and specific parts of the spinal cord are involved.

Summary

Sacral neuromodulation appears to be clinically efficacious for patients with fecal incontinence. Overall, the published series demonstrate a high effectiveness with a median 90.8% successful rate in the medium term at the cost of a reduced morbidity. Again, the long-term results are encouraging with a 76–81% of positive results, with up to 42% achieving full continence and the majority experiencing improvement in symptoms.

The possibility to select patients on the basis of a preliminary PNE makes, till now, sacral neuromodulation a unique technique in the spectrum of the possible treatments for fecal incontinence.

Given the low morbidity, reversibility, and minimal invasiveness of this procedure, the results provided by SNM therapy supersedes other surgical interventions for FI.

References

1. Bradley WE, Timm GW, Chou SN. A decade of experience with electronic stimulation of the micturition reflex. Urol Int. 1971;26:283–302.
2. Held T, Agrawal G, Kantrowitz A. Studies in stimulation of the bladder and its motor nerves. Surgery. 1966;60:848–56.
3. Jonas U, Jones LW, Tanagho EA. Spinal cord versus detrusor stimulation. A comparative study in six acute dogs. Investig Urol. 1975;13:171–4.
4. Ingersoll EH, Jones LL, Hegre ES. Effect on urinary bladder of unilateral stimulation of pelvic nerves in the dog. Am J Physiol. 1957;189:167–71.
5. Holmquist B. Electromicturition by pelvic nerve stimulation in dogs. Scand J Urol Nephrol. 1968;2:1–27.
6. Heine JP, Schmidt RA, Tanagho EA. Intraspinal sacral root stimulation for controlled micturition. Invest Urol. 1977;15(1):78–82.
7. Schmidt RA, Bruschini H, Tanagho EA. Urinary bladder and sphincter responses to stimulation of dorsal and ventral sacral roots. Invest Urol. 1979;16:300–4.
8. Tanagho EA, Schmidt RA. Bladder pacemaker: scientific basis and clinical future. Urology. 1982;20:614–9.
9. Matzel KE, Stadelmaier U, Hohenfellner M, Gall FP. Electrical stimulation of sacral spinal nerves for treatment of faecal incontinence. Lancet. 1995;346:1124–7.
10. NICE. Sacral nerve stimulation for faecal incontinence – interventional procedure guidance 99. London: National Institute for Clinical Excellence; 2004.
11. NICE. The management of faecal incontinence in adults – guideline CG49. London: National Institute for Clinical Excellence; 2007.
12. Malouf AJ, Wiesel PH, Nicholls T, Nicholls RJ, Kamm MA. Short-term effects of sacral nerve stimulation for idiopathic slow transit constipation. World J Surg. 2002;26:166–70.
13. Ganio E, Masin A, Ratto C, et al. Short-term sacral nerve stimulation for functional anorectal and urinary disturbances: results in 40 patients: evaluation of a new option for anorectal functional disorders. Dis Colon Rectum. 2001;44:1261–7.
14. Matzel KE. Sacral neurostimulation: principles and its role in the treatment of feacal incontinence. In: Proceeding of the 6th Biennial International Meeting of Coloproctology; 2000. p. 101–3.
15. Fowler CJ, Swinn MJ, Goodwin RJ, Oliver S, Craggs M. Studies of the latency of pelvic floor contraction during peripheral nerve evaluation show the muscle response is reflexly mediated. J Urol. 2000;163(3):881–3.
16. Scheepens WA, Weil EH, van Koeveringe GA, Rohrmann D, Hedlund HE, Schurch B, Ostardo E, Pastorello M, Ratto C, Nordling J, van Kerrebroeck PE. Buttock placement of the implantable pulse generator: a new implantation technique for sacral neuromodulation–a multicenter study. Eur Urol. 2001;40(4):434–8.
17. Spinelli M, Giardiello G, Arduini A, van den Hombergh U. New percutaneous technique of sacral nerve stimulation has high initial success rate: preliminary results. Eur Urol. 2003;43(1):70–4.
18. Koch SM, van Gemert WG, Baeten CG. Determination of therapeutic threshold in sacral nerve modulation for faecal incontinence. Br J Surg. 2005;92:83–7.
19. Medtronic Reference Manual. System eligibility, battery longevity, specifications. Minneapolis, MN: Medtronic Inc.; 2006.
20. Spinelli M, Sievert KD. Latest technologic and surgical developments in using InterStim Therapy for sacral neuromodulation: impact on treatment success and safety. Eur Urol. 2008;54:1287–96.
21. Ganio E, Luc AR, Clerico G, Trompetto M. Sacral nerve stimulation for treatment of fecal incontinence: a novel approach for intractable fecal incontinence. Dis Colon Rectum. 2001;44(5):619–29.
22. Michot F., Leroi AM., Grise P., Denis P. Effect of sacral nerve stimulation on pelvic floor function in patients with faecal incontinence. Acts from 6th Biennial Course – International Meeting of coloproctology, Saint Vincent 5–7/4/2000, pp. 135–137.
23. Melenhorst J, Koch SM, Uludag O, et al. Is a morphologically intact anal sphincter necessary for success with sacral nerve modulation in patients with faecal incontinence? Color Dis. 2008;10:257–62.
24. Chan MK, Tjandra JJ. Sacral nerve stimulation for fecal incontinence: external anal sphincter defect vs. intact anal sphincter. Dis Colon Rectum. 2008;51:1015–25.
25. Dudding TC, Parés D, Vaizey CJ, et al. Sacral nerve stimulation for the treatment of faecal incontinence related to dysfunction of the internal anal sphincter. Int J Color Dis. 2010;25:625–30.
26. Shaker H, Wang Y, Loung D, Balbaa L, Fehlings MG, Hassouna MM. Role of C-afferent fibres in the mechanism of action of sacral nerve root neuromodulation in chronic spinal cord injury. BJU Int. 2000;85(7):905–10.
27. Winning AJ, Hamilton RD, Shea SA, Guz A. Respiratory and cardiovascular effects of central and peripheral intravenous injections of capsaicin in man: evidence for pulmonary chemosensitivity. Clin Sci (Lond). 1986;71:519–26.
28. Black PH. Stress and the inflammatory response: a review of neurogenic inflammation. Brain Behav Immun. 2002;16(6):622–53.
29. Malouf AJ, Vaizey CJ, Nicholls RJ, Kamm MA. Permanent sacral nerve stimulation for fecal incontinence. Ann Surg. 2000;232:143–8.
30. Rosen HR, Urbarz C, Holzer B, Novi G, Schiessel R. Sacral nerve stimulation as a treatment for fecal incontinence. Gastroenterology. 2001;121:536–41.
31. Kenefick NJ, Vaizey CJ, Cohen RC, Nicholls RJ, Kamm MA. Medium-term results of permanent sacral nerve stimulation for faecal incontinence. Br J Surg. 2002;89:896–901.
32. Ganio E., Realis Luc A., Ratto C., Doglietto GB, Masin A., Dodi G., Altomare D.F., Memeo V., Ripetti

V., Arullani A., Falletto E., Gaetini A., Scarpino O., Saba V., Infantino A., La Manna S., Ferulano, Villani R.. Sacral nerve modulation for fecal incontinence. Functional results and assessment of the quality of life. Acts from 13th Annual Colorectal Disease – Fort Lauderdale, FL; February 14–16, 2002.
33. Matzel KE, Bittorf B, Stadelmaier U, Hohenberger W. Sacral nerve stimulation in the treatment of faecal incontinence. Chirurg. 2003;74(1):26–32.
34. Mirbagheri N, Sivakumaran Y, Nassar N, Gladman MA. Systematic review of the impact of sacral neuromodulation on clinical symptoms and gastrointestinal physiology. ANZ J Surg. 2016;86(4):232–6. https://doi.org/10.1111/ans.13257.. Epub 2015 Aug 5
35. Leroi A-M, Parc Y, Lehur P-A, et al. Efficacy of sacral nerve stimulation for fecal incontinence. Ann Surg. 2005;242:662–9.
36. Leroi AM, Michot F, Grise P, Denis P. Effect of sacral nerve stimulation in patients with fecal and urinary incontinence. Dis Colon Rectum. 2001;44:779–89.
37. Altomare DF, Ratto C, Ganio E, Lolli P, Masin A, Villani RD. Long-term outcome of sacral nerve stimulation for fecal incontinence. Dis Colon Rectum. 2009;52:11–7.
38. Oom DMJ, Steensma AB, van Lanschot JJB, Schouten WR. Is sacral neuromodulation for fecal incontinence worthwhile in patients with associated pelvic floor injury? Dis Colon Rectum. 2010;53:422–7.
39. Boyle DJ, Murphy J, Gooneratne ML, et al. Efficacy of sacral nerve stimulation for the treatment of fecal incontinence. Dis Colon Rectum. 2011;54:1271–8.
40. Hull T, Giese C, Wexner SD, et al. Long-term durability of sacral nerve stimulation therapy for chronic fecal incontinence. Dis Colon Rectum. 2013;56:234–45.
41. Oz-Duyos AM, Navarro-Luna A, Brosa M, Pando JA, Sitges-Serra A, Marco-Molina C. Clinical and cost effectiveness of sacral nerve stimulation for faecal incontinence. Br J Surg. 2008;95:1037–43.
42. Matzel KE, Kamm MA, Stösser M, et al. Sacral spinal nerve stimulation for faecal incontinence: multicentre study. Lancet. 2004;363:1270–6.
43. Jarrett ME, Matzel KE, Stösser M, Baeten CG, Kamm MA. Sacral nerve stimulation for fecal incontinence following surgery for rectal prolapse repair: a multicenter study. Dis Colon Rectum. 2005;48:1243–8.
44. Tjandra JJ, Chan MK, Yeh CH, Murray-Green C. Sacral nerve stimulation is more effective than optimal medical therapy for severe fecal incontinence: a randomized, controlled study. Dis Colon Rectum. 2008;51:494–502.
45. George AT, Kalmar K, Panarese A, Dudding TC, Nicholls RJ, Vaizey CJ. Long-term outcomes of sacral nerve stimulation for fecal incontinence. Dis Colon Rectum. 2012;55:302–6.
46. Santoro GA, Infantino A, Cancian L, Battistella G, Di Falco G. Sacral nerve stimulation for fecal incontinence related to external sphincter atrophy. Dis Colon Rectum. 2012;55:797–805.
47. Kenefick NJ. Sacral nerve neuromodulation for the treatment of lower bowel motility disorders. Ann R Coll Surg Engl. 2006;88:617–23.
48. Vaizey CJ, Kamm MA, Turner IC, Nicholls RJ, Woloszko J. Effects of short term sacral nerve stimulation on anal and rectal function in patients with anal incontinence. Gut. 1999;44:407–12.
49. Leroi AM, Parc Y, Lehur PA, Mion F, Barth X, Rullier E, Bresler L, Portier G, Michot F, Study Group. Efficacy of sacral nerve stimulation for fecal incontinence: results of a multicenter double-blind crossover study. Ann Surg. 2005;242(5):662–9.
50. Duelund-Jakobsen J, Dudding T, Bradshaw E, Buntzen S, Lundby L, Laurberg S, Vaizey C. Randomized double-blind crossover study of alternative stimulator settings in sacral nerve stimulation for faecal incontinence. Br J Surg. 2012;99(10):1445–52.
51. Kahlke V, Topic H, Peleikis H, Jongen J. Sacral nerve modulation for fecal incontinence: results of a prospective single-center randomized crossover study. Dis Colon Rectum. 2015;58(2):235–40.
52. Altomare DF, Giuratrabocchetta S, Knowles CH, Muñoz Duyos A, Robert-Yap J, Matzel KE, European SNS Outcome Study Group. Long-term outcomes of sacral nerve stimulation for faecal incontinence. Br J Surg. 2015;102(4):407–15. https://doi.org/10.1002/bjs.9740.. Epub 2015 Feb 2
53. Ratto C, Litta F, Parello A, Donisi L, De Simone V, Zaccone G. Sacral nerve stimulation in faecal incontinence associated with an anal sphincter lesion: a systematic review. Colorectal Dis. 2012;14(6):e297–304.
54. Ratto C, Litta F, Parello A, Donisi L, Doglietto GB. Sacral nerve stimulation is a valid approach in fecal incontinence due to sphincter lesions when compared to sphincter repair. Dis Colon Rectum. 2010;53(3):264–72. https://doi.org/10.1007/DCR.0b013e3181c7642c.
55. Leroi AM, Michot F, Grise P, Denis P. Effect of sacral nerve stimulation in patients with fecal and urinary incontinence. Dis Colon Rectum. 2001;44(6):779–89.
56. Conaghan P, Farouk R. Sacral nerve stimulation can be successful in patients with ultrasound evidence of external anal sphincter disruption. Dis Colon Rectum. 2005;48(8):1610–4.
57. Melenhorst J, Koch SM, Uludag O, van Gemert WG, Baeten CG. Is a morphologically intact anal sphincter necessary for success with sacral nerve modulation in patients with faecal incontinence? Color Dis. 2008;10(3):257–62. Epub 2007 Oct 19.
58. Chan MK, Tjandra JJ. Sacral nerve stimulation for fecal incontinence: external anal sphincter defect vs. intact anal sphincter Dis Colon Rectum. 2008;51(7):1015–24; discussion 1024–5. https://doi.org/10.1007/s10350-008-9326-0. Epub 2008 May 17.
59. Dudding TC, Parés D, Vaizey CJ, Kamm MA. Sacral nerve stimulation for the treatment of faecal incontinence related to dysfunction of the internal anal sphincter. Int J Color Dis. 2010;25(5):625–30. https://doi.org/10.1007/s00384-010-0880-2.. Epub 2010 Feb 2
60. Matzel KE. Sacral nerve stimulation for faecal incontinence: its role in the treatment algorithm. Colorectal Dis. 2011;13 Suppl 2:10–4. https://doi.org/10.1111/j.1463-1318.2010.02519.x.

61. Drossman DA. The functional gastrointestinal disorders and the Rome III process. Gastroenterology. 2006;130:1377–90.
62. Bharucha AE, Wald A, Enck P, et al. Functional anorectal disorders. Gastroenterology. 2006;130:1510–8.
63. Patton V, Stewart P, Lubowski DZ, Cook IJ, Dinning PG. Sacral nerve stimulation fails to offer long-term benefit in patients with slow-transit constipation. Dis Colon Rectum. 2016;59(9):878–85.
64. Hosseinzadeh ST, Sara Poorsaadati S, et al. Psychological disorders in patients with chronic constipation. Gastroenterol Hepatol Bed Bench. 2011;4(3):159–63.
65. Govaert B, Maeda Y, Alberga J, Buntzen S, Laurberg S, Baeten CG. Medium-term outcome of sacral nerve modulation for constipation. Dis Colon Rectum. 2012;55:26–31.
66. Ratto C, Ganio E, Naldini G. Long-term results following sacral nerve stimulation for chronic constipation. Colorect Dis. 2015;17:320–8.
67. Maeda Y, Kamm MA, Vaizey CJ, Matzel KE, Johansson C, Rosen H, Baeten CG, Laurberg S. Long-term outcome of sacral neuromodulation for chronic refractory constipation. Tech Coloproctol. 2017;21(4):277–86. https://doi.org/10.1007/s10151-017-1613-0. Epub 2017 Apr 20.
68. Eftaiha SM, Balachandran B, Marecik SJ, Mellgren A, Nordenstam J, Melich G, Prasad LM, Park JJ. Sacral nerve stimulation can be an effective treatment for low anterior resection syndrome. Colorectal Dis. 2017;19(10):927–33. https://doi.org/10.1111/codi.13701.
69. Ramage L, Qiu S, Kontovounisios C, Tekkis P, Rasheed S, Tan E. A systematic review of sacral nerve stimulation for low anterior resection syndrome. Colorectal Dis. 2015;17(9):762–71. https://doi.org/10.1111/codi.12968.
70. Bielefeldt K. Adverse events of sacral neuromodulation for fecal incontinence reported to the federal drug administration. World J Gastrointest Pharmacol Ther. 2016;7(2):294–305.
71. Hijaz A, Vasavada SP, Daneshgari F, et al. Complications and troubleshooting of two stage sacral neuromodulation therapy: a single institution experience. Urology. 2006;68:533–7.
72. Zbar AP. Sacral neuromodulation and peripheral nerve stimulation in patients with anal incontinence: an overview of techniques, complications and troubleshooting. Gastroenterol Rep. 2014;2:112–20.
73. Uludağ Ö, Morren GL, Dejong CHC, Baeten CGMI. The effect of sacral neuromodulation on the rectum. Br J Surg. 2005;92(8):1017–23.
74. Abdel-Halim MR, Crosbie J, Engledow A, Windsor A, Cohen CR, Emmanuel AV. Temporary sacral nerve stimulation alters rectal sensory function: a physiological study. Dis Colon Rectum. 2011;54:1134–40.
75. Roman S, Tatagiba T, Damon H, Barth X, Mion F. Sacral nerve stimulation and rectal function: results of a prospective study in faecal incontinence. Neurogastroenterol Motil. 2008;20:1127–31.
76. Michelsen HB, Buntzen S, Krogh K, Laurberg S. Rectal volume tolerability and anal pressures in patients with fecal incontinence treated with sacral nerve stimulation. Dis Colon Rectum. 2006;49:1039–44.
77. Michelsen HB, Maeda Y, Lundby L, Krogh K, Buntzen S, Laurberg S. Retention test in sacral nerve stimulation for fecal incontinence. Dis Colon Rectum. 2009;52:1864–8.
78. Michelsen HB, Worsøe J, Krogh K, et al. Rectal motility after sacral nerve stimulation for faecal incontinence. Neurogastroenterol Motil. 2010;22:36–41.. e6
79. Gourcerol G, Gallas S, Michot F, Denis P, Leroi A-M. Sacral nerve stimulation in fecal incontinence: are there factors associated with success? Dis Colon Rectum. 2007;50:3–12.
80. Maeda Y, Lundby L, Buntzen S, Laurberg S. Outcome of sacral nerve stimulation for fecal incontinence at 5 years. Ann Surg. 2014;259:1126–31.
81. Dudding TC, Parés D, Vaizey CJ, Kamm MA. Predictive factors for successful sacral nerve stimulation in the treatment of faecal incontinence: a 10-year cohort analysis. Color Dis. 2008;10:249–56.
82. Govaert B, Melenhorst J, Nieman FH, Bols EM, van Gemert WG, Baeten CG. Factors associated with percutaneous nerve evaluation and permanent sacral nerve modulation outcome in patients with fecal incontinence. Dis Colon Rectum. 2009;52:1688–94.
83. Schurch B, Reilly I, Reitz A, Curt A. Electrophysiological recordings during the peripheral nerve evaluation (PNE) test in complete spinal cord injury patients. World J Urol. 2003;20(6):319–22.. Epub 2002 Oct 10
84. Chan CLH, Facer P, Davis JB, Smith GD, Egerton J, Bountra C, Williams NS, Anand P. Sensory fibres expressing capsaicin receptor TRPV1 in patients with rectal hypersensitivity and faecal urgency. Lancet. 2003;361:385–91.
85. Hamdy S, Enck P, Aziz Q, Rothwell JC, Uengoergil S, Hobson A, Thompson D. Spinal and pudendal nerve modulation of human corticoanal motor pathways. Am J Physiol. 1998;274:G419–23.
86. Malaguti S, Spinelli M, Giardiello G, Lazzeri M, Van Den Hombergh U. Neurophysiological evidence may predict the outcome of sacral neuromodulation. J Urol. 2003;170(6 Pt 1):2323–6.
87. Giani I, Novelli E, Martina S, Clerico G, Luc AR, Trompetto M, Malaguti S, Nicholls J, Ganio E. The effect of sacral nerve modulation on cerebral evoked potential latency in fecal incontinence and constipation. Ann Surg. 2011;254(1):90–6.
88. Lundby L, Møller A, Buntzen S, Krogh K, Vang K, Gjedde A, Laurberg S. Relief of fecal incontinence by sacral nerve stimulation linked to focal brain activation. Dis Colon Rectum. 2011;54(3):318–23.
89. Janssen PTJ, Komen N, Melenhorst J, Bouvy ND, Jahanshahi A, Temel Y, Breukink SO. Sacral neuromodulation for fecal incontinence: a review of the central mechanisms of action. J Clin Gastroenterol. 2017;51(8):669–76.

Sacral Neuromodulation for Anorectal Dysfunction

37

Klaus E. Matzel and Birgit Bittorf

Background of Sacral Neuromodulation for Anorectal Dysfunction

The concept of recruiting residual function of the anorectal continence organ by stimulation of its peripheral nerve supply, the sacral spinal nerves, was introduced in 1994 [1].

Observation in patients treated for urinary dysfunction with sacral nerve stimulation revealed a potentially therapeutic effect on anorectal continence function. Ever since, the understanding of underlying mode of action has increased. Initially, the idea was an efferent nerve stimulating effect: today, it is appreciated that the effect of sacral spinal nerve stimulation (SNS)/sacral neuromodulation (SNM) is not confined to efferent nerves but also affects afferent nerve fibers, and it is not limited to the peripheral nerve system. As the mode of action is manyfold, there are no distinct physiological and morphological criteria, which allow to predict the therapeutic potential of the therapy in an individual patient. Based on this concept and the advantage of a testing phase to evaluate the therapeutic potential of SNM, the spectrum of application broadened over the years, even beyond the field of incontinence [2].

K. E. Matzel (✉) · B. Bittorf
University Hospital Erlangen, Department of Surgery, Section of Coloproctology, Erlangen, Germany
e-mail: Klaus.Matzel@uk-erlangen.de

SNM Technique

Patient selection is based on the outcome of a test stimulation phase: no clinical or physiologic predictor of success of chronic stimulation exists, and thus decision-making for implantation of a permanent device is based solely on the outcome of temporary test stimulation, usually of 2 weeks duration. Prerequisites for the test stimulation are residual sphincter function, an existing neuromuscular connection to the sphincter (tested by observation of voluntary squeeze or reflex activity after pinprick), and accessibility of the target sacral spinal nerves S3 and S4. Thus, the spectrum of application is not limited to specific physiological or morphological conditions.

The stimulation system consists of a fully implantable electrode placed close to a target nerve at the level of the sacral spinal nerves, most commonly S3 or S4, connected to an impulse generator placed in a subcutaneous pocket, which can be programmed and activated via telemetry.

The operative technique and the process of patient selection for permanent therapeutic SNM are standardized. With so-called percutaneous nerve evaluation (PNE), the sacral spinal nerves S3 and S4 are stimulated with needle electrodes placed through the dorsal sacral foramen. This aims to determine the single sacral spinal nerve functionally relevant to the innervation of the striated pelvic floor and anal sphincter muscles and to demonstrate whether nerve stimulation can induce muscular contraction.

L. C. C. Oliveira (ed.), *Anorectal Physiology*, https://doi.org/10.1007/978-3-030-43811-1_37

If PNE is successful, it is followed by the placement of one or more electrodes in the proximity of one or more target spinal nerves. Two techniques are available: temporary electrodes or electrodes that can remain in place for chronic stimulation if this phase is successful. The latter are quadripolar, so-called "tined lead" electrodes, and are placed with fluoroscopy guidance [3]. Both types of electrodes are connected to an external pulse generator for this test stimulation period.

If tined leads have been used and results indicate clinical efficacy, a pulse generator (IPG) is placed, usually in a subcutaneous pocket in the gluteal area. If temporary electrodes have been used, the complete neurostimulation system – electrode and pulse generator – needs to be implanted. Stimulation usually starts early after the implantation of the IPG, which is programmed by telemetry. The chronic stimulation pattern is standardized: 15 Hz, 210 μsec, and continuous or on/off cycle 5 sec/1 sec; voltage is adapted to the patients' perception of the stimulation in the anal and perineal region. The IPG can be deactivated and the intensity of stimulation changed within a preset range by the patient with a handheld device, the so-called patient programmer. During the PNE phase, bowel habits are documented with standardized bowel diaries (which are also used for follow-up) and then compared with pretest function. Commonly, a 50% improvement in symptoms – episodes of incontinence or days with incontinence episodes – is considered an indication for permanent stimulation.

Indication

Fecal Incontinence

For fecal incontinence (FI), commonly, a 50% improvement in symptoms with incontinence episodes is considered an indication for permanent stimulation. The predictive value of a positive test result is high: in approximately 80% of patients, the outcome of the test stimulation is at least equaled with permanent stimulation [2, 4]. The relevance of a false-negative test stimulation is unknown.

With the help of the highly predictive test stimulation, the spectrum of indications has been continuously expanded. Initially, the technique was confined to patients presenting with a weak, but morphologically intact, striated muscle pelvic floor and anal sphincter. Today, SNM is successfully applied to a wide etiologic spectrum [2, 3] such as weak external anal sphincter (with or without a deficit or defect of the smooth muscle internal anal sphincter), structuraö defects of the external anal sphincter of up to 180° [5]; fecal incontinence with or without urinary incontinence, low anterior resection syndrome [6] and neurogenic FI [7].

In recent years, the outcome has been increasingly presented on the basis of an intention-to-treat analysis (ITT). With this approach, clinically test stimulations are considered part of the treatment – not part of the diagnostic workup – and are consequently counted as failures if negative. Failure of the test stimulation, also dependent on patient selection, can reach up to 27% [2].

Constipation

As for incontinence, a symptom alleviation of 50% in constipation is commonly accepted to indicated chronic stimulation with a fully implantable neurostimulation system. Data of SNM for constipation are less robust than for incontinence treatment. Many series report a symptom improvement, which is less than in FI; however, outcome has been questioned recently [8]. A recent single report describes symptom improvement in a distinct group of patients presenting with constipation because of rectal hyposensitivity during test stimulation [9].

Contraindications

Contraindications to SNM are pathological conditions of the sacrum preventing adequate electrode placement, skin disease at the area of implantation, severe anal sphincter damage, trauma sequelae with micturition disorders or low bladder capacity, pregnancy, bleeding risk, psychological instability, low mental capacity, the presence of a cardiac pacemaker (although compatibility can be assessed) or implantable defibrillator, and the need for MRI (except head coil).

Morbidity of SNM is low, and severe complications are rare: device removal occurs in around 3%, and the overall complication rate ranges around 15% in patients with permanent implants [10, 11]. In a collective of 120 patients studied under a strict protocol, the cumulative revision rate at 5 years was 24%; after 5 years, the system was still in use in 81% [12].

The therapy requires maintenance: the IPG needs to be exchanged once the battery is depleted, and a substantial proportion of patients require repeated adjustment of the stimulation parameters. Indeed, a study revealed that 47% of the follow-up budget was used for 27% of the patients – those patients with suboptimal outcome [13].

Results

Fecal Incontinence

Usually, outcome is reported describing frequencies of incontinence episodes and/or applying incontinence scores such as the Cleveland Clinic Florida Fecal Incontinence Score (CCF-FIS) and quality of life scores. The reproducibility of this technique's clinical efficacy has been demonstrated in multiple studies: with chronic SNM, the frequency of incontinence episodes is reduced (Tables 37.1, 37.2, and 37.3), the CCF-FIS is reduced (Table 37.2), the ability to postpone bowel emptying is increased, and the quality of life is improved [10]. Long-term follow-up shows sustained efficacy: a median of 36% (4–52) of patients with chronic SNM experiences 100% symptom improvement, and 76% (21–96), a 50% improvement [2] if per-protocol analysis is applied (evaluating only those patients with a permanent device implanted based on the positive outcome of a test phase). Overall outcome has been seen to be poorer in patients with an underlying high-grade internal rectal intussusception [14], but no other correlation with morphological and physiological conditions or demographic features has been convincingly demonstrated [2].

In recent years, the outcome has been increasingly presented on the basis of an intention-to-treat analysis (ITT). With this approach, clinically unsuccessful test stimulations are considered part of the treatment – not part of the diagnostic workup – and are consequently counted as failures (Table 37.1). Failure of the test stimulation, also dependent on patient selection, can reach up to 27% [2].

Table 37.1 Chronic sacral nerve neuromodulation (SNM) for fecal incontinence (FI): >50% improvement of incontinence episodes/week, studies with at least 50 patients

Author	Year	Patients (n) (baseline)	Patients (n) (follow-up)	Median follow-up (month)	>50% improvement incontinence episodes/week	Intention-to-treat: 50% improvement incontinence episodes/week
Melenhorst et al.	2007	100	100	26[b]	79	59
Dudding et al.	2008	51	48	24	65	52
Tjandra et al.	2008	53	53	12[a]	71	63
Govaert et al.	2009	173	169	35[b]	77	53
Hollingshead et al.	2011	86	18	60[a]	83	n.a.
Uludag et al.	2011	50	50	85	84	n.a.
Duelund-Jakobsen et al.	2012	158	91	46	75	n.a.
Hull et al.	2013	120	76	60[a]	89	53
Altomare et al.	2015	272	228	84	78	50

Modified after Thin et al. [17]. [a]values at specific time point; [b]mean; n.a. not available

Table 37.2 Chronic sacral neuromodulation (SNM) for fecal incontinence (FI): Cleveland Clinic Incontinence Score, studies with at least 50 patients

Author	Year	Patients (n) (baseline)	Patients (n) (follow-up)	Median follow-up (months)	Median score baseline (range)	Median score follow-up (range)	*p*-value
Tjandra et al.	2008	53	53	12[a]	16 (1)[b]	1 (2)[b]	<0.001
Altomare et al.	2009	60	52	74[b]	15 (4)[b]	5 (5)[b]	<0.001
Brouwer et al.	2010	55	13	48[a]	15 (13–18)	6 (2–8)	0.008
Faucheron et al.	2010	87	87	45	13 (6–19)[b]	8 (1–17)[b]	n.a.
Michelsen et al.	2010	126	10	72[a]	20 (12–20)	7 (2–11)	<0.001
Gallas et al.	2011	200	54	24[a]	14 (2–20)	7 (0–19)	0.001
Lim et al.	2011	53	41	51[b]	12 (9–15)	8 (5–11)	0.001
Wong et al.	2011	61	61	31	14 (n.a.)	8 (n.a.)	n.a.
Faucheron et al.	2012	57	42	63	14 (4–19)	7 (1–16)	<0.001
Damon et al.	2013	102	101	48[b]	14 (3)	9 (1)	<0.0001
Maeda et al.	2014	108	101	60[a]	16 (6–20)	8 (0–19)	<0.0001
Altomare et al.	2015	272	228	84	16 (13–18)	7 (4–12)	<0.001
Duelund-Jakobsen et al.	2016	164	n.a.	22	15 (3–20)	9 (0–20)	<0.001

Modified after Thin et al. [17]. [a]values at specific time point; [b]mean; n.a. not available

Table 37.3 Chronic sacral neuromodulation (SNM) for fecal incontinence (FI): incontinence episodes, studies with at least 50 patients

Author	Year	Patients (n) PNE	Patients (n) (implants)	Patients (n) (follow-up)	Median follow-up (months)	Incontinence episodes/ week median (range)		*P* value
						Baseline	Last follow-up	
Uludag et al.	2004	63	50 (79%)	6	24[a]	8 (n.a.)	1 (n.a.)	n.s.
Melenhorst et al.	2007	134	100 (75%)	6	60[a]	10 (n.a.)[b]	2 (n.a.)[b]	<0.001
Dudding et al.	2008	60	51 (85%)	48	24	6 (0–81)	1 (0–59)	n.a.
Tjandra et al.	2008	60	53 (88%)	53	12[a]	10 (13)[b]	3 (10)[b]	<0.001
Altomare et al.	2009	94	60 (64%)	52	74[b]	4 (n.a.)[b]	1 (n.a.)[b]	0.004
Michelsen et al.	2010	167	126 (74%)	49	12[a]	8 (n.a.)	1 (n.a.)	<0.001
Hollingshead et al.	2011	113	86(76%)	86	33	9 (7)[b]	1 (2) [b]	<0.001
Uludag et al.	2011	n.a.	50	n.a.	60	8 (n.a.)	0 (n.a.)	<0.002
Duelund-Jakobsen et al.	2012	n.a.	147	147	46	6 (n.a.)	1 (n.a.)	<0.001
Hull et al.	2013	133	120 (90%)	76	>60	9 (n.a.)	2 (n.a.)	<0.0001
Altomare et al.	2015	407	272 (67%)	228	84	7 (4–11)	0.3 (0–3)	<0.001
Janssen et al.	2017	374	325 (87%)	?	7.1 years	5 (n.a.)[b]	1(n.a.)[b]	<0.001

Modified after Thin et al. [17]. [a]values at specific time point; [b]mean; n.a. not available

Constipation

Like in incontinence, the efficacy of SNN in the treatment of constipation is monitored by using symptom and quality of life scores. Existing studies include patients with heterogeneous causes of constipation. This prevents firm conclusions. In most studies, a symptom improvement is noted. Outcome is poorer when compared to SNM for FI (Table 37.4). While failure rate is higher, the general risk of complications is not different from other indications for SNM.

Table 37.4 Sacral neuromodulation (SNM) for constipation

Author	Year	Patients (*n*)	Follow-up	*n* Temporary	*n* Permanent	Improvement (intention-to-treat: %)
Kenefick et al.	2002	4	8 months (1–11)	ns	4	3/ns
Kenefick et al.	2002	2	12 months	2	2	2/2
Holzer et al.	2008	19	11 months (2–20)	19	8	8/19 (42%)
Vitton et al.	2009	6	2–50 weeks	6	5	0/6 (0%)
Kamm et al.	2010	62	28 months (1–55)	62	45	39/62 (63%)
Maeda et al.	2010	70	28 months (0–70)	70	38	35/38 (54%)
Naldini et al.	2010	15	42 months (24–60)	15	9	6/9
Carriero et al.	2010	13	22 months (12–26)	13	11	6/11
Sharma et al.	2011	21	38 months (18–62)	21	11	10/21 (48%)
Govaert et al.	2012	117	37 months (4–92)	117	68	61/117 (52%)
Knowles et al.	2012	13	19 months	13	11	9/13 (69%)
Ortiz et al.	2012	48	26 months (6–96)	48	23	14/48 (29%)
Graf et al.	2015	44	24 months (4–81)	44	15	5/44 (11%)
Ratto et al.	2015	61	51 months (±15)	61	42	20/61 (33%)
Patton et al.	2016	53	24 months	Ns	53	3/53 (ns)
Zerbib et al.	2017	36	12 months	36	20	11/36 (31%)
Maeda et al.	2017	62	60 months	62	45	14/62 (23%)

Mechanism of Action

The mechanism of action is complex and multifactorial: the effect of SNM is not limited to the anorectal continence organ and the large bowel, affecting the somatomotor, somatosensory, and autonomic nervous systems; it also appears to affect the central nervous system controlling bowel and sphincter activity [15].

Role in the Current Treatment Algorithm

SNM is a surgical therapy. Surgery for FI should only be considered if conservative means do not result in adequate symptom relief. The role of SNM in the evidence-based surgical treatment algorithm of FI is central (Fig. 37.1) [16]. SNM may be used as a singular treatment modality, but it also can be considered as part of a therapy making use of multiple treatments option, e.g., SNM after functional insufficient sphincter repair. The role of SNM in the treatment algorithm is not static. Recent developments like injectable, posterior tibial nerve stimulation, and Gatekeeper/Sphinkeeper challenge its role. The conceptual advantages of SNM are test stimulation, limited invasiveness, reversibility, high patient adherence to therapy, and sustainable long-term results.

In the context of surgical options for constipation, the role of SNM is less defined. Even though it is controversial, it may offer an alternative to much more invasive, resective surgical interventions in an individual patient. When compared with other, mostly resective treatment modalities, it is expected that the advantage of being not very invasive and of being reversible will determine the role of SNM in the therapeutic algorithm of constipation, despite the fact that the outcome is only moderate.

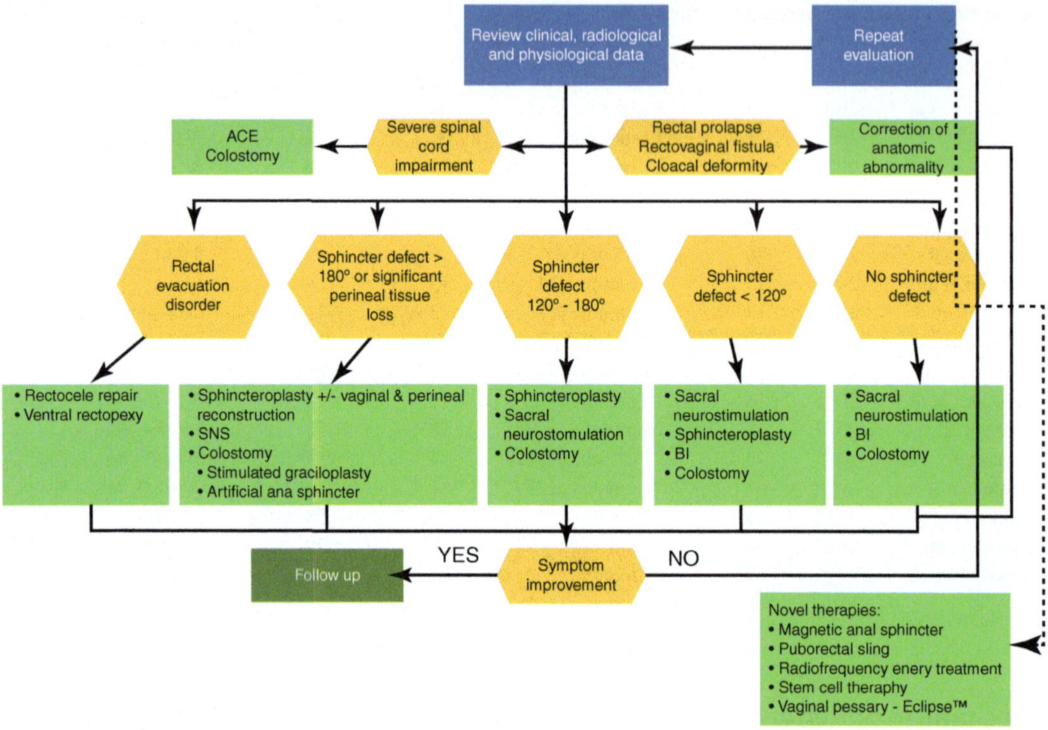

Fig. 37.1 International consultation on fecal incontinence [16]

Summary

SNM for FI should only be considered if conservative means do not result in adequate symptom relief. SNM may be used as a singular treatment modality, but it also can be considered as part of a therapy making use of multiple treatments options. Conceptual advantages of SNM are test stimulation, limited invasiveness, reversibility, high patient adherence to therapy, and sustainable long-term results. In the context of surgical options for constipation, the role of SNM is less defined. Although it is controversial, it may offer an alternative to much more invasive, resective surgical interventions in an individual patient, when compared with other, mostly resective treatment modalities.

References

1. Matzel KE, Stadelmaier U, Hohenfellner M, Gall FP. Electrical stimulation of sacral spinal nerves for treatment of faecal incontinence. Lancet. 1995;346:1124–7.
2. Altomare DF, Giuratrabocchetta S, Knowles CH, et al. Long-term outcomes of sacral nerve stimulation for faecal incontinence. Br J Surg. 2015;102:407–15.
3. Matzel KE, Chartier-Kastler E, Knowles CH, et al. Sacral neuromodulation: standardized electrode placement technique. Neuromodulation. 2017;20:816–24.
4. Matzel KE, Stadelmaier U, Hohenberger W. Innovations in fecal incontinence: sacral nerve stimulation. Dis Colon Rectum. 2004;47:1720–8.
5. Boyle DJ, Knowles CH, Lunniss PJ, et al. Efficacy of sacral nerve stimulation for fecal incontinence in patients with anal sphincter defects. Dis Colon Rectum. 2009;52:1234–3129.
6. Schwandner O. Sacral neuromodulation for fecal incontinence and "low anterior resection syndrome"

following neoadjuvant therapy for rectal cancer. Int J Color Dis. 2013;28:665–9.
7. Matzel KE. Sacral nerve stimulation for faecal incontinence: its role in the treatment algorithm. Colorectal. 2011;13(Suppl 2):10–4.
8. Thaha MA, Abukar AA, Thin NN, et al. Sacral nerve stimulation for faecal incontinence and constipation in adults. Cochrane Database Syst Rev. 2015; 24:8.
9. Knowles CH, Thin N, Gill K, et al. Prospective randomized double-blind study of temporary sacral nerve stimulation in patients with rectal evacuatory dysfunction and rectal hyposensitivity. Ann Surg. 2012;255:643–9.
10. Tan E, Ngo NT, Darzi A, et al. Meta-analysis: sacral nerve stimulation versus conservative therapy in the treatment of faecal incontinence. Int J Color Dis. 2011;26:275–94.
11. Goldman HB, Lloyd JC, Noblett KL, et al. International Continence Society best practice statement for use of sacral neuromodulation. Neurourol Urodyn. 2018;37:1823–48.
12. Hull T, Giese C, Wexner SD, et al. Long-term durability of sacral nerve stimulation therapy for chronic fecal incontinence. Dis Colon Rectum. 2013;56:234–45.
13. McMullin CM, Jadav AM, et al. Resource implications of running a sacral neuromodulation service: a 10-year experience. Colorectal Dis. 2014;16:719–22.
14. Prapasrivorakul S, Gosselink MP, Gorissen KJ, et al. Sacral neuromodulation for faecal incontinence: is the outcome compromised in patients with high-grade internal rectal prolapse? Int J Color Dis. 2015;30:229–34.
15. Carrington EV, Evers J, Grossi U, et al. A systematic review of sacral nerve stimulation mechanisms in the treatment of fecal incontinence and constipation. Neurogastroenterol Motil. 2014;26:1222–37.
16. Madoff R, Laurberg S, Lehur PA et al. Surgery for fecal incontinence International Consultation on Incontinence, Japan; 2016.
17. Thin NN, Horrocks EJ, Hotouras A, et al. Systematic review of the clinical effectiveness of neuromodulation in the treatment of faecal incontinence. Br J Surg. 2013;100:1430–47.

Technical Aspects of Sacral Neuromodulation

38

María Margarita Murphy

Introduction

Sacral neuromodulation has become an essential technique in the treatment of fecal incontinence. The surgically implanted system consists of a neurostimulator and the electrical lead that is implanted in the S3 foramen. The technique for placement of the lead is, without a doubt, the single most important step that will determine the level of success of the therapy. Attention to detail and, at times, patience will be most important to achieve the best results.

Stages of Sacral Neuromodulation

The procedure consists of a two-staged approach; the first stage or evaluation period serves to determine if the therapy will be successful and, thereby, identify eligible patients for implantation of the neurostimulator. Success is defined as an improvement in number and/or type of fecal incontinence events of at least 50%. Urgency, which has been identified as one of the most bothersome symptoms in some patients, can also be used to determine success of the evaluation. In the United States and some European countries, the percutaneous nerve evaluation (PNE) or *simple evaluation* is available. This procedure is performed with a non-tined lead designed to stay in the patient for no longer than 7 days. Unfortunately, the lead currently available for this test easily migrates out of position significantly decreasing the accuracy of the trial.

The *advanced evaluation* has been shown to be a more sensitive screening method [1, 2]. It also allows for a longer evaluation period that might be necessary in those patients whose symptoms are not as frequent. For this evaluation, a quadripolar tined lead is placed. This lead has four electrodes that will allow for more programming options and has tines to better anchor it to the subcutaneous tissues. If the evaluation is successful, this is the lead that will remain permanently in the patient as part of the implanted system. This test should last up to 14 days which should be enough time to demonstrate patient eligibility for the second-stage or long-term therapy [3]. During this second procedure, the permanent implantable neurostimulator is connected to the previously placed lead and implanted in a subcutaneous pocket in the buttock.

Advanced Evaluation or First Stage

Positioning and Preparation

Preoperatively, the patient should have completed baseline diaries documenting number of episodes of incontinence per day, severity of episodes, quality of the stools using the Bristol scale, and

M. M. Murphy (✉)
East Cooper Regional Medical Center, Department of Colon and Rectal Surgery, Mount Pleasant, SC, USA

L. C. C. Oliveira (ed.), *Anorectal Physiology*, https://doi.org/10.1007/978-3-030-43811-1_38

degree of urgency. I most commonly have the patient complete at least 14 consecutive days in their diary.

Before the start of the procedure, a prophylactic dose of antibiotic is given. Some authors follow the recommendations in their respective hospitals for hip replacement antibiotic prophylaxis. I administer 1 gram of vancomycin I.V. [4] within 60 minutes of surgical incision.

The placement of the lead can be done with local anesthetic only, allowing evaluation of sensory responses, or under sedation. Avoid paralytics or muscle relaxants to allow evaluation of the motor responses. The patient is placed in a prone position, with pillows under the hips and the shins for comfort. The lower back, buttocks, and upper thighs are prepped with an antiseptic solution. I use chlorhexidine gluconate which has been suggested to be better than iodine in preventing surgical site infections [5, 6].

Identifying the Third Sacral Foramen

The lead should be inserted in the third sacral foramen for best results. The second sacral foramen should be avoided as it will cause significant leg pain [7], and insertion in the fourth sacral foramen may not be as affective.

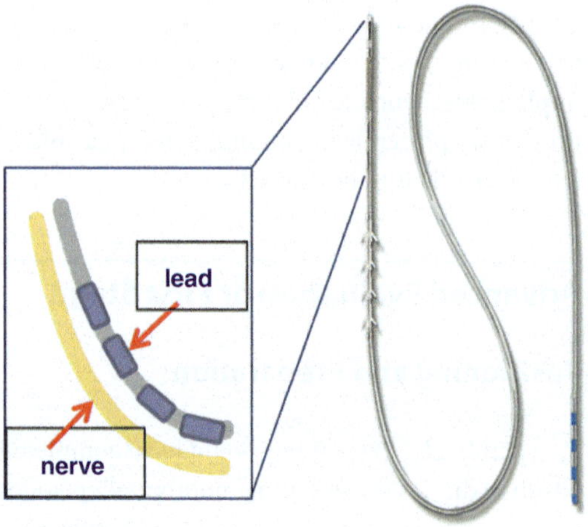

Fig. 38.1 The lead must follow the nerve's path, closely parallel to it. (Courtesy of Medtronic, Inc)

The ideal lead must follow a path parallel and very close to that of the S3 nerve as shown in Fig. 38.1. This optimal placement will allow all four electrodes to produce a motor and a sensory response, and each of those responses should be obtained with an electrical stimulation at very low voltages.

Anatomically, the third sacral nerve enters the foramen through its most medial and cephalad quadrant and exits the foramen on the anterior surface of the sacrum just above the bony hillock; the nerve then follows a medial to lateral path. Accordingly, the lead has to enter the foramen through the most cephalad and medial point of the foramen [7]. To that end, an anteroposterior (AP) fluoroscopic view is obtained to identify the medial edge of the sacral foramina bilaterally, and a corresponding line is marked over the patient's skin (Fig. 38.2).

Lead aprons are required for personal radiation protection. They should be worn by the surgeon and all the occupants of the room. It is the practice of the author to also protect the hands by wearing lead-impregnated gloves [8]. A cross-table or lateral fluoroscopic view is now obtained to locate the S3 foramen. S2 is adjacent to where the sacroiliac joint fuses, and S3 will be the first below it, characterized by an obvious anterior hump or hillock. This hillock will be an important landmark as the aim is to have the needle enter the foramen at approximately 0.5–1 cm above it. The trajectory of the needle should be perfectly parallel to the fusion line of the S2 and S3 vertebral bodies. To identify the entrance point of the needle, a hemostat is used. The point of entry is found on the previously marked line that identifies the medial edge of the foramina. The tip of the hemostat is placed over that line, and with the help of the fluoroscopy, the site is chosen based on an estimation of a line between the skin and the point 1 cm above the hillock paralleling the fusion line. This will allow for the correct angle of insertion in relation with the skin to be determined for each patient, regardless of their body habitus (Fig. 38.3).

Once identified, the point of entry is anesthetized with a mixture of 1% lidocaine and 0.25% bupivacaine. The subcutaneous tissue is infiltrated.

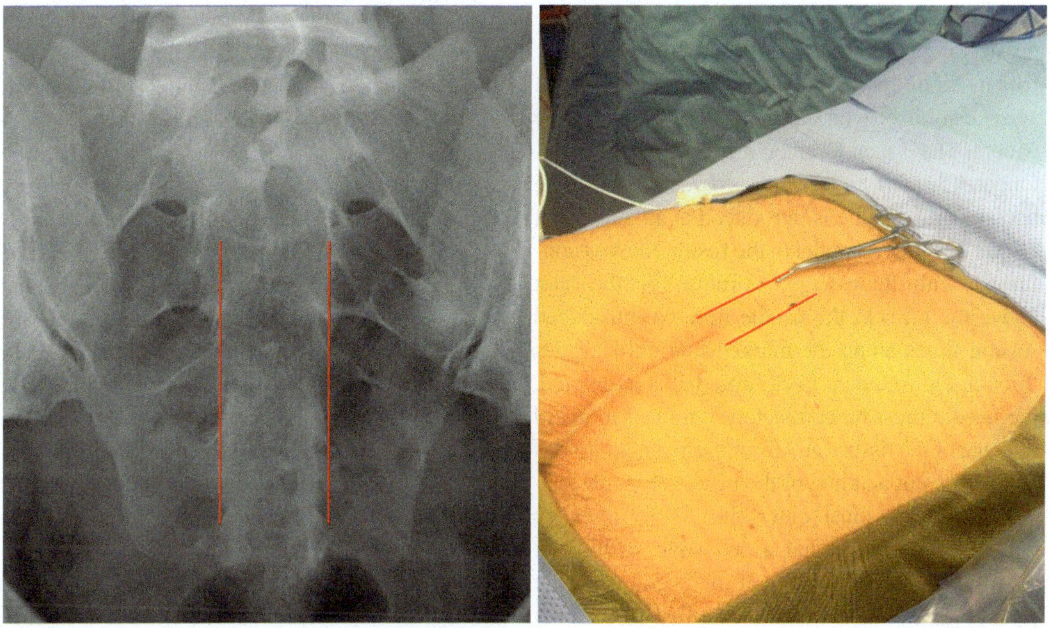

Fig. 38.2 A line marking the medial edge of the foramen is drawn over the patient's skin. (Courtesy of Medtronic, Inc)

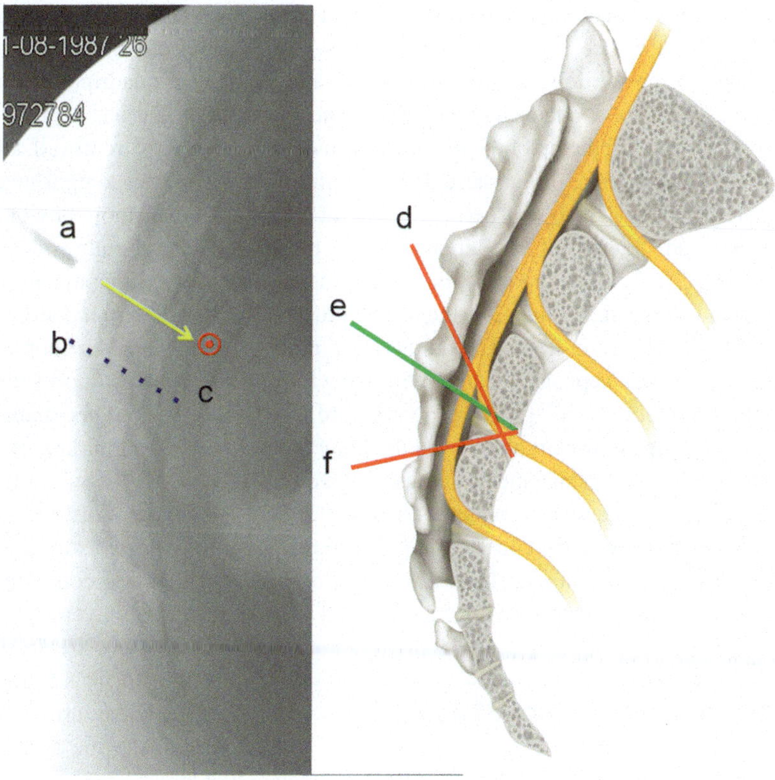

Fig. 38.3 The hemostat (a) is used to determine the entry site for the needle, which must follow the fusion line (b) aiming for a spot 1 cm above from the hillock (c). This will allow for the optimal placement of the needle (e) at the most medial and cephalad spot on the S3 foramen; d and f show incorrect needle placements. (Courtesy of Medtronic, Inc)

I avoid injecting on the sacral periosteum given the possibility of anesthetizing the nerve. To allow for patient comfort, I routinely have the help of an anesthesiologist who administers propofol to the patient. The 3-inch foramen needle is then used. A longer (5-inch) needle is provided for obese patients. The needle is introduced at an angle that allows it to enter parallel to the fusion lines as confirmed with fluoroscopy. Aiming for the spot above the hillock, the needle may be introduced several times along the marked line until it goes through the foramen. Each time this is attempted, the needle must be completely removed to avoid damage of the tissues and formation of hematomas and to prevent bending of the needle (Fig. 38.3).

The needle must enter the S3 foramen as cephalad and as medial as possible within the foramen. The introduction of the needle must be parallel to the central axis of the body, such that an AP view would show the needle as a perfectly straight line going up and down at the most medial edge of the foramina.

A grounding pad is placed on the patient's foot by an assistant and connected to an external stimulation device. This device in turn is connected to the sterile test stimulation cable and J hook. This hook is attached to the foramen needle for testing. With the tip of the needle just anterior to the anterior edge of the sacrum, stimulation is started. The goal is to see "bellowing" of the buttocks and flexing of the great toe. If these responses are not strong or require a stimulation with a voltage higher than 2 V, the needle must be removed and repositioned, usually to a more medial or more cephalad location within the foramen. Sometimes, if good responses are not obtained, one may start all over on the other side of the midline. Special attention must be given to the calf; ask an assistant to put his/her hand over it to ensure there is no contraction of those muscles and/or leg rotation. These motor responses would indicate the needle is in S2 (Table 38.1).

Introducing the Tined Lead

The next steps follow standard Seldinger technique. When satisfied with the needle positioning, a small nick is done on the skin alongside the needle with an 11 blade. The inner stylet of the needle is removed, and the directional guide wire is passed to the mark corresponding to the needle length used (3 in. or 5 in.). The needle is removed, leaving the guide wire in place. The introducer sheath is then introduced over the wire with controlled pressure under fluoroscopic surveillance. It is very important that the radiopaque marker at the tip of the sheath, seen as a transverse line, does not go pass two-thirds of the distance between the posterior and anterior edges of the sacrum. This is critical to ensure the lead follows a path parallel to the nerve. Once the sheath has been placed at the proper depth, the introducer stylet and directional guide wire are removed.

The quadripolar lead comes preloaded with a stylet that has a straight and stiff tip. This stylet should be removed and exchanged for the one with a curved tip found in the kit. This allows the lead to be more flexible and to follow the nerve in its natural path as it turns from medial to lateral when exiting the foramen. This has shown to allow better responses at lower amplitudes [9].

The lead is then passed through the introducer under fluoroscopy. The lead has two radiopaque markers shaped as dots in contrast with the flatter marker of the introducer. The lead times are between these two round markers. When the short line representing the tip of the introducer sheath is distal to the two markers, the tines are completely covered by the sheath and have not yet been deployed. This allows the lead to move freely until the most appropriate location for it is found. The four electrodes are identified as zero to three with zero being the most distal one at the tip of the lead and three the most proximal one. The lead is advanced until all the electrodes are anterior to the anterior edge of the bone. As it enters, the lead

Table 38.1 Sacral root responses

	Response		Sensation
S2	"Clamp" of anal sphincter	Leg/hip rotation, contraction of calf	Generally none
S3	"Bellows" of perineum	Flexing great toe, sometimes other toes	Pulling in rectum, extending to scrotum or labia
S4	"Bellows"	None	Pulling in rectum

should easily slide in and smoothly curve downward on the lateral view (Fig. 38.4).

On the AP view, it should be seen to curve medial to lateral giving it a "hockey stick" shape. If the lead is not pointing in a "down and out" position, it should be repositioned by pulling it out, slightly turning it, and advancing it once more under fluoroscopy. If after several attempts the appropriate curve is not seen, the whole procedure may need to be restarted with the foramen needle.

Once the lead is positioned, each of the four electrodes is tested documenting the amplitude threshold at which the appropriate motor responses are seen. If the amplitudes are low (less than 2 V) and are the same or very similar on all electrodes, it is indicative that the lead is in fact parallel to the nerve. I personally aim to have all the responses at an amplitude of 1 V. This precise positioning will allow longer battery life and more options for reprogramming and better symptom resolution. If the responses are not seen on all electrodes, the lead may need repositioning. If responses are seen only in the two most distal electrodes, the lead may need to be pulled out, some under fluoroscopic surveillance and vice versa. If there is only bellows but no toe flexion, responses consisting with the stimulation of S4, this is indicative of the lead curving down too much and stimulating the S4 nerve root. If there is mostly foot response as opposed to bellows or if the foot response happens before the perineal response, this is indicative of stimulation of the S2 root and can be seen if the lead is curving upward or not curving at all. Retracting the introducer sheath or advancing it under fluoroscopy may change the curvature of the lead. At times, advancing the sheath may be difficult without the stylet. If the stylet is replaced and the sheath advanced, it is very important to prevent the radiopaque marker from advancing too much as described above. If the marker passes beyond the anterior surface of the sacrum or even gets close to it, the hard-stiff stylet will create a false path for the lead to follow. This will prevent the lead from curving when introduced. The lead will thus cross over the nerve and not go parallel to it. Once again, if the path and responses are not optimal, it may be necessary to remove everything and start with the foramen needle again or switch to the other side of the midline.

Once the ideal position is achieved, the tines will have to be deployed to secure the lead at that location. This process, done under fluoroscopy, may be challenging as it is sometimes difficult to maintain the lead in the chosen position. The introducer sheath is slowly retracted over the lead. As the introducer radiopaque marker is seen moving upward, meaning toward the skin, the tines are being deployed. Once the sheath marker, seen as a radiopaque line, passes the most proximal round marker of the lead, the tines have been fully deployed. This is the "point of no return" as the sheath cannot be replaced over the tines once more (Fig. 38.5).

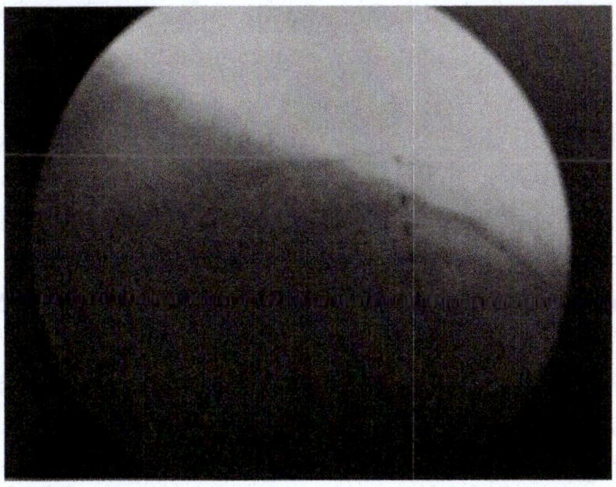

Fig. 38.4 Lateral view of lead. (Courtesy of Medtronic, Inc)

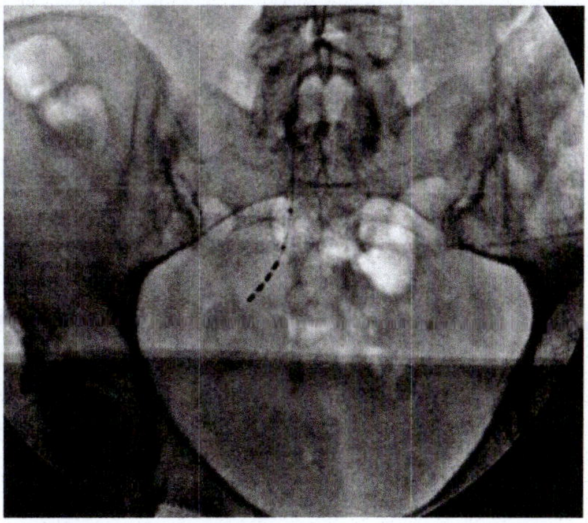

Fig. 38.5 AP view of the lead

Tunneling of the Lead

The site where the future implantable neurostimulator will remain is chosen. The site should be across the midline from the lead insertion site at a level above it but below and medial to the iliac crest. The exposed lead should be used as a guide such that the distance is appropriate for it to reach the pocket without having too much extra lead that will need to be wrapped around the device. With a ruler and marker, I draw a line at that location measuring 4.5 cm. The area is injected generously with local anesthetic. The incision is done over that mark and carried down through the subcutaneous tissues and Scarpa's fascia with the cautery. At that level, the pocket is created. The tunneling device is then used to pass the lead from the sacral insertion point to the pocket. It is important to first introduce the tunneling device as close as possible to the lead without damaging it, directing the device toward the bone, perpendicular to the skin, for approximately 1.5 cm. This will allow the lead to be tunneled deeply and not be trapped superficially within the skin. Then the tunneling device is directed toward the previously created pocket. The tip of the tunneling device is unscrewed and removed, and the introducer is removed leaving the plastic sheath or straw surrounding it in place. The lead is passed through the straw into the pocket and the straw is removed. A second tunnel is required for the percutaneous extension cable. A new plastic straw is loaded onto the tunneling device. Traditionally, this second tunnel is done following a path crossing the midline again and more cephalad than the pocket site. It is the preference of the author to create this tunnel downward and anterior toward the ipsilateral hip. This position allows the patient to be able to visualize and better control the exit site of the percutaneous extension cable, allows laying on the back without laying on the exit site, and avoids passing a sharp object over the vertebral bodies again. Once more, the tip of the device is unscrewed and the metal portion removed leaving the plastic sheath in place; the extension cable is passed through this second straw which is then removed. At the subcutaneous pocket, the lead must be connected to the percutaneous extension cable. It is important to clean bodily fluids from the lead and lead extension connectors. The plastic booth included in the kit is slid over the end of the lead. The lead is then introduced into the percutaneous extension connector, and each of the four screws is tightened by turning the wrench tool clockwise until a click is heard. The plastic booth is then slid over the connection and secured in place with 2-0 PDS at both ends. The connection is then introduced into the pocket which is now closed in layers, using 2-0 Vicryl for the thin fascial layer, and the skin is approximated with a 4-0 monocryl subcuticular running stitch. A simple stitch is used to approximate the edges of the sacral incision site. Sterile, waterproof dressings are placed. The author traditionally uses a thick layer of a topical skin glue over both incisions. This is then covered by adhesive skin bandages and a transparent adhesive film dressing over it.

Finally, the percutaneous extension pin connector is introduced into the twist lock of the screening cable, and the locking mechanism is further secured by taping over it. The screening cable is plugged into the external test stimulator box. Where available, the Verify™ external neurostimulator (Medtronic, Minneapolis, MN) can be used. This is a disposable, single-use device equipped with Bluetooth wireless technology that gives more freedom and comfort to the patient.

Postoperative Care

The patient is sent home with antibiotics. It is the choice of the author to use sulfamethoxazole/trimethoprim 800 mg/160 mg by mouth every 12 hours for 10 days. Very specific instructions are given to the patient to limit any activities that may displace the lead, e.g., bending over and reaching up or forward. The patient should shower every day, keeping the occlusive dressings intact. The patient is instructed to unplug the connector from the box and cover it with plastic secured with a rubber band. After showering, the dressings are patted dry, and the cable is reconnected to the external neurostimulator. The patient is directed to complete the fecal incontinence diaries once more. Every day, the patient is followed up via telephone to ensure there are no technical difficulties and to evaluate progress. On postoperative day 11, the patient is seen at the office for a wound check and to compare the current diaries with those completed at baseline. If there is an improvement of more than 50%, the trial is considered to be successful and the patient will be taken back to the operating room for the second stage on postoperative day 14. Alternatively, if the test failed, the patient will be taken to the operating room for removal of the lead.

Implantation of the Internal Neurostimulator or Second Stage

Perioperative management of the patient, type of anesthesia, and skin preparation are the same as for the first stage. The buttock incision is reopened being careful to protect the lead at all times. The connection components are located and brought up to the field. The suture securing the protective boot is cut and the boot removed exposing the screws. The four screws are loosened by turning the wrench counterclockwise. The lead is taken out form the percutaneous extension. The connector portion of the extension is cut at the level of the wire itself. This allows an assistant to pull the non-sterile end of the percutaneous extension where it exits the skin, thus removing the extension from the patient's body. Once again, the lead connectors are cleaned of any bodily fluids. The lead is slid into the implantable neurostimulator making sure the last blue connector is seen all the way to the end. The set screw is tightened until hearing a click. The device is then introduced into the already made subcutaneous pocket making sure the side with the writing is face up. Any extra lead is placed around and behind the neurostimulator making sure it is not bent. The wound and dressings are handled in the same manner as the first stage. Once more, the patient is discharged home on PO antibiotics per the author's preference.

Removal of the Tined Lead

If during the trial symptoms fail to improve by more than 50%, the tined lead and the percutaneous extension should be removed. The procedure starts in the same way as described above for the stage, and the percutaneous extension is removed in the same way. The lead should never be pulled from the pocket because of the high risk of it breaking with subsequent retention of a portion within the sacrum. Rather, the insertion site over the sacrum is reopened, and the lead is identified at that site. The distal portion is pulled from the subcutaneous pocket to this incision. The lead is clamped with a hemostat that will be used as a handle. The external portion of the lead is wound around the hemostat that is parallel to the skin. With constant careful tension, the lead is pulled following the same angle that was used to insert it at approximately 60 degrees from the skin. Once removed, the wounds are once again closed with absorbable sutures.

Complications

Infection of the implanted system is a possible complication that is not as common as initially thought [10] and can be prevented with good technique and antibiotic prophylaxis [4]. There are multiple reports on the literature of infections responding to antibiotic treatments, but if there is no response, the system will need to be explanted.

Another common complication is pain, either around the implanted neurostimulator or in the leg or vagina. The vast majority of these cases respond to reprogramming. If there is no improvement and the pain is associated with palpation of the device, an implant revision may be required. It may be as simple as reopening the subcutaneous pocket and repositioning of the neurostimulator at a deeper level. Less often, reimplantation of the whole system is needed. If the patient experiences discomfort going down the leg early in the postoperative course, it must be assumed that the lead was placed on the sacral foramina for the second sacral root, or the lead was placed with an angulation that allows for stimulation of the S2 nerve. Regardless, the system will need to be removed and replaced on the other side [11–13]. Lead migration is possible if the patient falls or was too active in the immediate postoperative period. If that were to happen, the system will need to be replaced.

It is important to mention that the use of magnetic resonance imaging (MRI) on patients who have a sacral nerve stimulator is not approved. The only exception remains imaging of the head only (the implant manufacturer must be consulted prior to allowing the patient to proceed with the imaging evaluation). The heating of the lead when exposed to the MRI may cause minor to very significant damage including severe burning of the neighboring nerve and other tissues. Future advances and research in this area may change the future recommendations. At this time during the preoperative visit, the patient must be counseled on the need to avoid MRIs if they undergo implantation of the sacral neuromodulator [4].

Summary

Sacral neuromodulation is the most effective therapy available for severe fecal incontinence. A meticulous surgical technique is necessary for the success of this therapy and to prevent complications.

References

1. Leong RK, De Wachter SG, Nieman FH, de Bie RA, van Kerrebroeck PE. PNE versus 1st stage tined lead procedure: a direct comparison to select the most sensitive test method to identify patients suitable for sacral neuromodulation therapy. Neurourol Urodyn. 2011;30:1249–52.
2. Janknegt RA, Weil EH, Eerdmans PH. Improving neuromodulation technique for refractory voiding dysfunctions: two-stage implant. Urology. 1997;49:358–62.
3. Baxter C, Kim JH. Contrasting the percutaneous nerve evaluation versus staged implantation in sacral neuromodulation. Curr Urol Rep. 2010;11:310–4.
4. Coffey RJ, Kalin R, Olsen JM. Magnetic resonance imaging conditionally safe neurostimulation leads: investigation of the maximum safe lead tip temperature. Neurosurgery. 2013;74(2)
5. Levin I, Amer-Alshiek J, Avni A, Lessing JB, Satel A, Almog B. Chlorhexidine and alcohol versus povidone-iodine for antisepsis in gynecological surgery. J Womens Health (Larchmt). 2011;20:321–4.
6. Noorani A, Rabey N, Walsh SR, Davies RJ. Systematic review and meta-analysis of preoperative antisepsis with chlorhexidine versus povidone-iodine in clean-contaminated surgery. Br J Surg. 2010;97:1614–20.
7. Swinn MJ, Schott GD, Oliver SE, Kitchen ND, Fowler CJ. Leg pain after sacral neuromodulation: anatomical considerations. BJU Int. 1999;84:1113–5.
8. Kim AN, Chang YJ, Cheon BK, Kim JH. How effective are radiation reducing gloves in C-arm fluoroscopy-guided pain interventions? Korean J Pain. 2014;27:145–51.
9. Jacobs SA, Lane FL, Osann KE, Noblett KL. Randomized prospective crossover study of interstim lead wire placement with curved versus straight stylet. Neurourol Urodyn. 2014;33:488–92.
10. Wexner SD, Hull T, Edden Y, et al. Infection rates in a large investigational trial of sacral nerve stimulation for fecal incontinence. J Gastrointest Surg. 2010;14:1081–9.
11. Maeda Y, Lundby L, Buntzen S, Laurberg S. Outcome of sacral nerve stimulation for fecal incontinence at 5 years. Ann Surg. 2014;259:1126–31.
12. George AT, Kalmar K, Panarese A, Dudding TC, Nicholls RJ, Vaizey CJ. Long-term outcomes of sacral nerve stimulation for fecal incontinence. Dis Colon Rectum. 2012;55:302–6.
13. Siegel SW, Catanzaro F, Dijkema HE, et al. Long-term results of a multicenter study on sacral nerve stimulation for treatment of urinary urge incontinence, urgency-frequency, and retention. Urology. 2000;56:87–91.

Posterior Tibial Nerve Stimulation

39

Charles H. Knowles

Introduction

The benefits of neuromodulation in the treatment of fecal incontinence are well proven (see chapter on SNS). Further, SNS is undoubtedly a safe and reversible therapy which sets it aside from surgery on the anal sphincter itself. Nevertheless, SNS is not successful in all patients, and there are others for whom this option is not available (due to co-morbidities, patient choice or local expertise). SNS also requires two operations that, despite advances in technology and technique, may still lead to complications [1, 2]. Although it is cost-effective compared to other surgical options, SNS does have high equipment costs (approx. £10,000 pp) and costs associated with ongoing management. On this basis, other opportunities to modulate neuromuscular functions relevant to the pelvic organs are attractive, especially if these were cheaper and less invasive.

History

The concept of modulating sacral nerve activity without the need for a permanent surgically implanted device was first described in 1983 by McGuire et al. in patients with urinary incontinence. They used a transcutaneous electrode over the tibial nerve, producing data suggesting long-term effectiveness [3]. The method was adjusted by Stoller in 1999, through the use of a percutaneous needle with a ground electrode on the ipsilateral extremity [4]. In 2003, Shafik proposed using PTNS for FI and reported a 78% functional success in 32 patients [5]. In the intervening period, TNS (especially PTNS) has become well established as a treatment of overactive bladder where its use is supported by pivotal trials [6].

Methods of Charge Delivery

As with sacral neuromodulation (SNM), the fundamental requirement of stimulation is the creation of a negative extracellular charge field in proximity to axons of the peripheral nerve in question. For the tibial nerve, this can now be achieved in three ways:

- Percutaneous tibial nerve stimulation (PTNS) (Fig. 39.1): Electrical stimulation is delivered via a needle placed adjacent to the tibial nerve just above the ankle. This focusses charge on the nerve, therefore limiting stimulation effects on the skin and thus, in theory, increasing potential charge delivery compared to transcutaneous stimulation. There are now several commercially available systems of which the main players are the Urgent® PC system (Cogentix Medical; previously

C. H. Knowles (✉)
National Bowel Research Centre, Blizard Institute, Barts and the London School of Medicine and Dentistry, London, UK
e-mail: c.h.knowles@qmul.ac.uk

L. C. C. Oliveira (ed.), *Anorectal Physiology*, https://doi.org/10.1007/978-3-030-43811-1_39

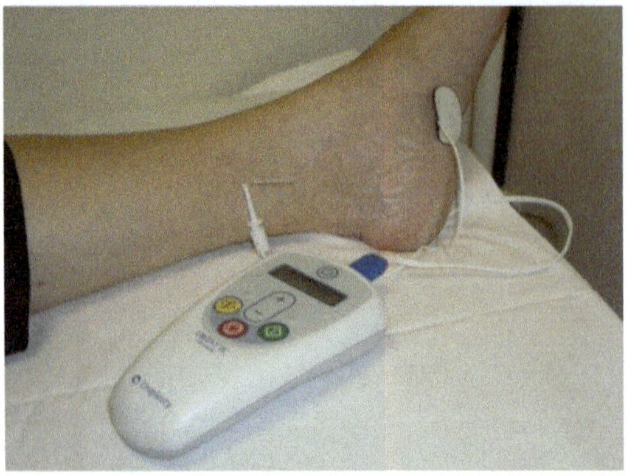

Fig. 39.1 Percutaneous tibial nerve stimulation using a commercially available device

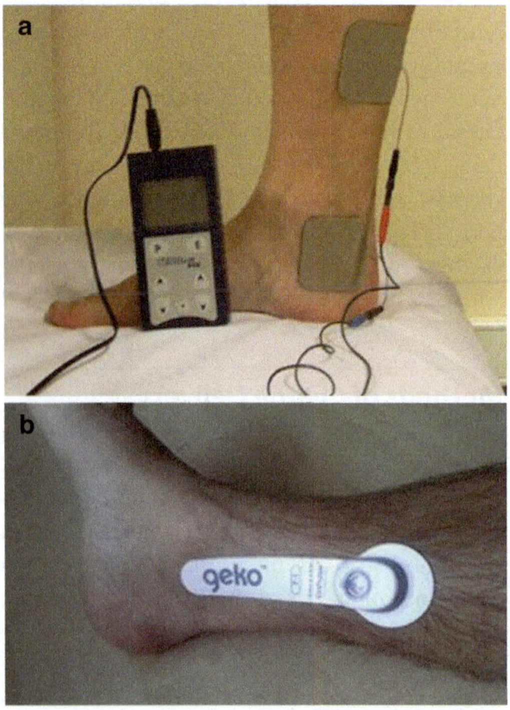

Fig. 39.2 (a) Transcutaneous tibial nerve stimulation using a standard TENS machine; (b) ambulatory adhesive device

Uroplasty Ltd., Minnetonka, MN, USA) and percutaneous tibial neuromodulation system (Medtronic Inc., Minneapolis, MN). Treatment is typically delivered as twelve 30-minute treatments, given usually weekly for 12 weeks or sometimes twice weekly for 6 weeks.

- Transcutaneous tibial nerve stimulation (TTNS) (Fig. 39.2): Electrical stimulation is delivered via two pad electrodes placed over the tibial nerve just above the ankle. This is usually delivered via a TENS (transcutaneous electrical nerve stimulation) machine (Figure 39.2A) although other ambulatory devices are under investigation, e.g. Geko (Firstkind Medical, UK) [7] (Figure 39.2B). Treatment regimens vary considerably though administration is usually in 20- to 30-minute sessions daily over a period of weeks or months.
- Implantable tibial nerve stimulation: This recent innovation utilises a miniature rechargeable device (BlueWind Medical, Herzliya, Israel) that is surgically implanted adjacent to the tibial nerve at the ankle. This device has thus far only been used for urinary indications (overactive bladder) [8] and will not be discussed further.

Physiological Effects of Tibial Nerve Stimulation

The mechanism of action of TNS in the treatment of FI is not well studied. TNS was borne from the hypothesis that the shared origin of innervation between pudendal and posterior tibial nerves (sacral nerve roots) could result in similar effects on the bladder and anorectum (innervated from S2 to S4) as achieved by SNM (Fig. 39.3).

The sticking point is that the mechanism of action of SNM is not entirely certain (see Chap. 38) and that, as for SNM, there has been an over-dependence on unblinded observations made of end organ effects based on what it is possible to measure rather than what might be relevant to mechanistic understanding. Thus, studies have reported end organ effects of TNS on various aspects of anorectal physiology, most commonly anorectal manometry [5, 9–16]. Results showed one with no change [15], whilst others showed an improvement in maximum squeeze pressure [9, 11, 16] or improvements in both maximum resting and maximum squeeze pressures [12, 14]. Two randomised studies (with sham-treated arms) showed improvements in manometry in both the sham and active intervention arms [11, 16]. Regardless, it is very difficult to countenance how TNS could possibly have a direct effect on motor

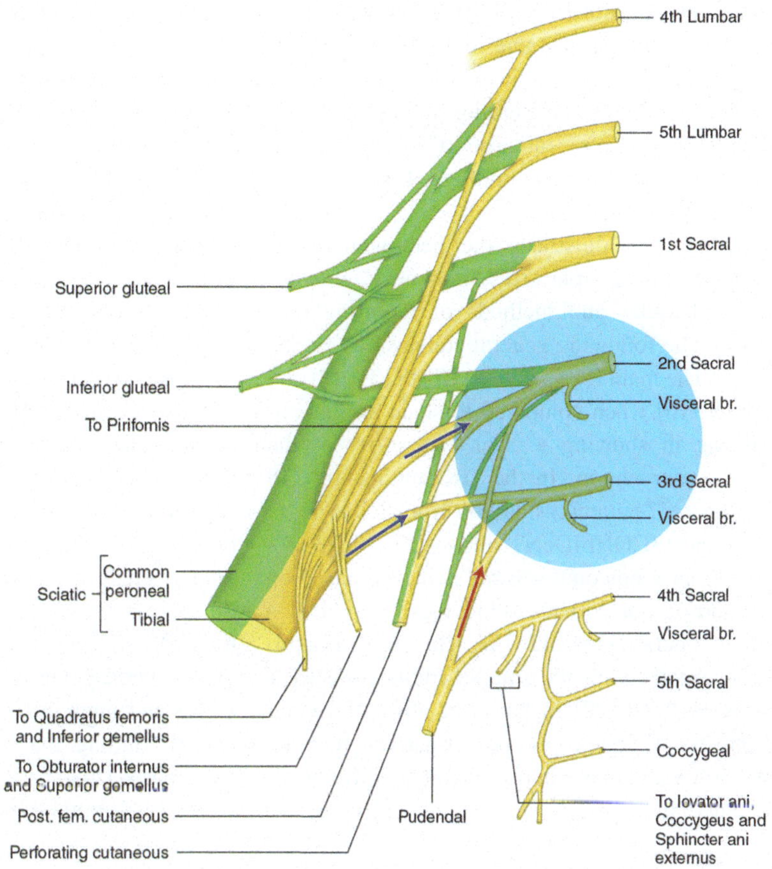

Fig. 39.3 Anatomy of the sacral plexus. The figure shows the confluence of axons derived from the tibial nerve (blue arrows) and pudendal nerve (red arrow)

function of the anus when one considers (anatomically) that convergence of pathways can only occur in the dorsal lamina of the spinal cord, i.e. a point where gating of afferent input occurs rather than motor outflow. Of course, long-term modulation could have an effect on the CNS that eventually led to improved higher motor control, but both this and the possibility of spinal reflex modulation have never been demonstrated in health or disease by any therapy.

An afferent effect of TNS is supported by experimental studies. A study of the effect of electrical stimulation over the tibial nerve in the rat demonstrated an increase in the peak amplitude of primary cortical evoked potentials by 45.1%, findings homologous to those with acute S1 nerve stimulation [17]. This was supported by a clinical study for the treatment of overactive bladder where treatment was associated with an increase in long latency somatosensory-evoked potentials, whereas placebo was not [18].

Clinical Efficacy and Effectiveness of TNS

General Overview

Evidence pertaining to the efficacy and effectiveness of PTNS for the treatment of fecal incontinence derives mainly from observational data and a small number of randomised trials. Broadly, this evidence provides a dichotomy of findings with optimistic results from observational studies tempered by less encouraging findings from controlled trials. The data have been presented on this basis.

Randomised Controlled Trials (RCTS)

RCTs have been performed for both PTNS and TTNS and include a large study of TTNS vs. sham [16], a small study of PTNS vs. TTNS vs.

sham [11], a pilot RCT of PTNS vs. SNS [19] and two randomised studies comparing different frequency of treatments (in effect doses) of TTNS [7, 20]. For PTNS, these earlier studies have been surpassed by two recent large RCTs of PTNS vs. sham [21, 22]. The main findings of these studies are shown in Table 39.1.

Table 39.1 illustrates the heterogeneity of comparisons, sample sizes, treatment intervention schedules and methods of reporting outcome. The following general conclusions can be made. The main RCTs of both PTNS [21, 22] and TTNS [16] when compared to sham are unconvincing in showing a clear superiority of the active intervention. In the UK multicentre (18 centres, 227 patients) study by Knowles et al., in the Lancet (CONFIDeNT study) [21] (Fig. 39.4), the primary outcome was not achieved based on proportions of patients achieving the goal of a 50% reduction in weekly FIE. This study triangulates exactly with the smaller multicentre RCT from Holland [22] in terms of absolute reductions in FIE and proportions of patients meeting the 50% reduction criteria. Analytical differences resulted in differing conclusions of these two papers with the former concluding a negative outcome vs. the latter concluding a tentative positive outcome. Clearly any outcome that is dependent on the vagaries of differing analyses must be treated with caution; however, both did show a significant effect on secondary outcomes, notably reduction in absolute FIE with an emphasis on reduction in urge FIE. The single blind study of Thin et al. [19], although a pilot, can be interpreted as showing that the small effect size achieved by PTNS is probably less than that of SNS, especially in those patients progressing to permanent implantation. It is simply impossible to conclude whether TTNS has a similar effect size to PTNS although data would tend not to support this [11, 16], i.e. the efficacy of TTNS is probably less than PTNS. Other studies are simply too small to be meaningful. The final study in the table is a post hoc analysis of 93 patients from the CONFIDeNT data set [23]. This paper analysed several potential predictors of response to PTNS and sham in 205 patients from the original cohort, finding that, contrary to the proposed hypothesis (that patients with isolated urge FI would preferentially benefit), the only (and very)

Table 39.1 Main findings of studies of TNS using randomised designs

Reference	Comparison	Numbers[a]	Treatment regimen	FIE/week (means)	>50%[b]	Incontinence score[c]
Leroi et al. [16]	TTNS vs. sham	68 vs. 63	3 months daily	−0.7 vs. −1.3	0 vs. 0	−3 vs. −2
George et al. [11]	PTNS vs. TTNS vs. sham	11 vs. 8 vs. 8	1.5 months twice weekly	−6.4 vs. −2.3 vs. −1.8	82 vs. 45 vs. 13	NR
Thomas et al. [20]	TTNS daily vs. TTNS weekly	14 vs. 12	1.5 twice weekly	−1.5 vs. −3.5	21 vs. 0	NR
Rimmer et al. [7]	Ambulatory TTNS daily 4 h vs. 1 h	22 vs. 21	6 weeks daily	−1.7 vs. −2.7	NR	−2.0 vs. −2.1
Thin et al. [19]	PTNS vs. SNS	16 vs. 15	12 weeks, weekly vs. standard 2 stage SNS	−4.3 vs. -9.0	44 vs. 75	-3.2 vs. -7.2
Knowles et al. [21]	PTNS vs. sham	115 vs. 112	12 weeks, weekly	−2.3 vs. sham	38 vs. 31	No difference[d]
Van der Wilt et al. [22]	PTNS vs. sham	29 vs. 30	6 weeks, twice weekly	−1.7 vs. sham	37 vs. 27	−3.2 vs. −1.8
Horrocks et al. [23]	PTNS vs. sham (stratified)[e]	93 evenly distributed	12 weeks, weekly	NR	48.9 vs. 18.2	NR

KEY: *FIE* fecal incontinence episodes, *NR* not reported, [a]numbers completing study and outcome assessments; [b]proportion of patients achieving greater than 50% reduction in FI episodes, [c]Cleveland Clinic Florida Fecal Incontinence Score (CCF-FIS) unless stated, [d]St Marks Continence Score; [e]represents a selected cohort and post hoc analysis of data presented in [23] based on exclusion of 112 patients with obstructed defaecation symptoms; TNS = tibial nerve stimulation

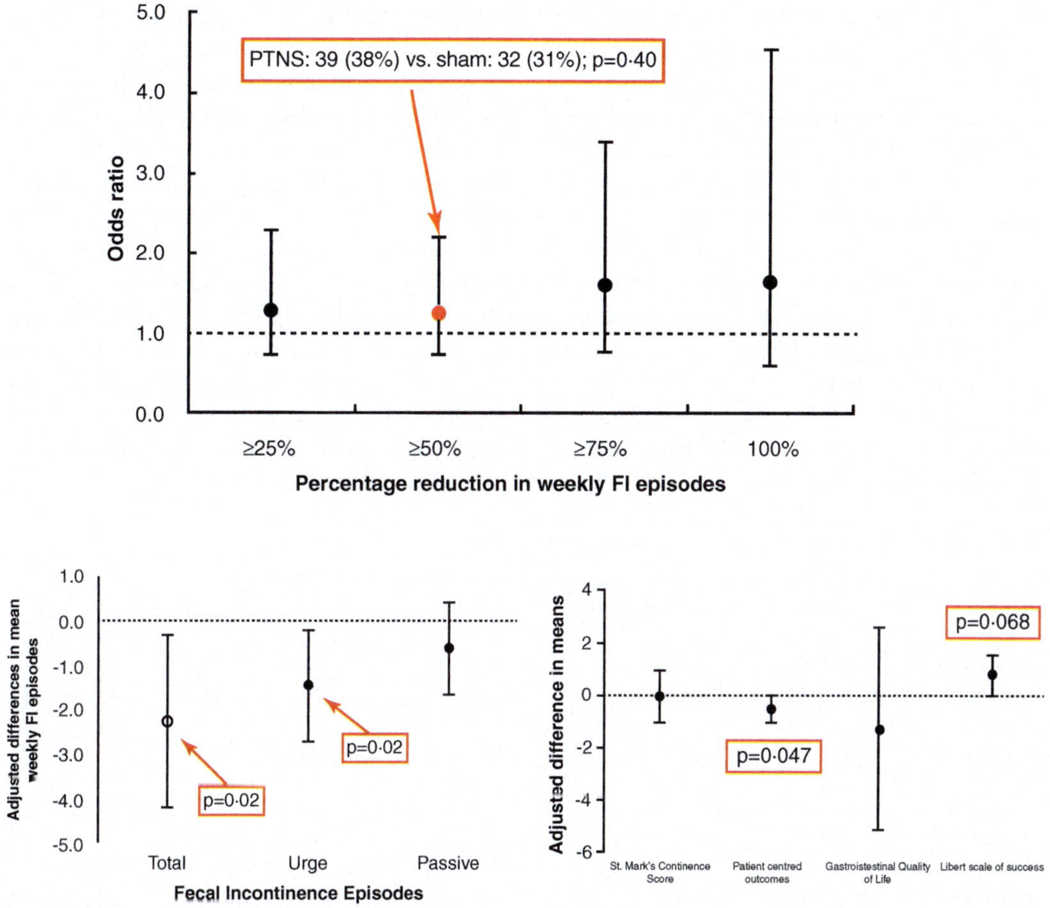

Fig. 39.4 Main findings from National Institute of Health-funded CONFIDeNT study [21]. (**a**) shows odds ratios for PTNS vs. sham of achieving four thresholds of treatment success based on percentage reductions in weekly FIE; (**b**. **c**) show secondary outcomes. (Reused with permission © Elsevier)

significant determinant of outcome was the presence of symptoms of obstructed defaecation. Patients with such symptoms were less likely to respond to PTNS (OR 0.38; CI 0.16–0.91) and more likely to respond to sham (OR 3.45; CI 1.31–9.21). Excluding these patients (112 of the 205) and re-performing the original analysis of primary outcome resulted in a profoundly different study interpretation (48.9 PTNS and 18.2% sham; multivariable OR 4.71; CI 1.71–12.93).

Observational Studies

A review up to 2015 [24] included a total of 20 observational studies (published between 2003 and 2014). These include ten case series of PTNS [5, 9, 10, 12–14, 25–29], a comparative case-matched study of PTNS vs. SNS [29], a prospective clinical audit of SNS and PTNS [30] and five case series of TTNS [15, 20, 31–33]. Aside from the inherent problems of internal validity of observational data, the quality of these studies is not high on formal scoring (NICE quality scores 3 to 6 from total of 8) [24]. Data from these studies are summarised in Table 39.2.

So how do these data inform the reader? The obvious answer to this question is 'not a lot' when putting together heterogeneity of treatment indication, outcomes and intervention with the findings from RCTs that largely refute any significant benefit of PTNS or TTNS over sham

(Fig. 39.5). This would, however, be a simplification for several reasons. First, and perhaps most obvious, is that in clinical practice (that observed), the therapy does appear to have benefit, and this observation is borne out by reports of patient experience [34, 35]. If this benefit (or much of it) does not exceed sham electrical stimulation (as demonstrated by RCTs), then the question is posed whether TNS is an adjunct to a broader, perhaps more holistic approach, to managing patient symptoms that includes other factors within a 'complex' intervention (this is discussed below). The second point relates to outcomes. It is possible that the basic outcome measures beloved of clinical trials in FI, e.g. reductions in FIE and proportion of patients achieving a fixed quantitative reduction, commonly and arbitrarily chosen as 50%, do not truly reflect the benefit of TNS. This issue is not limited to TNS and is also a problem with SNS where it is generally accepted that the restoration of conscious urge and thus advanced warning of rectal filling are very commonly reported as the main subjective improvement by patients. Studies such as that by Hotouras et al. [28] measured this 'deferment time' finding very significant improvements that mirror those measured for SNS [36]. Deferment time was not selected as an outcome in the RCTs, and this may have been an oversight.

Table 39.2 Main findings of case series of PTNS or TTNS

Reference	N starting (N F up)	Treatment regimen: N sessions (N weeks)	Top-ups[b]	[a]Follow-up (months)	Change in FIE	> 50% [c]	Change in CCF-FIS
PTNS							
Shafik et al. [5]	32 (32)	14 (4)	8	22	NR	NR	13 to 9
De la Portilla et al. [9]	16 (11)	16 (11)	8	14	NR	NR	13 to 9
Govaert et al. [25]	22 (16)	22 (16)	NR	12	7 to 1	59	12 to 6
Boyle et al. [10]	31 (30)	12 (12)	3	5	4 to 0	71	13 to 7
Findlay et al. [26]	13 (13)	12 (12)	0	4	NR	NR	NR
Hotouras et al. [27]	88 (88)	12 (12)	0	3	5 to 1	NR	12 to 9
Arroyo et al. [12]	16 (15)	12 (12)	12	6	NR	NR	10 to 5
Al Asari et al. [29]	21 (21)	12 (6)	> 3	12	NR	NR	15 to 9
Hotouras et al. [28]	146 (128)	12 (12)	0	3	4 to 1	NR	12 to 10
De la Portilla et al. [13]	30 (21)	12 (12)	12	27	NR	NR	14 to 9
Hotouras et al. [30]	115 (103)	12 (12)	6	29	5 to 1	52	12 to 10
Lopez Delgado et al. [14]	24 (19)	12 (12)	6	6	NR	NR	15 to 10
Median (range)	**554 (486)**	**12 (12–14)**	**6**	**6 (3–29)**	**5 to 1**	**52–71**	**−4 points**
TTNS							
Queralto et al. [15]	10 (10)	20 (4)	Y	4	NR	NR	13 to 1.5
Vitton et al. [32]	12 (9)	90 (12)	N	3	NR	NR	14 to 13
Vitton et al. [31]	24 (22)	90 (12)	N	15	NR	NR	12 to 10
Eleouet et al. [33]	32 (30)	56 (4)	N	6	NR	NR	15 to 11
Median (range)	**78 (71)**	**–**	**–**	**5**	**–**	**–**	**−3 points**

N number, *FU* follow up, [b]includes only longest follow-up where multiple time points were documented, [a]from start of treatment, *FIE* fecal incontinence episodes, *NR* not reported, [c]proportion of patients achieving greater than 50% reduction in FI episodes, *CCF-FIS* Cleveland Clinic Florida Fecal Incontinence Score, *PTNS* Percutaneous tibial nerve stimulation, *TTNS* Transcutaneous tibial nerve stimulation

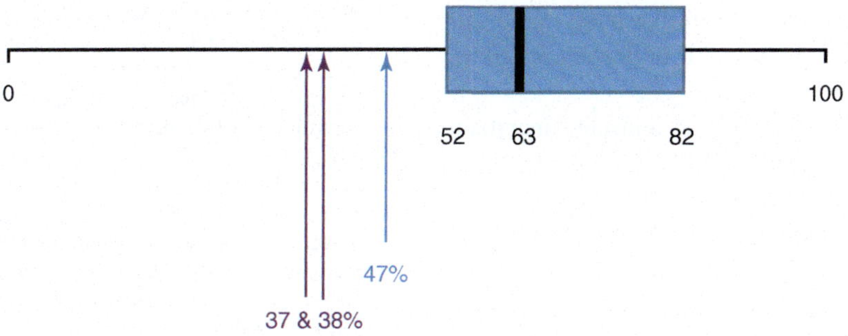

Fig. 39.5 Schematic representation of proportions of patients achieving 50% reduction in weekly FIE based on different study designs. The solid box shows the outcomes from pooled observational studies; the blue arrow shows the outcome from the single blind RCT of Thin et al. [19]; the two blue arrows show the outcomes from sham-controlled RCTs (37% van de Wilt et al. [22] and 38% Knowles et al. [21]). The influence of trial design is clearly observed

Table 39.3 Main continence requirements and goals of outpatient therapies

Continence requirements	Goals of biofeedback therapy	Goals of pelvic floor physiotherapy	Goals of tibial nerve stimulation
Increased abdominal and pelvic pressure	Abdominal wall training (avoidance of abdominal wall contraction)	Not routinely addressed	Not addressed
External anal sphincter and pelvic floor strength	Muscle training	Muscle training[a]	Not addressed
Conscious perception of rectal filling	Sensory training to recognise smaller distensions	Not addressed	May be modulated
Perception of rectal fullness	Urge resistance training by rectal desensitisation	Not addressed	May be modulated
Adequate IAS resting pressure	Not addressed	Not addressed	Not addressed

[a]May include adjuncts such as neuromuscular (galvanic) stimulation. *IAS* internal anal sphincter

Role of TNS in Current Clinical Practice

The above discussion of the evidence for various forms of TNS leaves the clinical community (and funders thereof) at a position of uncertainty. It is, however, important to appreciate that many complex interventions (e.g. behavioural interventions, surgery, etc.) that are generally considered to be of benefit have not (or would not) pass muster in high-quality clinical trials. This should not necessarily doom them, especially when they can be demonstrated to be very safe and highly acceptable to patients (as for TTNS and PTNS). The following points represent the author's view:

- TNS is well liked by patients and provides a measurable benefit in clinical practice (based on observational studies).
- This benefit is in part mediated by the wider interaction of the patient with the practitioner and other subtle interventions that were rightly excluded or controlled for in RCTs (as part of experimental design).
- As such, TNS can be viewed as an adjunctive therapy to holistic management of the patient, i.e. in a package of care for FI.
- The results of the post hoc analysis of CONFIDeNT [23] make a compelling argument to offer TNS as an adjunctive therapy only in those patients who do not have concomitant symptoms of obstructed defaecation (this is about 50% of patients in a general pelvic floor clinic). The benefit in these patients will mainly be observed as a decrease in urge FIE with improvement in deferment time.
- Particular attention should be paid to considering the role of TNS vs. other treatments that modulate defaecatory function (Table 39.3).

This places TNS at a similar point in the pathway of care as forms of biofeedback and pelvic floor muscle training.
- It is unlikely that the current uncertainty regarding evidence will be resolved by further RCTs although a CONFIDeNT-type trial could be considered with revised selection criteria.

Summary

This chapter makes clear that TNS is a safe and popular therapy although the current evidence base for its use is uncertain. It should probably be considered as an adjunct of other specialist conservative measures for FI and offered only to suitable patients.

References

1. Maeda Y, Lundby L, Buntzen S, et al. Suboptimal outcome following sacral nerve stimulation for fecal incontinence. Br J Surg. 2011;98:140–7.
2. Dudding TC, Meng Lee E, Faiz O, et al. Economic evaluation of sacral nerve stimulation for fecal incontinence. Br J Surg. 2008;95:1155–63.
3. McGuire EJ, Zhang SC, Horwinski ER, et al. Treatment of motor and sensory detrusor instability by electrical stimulation. J Urol. 1983;129:78–9.
4. Stoller ML. Afferent nerve stimulation for pelvic floor dysfunction. Eur Urol. 1999;35:16.
5. Shafik A, Ahmed I, El-Sibai O, et al. Percutaneous peripheral neuromodulation in the treatment of fecal incontinence. Eur Surg Res. 2003;35:103–7.
6. Peters KM, Carrico DJ, Perez-Marrero RA, et al. Randomized trial of percutaneous tibial nerve stimulation versus Sham efficacy in the treatment of overactive bladder syndrome: results from the SUmiT trial. J Urol. 2010;183:1438–43.
7. Rimmer CJ, Knowles CH, Lamparelli M, et al. Short-term outcomes of a randomized pilot trial of 2 treatment regimens of transcutaneous tibial nerve stimulation for fecal incontinence. Dis Colon Rectum. 2015;58:974–82.
8. van Breda HMK, Martens FMJ, Tromp J, et al. A new implanted posterior tibial nerve stimulator for the treatment of overactive bladder syndrome: 3-month results of a novel therapy at a single center. J Urol. 2017;198:205–10.
9. de la Portilla F, Rada R, Vega J, et al. Evaluation of the use of posterior tibial nerve stimulation for the treatment of fecal incontinence: preliminary results of a prospective study. Dis Colon Rectum. 2009;52:1427–33.
10. Boyle DJ, Prosser K, Allison ME, et al. Percutaneous tibial nerve stimulation for the treatment of urge fecal incontinence. Dis Colon Rectum. 2010;53:432–7.
11. George AT, Kalmar K, Sala S, et al. Randomized controlled trial of percutaneous versus transcutaneous posterior tibial nerve stimulation in fecal incontinence. Br J Surg. 2013;100:330–8.
12. Arroyo A, Parra P, Lopez A, et al. Percutaneous posterior tibial nerve stimulation (PPTNS) in fecal incontinence associated with an anal sphincter lesion: results of a prospective study. Int J Surg. 2014;12:146–9.
13. de la Portilla F, Laporte M, Maestre MV, et al. Percutaneous neuromodulation of the posterior tibial nerve for the treatment of fecal incontinence - midterm results: is retreatment required? Color Dis. 2014;16:304–10.
14. Lopez-Delgado A, Arroyo A, Ruiz-Tovar J, et al. Effect on anal pressure of percutaneous posterior tibial nerve stimulation for fecal incontinence. Color Dis. 2014;16:533–7.
15. Queralto M, Portier G, Cabarrot PH, et al. Preliminary results of peripheral transcutaneous neuromodulation in the treatment of idiopathic fecal incontinence. Int J Color Dis. 2006;21:670–2.
16. Leroi AM, Siproudhis L, Etienney I, et al. Transcutaneous electrical tibial nerve stimulation in the treatment of fecal incontinence: a randomized trial (CONSORT 1a). Am J Gastroenterol. 2012;107:1888–96.
17. Griffin KM, Pickering M, O'Herlihy C, et al. Sacral nerve stimulation increases activation of the primary somatosensory cortex by anal canal stimulation in an experimental model. Br J Surg. 2011;98:1160–9.
18. Finazzi-Agro E, Rocchi C, Pachatz C, et al. Percutaneous tibial nerve stimulation produces effects on brain activity: study on the modifications of the long latency somatosensory evoked potentials. Neurourol Urodyn. 2009;28:320–4.
19. Thin NN, Taylor SJ, Bremner SA, et al. Randomized clinical trial of sacral versus percutaneous tibial nerve stimulation in patients with fecal incontinence. Br J Surg. 2015;102:349–58.
20. Thomas GP, Dudding TC, Nicholls RJ, et al. Bilateral transcutaneous posterior tibial nerve stimulation for the treatment of fecal incontinence. Dis Colon Rectum. 2013;56:1075–9.
21. Knowles CH, Horrocks EJ, Bremner SA, et al. Percutaneous tibial nerve stimulation versus sham electrical stimulation for the treatment of fecal incontinence in adults (CONFIDeNT): a double-blind, multicentre, pragmatic, parallel-group, randomised controlled trial. Lancet. 2015;386:1640–8.
22. van der Wilt AA, Giuliani G, Kubis C, et al. Randomized clinical trial of percutaneous tibial nerve stimulation versus sham electrical stimulation in patients with fecal incontinence. Br J Surg. 2017;104:1167–76.
23. Horrocks EJ, Chadi SA, Stevens NJ, et al. Factors associated with efficacy of percutaneous tibial nerve

stimulation for fecal incontinence, based on post-hoc analysis of data from a randomized trial. Clin Gastroenterol Hepatol. 2017;15:1915–21.. e2

24. Horrocks, E.J. Fecal incontinence and the role of percutaneous tibial nerve stimulation. PhD thesis, University of London 2016.
25. Govaert B, Pares D, Delgado-Aros S, et al. A prospective multicentre study to investigate percutaneous tibial nerve stimulation for the treatment of fecal incontinence. Color Dis. 2010;12:1236–41.
26. Findlay JM, Yeung JM, Robinson R, et al. Peripheral neuromodulation via posterior tibial nerve stimulation – a potential treatment for fecal incontinence? Ann R Coll Surg Engl. 2010;92:385–90.
27. Hotouras A, Thaha MA, Boyle DJ, et al. Short-term outcome following percutaneous tibial nerve stimulation for fecal incontinence: a single-centre prospective study. Color Dis. 2012;14:1101–5.
28. Hotouras A, Murphy J, Walsh U, et al. Outcome of percutaneous tibial nerve stimulation (PTNS) for fecal incontinence: a prospective cohort study. Ann Surg. 2014;259:939.43.
29. Al Asari S, Meurette G, Mantoo S, et al. Percutaneous tibial nerve stimulation vs sacral nerve stimulation for fecal incontinence: a comparative case-matched study. Color Dis. 2014;16:O393–9.
30. Hotouras A, Murphy J, Allison M, et al. Prospective clinical audit of two neuromodulatory treatments for fecal incontinence: sacral nerve stimulation (SNS) and percutaneous tibial nerve stimulation (PTNS). Surg Today. 2014;44:2124–30.
31. Vitton V, Damon H, Roman S, et al. Transcutaneous electrical posterior tibial nerve stimulation for fecal incontinence: effects on symptoms and quality of life. Int J Color Dis. 2010;25:1017–20.
32. Vitton V, Damon H, Roman S, et al. Transcutaneous posterior tibial nerve stimulation for fecal incontinence in inflammatory bowel disease patients: a therapeutic option? Inflamm Bowel Dis. 2009;15: 402–5.
33. Eleouet M, Siproudhis L, Guillou N, et al. Chronic posterior tibial nerve transcutaneous electrical nerve stimulation (TENS) to treat fecal incontinence (FI). Int J Color Dis. 2010;25:1127–32.
34. Allison M. Percutaneous tibial nerve stimulation for patients with fecal incontinence. Nurs Stand. 2011;25:44–8.
35. A needle in my ankle beat my incontinence. https://www.dailymail.co.uk/health/article-2107303/A-needle-ankle-beat-incontinence.html.
36. Boyle DJ, Knowles CH, Lunniss PJ, et al. Efficacy of sacral nerve stimulation for fecal incontinence in patients with anal sphincter defects. Dis Colon Rectum. 2009;52:1234–9.

Pelvic Floor Dysfunction

40

Lucia Camara Castro Oliveira, Virginia C. Roncatti, Rogerio de Fraga, and Paulo Palma

Introduction

Pelvic floor disorders are common among female patients, especially in those above 60 years of age. It occurs in one of every three women including dysfunction of urinary and bowel control [1]. In addition, patients can present with pelvic pain.

According to Peter Petros' Integral Theory [2], the pelvic floor should be considered as a unit, where the organs are connected and maintained with balance by the fascias and ligaments (see Chap. 3). For example, if you have a posterior injury, all vector forces will move to the front to maintain the balance of the traction exerted by the muscles and transmitted by the fascia, which is fixed to the organs. In the proposed algorithm by Petros, each compartment has structures that can be damaged. These lesions cause symptoms, and the symptoms can be used to guide the structures to be fixed or the lesions to be repaired (Fig. 40.1).

Pelvic floor integrity is responsible to the maintenance of the organs in place, with normal function. The pelvis consists of bone wall, musculature, fascia, ligaments, and organs. Thus, lesions in the pelvic floor can cause dysfunction that will result in organ prolapse, fecal and/or urinary incontinence, and constipation.

Etiology

Denervation of the pelvic floor resulting from chronic evacuation or pudendal nerve injury by stretching is the most common mechanism to explain weak and hypotonic muscles. Traumatic spine injuries can also contribute to pelvic floor failure. Obstetric injury associated with the effects of aging plays an important role in the etiology of urinary incontinence and prolapse in women [1].

In addition to the musculature, the organs of the pelvis are supported by a fibromuscular fascia known as endopelvic fascia. This fascia involves the organs in the pelvis, and their rupture can lead to the appearance of herniations of the uterus, bladder, or the rectum. However, the major supporting components of the pelvic floor are the levator ani muscles. The levator muscles consist of four striated muscles: pubococcygeal, puborectalis, iliococcygeus, and coccygeus muscle. The holes between these muscles, through which the

L. C. C. Oliveira
Director Department of Anorectal Physiology of Rio de Janeiro and CEPEMED, Rio de Janeiro, Brazil

V. C. Roncatti (✉)
Department of Urogynecology and Pelvic Reconstruction, Heliópolis Hospital, Sao Paulo, Brazil

R. de Fraga
Hospital das Clinicas UFPR, Department of Surgery, Curitiba, Brazil

P. Palma
UNICAMP, Department of Urology, Sao Paulo, Brazil

L. C. C. Oliveira (ed.), *Anorectal Physiology*, https://doi.org/10.1007/978-3-030-43811-1_40

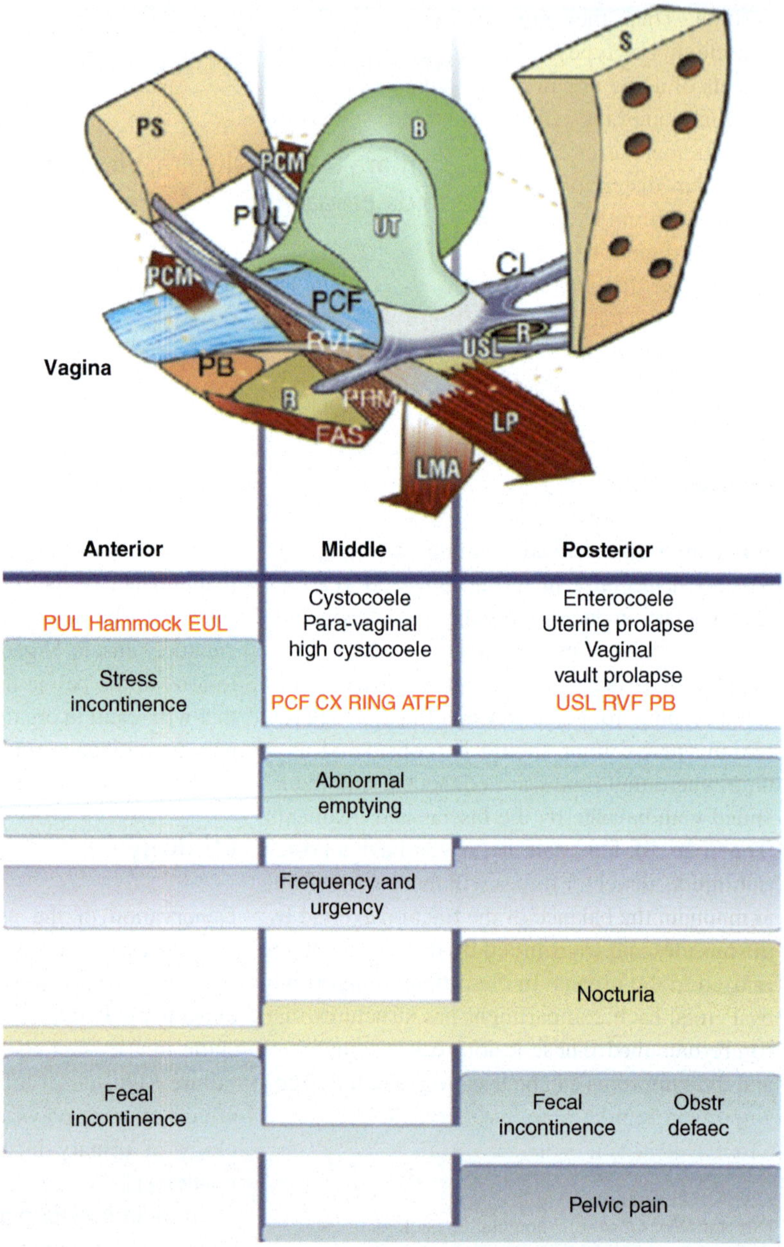

Fig. 40.1 Algorithm of symptoms

urethra and the vagina pass, are known as the urogenital hiatus, where the genital prolapses occur. Thus, it is possible to verify cystoceles, enteroceles, and rectoceles. As a consequence, protrusion of the bladder, bowel loops, uterus, and rectum can lead to urinary and fecal incontinence.

There is a high incidence of incontinent individuals with failure of the pelvic floor. In patients with anal incontinence, the association of urinary incontinence and genital prolapse may correspond to 50% and 22% of the cases, respectively [3].

In the elderly, there is some risk factors associated to urinary incontinence: cognitive impairment, diuretics and sedatives, diabetes, constipation, urinary infections, chronic cough, and cerebrovascular disease.

The pelvic floor muscles are in constant tonic status, except during bowel evacuation and micturition. Thus, they present a predominance of muscle fibers type I, resistant to fatigue. The effects of aging and hormonal changes influence the function of these muscles.

Extreme exercises, excessive stretching with muscular distention or any type of muscular trauma, especially during labor, can lead to weak and dysfunction of the pelvic floor [4]. Pathologies that affect innervation (e.g., multiple sclerosis and diabetes) may also be an important cause of pelvic floor failure, leading to severe double incontinence [5].

Estrogen has great influence on the pelvic floor. Although the most common symptoms of postmenopausal reduction in estrogen levels are related to vasodilation and emotional liability, urogenital atrophy occurs in up to 80% of women. The most frequent cases consist of vaginal dryness and dyspareunia, which are associated with devascularization of the vaginal mucosa. Because estrogen has a neurotrophic effect on nerve conduction, it is believed to exert influence on denervation and failure of the pelvic floor.

Pelvic organ prolapse occurs in one of every three women and has a multifactorial etiology. The main risk factors are obstetric injury and situations in which there is a chronic increase in intra-abdominal pressure, such as in patients with chronic obstructive diseases, severe constipation, and obesity and in individuals with occupations that require excessive weight lifting [6, 7]. Therefore, the inter-multidisciplinary approach for evaluation of these patients is considered important. Similarly, when treatment options are offered, it is noted that physiotherapy and biofeedback can be used for the treatment of different conditions, including anal and urinary incontinence and other disorders of defecation.

Clinical and Diagnostic Evaluation

The diagnosis of pelvic floor disorders begins with a careful clinical history, taking into account the patient's symptoms, clinical problems, and the physical history of emotional trauma. Many patients avoid discussing their problems with the family and with their doctors due to embarrassment and fear. A history of sexual abuse and other types of abuse is not uncommon among those patients.

Information about dietary habits and physical activities is obtained, and patients are advised on the type of evaluation and how they will be examined before any evaluation.

Thereafter, a thorough physical examination is performed, including assessment of the entire pelvic and perineal area, assessing the perineal body and performing a complete vaginal and anal evaluation. The strength of the anus muscles through the digital exam is obtained. The evaluation should be performed with the patient in different positions: sitting, standing, and lying down. Patients are asked to perform a Valsalva maneuver, and the perineum descend is evaluated; the patient is asked to make an effort to evacuate, and the presence of anismus or a hypertonic pelvic floor can be suspected.

The palpation of the sphincter muscles, with evaluation of the muscular strength and the capacity to sustain the contraction, is important. Rectal examination and vaginal evaluation are performed, alone or in combination, with the patient at rest and during the Valsalva maneuver [8].

In the evaluation of genital prolapses, the vaginal examination should be performed with the patient in the lithotomy or gynecological position (Fig. 40.2), in the Sims position (lateral decubitus with flexed legs), or in the sitting position.

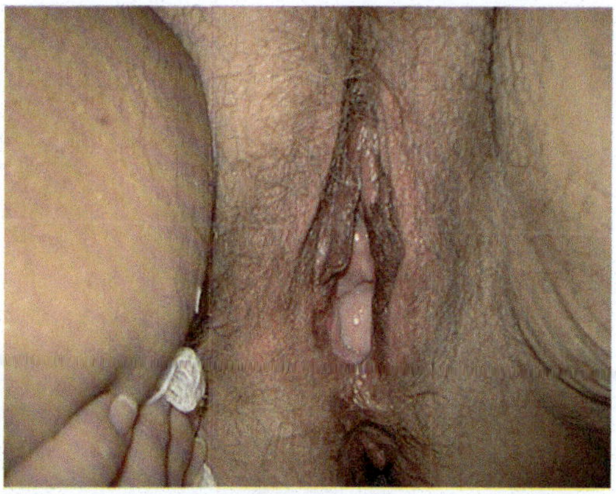

Fig. 40.2 Patient in the lithotomy (or gynecological) position

Figures 40.3 and 40.4 show a uterine and rectal prolapse, respectively. The rectal examination confirms the presence of a rectocele (Fig. 40.5). The International Continence Society (ICS) has developed a system for quantifying genital prolapse, known as the pelvic organ prolapse quantification (POP-Q) system, which allows the examiner to objectively quantify pelvic organ prolapse according to its height relative to the hymen [9, 10].

Patient assessment using validated questionnaires, especially the Organ Prolapse Questionnaire and Urinary Incontinence and Sexual Dysfunction Questionnaire, known as PSIQ-12, provides important information about sexual dysfunction.

Sexual dysfunction results from lack of orgasm, loss of sexual desire, and dyspareunia. Of women with pelvic pain, 76% may have some form of sexual dysfunction [11]. Spastic pelvic floor is also another aspect of pelvic floor dysfunction, which may manifest as pelvic pain, urinary tract symptoms, or defecation disorders [12].

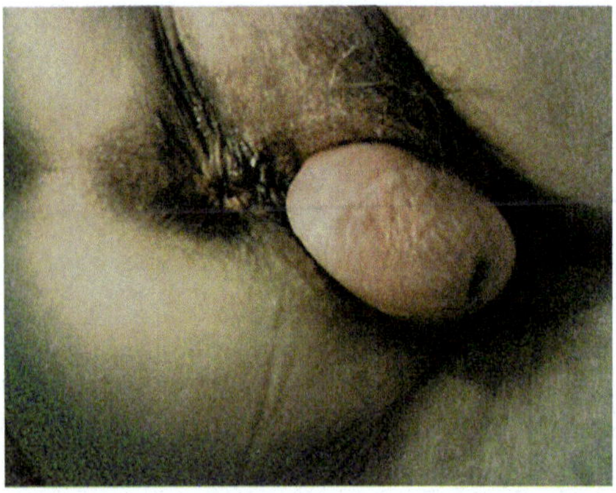

Fig. 40.3 Uterine prolapse

Pain is commonly reported as occurring in the perineum or in the labia majora. Urinary symptoms include difficulty in starting micturition, urinary urgency and frequency, and hyperalgesia of the urethra. Finally, the symptoms associated with defecation resemble those of defecation obstruction syndrome.

Diagnostic evaluation includes several methods, including urodynamic examination, defecography, anorectal manometry, electromyography (EMG), pudendal nerve latency (PNTML), and, more recently, defecation and echodefecography [12].

Defecography is a common study utilized to demonstrate the dynamics of defecation. Normally, the pelvic floor relaxes at the time of evacuation, allowing for the stretching of the rectum and opening of the anal canal to pass the fecal contents. These dynamic studies also allow us to assess whether the pelvic floor musculature is releasing properly for normal bowel movement or whether these muscles are flaccid, leading to incontinence. Figure 40.6 shows the paradoxical contraction of the puborectalis muscle during the evacuation phase, through defecography.

The defecography is also important to show the presence of internal invagination of the rectum, rectoceles, and paradoxical contraction of

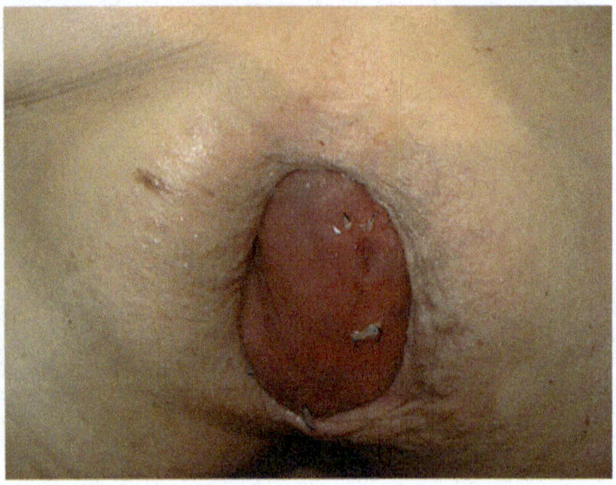

Fig. 40.4 Rectal prolapse

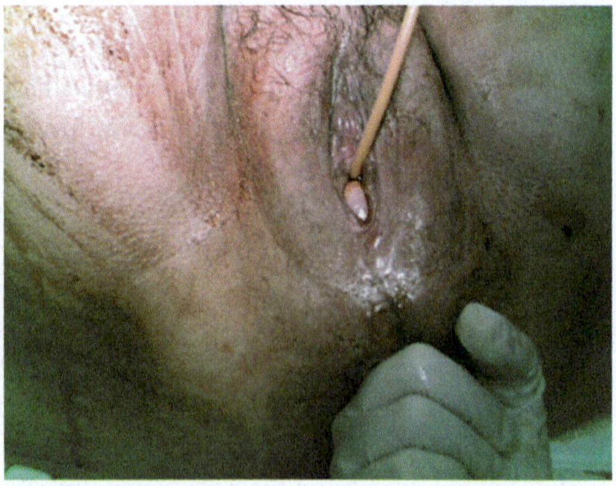

Fig. 40.5 Rectocele

the puborectalis muscle. EMG and PNTML may show pelvic floor dysfunction (Fig. 40.7). Other important exams for assessing pelvic floor dysfunction are anorectal manometry and dynamic echodefecography. Manometry evaluates anal sphincter strength, loss of muscle tone, loss of rectal sensitivity, and rectal capacity. Magnetic resonance imaging with defecography also allows the identification of urinary tract–related disorders (e.g., cystocele and enterocele), as well as adequate evaluation of the pelvic floor musculature (Fig. 40.8). This dynamic method may also show the relationship between the different organs constituting the three pelvic compartments, allowing a broad view of their dysfunctions (Figs. 40.9 and 40.10).

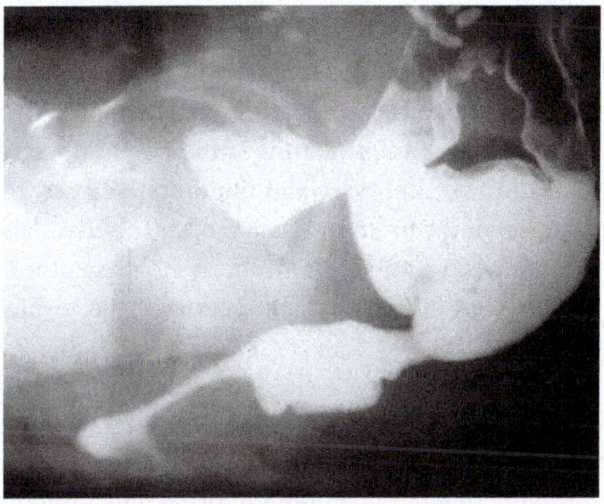

Fig. 40.6 Paradoxical contraction of the puborectalis muscle during the evacuation phase, through defecography

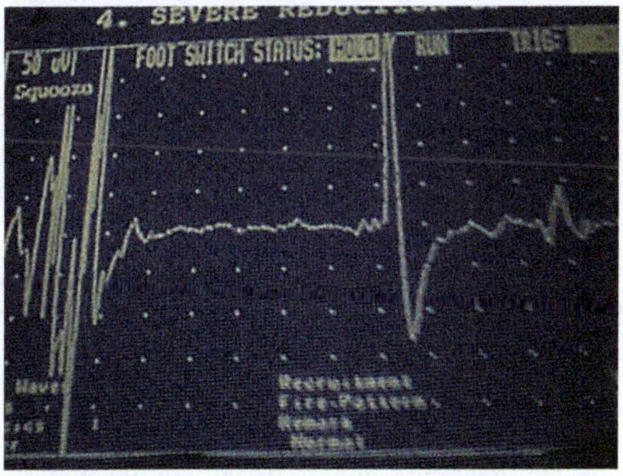

Fig. 40.7 EMG showing pelvic floor dysfunction

Algorithm of Symptoms According to the Integral Theory

According to the Integral Theory [2], stress urinary incontinence and urgency may result from damaged ligaments and fascias related to the urethra and vagina. Subsequently, pelvic pain, voiding, and anal symptoms were added. The systematization of these symptoms is shown in Fig. 40.1 (see also Chap. 3).

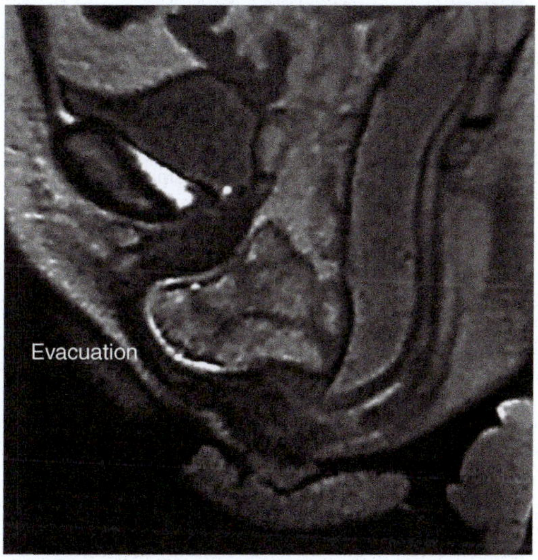

Fig. 40.8 Defecography showing the enterocele

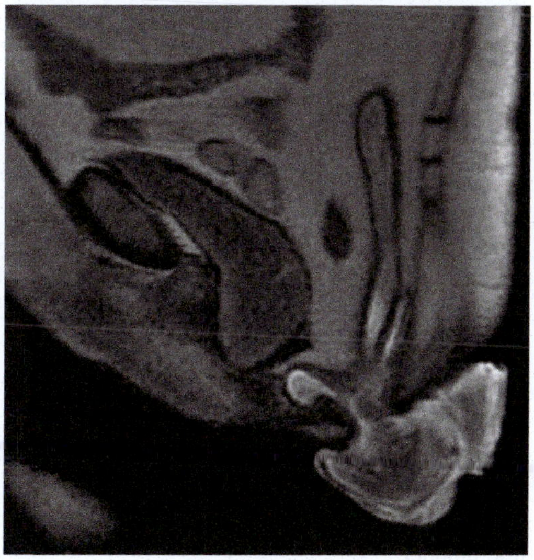

Fig. 40.9 Defecography showing cystocele and rectocele

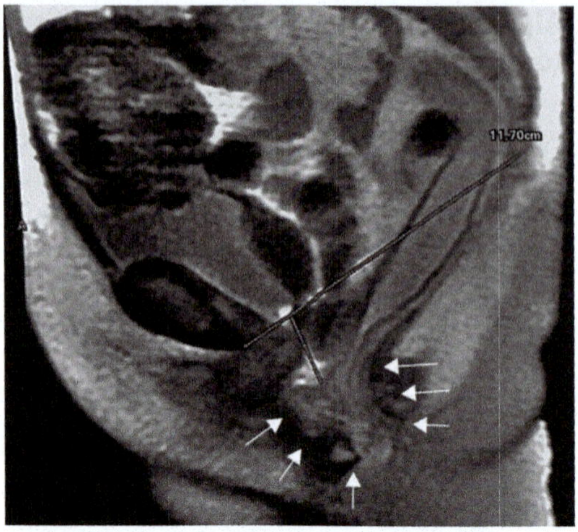

Fig. 40.10 Defecography showing cystocele, perineal descent, enterocele, and internal invagination

Notice that these compartments are different from the levels of support proposed by DeLancey [13]. The anterior compartment includes the pubourethral and urethropelvic ligaments, both level III according to DeLancey. The median compartment is composed by the pubocervical fascia and its insertions (level II), and the posterior compartment includes the complex cardinal sacrouterine ligaments, rectovaginal fascia, and perineal body, respectively, levels I, II, and III according to DeLancey [13].

Symptoms from the Anterior Compartment

The anterior compartment defects produce stress urinary incontinence, urgency, and anal incontinence. Urinary incontinence is easy to understand since damage to the elements of support of the urethra impairs the proper coaptation of the urethral lumen.

The urgency is derived from the urethral afferent nerve stimulation, since inadequate coaptation allows for the presence of urine in the proximal urethra, and this is interpreted as maximum bladder capacity, even with small volume in the bladder.

More difficult is to correlate anal incontinence, but remember that the urethral support ligaments help to support the insertion of the puborectalis muscle. In order to be effective, the muscular contraction needs good contra traction at its insertion. Therefore, lesion of the urethral ligaments may impair the tonus of the puborectalis muscle, diminishing the anorectal angulation, and may lead to anal incontinence.

Symptoms from the Middle Compartment

The middle compartment refers to the pubocervical fascia and its insertion at the arcus tendineus fasciae pelvis (ATFP) and pericervical ring. Damage to these structures may cause urgency, voiding dysfunction, nocturia, and pelvic pain. The symptoms of urgency and frequency are due to overstimulation of the afference of the bladder. The inhibition of the micturition reflex depends on the simultaneous and coordinated contraction of the striated musculature of the pelvic floor so that the mid and proximal urethra are closed by traction of the urethral ligaments and also by suppressing the stimulation of the stretch receptor in the trigone due to the contraction of the levator plateau.

When a tear in the fascia is present or even a damage to the pericervical ring, the force transmission will be inadequate and the hydrostatic pressure due to the column of urine will continue to be exerted on the receptor in the trigone. The brain will interpret this signal as a full bladder, even at small volumes.

Voiding symptoms such as low flow, hesitancy, dribbling, and incomplete voiding are due to the impaired transmission of the force generated by the levator plateau to the trigone. As described previously, the posterior force is needed for the urethral opening during the voiding phase.

Symptoms from the Posterior Compartment

In the posterior compartment are located the complex cardinal sacrouterine ligaments, rectovaginal fascia, and perineal body so that this compartment correlates with all symptoms derived from pelvic support damage to some extent.

- Urgency: May be due to lesion of the cardinal sacrouterine complex that prevents the tension due to the contraction of the levator plateau to be transmitted to the pericervical ring and rectovaginal fascia.
- Nocturia related to incomplete voiding and urine residuals.
- Pelvic pain: The nerve fibers that travel along the sacrouterine cardinal complex are distended by the gravity (G), producing pelvic pain.

The Aging Bladder

Urogenital tract, including the urinary bladder, is also affected by aging mechanisms. This organ must play the role of storing socially adequate and adding volumes of urine keeping its pressure constant (characteristic denominated compliance) and finally be emptied voluntarily under low pressure and low resistance to urinary flow [14, 15].

The bladder is a hollow organ, and the structure of its wall consists in four distinct parts: (a) an inner layer, delineated by the urothelium, and (b) the lamina propria, composed mainly by irregular and dense connective tissue, fibroblasts, and a thin layer of muscularis propria surrounded by (c) an external layer of smooth muscle (detrusor muscle), with fibers well organized in bundles and connective tissue rich in collagen and elastin involving its perimysium and (d) an adventitious external layer of loose connective tissue [16, 17].

The many components of the bladder play specialized and integrated roles for the correct performance of functions. The connective tissue and the collagen are intrinsically connected to the elastic properties of the bladder and adaptation of bladder to different pathophysiological situations, with implications in basic properties of the bladder like accommodation of volume and compliance [18, 19].

Some experimental studies are trying to correlate morphological, structural, and histological alterations in urinary bladder to events related to aging demonstrating that the aged urinary bladder shows functional alterations like reduced complacency, increase of post-volume residues as well as detrusor hyperactivity, impaired contractility, or the combination of both [20–22].

Although it is recognized functional alterations of urinary bladder in aging as well as the importance of connective tissue [17] and collagen to normal bladder functions, little is known about which are the structural alterations suffered by bladder in the aging process along with the correlation between these alterations to clinical and functional events related to the urinary bladder aging.

Clinical urodynamic studies have demonstrated advancing age to be associated with a reduced bladder capacity, an increase in uninhibited contractions, decreased urinary flow rate, and increased postvoid residual urine volume [23, 24]. The aging bladder specifically may be described as manifesting detrusor overactivity, impaired contractility, or a combination of both [20]. The important aspects of the aging bladder patient were discussed during the US National Health Consensus Conference in 1987 (Table 40.1).

We conclude that bladder aging should be taken into consideration when evaluating a patient. Some changes are structural and can impact assessment and treatment. It is a neces-

Table 40.1 The aging patient

Urogenital changes
Dryness
Pallor
Decreased rugation
Mucosal thinning
Inflammation with/without discharge
Decreased caliber and depth
Increased pH
Multidisciplinary evaluation
Multiple problems uncovered, described, and explained
Resources and strengths catalogued
Need for services assessed
Plan focused on problems
US National Health Consensus Conference, 1987

sary competence for the professional that treats elderly patients to act in a suitable way to the functional parameters.

Treatment

The treatment of pelvic floor dysfunctions can be divided into surgical and nonsurgical [25]. Nonsurgical treatment contemplates general measures and rehabilitation of the entire pelvic floor musculature, as well as the use of pessaries in the case of prolapses. Indications for the different options should be individualized, depending on the diagnosis, as well as general measures that would reduce pressure on the pelvic floor.

In general, the following measures are indicated:

1. Weight control: Body mass index (BMI) of 30 or above is an independent factor for urinary incontinence.
2. Avoid smoking since this is associated with urinary incontinence and the occurrence of prolapses.
3. Avoid weight lifting or strenuous exercise, especially in patients with urinary incontinence and genital prolapses.
4. Avoid chronic effort to evacuate through the use of fibers and lubricants and rehabilitation of the pelvic floor.
5. Rehabilitation of the pelvic floor through a coordinated program of exercises and different techniques supervised by an integrated team.
6. Hormone therapy in cases of atrophy caused by age.

Surgical treatment consists of correcting the existing specific defect using tissue from the patient or from synthetic or biological implants to replace the injured tissue. An existing option for elderly patients with no desire for sexual intercourse is obliterative procedures (colpocleisis) [26].

Pelvic Floor Rehabilitation

One of the most important measures in the treatment of pelvic floor dysfunction is the rehabilitation of the musculature by techniques of biofeedback, pelvic exercises, and electrostimulation, all of which have already been discussed in previous chapters. In addition to the different methods utilized, behavioral interventions such as diaries, pelvic muscle exercises, and weight and fluid intake are incorporated into the rehabilitation program for patients with urinary incontinence. Patients should receive guidance on the anatomy and functioning of the pelvic organs, the objectives of rehabilitation should be traced, and the professional must motivate them so that the results are successful [27, 28].

Hormonal Replacement

Hormonal replacement has been indicated for women with incontinence, especially after menopause [29], but evidences are not conclusive and reliable. Replacement is indicated for incontinent women who do not present contraindications to hormone therapy and who do not show signs of vaginal atrophy [30].

Medical Treatment

Overactive Bladder

Overactive bladder is defined as a symptom of urinary urgency associated with incontinence and, usually, nocturia and increased frequency [31]. It can be treated through behavioral techniques, including biofeedback and vesical exercise training, or medication (Table 40.2).

Table 40.2 Medical treatment for overactive bladder

Mixed mechanism
Oxybutynin
Propiverine
Muscarinic agents
Tolterodine
Trospium
Solifenacin
Darifenacin
Antidepressants
Imipramine
Diuretics
Desmopressin

Summary

Pelvic floor dysfunction and symptoms may have the same etiopathology. There is clinical correlation between pelvic floor muscle and components of the endopelvic fascia, and therefore, symptoms are generally associated. Therefore, the integration between the gynecologist, urologist, proctologist, physiotherapist, psychologist, and the psychiatrist is crucial to address these disorders, which often affect the middle, anterior, and posterior compartments.

Acknowledgment In memory of Sandra Sucar, MD.

References

1. Olsen AL, Smith VJ, Bergstrom JO. Epidemiology of surgically managed pelvic organ prolapse and urinary incontinence. Obstet Gynecol. 1997;89(4):501–6.
2. Petros PE. Cure of urinary and fecal incontinence by pelvic ligament reconstruction suggests a connective tissue etiology for both. Int J Urogynecol. 1999;10:356–60.
3. Dimpfl T, Jaeger C, Mueller-Felber W, et al. Myogenic changes of the levator ani muscle in premenopausal women: the impact of vaginal delivery and age. Neurourol Urodyn. 1998;17(3):197–205.
4. Hagen S, Stark D, Maher C, Adams E. Conservative management of pelvic organ prolapse in women. Cochrane Database Syst Rev. 2006;4:CD003882.
5. Lieber RL, Friden J. Mechanism of muscle injury gleaned from animal models. Am J Phys Med Rehabil. 2008;81:S70–9.
6. Kapoor DS, Thakar R, Sultan AH. Combined urinary and faecal incontinence. Int Urogynecol J Pelvic Floor Dysfunct. 2005;16(4):321–8.
7. Bump R, Norton P. Epidemiology and natural history of pelvic floor dysfunction. Obstet Gynecol Clin N Am. 1998;25(4):723–46.
8. Spence-Jones C, Kamm M, Henry M, Hudson C. Bowel dysfunction: a pathogenic factor in uterovaginal prolapse and urinary stress incontinence. Br J Obstet Gynecol. 1994;101(20):147–52.
9. Madhu C, Swift S, Moloney-Geany S, Drake MJ. How to use the Pelvic Organ Prolapse Quantification (POP-Q) system?. Neurourol Urodyn. 2018;37(S6):S39–43.
10. Espuña PM. Sexual health in women with pelvic floor disorders: measuring the sexual activity and function with questionnaires – a summary. Int Urogynecol J Pelvic Floor Dysfunct. 2009;20(Suppl 1):S65–71.
11. Verbeek M, Hayward L. Pelvic Floor Dysfunction And Its Effect On Quality Of Sexual Life. Sex Med Rev. 2019;7(4):559–64.
12. Cavallaro PM, Staller K, Savitt LR, et al. The Contributions of Internal Intussusception, Irritable Bowel Syndrome, and Pelvic Floor Dyssynergia to Obstructed Defecation Syndrome. Dis Colon Rectum. 2019;62(1):56–62.
13. DeLancey JO. Anatomic aspects of vaginal eversion after hysterectomy. Am J Obstet Gynecol. 1992;166(6 Pt 1):1717–24; discussion 1724–8
14. Kaplan EP, Richier JC, Howard PS, Ewalt DH, Lin VK. Type III collagen messenger RNA is modulated in non-compliant human bladder tissue. J Urol. 1997;157:2366–9.
15. Chang SL, Howard PS, Koo HP, Macarak EJ. Role of type III collagen in bladder filling. Neurourol Urodyn. 1998;17(2):135–45.
16. Deveaud CM, Macarak EJ, Kucich U, Ewalt DH, Abrams WR, Howard PS. Molecular analysis of collagens in bladder fibrosis. J Urol. 1998;160(4):1518–27.
17. Andersson KE, Arner A. Urinary bladder contraction and relaxation: physiology and pathophysiology. Physiol Rev. 2004;84(3):935–86.
18. Macarak EJ, Howard PS. The role of collagen in bladder filling. Adv Exp Med Biol. 1999;462:215–23.
19. Stevenson K, Kucich U, Whitbeck C, Levin RM, Howard PS. Functional changes in bladder tissue from type III collagen-deficient mice. Mol Cell Biochem. 2006;283(1–2):107–14.
20. Resnick NM, Yalla SV. Detrusor hyperactivity with impaired contractile function. An unrecognized but common cause of incontinence in elderly patients. JAMA. 1987;257(22):3076–81.
21. Hashim H, Abrams P. Overactive bladder: an update. Curr Opin Urol. 2007;17:231.
22. Zeni M, Melchioretto EF, Veronez DAL, et al. Stereological Study of collagen in the aging of urinary bladder in rats. Int J Surg Med. 2018;4(2):78–84.
23. Diokno AC, Brown MB, Goldstein NG, Herzog AR. Urinary flow rates and voiding pressures in elderly men living in a community. J Urol. 1994;151:1550–3.
24. Madersbacher S, Pycha A, Schatzl G, et al. The aging lower urinary tract: a comparative urodynamic study of men and women. Urology. 1998;51:206–12.
25. Davilla GW, Ghoniem GM, Wexner SD. Pelvic floor dysfunction. A multidisciplinary approach. London: Springer; 2008.
26. Jones K, Wang G, Romano R, St Marie P, Harmanli O. Colpocleisis: a survey of current practice patterns. Female Pelvic Med Reconstr Surg. 2017;23(4):276–80.
27. Berghmans B. Pelvic floor muscle training: what is important? A mini-review. Obstet Gynecil Int J. 2017;6(4):00214.
28. Schmidt RA, Vapnek JM. Pelvic floor behavior and interstitial cystitis. Sem Urol. 1991;9(2):154–9. Bo K. Pelvic floor muscle exercise for the treatment of stress urinary incontinence: an exercise physiology perspective. Int Urogynecol J. 1995;6:282–91.
29. Donnelly V, O'Connell PR, O'Herlihy C. The influence of oestrogen replacement on faecal incontinence

in post-menopausal women. Br J Obstet Gynaecol. 1997;104:311–5.
30. Haadem K, Ling L, Ferno M, Graffner H. Oestrogen receptors in the external anal sphincter. Am J Obstet Gynecol. 1991;164:609–10.
31. van Kerrebroeck P, Abrams P, Lange R, et al. Duloxetine versus placebo in the treatment of European and Canadian women with stress urinary incontinence. BJOG. 2004;111(3):249–57.

Index

L. C. C. Oliveira (ed.), *Anorectal Physiology*, https://doi.org/10.1007/978-3-030-43811-1

D

M

Q

R

S

The manufacturer's authorised representative in the EU is Springer Nature Customer Service Centre GmbH, Europaplatz 3, 69115 Heidelberg, Germany. If you have any concerns regarding our products, please contact ProductSafety@springernature.com

Printed and bound by CPI Group (UK) Ltd, Croydon, CR0 4YY
01/10/2026
02226525-0004